CORE TOPICS

FOURTH EDITION

Discover Biology

CORE TOPICS

FOURTH EDITION

Discover Biology

Michael L. Cain
Bowdoin College

Carol Kaesuk Yoon
Bellingham, Washington

W. W. NORTON & COMPANY
NEW YORK • LONDON

Anu Singh-Cundy
Western Washington University

Copyright © 2009, 2007 by W. W. Norton & Company, Inc.

Copyright © 2002, 2000 by Sinauer Associates, Inc.

All rights reserved

PRINTED IN THE UNITED STATES OF AMERICA

Fourth Edition

Composition by TSI Graphics

Manufacturing by RR Donnelley

Illustrations for the Fourth Edition by Dragonfly Media Group

Editor: Michael Wright

Senior Production Manager: Christopher Granville

Cover and Book Design: Rubina Yeh

Development Editor: Richard Morel

Associate Managing Editor: Kim Yi

Managing Editor, College: Marian Johnson

Marketing Manager: Betsy Twitchell

Copy Editors: Stephanie Hiebert, Philippa Solomon

Electronic Media Editor: Bonnie Melton

Print Ancillary Editor: Matthew Freeman

Editorial Assistant: Jennifer Cantelmi

Photo Researchers: Jane Sanders Miller, Stephanie Romeo

Library of Congress Cataloging-in-Publication Data

Cain, Michael L. (Michael Lee), 1956–

    Discover biology : core topics / Michael L. Cain, Carol Kaesuk Yoon, Anu Singh-Cundy.—4th ed.

      p. cm.

    Includes index.

    ISBN 978-0-393-93161-7 (pbk.)

1.   Biology.   I. Yoon, Carol Kaesuk.   II. Singh-Cundy, Anu.   III. Title.

QH308.2.C326 2009

570—dc22

2008049989

W. W. Norton & Company, Inc. 500 Fifth Avenue, New York, N. Y. 10110

www.wwnorton.com

W. W. Norton & Company Ltd., Castle House, 75/76 Wells Street, London WIT 3QT

3 4 5 6 7 8 9 0

Welcoming Anu Singh-Cundy

One of the most important changes in this edition of *Discover Biology* is the addition of Anu Singh-Cundy of Western Washington University to the author team. She has drawn on her research background and extensive experience teaching nonmajors to thoroughly revise the chapters on cells, genetics, and anatomy and physiology. We are delighted to welcome Anu. Here, she comments on her involvement in the Fourth Edition.

What appealed to you about working on this textbook?

Teaching nonmajors presents unique challenges: how to allay the fear that many students have of science, how to stimulate interest, and how to show that what they're learning will both enrich them intellectually and have relevance to their daily lives. I've grappled with these issues, discussed them with colleagues across the country, and tested various solutions in class. *Discover Biology* features many innovations that are part of the solution—like "Biology in the News" and "Biology Matters." The opportunity to build on those strengths and implement new ideas drew me to this project.

What were your most significant contributions to the new edition?

The chapters on cell biology, genetics, and organismal biology were my special focus. Many students are intimidated or bored by chemistry, so I tried to make the material friendlier by highlighting chemistry in our everyday lives. For example, we discuss the nutritional aspects of saturated fats and trans fats in the section on lipids. Mitosis is spiced up with a discussion of stem cells, including the breathtaking developments in the area of induced pluripotent stem cells. In genetics, we emphasize the influence of both nature and nurture, with concrete examples that range from human health to limb deformities in frogs.

I put a high premium on the precision and clarity of terms and concepts in a textbook. As developing learners, many students are uncomfortable with "fuzzy" definitions. So, what's a salt and how's that different from a molecule? What exactly is an organic molecule? Genetic trait, phenotype—what's the difference? We nailed down these concepts with clarity and scientific accuracy. We resisted the tendency to bend the science in the pursuit of simplification, while limiting specialized terminology to that which is needed to understand key ideas in biology. Our treatment isn't breezy and superficial, nor is it a dry listing of factoids. I think our attention to these and other aspects of pedagogy and style will make the book a better teaching tool.

What do you want students to take away from *Discover Biology*?

I'm confident the book will support instructors' efforts to engage their students and to demonstrate the value of scientific literacy. Most students will be able to appreciate that what they're learning contributes toward making them critical thinkers, wise consumers, informed voters, and good stewards of our world.

I also like to think that the book can help instructors bring out the *biophilia* in their students, fostering a lifelong curiosity about the natural world and some measure of the joy and inner sustenance we biologists find in it. I hope students who have read this textbook will not be intimidated by science, and will find that they have the wherewithal to understand the scientific issues of the day and participate self-assuredly in any discussion of the societal aspects of those issues.

About the Authors

Michael L. Cain received his PhD in ecology and evolutionary biology from Cornell University and did postdoctoral research at Washington University in molecular genetics. He taught introductory biology and a broad range of other biology courses at New Mexico State University and the Rose-Hulman Institute of Technology for 13 years. He is now writing full-time and is affiliated with Bowdoin College in Maine. Dr. Cain has published dozens of scientific articles on such topics as genetic variation in plants, insect foraging behavior, long-distance seed dispersal, and factors that promote speciation in crickets. Dr. Cain is the recipient of numerous fellowships, grants, and awards, including the Pew Charitable Trust Teacher-Scholar Fellowship and research grants from the National Science Foundation.

Carol Kaesuk Yoon received her PhD from Cornell University and has been writing about biology for the *New York Times* since 1992. Recent stories covered the sensory capabilities of plants and the field of "evo-devo," or evolution and development. Her articles have also appeared in *Science*, the *Washington Post*, and the *Los Angeles Times*. Dr. Yoon has taught writing as a visiting scholar at Cornell University's John S. Knight Writing Program, working with professors to help teach critical thinking in biology classes. She has also served as a science education consultant to Microsoft.

Anu Singh-Cundy received her PhD from Cornell University and did postdoctoral research in cell and molecular biology at Penn State. She is an associate professor at Western Washington University, where she teaches a variety of undergraduate and graduate courses, including organismal biology, cell biology, plant developmental biology, and plant biochemistry. She has taught introductory biology to nonmajors for over 12 years and is recognized for her pedagogical innovations that communicate biological principles in a manner that engages the nonscience student and emphasizes the relevance of biology in everyday life. Her research focuses on cell–cell communication in plants, especially self-incompatibility and other pollen–pistil interactions. She has published over a dozen research articles and has received several awards and grants, including a grant from the National Science Foundation.

Preface

Using stem cells to repair or replace damaged organs. Discovering every last branch on the vast evolutionary tree of life. Seeking a cure for cancer. These are but a few of the many reasons why biology is a gripping subject for us and, we hope, for students using this book. These topics are simultaneously intensely interesting and critically important. Because the scientific understanding of fundamental biological principles is growing by leaps and bounds, this is an exciting time to write, teach, and learn about all areas of the biological sciences.

Just consider that researchers are beginning to unlock the secrets of the human genome, allowing us to explore in unprecedented detail both the causes of and potential treatments for a host of human genetic disorders. And ecologists have begun to make real progress in understanding and predicting events that affect all of the world's ecosystems, such as the transport of pollutants and the contribution of human activities to global warming.

Discoveries such as these, along with a host of others, mean that biology is the subject of news stories on a daily basis. Recently, for example, the news has focused on both the disturbing (germs resistant to much of our antibiotic arsenal) and the exciting (the promise of biofuels made from algae). As these and other news reports show, biology carries important ramifications not only for individuals, but for all human societies, and it touches our lives every day, as well as in long-term ways that we are only beginning to comprehend.

Our Goals: Engage, Inform, Apply

The very things that make biology so interesting—the rapid pace of new discoveries and the many applications of these discoveries by human societies—can make it a difficult subject to teach and to learn. The problem is only exacerbated by the wide variation in background and interests that nonmajors bring to the course. When we set out to write the Fourth Edition of *Discover Biology*, we asked ourselves, "How can we convey the excitement, breadth, and relevance of biology to this varied group of students without burying them in an avalanche of information?"

We answered this question in several ways. In considering which topics and details to include from the vast group of possibilities, our goals were to

- Pique students' interest with engaging "real-world" examples

- Highlight fundamental concepts

- Demonstrate the application of fundamental concepts

We also responded to this question by writing clear, streamlined chapters and by emphasizing the development of biological literacy to provide students with tools that will serve them well regardless of the educational or career path they follow.

As described in detail in the guided tour that follows, to put our views into practice we begin all our chapters with a single main message, a small set of key concepts, and an opening vignette to pique students' interest as they start reading. In the body of the chapter, we develop the main message by illustrating key concepts with both applied topics and compelling examples from the natural world. We end the chapter by revisiting the opening vignette so that students might see how their newly acquired knowledge immediately empowers them to understand "real-world" issues in a more sophisticated manner.

Guided Tour of the Book

Key Features and goals of the text include:

A Chapter Organization That Promotes Scientific Literacy

Every chapter in *Discover Biology* is designed to engage the interest of students, highlight fundamental concepts, and teach students how to apply concepts for themselves.

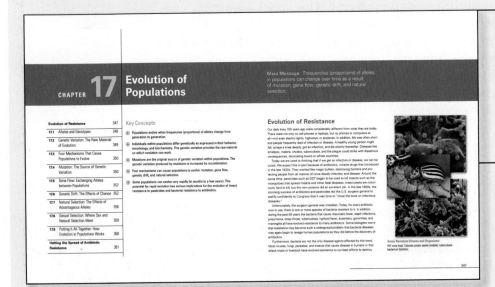

Each chapter is structured around a **Main Message** (the central idea of the chapter) and the **Key Concepts** that support that message (the important information that we will be sure students master by the end of the chapter). The Main Message and Key Concepts, along with a **Chapter Outline**, are listed on the chapter opening pages for easy reference while you are reading.

Each chapter opens with a contemporary **issue** that engages student interest and helps answer the all-too-common question, Why should I care about biology?

That **issue** is revisited at the end of the chapter, demonstrating how students can apply their newly acquired knowledge to gain a more sophisticated understanding of the topic.

Artwork Presented at the Right Level of Detail

The art program offers exceptional clarity. We avoid presenting overly complex figures and captions, preferring to focus on key concepts.

Many of the figures include **concise bubble captions** that guide students through the figure's most important features.

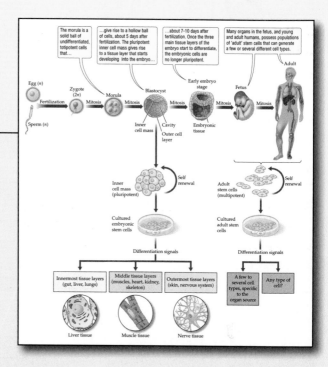

Empowering Students to Make Informed Choices in the "Real" World

"Biology Matters" boxes on topics such as human health and the environment connect chapter content to information of practical importance, with the goal of raising student awareness of the opportunities they have to make choices in their own lives.

Biology Matters

You'll Just Have to Tough It Out

Being sick is never enjoyable, but the more often you take antibiotics, the more likely they won't work for you—or others—in the future. Every time we use antibiotics, we are applying selection pressure: killing off any nonresistant bacteria and giving an advantage to any resistant bacteria. That is, we are actually helping to speed the evolution of resistance to those very antibiotics. And remember, antibiotics

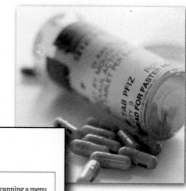

lasts only a couple of days to a week.
the flu may continue for several
tter while you are sick,

of fluids.
est.
fier—an electric device that puts
air.

tor if
ns get worse.
ns last a long time.
a little better, you develop signs of
s problem. Some of these signs are
ng, high fever, shaking, chills, and

ent of Health and Human Services.

Biology Matters

What's on Your Plate?

Humans eat many different kinds of species, but not all meals have the same impact on the planet's biodiversity. In particular, there are a number of fish and other marine species that humans consume that are being overharvested and are in rapid decline. Eating them only pushes them into steeper declines and could push them into extinction. In addition, some species, like salmon, are raised on aquatic farms. But while that might seem like the perfect solution to overfishing wild species, these farms sometimes can create high levels of ocean pollution, threatening other species.

So when we are shopping at the grocery store or scanning a menu at the restaurant, how can we decide what might be both delicious and harmless to eat? Various conservation organizations have put together lists of seafood that are best to eat and best to avoid, often printed on handy wallet-sized cards. Below are some of the recommendations from the wallet card of the marine conservation organization known as Blue Ocean Institute.

Just by choosing wisely, the next time you crave seafood, you can help preserve the planet's biodiversity.

SOURCE: The Blue Ocean Institute, "Guide to Ocean Friendly Seafood," September 2007.

ENJOY	BE CAREFUL	AVOID
Arctic Char	Crabs (Blue, Snow, and Tanner)	Chilean Sea Bass
Clams, Mussels, and Oysters (farmed)	Monkfish	Cod (Atlantic)
Mackerel	Rainbow Trout (farmed)	Halibut (Atlantic)
Mahi-Mahi (pole- and troll-caught)	Sea Scallops	Salmon (farmed)
Salmon (wild Alaskan)	Swordfish	Sharks
Striped Bass	Tuna: Albacore, Bigeye, Yellowfin, and Skipjack (canned or longline-caught)	Shrimp (imported)
Tilapia (U.S. farmed)		Tuna: Bluefin (Atlantic)

Highlighting the Process of Science and Biology in Everyday Life

"Science Toolkit" boxes give students insight into the methods, tools, and ideas that scientists employ in their research.

Science Toolkit

Tossing Coins and Crossing Plants: Probability and the Design of Experiments

The probability of an event is the chance that the event will occur. For example, there is a probability of 0.5 that a fair coin will turn up "heads" when it is tossed. A probability of 0.5 is the same thing as a 50 percent chance. As an illustration, consider a coin-tossing experiment. If you toss a penny only a few times, the observed percentage of heads may differ greatly from 50 percent. For example, if you tossed a coin only 10 times, you might get 70 percent (7) heads. However, if you tossed a coin 10,000 times, it would be very unusual to get 70 percent (7,000) heads.

Each toss of a coin is an independent event, in the sense that the outcome of one toss does not affect the outcome of the next toss. For a series of independent events, we can estimate the probability of each of the possible outcomes. Suppose we toss a coin 10,000 times and get 5,046 heads. From these results, it would be reasonable to estimate the chance of getting heads on the next toss as 50 percent, the percentage we expect from that coin. When only a small number of events are observed, our estimates of the underlying probabilities are less likely to be accurate, as when we toss a coin only a few times and get 7 of 10 tosses coming up heads.

How does all this relate to the design of a scientific experiment? Consider Mendel's work on peas. When Mendel crossed heterozygous plants (for example, *Pp* individuals) with each other, the offspring always had a phenotypic ratio close to 3:1. The reason for this consistent ratio is that the chances of getting offspring with *PP*, *Pp*, and *pp* genotypes are 25, 50, and 25 percent, respectively (see Figure 10.6). Because the 25 percent *PP* individuals and the 50 percent *Pp* individuals have the same phenotype, 75 percent of the ... 3:1 phenotypic

... e percentages of ... d have observed. ... and all egg cells ... tion. When large ... s assumption is ... ilures of the dif- ... one another. But ... f offspring in his ... red greatly from ... ght not have dis- ... w of segregation

What was true for Mendel's experiments is true for scientific experiments in general: chance events not under the control of the scientist can affect the outcome of an experiment, so large amounts of data are needed to compensate for the effect of chance events. This is true of any genetic cross in which the researcher has no control over which sperm and which egg fuse to produce an offspring, but it is also true in many other settings.

For example, a scientist wanting to know if pesticides were producing deformities in frogs would need to collect a lot of data to isolate the effects of chance events. One way scientists have investigated the impact of agricultural pesticides on frogs is to record the frequency of frogs with extra or missing legs in farm ponds that receive pesticides from nearby fields and compare that with data from ponds without pesticides. The researchers would have to collect data from as many ponds as possible to minimize the chance that some other factor, such as another pollutant or another organism found only in some ponds, affects the results one way or another. In one such study, although pesticides were not the direct cause of the deformities (a parasite was), the addition of pesticides weakened the frogs' immune systems, making it more likely that individuals would succumb to the parasite and develop deformities. (For an expanded discussion on deformities in frog populations, see Section 10.6.)

A Deformed Frog

Biology on the Job

Investing in Biotech

Biotechnology or "biotech" refers to the use of biology to make many different types of products, including agricultural products, DNA fingerprinting kits, and medicinal drugs. As techniques for working with DNA and proteins improved in the 1960s and 1970s, new companies using these innovations sprang up around the world. Here we interview Russell T. Ray, a specialist in biotechnology companies and a managing partner for HLM Venture Partners, a venture capital firm.

What do you do during a typical day at work? I arrive at the office around 7:30 AM and spend the next several hours responding to e-mail messages from coworkers and from companies we might invest in. I often meet with the managers of companies seeking to raise investment capital. I spend a few hours each day reviewing the scientific literature and reading health care newsletters, many of which are published daily. I also call colleagues at other venture capital firms to compare notes on what they are seeing in the way of interesting investments, and to share information on projects that we might work on together.

How did you become interested in investing in biotechnology companies? Almost by chance—in my first year in Merrill Lynch's investment banking department, I was selected to work on a financing project for a plant biotechnology company. Prior to that, I had completed an MBA in finance, as well as a master of science in biology that focused on the territorial behavior in a species of Costa Rican hummingbird. My scientific training helped me to understand the newly emerging field of biotechnology. I have now been working in this area for 23 years and I still love it!

What do you look for in a successful business plan or idea? In evaluating a biotechnology or medical device company, we focus on several areas:

- *Intellectual property.* Does the company have patents that protect its technology and science from other companies? If not, this is usually a deal killer.
- *Stage of development.* For example, does a pharmaceutical company have drug candidates that have successfully passed through the early stages of development? If not, we are unlikely to invest due to the high risk of failure associated with products that have not progressed from animal testing to human clinical testing.

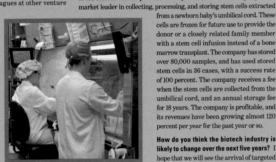

Technicians at a CBR System Cord blood storage facility in Tucson, Arizona.

- *Market analysis.* How big is the potential market? What are the competing products?
- *Quality of management.* Has management successfully done the job before? Who are the people behind the science?

Can you describe an example of an interesting company or a novel idea in the biotech industry? Take Cbr Systems, Inc. This company is the market leader in collecting, processing, and storing stem cells extracted from a newborn baby's umbilical cord. These cells are frozen for future use to provide the donor or a closely related family member with a stem cell infusion instead of a bone marrow transplant. The company has stored over 80,000 samples, and has used stored stem cells in 36 cases, with a success rate of 100 percent. The company receives a fee when the stem cells are collected from the umbilical cord, and an annual storage fee for 18 years. The company is profitable, and its revenues have been growing almost 120 percent per year for the past year or so.

How do you think the biotech industry is likely to change over the next five years? I hope that we will see the arrival of targeted medicines that match the patient's unique genotype.

What do you enjoy most about your job? I really enjoy working with bright entrepreneurs whose companies are on the cutting edge of science and medical technology. Each project I work on exposes me to new areas of science, so I'm constantly challenged to expand my horizons by learning more—something I love to do.

The Job in Context

Russell Ray and others like him straddle two worlds—they must understand both science and business. They invest in companies that take the information presented in this unit and apply it to solve real-world problems.

276 Unit 3 Genetics

"Biology on the Job" boxes examine how biological literacy can help in "nonbiological" careers such as crime-scene investigation and biotech investing.

Promoting Long-Term Retention of Key Concepts and a Deep Understanding of Terminology

Concept Check

1. What are allele frequencies, and what roles do they play in micro-evolution?

2. In addition to mutation, what are other sources of genetic variation in populations?

Concept Check Answers

1. Allele frequencies are the proportion or percentage of an allele (for example, A or a) in a population. Changes in allele frequencies over time constitute microevolution.
2. Recombination, in the form of crossing-over, independent assortment of chromosomes, and fertilization.

In-text pronunciation guides give students the confidence to incorporate new biological terms into their vocabulary.

ments, making combating the virus difficult because it is a "moving target."

In another example, genetic evidence indicates that the resistance of the mosquito ***Culex pipiens*** [*KYOO*-leks *PIP*-yenz] to organophosphate pesticides was caused by a single mutation that occurred in the 1960s. Since that time, mosquitoes blown by storms or moved accidentally by people have carried the initially rare mutant

Helpful to know

The word "antibiotic" has roots in Greek words meaning "against life." Because antibiotics work only on bacteria—and not against a wide range of disease agents, as many people think—we might do better to refer to them by the more accurate term "antibacterials," to avoid confusing them with substances used to kill fungi ("antifungals") and viruses ("antivirals").

Recognizing that mastering scientific terminology is a key part of true scientific literacy, *Discover Biology* includes **Helpful to Know** marginal annotations. These encourage deep understanding of vocabulary that will serve students through their next exam and beyond.

Empowering Students to Be Scientifically Literate Consumers of the News

As a nonmajors textbook, *Discover Biology* shows students how to actively apply core biological concepts to our rapidly changing world. Actual news stories in the textbook, on an instructor's DVD, and on the student website show students how their coursework helps them to make sense of current events.

Biology in the News

Antibiotic Runoff

One of the persistent problems of industrial agriculture is the inappropriate use of antibiotics. It's one thing to give antibiotics to individual animals, case by case, the way we treat humans. But it's a common practice in the confinement hog industry to give antibiotics to the whole herd, to enhance growth and to fight off the risk of disease, which is increased by keeping so many animals in such close quarters. This is an ideal way to create organisms resistant to the drugs. That poses a risk to us all.

A recent study by the University of Illinois makes the risk even more apparent. Studying the groundwater around two confinement hog farms, scientists had identified the presence of several transferable genes that confer antibiotic resistance, specifically to tetracycline. There is the very real chance that in such a rich bacterial soup these genes might move from organism to organism, carrying the ability to resist tetracycline with them. And because the resistant genes were found in groundwater, they are already at large in the environment.

There are two interdependent solutions to this problem, and hog producers should embrace them both. The first solution—the least likely to be acceptable in the hog industry—is to ban wholesale, herdwide use of antibiotics. The second solution is to continue to tighten the regulations and the monitoring of manure containment systems. The trouble is that there is no such thing as perfect containment.

The consumer, of course, has the choice to buy pork that doesn't come from factory farms. The justification for that kind of farming has always been efficiency, and yet, as so often happens in agriculture, the argument breaks down once you look at all the side effects. The trouble with factory farms is that they are raising more than pigs. They are raising drug-resistant bugs as well.

Although we tend to think of antibiotics as human medications, according to the Union of Concerned Scientists 70 percent of all antibiotics and related drugs used in the United States are given to livestock. An estimated 24 million pounds of antibiotics are used on livestock—pigs, cows, and chickens—each year.

In every exposure of bacteria to antibiotics—whether in humans or livestock—natural selection can weed out nonresistant bacteria and give an advantage to resistant bacteria. That is, every time antibiotics are used, there is the potential for antibiotic-resistant alleles to increase in frequency. Farmers use antibiotics to improve efficiency and cut costs because then they can raise more animals in closer quarters—conditions that, without widespread antibiotic use, would promote disease. In fact, however, this approach may be costing us all quite a lot in actual dollars. Treating antibiotic-resistant infections already costs billions of dollars each year in the United States alone.

One potentially huge cost described in this editorial could be incurred if resistance genes move from one kind of bacterium to another (in horizontal gene transfer). During such gene transfer, genes move not from one individual to its offspring, but from one individual to another individual of a different species. Such transfer is very common among bacteria. The real risk is movement of the resistance genes to a dangerous disease species—making the disease that much more difficult to treat. The groundwater around hog farms that contains such "transferable genes" could carry the organisms and their resistance genes anywhere.

Evaluating the News

1. How can raising healthier pigs be unhealthful for humans? Explain how herdwide applications of antibiotics could lead to increased antibiotic resistance in other kinds of bacteria, including bacteria that could cause disease in humans, via horizontal gene transfer.

2. Next time you're in a grocery store, check whether the packages of meat are from animals raised with antibiotics or without. (*Hint:* If they were raised without the use of antibiotics, the package will say so; otherwise, it will say nothing about antibiotics.) Compare the prices of meats raised with and without antibiotics; often those marked antibiotic-free are more expensive. Is it ethical for factory farms to use large-scale antibiotic treatment? On the other hand, would it be fair to farmers and consumers to outlaw such antibiotic use, if it meant that meats became more expensive?

3. A recent study found that a particular drug-resistant bacterium—methicillin-resistant *Staphylococcus aureus*, or MRSA—kills more Americans each year than even the AIDS virus. Common antibiotics, like methicillin and amoxicillin, are no longer effective against MRSA. Doctors are now using uncommon antibiotics to treat MRSA. How is natural selection likely to affect the effectiveness of those more uncommon antibiotics over time? If each use of an antibiotic helps promote resistance, by creating natural selection for resistant bacteria, what could doctors do to try to slow the evolution of additional resistance in MRSA?

SOURCE: *New York Times* editorial, September 18, 2007.

"Biology in the News" features are at the end of every chapter. An author-guided tour of a recent, actual news article demonstrates how mainstream news stories relate to the chapter's topics and how a mastery of the chapter's topics allows for a deeper understanding of the issues.

New to the Fourth Edition: The first **Evaluating the News** question in every chapter tests students' objective understanding of chapter material.

The second and third questions challenge students to use their biological knowledge by thinking about the ethical, societal, and political questions raised by the news stories.

New to the Fourth Edition:

The **INSTRUCTOR'S TEST BANK** now includes questions on all "Biology in the News" features so that instructors can more easily assign them as required reading.

Discover Biology in the News

Chapter 5: Cell Structure and Compartments

What's News?

T T ☒ 🖳

Food-Borne Illness-Causing Bacteria May Find Protection in Protozoa
Applied and Environmental Microbiology

Neighbors can sometimes be a bad influence on each other — especially when they are one-celled microscopic animals called protozoa and certain bacteria that are commonly linked to food-borne illnesses, like Salmonella.

» Read the full article

Active Reading Questions

1. This article reports research indicating that some protozoa consume bacteria and concentrate them into vesicles that are then released and may enhance the bacteria's ability to cause illness if they are consumed on produce. Both the bacteria and the protozoa are single-celled organisms, but they are very different cell types. Which of the following explains why bacteria and protozoa are so different?

   ○ a) Bacteria and protozoa are actually similar in many ways; both are prokaryotes

   ○ b) Bacteria are prokaryotes that do not contain membrane-bound structures; protozoa are eukaryotes and have membrane-bound structures like vesicles.

   ○ c) Bacteria are plant cells and protozoa are animal cells.

   ○ d) Bacteria can cause disease, and protozoa cannot

"What's News?" exercises on StudySpace (wwnorton.com/studyspace) and additional current news articles give students more opportunities to see how biology affects their everyday lives.

Discover Biology in the News **Instructor's Video DVD** presents 60 real, short news clips ideally suited for use in lecture. The clips are carefully selected for relevance to the key concepts presented in the text.

FOURTH EDITION
Discover Biology in the News
INSTRUCTOR DVD

Empowering Students to Understand Vital Issues Such as Cancer and Sustainability

INTERLUDE E Applying What We Learned

Building a Sustainable Society

Main Message: Aware that the current human impact on the global environment cannot be sustained, many individuals, corporations, and governments are taking innovative actions to help build a sustainable society.

The State of the World

Each year, representatives of nations and corporations give speeches and produce reports that summarize what they've done in the past year and where they are headed for the upcoming year. If such an update could be provided for Earth, it would tell us how the planet's air, water, soil, and living organisms changed in the previous year. No one makes anything close to a complete version of such a report, and indeed, no one could; we do not even know how many species there are on Earth, let alone the current status of each of those species and the environments in which they live.

Although we cannot give a complete "State of the World" report, we do know how some pieces of the planetary puzzle are changing over time. Some of that news is good. As described in Unit 5, populations of some endangered species (such as the bald eagle) are increasing in size, sulfur emissions that cause acid rain have decreased by about 40 percent in the United States after peaking to their highest levels in 1974, and the ozone layer is showing early signs of recovery. Other news is bad. Nitrogen pollution continues to have negative effects on ecosystems worldwide, populations of many species are in serious decline, and global CO_2 levels continue to rise rapidly.

In addition to being able to provide particular bits of good and bad news, we know enough about Earth to make an overall assessment of our effect on the planet (Figure E.1). Unfortunately, that assessment indicates that the current human impact on the biosphere is not sustainable. As we'll see, people are using and damaging many of Earth's resources more rapidly than they can be renewed.

Although scientific evidence indicates that our current impact is not sustainable, there are many hopeful signs for the future. Five aspects of human society—education, individual action, research, government, and business—have already begun to contribute to the formation of a sustainable society. In this essay we first describe some of the evidence suggesting that the current human impact on the biosphere is not sustainable. With that material as background, we turn to our main focus: sources of hope for the future, and case studies that provide clues as to how to build a sustainable society.

Figure E.1 Assessing the State of the World
New tools enable us to monitor Earth's vital signs in unprecedented detail, as in this computer image made using four different types of satellite data. Fires over land are shown in red. The large plume that extends from Africa over the Atlantic (and ranges in color from red to orange to yellow to green) was caused by the burning of vegetation and by windblown dust.

The Current Human Impact Is Not Sustainable

Many different lines of evidence suggest that the current human impact on the biosphere is not **sustainable**. What does this mean? An action or process is sustainable if it can be continued indefinitely without resources being used up or serious damage being caused to the environment. To begin with a simple example, modern societies depend on fossil fuels such as oil and natural gas to power our vehicles, heat our homes, and generate electricity. Although fossil fuels provide abundant energy now, our use of these fuels is not sustainable: they are not renewable, and hence supplies will run out, perhaps sooner rather than later (Figure E.2). Already, the volume of new sources of oil

discovered worldwide has dropped steadily from over 200 billion barrels during the period from 1960 to 1965, to less than 30 billion barrels during 1995 to 2000. In 2007, the world used about 31 billion barrels of oil, but only 5 billion barrels of new oil were discovered in that year.

Actions that cause serious damage to the environment are also considered unsustainable, in part because our economies depend on clean air, clean water, and healthy soils. But what constitutes "serious" environmental damage? One way to tell if an action causes serious damage is to see whether it disrupts important features of an ecosystem. As we have seen, many human actions have such effects. Human inputs to the global nitrogen and sulfur cycles, for example, now exceed all natural inputs combined (see Chapters 24 and 25). Such changes to the world's nutrient cycles

Figure E.2 Running Out of Oil
Many experts predict that the annual global production of oil will peak, then decline, sometime before 2020.

INTERLUDE B Applying What We Learned

Cancer: Cell Division Out of Control

Main Message: Cancer refers to the unrestrained growth and movement of abnormal cells within the body of a multicellular organism.

The Battle against Rogue Cells

Cancer represents a breakdown in the cooperative functioning of the cells in a multicellular organism. Every cancer begins with a single rogue cell that starts dividing with wild abandon, giving rise to a mass of abnormal cells. With the passage of time, the descendents of these abnormal cells adopt yet more malicious behaviors. They change shape, increase in size, quit their normal job, and sprawl over their neighbors (see Figure B.1). At their very worst, these highly abnormal cells break loose from their neighborhood, invade the bloodstream, and disperse to set up satellite colonies in other parts of the body. Wherever they establish themselves, they multiply furiously, overrunning their neighbors and monopolizing oxygen and nutrients to the extent that they starve the normal cells in the vicinity. The ability to spread through the body identifies these rogue cells as cancer cells, also known as malignant cells. Without restraints on their growth and migration, cancer cells take over, steadily destroying tissues, organs, and organ systems, until the body can no longer survive.

Cancer accounts for more than 500,000 deaths in the United States each year. While the past decade has seen improvements in treatment and prevention, more than a million Americans are diagnosed with some form of cancer every year. The National Cancer Institute estimates that the collective price tag for the various forms of cancer is more than $100 billion per year.

About 30 years ago, President Richard Nixon declared a war on cancer in the United States by making anticancer research a high priority. Since then, some major victories have been won, thanks to improvements in radiation and drug therapies. Whereas in the early twentieth century very few individuals survived cancer, today roughly 40 percent of patients are alive 5 years after treatment is begun. Nevertheless, the war against cancer is far from over, and the need for powerful new treatments to stop or kill the malignant cells is as urgent as ever. The greatest challenge in battling cancer is the selective destruction of rogue cells, while sparing healthy cells in the process. The standard war plan today relies on high-energy radiation, high doses of chemical poisons (chemotherapy), or both in sequence, to

disable any and all rapidly dividing cells. The devastating side effects of radiation therapy and chemotherapy are terrible precisely because this all-out assault also kills many innocent bystanders, cells necessary for the normal functions of the human body.

The good news is that discoveries in basic cell biology, and the large investment in cancer research, has produced a variety of innovative strategies for destroying malignant cells selectively. One inventive method uses genetically engineered viruses that infect and destroy specific types of cancer cells, while leaving normal cells alone. Another line of attack seeks to disable proteins that give cancer cells their unusual immortality, the capacity to divide indefinitely as long as oxygen and nutrients are available. A particularly promising approach is to choke off the supply of oxygen and nutrients to a cancerous mass by blocking the growth of blood vessels in its immediate vicinity. These and several other novel anticancer strategies are now in clinical trials. The early results are encouraging, although a lot of groundwork will have to be done before any of them become a routine treatment for any type of cancer.

The biggest lesson learned from the past three decades of cancer research is a surprisingly simple one: instead of dealing with them only when they become monsters, we should try to reduce the odds that our cells will start to progress toward cancer in the first place. It is now abundantly clear that environmental factors, including lifestyle choices, have a very large impact on a person's risk of developing cancer. The link between cancer and tobacco use is the most dramatic illustration of how chemical exposure can transform healthy cells into dangerous ones. Lung

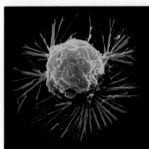

Figure B.1 A Home-Grown Monster
This color-enhanced photograph, captured with a scanning electron microscope, shows a breast cancer cell. A lab technician can recognize it as a cancer cell because of its large size, abnormally rounded shape, and altered cell surface. The spikes are membrane projections that help the cell creep about within the body. The ability to spread from one part of the body to another through the bloodstream is a hallmark of cancer cells.

B2

Each unit in *Discover Biology* culminates in an optional **Interlude chapter** on relevant topics. These Interludes give instructors the opportunity to introduce their students to an extended application of the preceding unit's topics.

Chapter Reviews That Test Understanding of the Key Concepts

The **Chapter Review** section covers the main factual and conceptual points of the chapter. This section is designed to help students review and assess their comprehension of the chapter material.

The **Summary** presents all of the major points introduced in the chapter, organized by chapter section.

Chapter Review

Summary

13.1 How Genes Work

- Genes code for RNA molecules. Protein-coding genes code for messenger RNA (mRNA), which contains instructions for building a protein.
- An RNA molecule consists of a single strand of nucleotides. Each nucleotide is composed of the sugar ribose, a phosphate group, and one of four nitrogen-containing bases. The bases found in RNA—adenine (A), cytosine (C), guanine (G), and uracil (U)—are the same as those in DNA except that uracil replaces thymine (T).
- At least three types of RNA (mRNA, rRNA, and tRNA) and many enzymes and other proteins participate in the manufacture of proteins.

13.2 How Genes Guide the Manufacture of Proteins

- In both prokaryotes and eukaryotes, protein synthesis requires two steps: transcription and translation.
- In transcription, an RNA molecule is made using the DNA sequence of the gene.
- In translation, ribosomes, mRNA, and tRNA molecules together direct protein synthesis. Ribosomes consist of rRNA and certain proteins.
- In eukaryotes, information for the synthesis of a protein is transmitted from the gene, located in the nucleus, to the site of protein synthesis, the ribosomes, located in the cytoplasm.

13.3 Gene Transcription: Information Flow from DNA to RNA

- During transcription, one strand of the gene's DNA serves as a template for synthesizing many copies of mRNA.
- The key enzyme in transcription is RNA polymerase.
- Each gene has a sequence (a promoter) at which RNA polymerase begins transcription. In prokaryotic genes, transcription stops at a special sequence called a terminator. Transcription termination in eukaryotes is more complicated.
- The mRNA molecule is constructed according to specific base-pairing rules: A, U, C, and G in mRNA pair with T, A, G, and C, respectively, in the template strand of DNA.
- In eukaryotes, most genes contain internal noncoding sequences of DNA (introns) that must be removed from the initial mRNA product while it is still in the nucleus. The remaining segments of mRNA correspond to exons, the DNA sequences of the gene that code for the protein.

13.4 The Genetic Code

- The information in a gene is encoded in its sequence of bases.
- The information encoded by an mRNA is read in sets of three bases; each three-base sequence is a codon. Of the 64 possible codons, most specify a particular amino acid, but certain codons signal the start or stop of translation. The information specified by each of the codons is collectively called the genetic code.

- When reading the genetic code, ribosomes begin at a fixed starting point on the mRNA (the start codon) and stop reading the code when any one of the three stop codons is encountered.
- The genetic code is unambiguous (each codon specifies no more than one amino acid), redundant (several codons specify the same amino acid), and nearly universal (used in almost all organisms on Earth).

13.5 Translation: Information Flow from mRNA to Protein

- In translation, the information in the sequence of bases in mRNA determines the sequence of amino acids in a protein.
- Translation occurs at ribosomes, which are composed of rRNA and more than 50 different proteins.
- Each transfer RNA (tRNA) molecule specializes in carrying a particular amino acid and has a three-base sequence, called the anticodon, that recognizes and pairs with a specific codon in the mRNA. A specific tRNA attaches to the ribosome–mRNA complex when its anticodon is able to bind a codon in the mRNA by complementary base pairing. In this way, a specific tRNA delivers a specific amino acid based on the codon message present in the mRNA.
- The ribosome makes the covalent bonds that link the amino acids together into a protein. It holds the mRNA and tRNA in an orientation that allows the amino acid carried by the tRNA to be covalently bonded to the growing polypeptide.

13.6 The Effect of Mutations on Protein Synthesis

- Many mutations are caused by the substitution, insertion, or deletion of a single base in a gene's DNA sequence.
- Insertion or deletion of a single base causes a genetic frameshift, resulting in a different sequence of amino acids in the gene's protein product.
- Mutations involving insertions or deletions of a series of bases are also common.
- Mutations can change the function of a gene's protein product, especially when a frameshift is involved or when the mutation affects a protein's binding site. Mutations causing frameshifts usually destroy the protein's function.
- Other mutations have neutral effects, and a few even have beneficial effects.

13.7 Putting It All Together: From Gene to Phenotype

- Some genes code for RNA only, but a large number encode proteins.
- Proteins are essential to life. In conjunction with the environment, they determine an organism's phenotype.
- Because genes guide the production of proteins, they play a key role in determining an organism's phenotype.

⊛ Review and Application of Key Concepts

1. What is a gene? In general terms, how does a gene store the information it contains?

2. Discuss the different products specified by genes. What are the function(s) of each of these products?

3. Describe the flow of genetic information from gene to phenotype.

4. How is the information contained in a gene transferred to another molecule? Why must the molecule that "carries" the information stored in the gene be transported out of the nucleus?

5. What is RNA splicing? Does it occur in both eukaryotes and prokaryotes? Explain your answer.

6. Describe the roles played by rRNA, tRNA, and mRNA in translation.

7. Scientists have discovered a mutation in a gene encoding a tRNA molecule that appears to be responsible for a series of human metabolic disorders. The mutation occurred at a base located immediately next to the anticodon of the tRNA, a change that destabilized the ability of the tRNA anticodon to bind to the correct mRNA codon. Why might a single-base mutation of this nature result in a series of metabolic disorders?

8. Write a paragraph explaining to someone with little background in biology what a mutation is and why mutations can affect protein function.

Key Terms

anticodon (p. 275)
codon (p. 273)
deletion (p. 278)
exon (p. 272)
frameshift (p. 278)
gene (p. 268)
genetic code (p. 273)
insertion (p. 278)
intron (p. 272)
messenger RNA (mRNA) (p. 269)
point mutation (p. 277)
promoter (p. 271)
ribosomal RNA (rRNA) (p. 269)
RNA polymerase (p. 271)
RNA splicing (p. 272)
start codon (p. 273)
stop codon (p. 273)
substitution (p. 277)
template strand (p. 271)
terminator (p. 272)
transcription (p. 270)
transfer RNA (tRNA) (p. 269)
translation (p. 270)

Self-Quiz

1. For genes that specify a protein, what molecule carries information from the gene to the ribosome?
   a. DNA
   b. mRNA
   c. tRNA
   d. rRNA

2. During translation, each amino acid in the growing protein chain is specified by how many nitrogenous bases in mRNA?
   a. one
   b. two
   c. three
   d. four

3. Which molecule delivers the amino acid specified by a codon to the ribosome?
   a. rRNA
   b. tRNA
   c. anticodon
   d. DNA

4. During transcription, which molecule or molecules are produced?
   a. mRNA
   b. rRNA
   c. tRNA
   d. all of the above

5. A portion of the template strand of a gene has the base sequence CGGATAGGGTAT. What is the sequence of amino acids specified by this DNA sequence? (Use the information in Figure 13.5 and assume that the corresponding mRNA sequence will be read from left to right.)
   a. alanine–tyrosine–proline–isoleucine
   b. arginine–tyrosine–tryptophan–isoleucine
   c. arginine–isoleucine–glycine–tyrosine
   d. none of the above

6. Which of these is responsible for creating the covalent bonds that link the amino acids of a protein together in the order specified by the gene that encodes that protein?
   a. tRNA
   b. mRNA
   c. rRNA
   d. ribosome

7. A mutation occurs in which the fourth, fifth, and sixth bases are deleted from a gene that encodes a protein with 57 amino acids. Which of these would be expected to happen?
   a. The resulting frameshift will prevent protein synthesis.
   b. A protein with 56 amino acids will be constructed.
   c. A protein that differs from the original one—but still has 57 amino acids—will be constructed.
   d. A protein with 54 amino acids will be constructed.

8. Most eukaryotic genes contain one or more segments that are transcribed but not translated. Each such segment is known as
   a. a start codon.
   b. a promoter.
   c. an intron.
   d. an exon.

Review and Application of Key Concepts and **Self-Quiz** questions help students determine whether they have developed a deep understanding of the key concepts that were listed at the beginning of the chapter. Answers to all Self-Quiz questions and most Review and Application questions are in the back of the text.

The **Key Terms** list is a helpful checkpoint for reviewing the material. Page numbers are included so that students can easily return to the page where each term was introduced.

Instructor Supplements

Instructor Resource CD-ROMs

These CDs include the following:

- **Drawn art and photographs.** All drawn art, with and without labels, and most photographs in JPEG and PowerPoint formats.
- **Process animations and activities.** Developed specifically for *Discover Biology,* over 200 animations and activities bring key figures from the text to life and aid students' review of difficult concepts. The animations can be enlarged to full-screen view and include VCR-like controls that make it easy to control the pace of animation during lecture.
- **Editable PowerPoint lectures.** Lecture outlines for every chapter integrate art, notes, animations, and in-class quizzes ideally suited as "clicker" questions.
- **Test Bank file in WebCT, Blackboard, Word, and ExamView formats.**

Everything on the CDs are also available for download at **wwnorton.com/instructors.**

Test Bank

By Jennifer Schramm (Chemeketa Community College) and Stephen Lebsack (Linn-Benton Community College)

Thoroughly revised and expanded for the Fourth Edition, each chapter of the Test Bank consists of three question types classified according to a taxonomy of educational objectives:

1. **Factual** questions test students' basic understanding of facts and concepts.
2. **Applied** questions require students to apply knowledge in the solution of a problem.
3. **Conceptual** questions require students to engage in qualitative reasoning and to explain why things are as they are.

Questions are further classified by section and difficulty, making it easy to construct tests and quizzes that are meaningful and diagnostic. The question types are multiple-choice, completion, and true–false.

    New to this edition are figure interpretation questions and questions about all **"Biology in the News"** features.

    The Test Bank is available in print and ExamView formats.

Discover Biology in the News Instructor's DVD

This DVD features 60 real news clips for use in lectures. Carefully selected for currency and relevance, these news topics serve as an ideal complement to the text.

Coursepacks

Coursepacks include chapter overviews, PowerPoint lecture outlines, animations and flash cards, Test Bank questions, and a glossary. Available at **wwnorton.com/instructors**, coursepacks are offered in Blackboard CE 4 and 6 (formerly, WebCT CE 4 and 6), Blackboard Vista (formerly, WebCT Vista), and other Blackboard formats.

Norton Gradebook

Norton Gradebook is an online resource (**wwnorton.com/nrl/gradebook**) that enables instructors and students to store and track their online quiz results. The results from each quiz that students take at **discoverbiology.com** are uploaded to the password-protected Gradebook, where instructors can access and sort them by section, book, chapter, student name, and date. Students can access the Gradebook to review their personal results. Results can easily be downloaded to instructors' desktops and imported into course management systems. Registration for the Gradebook is instant, and no setup is required.

Biology Video Library

This extensive video library, available to qualified adopters of *Discover Biology,* Fourth Edition, includes content corresponding to each unit of the textbook.

Overhead Transparencies

A full-color transparency set includes all drawn art and tables from the textbook in a relabeled and resized format that is optimal for projection.

Student Supplements

Student Website

The free and open student website for the Fourth Edition of *Discover Biology* does not require a password! **Discoverbiology.com** offers a complete study plan that will improve students' conceptual understanding of the reading and help them develop biological literacy.

- **Animations and activities** include mini-lectures that offer step-by-step explanations of key concepts, drag-and-drop labeling exercises, and simulated experiments.

- **Self-test quizzes** provide students with an opportunity to assess their understanding. Quiz results are connected to Norton's online Gradebook, making it easy for instructors to collect scores and enabling students to track their own progress.

- **"What's News" exercises** are critical-thinking exercises that challenge students to apply the topics they're learning about in class to real news stories.

- **"Biology in the News"** weekly updates highlight relevant news stories and encourage students to apply what they learn in class to their daily lives.

- **Key-figure quizzes** test students' understanding of key figures from the book.

- **Quiz+** offers more than a list of correct answers; it generates a custom study plan and assembles a resource of online materials that guide students to the correct answers.

- **Chapter overviews** summarize key concepts and prompt students to explore animations and activities that will enhance their understanding.

- A **glossary** includes audio pronunciations for over 100 difficult-to-pronounce words.

- **Vocabulary flash cards** make it easy for students to learn and remember new terms.

Study Guide

By David Demers (Sacred Heart University), Edward Dzialowski (University of North Texas) and Stephen Lebsack (Linn-Benton Community College)

The Study Guide offers a study plan and practice questions for each chapter.

Art Notebook

The Art Notebook is an invaluable tool for taking effective notes during lecture. All of the drawn art from the textbook is reproduced in color, with ample space for taking notes. This resource allows the student to focus on the lecture and not worry about trying to copy drawings from the blackboard or projector.

Lab Manual

By Tara Scully, The George Washington University

Written specifically to accompany the textbook, *Discovering Biology in the Lab* targets the nonscience major by emphasizing core concepts, critical thinking, and relevance.

- **Applications** include antibiotic resistance (in one of the evolution labs), birth control (in the mammalian reproduction lab), nutrition (in the macromolecules lab), and global change (in one of the ecology labs).

- **Concept Checks** after each activity teach students to connect what they're doing in the lab to key biological concepts. Use of many of the figures from *Discover Biology* further connect the lab manual to the textbook.

Ebook

Ebook chapters are available for as little as $1 per chapter. See **nortonebooks.com** for details.

Acknowledgments

We would like to thank the many people who provided critical commentary on our revisions. Their insights helped us make *Discover Biology* a better book.

Reviewers of the Fourth Edition

Laura Ambrose, University of Regina
Angelika M. Antoni, Kutztown University of Pennsylvania
Idelisa Ayala, Broward College
Neil R. Baker, Ohio State University
Christine Barrow, Prince George's Community College
Janice M. Bonner, College of Notre Dame of Maryland
Randy Brewton, University of Tennessee/Knoxville
Peggy Brickman, University of Georgia
Art Buikema, Virginia Tech University
Wilbert Butler, Jr., Tallahassee Community College
Kelly Cartwright, College of Lake County
Francie Cuffney, Meredith College
Kathleen Curran, Wesley College
Garry Davies, University of Alaska/Anchorage
Kathleen DeCicco-Skinner, American University
Lisa J. Delissio, Salem State College
Christian d'Orgeix, Virginia State University
Jean Engohang-Ndong, Brigham Young University/Hawaii
Richard Farrar, Idaho State University
Tracy M. Felton, Union County College
Ryan Fisher, Salem Sate College
Susan Fisher, Ohio State University
Kathy Gallucci, Elon University
Gail Gasparich, Towson University
Beverly Glover, Western Oklahoma State College
Tamar Goulet, University of Mississippi
Nancy Holcroft-Benson, Johnson County Community College
Thomas Horvath, SUNY College at Oneonta
James L. Hulbert, Rollins College
Karel Jacobs, Chicago State University
Robert M. Jonas, Texas Lutheran University
Arnold Karpoff, University of Louisville
Will Kopachik, Michigan State University
Olga Kopp, Utah Valley University
Erica Kosal, North Carolina Wesleyan College
Shawn Lester, Montgomery College
Lee Likins, University of Missouri/Kansas City
Melanie Loo, California State University/Sacramento
David Loring, Johnson County Community College
Lisa Maranto, Prince George's Community College
Quintece Miel McCrary, University of Maryland/Eastern Shore
Dorian McMillan, College of Charleston
Alexie McNerthney, Portland Community College

Susan Meacham, University of Nevada, Las Vegas
Steven T. Mezik, Herkimer County Community College
Jonas Okeagu, Fayetteville State University
Alexander E. Olvido, Longwood University
Murali T. Panen, Luzerne County Community College
Brian Perkins, Texas A&M University
Jim Price, Utah Valley University
Barbara Rundell, College of DuPage
Jennifer Schramm, Chemeketa Community College
John Richard Schrock, Emporia State University
Tara A. Scully, The George Washington University
Marieken Shaner, University of New Mexico
William Shear, Hampden-Sydney College
Jennie Skillen, College of Southern Nevada
Julie Smit, University of Windsor
James Smith, Montgomery College
Ruth Sporer, Rutgers
Jim Stegge, Rochester Community and Technical College
Richard Stevens, Monroe Community College
Josephine Taylor, Stephen F. Austin State University
Doug Ure, Chemeketa Community College
Rani Vajravelu, University of Central Florida
Cindy White, University of Northern Colorado
Daniel Williams, Winston-Salem State University
Elizabeth Willott, University of Arizona
Silvia Wozniak, Winthrop University
Carolyn A. Zanta, Clarkson University

Reviewers of Past Editions

Michael Abruzzo, California State University/Chico
James Agee, University of Washington
Marjay Anderson, Howard University
Caryn Babaian, Bucks County College
Sarah Barlow, Middle Tennessee State University
Gregory Beaulieu, University of Victoria
Craig Benkman, New Mexico State University
Elizabeth Bennett, Georgia College and State University
Stewart Berlocher, University of Illinois/Urbana
Robert Bernatzky, University of Massachusetts/Amherst
Nancy Berner, University of the South
Juan Bouzat, University of Illinois/Urbana
Bryan Brendley, Gannon University
Sarah Bruce, Towson University
Neil Buckley, State University of New York/Plattsburgh
John Burk, Smith College

Kathleen Burt-Utley, University of New Orleans
David Byres, Florida Community College/Jacksonville—South Campus
Naomi Cappuccino, Carleton University
Heather Vance Chalcraft, East Carolina University
Van Christman, Ricks College
Jerry Cook, Sam Houston State University
Judith D'Aleo, Plymouth State University
Vern Damsteegt, Montgomery College
Paul da Silva, College of Marin
Sandra Davis, University of Louisiana/Monroe
Véronique Delesalle, Gettysburg College
Pablo Delis, Hillsborough Community College
Alan de Queiroz, University of Colorado
Jean de Saix, University of North Carolina/Chapel Hill
Joseph Dickinson, University of Utah
Gregg Dieringer, Northwest Missouri State University
Deborah Donovan, Western Washington University
Harold Dowse, University of Maine
John Edwards, University of Washington
Jonathon Evans, University of the South
William Ezell, University of North Carolina/Pemberton
Deborah Fahey, Wheaton College
Marion Fass, Beloit College
Richard Finnell, Texas A&M University
April Fong, Portland Community College
Wendy Garrison, University of Mississippi
Aiah A. Gbakima, Morgan State University
Dennis Gemmell, Kingsborough Community College
Alexandros Georgakilas, East Carolina University
Kajal Ghoshroy, Museum of Natural History/Las Cruces
Jack Goldberg, University of California/Davis
Andrew Goliszek, North Carolina Agricultural and Technological State University
Glenn Gorelick, Citrus College
Bill Grant, North Carolina State University
Harry W. Greene, Cornell University
Laura Haas, New Mexico State University
Barbara Hager, Cazenovia College
Blanche Haning, University of North Carolina/Chapel Hill
Robert Harms, St. Louis Community College/Meramec
Chris Haynes, Shelton State Community College
Thomas Hemmerly, Middle Tennessee State University
Tom Horvath, SUNY College at Oneonta
Daniel J. Howard, New Mexico State University
Laura F. Huenneke, New Mexico State University
Paul Kasello, Virginia State University

Laura Katz, Smith College

Andrew Keth, Clarion University of Pennsylvania

Tasneem Khaleel, Montana State University

John Knesel, University of Louisiana/Monroe

Hans Landel, North Seattle Community College

Allen Landwer, Hardin-Simmons University

Katherine C. Larson, University of Central Arkansas

Harvey Liftin, Broward County Community College

Craig Longtine, North Hennepin Community College

Kenneth Lopez, New Mexico State University

Ann S. Lumsden, Florida State University

Blasé Maffia, University of Miami

Patricia Mancini, Bridgewater State College

Roy Mason, Mount San Jacinto College

Joyce Maxwell, California State University/Northridge

Phillip McClean, North Dakota State University

Amy McCune, Cornell University

Bruce McKee, University of Tennessee

Bob McMaster, Holyoke Community College

Gretchen Meyer, Williams College

Brook Milligan, New Mexico State University

Ali Mohamed, Virginia State University

Daniela Monk, Washington State University

Brenda Moore, Truman State University

Ruth S. Moseley, S.D. Bishop Community College

Jon Nickles, University of Alaska/Anchorage

Benjamin Normark, University of Massachusetts/Amherst

Douglas Oba, University of Wisconsin/Marshfield

Mary O'Connell, New Mexico State University

Marcy Osgood, University of Michigan

Donald Padgett, Bridgewater State College

Penelope Padgett, University of North Carolina/Chapel Hill

Kevin Padian, University of California/Berkeley

Brian Palestis, Wagner College

John Palka, University of Washington

Anthony Palombella, Longwood College

Snehlata Pandey, Hampton University

Robert Patterson, North Carolina State University

Nancy Pelaez, California State University/Fullerton

Pat Pendarvis, Southeastern Louisiana University

Patrick Pfaffle, Carthage College

Massimo Pigliucci, University of Tennessee

Jeffrey Podos, University of Massachusetts/Amherst

Robert Pozos, San Diego State University

Ralph Preszler, New Mexico State University

Jerry Purcell, Alamo Community College

Richard Ring, University of Victoria

Ron Ruppert, Cuesta College

Lynette Rushton, South Puget Sound Community College

Shamili Sandiford, College of DuPage

Barbara Schaal, Washington University

Kurt Schwenk, University of Connecticut

Harlan Scott, Howard Payne University

Erik Scully, Towson University

David Secord, University of Washington

Cara Shillington, Eastern Michigan University

Barbara Shipes, Hampton University

Mark Shotwell, Slippery Rock University

Shaukat Siddiqi, Virginia State University

Donald Slish, State University of New York/Plattsburgh

Philip Snider, University of Houston

Julie Snyder, Hudson High School

Ruth Sporer, Rutgers University/Camden

Neal Stewart, University of North Carolina/Greensboro

Tim Stewart, Longwood College

Bethany Stone, University of Missouri

Nancy Stotz, New Mexico State University

Steven Strain, Slippery Rock University

Allan Strand, College of Charleston

Marshall Sundberg, Emporia State University

Alana Synhoff, Florida Community College

Joyce Tamashiro, University of Puget Sound

Steve Tanner, University of Missouri

John Trimble, Saint Francis College

Mary Tyler, University of Maine

Roy Van Driesche, University of Massachusetts/Amherst

Cheryl Vaughan, Harvard University

John Vaughan, St. Petersburg College

William Velhagen, Longwood College

Mary Vetter, Luther College

Alain Viel, Harvard Medical School

Carol Wake, South Dakota State University

Jerry Waldvogel, Clemson University

Elsbeth Walker, University of Massachusetts/Amherst

Daniel Wang, University of Miami

Stephen Warburton, New Mexico State University

Carol Weaver, Union University

Paul Webb, University of Michigan

Peter Wilkin, Purdue University North Central

Peter Wimberger, University of Puget Sound

Allan Wolfe, Lebanon Valley College

David Woodruff, University of California/San Diego

Louise Wootton, Georgian Court University

Robin Wright, University of Washington

Thanks to the *Discover Biology* Team

Putting together an introductory biology textbook is no small task, and the work to complete this Fourth Edition of *Discover Biology* was no exception. The authors would like to thank the many editors, researchers, and assistants at W. W. Norton who helped shepherd this book through the significant revisions in text, photos, and artwork that you see here. In particular, we'd like to thank Mike Wright for his leadership and vision, and his insights about biology textbooks and textbook publishing in general. Mike, who oversaw the revision, production, and publication of the Fourth Edition, has been thoroughly engaged at every step of the project, from the gathering of the first reviews to the choosing of the last pieces of art. Thanks to our eagle-eyed copy editors, Stephanie Hiebert and Philippa Solomon, both of whom are superbly meticulous, perceptive, and skillful wordsmiths. Our developmental editor, Dick Morel, greatly improved the readability and accessibility of this book. Thanks also to Kim Yi for seamlessly coordinating the movement and synthesis of the innumerable parts of this book. Her patience and firmness, as well as unflappability in the face of tight production deadlines, are much appreciated. Our thanks also to Chris Granville for skillfully overseeing the final assembly into a tangible, beautiful book. With Betsy Twitchell's tireless advocacy of this book in the marketplace, we're confident that it will reach as wide an audience as possible. Matthew Freeman progressed from assisting us in all aspects of this project to managing the improvement of the book's print ancillaries, most notably the test bank. Jennifer Cantelmi deserves thanks for making sure that the many parallel tracks of reviewing, revising, and correcting eventually converged at the right time and place. Finally, we would like to thank our families for support during the long process that is a textbook revision, including Merrill Peterson, Emiko Peterson-Yoon, Erik Peterson-Yoon, and June Ginoza Yoon; and Don, Ryan, and Erika Singh-Cundy.

Contents

Unit 1 The Diversity of Life

Unit 2 Cells: The Basic Units of Life

CORE TOPICS

FOURTH
EDITION

Discover Biology

CHAPTER 1

The Nature of Science and the Characteristics of Life

Key Concepts

◉ To investigate the natural world, scientists use the scientific method: making an observation, forming a hypothesis to explain that observation, generating predictions from the hypothesis, and testing those predictions. When an experiment upholds the predictions, a hypothesis gains strength and support. When the predictions are not upheld, a hypothesis may be discarded or modified. A scientist at this point may also choose to go back and reexamine or reconsider the initial observations, predictions, and tests.

◉ All living organisms are thought to have descended from a single common ancestor. Therefore they share certain characteristics: they are composed of cells, reproduce using DNA, grow and develop, actively take in energy from their environment, sense and respond to their environment, maintain constant internal conditions, and evolve.

◉ Viruses challenge descriptions of the boundary between living and nonliving because they exhibit some characteristics of living organisms but lack others. In this book we will follow the lead of most biologists and consider viruses nonliving.

◉ Living organisms are just one part of the biological hierarchy, which ranges in scale from molecules at the lowest level to cells, tissues, organs, organ systems, individuals (living organisms), populations, communities, ecosystems, biomes, and finally the biosphere.

◉ Energy maintains a characteristic flow through biological systems. In most ecosystems, energy goes from the sun to producers (such as plants) to consumers and decomposers (such as animals and fungi that consume plants and other organisms). Food webs depict the complex relationships between organisms that eat and those that are eaten.

They're Alive! Or Are They?

Minuscule and mysterious, they had remained hidden since time immemorial, buried deep within Earth, 3 miles beneath the ocean floor off the coast of Western Australia. Then, several years ago, a team of scientists spotted the tiny oddities while using ultra-high-powered microscopes to study ancient rocks retrieved from an oil-drilling site. Researchers dubbed the miniature things "nanobes" because they are so small that they measure only billionths of a meter (nanometers), with a million or so able to fit comfortably in the dot over this "i." The discovering scientists proclaimed these tiny creatures to be the world's smallest living organisms, thereby causing a controversy.

Nanobes look much like other, larger living organisms, with shapes similar to small molds, and they grow when brought into the laboratory. In fact, they grow so quickly that within weeks they can go from being visible only with the world's most powerful microscopes to forming fast-expanding threadlike mats that are easily visible to the naked eye.

Some have hailed this work as a fundamental and important discovery, even suggesting that nanobes are likely to be widespread—possibly the most abundant form of life in and on Earth. These scientists suggest that nanobes could be critical in the decomposition of Earth's rocks and the formation of soils—fundamental chemical processes that shape the very Earth we stand on. Even more recently, scientists have reported finding these minuscule forms in eggshells, blood, and decaying leaves.

Although nanobes are indisputably tiny, scientists cannot seem to agree on whether they are the world's smallest organisms. That is because there still is no agreement on whether nanobes are actually alive.

Isn't it a simple matter to tell living from nonliving? Any schoolchild can distinguish between an inanimate stone and the living being who skips it across a pond. **Are nanobes living things or not? Why can't scientists agree on whether something is alive? What, in fact, is life?** In this chapter we examine these questions by considering what the characteristics of life are and how biologists study life.

Nanobes: A Tempest in a Teapot
Are these nanobes the smallest living organisms ever discovered? Or are they not alive at all?

What is life? This deceptively simple question is, in many ways, one of the most profound. It underlies medical controversies ranging from abortion and determining when life begins, to the right to die and when life ends. The same question reaches deep into the sciences as researchers seek to understand when life on our own planet first took hold and whether life has ever existed on other planets. One of the sciences asking these questions is **biology**, which is the scientific study of life, and the subject of this book. The goal of biology is to improve our understanding of living organisms, from microscopic bacteria to giant redwood trees to humans.

In this chapter we begin with an exploration of science and how scientists ask and seek answers to questions about living organisms. Then we address the question, "What is life?" We will see that all living things, diverse though they are, share characteristics that are common to all of them, and that all living organisms are part of a greater biological hierarchy of life. We will also see how organisms play different roles in biological systems, through which energy flows. We close by returning to the question asked at the outset: whether nanobes are living or not.

Figure 1.1 You Can Hear Me Now
As remote as science can sometimes seem, it constantly affects how we live our lives. The scientific process provides an ever-deepening knowledge of the physical world—knowledge that results in new ideas, new technologies, and new products, some of which we quickly grow to adore.

1.1 Asking Questions, Testing Answers: The Work of Science

Science is a method of inquiry, a way of discovering truths about the natural world. Because it is such a powerful way of understanding nature, science holds a central place in modern society. For scientists and nonscientists alike, knowing how nature works can be exciting and fulfilling. In addition, applications of scientific knowledge influence all aspects of modern life. Every time we take medicine, instant-message a friend, or run on a treadmill, we are enjoying the benefits of science.

Few of us, however, have a good picture of how science works, how it generates knowledge, and what its powers and limitations are. Meanwhile, science plays an increasingly important role in decisions made by society as a whole, as well as in personal decisions made by individuals. As a society, for example, we must evaluate the discoveries made by scientists when making decisions about global warming, the courtroom use of DNA fingerprinting, and even whether teachers in our public schools are able to provide their students with the most current scientific knowledge about evolution. As individuals, we must evaluate daily the reports of scientific studies that we see in the news. Should we avoid genetically engineered foods, or are they safe? Is drinking red wine really good for the heart? Is it okay to use a cell phone for hours at a time, or could it cause damage to the body? (Figure 1.1). To make good decisions on these and many other issues, all of us—not just scientists—benefit from understanding how the scientific process works.

Scientists use the scientific method

To study the natural world, scientists follow a series of logical steps known as the **scientific method** (Figure 1.2). Many people assume that because it is logical, the scientific method is a mechanical process carried out by adherence to a rigid set of rules but, as we shall see, luck and imagination play an important role as well.

The scientific method begins with **observations**. A scientist can make observations of the natural world in many different ways: by looking through a microscope, diving to the ocean floor, or walking through a meadow. The observation that started Dr. JoAnn M. Burkholder on her line of inquiry is one that many people made in North Carolina in the 1990s. Dr. Burkholder, a biologist at North Carolina State University, observed that huge numbers of fish were periodically being killed in myste-

Dr. JoAnn Burkholder

rious die-offs; their bodies, covered with bleeding sores, were found floating by the millions in the region's estuaries, where the rivers meet the sea (Figure 1.3). Though this observation is particularly dramatic, it shares a key feature with *all* observations that begin the scientific method: it strikes biologists as curious or interesting, making them want to know, why is this happening?

After observation, the next step in the scientific method is the creative process of generating a **hypothesis**, or an explanation that identifies natural causes for the observation. Some people describe a hypothesis as an educated guess. Generating a hypothesis to explain an observation is not always easy. For some

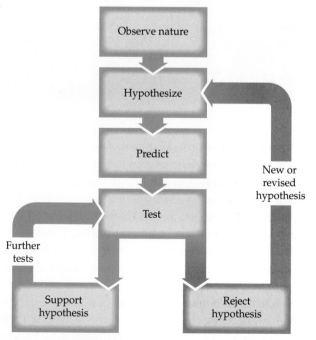

Figure 1.2 The Scientific Method

time, researchers were thoroughly stumped by the North Carolina fish kills, unable to come up with explanations as to what might be causing them. As it turned out, several years earlier, colleagues of Dr. Burkholder had been dismayed to find that their laboratory fish were dying suddenly after exposure to local river water. When Dr. Burkholder looked into their problem, she found that a kind of microscopic organism called a protist greatly increased in numbers in the laboratory aquariums just before the fish died and decreased in numbers unless live fish were added. (Chapter 3 will describe protists in more detail.) When the fish die-offs began happening in the wild, Dr. Burkholder came up with the hypothesis that the same tiny protist, known by the name *Pfiesteria* [fih-STEER-ee-ah], that appeared to have killed the laboratory fish was also causing the fish die-offs in local rivers.

Dr. Burkholder then used her hypothesis to generate **predictions** that she could test. She asked herself, "If my hypothesis is correct and *Pfiesteria* is causing fish die-offs in local rivers, what else can I expect to happen?" Her first prediction was that *Pfiesteria* would be found in abundance in the river water during times when fish die-offs were happening and would not be found there when fish die-offs were not happening. Her second prediction was that the same *Pfiesteria* would be capable of killing healthy fish if introduced into the aquariums housing fish in her laboratory.

The next step in the scientific process is testing the predictions of a hypothesis. One way that scientists test their predictions is by devising and conducting **experiments**—controlled, repeated manipulations of nature. Another way is by making further observations.

What, then, did Dr. Burkholder and her colleagues find when they began to test their predictions? First, they observed that *Pfiesteria* could indeed be found swarming in river regions where fish were dying, but not in those same regions when fish were not dying—upholding her first prediction. Then, in laboratory experiments, the researchers exposed fishes of many different kinds to the protist. The fish were quickly killed, upholding her second prediction.

When scientists test a prediction of a hypothesis and find it upheld, the hypothesis is said to be supported. They can become more confident in the hypothesis. However, they cannot say that the hypothesis has been

See the flowchart of the scientific method in action. 1.1

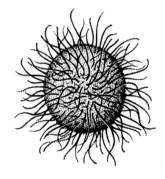

Pfiesteria

(a)

Figure 1.3 The Mystery of the Dead Fish
(*a*) Massive fish die-offs were seen in many North Carolina rivers in the 1990s. Approximately 1 million fish were affected, as this photograph taken in 1991 of Blount Bay, in the Pamlico Estuary of North Carolina, illustrates. (*b*) Scientists observed bloody sores on the bodies of the dead fish.

(b)

proved true. Proving a hypothesis true is not possible, because the hypothesis could always fail when subjected to a different test. The more evidence that supports a hypothesis, however, the stronger that hypothesis is deemed to be. When a prediction is not upheld, the hypothesis is reexamined and changed, or discarded. Sometimes, especially when experimental results are particularly surprising or confounding, scientists not only reconsider the hypothesis but may go back and reexamine their experiments, their predictions, and even their initial observations. In Dr. Burkholder's case, both observation and experiments upheld the predictions, providing strong support for the hypothesis that *Pfiesteria* is the culprit behind the massive fish kills.

Like all scientific studies, Dr. Burkholder's work raises as many questions as answers, and her studies have generated much subsequent research. Scientists still don't know exactly how *Pfiesteria* kills fish—by means of a toxin (a harmful or poisonous substance), by attacking the fish physically, or by some other method. Dr. Burkholder

hypothesized a toxin as the agent of destruction. Recent studies from other laboratories have found no evidence of a toxin coming from fish-killing strains of *Pfiesteria*. However, Dr. Burkholder maintains that these other labs tested the wrong strains of *Pfiesteria* and, in addition, handled them improperly. In other words, she questions whether their experiments were really a proper test. This kind of disagreement is not unusual at all and is a good example of how scientific progress is made.

What will happen next? Researchers will continue their work using the scientific method—testing their hypotheses about toxins versus physical attacks, and keeping track as experiments agree with or contradict their predictions. Along the way, hypotheses will be supported, modified, or discarded. Slowly, through this self-correcting process, scientists come ever closer to a more accurate understanding of a biological phenomenon.

In the meantime, debate about *Pfiesteria* is likely to remain heated. Some researchers hypothesize that *Pfiesteria*, and possibly toxins produced by *Pfiesteria*, is also causing human health problems in the study area. And some biologists hypothesize that the fish die-offs are ultimately caused by big businesses—including factories and pig farms—that dump waste into waterways, causing outbreaks of killer strains of *Pfiesteria*.

As we have seen, scientists make observations, develop a hypothesis, make and test its predictions, then continue to test, and perhaps change or discard hypotheses. Together, these steps make up the basics of the scientific method. Dr. Burkholder's studies are just the first step toward finding a thorough answer to why fish are dying, what the consequences are for humans, and what can be done about it. It is the accumulation of studies like Dr. Burkholder's that expands our base of scientific knowledge.

The terms "theory" and "fact" are used differently in science

Outside of science, people often use the word "theory" to mean an unproved explanation. If something unusual occurs, someone might say, "I have a theory about how that happened." The theory could be anything from a wild guess to a well-considered explanation, but either way, it is "just a theory." Scientists use the word "theory" to mean something very different. If an idea is merely one of many plausible explanations, it is a hypothesis. In science, an idea is elevated to the status of theory only if it is a major idea that explains a wide variety of related phenomena and has been supported by many different observations and experiments.

One example is the theory of global warming. This theory says that Earth's climate is warming and that this warming is due to human activities. The vast majority of scientists agree that this theory is well supported by many scientific experiments and observations. Dr. Burkholder's work is part of the body of evidence supporting the theory. How does her research fit in? Toxin-producing organisms, like *Pfiesteria*, are known to increase rapidly in number when waters warm. Dr. Burkholder's work on the cause of increasingly frequent fish kills is important as one of many pieces of evidence from around the world that such toxin-producing organisms—and ocean temperatures—are indeed on the rise.

What about the term "facts"? In casual conversation, we typically use the term to mean things that are known to be true, for example, as opposed to things that are "just a theory." Scientists typically use the term "fact" to mean a direct and repeatable observation. An example of such an observation might be that an apple, when dropped, falls to the ground—not up into the sky. Another fact is that fish were dying in huge numbers in North Carolina estuaries in the 1990s. So how do facts in science relate to theories in science? In science, a theory, such as the theory of global warming, is not "just a theory" as in casual conversation. Instead, its status as a theory is based on decades of accumulated facts—confirmed observations—that Earth's climate does appear to be warming.

The scientific method has limits

As powerful as the scientific method is, it is restricted to seeking natural causes to explain the workings of the natural world. So there are areas of inquiry that science cannot address. The scientific method cannot tell us, for example, what is morally right or wrong. Science can inform us about how men and women differ physically, but it cannot tell us what the morally correct way is to act on that information. Science cannot speak to the existence of a God or any supernatural being. Nor can science tell us what is beautiful or ugly, which poems are lyrical, or which paintings most inspiring. So although science can exist comfortably alongside different belief systems—religious, political, and personal—it cannot answer all their questions.

1.2 The Characteristics of Living Organisms

Though many have tried, no scientists have been able to produce a simple, single-sentence, airtight definition of life that encompasses the great diversity of living forms—from massive redwoods to microscopic bacteria and everything in between. But all living organisms are thought to be the descendants of a single common ancestor that arose billions of years ago (see the box on page 16). So, not surprisingly, living things share certain features that characterize life, though they do not define it. The shared features (Table 1.1) recognized by biologists as the key hallmarks of life are detailed in the discussion that follows.

Learn more about the characteristics all life shares.

▶ ❙❙
1.2

Table 1.1

The Shared Characteristics of Life

Living organisms

- are composed of cells
- reproduce using DNA
- grow and develop
- actively take in energy from their environment
- sense their environment and respond to it
- maintain constant internal conditions
- can evolve as groups

Living organisms are composed of cells

The first organisms were single cells that existed billions of years ago. The **cell** remains the smallest and most basic unit of life, the fundamental building block of all living things. Cells are tiny, self-contained units enclosed by a membrane. All organisms are made of cells. The simplest of organisms, such as bacteria, are made up of just a single cell.

Larger organisms, such as monkeys and oak trees, are made up of many different kinds of specialized cells and are known as **multicellular organisms**. Our bodies, for example, are composed of trillions of cells with specialized functions, including skin cells, muscle cells, cells that fight disease, and brain cells (Figure 1.4).

Another way to think of cells as the fundamental unit of life is to think about how all living organisms are organized. In human bodies, for example, cells are grouped into

Figure 1.4 The Basic Building Block of Life: The Cell

Humans, like all organisms, are composed of cells. The human body is composed of many types of cells.
(*a*) Red blood cells move oxygen around the body. (*b*) These cells are part of the immune system, which fights off diseases that invade the body. (*c*) This cross-section of intestinal tissue shows some of the variety of cells involved in the absorption of nutrients that takes place there. (*d*) A brain cell. (*e*) Muscle cells. (*f*) Skin cells.

tissues that are organized into internal organs. For the body to function properly, those organs must not only be present but also be arranged in a particular way spatially, with the stomach in its proper place, the brain positioned just so, and so on. Likewise, close examination of a flower reveals that the parts are far from randomly organized. A living organism must maintain its characteristic spatial organization to function properly. And the cell is the fundamental and most basic level of that organization.

Living organisms reproduce themselves via DNA

One of the key characteristics of living organisms is that they can reproduce. Single-celled organisms, such as bacteria, can reproduce by dividing into two new genetically identical copies of themselves. In contrast, multicellular organisms can reproduce in a variety of ways. Humans and other mammals, for example, can reproduce by having sex, which allows a specialized reproductive cell in the male, called a sperm, to fertilize (join with) a female's specialized reproductive cell, an egg; a certain amount of time later—9 months in humans—the female gives birth to young. Some plants reproduce using a form of sex in which their flowers exchange pollen (the equivalent of sperm), which fertilizes cells equivalent to eggs inside the flowers. The plants then produce seeds, which develop into young plants. Some multicellular organisms can also reproduce without sex—for example, by simply budding off new individuals, as sponges and some plants do. Familiar houseplants, such as "mother of millions" and spider plants, can produce tiny plantlets that drop off of the parent plant and take root independently.

Whether organisms produce seeds, lay eggs, give birth, or just split in two, they all reproduce using a molecule known as **DNA (deoxyribonucleic** [dee-*OX*-ee-*RYE*-boh-noo-*CLAY*-ic] **acid**). DNA is the hereditary, or genetic, material that transfers information from parents to offspring. Briefly, the DNA molecule can be thought of as a blueprint or set of instructions for building an organism. A DNA molecule is shaped like a ladder that is twisted into a spiral along its length—a form known as the double helix (Figure 1.5).

This molecule contains a wealth of information—all the information necessary for an organism to create more cells or to grow from a fertilized egg into a complex multicellular organism that will eventually produce its own offspring. DNA is stored in every cell of every living organism. Life, no matter how simple or how complex, uses this inherited blueprint. We will discuss DNA in detail in Unit 3.

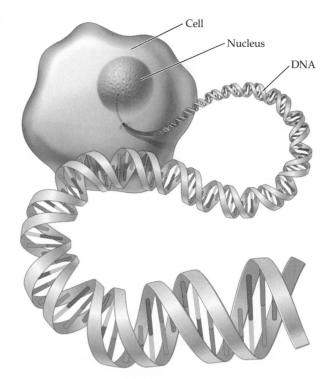

Figure 1.5 The DNA Molecule: A Blueprint for Life
DNA is a hereditary blueprint found in the cells of every living organism. DNA provides a set of instructions by which an individual organism can grow and develop, and that it can pass on to its own young so that they can grow and develop.

Living organisms grow and develop

Using DNA as their blueprint, organisms come into being by building themselves anew every generation—a process known as **development**. For example, a human sperm and a human egg fuse to form a single cell that grows and develops inside a woman's body into a living, breathing baby, who then grows and develops into an adult (Figure 1.6). All organisms go through some process of development in which one organism arises from another organism's cell or cells, whether that organism completes its development by splitting off as a single cell or grows into something as complicated as a multicellular cactus or octopus.

Living organisms actively take in energy from their environment

All organisms need energy to persist. Organisms use a wide variety of methods to capture this energy from their environment.

Plants are among the organisms that can capture the energy of sunlight through a chemical process known as photosynthesis, by which they produce

(a)

Acorn   Seedling   Sapling   Oak tree

(b)

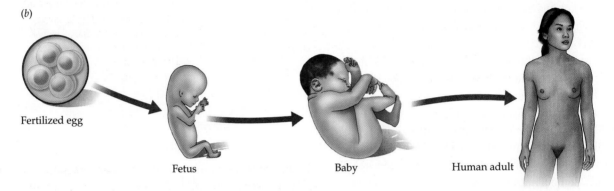

Fertilized egg   Fetus   Baby   Human adult

Figure 1.6 Growing Up
All living organisms develop. (a) An oak develops through several stages, from acorn to seedling to sapling to full-grown adult oak. (b) In humans, after a sperm fertilizes (fuses with) an egg, the resulting cell eventually develops into an adult.

sugars and starches. (We will discuss photosynthesis in detail in Chapter 8.) Some bacteria can also capture energy from the sun through photosynthesis. In addition, bacteria can harness energy from chemical sources such as iron or ammonia through an entirely different chemical reaction. And many organisms, including animals like ourselves, fungi (mushroom-producing organisms and their kin), and certain single-celled organisms, gather energy by consuming other organisms (Figure 1.7).

Living organisms sense their environment and respond to it

Living organisms sense many aspects of their external environment, from the direction of sunlight (as many plants can do; Figure 1.8) to the presence of food and mates. Like humans, many animals can smell, hear, taste, touch, and see the environments around them. Some organisms can sense things that humans cannot, such as ultraviolet and infrared light, electrical fields, and ultrasonic sounds. Some bacteria can even act like a living compass, sensing which direction is north and which direction is up or down by means of magnetic

Figure 1.7 Finding the Energy
Whereas plants can capture energy from sunlight through photosynthesis, animals must get their energy by eating other organisms. This green tree python is ingesting a source of energy that it has captured.

Figure 1.8 Here Comes the Sun
All living organisms must be able to sense and respond to stimuli in their environment. These sunflowers have all detected sunshine, turning toward its light and warmth.

particles within them. All organisms gather information about their internal and external environment by sensing it, then respond appropriately for their continued well-being.

Figure 1.9 Do Sweat It
Organisms must expend energy to maintain consistent internal conditions; for example, humans maintain a steady internal body temperature of about 98.6°F. So when internal temperatures rise, the human body will break a sweat over it—one that begins cooling a person right back down.

Living organisms actively maintain constant internal conditions

Living organisms do not only sense and respond to external conditions; they sense and respond to their internal conditions as well. In fact, organisms maintain remarkably constant internal conditions—a process known as **homeostasis** (Figure 1.9). An example of homeostasis in humans is the maintenance of a fairly constant internal body temperature of 98.6°F. When heat or cold threatens to alter our inner temperature, our bodies quickly respond—for example, by sweating to cool or by shivering to heat.

Groups of living organisms can evolve

An individual organism can physically change over its life—for example, a seedling can grow into a towering tree or a person can dye his hair—but such change is not evolution. Instead, **evolution** is change in groups of organisms over time—and not just change in how those organisms look or act, but change in their genetic material or DNA.

A **species** consists of all organisms that can interbreed to produce fertile offspring (that is, offspring that can themselves reproduce), but that do not, or cannot, breed with other organisms. For example, all mountain lions comprise a single species, as do all monarch butterflies, and all Douglas fir trees. **Populations** are groups of organisms within a species that live and interact with one another, like the mountain lions of one particular mountain range. Both populations and

Biology Matters

What's for Dinner?

All organisms must capture energy to survive, develop, and grow, and humans are no different. Most animals, however, eat only one sort of food—for example, only meat. In fact, many insects go an entire lifetime eating just one particular species of plant. For humans, that would be like going from birth to death eating only corn. But even to obtain such dull diets, animals use a lot of energy, chasing down prey, sucking juices out of plants, and so on. Humans, on the other hand, have the luxury of not only thriving on a huge variety of kinds of organisms as food, but of being able to acquire nutrients as easily as unwrapping a candy bar. No wonder so many Americans and others throughout the developed world are suffering from obesity, or being unhealthily overweight.

According to recent data reported by the National Center for Health Statistics,

- 64 percent of adults age 20 years and over are overweight or obese.
- 30 percent of adults age 20 years and over are obese.
- 15 percent of adolescents age 12–19 are overweight.
- 15 percent of children age 6–11 years are overweight.

Interestingly, many people find that it's easiest to lose the extra weight by severely restricting their diets—on the order of what other animal species do every day. Popular diets have often allowed only a single food—for example, only grapefruit or only rice. More recently, diets like the Atkins diet have become very popular. They severely limit or forbid foods that contain compounds known as carbohydrates (which we'll learn about in detail in Chapter 4) found in cakes, breads, even fruits. The attraction is that Atkins dieters can eat as much as they like of high-fat foods like steak and cheese.

Many people are fans of these diets. Yet some nutritionists say that such diets are too limited to provide sufficient nutrients and energy for the human body, sparking a reassessment of long-held notions about nutrition. Even the government has reexamined its approach, replacing its traditional, one-size-fits-all food pyramid with 12 different pyramids tailored to varying lifestyles. All, however, recommend a broad variety of foods and suggest that most people need to eat more dark green and orange vegetables, legumes, fruits, whole grains, and low-fat milk products.

All food pyramids encourage people to consume

- 2 cups of fruit and 2.5 cups of vegetables per day
- 3 or more ounce-equivalents per day of whole-grain products
- 3 cups per day of fat-free or low-fat milk or equivalent milk products

And they limit the consumption of other foods (despite their tastiness) as follows:

- Keep total fat intake between 20 percent and 35 percent of calories, with most fats coming from sources of polyunsaturated and monounsaturated fatty acids, such as fish, nuts, and vegetable oils.
- Consume less than 2,300 milligrams of sodium per day (approximately 1 teaspoon of salt).
- Limit alcoholic beverages to one drink per day for women and up to two drinks per day for men.

Source: *Dietary Guidelines for Americans, 2005.*

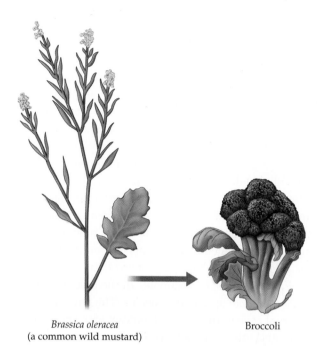

Brassica oleracea
(a common wild mustard)

Broccoli

Figure 1.10 Groups of Living Organisms Can Evolve
Species and populations of living organisms can change over time—that is, evolve. And many of our favorite foods (as well as some of our least favorite) are products of just such evolutionary change that has been controlled by humans, through a process known as artificial selection. For example, plant breeders used the techniques of artificial selection (discussed in more detail in Unit 4) to cause the wild mustard plant to change over time, or evolve, into broccoli.

species can evolve. (We will discuss evolution in detail in Unit 4.)

Pronghorn antelope, for example, are the fastest-running creatures in North America. Over time, pronghorn as a species evolved to be faster runners because only the fastest pronghorn could outrace their predators to survive and then reproduce. Slower pronghorn were eaten. The young of the survivors tended to be speedy themselves because they inherited their DNA—the blueprint for building a speedy antelope—from their speedy parents. In the same way, a species of plant can evolve to have a different shape or a different number of flowers (Figure 1.10).

A nineteenth-century Englishman named Charles Darwin, considered the father of evolutionary biology and one of the great thinkers of all time, made the

Charles Darwin

first convincing argument for evolution in his book *The Origin of Species*. As Darwin explained, in the struggle to survive and the contest to reproduce, characteristics of species—such as the speed at which a pronghorn can run—tend to change over time. Features that are advantageous in the struggle to survive and to reproduce are known as **adaptations**. Because evolutionary change can explain so many of the features of living organisms—from how fast pronghorn run to the sharpness of a shark's tooth—evolution is considered the central, unifying theme in biology.

1.3 Viruses: A Nasty Puzzle

Many people are laid up each winter by the influenza virus. Flu sufferers cough and sneeze, have a fever, and ache all over. The reason is that the influenza virus is infecting the body's cells, reproducing throughout the nose, throat, and lungs. In response, specialized defensive cells of the immune system are fighting back. One way is to attack and destroy cells infected by the virus. Another is to turn up body temperature to hinder viral reproduction—hence the high fever.

As this microscopic virus proliferates throughout the body's cells—whether for days or for weeks—it is evolving rapidly. In fact, viruses evolve so quickly that they are able to evade many defenses, which makes them difficult for medicines or the body to fight. Many different viruses besides influenza—the virus that causes AIDS, for example—affect people, and still other viruses infect all the other forms of life.

Viruses are such powerful foes that they certainly seem alive. In fact, however, viruses are hard to characterize. Like living organisms, viruses can contain DNA, reproduce, and evolve. Yet all viruses, including the influenza virus, lack some of the key characteristics of life. For one thing, viruses are not made up of cells. A virus is much simpler than a cell, usually consisting of just a strand of hereditary blueprint—that is, a small piece of genetic material like DNA—that is wrapped in a coat of molecules known as proteins. Another difference is that viruses lack all the structures within cells that are necessary to perform nearly all the activities that living organisms do, including homeostasis, reproduction, and energy collection. To accomplish these tasks, viruses make the cells of other organisms do their work for them. They accomplish this by invading those cells, just as the influenza virus invades the cells of the body. A third unusual feature

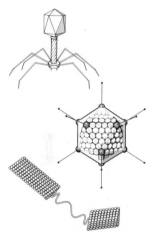

Three different viruses

of viruses, unlike living organisms, is that the genetic material they pass from one generation to the next is not always DNA. Some viruses use a related molecule, known as **RNA**, or **ribonucleic acid**. (We will discuss RNA in Chapter 13.)

Viruses defy easy definition. But though some scientists consider viruses to be a very simplified form of life and others say they cannot be defined, we follow the lead of most and consider viruses to be nonliving.

Concept Check

1. What is a hypothesis, and what is its role in the scientific method?
2. In everyday language, we use the word "theory" to mean an entirely unproved explanation. What do scientists mean when they say an idea is a theory?
3. In summer heat or winter cold, the human body maintains an internal temperature of 98.6°F. Which of the shared characteristics of living things does this phenomenon demonstrate?
4. Why is a virus considered nonliving?

Explore the biological hierarchy.

1.3

1.4 The Biological Hierarchy

Biologists find it useful to organize life into a **biological hierarchy** (Figure 1.11). The hierarchy has many levels of organization, ranging from molecules at the lowest level up to the entire biosphere at the highest level. In scale, the hierarchy ranges from one-millionth of a meter (the approximate length of a molecule) to 12 million meters (the diameter of Earth).

Levels in the biological hierarchy provide biologists with a framework for understanding life

At its lowest level, the hierarchy begins with the **molecules** found primarily in living organisms. An example of such a molecule is DNA, the hereditary material that carries the blueprint for building an organism. Many such specialized molecules are organized and are the working parts and machinery, so to speak, in the next level of the hierarchy—namely, cells, the basic unit of life. As mentioned earlier, some organisms, such as bacteria, consist of only a single cell. Multicellular organisms have **tissues**, which are specialized, coordinated collections of cells that perform particular functions in the body, such as muscle or nerve tissues. Sometimes tissues are

organized into **organs**, which are body parts composed of different tissues that are organized to carry out specialized functions, such as hearts and brains. Groups of organs can work together as **organ systems**; for example, the stomach, liver, and intestines are parts of the organ system known as the digestive system. Groups of organ systems work together for the benefit of a single **individual**.

There are also several levels in the hierarchy above the individual. Each individual is a member of a population, a group of organisms of the same species living and interacting in the same area. Earlier we mentioned that a population can consist of all the mountain lions in a particular mountain range. Different populations that can interbreed are considered to be members of the same species; for example, all the mountain lions on Earth make up a single species. The populations of different species that live and interact with one another in a particular area are known as a **community**—for example, the community of insect species living in a forest or the community of plant species on a mountainside.

Communities, along with the physical environments they inhabit, are known as **ecosystems**; for example, a river ecosystem includes the river itself, as well as the communities of organisms living in it. At the next level are **biomes**, large regions of the world that are defined on land by the plants that grow there (for example, the arctic tundra) and in water by the physical characteristics of the environment (for example, coral reefs). Finally, each biome is part of the one **biosphere**, which is defined as all the world's living organisms and the places where they live.

Knowing the biological hierarchy can be useful in everyday life

Few people are familiar with the concept of the biological hierarchy, so it might at first seem like a scientific idea important only to biologists. However, many of us interact with aspects of the biological hierarchy every day—whether at home, at work, shopping for food, or getting health care, as the following examples show.

- When we take prescription drugs, we are using particular *molecules* that interact with very specific natural molecules in our bodies, as well as with any other drugs and vitamins we may be taking at the same time.

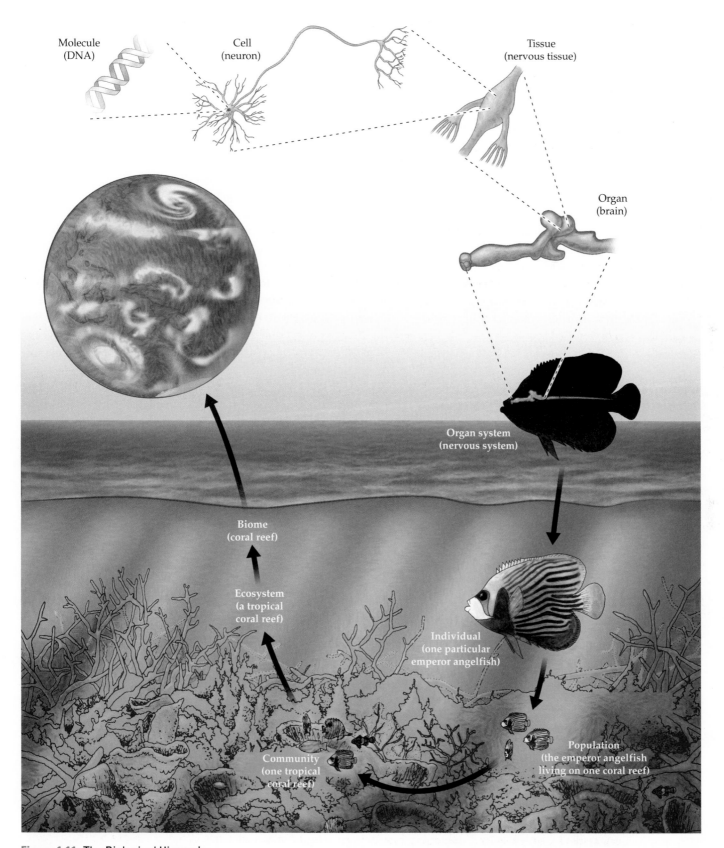

Molecule
(DNA)

Cell
(neuron)

Tissue
(nervous tissue)

Organ
(brain)

Organ system
(nervous system)

Biome
(coral reef)

Ecosystem
(a tropical
coral reef)

Individual
(one particular
emperor angelfish)

Population
(the emperor angelfish
living on one coral reef)

Community
(one tropical
coral reef)

Figure 1.11 The Biological Hierarchy

Levels of biological organization can be traced from molecules found in organisms (such as DNA) all the way up through levels of increasing organization to the biosphere, which includes all living organisms and the places where they live. Aquarium enthusiasts may recognize the emperor angelfish pictured in the hierarchy.

Science Toolkit

Thinking outside the Box: Thinking inside the Soup

At the time Earth formed, 4.6 billion years ago, the planet was lifeless. The first sure signs of cellular life can be found in the fossil record as early as 3.5 billion years ago. But what happened during the intervening 1.1 billion years—how life arose from nonlife—remains one of the most fascinating questions in science.

How do scientists devise hypotheses and experimental tests to explain an event—the origin of life—thought to have happened billions of years ago? This is one of the areas of biological research that not only benefits from thinking outside the box, but requires it. So what have scientists come up with? One of the many competing hypotheses suggests that life originated in hot springs deep on the ocean bottom. Another hypothesis proposes that life's birthplace lies near underwater volcanoes or in hot-water geysers on land, such as Yellowstone's Old Faithful. Still another suggests that life, or its building blocks, did not arise on Earth at all, but arrived here from another planet (such as Mars) by traveling through space on an asteroid or meteorite. All of these competing hypotheses have their strengths and weaknesses.

The best-known and longest-standing hypothesis, however, is the "soup" hypothesis suggesting that life originated in the primordial soup of Earth's ancient seas. But how could testable predictions be generated from this hypothesis? In 1953, Stanley L. Miller, while still a student, came up with a way. He realized that one prediction of the hypothesis was that a re-creation of Earth's ancient seas should re-create the building blocks required to make a living cell. Miller attempted to test this prediction by simulating the conditions—hot seas and lightning-filled skies—of early Earth.

To make his "primordial soup," Miller began with water, which he kept boiling. The water vapor rose into the simulated atmosphere, which contained the gases methane, ammonia, and hydrogen—some of the gases that Earth's early atmosphere is thought to have harbored, and the very sort that could have belched forth from ancient volcanoes. Miller added an electrical spark to simulate lightning. He then cooled the sparking vapors until they turned to liquid, and the resulting liquid, like rain falling back to the seas, returned to the boiling water. Miller allowed his apparatus—a closed system in which the water continuously recycled between boiling seas and sparking skies—to "cook" for a week.

The whole situation was absurd—not least of all the idea that a person could simply cook up life's crucial components as Earth had done so long ago from such a seemingly simple recipe—except that Miller's experiment confirmed the prediction of the soup hypothesis. When he examined the contents of the soup at the end of 7 days, Miller discovered that from water and simple gases alone, he had created an array of molecules critical to the origin of life, including two amino acids, the building blocks of molecules known as proteins. We know today that Earth's original conditions weren't exactly like those in Miller's experiment, but since then, other such soup experiments attempting to re-create alternate versions of Earth's early conditions have likewise produced key biological molecules, including all the common amino acids, as well as sugars, lipids (fats), and the basic building blocks of the genetic material, DNA and RNA.

As is typical in the scientific process, new discoveries lead to new questions: Once a soup of critical molecules had developed, how did those molecules get organized into larger molecules and into the first cells able to gather energy and reproduce? Did the first cells float freely in such a soup, or did they not become cells until their parts fell out of the soup onto a surface, such as the ocean bottom?

How Earth made the crucial leap from barren stone to cradle of life remains one of the most difficult questions facing science—but as Miller showed, it is also the perfect subject on which scientists can continue to hone their creative talents.

The Primordial Soup of Life

Stanley Miller cooked up a brew inside his apparatus (shown here) that contained many of the molecules necessary to life. His experiment supports the hypothesis that life could have originated in the early seas under lightning-filled skies.

In the icy waters off the New England coast, fishermen clean their haul of Atlantic cod. The capture of this tasty fish is regulated to protect its remaining populations. From the fishermen's point of view, the biological level of the individual is what matters, since each fish, quite literally, counts.

- When we drink water from a household tap that is fed by a municipal water supply, we benefit from monitoring by water-quality engineers, who check for harmful organisms that consist of single *cells*, such as toxic *E. coli* [ee *кон*-lye] bacteria and *Giardia* [jee-*ahr*-dee-uh] (a protist), both of which can cause illness in people.
- If we get a professional massage, we trust that the massage therapist has been trained to know that the body has various kinds of *tissues*. With expert hands, the therapist uses techniques to relax certain tissues, particularly muscles and tendons.
- Savvy grocery shoppers are keenly aware that *organs* are made of specialized—and different-tasting—tissues, some days shopping for liver to cook up with onions while carefully avoiding the packages of brain or tongue. Other organs make their way onto the Thanksgiving dinner table: giblet gravy is made from some of the turkey's organs—such as heart and liver—which are wrapped in paper and packed inside most turkeys by supermarket butchers.
- Medical doctors are well versed in many levels of the biological hierarchy. When we visit a medical special-

ist, it is often to take advantage of that practitioner's knowledge of the ailments of a particular *organ system*. For example, specialists dealing with the nervous organ system are called neurologists, and those dealing with women's reproductive organ systems are called gynecologists.

- *Individual organisms* are the most familiar level of the hierarchy because we deal with individuals every day. Each member of a household is an individual, and so is each pet and each separately potted houseplant. When the hostess in a restaurant asks us, "How many for dinner?" we know she's asking how many individuals are in our party.
- Many fish species are declining in number around the world's oceans. As a result, a number of nations limit how many fish can be taken annually from certain *populations* (one example is Atlantic cod in the once-rich Georges Bank region off the coast of New England, as in Figure 1.12). Population size and regulation in turn affect which fishes are available in the supermarket and how expensive they are.
- People who enjoy honey know that bees collecting pollen from different *communities* of flowers—say, a meadow of blooming clover and alfalfa or a field full of

wild buckwheat—will produce very different-tasting honey.

- When we take a walk through the woods, we are encountering a forest *ecosystem*, though we may not be able to sense all the many organisms within that ecosystem.
- Any zoo or aquarium is likely to have displays on *biomes* such as tundras and coral reefs. Natural-history museums often illustrate biomes as dioramas.
- Only a small number of people—the men and women who fly in space—have so far been in a position to see the entire *biosphere*. Astronauts often say they have been forever changed by the awe-inspiring sight of that familiar blue and white orb, full of life, that is our biosphere.

1.5 Energy Flow through Biological Systems

In most biological systems on Earth, the original source of energy is the sun. Energy flows first from the sun to plants and other photosynthesizing organ-

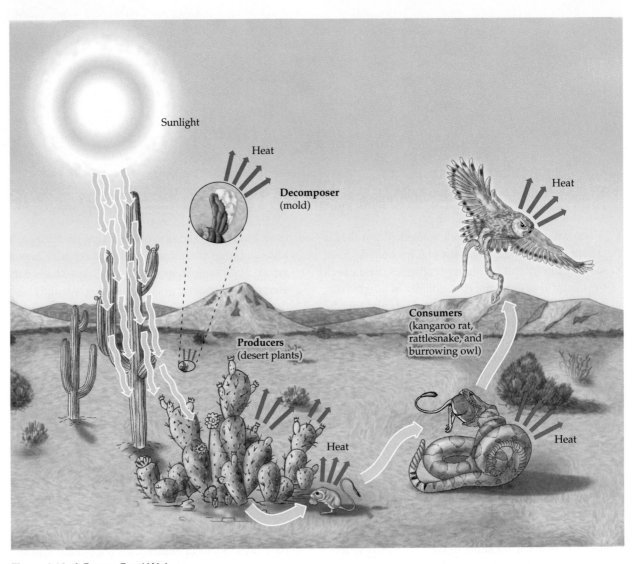

Figure 1.13 A Desert Food Web

The energy of sunlight is captured by producers, in this case desert plants. Energy then flows to consumers, such as a fruit-eating kangaroo rat. These consumers are then eaten by other consumers, such as snakes, which can then be eaten by other consumers in turn. Decomposers, such as the mold on a cactus pad that has fallen off the plant, likewise get their energy either from producers or from organisms whose energy ultimately derives from producers. All organisms—producers, decomposers, and consumers—give off energy in the form of heat (though it can be very slight). Energy flows from the sun to producers and then to consumers and decomposers throughout this food web.

isms, such as green algae. These organisms are called **producers** because they transform the sun's energy (in the form of light) into chemical energy in the form of sugars and starches. That energy is used by plants and can be harvested by **consumers**, organisms that eat either producers or other organisms whose energy ultimately derives from producers. Animals are a familiar example of consumers. There are also **decomposers**, organisms that derive their sustenance from dead organisms or cast-off parts of living organisms. In the process they return nutrients to the biosphere. Many fungi are decomposers. Energy flows almost entirely in one direction through the biosphere: from the sun to producers and then to consumers and decomposers, and all these organisms give off energy as heat. A depiction of producers, consumers, and decomposers that illustrates who eats whom is known as a **food web** (Figure 1.13).

Most, but not all, biological systems use the sun as the original source of energy, however. The deep-sea floor, far from the sun's light, contains entire ecosystems that depend instead on the energy that bacteria harness from nonliving materials, such as iron or ammonia.

Concept Check

1. Unscramble this scrambled biological hierarchy: community, organ system, ecosystem, tissue, individual, biosphere, organ, cell, biome, population, molecule.
2. What provides the energy used by consumers in most biological systems?

Applying What We Learned

Are Nanobes Indeed the Smallest Form of Life?

Now that we have a list of shared characteristics of living things, how do nanobes stack up? Measuring from 20 to 150 nanometers across, are they the smallest living things?

As we have seen, all living organisms are composed of cells. Scientists examining the structure of nanobes report that they appear to be highly structured and have very much the "look" of a complex microscopic organism—appearing as spheres, chains of beads, or bean- or sausagelike shapes, as many microscopic organisms do—as opposed to looking like an array of lifeless crystals or disorganized nonliving material. Scientists say that the nanobes also appear to have a cell-like structure, with a cell-like outer surface, or membrane, encircling the nanobe. So far, so good.

Living organisms also contain DNA. What about nanobes? In recent studies, scientists found that applying chemical stains for DNA to nanobes reveals the presence of DNA. Researchers intend to recover and study that DNA in greater detail.

Living organisms don't just *contain* DNA, however; they reproduce using that DNA. Has anyone seen a nanobe reproduce? Not yet. We also know that living organisms grow and develop. But although some scien-

tists, after seeing the enlargement of nanobe colonies in the laboratory, have concluded that nanobes can grow, others say that the evidence is too weak. Some skeptics suggest that nanobes may be nonliving crystals that increase in size under the proper conditions, similar to the way drying salt water can create a growing crust of salt crystals. Nanobes, they suggest, merely mimic the growth of a living organism.

Living organisms capture energy from their environment, and they sense and respond to their environment. They maintain constant internal conditions. Populations of living things can evolve. And nanobes? Scientists still have not been able to determine whether nanobes do any of these things.

The vast majority of scientists agree that the jury is still out on nanobes. More hypotheses will be generated and more experiments done, since much more remains to be learned about them, but so far, as far as most scientists are concerned, nanobes have not quite proved that they are among the ranks of the living. Still, however scientists choose to define such entities—whether as living or nonliving or somewhere in between—both nanobes and viruses remind us how varied and diverse life on Earth can be.

Are Nanobacteria Making Us Ill?

By Amit Asaravala

Olavi Kajander didn't mean to discover the mysterious particles that have been called the most primitive organisms on Earth and that could be responsible for a series of painful and sometimes fatal illnesses.

He was simply trying to find out why certain cultures of mammalian cells in his lab would die no matter how carefully he prepared them.

So the Finnish biochemist and his colleagues slipped some of their old cultures under an electron microscope one day in 1988 and took a closer look. That's when they saw the particles. Like bacteria but an astonishing 100 times smaller, they seemed to be thriving inside the dying cells.

Believing them to be a possible new form of life, Kajander named the particles "nanobacteria," published a paper outlining his findings and spurred one of the biggest controversies in modern microbiology.

At the heart of the debate is the question of whether nanobacteria could actually be a new form of life . . . [The question has become more urgent, though, because of a] steadily increasing number of studies linking nanobacteria to serious health problems, including kidney stones, aneurysms and ovarian cancer.

"It's all pretty exciting stuff," said David McKay, chief scientist for astrobiology at NASA's Johnson Space Center.

Kajander stumbled upon his discovery simply by trying to grow a culture of cells from mammals. Cells of bacteria can be kept alive in the laboratory, reproducing to form populations called cultures. Likewise, a single cell—like one skin cell—can be taken from a multicellular organism like a human, dog, or other mammal and grown in laboratory cultures. But Dr. Kajander's cultures kept dying inexplicably, until he saw that they appeared to be infected with nanobacteria, also known as nanobes or nano-organisms.

This discovery might have remained an isolated curiosity—except that other researchers began finding nanobacteria associated with human diseases. Nanobes, they reported, were to be found inside of kidney stones, calcified arteries, and the tissue of cancer patients.

If nanobacteria are playing a role in these diseases, as some research suggests, then understanding what nanobacteria really are—living or nonliving—could be important in prevention and cures.

Let's look at where the scientific method has taken biologists so far in understanding nanobes and where it might take them next. One hypothesis is "nanobes are alive." One prediction, then, is that nanobes will exhibit the characteristics of living things.

Our checklist of characteristics of living things tells us that they are made of cells, and nanobes appear to be enclosed in a cell membrane. But why isn't that fact compelling evidence to many scientists that nanobes are living cells? The reason is that many other pieces, the working machinery inside of a cell, must also be present for a thing to be truly alive—lots of biological molecules, like proteins, carbohydrates, and many more. Remember, living things grow, reproduce, and maintain constant internal conditions. They require a lot of moving parts and pieces inside a cell to keep those operations going. For example, one important piece of cellular machinery is the ribosome.

(We will learn about ribosomes in detail in Chapter 5.) For now, suffice it to say that the ribosome makes proteins. A typical cell contains several hundred thousand ribosomes, each 25 to 30 nanometers wide. Yet some nanobes measure just 20 nanometers across. And ribosomes are just one of many elements needed by a working, living cell. Some scientists have rejected the idea that nanobes are alive, pointing out that not even a single ribosome would fit inside of one.

Life is both diverse and complex, and many aspects beyond the bare bones of our checklist of characteristics—from DNA to ribosomes and beyond—are required for the maintenance and reproduction of life.

Evaluating the News

1. How does the size of a ribosome compare with the size of a nanobe, and why does it matter?

2. The person quoted in the news story is from NASA. Why would space scientists be interested in what's inside of kidney stones? Why would these explorers of outer space care about nanobes and how to determine whether a newly discovered form is living or nonliving?

3. Some of the researchers studying nanobes have started a business, called Nanobac Life Sciences, which performs research aimed at detecting nanobes and treating disease symptoms caused by nanobes. These scientists have made the already controversial nanobes even more controversial; some say their research is now suspect because they have a direct financial interest in a belief in nanobes. Are you less likely to believe evidence they might produce that nanobes are alive? Should you be?

SOURCE: Wired.com, March 14, 2005.

Chapter Review

Summary

1.1 Asking Questions, Testing Answers: The Work of Science

- Science is a rational way of discovering truths about the natural world.
- To answer questions about the natural world, scientists use the scientific method, which has four steps: (1) make observations, (2) devise a hypothesis to explain the observations, (3) generate predictions from that hypothesis, and (4) test those predictions.
- Scientists can test their hypotheses by making further observations or by performing experiments (controlled, repeated manipulations of nature) that will either uphold or not uphold the predictions.
- A hypothesis cannot be proved true or false. It can only be upheld or not upheld. If the predictions of a hypothesis are not upheld, the hypothesis is rejected or modified. If the predictions are upheld, the hypothesis is supported. Even if one prediction is upheld, other predictions of the same hypothesis may not be upheld.
- A scientific theory is a major idea that has been supported by many different observations and experiments.
- The scientific method can explore only natural causes for natural phenomena. It is not a useful way of addressing questions that cannot be tested, such as moral questions or questions that deal with the supernatural.

1.2 The Characteristics of Living Organisms

- The great diversity of life on Earth is unified by a set of shared characteristics.
- Living organisms are built of cells, the building blocks of living organisms. Some organisms are single-celled, composed of one building block; some are multicellular, composed of many.
- Living organisms reproduce using DNA.
- Living organisms grow and change during the course of their lives in a process called development.
- Living organisms take in energy from their environment.
- Living organisms sense and respond to their environment.
- Living organisms actively maintain constant internal conditions.
- Groups of living organisms, such as species and populations, can change over time, in a process known as evolution. Evolutionary changes are not merely visible changes but changes in the genetic material or DNA of a species or population. Adaptations are features that have evolved to allow organisms to survive and reproduce better. Evolution is the central theme in biology.

1.3 Viruses: A Nasty Puzzle

- Viruses share some of the characteristics of living organisms: they evolve and reproduce.

- Viruses lack some of the characteristics of living organisms: they are not made of cells, and they lack the structures necessary to perform activities essential to life, such as collecting energy and maintaining constant internal conditions. Some viruses lack DNA.
- Because viruses exhibit only some of the characteristics of living organisms, most biologists consider them to be nonliving. Viruses also illustrate how difficult it is to define life precisely.

1.4 The Biological Hierarchy

- Living organisms are part of a biological hierarchy.
- Molecules are at the smallest level of the hierarchy and include such things as DNA. Specialized molecules can be organized into cells, the basic unit of life.
- Cells are organized into tissues, specialized tissues are organized into organs, and organs working together make up an organ system. Organ systems work in concert for the next level of the hierarchy, whole individuals.
- The individuals of a given species in a particular area constitute populations. Populations of different species in an area make up communities. Communities, along with the physical habitat they live in, constitute ecosystems.
- Ecosystems make up biomes, large regions of the world that are defined on land by the kinds of plants that grow there and in water by the physical characteristics of the environment. All the biomes on Earth make up our one single biosphere.
- The biological hierarchy can be seen and recognized by people in their everyday lives.

1.5 Energy Flow through Biological Systems

- Biological systems are made up of three groups of organisms: (1) producers, such as plants, which create chemical energy in the form of sugars and starches from light; (2) consumers, such as animals, which eat producers or other organisms whose energy ultimately derives from producers; and (3) decomposers, such as fungi and certain bacteria, which eat decomposed organisms or cast-off parts of living organisms, releasing nutrients back to the biosphere.
- Most biological systems are fueled by energy from the sun. In these systems, energy flows in one direction: from sun to producers to consumers and decomposers. However, some deep-sea biological systems are fueled by energy that bacteria extract from nonliving materials such as iron or ammonia.
- We can diagram the relationships, in terms of who is eating whom, among producers, consumers, and decomposers in an ecosystem using a food web.

Review and Application of Key Concepts

1. Describe one observation, one hypothesis, and one experiment from Dr. Burkholder's work.

2. Imagine you are a scientist who thinks the observations of nanobes favor the hypothesis that nanobes are living organisms. If they are living, what is one testable prediction you could make about them? For some ideas, try to finish the following sentence: "If nanobes are living things, then it should also be true that . . . "

3. Some people avoid eating beef because they fear contracting mad cow disease, which is thought to be caused by a class of molecules known as prions [PREE-onz]. These poorly understood molecules are a kind of protein that is able to infect the nervous system, eventually creating holes in the brain tissue, leading to loss of mental and physical function and eventually to death. Prions are thought to spread their effects through the body rather like "falling dominoes"—by causing other normal molecules to mimic the prion's shape and thus become infectious molecules themselves. Like a virus, a prion can cause disease. A prion can multiply or reproduce itself by making other molecules mimic its shape and behavior. Prions can spread from one part of the body to another. Are prions alive? Why or why not?

4. What are the levels of the biological hierarchy? Arrange them in their proper relationship with respect to one another, from smallest to largest. Give an example for each level that you know from your own experience.

5. Describe the flow of energy through a biological system that includes grasses, lions, sunshine, antelope, and ticks.

6. For the biological system described in question 5, (a) identify which are producers and which are consumers, and (b) diagram the food web of the system.

7. Choose one of these two hypotheses:
   I. Viruses are living organisms.
   II. Viruses are not living organisms.

Given what you know about the characteristics of life, (a) state some testable predictions about viruses based on the hypothesis you chose; and (b) propose an experiment to test one of your predictions. (For example, if you chose hypothesis I, then one prediction might be, "Viruses require energy." An experiment to test this prediction might be described like this: "I will provide one group of viruses with a lot of energy in the form of nutrients, heat, and light, and another group of viruses with no energy. If the first group of viruses multiplies and the second group does not, these observations will support my prediction and hypothesis".) In dreaming up your experiment, don't hold back. Remember—science thrives on creativity.

Key Terms

adaptation (p. 13)
biological hierarchy (p. 14)
biology (p. 4)
biome (p. 14)
biosphere (p. 14)
cell (p. 8)
community (p. 14)
consumer (p. 19)
decomposer (p. 19)
development (p. 9)
DNA (deoxyribonucleic acid) (p. 9)
ecosystem (p. 14)
evolution (p. 11)
experiment (p. 5)
food web (p. 19)
homeostasis (p. 11)
hypothesis (p. 5)
individual (p. 14)
molecule (p. 14)
multicellular organism (p. 8)
observation (p. 4)
organ (p. 14)
organ system (p. 14)
population (p. 11)
prediction (p. 5)
producer (p. 19)
RNA (ribonucleic acid) (p. 14)
science (p. 4)
scientific method (p. 4)
species (p. 11)
tissue (p. 14)
virus (p. 13)

Self-Quiz

1. Which of the following is *not* an element of the scientific method?
   a. observation
   b. religious belief
   c. experiment
   d. hypothesis

2. A hypothesis is
   a. an educated guess explaining an observation.
   b. a prediction based on an observation.
   c. a test of a prediction.
   d. an experiment that works well.

3. Which of the following are both shared characteristics of life?
   a. the abilities to move and to reproduce
   b. the abilities to reproduce using RNA and to capture energy directly from the sun
   c. the abilities to reproduce and to sense the environment
   d. the abilities to sense the environment and to capture energy by eating

4. Viruses
   a. consist entirely of proteins.
   b. contain no hereditary material, such as DNA.
   c. do not affect humans.
   d. are considered by most to be nonliving.

5. Which of the following is the basic unit of life?
   a. plants
   b. DNA
   c. the cell
   d. biomes

6. Which of the following can reproduce without its own DNA?
   a. a human being
   b. a virus
   c. a single-celled organism
   d. none of the above

7. Which of the following is a multicellular organism?
   a. a beetle
   b. a brain
   c. a bacterium
   d. a forest

8. The energy in biological systems can originate from
   a. the sun and the moon.
   b. the sun only.
   c. plants and algae.
   d. the sun, as well as nonliving materials such as iron.

CHAPTER 2

Organizing the Diversity of Life

Key Concepts

- Biologists examine all aspects of an organism's biology—its body structure, its behavior, its DNA—to look for inherited similarities shared with other organisms. Using that information, scientists form a hypothesis of the evolutionary relationships of those organisms and depict them in an evolutionary tree.

- The most informative similarities for determining evolutionary relationships are shared derived features, which are shared by a group of organisms because the features arose in their most recent common ancestor.

- Evolutionary trees can be used to predict attributes, such as behaviors, of organisms on the tree.

- The Linnaean hierarchy is a system for classifying all of life. It ranges from species at the lowest category level, up through genera, families, orders, classes, phyla, and kingdoms. Biologists also widely recognize the three-domain system as the most basic division of living organisms.

- Biologists are making dramatic progress in understanding the evolutionary relationships of the world's many organisms. In part, this progress is due to modern DNA studies, which can reveal some surprising relationships.

The Iceman Cometh

On a September day in 1991, hikers high in the Alps near the border between Austria and Italy came across a remarkable sight: the body of an ancient, mummified man in the melting glacial ice. Instantly dubbed the Iceman, the well-preserved body appeared to be a prehistoric hiker dating to the Stone Age, possibly a shepherd or traveler, who had died on the mountainside, his body captured by the ice, which had preserved it for some 5,000 years.

Extremely well preserved—right down to his underwear—this one-and-a-half-meter-tall (about 5-foot) visitor from another time promised a tantalizing peek into the past. Sporting tattoos on his body, he wore a loincloth, a leather belt and pouch, leggings and a jacket made of animal skin, a cape of woven grasses, and calfskin shoes lined with grass. He carried a bow and arrows, an axe, knives, and two pieces of birch fungus, perhaps a kind of prehistoric penicillin.

Perfectly suited for scientific examination, this first-ever corpse from the Stone Age, found right in the path of hikers, seemed too good to be true. Researchers began to fear that the Iceman might be an elaborate hoax. People speculated that the body was a transplanted Egyptian or pre-Columbian American mummy. **How could biologists determine the true identity of this long-lost wanderer?** To know who the Iceman was, scientists realized they would have to figure out who his closest relatives had been. **Was he most closely related to Columbian Indians? Or were his closest living relatives Egyptians or northern Europeans?** To solve the mystery, scientists would have to place him in the human family tree. How scientists identify organisms—that is, how they determine their proper place on the family tree that includes all living things, the evolutionary tree of life—is the topic of this chapter.

A Stone Age Mummy
Known as the Iceman, this mummy was discovered in melting ice by hikers in the Alps in 1991, some 5,000 years after he died.

iologists often come across mysterious, previously unknown organisms. When they find a puzzling specimen, the first question they ask is, "What is this?" As with the Iceman, we can answer this question only by placing the organism into a kind of family tree of living things, known as an evolutionary tree.

In this chapter we begin by examining how biologists determine evolutionary relationships among organisms and how they depict them on evolutionary trees. Next we examine the grand classification scheme known as the Linnaean hierarchy. Finally we look at some of the more interesting and important branches on the evolutionary tree of life that scientists are working to understand. We close with new information that scientists are uncovering about the Iceman and his place in the human family.

2.1 Evolutionary Relationships Are Depicted in Evolutionary Trees

Genealogists and genealogy enthusiasts gather information to show how members of a human extended family are related. They look from one generation to the next (who are my great-grandparents? my great-great-great-grandparents?) and within a given generation as well (who is whose sister or brother?). Once genealogists know how members of a group are related, they depict that information in a family tree (Figure 2.1*a*). In the same way, **systematists** [*siss*-tuh-muh-*TISTS*]—biologists who study the relationships among groups of different organisms—depict those evolutionary relationships in **evolutionary trees** (Figure 2.1*b*). Taken together, the many evolutionary trees of all the variety of living organisms make up the evolutionary tree of all life. But although a family tree can record known family relationships, an evolutionary tree is quite different.

An evolutionary tree is not a recording of facts, but a scientific hypothesis. It is a hypothesis because the evolutionary relationships depicted on an evolutionary tree are a scientist's best educated guess at the relationships, based on all previous studies. Like all hypotheses, an evolutionary tree generates predictions. And every time a new study is undertaken, it can support or fail to support those predictions. So as new information is gathered, evolutionary trees—like all hypotheses—can and do change.

To place a new specimen on the tree of life, biologists must ask themselves a series of questions. First, is this mystery organism a member of a species that is already named and placed on the tree of life? This is not a trivial question. Organisms that appear unique sometimes turn out to be a different form—even a different life stage or a different sex—of a species already known to exist. Conversely, organisms that look outwardly identical to known species can turn out to be quite distinct in terms of their DNA, thereby signaling to a biologist that the two specimens may be from different species.

If the mystery organism is not a member of a known species, then what are its closest relatives? If a plant, is it a close relative of an oak tree or a bluebell? If an animal, is it more closely related to a guppy or to an eagle?

Once scientists determine where an organism should sit in the tree of life, they can name it. So, whether scientists are demystifying mummies or collecting new species from the rainforest, their first question always is, "Where does this organism fit in the evolutionary tree of life?"

Groups are related because they evolved from common ancestors

Let's look more closely at the process of constructing an evolutionary tree. How *do* biologists assess which groups of organisms are more closely or more distantly related to one another?

As anyone who has been to a family reunion knows, the more closely related two people are, the more similar they tend to be, in the way they look and sometimes even in the way they act—laughing or walking in just the same way. Close relatives tend to be similar even in how their bodies work, often exhibiting the same physical strengths or vulnerabilities to the same kinds of illnesses. In fact, the similarities extend right down to the level of a person's genetic material (DNA), and for good reason.

Recall that DNA is the molecule passed from one generation to the next, the genetic blueprint for the development of an individual. We inherited these blueprints for body structures and behaviors from our parents, who inherited the blueprints from their parents, and so on. As a result, we exhibit many of the characteristics that our relatives exhibit. In the same way, closely related groups of organisms (for example, several species that arose from the same ancestor) tend to resemble one another.

Unit 4 discusses evolution in detail. For now, it is enough to know that over time, populations within a spe-

(*a*) Family tree of Britain's royalty.

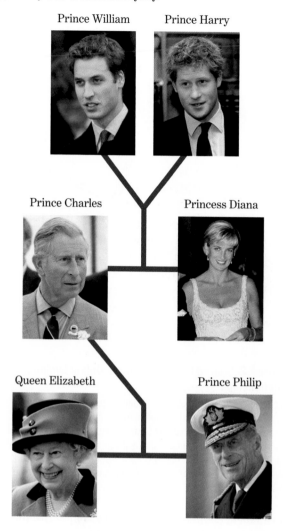

Prince William | Prince Harry

Prince Charles | Princess Diana

Queen Elizabeth | Prince Philip

(*b*) Evolutionary tree of the swallowtail butterfly.

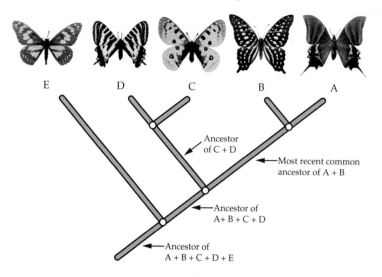

E D C B A

Ancestor of C + D

Most recent common ancestor of A + B

Ancestor of A + B + C + D

Ancestor of A + B + C + D + E

Figure 2.1 **Family Trees versus Evolutionary Trees**
(*a*) This family tree shows the relationships of several familiar members of Britain's royal family, including Queen Elizabeth and Prince Philip; their son Charles; his first wife (the late Princess Diana); and Charles and Diana's two sons, William and Harry. (*b*) An evolutionary tree is a hypothesis—a best educated guess—of evolutionary relationships based on scientific studies. In this case the evolutionary relationships hypothesized are among swallowtail butterfly species. This tree hypothesizes that butterfly groups A and B are each other's closest relatives. They descended or evolved from a most recent common ancestor, as shown. The hollow circle connecting the two branches leading each to A and B represents the moment when the ancestor split, or evolved into two separate groups A and B.

cies can evolve to become different enough to form new species. That is, an ancestor species can give rise to new species that are its descendants. On evolutionary trees, these descendants are depicted as the tips of branches (see Figure 2.1*b*). When we trace downward from the tips of any two branches to the point at which they meet (on Figure 2.1*b*, that point is a hollow circle), we are really tracing backward through time to the moment when the ancestor species split into those two distinct **lineages**. The ancestor (depicted as the line leading up to the circle) is called the **most recent common ancestor** of the two lineages, because it is the most recent ancestral group from which the descendants arose. The farther down the tree we go, the farther back into history we are delving. Note that any two groups may have many common ancestors, but only one most recent common ancestor.

Shared derived features help identify the most recent common ancestor

Descendants share features because they share an ancestor. In human families, a father's distinctive nose may be seen on the faces of all his children. Similarly, when biologists compare any groups of organisms—for example, the organisms in Figure 2.2—they find many characteristics in common among them. All these animals eat food to survive. Some have four legs, others have fins, some have hands. All these animals are built of cells. All these animals can move, and so on. How do biologists decide which features to compare? It turns out that not all similarities between groups are equally useful for understanding relationships. In fact, most similarities are not useful for determining evolutionary relationships.

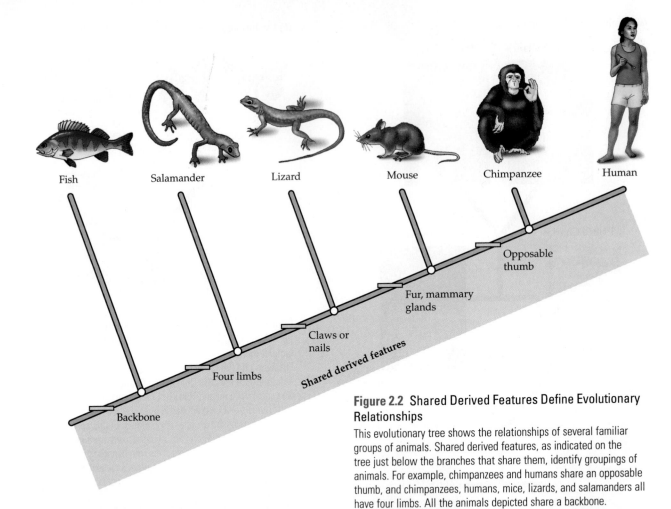

Fish Salamander Lizard Mouse Chimpanzee Human

Opposable thumb

Fur, mammary glands

Claws or nails

Shared derived features

Four limbs

Backbone

Figure 2.2 Shared Derived Features Define Evolutionary Relationships
This evolutionary tree shows the relationships of several familiar groups of animals. Shared derived features, as indicated on the tree just below the branches that share them, identify groupings of animals. For example, chimpanzees and humans share an opposable thumb, and chimpanzees, humans, mice, lizards, and salamanders all have four limbs. All the animals depicted share a backbone.

Test your knowledge of these derived characteristics. ▶ ❙❙ 2.1

The most useful are the features that originated in a group's most recent common ancestor. Those features are then shared by the descendant species, having been passed down, or derived, from the most recent common ancestor. Known as **shared derived features**, these key features mark a group as a set of close relatives. Groups that are not as closely related—that is, not descended from the same most recent common ancestor—do not display those shared derived features. Instead, they display a different set of shared derived features that is unique to themselves and their close relatives.

Figure 2.2 depicts the relationships of some familiar animals, identifying some of the shared derived features that define these groups. For example, at the upper right-hand corner of the tree, the first shared feature is opposable thumbs (thumbs that are capable of being placed against one or more of the other fingers on the hand). This is one of the many features that set chimpanzees and humans apart from the other animals on the tree; no other organisms on the tree share this feature. Most likely we share it with chimpanzees because our

most recent common ancestor had an opposable thumb, making it one piece of evidence that humans and chimpanzees share a more recent common ancestry with each other than with the other animals on the tree.

Convergent features are similarities that do not indicate relatedness

Some similarities do not indicate close evolutionary relationships and can be very misleading. For example, sometimes very distantly related organisms share features because they just happened to evolve the same feature independently. Such **convergent features** can be misleading. For example, the panda, like the human and the chimpanzee, has an opposable thumb (Figure 2.3). Should pandas be placed beside humans and chimps on an evolutionary tree? The answer is a clear no, based on everything else scientists know about humans, chimpanzees, and pandas. Therefore, we must share an opposable thumb with pandas—our distant relations—simply because pandas evolved this useful feature independently.

(a)  (b)  (c)

Figure 2.3 Misleading Convergent Features

A panda's opposable thumb (*a*), which helps it grasp the bamboo shoots it loves so much, would seem to make it a very close relative of humans and chimpanzees. But in actuality its "thumb" is an enlarged wrist bone that evolved separately. Notice that humans (*b*) and chimps (*c*) have four digits plus an opposable thumb; the panda has five digits plus an opposable thumb.

Though it might seem simple to separate similarities that are convergent features from those that are true shared derived features, in practice it can be quite difficult and it remains one of the major challenges for systematists.

DNA comparison is a powerful new way to study evolutionary relationships

Traditionally, biologists looking for similarities that suggest close evolutionary relationships have examined structural characteristics: the number of legs, the arrangement of petals in a flower, the anatomy of an animal's heart, and so on. In recent years, however, researchers have begun studying other features, including behaviors. Perhaps the most powerful new feature being studied by systematists is an organism's DNA.

All living organisms use DNA as their hereditary material. Therefore, by investigating shared derived features of organisms' DNA, biologists have been able to study the relationships of many different groups that were difficult or impossible to study before. An example is the relationships among major groups such as bacteria, plants, and animals—groups whose bodies are so different that they are often impossible to compare on

the basis of structural features alone. These relationships are described in Sections 2.3 and 2.4. Researchers are also learning more about groups that were difficult to study previously because their structures are so simple. For example, certain kinds of worms—in particular, those that live as parasites inside other organisms—are difficult to compare because they are often tiny, smooth, and nearly featureless. Systematists can even unravel criminal mysteries using DNA (see the box on page 35).

2.2 Using Evolutionary Trees to Predict the Biology of Organisms

Because evolutionary trees are hypotheses of the relationships of organisms, they can generate predictions. That's because close relatives can be predicted to share many features passed down by their most recent common ancestor.

As surprising as it might seem, there is now overwhelming evidence from living and fossil animals

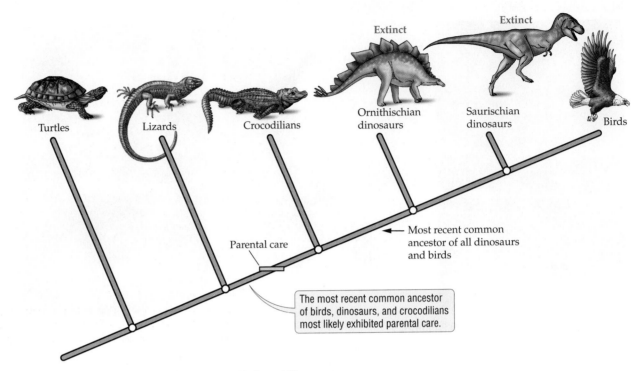

Extinct Extinct

Turtles   Lizards   Crocodilians   Ornithischian   Saurischian   Birds
                                       dinosaurs   dinosaurs

Parental care

Most recent common
ancestor of all dinosaurs
and birds

The most recent common ancestor
of birds, dinosaurs, and crocodilians
most likely exhibited parental care.

Figure 2.4 The Close Relationship between Birds and Dinosaurs
On this evolutionary tree, birds and the two lineages of dinosaurs—ornithischians [ore-nih-THISS-shee-unz]
(plant eaters such as *Stegosaurus*) and saurischians [sore-ISS-shee-unz] (animals such as *Tyrannosaurus*)—
share the most recent common ancestor. The next-closest relatives of this group are the crocodilians,
including crocodiles and alligators.

that the closest relatives of birds are the extinct creatures we know as dinosaurs. Of the animals shown in Figure 2.4, the next-closest relatives of birds are crocodiles and alligators—a group known as crocodilians. Knowing the relationships among these groups has made it possible for biologists to do the seemingly impossible: make predictions about the behavior of long-extinct dinosaurs.

Crocodilians and birds are known to be dutiful parents. They both build nests and defend their eggs and young. Scientists reasoned that if crocodilians and birds both exhibit extensive and complex parental care, then their most recent common ancestor probably exhibited this behavior as well. Because dinosaurs share this same common ancestor, scientists were able to predict that these long-gone and unobservable creatures had tended their eggs and hatchlings too—a shocking notion for creatures with a reputation for being big, vicious, and pea brained.

Recently, researchers confirmed that dinosaurs had exhibited parental care. These researchers discovered the fossil of a dinosaur that had died 80 million years ago while sitting on a nest of eggs in what today is the Gobi Desert in Mongolia. At the time this dinosaur was originally unearthed—in 1923—the idea that it might be sitting on a nest of its own eggs was so unthinkable that the assumption was that it had been attacking and eating the eggs. Hence, its discoverers gave it the name *Oviraptor* [OH-vee-rap-ter], which means "egg seizer." Not knowing then that birds and dinosaurs are close relatives, biologists did not expect the two groups to show similar behaviors. But in 1994, a new evolutionary tree—a revised hypothesis—for the evolutionary relationships of dinosaurs, birds, and crocodilians made *Oviraptor*'s intentions toward the eggs clear. Rather than attacking the eggs, *Oviraptor* appears to have died in a sandstorm protecting its nest, its limbs encircling its unhatched young in a posture as protective as that of any bird (Figure 2.5).

Concept Check

1. What is an evolutionary tree?
2. What features are most informative in determining which groups of organisms are most closely related?

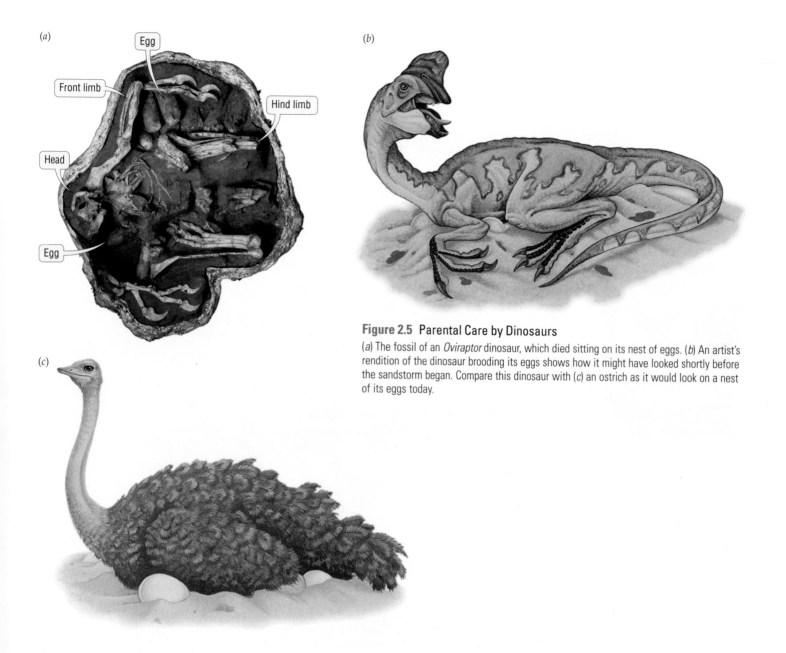

(a)

Egg

Front limb

Head

Egg

Hind limb

(b)

(c)

Figure 2.5 Parental Care by Dinosaurs

(*a*) The fossil of an *Oviraptor* dinosaur, which died sitting on its nest of eggs. (*b*) An artist's rendition of the dinosaur brooding its eggs shows how it might have looked shortly before the sandstorm began. Compare this dinosaur with (*c*) an ostrich as it would look on a nest of its eggs today.

2.3 A Classification System for Organizing Life: The Linnaean Hierarchy and Beyond

In addition to using evolutionary trees to show relatedness, biologists use a hierarchical classification system. The system was developed in the 1700s by the father of modern scientific naming, a Swedish biologist named Carolus Linnaeus [lih-*NEE*-us], and is known as the **Linnaean** [lih-*NEE*-un] **hierarchy** (Figure 2.6). The spe-

cies is the smallest unit (lowest level) of classification. Closely related species are grouped together to form a **genus** (plural "genera"). Using these two categories in the hierarchy, every species is given a unique, two-word Latin name called its **scientific name.** The first word identifies the genus to which the organism belongs, and the second defines the species. For example, humans are called *Homo sapiens*: *Homo* ("man") is the genus to which we belong, and *sapiens* [*SAY*-pee-enz] ("wise") is our species name. We are the only living species in our genus. Other species in our genus include *Homo erectus* ("upright man") and *Homo habilis* ("handy man"), both of which are extinct.

Homo habilis

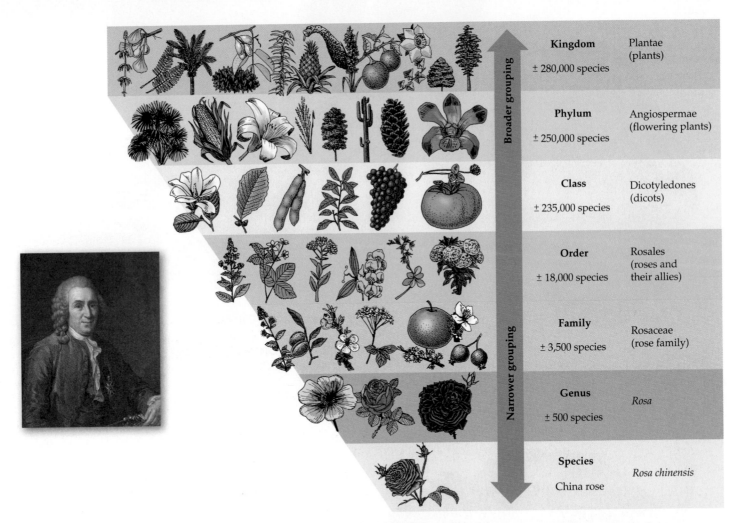

Kingdom ± 280,000 species	Plantae (plants)	
Phylum ± 250,000 species	Angiospermae (flowering plants)	
Class ± 235,000 species	Dicotyledones (dicots)	
Order ± 18,000 species	Rosales (roses and their allies)	
Family ± 3,500 species	Rosaceae (rose family)	
Genus ± 500 species	*Rosa*	
Species China rose	*Rosa chinensis*	

Broader grouping ↑ — Narrower grouping ↓

Figure 2.6 The Linnaean Hierarchy

The smallest unit of classification is the species (here, the China rose, whose scientific name is *Rosa chinensis*). This species belongs to the genus *Rosa*, which includes other roses. The genus *Rosa* lies within the family Rosaceae [roze-*ACE*-ee-ee], which lies within the order Rosales [roze-*AH*-leez], within the class Dicotyledones [dye-kot-oh-*LEE*-duh-neez], within the phylum Angiospermae and the kingdom Plantae. We can use the same categories—from species to kingdom—to classify all organisms. This classification system was first devised by the Swedish naturalist Carolus Linnaeus (inset).

All organisms belong to higher levels in the hierarchy as well, even though the scientific name does not indicate those levels. Closely related genera are grouped together into a **family**. Closely related families are grouped into an **order**. Closely related orders are grouped into a **class**. Closely related classes are grouped into a **phylum** [*FYE*-lum] (plural "phyla"). Finally, closely related phyla are grouped together into a **kingdom**.

Systematists refer to groups of organisms at these various levels of classification as taxonomic groups, or more simply, as taxa (singular "**taxon**"). Using Figure 2.6 as an example, we see that the species *Rosa chinensis* is a taxon, but so are the higher levels of classification to which it belongs: Rosaceae, Rosales, Dicotyledones, and so on up to Plantae. Each of these is a taxon, or taxonomic group, to which the China rose belongs, along with all the other organisms grouped with it in that taxon.

As biologists have learned more about different organisms and as evolutionary trees have changed, the Linnaean hierarchy has continued to change as well. For example, Linnaeus originally described just two kingdoms: plants and animals. Today biologists recognize anywhere from 5 to 13 or more kingdoms and there continues to be vigorous debate about exactly

Review the Linnaean hierarchy. ▶❙❙ 2.2

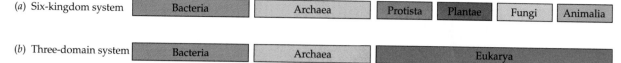

Figure 2.7 Kingdoms and Domains

This book employs both the widely used six-kingdom system (*a*) and the three-domain system (*b*). The domain Bacteria is equivalent to the kingdom Bacteria, and the domain Archaea is equivalent to the kingdom Archaea. The domain Eukarya encompasses four kingdoms in the six-kingdom scheme: Protista (protists, which include organisms such as amoebas and algae), Plantae (plants), Fungi (including yeasts and mushroom-producing species), and Animalia (animals).

how to organize life into a Linnaean hierarchy. In this book we adopt the widely used six-kingdom system (Figure 2.7*a*). Chapter 3 will describe these six kingdoms in detail.

In addition to the kingdoms designated by the Linnaean hierarchy, most biologists use an even higher level of organization called **domains** (Figure 2.7*b*). The three domains are **Bacteria** (which includes familiar disease-causing bacteria); **Archaea** [ahr-*KEE*-uh] (bacteria-like organisms best known for living in extremely harsh environments); and **Eukarya** [yoo-*KAIR*-ee-uh] (all the rest of the living organisms, from amoebas to plants to fungi to animals). The domains describe the most basic and ancient divisions among living organisms. DNA studies initially alerted biologists to the existence of the three domains and to the fact that the Archaea, which were once thought to be just more bacteria, are a separate and ancient group. When the three-domain scheme was first proposed, it was highly controversial, but it is now widely accepted.

2.4 Branches on the Tree of Life

The study of evolutionary relationships is one of the most quickly advancing areas of biology. This rapid advancement is due in part to new DNA studies, some of which have produced surprising results.

Unexpected evolutionary relationships exist among the plant, fungus, and animal kingdoms

As new studies show, an organism's DNA is an excellent place to look for shared derived features because all living organisms carry DNA. Even extremely distantly related groups, such as domains and kingdoms, can be compared to reconstruct the arrangement of their branches on the tree of life.

Figures 2.8 and 2.9 show trees encompassing all of life: Figure 2.8 depicts one hypothesis of the relation-

Helpful to know

An easy way to remember the levels of the Linnaean hierarchy is to memorize the following sentence, in which the first letter of each word stands for each descending level: *K*ing *P*hillip *C*leaned *O*ur *F*ilthy *G*ym *S*horts.

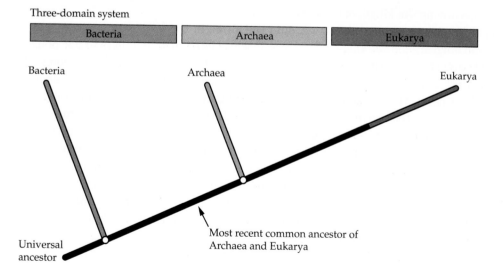

Figure 2.8 Evolutionary Tree of Domains

This tree shows one hypothesis of the relationships of the three domains. At the root of the tree is the universal ancestor, from which all living things descended. Of the three surviving lineages, the first split came between the Bacteria and the lineage that would give rise to the Archaea and Eukarya. The next split was between the Archaea and the Eukarya, making Archaea and Eukarya more closely related to each other than either group is to the Bacteria.

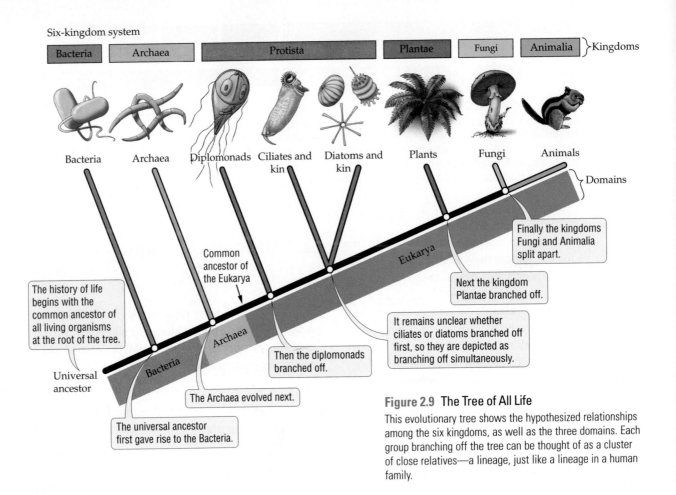

Six-kingdom system

| Bacteria | Archaea | Protista | Plantae | Fungi | Animalia | } Kingdoms |

The history of life begins with the common ancestor of all living organisms at the root of the tree.

Universal ancestor

The universal ancestor first gave rise to the Bacteria.

The Archaea evolved next.

Then the diplomonads branched off.

Common ancestor of the Eukarya

It remains unclear whether ciliates or diatoms branched off first, so they are depicted as branching off simultaneously.

Next the kingdom Plantae branched off.

Finally the kingdoms Fungi and Animalia split apart.

Domains

Figure 2.9 The Tree of All Life
This evolutionary tree shows the hypothesized relationships among the six kingdoms, as well as the three domains. Each group branching off the tree can be thought of as a cluster of close relatives—a lineage, just like a lineage in a human family.

ships of the three domains; Figure 2.9 provides the next level of detail by showing the hypothesized relationships among the kingdoms. Both of these trees are the subject of ongoing DNA comparison studies, and not all studies agree. Different studies support different evolutionary trees (or hypotheses).

In general, the relationships among the kingdoms in Figure 2.9 are not surprising. The most distant relatives of plants, animals, and fungi are the two kingdoms Bacteria and Archaea; bacteria and archaeans are single-celled organisms that, at best, seem like very distant cousins of ours. But notice the relationships among plants, fungi, and animals: the two most closely related groups among the three are fungi and animals. For decades, the fungi—which include yeasts, molds, and mushroom-producing organisms—were thought to be more closely related to plants. Unable to move and very unlike animals, these faceless organisms seem more akin to trees, shrubs, and mosses. As a result, it came as a huge surprise when recent

studies showed that fungi are actually more closely related to animals, including people, than to plants. That is, fungi share a more recent common ancestor with animals than with plants. Bringing this closer to home, the mushrooms on a pizza are more closely related to us than they are to the green peppers sitting next to them.

Could the slime in the bathroom shower really be more closely related to us than it is to our houseplants? How much do we really have in common with the likes of bread mold or yeast? A lot, it turns out. The finding that fungi and animals are close relatives solved the long-standing mystery of why doctors often have such a difficult time treating fungal infections—particularly internal infections, in which a fungus has begun living inside the human body. Because fungi and animals are such close relatives, their cells work very similarly. Thus, anything a doctor might use to kill off a fungus could kill, or nearly kill, the person as well.

Science Toolkit

The Guilty Dentist

Although biologists use the study of evolutionary relationships mainly to build evolutionary trees of the world's organisms, scientists are finding wider and wider uses for evolutionary trees. Scientists even used evolutionary trees to solve the mystery of whether a dentist infected his patients with HIV, the virus that causes acquired immunodeficiency syndrome (AIDS) in humans.

The dentist, who was himself infected with HIV, had been working, as usual, with his patients. Then some of his patients began turning up HIV positive. Against a backdrop of increasing public controversy over whether health care workers with HIV pose a risk to their patients, many people wanted to know: had this dentist transmitted the deadly virus to his patients? Adding confusion to the situation was the fact that some of the infected patients had lifestyles or habits that put them at risk for HIV infection by other means. Those on both sides of the issue—some saying the dentist should not be blamed, others saying he was a threat—argued heatedly without resolution.

To answer the question, researchers took advantage of the observation that HIV's genetic material, like that of most viruses, evolves

quickly, resulting in many different but closely related strains of HIV. The researchers created an evolutionary tree of the genetic material found in the viruses in the dentist, in each of the infected patients, and in local people who were also infected with HIV but were not patients of the dentist. If the dentist had infected his patients, then genetic material of the AIDS virus found in his patients should be more closely related to his viruses and less closely related to the viruses from other infected people.

The results showed that some of the dentist's patients probably were infected by him and that some probably were not. Two patients, X and Y, who could have been exposed to HIV as a result of their lifestyles or habits, were infected with AIDS viruses whose genetic material was not closely related to the viruses that the dentist had. However, five other patients—A, B, C, D, and E—were carrying AIDS viruses whose genetic material was very closely related to the AIDS viruses in the dentist. Moreover, those people were not at risk of contracting HIV by other means. So scientists concluded that the dentist had indeed infected at least five patients.

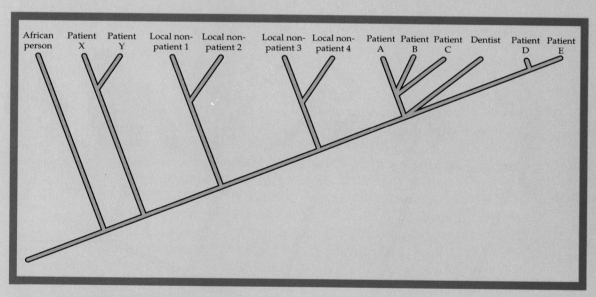

An Evolutionary Tree Solves the Whodunit

By examining the evolutionary relationships of the HIV (AIDS) viruses found in a number of infected people, including a dentist and his infected patients, researchers were able to solve the mystery of how the dentist's patients were likely to have been infected.

Figure 2.10 Astrophysicist Stephen Hawking

The brilliant scientist Stephen Hawking has suffered from Lou Gehrig's disease, or amyotrophic lateral sclerosis, for over 40 years. Despite its ravages to his body (he is paralyzed and cannot use his limbs or voice), he is responsible for breakthroughs in the fields of theoretical cosmology and quantum gravity. The study of yeasts with a similar disorder may provide hope to Hawking and the estimated 5,000 people in the United States who are diagnosed with Lou Gehrig's disease each year.

So similar are humans and fungi that they even share a similar disease. Lou Gehrig's disease is a fatal disease of humans in which the nervous system degenerates quickly (Figure 2.10). Yeasts get a similar "disorder." Thus, yeast is an excellent, if surprising, model for studying this dangerous human disease.

The primate evolutionary tree reveals the closest relative of humans

When we visit the ape house at a zoo, the striking similarities between the beings standing outside the cage looking in and those inside the cage looking out are obvious. These similarities are not surprising, given that humans, apes, and monkeys all belong to the order Primates. But which of our primate relatives is our clos-est relation? Over the years this question has generated intense interest and heated debate. Researchers have studied everything from bone structure to behavior to DNA in attempts to determine which primate is humankind's closest kin.

Though controversy remains, the emerging consensus suggests that our closest relative is the chimpanzee, a fellow tool user with whom we share a remarkable degree of similarity in DNA (Figure 2.11). More distantly related are gorillas, and beyond that orangutans, gibbons, monkeys (such as the spider monkey), and, most distantly, lemurs.

A surprising tangle can be found at the roots of the tree of life

One of the oddest and most recent surprises to come out of DNA studies is the hypothesis that, at its base, the tree of life might look a lot less like a tree than like a highly interconnected web. This idea arose when biologists began to find DNA in what seemed to be the "wrong" places. For example, scientists have found bacterial DNA in archaeans and in organisms in the Eukarya as well.

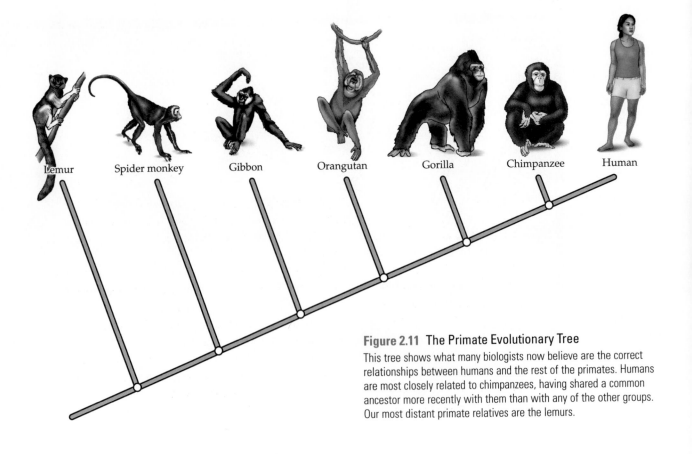

Lemur Spider monkey Gibbon Orangutan Gorilla Chimpanzee Human

Figure 2.11 The Primate Evolutionary Tree

This tree shows what many biologists now believe are the correct relationships between humans and the rest of the primates. Humans are most closely related to chimpanzees, having shared a common ancestor more recently with them than with any of the other groups. Our most distant primate relatives are the lemurs.

How could this happen if the three lineages—Bacteria, Archaea, and Eukarya—diverged long ago? Each lineage should have evolved to be quite distinct over time. Yet scientists have found that some organisms' DNA looks like a grab bag of DNA collected from across the tree of life.

One hypothesis to explain the grab bag comes from Dr. W. Ford Doolittle, a biochemist at Dalhousie University in Canada. He suggests that throughout the early history of life, both before and after the three major lineages evolved, organisms in those three lineages were freely exchanging many different genes. So in addition to genes being passed "vertically" (from one generation to the next), genes also appear to have been moving across from one lineage to another—a process known as **horizontal gene transfer** (Figure 2.12). In this view, the tree of life stands as it is, but with a lot of extra, unexpected movement of genes in an unexpected direction—horizontally. According to Doolittle's hypothesis, the base of the tree of life might be most accurately pictured as a loosely knit community of very primitive cells that were freely exchanging genes, creating the grab-bag mixtures of genes that persist in organisms across the tree of life even today. Still controversial, this startling new hypothesis remains a matter of active debate.

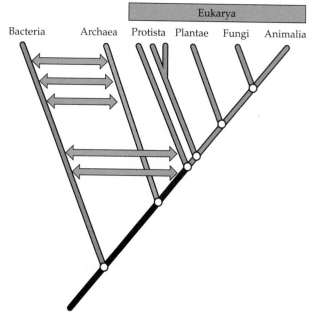

Figure 2.12 A Tangled Web at the Base of the Tree of Life?

In this representation of the evolutionary tree, green arrows represent multiple instances of horizontal gene transfer, in which genes moved among the lineages Bacteria, Archaea, and Eukarya, even after the three groups had become well established as separate lineages.

Concept Check

1. What is the Linnaean hierarchy, and how does it relate to the domain system for organizing life?
2. Are fungi more closely related to plants or animals?

Concept Check Answers

1. The Linnaean hierarchy is an organizational system for all living things, from species up to kingdoms. In the broader domain system, three domains encompass all six Linnaean kingdoms.
2. Animals, as has been shown by recent studies of DNA.

Applying What We Learned

And the True Identity of the Iceman Is …

Let's return to the mystery from the frozen Alps: who was the Iceman?

Scientists hoping to uncover his true identity looked for clues in the DNA still preserved in the Iceman's tissues. By comparing his DNA with the DNA of various groups of people around the world, they found that he shared key similarities—shared derived features in his DNA—with modern peoples of northern Europe. His DNA indicated that he was much less closely related to Egyptian or South American peoples, from whom a mummy could have been stolen and moved. Researchers solved the mystery by placing the Iceman in the tree of life—specifically, in that part of the tree that depicts

relationships among different groups of human beings. So the Iceman—dubbed Ötzi because he was found in the Ötzal Alps on the border of Austria and Italy—was no hoax after all.

Ongoing studies of Ötzi's remains continue to confirm that he lived in the region where he was found. They have also revealed details of what appear to have been the dramatic last hours of his life. In Ötzi's gut, scientists discovered ancient pollen (the sperm-like cells released by flowers) from the hop hornbeam tree. Because the hop hornbeam blooms at a specific time of year and is found only in the Schnals Valley to the south of the Ötzal Alps, scientists know that Ötzi

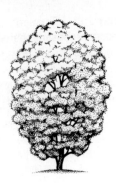

Hop hornbeam tree

had made his way out of the warm Schnals Valley and into the mountains on an early summer day. Scientists also know, from examining the minuscule remaining contents of his gut, that Ötzi had recently eaten a bit of tough, unleavened bread made from einkorn [*INE-korn*] wheat, one of the few grains known to have been domesticated by ancient Europeans at that time. And from a bit of bone and muscle fiber, researchers guess that the lucky Ötzi even had a bit of meat to chew on.

What, then, brought Ötzi to the mountains and his death, after his meal? Recent studies revealed the presence of blood from four other individuals—from two people on an arrow in his quiver, from a third person on his knife blade, and from the fourth on the back of his cloak. Scientists also found an arrow wound in Ötzi's back, a gash on his left hand, a slash on his right forearm, and three deep bruises on his side. Ötzi appears to have been in a fight for his life, attacked by several assailants and possibly carrying a wounded friend

on his back—hence the stain on his cloak. Scientists hypothesize that in attempting to escape his attackers, Ötzi fled into the mountains, where he was caught at 300 meters altitude by a snowstorm, which killed him. Or perhaps the struggle took place nearby and as the glacier continues to melt, more mummies may be revealed—perhaps that of Ötzi's killer or his friend. Betting on such possibilities, scientists have begun scouring the mountains for more mummies. If more are found, researchers will again start the process of answering the question, "Who are these people?" by placing them on the tree of life.

As unusual as Ötzi is, the way in which scientists determined who he was—by finding his correct place in the tree of all living things—is standard procedure for biologists. Whether a strange new monkey, frog, or flowering plant is discovered—or a dangerous new disease organism—knowing where an organism fits in the tree of life, as with all other biological problems in the real world, is the first order of business.

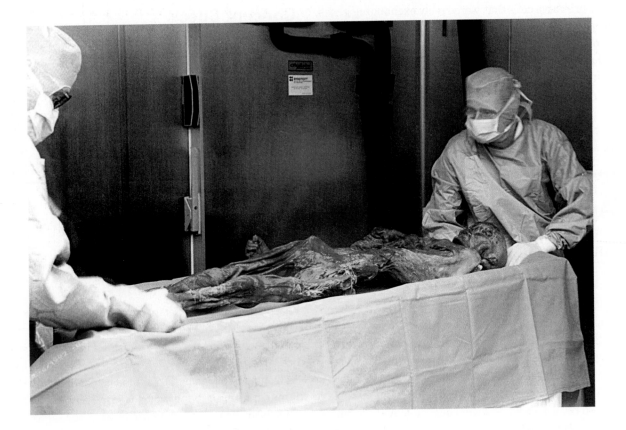

Sasquatch Sighting Reported in Yukon

It may have been a sasquatch, or perhaps some kind of bushman, but some people in the Yukon say they saw something larger than life pass by their house.

The report comes out of the community of Teslin, about 180 kilometres east of Whitehorse [in Canada's Yukon Territory]. Nine people there say a large, human-like figure covered in hair passed by the window of a house over the weekend.

Shortly afterwards, they say they saw it standing beside an abandoned car near some other homes.

Two men, Chuck Chouman and Trent Smarch, said they heard trees snapping and creaking as the mysterious figure went by, and they say there was no wind at the time. They said whatever they saw was three metres tall and moving fast, too fast for them to keep up with on foot.

And they say they have double proof this isn't just another tall tale: a huge footprint, twice the size of a human's, and a small patch of hair.

This isn't the first time there has been a reported sasquatch sighting in the area around Teslin.

In June 2004, a man and a woman reported seeing a tall figure completely covered in hair standing by the highway they were travelling on. They said as they looked back the figure crossed the highway in two giant strides and vanished into the bush.

For thousands of years sightings of the elusive, bipedal ape-man have been reported, throughout North America's Pacific northwest and as far away as China. The creature has also been a feature of aboriginal folklore.

As laughable as some people find the idea of Sasquatch or Bigfoot, it is an idea that never seems to die. In fact, throughout the history of science, what were considered unbelievable animals have been discovered again and again. For example, rhinoceros were once considered likely to be mythical, as were giant squid and the orangutan.

The sighting reported above took on extra urgency because it was the second of that summer. The same news agency, CBC News, had then reported that a man driving a ferryboat had seen, and recorded on videotape, a "tall, dark, humanoid figure moving on the riverbank." The creature appeared to be 3 meters tall and covered with hair.

In the most recent sighting, however, actual physical evidence was left behind—a clump of hair. And where there is hair, scientists can extract DNA. But what could scientists learn from such DNA? How could scientists tell whether this was a thoroughly new humanlike creature, a long-lost relative of our own species, or something else altogether? As with Ötzi, scientists could solve the mystery by finding where the organism best fit on the evolutionary tree of life.

Dave Coltman and Corey Davis, scientists at the University of Alberta in Edmonton, did just that. Had it been an entirely novel organism, these researchers would have searched for shared derived features in the mystery DNA to find where it best fit on the tree of life. But the case turned out to be much simpler. The DNA matched perfectly with an organism already in the evolutionary tree of life: the bison (sometimes referred to as buffalo). DNA is not the only evidence available, of course, and an initial study of the look, shape, and texture of the hair suggested it might be bison fur.

"If Sasquatch is indeed a primate," said Coltman, "then we would expect the sample to be closer to humans or chimpanzees or gorillas." Humans, chimps, and gorillas, remember, are species within the Linnaean hierarchy. All are part of the order—a grouping higher in the Linnaean hierarchy—known as primates.

The scientists noted that the hair recovered was not fresh, leaving open the possibility that a bison had left its hair long before, and that whatever walked past the window that summer night might still be a mystery.

Evaluating the News

1. Assuming that the hairs studied were indeed from the mystery creature, and according to the scientists' findings, which are more closely related—the mystery creature and a bison, or the mystery creature and a human? Draw an evolutionary tree showing the relationships between bison, mystery creature, and humans.
2. What evidence would it take to make systematists—the biologists who study evolutionary relationships—believe that Sasquatch is real?
3. Does the discovery that the hairs found were from a bison invalidate ancient stories of an elusive bipedal creature? Should it? Why or why not?

Source: Canadian Broadcasting Corporation News, July 13, 2005.

Chapter Review

Summary

2.1 Evolutionary Relationships Are Depicted in Evolutionary Trees

- Systematists depict hypothesized relationships among different organisms in evolutionary trees. To understand an organism's place in the diversity of life, a systematist studies the features it shares with other organisms to determine its closest relatives and thereby its place on the evolutionary tree of life.
- Like all hypotheses, evolutionary trees can generate predictions that can be tested, and like all hypotheses, evolutionary trees can be modified on the basis of new studies.
- Closely related groups of organisms share distinctive features that originated in their most recent common ancestor. Systematists use this particular set of features, known as shared derived features, to identify lineages of closely related organisms.
- Evolutionary trees depict the hypothesized evolutionary relationships of groups of organisms. The tips of branches represent existing groups of organisms. The point at which they branch off from one another—here denoted by a hollow circle—represents the moment when the most common recent ancestor split into the two descendant groups. And the line on a tree leading up to that point represents the most common recent ancestor.
- Convergent features are shared by distantly related organisms because the features evolved independently, not because they were inherited from the organisms' most recent common ancestor. Thus, they are not useful for determining relatedness.
- DNA comparison is a useful new tool for studying evolutionary relationships.

2.2 Using Evolutionary Trees to Predict the Biology of Organisms

- Evolutionary trees can predict features of organisms and can lead to surprising discoveries. An example is the realization that long-extinct dinosaurs exhibited parental care, protecting nests of their eggs, just as crocodilians and birds do today. All three groups share a common ancestor that probably exhibited parental care.

2.3 A Classification System for Organizing Life: The Linnaean Hierarchy and Beyond

- The Linnaean hierarchy is a classification system for organizing life forms. In this scheme, every species of organism has a two-part scientific name indicating its genus and species.

- The lowest level of the Linnaean hierarchy is the species. From there the hierarchy moves upward to genera, families, orders, classes, phyla, and kingdoms.
- Like evolutionary trees, the organization of life into the Linnaean hierarchy changes with new knowledge. There is still active debate over how many kingdoms life should be divided into. In this book we use the common six-kingdom scheme.
- Biologists use a level of classification above kingdoms, known as domains. The domains are Bacteria, Archaea, and Eukarya.

2.4 Branches on the Tree of Life

- DNA studies have increased our understanding of the tree of life. Some DNA studies provide interesting surprises.
- From DNA studies, biologists now know that the kingdom Fungi shares a more recent common ancestor with the kingdom Animalia than it does with the kingdom Plantae. This finding overturns the long-held assumption that fungi were more closely related to plants.
- Scientists have greatly improved our understanding of the primate evolutionary tree. There is growing agreement that chimpanzees are humans' closest primate relatives.
- DNA studies indicate that the base of the tree of life may look more like a complicated web than a simple, single root. The evidence for this web is the grab-bag nature of DNA in some organisms. Some biologists explain this pattern as a result of horizontal gene transfer among kingdoms early in the history of life.

◉ Review and Application of Key Concepts

1. How do biologists identify an unknown organism and place it on the tree of life?

2. How are family trees like evolutionary trees? How do they differ?

3. Define "shared derived features" and describe how they differ from other similarities between organisms. Why aren't the panda's thumb and the human thumb shared derived features?

4. How is an evolutionary tree a hypothesis?

5. Why did biologists think that dinosaurs might exhibit parental behavior? Use the tree in Figure 2.4 to explain.

6. It is well known that birds can sing, and that they sing often. Crocodilians are also known to make similar chirping vocalizations. Judging by the tree in Figure 2.4, what might scientists hypothesize about the singing abilities of dinosaurs? And why?

7. What are the levels of the Linnaean hierarchy? Which is the smallest grouping? Which is the largest grouping?

8. Why has the study of DNA made such revolutionary changes in our understanding of the evolutionary tree of life? To defend your statement, give one example of a change made by a DNA study.

Key Terms

Archaea (p. 33)
Bacteria (p. 33)
class (p. 32)
convergent features (p. 28)
domain (p. 33)
Eukarya (p. 33)
evolutionary tree (p. 26)
family (p. 32)
genus (p. 31)
horizontal gene transfer (p. 37)
kingdom (p. 32)

lineage (p. 27)
Linnaean hierarchy (p. 31)
most recent common ancestor
(p. 27)
order (p. 32)
phylum (p. 32)
scientific name (p. 31)
shared derived features (p. 28)
systematist (p. 26)
taxon (p. 32)

Self-Quiz

1. In Figure 2.1*b*, which two are most closely related?
   a. A and C   c. A and B
   b. D and E   d. B and D

2. The most powerful new tool being used by biologists to determine evolutionary relationships today is
   a. behavior.   c. DNA.
   b. the cell.   d. organs.

3. As depicted in Figure 2.11, the closest relative of humans is the
   a. chimpanzee.   c. orangutan.
   b. gorilla.   d. lemur.

4. Dinosaurs and crocodilians most likely exhibit similar parental behaviors because
   a. they can both be dangerous predators.
   b. they are ancient animals.
   c. they are both closely related to birds.
   d. they share a common ancestor that exhibited parental behaviors.

5. Which of the following can be concluded from Figure 2.9?
   a. Archaea and Bacteria are more closely related to each other than either is to a squirrel.
   b. Fungi, animals, and plants are equally closely related.
   c. Animals and fungi are more closely related to each other than either is to plants.
   d. Diplomonads are the most ancient group known.

6. Which of the following groupings list only domains?
   a. Eukarya, Bacteria, and Animalia
   b. Plantae, Protista, and Archaea
   c. Archaea, Bacteria, and Eukarya
   d. Bacteria, Archaea, and ciliates

7. Bacteria are
   a. a kingdom.
   b. a domain.
   c. part of the Eukarya.
   d. a kingdom and a domain.

8. The Linnaean hierarchy is organized as follows:
   a. species, order, class, genus, phylum, kingdom, class.
   b. species, genus, family, order, phylum, class, kingdom.
   c. species, class, order, phylum, kingdom, genus, family.
   d. species, genus, family, order, class, phylum, kingdom.

9. The point at which two lineages split off from one another (represented for example, in Figure 2.1*b* by a hollow circle) is
   a. a convergent feature.
   b. the lineage.
   c. the moment when the ancestor species split into two descendant species.
   d. the shared derived feature.

10. Convergent features
    a. are similarities that indicate close evolutionary relatedness.
    b. are shared derived features.
    c. are similarities that do not indicate close evolutionary relationship.
    d. are obvious differences between two species.

3 Major Groups of Living Organisms

Key Concepts

⊙ Biologists categorize living organisms in three major ways: (1) the domains; (2) the Linnaean hierarchy; and (3) the tree of life.

⊙ Organisms can also be categorized according to their cellular structure, as either prokaryotes or eukaryotes.

⊙ Bacteria and Archaea are prokaryotes. They are the most numerous in terms of numbers of individuals, the most widespread, and the most diverse in methods of obtaining nutrients. They can act as producers, consumers, and decomposers.

⊙ Protists are a diverse group of single-celled and multicellular organisms that can be animal-like, plantlike, or funguslike. One of the key innovations of the Protista was sexual reproduction. Some protists represent early stages in the evolution of multicellularity.

⊙ Plants pioneered living on land. Plants evolved numerous evolutionary innovations, including seeds and flowers. Plants are producers; they provide the nutrients that almost all land-living consumers eventually use.

⊙ Fungi have evolved a unique body plan that allows them to penetrate other organisms, digest their tissues externally for food, and then absorb that material. Fungi can be critical components of ecosystems, acting primarily as decomposers. Some fungi are mutualists with algae and plants. Some fungi are parasites.

⊙ Animals are multicellular and range in complexity from sponges to mammals. Some key evolutionary innovations of animals include specialized tissues, organs and organ systems, complete body cavities, and an astounding range of behaviors. Insects (an animal group) are the most species-rich group of all organisms. Animals act mostly as consumers but also as decomposers.

⊙ Viruses are not classified into any kingdom or domain.

Weird, Wild, and Wonderful

In December of 2006, a team of Japanese scientists hauled her out of the sea, a gargantuan beast 7 meters long—a writhing, muscular, sucker-covered, deep-red mystery of a creature—the first of her kind that had ever been seen alive. She was a giant squid, a bizarre animal with nightmarishly long tentacles adapted to terrorize the tasty creatures of the ocean bottom. Yet as huge as she was, she and her kind had remained a mystery, perhaps in part because giant squid are built to swim quickly not only toward their prey but, when necessary, out of danger and out of sight.

This monstrous denizen of the deep is only one of many amazing organisms on Earth. The Indonesian plant known as *Rafflesia* grows a flower that can stretch to a meter (over 3 feet) across—a blossom that smells like rotting flesh, the better to attract flies with. The peregrine falcon, once in danger of extinction, has evolved to be such a powerful flier that it can dive through the air at a speed of 250 kilometers (about 155 miles) an hour. Less flashy, but equally impressive, are single-celled Archaeans, bacteria-like organisms, that have evolved the ability to live in nearly boiling water.

It all began with that original living thing that arose more than 3.5 billion years ago: a simple, single cell in an ancient, primordial soup. From that first form, all these living things and so many more have descended, some having evolved advantageous new characteristics that allowed major groups of organisms to colonize not only the oceans, but rivers and lakes, as well as the land from mountains to deep valleys, to deserts, ice fields, and even the air.

What are the major groups of organisms on Earth, and what features have they evolved that have allowed them to spread, thrive, and succeed?

That's a Lot of Calamari
Scientists pull a 7-meter-long giant squid alive from the sea for the first time ever in December of 2006. Yet, as amazing as giant squid are, they are only one of many living things that have evolved amazing adaptations to thrive on Earth.

As we saw in Chapter 1, all living organisms share a basic set of characteristics. Life shares this set of common properties because all living organisms descended from a common ancestor, known as the universal ancestor.

Biodiversity is all the world's living things, in every size, shape, and form. This great diversity of life is still far from being completely known, counted, or named. Most biologists estimate that there are between 3 million and 30 million species on Earth. No chapter, including this one, could provide a comprehensive examination of all the world's species. Instead, here we examine the major groups of organisms. First we orient ourselves by looking at how biologists organize the major groups. Then we look at key features characterizing each group, especially features that allow that group to thrive—advantageous characteristics known as **adaptations**. In addition, we describe some of the more important and interesting members of each group. We close by considering the places left on Earth that are still unexplored and that could harbor unknown life, as well as unknown adaptations.

Explore the six kingdoms of life.

▶‖

3.1

3.1 Getting Oriented: How Biologists Organize and Classify the Major Groups

The major groups of organisms are the Bacteria (including familiar disease-causing bacteria, like *Chlamydia* which can cause a sexually transmitted disease); the Archaea (bacteria-like organisms, some of which can live in extreme environments); the Protista (a diverse group that includes amoebas and algae); the Plantae [*PLAN*-tee] (plants); the Fungi [*FUNJ*-eye] (mushrooms, molds, and yeasts); and the Animalia (animals).

All of life is classified in three major ways

In Chapter 2 we learned about three different ways of organizing living organisms: (1) life's three domains, (2) the Linnaean hierarchy, and (3) the evolutionary tree of life. Figure 3.1 shows how biologists organize the major groups of organisms in each of those three different systems. Recall

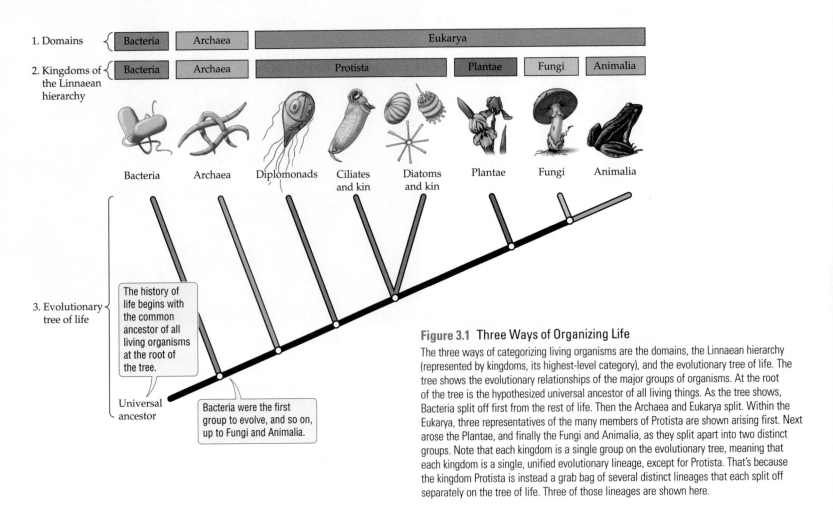

Figure 3.1 Three Ways of Organizing Life
The three ways of categorizing living organisms are the domains, the Linnaean hierarchy (represented by kingdoms, its highest-level category), and the evolutionary tree of life. The tree shows the evolutionary relationships of the major groups of organisms. At the root of the tree is the hypothesized universal ancestor of all living things. As the tree shows, Bacteria split off first from the rest of life. Then the Archaea and Eukarya split. Within the Eukarya, three representatives of the many members of Protista are shown arising first. Next arose the Plantae, and finally the Fungi and Animalia, as they split apart into two distinct groups. Note that each kingdom is a single group on the evolutionary tree, meaning that each kingdom is a single, unified evolutionary lineage, except for Protista. That's because the kingdom Protista is instead a grab bag of several distinct lineages that each split off separately on the tree of life. Three of those lineages are shown here.

that Bacteria and Archaea are each both a kingdom and a domain, whereas Eukarya is a single domain made up of four kingdoms.

To help us keep track of where each of the major groups is in each of the different ways of organizing the living world, a guidepost diagram will be provided in the margin at the beginning of each section where a new group is introduced. This diagram will highlight that group on the tree of life and identify which kingdom or domain it belongs to. In addition, for each major group a large overview figure shows an evolutionary tree of the subgroups within that major group, and a photo gallery illustrates some of the prominent subgroups. These overview figures provide an evolutionary framework for discussing each group's major **evolutionary innovations**—that is, the new adaptations that have allowed its members to live and reproduce successfully.

Organisms can also be identified as prokaryotes and eukaryotes

In addition to using the three systems just described, biologists refer to organisms as either prokaryotes or eukaryotes. **Eukaryotes** (organisms in the domain Eukarya) are distinguished from **prokaryotes** (the Bacteria and the Archaea) by the structure of their cells. Eukaryotes contain specialized compartments inside of each of their cells, known as **organelles**, compartments that perform specialized functions. For example, one important compartment in eukaryotic cells is the nucleus, which holds the genetic material (DNA). Although commonly used, the terms "eukaryote" and "prokaryote" simply refer to the structure of an organism's cells and are not part of the domain, kingdom, or tree-of-life classifications.

3.2 The Bacteria and Archaea: Tiny, Successful, and Abundant

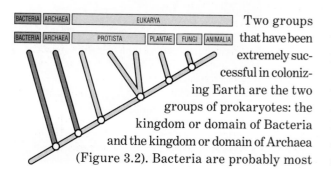

Two groups that have been extremely successful in colonizing Earth are the two groups of prokaryotes: the kingdom or domain of Bacteria and the kingdom or domain of Archaea (Figure 3.2). Bacteria are probably most familiar as single-celled disease-causing organisms such as *Streptococcus pneumoniae* [noo-*MOH*-nee-eye], which can cause pneumonia in humans. However, most bacteria are not harmful to humans. Discovered in the 1970s, the Archaea are single-celled bacteria-like organisms, some of which are **extremophiles** (literally, "lovers of the extreme"): they have been found thriving in boiling hot geysers, highly acidic waters, highly salted foods, and the freezing-cold seas off Antarctica. There is still uncertainty about this deepest part of the tree of life. We present one of the hypotheses here, depicting the Bacteria arising first and then the lineage that it split off itself splitting into the Archaea and the Eukarya.

Although they are distinct groups, Bacteria and Archaea are similar in both size (microscopic) and structure (single-celled); they are similar in many other ways as well. So it is appropriate that we begin our introduction to the major groups of life by describing these two groups at the same time.

Bacteria and Archaea are quite variable in shape, with some having shapes like spheres (called cocci [*KOCK*-eye]; singular "coccus"), rods (called bacilli [ba-*SILL*-eye]; singular "bacillus"), or corkscrews (called spirochetes [*SPY*-roh-keets]); however, they all share a basic structural plan (Figure 3.3). The picture of efficiency, these organisms are nearly always single-celled and small. They typically have much less DNA than the cells of organisms in the Eukarya have. Eukaryotic genetic material is often full of what appears to be extra DNA, known as junk DNA, that serves no clear function. In contrast, prokaryotic genetic material contains only DNA that is actively used for the survival and reproduction of the bacterial cell. Prokaryotic reproduction is similarly uncomplicated: prokaryotes typically reproduce by splitting in two—a process called fission.

Prokaryotes represent simplicity translated into success

When most people think of the living world, they tend to think in terms of butterflies, tigers, and orchids, but though we seldom notice microscopic organisms, the vast majority of life on Earth is in fact single-celled and prokaryotic. Although scientists are still struggling to assess the number of prokaryotic species, they estimate that the number of individual prokaryotes on Earth is about 5,000,000,000,000,000,000,000,000,000,000 (5×10^{30}). The success of prokaryotes is due, in part, to how quickly they reproduce. Overnight, a single bacterium of the common species *Escherichia coli* (usually referred

Figure 3.2 The Prokaryotes: Bacteria and Archaea

Prokaryotes are simple, microscopic, single-celled organisms that are the most ancient forms of life. In the evolutionary tree shown, scientists hypothesize that Bacteria branched off first, then the Archaea split off from what would become the rest of the living organisms, the Eukarya.

- Number of species discovered to date: ~4,800
- Functions within ecosystems: Producers, consumers, decomposers
- Economic uses: Many, including producing antibiotics, cleaning up oil spills, treating sewage
- Did you know? The number of bacteria in a person's digestive tract outnumber all the humans that have lived on Earth since the beginning of time.

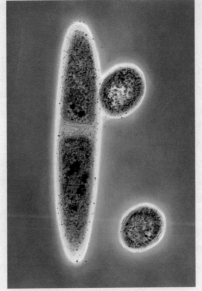
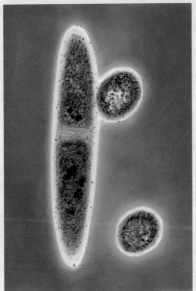

These archaeans, known as *Methanospirillum hungatii*, are shown in cross section (the two circular shapes) and as an elongated cell that is about to fission into two cells.

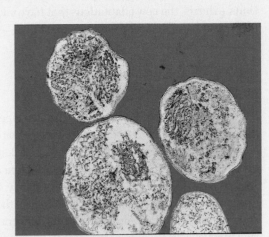

This bacterium, known as *Escherichia coli* (or, more commonly, *E. coli*), is usually a harmless inhabitant of the human gut. However, toxic strains can contaminate and multiply on foods, such as raw hamburger, and cause illness or death in humans who eat them.

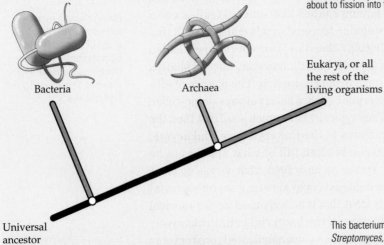

Bacteria

Archaea

Eukarya, or all the rest of the living organisms

Universal ancestor

This bacterium is a member of the genus *Streptomyces*, which produces the antibiotics streptomycin, erythromycin, and tetracycline.

The bacterium *Chlamydia trachomatis* causes the most common sexually transmitted disease, chlamydia.

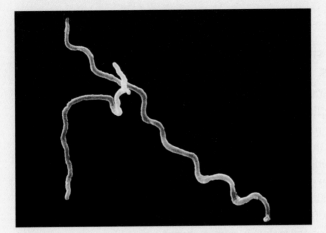

These bacteria, *Borrelia burgdorferi*, known as spirochetes because of their spiral-shaped cells, cause Lyme disease, which is transmitted to humans through a tick bite.

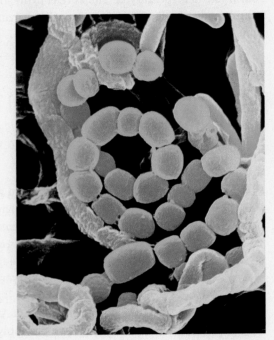

to by the abbreviated form *E. coli*), which normally lives harmlessly in the human gut (see Figure 3.2), can divide to produce a population of 16 million bacteria.

Prokaryotes are also the most widespread of organisms, able to live in many places where no other life can exist, such as the lightless ocean depths, the insides of boiling-hot geysers, and miles below Earth's surface. Because of their small size, prokaryotes also live in great numbers on and in other organisms. Scientists estimate that 1 square centimeter of healthy human skin is home to between 1,000 and 10,000 bacteria.

In addition, though many prokaryotes need the gas oxygen to survive—that is, they are **aerobes**, from *aero*, "air"; *bios*, "life"—many others can survive without oxygen. The latter are **anaerobes**, from *an*, "without." This ability to exist in both oxygen-rich and oxygen-free environments also increases the number of habitats in which prokaryotes can persist. But the real key to the success of these groups is the great diversity of ways in which they obtain and use nutrients.

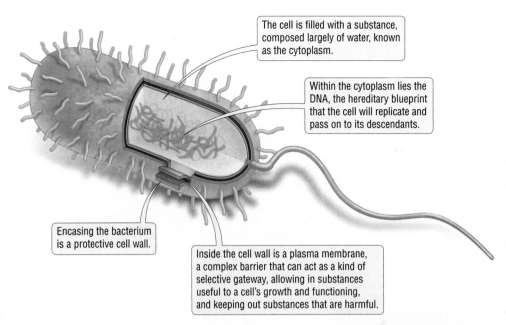

The cell is filled with a substance, composed largely of water, known as the cytoplasm.

Within the cytoplasm lies the DNA, the hereditary blueprint that the cell will replicate and pass on to its descendants.

Encasing the bacterium is a protective cell wall.

Inside the cell wall is a plasma membrane, a complex barrier that can act as a kind of selective gateway, allowing in substances useful to a cell's growth and functioning, and keeping out substances that are harmful.

Figure 3.3 The Basic Structure of the Prokaryotic Cell
Prokaryotic cells tend to be about 10 times smaller than the cells of organisms in the Eukarya, and have much less DNA.

Get to know the basic parts of a bacterial cell.
3.2

Prokaryotes exhibit unmatched diversity in methods of obtaining nutrients

All organisms require nutrients to survive, grow, and reproduce. Nutrients provide energy to organisms. Nutrients also provide key chemical elements, like carbon. Carbon is the chemical building block used to make critical molecules for life, such as proteins and DNA.

Prokaryotes are distinguished by having more diverse methods of obtaining nutrients than any group of eukaryotes or even all the eukaryotes combined—that is, all the other living things on Earth.

When humans and other animals need nutrients, we consume other species. Our bodies then break down the tissues of the other species, from which we get both energy (in the form of chemical bonds, which we'll learn more about in Unit 2) and carbon (in the form of carbon-containing molecules). In fact, many organisms, including all animals, all fungi, and some protists, get their nutrients by consuming other organisms—or, in humans, by what we call eating. Prokaryotes can do the same. Some prokaryotes can even get carbon nutrients from carbon-containing compounds (such as pesticides) that are not parts of other organisms. These organisms are all known as **chemoheterotrophs**. The first part of the term—*chemo*—denotes where the organism gets its energy, in this case from chemical compounds, like sugars or carbohydrates. The second part of the term—*hetero*—means where the organism gets its carbon, *hetero* meaning "other" and here denoting that

the carbon comes from molecules taken most often from other organisms. The term *troph* means "to eat."

Plants and the remaining protists, in contrast, harness the energy of sunlight and gather carbon from carbon dioxide, the gas in the air that humans and other animals exhale. They use sunlight and carbon dioxide during photosynthesis to produce sugars. Some prokaryotes carry out photosynthesis as well. Cyanobacteria [sye-*AN*-oh- . . .], the green slime more commonly known as pond scum, are a familiar example of prokaryotes that use the energy of sunlight and get their carbon from carbon dioxide (Figure 3.4). (In fact, as we will see in Chapter 5, photosynthesis was originally an evolutionary innovation of cyanobacteria

Figure 3.4 Pond Scum: Bacteria That Photosynthesize
These photosynthetic bacteria, called cyanobacteria or blue-green algae, can be found growing as slimy mats on freshwater ponds.

Biology on the Job

A Passion for Life

Dr. Niles Eldredge grew up to get a job that is the stuff of dreams for many children (spending all day in one of the world's greatest natural-history museums) and the stuff of nightmares for other people (stuck all day in the back hallways of a museum full of dead animals and pressed plants). Dr. Eldredge's job has him immersed every day in exactly what this chapter talks about: the diversity of living things. He studies diversity in the science he carries out, and he handles the diversity of life in his job as a curator taking care of the museum's collections. He is even quite literally immersed in diversity as he walks past the glorious displays of the living world for which the American Museum of Natural History (AMNH) is world famous.

What is your official title? Curator in the Division of Paleontology at the American Museum of Natural History in New York City. I was also Curator-in-Chief of the Hall of Biodiversity—with overall responsibility for the scientific content—as we were developing the hall.

What would you say is the goal of AMNH's Hall of Biodiversity exhibit, or of biodiversity research in general? Our goal was to impress upon the visitor that life is beautiful, rich, vibrant, and still of great importance to humanity, and very much under threat—even though many of us live in cities and are not bombarded with life's beauties and richness every day (though there is still a lot of biodiversity in cities!).

When did biodiversity become a separate field of scientific study? In a sense, biodiversity has always been a focus of biology—but only in the past 15 years or so have scientists joined others in seeing the rapidly accelerating loss of species as the world's ecosystems continue to be degraded at an alarming pace.

What skills or personal characteristics do you think have been most beneficial to you in your job as a curator? Profound, unending curiosity and a passion for the history of life.

What do you enjoy most about your job? Thinking.

The Job in Context

Dr. Eldredge says that unending curiosity and a passion for the entire history of living organisms are crucial for doing a job that focuses on biodiversity, the diversity of all living things. But maybe good old-fashioned fun—the wonder of peering at animals, the fun of collecting wild plants, examining stony fossils, playing with ideas, roaming the wild outdoors—is equally important. For what comes across in the Hall of Biodiversity—which, by definition, covers all the living things we cover in this long chapter—is not just the work of science, but the wonder of the living world, as well as the many, many more organisms that once lived but went extinct.

It is interesting, too, that although most people do not think of museums as dynamic scientific centers, they are exactly that, with researchers like Dr. Eldredge seeking out and documenting the many fantastic organisms that live now and lived earlier on Earth. The creation of the evolutionary trees illustrated in this chapter, the estimates of numbers of species in different groups, and the sorting of organisms into kingdoms and domains—all these activities are the work of people like Dr. Eldredge. And, as he has pointed out, these researchers have become responsible not only for documenting what lives on Earth, but also for sending out a warning call that many of these organisms are in need of saving as well.

Figure 3.5 Curious Appetites
The crusty orange and yellow puddle is a colony of the organism known as *Sulfolobus* [sul-*FALL*-uh-buss], an archaean that gets its carbon from carbon dioxide, as plants do. This archaean gets its energy, however, in an unusual way—not by harnessing sunlight (as plants do), or by eating other organisms (as animals do), but by chemically processing inorganic materials (metal compounds) such as iron. This chemoautotroph is living in a volcanic vent in Japan.

that has since been successfully exploited by organisms like green algae and plants, which evolved to employ it as a key adaptation.) All these organisms are producers. They produce sugar and starches, the energy-containing compounds that all other organisms are ultimately dependent upon. They are known as **photoautotrophs**. The first part of the term—*photo*—indicates how the organism obtains energy, *photo* meaning "light," as in the energy of sunlight. The second part of the term—*auto*—means "self" and here indicates that the organism is not dependent on other organisms for its carbon but uses carbon dioxide.

The Eukarya are all either chemoheterotrophs or photoautotrophs.

Prokaryotes, however, show the most evolutionary innovations in their means of obtaining nutrients. Some are **photoheterotrophs**, meaning they use light as an energy source (as do plants) but derive their carbon from carbon-containing compounds (as opposed to deriving it from carbon dioxide as plants do). Finally, and most curiously, some prokaryotes get their energy not from light but from chemicals,

including such unlikely materials as iron and ammonia (Figure 3.5)—basically by eating rocks! These prokaryotes use carbon dioxide in the air as a carbon source. Given the rules for naming illustrated thus far, can you guess what they are called? They are known as **chemoautotrophs**.

This unrivaled diversity in modes of obtaining nutrients is another reason that prokaryotes are able to live in so many places and claim the title of "the most widespread organisms on Earth."

Prokaryotes can thrive in extreme environments

Prokaryotes can live in many places where no other organisms can live. Although some of the Bacteria thrive in unusual environments, Archaea is the group best known for the extreme lifestyles of some of its members. Some are extreme thermophiles (*thermo*, "heat"; *phile*, "lover") that live in geysers and hot springs. The cells of most organisms cannot function at such high temperatures, but thermophiles have come up with evolutionary innovations—for example, proteins that are not destroyed at high heat—that allow them to succeed where others cannot. Others are extreme halophiles (*halo*, "salt"), thriving in very salty, high-sodium environments where nothing else can live—for example, in the Dead Sea and on fish and meat that have been heavily salted to keep most bacteria away (Figure 3.6).

Figure 3.6 Better Than a Bag of Chips
For those who love salt, such as archaeans that are extreme halophiles, nothing beats a salt farm. Here in Thailand, seawater is evaporated, and vast piles of dried salt are accumulated, creating an environment that only an archaean could love. Inset photo shows a halophilic archaean species in the genus *Halobacterium*.

Not all archaeans, however, are so remote from daily experience. Members of one group, the methanogens (*methano*, "methane"; *gen*, "producer"), inhabit animal guts and produce the methane gas in such things as human flatulence (intestinal gas) and cow burps.

Prokaryotes play important roles in the biosphere and in human society

Because of their wide range of evolutionary innovations, in their modes of obtaining nutrients, in particular, prokaryotes play numerous and important roles in ecosystems and in human society. Bacteria like cyanobacteria can act as producers. Bacteria can be consumers, like the bacteria that eat our food in a process we typically refer to as rotting. Bacteria can be decomposers, like oil-eating bacteria used to clean up ocean oil spills or the bacteria that live on sewage, decomposing human waste so that it can be safely, usefully returned to the environment. Bacteria can directly aid plants as well. Plants require the chemical nitrate, which they cannot make themselves. For this they depend on bacteria that can take nitrogen, a gas in the air, and convert it to nitrate. Without these bacteria there would be no plant life, and without plant life there would be no life on land.

Of course, not all prokaryotes are helpful. Bacteria cause terrible diseases; some are the source of nightmares, such as the flesh-eating bacteria able to destroy human flesh at frightening rates. With their ability to use almost anything as food, bacteria also attack crops, stored foods, and domesticated livestock.

Bacteria and Archaea exhibit key differences

Though similar in many ways, the Bacteria and Archaea are distinct groups. Much of archaean DNA is unique to the Archaea and not found in the Bacteria. The Archaea and Bacteria also differ in how their metabolism (cellular chemistry) runs. In addition, the cells of the two groups show key structural differences: most prokaryotic cells have both a cell wall and a plasma membrane (see Figure 3.3), but some molecules in those structures differ between the two groups.

Concept Check

1. Which of the six kingdoms in the Linnaean hierarchy is *not* represented by only one branch in Figure 3.1, and why?
2. What characteristics of prokaryotes make them so numerous and so successful?

3.3 The Protista: A Window into the Early Evolution of the Eukarya

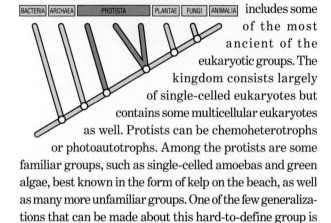

Protista includes some of the most ancient of the eukaryotic groups. The kingdom consists largely of single-celled eukaryotes but contains some multicellular eukaryotes as well. Protists can be chemoheterotrophs or photoautotrophs. Among the protists are some familiar groups, such as single-celled amoebas and green algae, best known in the form of kelp on the beach, as well as many more unfamiliar groups. One of the few generalizations that can be made about this hard-to-define group is that its members are diverse in size, shape, and lifestyle.

Much remains unclear about the evolutionary relationships of the protists to one another and to other living organisms. As a result, there are a number of competing hypotheses, and therefore different evolutionary trees, of the relationships of the protists. Figure 3.7 presents one hypothesis for the evolutionary tree of a few of the major groups of protists. Diplomonads are shown branching off first. Then, three major groups—one including dinoflagellates, apicomplexans [ay-pee-kom-*PLEX*-unz], and ciliates; another including diatoms and water molds; and the last, comprising the green algae—are shown splitting off at the same time. That's because scientists still do not know which of these three groups branched off first.

One thing that's clear is that the protists do not make up just one branch on the evolutionary tree of life. In other words, protists are a kind of evolutionary grab bag of different evolutionary lineages, whose actual evolutionary relationships remain to be worked out. In part because they are not a single evolutionary lineage, the protists show great variety in their lifestyles. Some are plantlike; for example, green algae can photosynthesize, and plants are thought to have evolved from them. Some are animal-like, such as the amoebas, moving and hunting for food. Still others, such as the so-called slime molds, act more like fungi.

Protists provide insight into the early evolution of multicellularity

Some groups of protists are of interest to biologists because they have evolved from living as single-celled creatures to forming multicellular associations that

Figure 3.7 The Protista

The protists form a diverse group of single-celled and multicellular organisms. The evolutionary relationships among protist groups are poorly understood. The tree shown is just one of a number of competing hypotheses for how protists evolved. Even here, dinoflagellates, apicomplexans, and ciliates are shown branching off the tree simultaneously, to indicate that the order in which these groups evolved is still unknown.

- Number of species discovered to date: ~30,000
- Functions within ecosystems: Producers, consumers, decomposers
- Economic uses: Kelp, a multicellular alga, is raised for food. However, protists are best known for the damage they do, causing red tides and diseases such as amoebic dysentery.
- Did you know? Malaria, the second most deadly disease after AIDS, is caused by *Plasmodium*, a protist.

Seaweeds, such as this sea lettuce, are among the most familiar protists, green algae, seen along coastal shores.

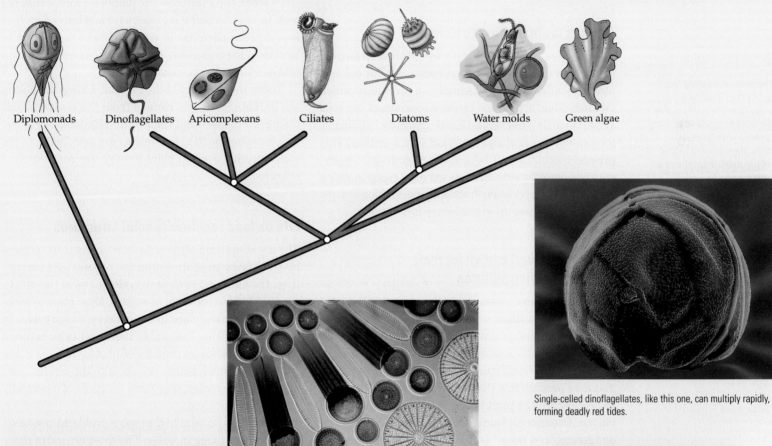

Diplomonads   Dinoflagellates   Apicomplexans   Ciliates   Diatoms   Water molds   Green algae

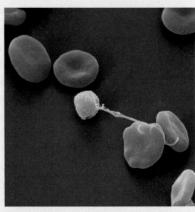

Single-celled dinoflagellates, like this one, can multiply rapidly, forming deadly red tides.

Diatoms are important producers in freshwater and marine environments. Their glasslike outer shells are microscopic works of art.

Animal-like organisms such as this *Paramecium*, a ciliate, swim using tiny hairlike structures called cilia.

An apicomplexan surrounded by red blood cells. This protist parasite, *Plasmodium*, causes malaria.

function to varying degrees like more complex multicellular individuals. Among the more interesting of these experiments in the evolution of multicellularity are the slime molds, protists that were originally mistaken for molds (which are fungi, belonging to another kingdom entirely). Commonly found on rotting vegetation, slime molds are protists that eat bacteria and live their lives in two phases: as independent, single-celled creatures and as members of a multicellular body. Like other protists that can live as either single-celled or multicellular organisms, slime molds are studied by biologists who hope to gain insight into how that key evolutionary innovation, multicellularity, first arose.

Protists were the first to have sex

Prokaryotes reproduce simply by splitting in two—a form of asexual (nonsexual) reproduction. But eukaryotes typically reproduce sexually—a process in which two individual organisms produce specialized sex cells known as male gametes and female gametes (in humans, for example, sperm and egg) that fuse together. This process combines the DNA contributions from two parents into one offspring. Protists were the group in which this form of reproduction first appeared, making sex one of its most noteworthy evolutionary innovations.

Protists are well known for their disease-causing abilities

Protists, being so diverse, can be found playing the roles of producer, consumer, or decomposer.

Although most protists are harmless, many of the best-known protists cause diseases. One example is the dinoflagellates (see Figure 3.7), a group of microscopic plantlike protists that live in the ocean and sometimes experience huge population explosions, known as blooms. Occasional blooms of toxic dinoflagellates cause dangerous "red tides." During red tides, shellfish that have eaten these toxic dinoflagellates are, in turn, poisonous to humans. The animal-like protist *Plasmodium* causes malaria, which kills millions of people around the world each year—more than any other disease except AIDS. Finally, protists left their mark on human history forever when one of them (often mistakenly referred to as a fungus) attacked potato crops in Ireland in the 1800s, causing the disease known as potato blight. The resulting widespread loss of potato crops caused a devastating famine and a major emigration of Irish people to the United States in the 1840s. Many of our family histories were altered by a protist.

Test your knowledge of the properties of multicellular eukaryotes.

▶❚❚

3.3

3.4 The Plantae: Pioneers of Life on Land

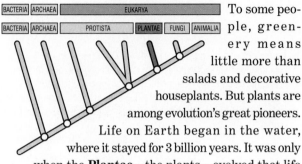

To some people, greenery means little more than salads and decorative houseplants. But plants are among evolution's great pioneers. Life on Earth began in the water, where it stayed for 3 billion years. It was only when the **Plantae**—the plants—evolved that life took to land. In doing so, plants turned barren ground into a green paradise in which a whole new world of land-dwelling organisms, including ourselves, could then evolve. Plants are photoautotrophs.

Today the diversity of the Plantae, a kingdom within the domain Eukarya, ranges from the most ancient lineages—mosses and their close relatives—to ferns, which evolved next; to gymnosperms; and finally, to the most recently evolved plant lineage, the angiosperms (Figure 3.8).

Life on land requires special structures

The key adaptation of plants is their ability to use chloroplasts in order to photosynthesize—to use light (energy from the sun) and carbon dioxide (a gas in the air) to produce food in the form of sugars. Most photosynthesis in plants takes place in their leaves, which typically grow in ways that maximize their ability to capture sunlight. A useful by-product of photosynthesis is the gas oxygen, which plants release into the air. Because plants are producers, they form the basis of essentially all food webs on land.

Organisms on land had to solve problems not faced by organisms living in water. The most crucial of these was how to obtain and conserve water. One of the evolutionary innovations allowing plants to do this is the **root system**, a collection of fingerlike growths that absorb water and nutrients from the soil. Another is the waxy covering over stem and leaves, known as the **cuticle**, which prevents plant tissues from drying out even when exposed to sun and air. A second challenge of life on land was gravity. Whereas plants can float in water, they cannot "float" in the air. But plant cells have rigid cell walls composed of a substance known as cellulose. Cell walls add to the rigidity and structural strength that a plant needs to grow up into the air.

Figure 3.8 The Plantae

Plants are multicellular organisms that make their living by photosynthesis. They are a diverse group that pioneered life on land.

- Number of species discovered to date: ~250,000
- Functions within ecosystems: Producers
- Economic uses: Flowering plants provide all our crops: corn, tomatoes, rice, and so on. Fir trees and other conifers provide most of our wood and paper. Plants also produce important chemicals, such as morphine, caffeine, and menthol.
- Did you know? Of the 250,000 species of plants, at least 30,000 have edible parts. In spite of this abundance of potential foods, just three species—corn, wheat, and rice—provide most of the food that the world's human populations eat.

Ferns and their close relatives evolved vascular systems that allowed them to grow to greater heights. This Ama'uma'u fern grows only in Hawaii.

Giant sequoia, like this huge tree, are conifers; the most familiar gymnosperms, sequoia are important wood and paper producers.

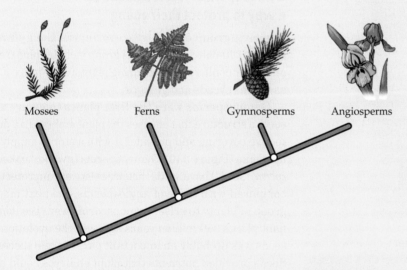

Mosses   Ferns   Gymnosperms   Angiosperms

Mosses and their close relatives, the most ancient group of plants, do not have vascular systems and so cannot grow more than a few inches high.

The orchids, the most species-rich family of angiosperms in the world, also produce some of the world's most beautiful flowers.

On Mount Sago in Sumatra, the angiosperm *Rafflesia arnoldii* produces the world's largest blossoms, measuring as much as 1 meter across.

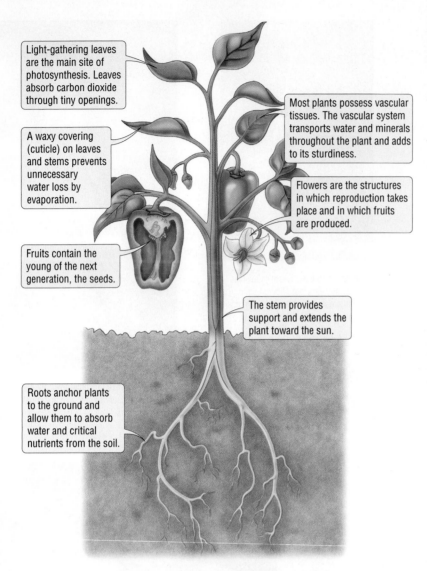

Light-gathering leaves are the main site of photosynthesis. Leaves absorb carbon dioxide through tiny openings.

A waxy covering (cuticle) on leaves and stems prevents unnecessary water loss by evaporation.

Fruits contain the young of the next generation, the seeds.

Roots anchor plants to the ground and allow them to absorb water and critical nutrients from the soil.

Most plants possess vascular tissues. The vascular system transports water and minerals throughout the plant and adds to its sturdiness.

Flowers are the structures in which reproduction takes place and in which fruits are produced.

The stem provides support and extends the plant toward the sun.

Figure 3.9 The Basic Structures of a Plant

Shown here is a familiar garden vegetable, a pepper plant. Because it is a member of the angiosperms, the last of the major plant groups to evolve, this one plant illustrates all the evolutionary innovations that distinguish plants.

In addition to the features just mentioned, three other major evolutionary innovations were critical to plants' highly successful colonization of land: vascular systems, seeds, and flowers. Each of these innovations marks the rise of a separate major plant group. Figure 3.9 shows the basic structures of a plant.

Vascular systems allow ferns and their allies to grow to great heights

Early in their evolution, plants grew close to the ground. Mosses and their close relatives, which make up the most ancient plant lineage, represent those early days

in the history of plants. These plants rely on each cell's having access to water. The innermost cells of their bodies receive water only after it has managed to pass through every cell between them and the outermost layer. Because such movement of water from cell to cell is relatively inefficient, these plants cannot grow to great heights or sizes.

Ferns and their close relatives, the next major plant group to arise, were able to grow taller because they evolved **vascular systems**—networks of specialized tissues that extend from the roots throughout the bodies of plants (see Figure 3.9). Vascular tissues can much more efficiently transport fluids and nutrients, much as the human circulatory system of veins and arteries transports blood. All plants, except mosses and their close relatives, have vascular systems.

Gymnosperms evolved seeds as a way to protect their young

The next group of plants to evolve was the **gymnosperms** [JIM-noh-spermz]. This group includes pine trees and other conifers (cone-bearing plants; see Figure 3.8), as well as cycads and ginkgos.

Gymnosperms were the first plants to evolve the **seed**, a structure that encases the plant embryo in a protective covering and provides it with a stored supply of nutrients (Figure 3.10). Gymnosperms (*gymno*, "naked"; *sperm*, "seed") have seeds that are relatively unprotected compared with those of angiosperms, the next major group of plants to arise. Gymnosperms were the dominant plants 250 million years ago, and the evolution of seeds was probably an important part of their success. Seeds provided nutrients that plant embryos could use to grow before they were able to produce their own food via photosynthesis. Seeds also provided embryos with protection from drying or rotting and from attack by predators.

Angiosperms produced the world's first flowers

Although typically we think of flowers when we think of plants, flowering plants are a relatively recent development in the history of life. Today the flowering plants, the **angiosperms** [AN-jee-oh-spermz], are the most dominant and diverse group of plants on the planet, including orchids, grasses, corn plants, and apple and maple trees. Angiosperms produce seeds that are well protected (*angio* means "vessel," referring to the tissues that encase the plant's embryo).

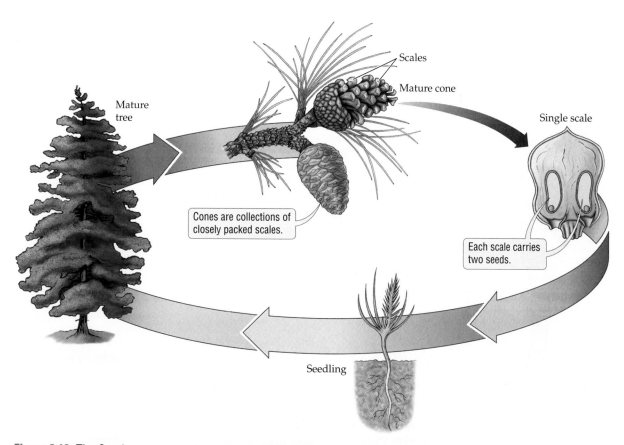

Scales

Mature cone

Single scale

Mature tree

Cones are collections of closely packed scales.

Each scale carries two seeds.

Seedling

Figure 3.10 The Seed

The first plants with seeds were the gymnosperms, which include conifers such as pine trees. Modern descendants of the first seeds can be found in pine cones.

Highly diverse in size and shape, angiosperms live in a wide range of habitats—from mountaintops to deserts to salty marshes and fresh water. Almost any plant we can think of that is not a moss, a fern, or a cone-producing tree is an angiosperm. The key evolutionary innovation of angiosperms and the key to their success is the **flower**. Flowers are specialized structures for sexual reproduction. They are the site of fertilization—the process in which male and the female gametes meet (Figure 3.11).

Pollen grains are most familiar as the irritating dust-like particles that some plants release into the air in spring and summer and can cause allergies. But pollen grains are not just nasal irritants. When pollen reaches the female reproductive parts of a flower (a process known as pollination), it can produce sperm—male gametes—that fuse with the plant's female gametes (a process known as fertilization). Some plants have pollen that is transported from flower to flower by wind (this kind can cause allergies). Some pollen is transported by animals, like bees. In this way, wind or animal pollina-

tors can facilitate sexual reproduction between even very distant plants.

Many flowers have evolutionary innovations that increase their chances of having their pollen moved or of receiving pollen from another plant. For example, some flowers provide food, such as the sugary liquid known as nectar, that attracts animals that encounter pollen and then move on, pollinating other flowers. Honeybees, for example, collect nectar, which they use to make honey, and they collect pollen to eat—but in the process of their foraging, they end up pollinating many flowers. Other flowers simply smell like food—like the giant *Rafflesia* flower, which smells like rotting flesh—to attract the flies that pollinate it.

Angiosperms have also evolved a variety of ways to distribute their seeds to distant places in order to get their young off to a good start. One of these is the use of tasty fruits that attract animals. As the embryos of angiosperms are developing, the surrounding ovary develops into a ripened fruit (see

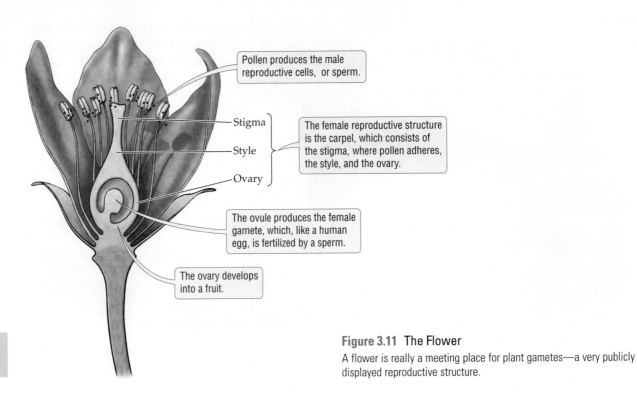

Pollen produces the male reproductive cells, or sperm.

Stigma

Style

Ovary

The female reproductive structure is the carpel, which consists of the stigma, where pollen adheres, the style, and the ovary.

The ovule produces the female gamete, which, like a human egg, is fertilized by a sperm.

The ovary develops into a fruit.

Test your knowledge of the parts of a flower. ▶❙❙ 3.4

Figure 3.11 The Flower
A flower is really a meeting place for plant gametes—a very publicly displayed reproductive structure.

Figure 3.11). Very fleshy, tasty, aromatic fruits can attract animals that eat them and excrete the seeds in their feces. These nutrient-rich wastes provide a good place for the seeds to begin life, often far from their parent plant, where they will not compete with that parent for water, nutrients, or light. Other plant seeds have evolved other innovations for travel (Figure 3.12).

(a)

(b)

Figure 3.12 Getting Around
Plants have evolved many ways of spreading to new areas. (a) A palm tree seed in a coconut fruit can float for hundreds of miles until it reaches a new beach where it can take root and grow. (b) Some seeds have wings (for example, maple "keys") or other structures (such as milkweed fluff, shown here) that allow them to be carried by the wind, sometimes over great distances.

Plants are the basis of land ecosystems and provide many valuable products

It is difficult to overstate the significance of plants. They play the role of producers. As photosynthesizing organisms, plants use sunlight and carbon dioxide to make sugars, food that they and the organisms that eat them can use. Nearly all organisms on land ultimately depend on plants for food, either directly by eating plants or indirectly by eating other organisms (such as animals) that eat plants or that eat other organisms that eat plants, and so on. Many organisms live on or in plants, or on or in soils largely made up of decomposed plants. Many plants live in water as well, some familiar ones being water lilies and duckweed. And just as on land, in the water plants can be key sources of food for other organisms.

Flowering plants provide humans with materials such as cotton for clothing and with pharmaceuticals such as morphine. Essentially all agricultural crops are flowering plants, and the entire floral industry rests on the reproductive structures of angiosperms. Gymnosperms such as pines, spruces, and firs are the basis of forestry industries, providing wood and paper.

As valuable as plants are when harvested, they are also valuable when left in nature. By soaking up rainwater in their roots and other tissues, for example, plants prevent runoff and erosion that can contaminate streams. Plants also recycle carbon dioxide and produce the gas we breathe, oxygen.

Concept Check

1. Why are protists said to have had sex first?
2. Why are protists so diverse in so many ways?
3. Name three key evolutionary innovations of plants.
4. Why are plants so important in the colonization of land and in land-based food webs?

3.5 The Fungi: A World of Decomposers

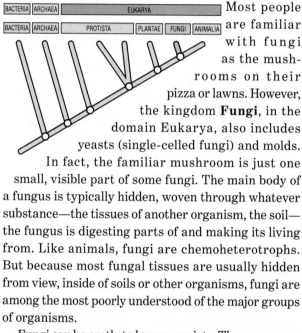

Most people are familiar with fungi as the mushrooms on their pizza or lawns. However, the kingdom **Fungi**, in the domain Eukarya, also includes yeasts (single-celled fungi) and molds. In fact, the familiar mushroom is just one small, visible part of some fungi. The main body of a fungus is typically hidden, woven through whatever substance—the tissues of another organism, the soil—the fungus is digesting parts of and making its living from. Like animals, fungi are chemoheterotrophs. But because most fungal tissues are usually hidden from view, inside of soils or other organisms, fungi are among the most poorly understood of the major groups of organisms.

Fungi can be costly to human society. They can cause diseases, contaminate crops, rot food, and force us to clean our bathrooms more often than we like. Other fungi are beneficial, providing us with pharmaceuticals, including antibiotics such as penicillin. Yeasts such as *Saccharomyces cerevisiae* [*SAK*-uh-roh-*MYE*-seez sair-uh-*VEE*-see-eye] can feed on sugars and produce two important products: alcohol and the gas carbon dioxide—both crucial to the rising of bread, the brewing of beer, and the fermenting of wine. Fungi also provide highly sought-after delicacies such as truffles, whose underground growing locations can be found only by specially trained dogs or pigs.

As Figure 3.13 on page 58 shows, the fungi are divided into three distinct groups: zygomycetes, which evolved first; ascomycetes; and basidiomycetes. Each group differs in—and is named for—its unique reproductive structures.

Fungi play several roles in terrestrial ecosystems. Many are decomposers. Playing the role of garbage processor and recycler, these fungi speed the return of the nutrients in dead and dying organisms to the ecosystem. Some fungi are **parasites** (organisms that live in or on other organisms and harm them); others are **mutualists** (organisms that benefit from, and provide benefits to, the organisms they associate with).

Fungi have evolved a structure that makes them highly efficient decomposers

One of the key evolutionary innovations of the Fungi is their body form. The main body of a typical fungus is a **mycelium** [my-*SEE*-lee-um] (plural "mycelia"), which is a mat of threadlike projections called **hyphae** ([*HYE*-fee]; singular "hypha"). The mycelium typically grows hidden within the soil or the tissues of the organism that the fungus is decomposing (Figure 3.14).

Test your knowledge of how hyphae help fungi grow. 3.5

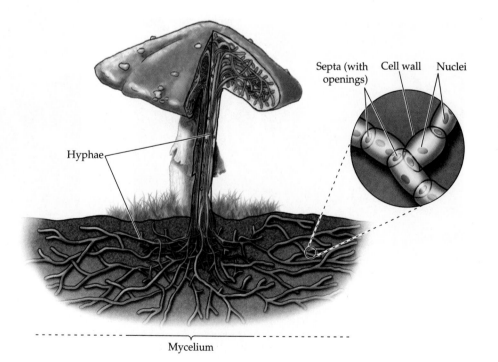

Septa (with openings)   Cell wall   Nuclei

Hyphae

Mycelium

Figure 3.14 The Basic Structures of a Fungus
Mats of hyphae, known collectively as a mycelium, form the main body of a fungus. Hyphae are composed of cell-like compartments separated by septa. Openings in the septa allow organelles to move from one compartment to another. Unlike plant cell walls, the fungal cell walls encasing the hyphae are composed of chitin, the same material that makes up the hard outer skeleton of insects.

Figure 3.13 The Fungi

Fungi are most familiar to us as mushrooms, but the main bodies of such fungi typically lie hidden underground or in another organism's tissues. Some fungi are decomposers, breaking down dead and dying organisms. Others are parasites, living on or in other organisms and harming them. Others are mutualists, living with other organisms to their mutual benefit. The three major groups of fungi are shown in the evolutionary tree.

- Number of species discovered to date: ~70,000
- Function within ecosystems: Decomposers and consumers
- Economic uses: Mushrooms are used for food, yeasts are used for producing alcoholic beverages and bread. Some fungi also produce antibiotics, drugs that help fight bacterial infections.
- Did you know? Highly sought-after mushrooms known as truffles can sell for $600 a pound.

These foul-smelling basidiomycetes, known as stinkhorn mushrooms, attract flies, which get covered with their sticky spores and then scatter the spores as they fly to other locations.

Zygomycetes   Ascomycetes   Basidiomycetes

Pilobolus, a zygomycete that lives on dung, can shoot its spores out at an initial speed of 50 kilometers per hour.

This ascomycete, *Penicillium roqueforti*, is used in the making of Roquefort cheese, producing its characteristic blue veins. It is a relative of the original species that produced the antibiotic penicillin, a drug that fights bacterial infections and has saved the lives of countless people.

Figure 3.15 Fungi Spread via Spores
This puffball fungus is expelling a cloud of spores into the air.

Hyphae are composed of cell-like compartments encased in a cell wall. Unlike the cells in other multicellular organisms, the cell-like compartments of fungi are not completely encased by cell membranes. These compartments are instead only incompletely separated by a structure called a septum (plural "septa"). Openings in the septum allow organelles—including nuclei—to pass from one compartment to another.

Like animals, and all chemoheterotrophs, fungi rely entirely on other organisms for both energy and nutrients. Most animals ingest food and then release digestive juices and proteins into a stomach to digest it. In contrast, fungi digest their food externally. They release special digestive proteins to break down the tissues or other substances through which they grow. The hyphae then absorb the nutrients for the fungus to use.

The ability of fungi to grow through tissues makes them well suited to the roles of consumer and decomposer. In fact, fungi are among the most important groups of decomposers, recycling a large proportion of the dead and dying organisms on land.

Fungi have unique ways of reproducing

Fungi are characterized by complex mating systems. For starters, they are divided not into male and female sexes, but into a variety of mating types. Each type can mate successfully only with a different mating type. Another, more familiar aspect of fungal reproduction is **spores**, reproductive cells that typically are encased in a protective coating that shields them from drying or rotting. Known to most of us as the green powdery dust on old bread, fruit, and cheese, fungal spores, like plant seeds, are scattered into the world by wind, water, and animals (Figure 3.15). Once carried to new locales, spores can begin growing as new, separate individuals.

The same characteristics that make fungi good decomposers make them dangerous parasites

Some fungi are parasites. Parasitic fungi grow their hyphae through the tissues of living organisms, causing diseases in animals (including humans) and plants (including crops).

In humans, fungi can cause mild diseases such as athlete's foot. Fungal diseases can also be deadly, like the pneumonia caused by the fungus *Pneumocystis carinii* [*NOO*-moh-*SISS*-tiss kuh-*REE*-nee], the leading killer of people suffering from AIDS. Fungi attack plants too. *Ceratocystis ulmi* [*SAIR*-uh-toh-*SISS*-tiss *OOL*-mee] causes Dutch elm disease, which has nearly eliminated the elm trees that once formed arching canopies over streets all across the United States. Rusts and smuts are fungi that attack crops. Still other fungi are specialized for eating insects, and biologists are trying to use these fungi to kill off insects that are crop pests (Figure 3.16).

Check out the unusual way that fungi digest food externally. ▶❚❚ 3.6

Figure 3.16 Fungal Parasites
Some fungi are parasites, making their living by attacking the tissues of other living organisms. This beetle, a weevil in Ecuador, has been killed by a *Cordyceps* [*KORE*-duh-seps] fungus, the stalks of which are growing out of its back.

Some fungi live in beneficial associations with other species

Some fungi are mutualists, living in association with other organisms to their mutual benefit. One broad group of mutualists—found in all three groups of fungi (zygomycetes, ascomycetes, and basidiomycetes)—is known as **mycorrhizal** [MYE-koh-RYE-zul] fungi. These species live in mutually beneficial associations with plants. The fungi form thick, spongy mats of mycelium on and in the plants' roots that help the plants absorb more water and nutrients. These fungi are critical to the survival of many plants. In fact, more than 95 percent of ferns (and their close relatives), gymnosperms, and angiosperms have mycorrhizal fungi living in association with their roots. For example, morels—a group of mushrooms highly prized as food by some—are the reproductive structures of mycorrhizal fungi.

Another familiar fungal association is the **lichen** [LYE-kin], consisting of lacy, orange or gray-green growths often seen on tree trunks or rocks. A lichen is an association of an alga (a photosynthetic protist, as we learned earlier) and a fungus (Figure 3.17). Both ascomycetes and basidiomycetes form lichens. The alga and fungus in a lichen grow with their tissues intimately entwined, allowing the fungus to receive sugars and other carbon compounds from the alga. In return, the fungus produces lichen acids, a mixture of chemicals that scientists believe may function to protect both the fungus and the alga from being eaten by predators.

Morel

Figure 3.17 Mutualist Fungi

A lichen consists of an alga and a fungus intimately entwined in a mutually beneficial association. Here lichens can be seen growing on the trunk of a maple tree.

3.6 The Animalia: Complex, Diverse, and Mobile

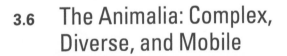
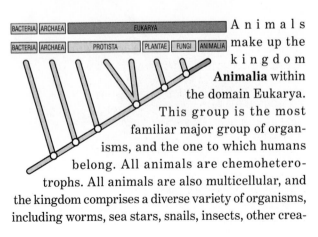

Animals make up the kingdom **Animalia** within the domain Eukarya. This group is the most familiar major group of organisms, and the one to which humans belong. All animals are chemoheterotrophs. All animals are also multicellular, and the kingdom comprises a diverse variety of organisms, including worms, sea stars, snails, insects, other creatures that are less obviously animal-like (for example, sponges and corals), and flashy beasts like Bengal tigers and us.

The sponges, the most ancient of animal lineages, were the first to branch off of the evolutionary tree (Figure 3.18 on pages 62–63). Next to evolve were the cnidarians [nye-DAIR-ee-unz] (including jellyfish, sea anemones, and corals), and then the flatworms. Protostomes were the next group to evolve, comprising more than 20 separate subgroups, including mollusks (such as snails and clams), annelids (segmented worms), and arthropods (including crustaceans, spiders, and insects)—the three shown in Figure 3.18. These three protostome groups are depicted branching off together because it is unclear which of them evolved from the others first. What is known is that they are part of a single lineage descending from an ancestor that branched off the tree after flatworms, but before echinoderms [ee-KYE-noh-dermz]. Next to evolve were the echinoderms (sea stars and the like) and the vertebrates (animals with backbones, such as fishes, birds, and humans)—both deuterostomes [DOO-ter-oh-stomez].

Like all fungi and some bacteria and protists, animals are chemoheterotrophs. They play the role of consumers by eating the tissues of other organisms, from which they derive both carbon and energy. Animals differ from fungi and plants in that animal cells do not have cell walls surrounding their plasma membranes. Typically mobile and often in search of either food or mates, animals have evolved a huge diversity in their ways of life.

Animals evolved true tissues

Sponges are among the simplest of animals. They represent a time in the evolution of animals before tissues—specialized, coordinated collections of cells—evolved. A sponge is a loose collection of cells (Figure 3.19). If it is put through a sieve and broken apart into individual cells, it will slowly reassemble as a whole sponge. Widespread and highly successful, sponges feed on amoebas and other tiny organisms in their aquatic environment, filtering a ton of water just to get enough food to grow an ounce.

An important evolutionary innovation of animals is the development of true tissues. One of the earliest animal groups to evolve true tissues was the cnidarians. Their name—Cnidaria—comes from the Greek word for "nettle," a stinging plant found on land. Cnidarians are characterized by stinging cells that they use to immobilize prey and to protect

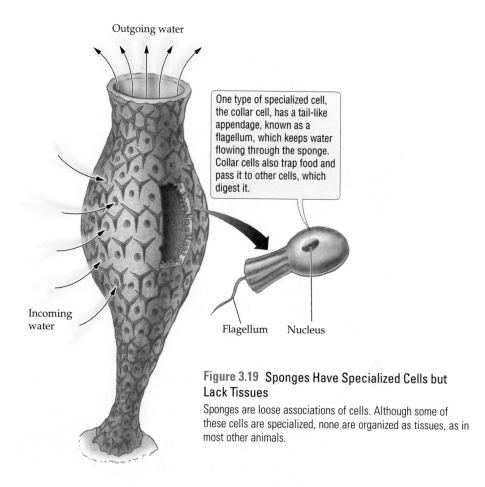

Outgoing water

One type of specialized cell, the collar cell, has a tail-like appendage, known as a flagellum, which keeps water flowing through the sponge. Collar cells also trap food and pass it to other cells, which digest it.

Incoming water

Flagellum   Nucleus

Figure 3.19 Sponges Have Specialized Cells but Lack Tissues
Sponges are loose associations of cells. Although some of these cells are specialized, none are organized as tissues, as in most other animals.

themselves from predators. Like other cnidarians, jellyfish (Figure 3.20) have specialized tissues, including musclelike tissues and digestive tissues. This specialization allows behavior—such as rapid swimming away from predators—that requires the coordination of many cells.

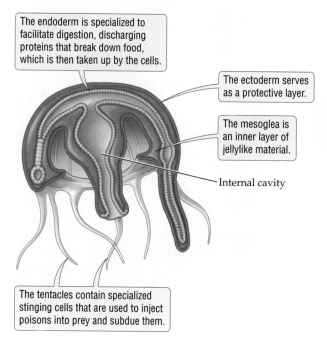

The endoderm is specialized to facilitate digestion, discharging proteins that break down food, which is then taken up by the cells.

The ectoderm serves as a protective layer.

The mesoglea is an inner layer of jellylike material.

Internal cavity

The tentacles contain specialized stinging cells that are used to inject poisons into prey and subdue them.

Figure 3.20 Jellyfish Have True Tissue Layers
Cnidarians (including jellyfish) were one of the earliest groups to evolve true tissues. These tissues include the ectoderm (*ecto*, "outer"; *derm*, "skin") and the endoderm (*endo*, "inner"). For clarity, these two layers are color-coded blue and yellow, respectively. Sandwiched between them is an inner (red) layer of secreted material known as the mesoglea (*meso*, "middle"; *glea*, "jelly"). The ectoderm coordinates with the endoderm to contract like muscle tissue does. Tentacles bring food into the internal cavity through a single opening, which serves as both a mouth and an anus.

Review the tissue layers of jellyfish. ▶ ‖
3.7

Figure 3.18 The Animalia

Animals are multicellular organisms that are typically mobile and display a wide variety of sizes and shapes. They range from sponges, which do not seem very animal-like, to more familiar forms such as elephants and whales.

- Number of species discovered to date: ~1 million
- Functions within ecosystems: Consumers and decomposers
- Economic uses: Humans use other animals as food, as workers, and as laboratory specimens.
- Did you know? Three-fourths of all known animal species are insects.

Flatworms, like this oceangoing flatworm from the West Coast of the United States, were among the earliest animals to evolve true organs and organ systems.

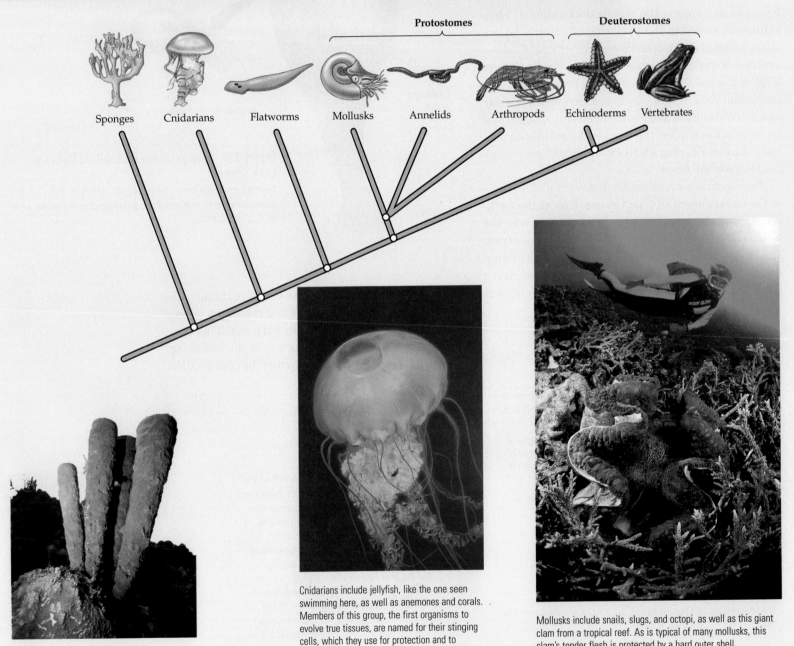

Protostomes

Deuterostomes

Sponges   Cnidarians   Flatworms   Mollusks   Annelids   Arthropods   Echinoderms   Vertebrates

Cnidarians include jellyfish, like the one seen swimming here, as well as anemones and corals. Members of this group, the first organisms to evolve true tissues, are named for their stinging cells, which they use for protection and to disable prey.

Mollusks include snails, slugs, and octopi, as well as this giant clam from a tropical reef. As is typical of many mollusks, this clam's tender flesh is protected by a hard outer shell.

Sponges are ancient aquatic animals. They have evolved some specialized cells, but no true tissues.

A key feature of annelids, also known as segmented worms, is segmentation. This body plan of repeating units can be seen as the series of distinct segments in this fire worm. The segmented body plan, which is also seen in arthropods and vertebrates, facilitated the evolution of many different body forms.

Echinoderms include sea stars, like the one from Indonesia shown here, and sea urchins. They are closely related to the vertebrates.

Arthropods include crustaceans, like lobsters and crabs, as well as millipedes, spiders, and the most species-rich of all groups, the insects. This *Morpho* butterfly, an inhabitant of the tropical rainforest, is one of the most spectacular insects on Earth.

Amphibians, slimy creatures that include frogs (like this poison arrow frog from Costa Rica) and salamanders, typically spend part of their lives in water and part on land.

Vertebrates are the animals that have backbones, including fishes, reptiles, amphibians, birds, and mammals. Shown here is a coral reef fish from Thailand. Fish were the earliest vertebrate animals.

Mammals are characterized by milk-producing mammary glands in females, as well as young that are born live (with the exception of the platypus and its relatives, known as monotremes, whose young are laid in an egg that later hatches). These kangaroos are mammals, as are bears, dogs, lions, and humans.

Primates include monkeys, apes, and humans. In this group we find our closest relative (the chimpanzee, shown here) and the gorilla, among others.

Animals evolved organs and organ systems

Two other key evolutionary innovations to arise in animals were organs and organ systems, which allowed animals to function even more efficiently. Recall that organs are body parts composed of different tissues organized to carry out specialized functions. Usually organs have a defined boundary and a characteristic size and shape; an example is the kidney.

Recall from Chapter 1 that an organ system is a collection of organs functioning together to perform a specialized task. The human digestive system, for example, is an organ system that includes the stomach as well as other digestive organs, such as the pancreas, liver, and intestines. Flatworms, a group of fairly simple wormlike animals, were one of the earliest groups to evolve true organs and organ systems, such as nervous and reproductive systems (Figure 3.21).

Animals evolved complete body cavities

Still later, animals evolved another innovation: a complete body cavity—an interior space of the body within which the organs reside. The evolution of a body cavity enabled an animals' internal organs to function and grow more freely, as they became separate from the body wall. In addition, the fluid that fills such a cavity

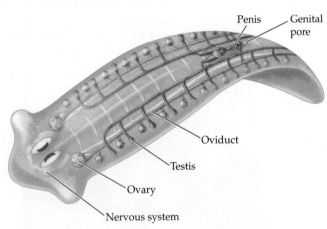

Figure 3.21 Flatworms Evolved Organs and Organ Systems

One of the several organ systems in the flatworm is the reproductive system. The flatworm is a hermaphrodite, meaning that it contains both male and female structures. For clarity, we have color-coded the female structures pink (ovary, oviduct, and genital pore) and the male structures blue (penis and testis).

provides padding and protection for the organs, as well as turgidity and support for the entire body. The two distinct evolutionary lineages that exhibit such cavities are the protostomes and the deuterostomes (see Figure 3.18). Protostomes include animals such as insects, worms, and snails. Deuterostomes include animals such as sea stars (echinoderms) and all the animals with backbones (vertebrates), such as humans, fishes, and birds.

The names for these two lineages refer to which of the two openings in the early embryo becomes the mouth. In **protostomes** (from *proto*, "first"; *stome*, "opening"), the mouth forms from the first opening to develop, and the anus forms elsewhere later. In **deuterostomes** (*deutero*, "second"), the first opening develops into the anus, and the second opening becomes the mouth.

Animal body forms exhibit variations on a few basic structures

Animals exhibit a great variety of shapes and sizes, much of which reflect variations on a few basic structures.

Arthropods (*arthro*, "jointed"; *pod*, "foot") have a hard outer skeleton called an **exoskeleton** (*exo*, "outer"), which is made of chitin [KYE-tin], the same material found in the cell walls of fungi.

One feature that has facilitated the evolution of arthropod bodies is their segmented body plan. Over time, individual body segments have evolved different combinations of legs, antennae, and other specialized appendages, resulting in a huge number of different types of animals, some of them extremely successful. Probably the best-known arthropod group is the **insects** (grasshoppers, beetles, butterflies, and ants, among others), which have 6 legs. Whereas prokaryotes dominate Earth in terms of sheer numbers of individuals, insects dominate in number of species, having many more species than any other group of organisms.

Other arthropod groups include the arachnids [uh-RACK-nidz] (spiders, scorpions, and ticks), which have 8 legs; the crustaceans (lobsters, shrimps, and crabs), which have 10 or more legs and live primarily in water; and millipedes and centipedes, which live on land and have many more legs—but less specialization—than the previously mentioned groups. Arthropods are a wonderful illustration of how evolution can take a basic structure and modify it to produce many variations over time (Figure 3.22). The evolution of just the last

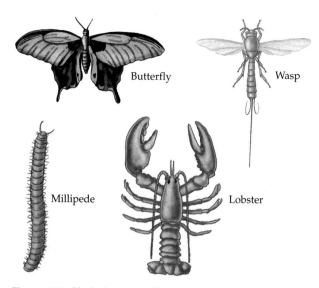

Figure 3.22 Variations on a Theme
From a simple segmented body plan, arthropods have evolved a huge diversity of forms and sizes. The millipede can be viewed as the simplest form of these segmented animals because all of its segments are similar. As segments have evolved and diversified, a variety of organisms have arisen, from lobster to swallowtail butterfly to parasitoid wasp.

segments (the rear end) of these animals shows that the evolutionary changes support a huge variety of shapes and lifestyles. The last segments have evolved into the delicate abdomen of a butterfly, the piercing abdomen of a wasp (which has a huge needlelike structure for inserting and laying eggs deep in another animal's body), and the delicious tail of the lobster.

Such segmentation can also be seen in the annelids (segmented worms; see Figure 3.18). This group includes the familiar earthworm, whose body is made up of a repeated series of segments (Figure 3.23a).

Vertebrates—animals with an internal backbone—are also built on a segmented body plan (Figure 3.23b). The major vertebrate groups include fishes, amphibians (frogs and salamanders), reptiles (snakes, lizards, turtles, and crocodiles), birds, and mammals (including humans and kangaroos). The front appendage of vertebrates is another good example of variations on a basic structure. This appendage has evolved as an arm in humans, a wing in birds, a flipper in whales, an almost nonexistent nub in snakes, and a front leg in salamanders and lizards. Like annelids and arthropods, vertebrates illustrate how a variety of very different forms can evolve from relatively few basic structures.

Animals exhibit an astounding variety of behaviors

Another fascinating characteristic of animals, and a key evolutionary innovation, is their ability to move, which allows for a wide range of behaviors. Animals

(a)

(b)

Figure 3.23 Many Animals Are Segmented
Segmentation, a body plan in which segments repeat and often can evolve independently of one another, is shown here in (a) an earthworm and (b) the "six-pack" of a familiar vertebrate.

have evolved varied ways to capture prey, eat prey, avoid being captured, attract mates, care for young, and migrate to new habitats. As we saw earlier, animals are quite useful to immobile organisms, such as plants, which have evolved ways to get animals to carry their pollen and seeds.

Animals play key roles in ecosystems and provide products for humans

Because they live by eating other organisms, and because most are mobile, animals play many roles in ecosystems. Most serve as consumers, preying on many species of plants and animals. Some animals, such as carrion beetles, serve as decomposers of dead animals. Animals also help spread plant pollen, seeds, and fungal spores. And they can be pests, carrying diseases; ticks, for example, spread the bacterial parasite that causes Lyme disease. Animals, especially insects, can also be crop pests, such as the tomato hornworm, a caterpillar that attacks tomatoes.

Domesticated animals provide humans with food and material for clothing, including feathers and leather. They provide transportation when we travel by horse or camel and even good company when we spend time with dogs or cats.

We humans, however, are the animal species having the greatest impact on life on Earth. Our rapidly growing population and our ability to drastically and rapidly modify Earth with cities, agriculture, and industries risk making the planet uninhabitable for ourselves and other species unless we take care (see the "Biology Matters," box below).

Biology Matters

What's on Your Plate?

Humans eat many different kinds of species, but not all meals have the same impact on the planet's biodiversity. In particular, there are a number of fish and other marine species that humans consume that are being overharvested and are in rapid decline. Eating them only pushes them into steeper declines and could push them into extinction. In addition, some species, like salmon, are raised on aquatic farms. But while that might seem like the perfect solution to overfishing wild species, these farms sometimes can create high levels of ocean pollution, threatening other species.

So when we are shopping at the grocery store or scanning a menu at the restaurant, how can we decide what might be both delicious and harmless to eat? Various conservation organizations have put together lists of seafood that are best to eat and best to avoid, often printed on handy wallet-sized cards. Below are some of the recommendations from the wallet card of the marine conservation organization known as Blue Ocean Institute.

Just by choosing wisely, the next time you crave seafood, you can help preserve the planet's biodiversity.

SOURCE: The Blue Ocean Institute, "Guide to Ocean Friendly Seafood," September 2007.

ENJOY	BE CAREFUL	AVOID
Arctic Char	Crabs (Blue, Snow, and Tanner)	Chilean Sea Bass
Clams, Mussels, and Oysters (farmed)	Monkfish	Cod (Atlantic)
Mackerel	Rainbow Trout (farmed)	Halibut (Atlantic)
Mahi-Mahi (pole- and troll-caught)	Sea Scallops	Salmon (farmed)
Salmon (wild Alaskan)	Swordfish	Sharks
Striped Bass	Tuna: Albacore, Bigeye, Yellowfin, and Skipjack (canned or longline-caught)	Shrimp (imported)
Tilapia (U.S. farmed)		Tuna: Bluefin (Atlantic)

3.7 Viruses: Sprouting Up All over the Evolutionary Tree?

As we saw in Chapter 2, viruses are not represented on the tree of life or classified into any kingdom or domain. That is because they are largely considered to be nonliving. A virus is simply some protein wrapped around a fragment of DNA or RNA.

However, a recent hypothesis about viruses suggests a way, albeit odd, to place viruses on the evolutionary tree of life. Some scientists hypothesize that, rather than branching off from any single point on the evolutionary tree of life, viruses have instead arisen from the DNA or RNA of many different organisms at many different times. In fact, this process may be ongoing, with new viruses appearing all the time. If this hypothesis is correct, then scientists need to put viruses here and there all over the tree of life. Viruses that arose from chickens would need to be put on a branch next to chickens, viruses that arose from oak trees would need to be put on an evolutionary branch next to oak trees, and so on. Though this idea is intriguing, so far biologists are not adding viruses to the tree of life.

Concept Check

1. What is a mycelium, and why does it make fungi so hard to notice?
2. What are mycorrhizal fungi, and what is their significance?
3. Name four evolutionary innovations of animals.
4. Why do some scientists think that viruses should be placed, in scattered fashion, all across the evolutionary tree of life?

Applying What We Learned

Wonderful Life

Life has evolved a diverse array of evolutionary innovations that have allowed many different forms of living things to colonize, exploit, and thrive on Earth, from archaeans to cacti to humans to giant squid.

We still have much to learn about the species that we already know exist. Every day biologists find that these organisms have powerful abilities and curious talents, odd habits and diseases, previously unimagined. Fish are found that can swim in freezing waters without ever suffering the cold, because their blood carries special proteins that act as an antifreeze. Familiar trees are discovered to have the ability to detect when other trees in the area are being attacked by plant eaters, causing them to increase the toxins in their leaves and thus affording them better protection. Bacteria are found to be able to withstand 10,000 times the radiation that would kill a human being. Any number of such evolutionary innovations have the potential to allow a group to survive when or where others will not, and multiply in new environments.

Beyond those organisms already known, many more are yet to be discovered—species that will undoubtedly be shown to harbor still more adaptations, and unimagined evolutionary innovations. The deep sea still harbors much unknown life, as do tropical treetops. But some researchers predict that the next great biological discoveries will be made not out in the great wilds, but in the uncharted regions within organisms. New microscopic organisms are being discovered in the moist, warm mouths of animals. Researchers say there are unexplored worlds of species in and on many organisms—such as the mysteries of a toucan's gut or the world of plants and animals living in a slow-moving sloth's hair.

We humans, too, are territory for exploration. The human body is normally host to millions of bacteria, both externally (on the skin) and internally (in the digestive system). In addition, we harbor hundreds of microscopic arthropods. Each of us is a densely populated community of living organisms, many still likely to harbor wondrous new ways of living, evolutionary innovations still undreamed of. Don't be surprised if biology's next great discovery is waiting not just *for* you, but quite literally *on* you.

Honeybees Vanish,
Leaving Keepers in Peril

By Alexei Barrionuevo

Visalia, Calif., Feb. 23 – David Bradshaw has endured countless stings during his life as a beekeeper, but he got the shock of his career when he opened his boxes last month and found half of his 100 million bees missing.

In 24 states throughout the country, beekeepers have gone through similar shocks as their bees have been disappearing inexplicably at an alarming rate, threatening not only their livelihoods but also the production of numerous crops, including California almonds, one of the nation's most profitable.

"I have never seen anything like it," Mr. Bradshaw, 50, said from an almond orchard here beginning to bloom. "Box after box after box are just empty. There's nobody home."

The sudden mysterious losses are highlighting the critical link that honeybees play in the long chain that gets fruit and vegetables to supermarkets and dinner tables across the country.

In a mystery worthy of Agatha Christie, bees are flying off in search of pollen and nectar and simply never returning to their colonies.

Researchers say the bees are presumably dying in the fields, perhaps becoming exhausted or simply disoriented and eventually falling victim to the cold.

"Every third bite we consume in our diet is dependent on a honeybee to pollinate that food," said Zac Browning, vice president of the American Beekeeping Federation.

One of the most surprising things to happen in recent years has been the sudden disappearance of millions upon millions of honeybees. Never returning to their hives, these bees pose a huge mystery for scientists who have no idea where the bees are going, how they are dying, or why. Their disappearance has created great public concern because of the critical role that honeybees play in agriculture.

As we learned earlier, angiosperms produce flowers that contain their sexual organs. For a plant to complete sexual reproduction or sex, pollen must be delivered to a flower, where it will release male gametes that can fuse with a female gamete in the process known as fertilization. For many plants, if there are no bees to move pollen from one flower to another, there is no chance for them to have sex.

Why are bees and plant sex so important? Because we love to eat the products of plant sex: the young of plants—their seeds (like almonds) or the fleshy, developed ovaries of plants around those seeds that we call fruits (like peaches). We love them so much that we are willing to pay quite a bit of money for them. In a recent study, scientists estimated that honeybees pollinate more than $14 billion worth of crops each year in the United States.

Evaluating the News

1. Draw a food web (see Chapter 1) that includes humans (the species *Homo sapiens*), honeybees (*Apis mellifera*), cows (*Bos taurus*), and three flowering plants that are pollinated by bees: tomatoes (*Lycopersicon esculentum*), wheat (*Triticum aestivum*), and alfalfa (*Medicago sativa*). Include in this human's diet a good hamburger—that is, a beef patty, a wheat bun, and a sliced tomato. Note also that cows eat alfalfa. Complete your food web by showing all the "eats" and " is eaten by" connections between species that you can. Now, what happens if honeybees aren't around to do their job? What happens to hamburgers and to humans that eat them?

2. Honeybees require a bacterial species—that is, a species from the kingdom Bacteria—in their guts to survive. Cows require an archaean species in their guts to process their food and survive. Look again at your food web. How many species go into the making of a hamburger? How many kingdoms? How many domains? Now name at least two evolutionary innovations that have enriched your diet.

3. Environmentalists tell us that biodiversity is important. If people like hamburgers, honey in tea, tomato soup, or nice buttery croissants with raspberry jam, explain why the diversity of life should matter to them.

Source: *New York Times*, February 27, 2007.

Chapter Review

Summary

3.1 Getting Oriented: How Biologists Organize and Classify the Major Groups

- The major groups of organisms are Bacteria, Archaea, Protista, Plantae, Fungi, and Animalia.
- Biologists organize all living organisms in one of three ways: into domains (Bacteria, Archaea, and Eukarya); into the Linnaean hierarchy, in which the six major groups are kingdoms; and onto the evolutionary tree of life.
- Each major group is characterized by evolutionary innovations that allowed the group to live and reproduce successfully.
- All organisms can also be classified according to cellular structure, as either prokaryotes (simple one-celled organisms without organelles) or eukaryotes (one-celled and multicellular organisms whose cells have organelles).

3.2 The Bacteria and Archaea: Tiny, Successful, and Abundant

- The two prokaryotic kingdoms and domains are Bacteria and Archaea. Both are microscopic, single-celled organisms. The two groups differ in significant ways: in their DNA, in components of their cell walls and plasma membranes, and in their metabolism.
- Prokaryotes can reproduce extremely rapidly and are the most numerous life forms on Earth. They also have the most widespread distribution. Some prokaryotes, including many archaeans, are extremophiles, thriving in very hot (thermophiles), very salty (halophiles), or other extreme environments.
- Prokaryotes exhibit unmatched diversity in methods of getting and using energy and nutrients. Only prokaryotes can be chemoheterotrophs, photoautotrophs, chemoautotrophs, or photoheterotrophs.
- Prokaryotes perform key tasks in ecosystems, including photosynthesizing, providing nitrate to plants, and decomposing dead organisms. Prokaryotes are useful to humanity in many ways (for example, in cleaning up oil spills and helping with our digestion), but they also cause deadly diseases.

3.3 The Protista: A Window into the Early Evolution of the Eukarya

- The Protista, the most ancient group of eukaryotes, is a highly diverse group. It includes plantlike, animal-like, and funguslike organisms. Protists can be photoautotrophs or chemoheterotrophs.
- The evolutionary relationships among members of the Protista remain poorly understood.
- Protists provide examples of the early evolution of multicellularity.
- A key evolutionary innovation of protists was sex or sexual reproduction.
- Although protists include many harmless or helpful organisms, they also include many disease-causing organisms.

3.4 The Plantae: Pioneers of Life on Land

- The Plantae (plants) are multicellular, photosynthesizing organisms. Plants are photoautotrophs. They pioneered life on land.
- Colonizing land presented two key problems: how to get and conserve water, and how to grow in the presence of gravity. Evolutionary innovations that helped plants meet these challenges include a root system, the cuticle, and cell walls stiffened with cellulose. Other important evolutionary innovations that allowed plants to thrive and diversify include vascular systems, seeds, and flowers.
- Ferns and their allies were the first plants to evolve vascular systems, allowing plants to grow to greater heights.
- Gymnosperms evolved the first seeds, which provide nutrients so that plant embryos can grow and develop before they are able to produce their own food via photosynthesis. Seeds also protect the plant embryo from drying, rotting, and attack by predators.
- Angiosperms evolved the first flowers, specialized reproductive structures that are the site of fertilization, the fusing of male and female gametes. Flowers have evolved numerous innovations to aid in pollen dispersal and seed dispersal.
- As producers, plants are the ultimate food source for nearly all land organisms and are therefore critical components of land-based food webs.
- Plants are harvested for food, clothing, pharmaceuticals, lumber, oxygen, and many other important products. Plants are also valuable when left in nature, where they provide oxygen and prevent erosion and water contamination, among other benefits.

3.5 The Fungi: A World of Decomposers

- Fungi have a unique body plan: a mat of hyphae called the mycelium, which grows as the hyphae penetrate, digest, and absorb food. This evolutionary innovation allows fungi to be successful as decomposers (breaking down dead and dying organisms) and consumers. Hyphae make fungi excellent parasites (extracting nutrients from living organisms without providing benefit in return) and mutualists (living in close association with other organisms for their mutual benefit). Fungi are chemoheterotrophs.
- Reproduction in fungi is unique to this group. Instead of sexes, there are many mating systems and mating types. Encased reproductive cells called spores are protected from harsh conditions and able to disperse long distances via animals, wind, and water before beginning their development. Fungi are divided into three groups—zygomycetes, ascomycetes, and basidiomycetes—based on reproductive structures specific to each group.
- There are two groups of mutualist fungi. Mycorrhizal fungi live in and on the roots of most plants. In lichens, fungi and algae live in close association with each other.
- Fungi cause dangerous diseases, but they also produce valuable products such as foods and pharmaceuticals.

3.6 The Animalia: Complex, Diverse, and Mobile

- The Animalia (animals) make a living by eating other organisms and are typically mobile. They are chemoheterotrophs.
- The bodies of sponges, the most primitive animal group, have clusters of specialized cells, but cnidarians were one of the first groups with the evolutionary innovation of true tissues (groups of specialized cells that are also coordinated).
- Flatworms were one of the earliest groups to exhibit the evolutionary innovation of true organs (different tissues organized for a specialized function, with a defined boundary and characteristic size and shape) and organ systems (groups of organs functioning together for specialized tasks).
- The evolutionary innovation of complete body cavities did not show up in animal bodies until the evolution of protostomes (mollusks, annelids, and arthropods) and deuterostomes (echinoderms and vertebrates). In protostome embryos, the mouth develops from the first opening that forms; in deuterostomes, the mouth develops from the second opening.
- Animals exhibit a great variety of forms and sizes, often variations on a single theme. One major theme is the segmented body plan, in which a given segment evolves different shapes in different animal groups in order to perform different functions. Annelids, arthropods, and vertebrates all exhibit variations on a basic segmented body plan.
- Insects are the most familiar of the arthropods, a group characterized by a hard outer skeleton known as an exoskeleton.
- Vertebrates (which include fishes, amphibians, reptiles, birds, and mammals) are distinguished by having an internal backbone.
- Animals' ability to move allows them a great variety of behaviors, including ways to capture and eat prey, avoid being captured, attract mates and care for young, and migrate.
- Animals play a variety of roles in ecosystems, serving mainly as consumers but sometimes as decomposers. Some spread diseases; others are crop pests. Animals also provide food, clothing, and other products to human society. Humans are the animals that have had the greatest impact on ecosystems worldwide.

3.7 Viruses: Sprouting Up All over the Evolutionary Tree?

- Viruses, widely considered to be nonliving, are not placed on the tree of life or in the kingdom or domain classification system.
- A new hypothesis suggests that viruses arose multiple times from different organisms across the evolutionary tree.

◉ Review and Application of Key Concepts

1. What are the three major systems used to categorize living organisms?

2. What two kingdoms make up the prokaryotes? Name three factors that contributed to the success of prokaryotes.

3. Why are slime molds of particular interest to biologists interested in the early evolution of eukaryotes?

4. Describe two evolutionary innovations of animals and why they were important. What animal groups exhibit these innovations?

5. To what kingdom and domain do viruses belong? Why?

6. Two of the major challenges facing plants when they colonized land were (a) obtaining and retaining water and (b) gravity. What evolutionary innovations did plants use to deal with these challenges?

7. Which kingdom(s) contain organisms that have chloroplasts in their cells? Why are chloroplasts significant?

8. Animals are typically mobile, whereas plants are not. In what ways have plants evolved to utilize the mobility of animals to aid in their survival and reproduction?

Key Terms

adaptation (p. 44)
aerobe (p. 47)
anaerobe (p. 47)
angiosperm (p. 54)
Animalia (p. 60)
arthropod (p. 64)
biodiversity (p. 44)
chemoautotroph (p. 49)
chemoheterotroph (p. 47)
cuticle (p. 52)
deuterostome (p. 64)
eukaryote (p. 45)
evolutionary innovation (p. 45)
exoskeleton (p. 64)
extremophile (p. 45)
flower (p. 55)
Fungi (p. 57)
gymnosperm (p. 54)
hypha (p. 57)

insect (p. 64)
lichen (p. 60)
mutualist (p. 57)
mycelium (p. 57)
mycorrhizal (p. 60)
organelle (p. 45)
photoautotroph (p. 49)
photoheterotroph (p. 49)
parasite (p. 57)
Plantae (p. 52)
prokaryote (p. 45)
Protista (p. 50)
protostome (p. 64)
root system (p. 52)
seed (p. 54)
spore (p. 59)
vascular system (p. 54)
vertebrate (p. 65)

Self-Quiz

1. Which group is the most abundant in number of individuals?
   a. insects
   b. eukaryotes
   c. protists
   d. prokaryotes

2. Eukaryotes differ from prokaryotes in which of the following ways?
   a. They do not have organelles in their cells, whereas prokaryotes do.
   b. They exhibit a much greater diversity of nutritional modes than prokaryotes do.
   c. They have organelles in their cells, but prokaryotes do not.
   d. They are more widespread than prokaryotes.

3. Which of the following groups contains organisms that represent early stages in the evolution of multicellularity?
   a. Plantae
   b. Protista
   c. Fungi
   d. Animalia

4. Which of the following groups was the first to succeed on land?
   a. plants
   b. animals
   c. angiosperms
   d. slime molds

5. Which of the following were key evolutionary innovations of the Plantae?
   a. seeds, organelles, flowers
   b. roots, cuticle, seeds, flowers
   c. roots, hyphae, flowers
   d. hyphae, cuticle, organelles

6. Fungi grow using
   a. hyphae.
   b. septa.
   c. basidiomycetes.
   d. complete body cavities.

7. Animals that are segmented include
   a. vertebrates and annelids.
   b. vertebrates and flatworms.
   c. sponges and humans.
   d. sponges and flatworms.

8. Mycorrhizal fungi are
   a. beneficial to plants because they help plants stay dry.
   b. harmful to plants because they secrete acids.
   c. beneficial to plants because they help plants take up water.
   d. beneficial to plants because they help plants form lichens.

9. An evolutionary innovation is
   a. an ancient feature exhibited by a group of organisms.
   b. a novel adaptation that helps a group to survive and succeed.
   c. a newly evolved species of living thing.
   d. an ancient adaptation.

10. Three evolutionary innovations important to the success of the kingdom Animalia are
    a. true tissues, organ systems, and mobility.
    b. true tissues, hyphae, and complete body cavities.
    c. protostomes, true tissues, and mobility.
    d. organ systems, anthers, and mobility.

Biodiversity and People

Where Have All the Frogs Gone?

In 1987, a spectacularly beautiful creature known as the golden toad could be found in abundance in the only spot in the world where it was known to live: a tropical forest known as Monteverde high in the mountains of Costa Rica. That year, hundreds of the brightly colored animals were seen (Figure A.1). The next year, just a few were found. Within a few years the golden toad had disappeared entirely, never to be seen again.

Although there is always a concern when a species plummets into extinction, most extinctions are easier to understand than is the loss of the golden toad. When a forest-dwelling bird goes extinct because its forest is cut down, there is no lingering mystery. But the golden toads were living in a pristine area far from deforestation or development. There was no obvious reason for these frogs to go extinct.

Since the time when the golden toad was last seen, biologists around the world have documented population declines of numerous amphibians (a class within a larger group, the vertebrate animals, that includes not only frogs but also toads and salamanders). Many of these declines have occurred in preserved areas. In the United States, for example, in and around Yosemite National Park, where frogs and toads were once abundant, many species have declined or disappeared.

Scientists currently believe that no single cause is responsible for these worldwide losses of amphibians. For some species living at high altitudes, increasing exposure to damaging ultraviolet light may be a problem as the protective ozone layer in the atmosphere thins. In many parts of the world, huge numbers of amphibians are being killed by fungal diseases. In other places, researchers report that

Figure A.1 Gone but Not Forgotten
Once abundant on a mountaintop in Costa Rica, the golden toad mysteriously went extinct in the late 1980s. Here the orange-colored male mates with the larger, and very differently colored, female.

pollution, in the form of artificial chemicals in the water, may be affecting processes of amphibian development and causing deformities. In some areas, parasites may be causing similar deformities in developing amphibians.

Whatever the problem in any particular region, amphibians are being lost around the world. Many scientists believe that these animals are more sensitive than other animals to environmental deterioration, and that—like canaries in a coal mine—the dying amphibians are warning signs of an increasingly poisoned environment. Meanwhile, many other species are also going extinct.

Does it really matter if the golden toad is lost forever? Or if any other particular species, or even many species, disappear? Everywhere we hear warnings about the loss of species around the globe. How extensive is this loss of biodiversity, what is causing it, and do these losses really matter? In this essay we look at how the numbers of species on Earth have changed over time, during mass extinctions in the past and in today's human-caused extinction. We also look at biodiversity and ask what value the totality of the world's species has for humanity.

How Many Species Are There on Earth?

To understand whether the current loss of species matters, we must begin with some idea of how many species have disappeared. That means knowing how many species there are and how many there have been on Earth. Surprisingly, in spite of intense worldwide interest, scientists do not know the exact number of species alive today. Estimates range widely, from 3 million to 100 million species. Most estimates, however, fall in the range of 3 million to 30 million.

Scientists use indirect methods to estimate total species numbers

So far, a total of about 1.5 million species have been collected, identified, named, and placed in the Linnaean hierarchy. Despite this massive cataloging effort, which has taken more than two centuries, some researchers believe that biologists have barely scratched the surface. Some estimates suggest that 90 percent or more of all living organisms remain to be identified and named by biologists.

How can scientists know how many species exist that they've never seen? Biologists use indirect methods to estimate how many species remain unknown. In 1952, for example, a researcher at the U.S. Department of Agriculture estimated, on the basis of the rate at which unknown insects were pouring into museums, that there were 10 million insect species in the world.

Thirty years later, Terry Erwin, an insect biologist at the Smithsonian Institution, shocked the world with his estimate that the arthropods alone (the group in the Linnaean hierarchy, known as a phylum, that includes insects, spiders, and crustaceans) numbered more than 30 million species. That is 20 times the number of all the

Here's how Erwin did the calculation: There are an estimated 50,000 tropical tree species. If the tree species that Erwin studied was typical, then the total number of beetle species living in tropical trees in the world should number 8 million (50,000 tree species × 160 specialist beetle species). Beetle species are thought to make up about 40 percent of all arthropod species. Eight million beetle species is 40 percent of 20 million. Therefore, the total number of arthropod species in the rainforest canopy should be 20 million. Many scientists believe that the total number of arthropod species in the canopy is double the number found in other parts of tropical forests, suggesting that there are another 10 million arthropod species in the rest of the tropical forests. Assuming these numbers, the total number of arthropod species in the tropics alone should be 30 million (20 million + 10 million). That means, of course, that there should be even more species of *all* kinds in the tropics, and more still when the entire world is taken into consideration.

Like all such estimates, this one is based on numerous assumptions, some of which could be wrong. A change in even one of these assumptions could drastically alter the final number. Though Erwin's calculation is among the most famous of these estimates, such indirect measures are typical of how the numbers of species yet to be discovered are determined, since it is impossible to count them directly. Scientists continue to argue over the exact figures and the assumptions on which they are based, but one thing is certain: the 1.5 million species discovered and named to date are far from the total number living in this world.

Figure A.2 Fogging to Count Species
Tropical biologist Terry Erwin fogs the canopy of a tree to collect its many insects. On the basis of his studies of insects from tropical treetops, Erwin estimated that, worldwide, there could be as many as 30 million species of tropical arthropods alone.

previously named species of all groups on Earth. Erwin believed that most of these arthropods were insects living in the tropical rainforest **canopy**—the nearly inaccessible habitat in the branches of rainforest trees.

Erwin based his estimate on actual species counts that he obtained by fogging (Figure A.2). In this method, first a biodegradable insecticide is blown high into the top of a single rainforest tree. Then the dead and dying insects that fall to the ground are collected and counted, and the number of different species is tallied. Erwin found more than 1,100 species of beetles alone living in the top of one particular tropical tree. He estimated that 160 of these species were likely to be **specialists**—that is, able to live only in this species of tree and not found anywhere else on Earth. From there, he was able to come up with a minimum estimate of the number of arthropod species in the world.

Some groups are well known and others are poorly studied

About half of the 1.5 million known species (about 750,000) are insects. All the remaining animals make up a mere 300,000 species or so (Figure A.3). The next-largest group is the plants, with about 250,000 known species. There are also approximately 69,000 named fungi, 85,000 protists, and some 4,800 prokaryotes, including both bacteria and archaeans.

Among these groups, some are very well studied because they are large or easy to capture or they are popular with biologists. Others are very poorly studied, often because they are microscopic or otherwise hard to collect and identify. The birds, for example, total 9,000 species and are among the best-studied organisms on Earth; relatively few new bird species are left to be discovered. Insects, on the other hand, remain poorly known; the majority, possibly the vast majority, are still undiscovered and unidentified. The result is that most of us tend to have a rather skewed view of the makeup of the living world, thinking of it as dominated by well-studied organisms like birds or mammals and tending to forget altogether the fact that there are many more species of, for example, insects (Figure A.4).

Whole kingdoms, such as Fungi, Bacteria, and Archaea, are also poorly known. In a single gram of Maine soil, scientists have estimated that there may be as many as 10,000 species of bacteria. That's about 5,000 more species than have so far been named by biologists. The studies forming the basis of this estimate examined the different types of DNA contained in the soil. Scientists estimated 10,000 different types, each of which probably represents a different species. There is an additional problem with poorly studied organisms: when little is known about a group of organisms, it can sometimes be difficult to determine whether two organisms from that group are members of the same species or two different species, making the tallying of total numbers of species even more complex.

Biologists continue to discover new species even in relatively well-known groups. For example, about 100 new fish species are discovered each year. And in 1992, although scientists claimed to be sure that all the large land mammals had already been accounted for, a large deerlike species was found in Vietnam. Then, 2 years later, the barking deer, another large deer species, emerged from the mountains there as well.

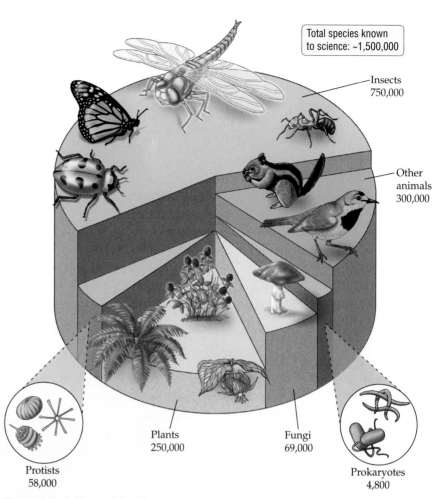

Total species known to science: ~1,500,000

Insects 750,000

Other animals 300,000

Plants 250,000

Fungi 69,000

Protists 58,000

Prokaryotes 4,800

Figure A.3 A Piece of the Pie

This pie chart breaks down all the known species on Earth into the major groups of organisms. Animals (particularly insects) and plants make up the vast majority of the known species, but many more species remain to be discovered.

The Beginnings of a Present-Day Mass Extinction

The history of life on Earth includes a handful of drastic events, known as **mass extinctions**, during which huge numbers of species went extinct. Today, even as biologists struggle to get a total species count, many biologists assert that we are on our way toward a new mass extinction. In fact, many biologists say that the ongoing extinction—if it continues unabated—will lead to the most rapid mass extinction in the history of Earth. As with the total number of species on the planet, extinction rates are estimates. But even using conservative calculations, species are being lost at a staggering pace. As we shall see, the cause of this mass extinction is clear: the activities of the ever-increasing number of people living on Earth.

The current mass extinction probably began with early humans

Increasing evidence suggests that humans have been driving species to extinction around the world for a very long time. The fossil record shows that at about the same time humans arrived in North America, Australia, Madagascar, and New Zealand, species of large animals (including mammoths, giant ground sloths, camels, horses, and saber-toothed cats) began to disappear. Although some people suggest that these species extinctions may also have been due to climate changes, the coincidence of these losses with the arrival of humans in three different parts of the world is striking and consistent. Of the genera of large mammals that roamed Earth 10,000 years ago, 73 percent are now extinct. (Remember from Chapter 2 that the genus is the next category above species in the Linnaean hierarchy.)

Barking deer

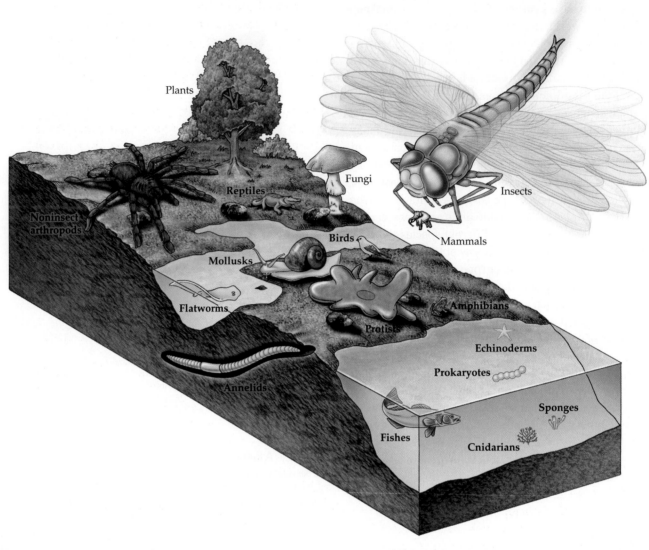

Figure A.4 A Species-Scape

In this strange world, each group of organisms is represented not by its true size, but by the number of known species in the group. As a representative of mammals, the elephant shown here is uncharacteristically tiny to reflect the relatively small number of mammalian species. The enormous size of the dragonfly holding the elephant reflects the vast number of insect species.

Giant ground sloth

Explore the diversity of species on Earth.

▶ ‖

A.1

There are several ways that humans probably affected the animal life around them. Many large mammals would have made a hearty meal for prehistoric hunters. A similar number of birds, particularly those that would have been the easiest of prey for humans, disappeared. One species lost, for example, was a flightless duck—literally, a sitting duck. Many large animals also may have suffered from having to compete with humans for the same prey. And in addition to predation and competition, humans had indirect effects. For example, numerous other animals that depended on large animals for their survival, such as vulture species that fed on their carcasses, went extinct as those animals declined in numbers.

Scientists have estimated the species losses occurring today

The most devastating and obvious losses of species today are occurring in tropical **rainforests**—wet, lush forests in the tropical regions of the world

Figure A.5 The Rainforest

Tropical rainforests, like this one in Hawaii, are typically home to numerous species not found in other habitats.

(Figure A.5). Known to harbor huge numbers of species found nowhere else, rainforests are being burned or cut at alarming rates. Direct observations of rainforest destruction and satellite photographs of Earth both indicate that more than half of the rainforest area that existed in prehistoric times has been destroyed. And scientists estimate that the rest is going fast. According to some estimates, an area of rainforest the size of a football field is completely destroyed every second, amounting to the destruction of an area the size of Florida each year. An area of the same size is simultaneously degraded, meaning that although trees are still left standing in these areas, many of the plants and animals and other organisms of the forest have been harvested, killed, or otherwise destroyed.

How do such large-scale losses of natural areas translate into numbers of species lost? According to Edward O. Wilson, a biologist at Harvard University, at this rate of rainforest destruction 27,000 additional species will be doomed to extinction each year—an average of 74 per day, or 3 every hour. And though rainforests are particularly rich in biodiversity, they are just one of many different **habitats**—characteristic places or types of environments in which different species can live.

Though estimating extinction rates for the world as a whole can be difficult, scientists have definitively documented the extinction, caused by humans, of many hundreds of particular species in the last few thousand years. Twenty percent of the freshwater fish species known to be alive in recent history either have gone extinct or are nearly extinct. One large-scale study showed that 20 percent of the world's species of birds that existed 2,000 years ago are no longer alive. Of the remaining bird species, 10 percent are estimated to be **endangered**—that is, in danger of extinction.

Although it may be tempting to assume that little of this extinction is happening close to home, evidence suggests otherwise (Figure A.6). As mentioned earlier, in the United States, frogs are disappearing in Yosemite National Park. In addition, some 200 plant species known to have existed in recent history have already gone extinct, and at least 600 more are predicted to disappear within the next few years. In North America overall, 29 percent of freshwater fishes and 20 percent of freshwater mussels are endangered or extinct.

Altogether, biologists have amassed a wealth of data on species already gone or on their way out, and even conservative analyses suggest that huge numbers of species have been, and continue to be, lost. Why are these species disappearing?

(a)

(b)

Figure A.6 The First Not to Last

(*a*) Once fairly common on the sand dunes of the San Francisco peninsula, the xerces blue butterfly has earned dubious fame as the first butterfly documented to have gone extinct in the United States because of human disturbance. Also known by the scientific name *Glaucopsyche xerces* [GLOU-koh-sye-kee ZER-seez], the xerces blue was last seen in the 1940s, and this photograph of a museum specimen is rare. Scientists say that the major cause of its extinction was probably destruction of its seaside habitat in the populous San Francisco area. (*b*) Researchers say that other factors may have contributed as well. For example, the xerces blue in its caterpillar phase was protected from predators by certain native ant species, which—also because of human disturbance—have declined in numbers or disappeared altogether. Here ants "milk" the caterpillar of another species of butterfly in a similar reciprocal arrangement. The ants drink a sugary substance released by the caterpillar and provide protection in return.

The Many Threats to Biodiversity

The remaining species of the world face continuing challenges to their survival. A number of human activities are threatening and destroying biodiversity around the globe and in our own backyards.

Habitat loss and deterioration are the biggest threats to biodiversity

Foremost among the direct threats to biodiversity is the destruction or deterioration of habitats. As human homes, farms, and industries spring up where natural areas once existed, habitats suited to nonhuman species continue to disappear or become radically altered. For many people the term "habitat loss" conjures up images of burning rainforest in the Amazon, but the problem is much more widespread and much closer to home.

Every time a suburban development of houses goes up where once there was a forest or field, habitat is destroyed (Figure A.7a). So widespread is the impact of growing human populations in urban and suburban areas that species are disappearing even from parks and reserves in heavily populated areas. For example, ecologists studied a large preserve in the midst of increasing suburban development outside Boston; there they found that 150 of the park's native plant species had disappeared. The immediate cause of the loss of species was most likely trampling and other disturbances, as more and more people—likely including many nature lovers—used the park (Figure A.7b). But the increasing number of homes in the area and the decreasing number of nearby natural areas—from which seeds could have come to repopulate the park—also played an important role in the loss of species. In addition, pollution, erosion, and other effects of human activities and human population growth are altering natural habitats to the point where many species can no longer inhabit them.

Introduced foreign species can wipe out native species

Another devastating problem is the introduction of non-native, or foreign, species—that is, species that do not naturally live in an area but are brought there on purpose or accidentally by humans. Researchers estimate that 50,000 such **introduced species** have entered the United States since Europeans arrived. Some of these introduced species, referred to as invasive species, are able to sweep through a landscape, wiping out native species as they go. The damage can happen directly, through their eating or parasitizing native species, or

(a)

(b)

Figure A.7 The Threats of Habitat Loss and Deterioration
(*a*) Habitat loss doesn't have to mean the burning of tropical rainforest in exotic locales. Here a development of homes and swimming pools has replaced what was at one time a natural landscape.
(*b*) Researchers discovered that 150 plant species have disappeared from the Middlesex Fells Reservation in Massachusetts, a 100-year-old preserve, despite the ban on development within the park. Many of the plants (which still exist in other locations) were suspected to have been lost as the result of increasing use of the park by people from the growing suburban area surrounding it.

it can happen indirectly, by their outcompeting native species for food, soil, light, and other resources.

In Africa's Lake Victoria, roughly 500 species of native cichlid [SICK-lid] fishes evolved in some 10,000 years, making this lake a trove of fish diversity. Fewer than half of those species now remain, however, and many of the surviving species are close to extinction. The Nile perch, introduced to the lake as a food fish for people, is responsible for the extinction of many of these species through predation. (Pollution and increased cloudiness of the water are other causes.)

In Guam, the brown tree snake (Figure A.8a), another invasive species, has drastically reduced the numbers of most of Guam's forest birds, leaving an eerie quiet where once the forest was noisy with tropical birdsong. The snake is thought to have been introduced accidentally, brought by U.S. military planes, from New Guinea, where it occurs naturally.

In Hawaii, introduced pigs that have escaped into and are living in the wild are devouring the native plant species. Domesticated cats and mongooses, also introduced species, have killed many of Hawaii's native birds, especially the ground-dwelling species whose nests are easy targets. Purple loosestrife, eucalyptus trees, and Scotch broom (Figure A.8b) are invasive plant species that are choking out native plants in various parts of the United States.

Climate changes also threaten species

Recent changes in climate, which the vast majority of scientists now agree are caused largely by human activities, constitute another threat that seems to be affecting many species. In Austria, for example, biologists have found whole communities of plants moving slowly up the Alps; apparently this movement is a response to global warming, as these plants are able to survive only at ever higher, cooler elevations. These plant communities are moving at an average rate of about a meter per decade. If the climate continues to warm, these alpine plants—which exist nowhere else in the world—will eventually run out of mountaintop and go extinct (Figure A.9). And in areas where organisms don't have cool mountains to climb, many species have begun moving to and living in increasingly northern latitudes to escape the heat.

Some threats are difficult to identify or define

Biologists now agree that many frogs, salamanders, and other amphibians—as well as many other kinds of organisms—are disappearing around the world. In

(a)

(b)

Figure A.8 The Threat of Introduced Species
Introduced species threaten native ecosystems everywhere. (a) The brown tree snake, which was inadvertently introduced to Guam, has not only decimated bird populations there but also disturbed human populations, climbing into babies' cribs and swimming up sewer drains to appear in people's toilets. (b) Scotch broom, brought from Europe to the United States, is driving out native plants across the country. This hardy plant has seedpods that explode to disperse its seeds. Spreading quickly, it can often be seen blooming along highways and other roadsides.

some cases, like that of the amphibians, the reason for the disappearances remains unclear. For some amphibians, pollution appears to be the culprit. In other cases, increased ultraviolet radiation seems to be killing frogs off. In still other instances, diseases are wiping out amphibians.

According to one study, climate change may be at the root of the problem, making amphibians susceptible to both ultraviolet light and disease. Scientists studying Western toads found evidence that changing climate patterns were resulting in less rain falling in the mountains of the northwestern United States. Frog and toad eggs developing in what are now sometimes very shallow pools of water are subjected to stronger, damaging ultraviolet light. These weakened eggs become vulnerable to disease, including infection by a deadly fungus.

Despite what scientists have learned about many of the specific cases of amphibian demise, it remains unclear why so many amphibians all around the world are dying off at the same time. Is one force weakening them all? Lacking an obvious "smoking gun," biologists

Figure A.9 Plants with Nowhere to Go
The alpine *Androsace* is part of a community of plant species that has been forced to move up the mountainsides of the Alps. On Mount Hohe Wilde in the Austrian Alps, this mountain wildflower has been migrating upward at a rate of about 2 meters per decade. Researchers say that if global warming continues, this species will soon run out of mountaintop to climb and will go extinct.

continue to confront a variety of potential threats without knowing which ones cause the biggest problems for the populations they study.

Human population growth underlies many, if not all, of the major threats to biodiversity

The biggest threat overall to nonhuman species is the growth of human populations. Many of the problems that we have mentioned here are the direct result of the increasing numbers of people on Earth. Our growth is what spurs continuing habitat deterioration as natural areas are converted to the farms, roads, and factories needed to support human life. As more people demand more cars, burn more gas and oil, and buy more paper, plastic, pesticides, herbicides, fertilizers, and food, more land must be devoted to making such items, and more pollution finds its way into the water and air. When people are finished using what they've bought, more waste crowds the landfills. And the effects of our growing population are further magnified by the fact that more resources are being used *per person* now than in the past.

All these changes alter the environment and hasten the demise of other species. In the search for solutions to these problems, we need to consider not only direct destruction (for example, deforestation) but also the indirect causes of biodiversity loss—in particular, the increasing use of resources by an ever-growing human population.

The factors we have discussed in this section are causing extinctions around the globe. Many more species have dwindled to dangerously low levels because of these same problems. Though such species continue to hang on, their low numbers make them much more susceptible to future extinction.

Mass Extinctions of the Past

The current mass extinction is not the first in the history of life on Earth. As described more fully in Chapter 19, since the time when life first took hold on this planet, more than 3.5 billion years ago, there have been five previous, well-documented mass extinctions. These mass extinctions took place about 440, 350, 250, 206, and 65 million years ago.

Unlike the current extinction, previous mass extinctions were not caused by humans (which had not yet evolved). Scientists have hypothesized other causes for the different extinctions, including climate change, increased volcanic activity, and reductions in sea level. One of the newest and best-supported explanations is that at least some mass extinctions were caused by the aftereffects of an extraterrestrial object—such as an asteroid—hitting Earth and filling the planet's atmosphere with a thick cloud of dust (Figure A.10).

Originally, the idea of an asteroid blasting the Earth and diminishing biodiversity was met with skepticism and ridicule. But no more. According to recent studies, extraterrestrial objects appear to have caused at least two of the five prior mass extinctions. One of these impacts occurred 65 million years ago, at the end of the Cretaceous period; it is the most famous of the five because it wiped out the last of the dinosaurs. But the largest of all prior extinctions occurred at the end of the Permian period, 250 million years ago; at that time, 80 percent to 90 percent of all marine species went extinct. With new evidence being revealed about the importance of extraterrestrial objects in mass extinctions, other extinctions may prove to have been caused by extraterrestrial impacts as well.

During these mass extinctions, certain groups of organisms disappeared while others survived, apparently unscathed. And after each mass extinction, species numbers rebounded, as whole new groups of organisms evolved and colonized the planet. For example, when dinosaurs were dominant, only a few kinds of small mammals were around. But when the dinosaurs went extinct, mammals evolved into many new species. These

Learn more about the rapid explosion of the human population in the last few hundred years.

►II
A.2

Biology Matters

Does What I Do Matter?

Why is it that we humans are so busy cutting down trees and clearing fields—doing in the biodiversity all around us—just to put up factories and shopping malls? Part of the reason is that we use so much stuff. For example, the average college student uses 500 disposable cups and 320 pounds of paper each year—think of all those coffees to go, the beer cups, notebooks, printer paper, and so on. Not only are wild places and wild organisms destroyed so that we can harvest the resources needed to make these and all the other things we buy, but wild places also end up being cleared so that there's a place to put all the garbage when we throw these things away. That same average college student also produces 640 pounds of solid waste each year, adding up to more than 200 million tons of waste generated yearly by college students alone. What can we do about it? At the University of Colorado at Boulder, a once-weekly collection of recyclables has kept about 40 percent of the school's disposable waste out of the landfill. Along similar lines, campuses offering reusable mugs and drink discounts have seen disposable waste decrease by as much as 30 percent.

Beyond such university-based efforts, more can be done. For instance, if all morning newspapers read around the country were recycled, 41,000 trees would be saved daily and 6 million tons of waste would never make its way to landfills. Americans throw away more than 25 *million* plastic beverage bottles every hour—an estimated 1 million to 6 million from bottled water alone—and discard enough aluminum to rebuild the nation's commercial airline fleet every 3 months.

By recycling a single ton of paper, we save

- 17 trees
- 6,953 gallons of water
- 463 gallons of oil
- 587 pounds of air pollution
- 3.06 cubic yards of landfill space
- 4,077 kilowatt-hours of energy

What has already been done is a good start. For example, 42 percent of all paper is now recycled. According to the U.S. Environmental Protection

Agency, 64 million tons of material avoided ending up in landfills and incinerators because of recycling and composting activities in 1999. Today, the United States recycles 28 percent of its waste—a rate that has almost doubled in the past 15 years.

Sources: www.columbia.edu/cu/cssn/greens/waste.html; www.depts.drew.edu/admfrm/recycling.html; www.epa.gov.

Figure A.10 Fact or Fiction: The End of the World

Though humans are responsible for the current mass extinction, in the past various natural occurrences have been the culprit. Current research suggests that asteroids striking the planet destroyed biodiversity on at least two occasions—a "Chicken Little" idea that initially sparked ridicule but is now widely accepted. This artist's conception depicts how such an impending event might look from space.

species began living in new habitats and exhibiting new habits. This great diversification of the mammals eventually resulted in the evolution of our own species.

If Earth has recovered from mass extinctions in the past, why should we be concerned about the current mass extinction? Won't Earth recover as it always has? The number of species on Earth would likely increase again as new species evolved, but those that are being lost would be gone forever. More important, recoveries from past mass extinctions have required many millions of years. Ocean reefs, for example, have been destroyed and have recovered from mass extinctions multiple times in the past 500 million years, but the recovery time was on the order of 5 million to 10 million years (Figure A.11). Meanwhile, life on Earth for the remaining species was very much changed with the loss of so many species that were parts of food webs—important producers, consumers, and decomposers. Are we really

Figure A.11 Starting from Scratch

Coral reefs have rebounded from mass extinctions in the past, although it took millions of years to achieve the level of biodiversity that now flourishes in some.

willing to wait—and can we survive to wait—long enough for species numbers to rebound?

The actions we take in the next 50 years could determine whether Earth's biodiversity is impoverished for millions of years. What kind of world can we, should we, and will we leave for our descendants?

The Importance of Biodiversity

Many people wonder whether the loss of one mouse species here or one beetle species there really makes a difference to humanity. To answer these questions, it helps to look at the situation from the perspective of biologists who have long wrestled with how to assess the value of diverse species within particular habitats. One question that biologists are particularly interested in answering is how, if at all, biodiversity affects the forests, wetlands, oceans, rivers, and other wild ecosystems of the world. Since the 1990s, researchers have been studying how biodiversity can contribute to the health and stability of ecosystems.

Biodiversity can improve the function of ecosystems

From tiny experimental ecosystems in an English laboratory to experimental prairies in the midwestern United States (Figure A.12), researchers have found that the more species an ecosystem has, the healthier it appears to be. Biodiversity provides ecosystems with several benefits.

For one thing, researchers looking at a variety of ecosystems have found that the more diverse ecosystems are, the higher their **productivity** is—that is, the higher is the actual mass of plant matter (leaves, stems, fruits) that they can produce. Why should biodiversity increase the productivity of an area? Different species are good at using differing resources; for example, some thrive in the sun-drenched portions of a habitat, whereas others can make the best use of the areas that lie mostly in the shade. The more different kinds of species there are, the more productively all parts and resources of a habitat can be utilized. The same goes for other resources, such as wet and soggy versus dry and parched soils. The more species there are in a habitat, the more likely it is that at least one species will be able to use a given resource productively.

Evidence also suggests that the more species there are in an ecosystem, the greater the resilience of that

Figure A.12 Testing the Importance of Biodiversity (*a*) The ecotron is a tiny experimental ecosystem created in England to test the importance of biodiversity under controlled laboratory conditions. These constructed ecosystems include grasses, wildflowers, snails, flies, and other organisms. (*b*) Much of the pioneering work on the importance of biodiversity has come from huge experiments in Minnesota prairies, in which scientists have created numerous plots containing different numbers of species.

ecosystem is. For example, the more species are present in a patch of prairie, the more easily that area can return to a healthy state following a drought. Scientists have also found that an increased diversity of species in an area leads to a lower incidence of disease and lower rates of invasion by introduced species.

Productivity, resilience, resistance to disease, and resistance to invasion are all elements of good ecosystem health. Furthermore, diversity can even lead to more diversity: researchers found that the more plant diversity there was in a plot of ground, the more species of insects were feeding on those plants and on one another. But why should people care whether ecosystems—such as forests, streams, marshes, and oceans—are healthy?

Biodiversity provides people with goods and services

Though we rarely think about it, the biosphere directly provides us with many goods and services. Even the most basic requirements of human life are provided by other organisms. Plants produce the oxygen we breathe. Every bit of food we eat is provided by other species. A wide variety of crops and livestock, such as tomatoes and cows, have been domesticated from wild species, forming the basis of our agricultural system.

In addition, in many societies, wild species provide important foodstuffs. Insects—especially crunchy, tasty ones such as grasshoppers and ants—are an important source of protein for many peoples around the world. In Central America, many people dine on the green iguana, a huge lizard that likes to sunbathe in treetops. Known

as the "chicken of the trees," this lizard has been a food source for 7,000 years. In some parts of South America, the guanaco (a relative of the llama) is an important source of meat and hides. Also in South America, the capybara, the world's largest rodent, is a prized source of meat. In Japan, many different kinds of algae are used to make food, from sushi wrappings to seaweed soup. Many different types of fish are used to top off those pieces of sushi as well.

Biodiversity also comes to our assistance when we are ill. One-fourth of all prescription drugs dispensed by pharmacies are extracted from plants (Figure A.13). Quinine, used as an antimalarial drug, comes from a plant called yellow cinchona [sing-*KOH*-nah]. Taxol, an important drug for treating cancer, comes from the Pacific yew tree. Bromelain [*BROH*-muh-lin], a substance that controls tissue inflammation, comes from pineapples. Nearly as many other drugs derive from animals, fungi, or microscopic organisms such as bacteria.

In addition to food and medicine, wild species provide many other useful products. Among these are chemicals used as glues, fragrances, pesticides, and flavorings. Many species in the kingdom Plantae provide us with shelter and furniture; examples include many of the world's trees, of course, but also various grasses (such as bamboo). Bacteria have turned out to be particularly useful to us, in part because of their great diversity and their ability to capture energy from so many different sources. Some bacteria, for example, can be used to "eat" oil spills and to clean up sewage. Bacteria also produce numerous extremely useful substances—including antibiotics, which they use to kill off competing bacteria and we use to rid our bodies of bacterial infections.

Capybara

(a)

(b)

(c)

(d)

Figure A.13 Biodiversity and Your Health
Many important and common drugs originally were found in (and sometimes continue to be produced by and harvested from) other species, most often plants. A look at some of the drugs in a medicine cabinet will bring this point home. (*a*) The Madagascar periwinkle was originally found only on Madagascar, the large island off the southeastern coast of Africa. This flowering plant is the source of vinblastine, an important anticancer drug. (*b*) The bark of the cinchona tree is the source of quinine, a drug widely used to treat malaria, a disease that strikes 350 million to 500 million people each year. (*c*) Though we may not think of caffeine as a drug, this powerful stimulant derived from coffee beans has energized and sustained humanity through long days and nights of study and work. (*d*) The wild yam is the source of diosgenin [dye-*oss*-juh-nin], the substance that is converted into progesterone for use in birth control pills.

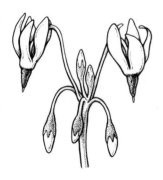

Shooting star

Whole ecosystems full of species can provide what are known as **ecosystem services**. For example, in coastal northern California many hillsides were once covered with towering redwood trees, whose needles and branches gathered and channeled fog, mist, and rain, bringing water onto and into the ground. Now that so many of these trees have been cut, the clouds full of fog, mist, and rain are much more likely to float through or burn off in the sun, the water evaporating into the air. Much less water ends up entering the ground and reservoirs, so less water is available to people living in the area. Without trees in the ground, hillsides can also easily erode and rivers become filled with sediments. Seemingly useless marshes full of reed and grass species can act as natural filters for the water moving through them, providing water-cleaning services to growing human populations.

Biodiversity also provides the world with aesthetic gifts. Scarlet macaws, parrot fishes, sea anemones, tulip trees, and shooting-star wildflowers are among the species that make it clear that if there is a value to beauty, then biodiversity is worth a lot. For many people, the existence of such a rich, living world goes beyond mere beauty to providing spiritual refreshment and rejuvenation.

Finally, consider the argument that biodiversity has value in its own right, without having to fulfill human needs and desires. This viewpoint assumes that wildlife has the right to exist unperturbed and not be destroyed by other species, such as our own.

What, then, of the golden toad now irretrievably extinct? We will never know whether the golden toad was providing important benefits to the mountain ecosystem where it lived—whether its activities helped other kinds of organisms to survive and reproduce, whether it played a key role in food webs, or whether it was important in altering the quality of the water or other aspects of its habitat. It was undoubtedly one of the flashiest and most stunning of toad species. And its bright colors—very similar to the warning colors that many toxic or bad-tasting organisms share—suggests that it might have had a chemical bounty that would interest pharmaceutical researchers. But this is just one more thing we will never know. All of this illustrates that there is one guaranteed loss every time a species goes extinct: the loss of the opportunity to ever learn more or to gain more—in any way—from the living species that disappeared.

Nature's bounty of life is great, but with so many species yet undiscovered, the vast majority of the living world's wealth remains untapped. If the sheer numbers of species yet to be discovered are any indication, much more awaits—beauty, food, shelter, medicine—if we can just find it before it disappears along with the golden toad and so many other species.

◉ Review and Discussion

1. How did Terry Erwin estimate the number of insect species in the world, and why are such estimates so difficult to do? Try to think of one other way that biologists could attempt to estimate the remaining number of unknown species on Earth. Be creative—fogging trees seemed crazy at first too.

2. Biologists don't know either the exact number of species on Earth or the exact rate at which species are going extinct. Given their lack of precise information, should we take seriously biologists' concerns about worldwide extinction of species? Why or why not?

3. How has biodiversity risen and fallen over time—when and in what way?

4. How has human population growth helped to spur threats to biodiversity?

5. Often the perceived needs of people and the preservation of endangered species come into conflict. If people continue to farm and build houses and shopping malls, many more habitats and species will disappear. But many people need and want such development. Do you think human beings have a right to such development? Do nonhuman species have a right to exist? If so, where do the rights of human beings end and the rights of other species begin?

6. In the past, when endangered species were threatened by a new building or other development, that development was often forbidden by the courts and the Endangered Species Act. Now, more often, when development conflicts with the survival of an endangered species, developers and conservationists compromise. Some land is used for building, and some is set aside for the endangered species. Why might such compromise be a good idea? Why not?

7. Every community has its conflicts between biodiversity and development. Consider the area where you live or study: Is there a habitat or environment that's endangered there because of human activity, or a way of life or profession endangered by environmental activism? Are loggers losing their jobs? Are the rivers becoming increasingly polluted with agricultural runoff? Together with one or two classmates, write a short letter to the editor of your local newspaper, arguing either for or against the importance of preserving biodiversity or human lifestyles; back up your statements with evidence wherever possible. Every voice counts, so send your letter in to the paper.

Key Terms

canopy (p. A3)
ecosystem services (p. A13)
endangered (p. A6)
habitat (p. A6)
introduced species (p. A7)

mass extinction (p. A4)
productivity (p. A11)
rainforest (p. A5)
specialist (p. A3)

4 Chemical Building Blocks

Key Concepts

- Living organisms are composed of atoms held together by chemical bonds. Four elements—oxygen, carbon, hydrogen, and nitrogen—account for 96 percent of the weight of a living cell.

- A molecule contains two or more atoms linked through covalent bonds; a covalent bond is the strongest chemical linkage that can form between two atoms.

- Some types of atoms interact through weaker linkages known as noncovalent bonds. Hydrogen bonds and ionic bonds are common noncovalent bonds in biologically important molecules.

- A chemical reaction occurs when chemical bonds between atoms are formed or reformed. To sustain life, thousands of different types of chemical reactions must occur inside even the simplest cell.

- The unique properties of water, the primary medium for life-supporting chemical reactions, have a profound influence on the chemistry of life.

- An acid releases hydrogen ions and a base absorbs them. Many chemical reactions within a cell are sensitive to the levels of acids and bases.

- Four main classes of molecules are common to all living organisms: carbohydrates, nucleic acids, proteins, and lipids. The functions of these critical molecules range from providing energy to storing hereditary information.

Looking for Life In the Goldilocks Zone

In every culture, and throughout recorded history, humans have gazed at the stars and wondered what secrets those distant points of light might hold. That fascination continues into modern times with the unabated popularity of sci-fi books and movies, and regular reports of UFO sightings and abduction by aliens.

Despite the half-wishful, half-fearful imaginings of popular culture, there is at present no scientific evidence for the existence of life beyond our blue planet. So far, the closest encounters have been entirely of the chemical kind. Earth-based instruments have detected chemical signatures that hint at the possibility of life on distant planets. The chemical hallmarks of life on earth have been found in material from deep space. Amino acids and simple sugars, building blocks of some of the key components of terrestrial life, have been identified in meteorites, for example.

About 20 light years away from us, in the constellation of Libra, is Gliese 581c, a planet with conditions so perfect for life that it has been dubbed the Goldilocks planet. With atmospheric temperatures that range from 0°C to 40°C, the planet is not too hot, not too cold, just the way Goldilocks in the children's story preferred her porridge. The Goldilocks planet has been a media sensation because liquid water, critical for life on earth, is predicted to exist in its reasonably pleasant climate. If space aliens have a chemistry similar to ours, we could run into a few on Gliese 581c. But 20 light years is a long way to travel in hopes of a chance encounter with an alien. **Are we ready to write off planets closer to home as dead zones for life because they don't fall into that Goldilocks zone? Do we even understand the Goldilocks zone, the range of conditions that can support life as we know it?** Astrobiologists, whose goal is to explore the origins, evolution, and distribution of life in the known universe, are turning to the extreme zones of our own planet for a deeper understanding of life's limits and therefore a more informed quest for extraterrestrial life. We shall take a look at extreme living on earth, but first we will survey the chemical basis of life, the primary focus of this chapter.

Looking for Life in All the Right Places
Artist's impression of Gliese 581c (bluish-gray sphere in the foreground), the most earthlike planet yet discovered. Five times more massive than the earth, the so-called Goldilocks planet is predicted to have liquid water and it could harbor life. The red sphere in the background is the planet's host star, Gliese 581; the two smaller spheres are sister planets.

For all its remarkable diversity, life as we know it is built from a rather limited variety of atoms. The fact that all cells share this limited range of atomic ingredients reminds us of the fundamental unity of all life on earth.

In this chapter we begin our exploration of the cellular complexity of living organisms by identifying the chemical components shared by all of them. We examine how assemblies of atoms that produce small molecules are linked yet again to create larger units, or macromolecules, to serve as structural material, fuel, and other functional components. Many of the topics introduced in this chapter serve as a foundation for the more in-depth investigation of life, at every level of the biological hierarchy, that follows in later chapters.

4.1 The Physical World Is Made Up of Atoms

The universe contains at least 92 different types of pure materials known as elements. An **element** has a distinctive set of physical and chemical properties and cannot be broken down to other substances by ordinary chemical methods. Each element is identified by a one- or two-letter symbol; for example, oxygen is identified as O, calcium as Ca. An element itself is composed of tiny units called atoms that are so small that more than a trillion of them could easily fit on the head of a pin. An **atom** is defined as the smallest unit of an element that still has the characteristic chemical properties of that element. Because there are 92 naturally occurring elements, there are also 92 different types of atoms. The uniqueness of an element comes entirely from the special characteristics of the atoms that compose it.

What makes the atoms of one element different from those of another? The answer lies in the specific combination of three atomic components. The first two components are electrically charged: **protons** have a positive charge (+), and **electrons** have a negative charge (–). As the name of the third component implies, **neutrons** lack an electrical charge and are therefore considered electrically neutral. These three components, especially the electrons, determine how atoms behave in the physical world and how they interact with one another.

A single atom has a dense central core, called the **nucleus** (plural "nuclei"), that contains one or more protons and is therefore positively charged. Except in the case of hydrogen atoms, the nucleus also contains one or more neutrons. One or more negatively charged electrons move in space around the nucleus (Figure 4.1). If a hydrogen nucleus were the size of a marble, the electron would move around it in a space as big as the Houston Astrodome. As a whole, the positive charge on the nucleus balances the negative charge on the electrons such that atoms are electrically neutral.

Describing atoms with numbers

The distinctive aspects of the atom of each element can be summarized by numbers that describe the atom's structure and mass. The number of protons found in an atom's nucleus is the **atomic number** of that particular element. Hydrogen, with its single proton, has an atomic number of 1, while carbon, with its six protons, has an atomic number of 6.

Elements can also be distinguished by their **atomic mass number**, which is the sum of an atom's protons and neutrons. The mass of any object is the amount of matter in it. On earth, mass and weight are equivalent quantities. The mass of an electron is only about 1/2,000 that of a proton or neutron, so the atomic mass number of an element is based on the total number of protons and neutrons contained in the nucleus of the atom. Hydrogen has a single proton and no neutrons, so the atomic mass number for hydrogen is the same as its atomic number, namely 1, written ^{1}H. In contrast, the nucleus of a carbon atom contains six protons and six neutrons, giving carbon an atomic mass number of 12 (^{12}C). A carbon atom has about 12 times as much mass as a hydrogen atom.

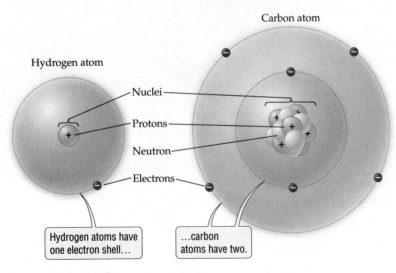

Carbon atom

Hydrogen atom

Nuclei

Protons

Neutron

Electrons

Hydrogen atoms have one electron shell...

...carbon atoms have two.

Figure 4.1 Atomic Structure
The electrons, protons, and nuclei of these hydrogen and carbon atoms are shown greatly enlarged in relation to the size of the whole atom. An electron shell is a simplified way of representing the space that electrons move in as they orbit the nucleus.

Some elements have variant forms of atoms that differ only in the number of neutrons in their nuclei

So far we have discussed elements as if each is found in only one form, but in the natural world some elements vary in the number of neutrons in their atomic nuclei. These different forms of an element are called **isotopes**. All isotopes of an element have the same number of protons (the same atomic number) and electrons. But they differ in the number of neutrons and therefore in their atomic mass numbers. For example, over 99 percent of the carbon atoms found in atmospheric carbon dioxide gas (CO_2) have an atomic mass number of 12 (^{12}C). However, a tiny fraction—about 1 percent—of those carbon atoms exist as an isotope containing eight neutrons instead of six. These eight neutrons, together with the six protons, give an atomic mass number of 14 (^{14}C). This isotope is therefore referred to as carbon-14.

Some isotopes have unstable nuclei that change (decay) into simpler forms, releasing high-energy radiation in the process. Isotopes that tend to decay in this manner are known as **radioisotopes** [RAY-dee-oh-EYE-so-topes]. The carbon-14 isotope of carbon has an unstable nucleus and is therefore a radioisotope. Although only a fraction of known isotopes are radioactive, some radioisotopes, such as those of carbon (carbon-14), phosphorus (phosphorus-32), and hydrogen (deuterium, ^2H, and tritium, ^3H), have important uses in both research and medicine.

The radiation given off by radioisotopes can be detected by a variety of methods, ranging from simple film exposure to the use of sophisticated scanning machines. The detectability of radioisotopes means that their location and quantity can be tracked fairly easily, a characteristic that makes them useful in medical diagnostics. For example, the thyroid gland takes up iodine for producing a special type of hormone required by the body. When a low dose of an iodine radioisotope (iodine-131) is administered to patients with thyroid disease, the uptake of the radioisotope by the thyroid allows physicians to visualize the gland with an imaging device (Figure 4.2). If a patient is found to be suffering from cancer of the thyroid, repeated doses of iodine-131 can be administered as a therapy, since the accumulation of radioactivity in the thyroid tends to kill the cancer cells.

Atoms can associate with other atoms through chemical bonds

The number of electrons in an atom, and the way in which those electrons are packed around the nucleus, is the main determinant of an atom's chemical behavior.

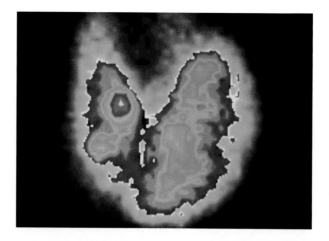

Figure 4.2 Radioisotopes Are Useful in Medical Imaging
Small amounts of radioactive iodine were given to the patient, whose thyroid gland is visualized in the gamma ray scan shown here. The scan shows a normal lobe (on the left) and an abnormal lobe (on the right) swollen to twice its usual size because of a goiter. Note the higher metabolic activity (in red) in the normal lobe.

Some elements are said to be chemically inert because their atoms tend not to interact with other atoms by losing, gaining, or sharing electrons. Most elements, however, and all elements that are biologically important, have atoms that are far more social. Such atoms have a tendency to donate electrons, accept electrons, and even share electrons, if the right type of atom becomes available under the right conditions. The attractive interaction that causes two atoms to associate with each other is known as a **chemical bond**.

When an atom loses one or more of its negatively charged electrons, it acquires an overall (net) positive charge. Likewise, when an atom gains one or more electrons, it becomes negatively charged. Atoms that become charged due to loss or gain of electrons are called **ions**. Ions with opposite charge experience an electrical attraction, and such ionic interactions have an important role in the chemistry of living systems. The chemical attraction between negatively charged and positively charged ions is called an **ionic bond**, and it constitutes one type of chemical bonding. A salt is made up of ions from at least two different elements; these ions are held together exclusively by the electrical attraction between them.

A **molecule** is an assemblage of atoms in which at least two of the atoms are linked through electron sharing. Electron sharing creates an exceptionally strong chemical bond known as a **covalent bond**. A molecule contains a minimum of two atoms, but larger molecules can be composed of millions of atoms. A molecule may consist of atoms of the same element (say, oxygen atoms only), or it may contain atoms from two or more elements (two hydrogen atoms and one oxygen atom in the water molecule, for example).

Any substance that contains atoms from two or more *different* elements, each in a precise ratio, is known as a **chemical compound**. A molecule of water, for example, is a compound of hydrogen and oxygen atoms. Chemists have developed a simple shorthand, known as a chemical formula, to represent the atomic composition of salts and molecules. The formulas use the letter symbol of each element and a subscript number to show how many atoms of that element are contained in the salt or molecule. For example, each molecule of water is made up of two hydrogen (H) atoms and one oxygen (O) atom, so the chemical formula for this molecule is H_2O. The same notation is used for more complex compounds, such as table sugar (sucrose), which has 12 carbons, 22 hydrogens, and 11 oxygens per molecule ($C_{12}H_{22}O_{11}$). For ionic compounds such as table salt, a similar method is used. Table salt has equal numbers of sodium ions (Na^+) and chloride ions (Cl^-). The molecular formula for salt is therefore NaCl.

4.2 Covalent Bonds Are the Strongest Chemical Bonds in Nature

As we mentioned earlier, a molecule contains at least two atoms held together by covalent bonds. One covalent bond represents the sharing of one pair of electrons between two atoms. The atoms can be of the same type (Figure 4.3a) or they can be atoms of two different elements (as, for example, in all the molecules in Figure 4.3b except hydrogen gas).

What would drive atoms to share electrons with one another? To answer that question, we must first examine how electrons are distributed in the space around the nucleus. The electrons of every atom move in volumes of space that can be visualized as concentric

(a)

(b)

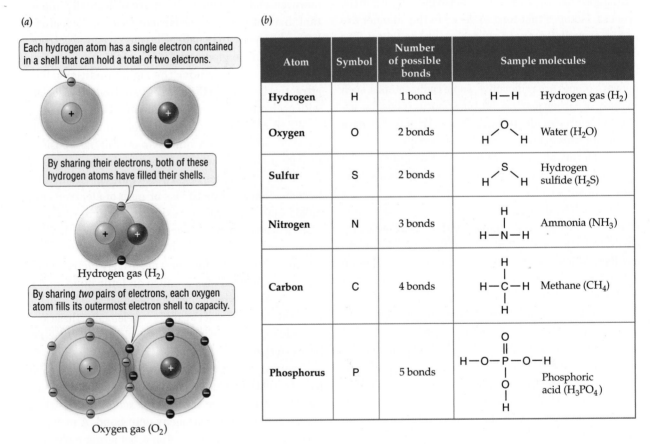

Atom	Symbol	Number of possible bonds	Sample molecules	
Hydrogen	H	1 bond	H—H	Hydrogen gas (H_2)
Oxygen	O	2 bonds		Water (H_2O)
Sulfur	S	2 bonds		Hydrogen sulfide (H_2S)
Nitrogen	N	3 bonds		Ammonia (NH_3)
Carbon	C	4 bonds		Methane (CH_4)
Phosphorus	P	5 bonds		Phosphoric acid (H_3PO_4)

Figure 4.3 Covalent Bonds

(a) Atoms in biologically important molecules have as many as four electron shells. The innermost holds a maximum of two electrons and the next three have a maximum capacity of eight, eighteen, and thirty-two. The number of covalent bonds that an atom can form depends on the number of electrons needed to fill its outermost shell to capacity. For example, two hydrogen atoms can share one pair of electrons, forming one covalent bond. An oxygen atom requires two additional electrons to fill its outermost shell and can therefore form two covalent bonds. A carbon atom has four electrons in its outermost shell and requires four more electrons to fill that shell; a carbon atom can therefore form a maximum of four covalent bonds (b).

layers, called shells, around the nucleus (see Figure 4.1). The maximum number of electrons in any shell is fixed. When all its shells are filled to capacity, an atom is in its most stable state. Atomic shells are filled starting from the innermost shell. The innermost shell can hold two electrons at most. The next shell outside it can hold a maximum of eight electrons.

Atoms that have unfilled outer shells can achieve a more stable state by reacting with one another in ways that will achieve maximum occupancy of the outermost shell. One way for an atom to achieve greater stability is to fill its outermost shell by sharing one or more of its outer shell electrons with a neighboring atom. Each atom in this arrangement contributes one electron to every pair of shared electrons. Each pair of electrons that is shared between the two atoms constitutes one covalent bond (Figure 4.3a).

The number of covalent bonds that an atom can form depends on the number of electrons needed to fill its outermost shell. Consider the electron sharing that occurs in the covalent bonds between hydrogen and oxygen in a water molecule. Hydrogen has one electron in its single shell, but that shell is filled only when it has two electrons. The inner shell of oxygen is filled, but its outer shell is not: it has six electrons, but can hold eight. This situation can be resolved by mutual borrowing, on a "time-sharing" basis, between two hydrogen atoms and an oxygen atom: each hydrogen atom shares its one electron with oxygen, and the oxygen atom shares two of its electrons, one with each of the hydrogen atoms. So the atoms contribute electrons in a way that makes the outer shells of all three atoms complete, at least on a shared basis. This kind of sharing requires an intimate association between the atoms, which is why covalent bonds are so strong.

Water and the natural gas called methane have the chemical formulas H_2O and CH_4, respectively. These formulas reveal the atomic components of each compound, but they say nothing about how the various atoms are bonded together and arranged in space. Another type of notation, known as a structural formula, is used to indicate both the atoms and the bonding arrangement in a molecule. As Figure 4.3b shows, a water molecule is held together by two covalent single bonds, each of which joins the oxygen atom with one of the two hydrogen atoms. Likewise, methane has four covalent bonds, each of which links the lone carbon atom to one of the four hydrogen atoms in the molecule.

Carbon, nitrogen, hydrogen, oxygen, phosphorus, and sulfur are among the most common elements found in living organisms, and their atoms all form covalent bonds. Combinations of these atoms are found in the molecules that make up living organisms.

4.3 Hydrogen Bonds and Ionic Bonds Are Individually Weak, but Collectively Potent

Atoms can be linked, not only by covalent bonds, but also via **noncovalent bonds**, which are based on other types of chemical interactions between atoms, such as the mutual attraction between atoms that have an opposite electrical charge. Noncovalent bonds help create the complex physical organization and diverse metabolic activities we associate with living organisms. They assume this important role despite the fact that noncovalent bonds are weaker than covalent bonds. Noncovalent bonds exert their substantial effects on the structure and function of living cells through their sheer number and dynamic nature.

The relative weakness of noncovalent bonds is a virtue in situations where they link many molecules with each other, or help establish the overall configuration of large molecules. Because noncovalent bonds can form, break, and re-form rapidly, they foster atomic associations that are readily altered. The liquidity of water at room temperature stems from the fact water molecules are constantly linking and de-linking as noncovalent bonds shift among molecules. Noncovalent bonds also allow large, complex molecules, such as proteins, to adjust to changes in their surroundings, a feature necessary for many biological processes. For example, when skin is pinched and then released, its ability to stretch and then spring back depends on the breaking and re-forming of a multitude of noncovalent bonds between many different classes of molecules, including those in an unusually "springy" protein known as elastin.

See how temperature affects molecular movement.

4.1

Hydrogen bonds are especially dynamic noncovalent bonds that are widely encountered in the chemistry of life

The hydrogen bond is one of the most important kinds of noncovalent bonds in the chemistry of living organisms. A hydrogen bond is about 20 times weaker than a covalent bond. A simple example of a hydrogen bond is the one that exists between water molecules in liquid

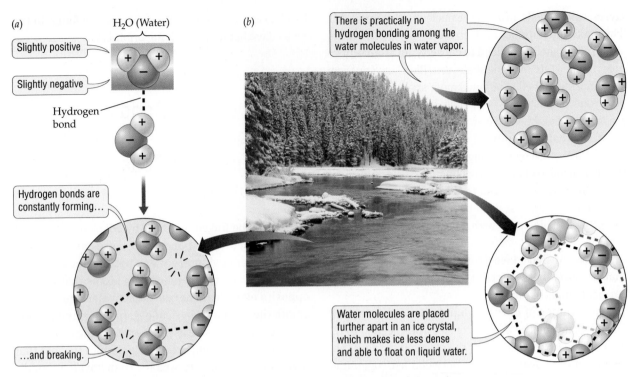

Figure 4.4 The Polarity of Water Leads to Hydrogen Bonding between Water Molecules

In liquid water and ice, water molecules interact with each other through hydrogen bonds. (*a*) Water is a polar molecule because it has two distinctly different ends, or poles: the oxygen end has a slight negative charge; the opposite end bears a slight positive charge. This unevenness in the distribution of electrical charge comes about because each pair of shared electrons is drawn more strongly toward the oxygen atom than toward the hydrogen atom. The weak attraction between a hydrogen atom with partial positive charge and the partially negative region of any polar molecule is known as a hydrogen bond. In the liquid state, hydrogen bonds are constantly formed and re-formed among water molecules. (*b*) When water freezes, the hydrogen bonds become more rigid as water molecules becomes stacked into a three-dimensional network of hexagons, forming an ice crystal. When water molecules gain enough energy, they enter their gas state, as water vapor. Because water molecules are spaced farther apart, and moving around much faster, they have little opportunity to form hydrogen bonds when they are in their gaseous state.

water or ice (Figure 4.4). Hydrogen bonding is an important attribute of water and is the basis of many of water's unique chemical properties. Most organisms consist of more than 70 percent water by weight, and nearly every chemical process associated with life occurs in water.

As we have seen, each molecule of water is made up of two hydrogen atoms and one oxygen atom held together by covalent bonds. The positive nucleus of the oxygen atom tends to attract the negative electrons more powerfully than the nuclei of the two hydrogen atoms do, causing each water molecule to have an uneven distribution of electrical charge. That is, the oxygen atom carries a slightly negative charge, while the hydrogen atoms in turn carry a slightly positive charge (Figure 4.4*a*). Molecules with an uneven distribution of electrical charge are called **polar molecules**.

Because opposite charges attract, the slightly positive hydrogen atoms of one water molecule are attracted to the slightly negative oxygen atom of a neighboring water molecule. This attraction forms a hydrogen bond between the two water molecules, which can also form hydrogen bonds with yet other neighboring water molecules (see Figure 4.4). The resulting dynamic interplay of hydrogen-bonded water molecules is responsible for the properties of liquid water (in particular, for making it liquid at room temperature). The more stable network of water molecules that forms in ice is also the result of hydrogen bonding (Figure 4.4*b*).

A **hydrogen bond** forms when a hydrogen atom with a partial positive charge interacts with a neighboring polar molecule that contains a partially negative atom. Water molecules can form hydrogen bonds with other polar compounds, with the result that these compounds dissolve in water. Such compounds are said to be **soluble** in water because they dissolve, that is, mix completely with the water. Ions also dissolve readily in water because water molecules can form a network, or hydration shell, around them. For example, when table salt is added to water, the solid crystals dissolve. The crystals break apart as the ions in the salt crystal are surrounded by water molecules

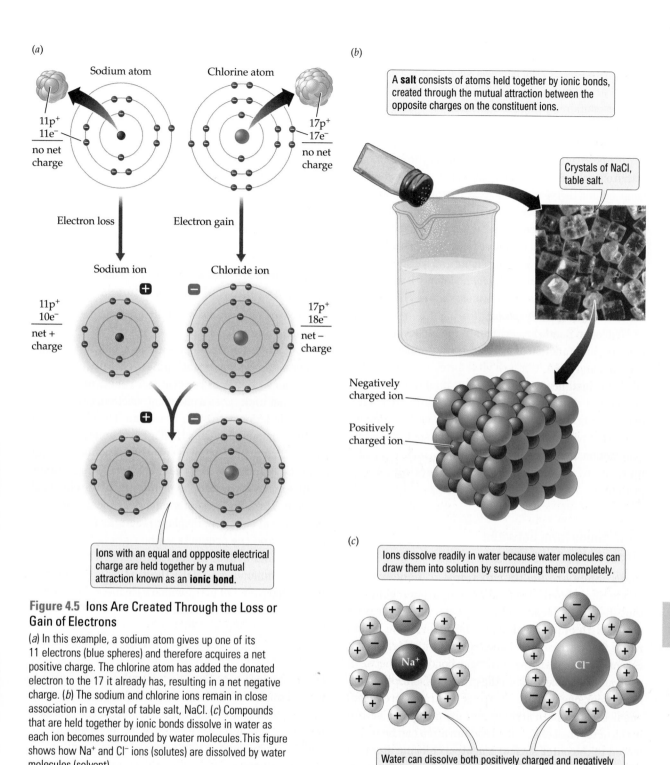

(a)

Sodium atom   Chlorine atom

11p⁺ / 11e⁻
no net charge

17p⁺ / 17e⁻
no net charge

Electron loss   Electron gain

Sodium ion   Chloride ion

11p⁺ / 10e⁻
net + charge

17p⁺ / 18e⁻
net – charge

Ions with an equal and oppposite electrical charge are held together by a mutual attraction known as an **ionic bond**.

Figure 4.5 Ions Are Created Through the Loss or Gain of Electrons

(a) In this example, a sodium atom gives up one of its 11 electrons (blue spheres) and therefore acquires a net positive charge. The chlorine atom has added the donated electron to the 17 it already has, resulting in a net negative charge. (b) The sodium and chlorine ions remain in close association in a crystal of table salt, NaCl. (c) Compounds that are held together by ionic bonds dissolve in water as each ion becomes surrounded by water molecules. This figure shows how Na⁺ and Cl⁻ ions (solutes) are dissolved by water molecules (solvent).

(b)

A **salt** consists of atoms held together by ionic bonds, created through the mutual attraction between the opposite charges on the constituent ions.

Crystals of NaCl, table salt.

Negatively charged ion

Positively charged ion

(c)

Ions dissolve readily in water because water molecules can draw them into solution by surrounding them completely.

Na⁺   Cl⁻

Water can dissolve both positively charged and negatively charged ions because water molecules can orient themselves in such a way that they expose to the ion either the partial negative charge on the oxygen atom, or the partial positive charge on the hydrogen atoms.

Explore the process of salt dissolving in water.
4.2

and scattered uniformly throughout the liquid (Figure 4.5). Since dissolved compounds abound in and around living cells, chemists and biologists use specific terms to describe these mixtures: a **solution** is any combination of a **solute** (a dissolved substance, such as salt)

and a **solvent** (in this case water), the fluid into which the salt has dissolved.

The polarity of water molecules also means that they will not associate with other molecules that are not charged—that is, with **nonpolar molecules**. When they are added to water, nonpolar molecules tend to gather among their own kind. This is exactly what happens when olive oil (composed of uncharged molecules) is added to vinegar, a watery solution containing charged molecules. To make a salad dressing, the two substances have to be shaken vigorously, otherwise the oil separates from the vinegar. The distribution of electrons among carbon atoms and hydrogen atoms within oil molecules is nearly equal, making these molecules nonpolar and therefore insoluble. Waxes are also nonpolar, and automobile enthusiasts wax their cars not just to make them look shiny, but also to repel water and reduce the risk that their hot rods will be marred by rust. Molecules that associate with water (such as sugar and salt) are called **hydrophilic** (*hydro*, "water"; *philic*, "loving"), while molecules that are excluded from water (such as oil and wax) are called **hydrophobic** (*phobic*, "fearing").

Hydrogen bonds are critical to the structure, organization, and function of many large molecules, such as proteins and DNA, that are vital for cell function. Our DNA, for instance, would fall apart if all the hydrogen bonds in it didn't exist. At the same time, hydrogen bonds facilitate dynamic associations between molecules. During rapid biological processes such as muscle contraction, hydrogen bonds are broken and re-formed moment by moment between interacting sets of proteins.

Ionic bonds form between atoms of opposite charge

Ionic bonds are another example of noncovalent bonds. Ionic bonds are similar to hydrogen bonds in that they form because of mutual attraction between atoms with opposite charges; however, ionic bonds result from attraction between species with a *net* electrical charge (ions), whereas hydrogen bonds form between neutral polar molecules. In the process of ionization (ion formation), one or more electrons are transferred from one neutral atom to another. For example, an electron in the outer shell of a neutral sodium atom can be transferred to the outer shell of a neutral chlorine atom, as shown in Figure 4.5a. This loss of an electron from the sodium atom and the gain of an electron by chlorine converts both neutral atoms into ions with a *net* charge that is equal and opposite. For maximum stability the ions created through such a charge transfer must remain closely associated. The mutual electrical attraction between these ions is known as an ionic bond. Charged atoms that are held together through ionic bonds are known as **salts**. In ionic solids such as crystals of table salt (NaCl), these ions are closely packed in an orderly pattern.

A grain of dry table salt consists of countless sodium ions (Na^+) linked by ionic bonds to chloride ions (Cl^-), as shown in Figure 4.5b. In the absence of water, these ions pack tightly to form the hard, three-dimensional structures we know as salt crystals. When salt is added to water, the water molecules, because they are polar, are attracted to and surround both types of charged ions in NaCl. This interaction with water breaks up and dissolves the salt crystals, scattering both positive sodium ions and negative chloride ions throughout the liquid (see Figure 4.5c).

Chemical reactions are characterized by rearrangement of chemical bonds

Many biological processes require atoms to break existing connections and form new ones with other atoms or groups of atoms. The process of breaking existing chemical bonds and creating new chemical bonds is known as a **chemical reaction**. A **reactant** is a substance that undergoes a chemical reaction, either alone or in conjunction with other reactants. The alteration of electron sharing patterns through a chemical reaction yields at least one chemical substance that is different from the reactant(s) and the newly formed substance or substances are called the **product(s)** of the chemical reaction. The standard notation for chemical reactions, the chemical equation, displays reactants to the left of an arrow and the products to the right of that arrow. Nitrogen and hydrogen, for example, can combine to produce ammonia gas (NH_3), the gas that is responsible for the sharp odor of many window cleaners. The chemical equation for this reaction is

$$3\,H_2 + N_2 \rightarrow 2\,NH_3$$

The arrow indicates that the molecules on the left side of the equation (the reactants, hydrogen molecules and nitrogen molecules) are converted to the product, ammonia. The numbers in front of the molecules define how many molecules participate in the reaction. In this case, three molecules of molecular hydrogen (H_2) combine with one molecule of molecular nitrogen (N_2) to produce two molecules of ammonia (note that "1" is generally omitted).

Chemical bonds in atoms are rearranged during a chemical reaction, but the process can neither create nor destroy atoms. Therefore, the reaction must begin and end with the same number of atoms of each element. In the example shown here, there are six hydrogen atoms for each pair of nitrogen atoms among the reactants, and all six hydrogen atoms and both the nitrogen atoms are accounted for in the two molecules of product (ammonia).

Concept Check

1. An atom of iron (Fe) contains 26 electrons and 30 neutrons. (a) How many protons does it have? (b) How many uncharged particles? (c) What is its atomic mass number?

2. What is a molecule? How many atoms of oxygen (symbol: O) are present in each molecule of table sugar (sucrose), the chemical formula of which is $C_{12}H_{22}O_{11}$?

3. Describe how ions are formed.

4. Why is it that oil and water don't mix?

4.4 The pH Scale Expresses the Acidity or Basicity of a Solution

All chemical reactions that support life occur in water. Some of the most important are those that involve two classes of compounds: acids and bases. An **acid** is a polar compound that dissolves in water and loses one or more hydrogen ions (H^+). These hydrogen ions tend to bond with the surrounding water molecules, forming positively charged hydronium ions (H_3O^+). Because the formation of hydronium ions is easily reversed, hydrogen ions are constantly being exchanged between water molecules and other molecules dissolved in water.

Bases are also polar compounds, but unlike acids, they *accept* hydrogen ions from their surroundings. Acids and bases interact with water molecules in different ways and have opposite effects on the amount of free hydrogen ions in water. A base can accept one H^+ ion from water, leaving one hydroxide ion (OH^-) behind.

The concentration of free hydrogen ions in water influences the chemical reactions of many other molecules. Hydrogen ion concentration can be expressed as a scale of numbers that conventionally ranges from 0 to 14, where 0 represents an extremely high concentration of free hydrogen ions and 14 represents the lowest. This scale is called the **pH** scale (Figure 4.6). Each unit of the scale represents a tenfold increase or decrease in the concentration of hydrogen ions in a water solution.

In pure water, the concentrations of free hydrogen ions and hydroxide ions are equal, and the pH is said to be neutral, or in the middle of the scale, at pH 7. The addition of acids to pure water raises the concentration

of free hydrogen ions making the solution more acidic. An acid pushes the pH below the neutral value of 7. Adding a base lowers the concentration of free hydrogen ions in the solution, making the resulting solution more basic. The addition of a base raises pH above 7 because a base removes hydrogen ions from water molecules, thereby creating an excess of hydroxide ions (OH^-) and a comparable deficit of hydrogen ions.

We have all encountered acidic and basic substances. The tartness of lemon juice is due to the acidity of the juice (about pH 2). Our stomach juices are able to break down food because they are very acidic (about pH 2). At this low pH, noncovalent bonds within and between molecules are disrupted by the high concentration of free hydrogen ions, and even the covalent bonds of some molecules are broken. At the other extreme, a very basic substance, such as oven cleaner (pH 13 or so), can also cause molecules to be disrupted or broken. This is why extremes of pH can be caustic, causing chemical burns on the skin. Figure 4.6 shows the approximate pH of some common substances.

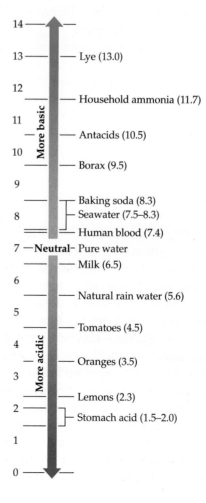

Learn more about acids, bases, and pH.
▶ ❙❙
4.3

Figure 4.6 The pH Scale

A pH of 7 means that a solution is neutral, neither basic nor acidic. Values below 7 indicate acidic solutions; the lower the value, the more acidic the solution. A solution with a pH of 3 is ten times more acidic than one with a pH of 4. Values above 7 indicate basic solutions; the higher the value, the more basic the solution. A solution with a pH of 10 is one hundred times more basic than a solution with a pH of 8.

Biology Matters

Elements 'R Us

Although there are traces of other atoms in our bodies, we are mostly made up of just eleven different elements. The chart that follows shows the percentage by weight of each element that is part of us. Much of the hydrogen and oxygen in our bodies is combined in the form of H_2O molecules, so that approximately 70 percent of our weight is water. Other trace elements make up about 0.01 percent of our weight. We need to eat, in part, to replace the elements that we naturally lose each day. We also need food energy to keep us going.

The nutrition label on a carton of whole milk specifies the amount of energy, expressed in calories, that is stored in the chemical bonds of one serving. The label also shows the types of molecules the milk contains: Fats and carbohydrates contain atoms of the elements carbon (C), hydrogen (H), and oxygen (O). Cholesterol contains the same three elements. Proteins contain nitrogen (N) and sulfur (S) in addition to C, H, and O. We can see that the elements sodium (Na), calcium (Ca), iron (Fe), and phosphorus (P) are included on the whole-milk label, as well. Vitamins also contain carbon, hydrogen, and nitrogen. So, whole milk is a source of at least nine of the eleven major elements that, combined in many different ways, make up our bodies.

Keeping an eye on food labels helps us better understand the amounts of energy and nutrients in our diets, and eating a variety of foods can help assure a healthy intake of elements. Potatoes and bananas are rich in potassium, an element important in nerve function. Dairy products are richer than most food in calcium, a major

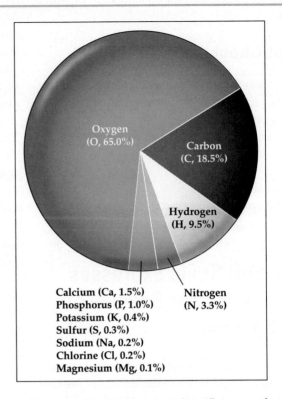

Oxygen (O, 65.0%)

Carbon (C, 18.5%)

Hydrogen (H, 9.5%)

Nitrogen (N, 3.3%)

Calcium (Ca, 1.5%)
Phosphorus (P, 1.0%)
Potassium (K, 0.4%)
Sulfur (S, 0.3%)
Sodium (Na, 0.2%)
Chlorine (Cl, 0.2%)
Magnesium (Mg, 0.1%)

component of bones and teeth. Yogurt and sunflower seeds are good sources of phosphorous, which forms the phosphate groups that link the building blocks of DNA and RNA. Processed foods often contain large amounts of sodium in the form of added salt. Deficiencies in sodium are rare, and experts warn against ingesting too much of this element, because excess sodium is associated with high blood pressure.

Buffers prevent large changes in pH

Most living systems function best at an internal pH that is close to neutral. Any change in pH to a value significantly below or above 7 adversely affects many biological processes. Because hydrogen ions can move so freely from one molecule to another during normal life processes, organisms must have ways of preventing dramatic changes in the pH levels of their internal environments. Substances called **buffers** meet this need by maintaining the concentration of hydrogen ions within narrow limits. They do so by releasing hydrogen ions when the surroundings become too basic (excessive OH^- ions, high pH) and accepting hydrogen ions when the surroundings become too acidic (excessive H^+ ions, low pH).

4.5 The Chemical Building Blocks of Living Systems

If we removed the water from any living organism, we would be left with four major classes of molecules, all of them critical for living cells: carbohydrates, nucleic acids, proteins, and lipids (informally known as "fats" and "oils"). Each of these biologically important molecules is built on a framework of covalently linked carbon atoms associated with hydrogen. Oxygen, nitrogen, phosphorus, and sulfur atoms are also found in some of these molecules.

Carbon is the predominant element in living systems partly because it can form large molecules that contain thousands of atoms. A single carbon atom can form strong covalent bonds with up to four other atoms. Even more importantly, carbon can bond to carbon, forming long chains, branched molecules, or even rings. The diversity of biological processes depends on the wide variety of small and large molecular structures that can be built from a carbon–carbon framework and a small handful of other types of atoms.

Molecules that include at least one carbon–hydrogen bond are referred to as **organic molecules**. The cell contains many types of small organic molecules, which contain up to 20 atoms or so. These include substances such as sugars and amino acids. Small organic molecules can link up via covalent bonds to create larger assemblies of atoms called **macromolecules** (*macro,* "large"), which include substances such as starch and proteins. Carbon compounds in living organisms often follow this principle of building very large macromolecules from smaller units. Small molecules that serve as repeating units in a macromolecule are called **monomers** (*mono,* "one"; *mer,* "part"). Many macromolecules in a cell are made up of hundreds of monomers covalently bonded to each other. Macromolecules that contain monomers as building blocks are called **polymers** (*poly,* "many") (Figure 4.7). Polymers account for most of an organism's dry weight (its weight after all water is removed) and are essential for every structure and chemical process that we associate with life.

In living organisms, fewer than 70 different biological monomers are combined in an endless variety of ways to produce polymers with many different properties. Polymers are therefore a step up from monomers in organizational complexity, and they have chemical properties that are not manifested in a monomer. Furthermore, the properties of organic polymers depend on the properties

(a)

Name	Structural formula	Notes
Methane	H–C–H (with H above and below)	Swamp gas.
Butane	Carbon atoms form chains of different lengths...	Component of natural gas; used for camp stove burners.
Isoprene	...that can branch...	Building block of natural rubber.
Benzene	...or form rings.	Important industrial solvent.

(b)

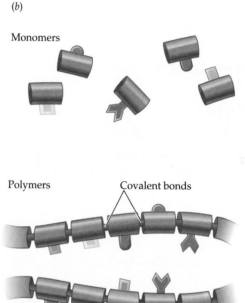

Monomers

Polymers   Covalent bonds

A limited set of monomers can make many kinds of polymers, each with different chemical characteristics.

Figure 4.7 Assembling Complex Structures from Smaller Components
In living organisms, most large and complex structures are built from smaller components. This principle applies even at the level of atoms and molecules. (*a*) A single carbon atom can form a total of four covalent bonds by sharing electrons with other atoms. Carbon atoms can bond with other carbon atoms to form a variety of structures, including unbranched or branched chains and one or more rings. (*b*) Just as an individual atom can link together to form molecules, building blocks formed by small molecules (monomers) can be joined to make larger assemblies, called polymers.

of attached clusters of atoms, called functional groups (Table 4.1). As the name implies, **functional groups** are clusters of covalently bonded atoms that have the same specific chemical properties no matter what molecule they are found in. Some functional groups help establish covalent linkages between monomers; others have more general effects on the chemical characteristics of a polymer. Four general classes of small organic molecules are critical for life in every organism—sugars, nucleotides, amino acids, and fatty acids—and the properties of each of these classes of molecules are strongly influenced by the specific functional groups found in it.

Carbohydrates provide energy and structural support for living organisms

Sugars are familiar to us as compounds that make foods taste sweet. Although not all sugars are perceived as sweet by human taste buds, most sugars are important food sources and serve as a major means of storing energy in living organisms. Sugars and their polymers are referred to as **carbohydrates**. The name comes from the ratio of C, H, and O in the compounds, in which, for each carbon atom (*carbo*), there are two hydrogens and one oxygen (corresponding to a molecule of water, a *hydrate*).

The simplest sugar molecules are called **monosaccharides** [*MAH*-noh-*SACK*-uh-ridez] (*mono,* "one"; *sacchar,* "sugar"). Like most carbohydrates, monosaccharides are made up of units containing carbon, hydrogen, and oxygen atoms in the ratio of 1:2:1, that is, one carbon atom to two hydrogen atoms to one oxygen atom. This ratio can also be expressed as the molecular formula $(CH_2O)_n$, with n ranging from 3 to 7. This means the many different types of monosaccharides found in nature have anywhere from 3 to 7 carbon atoms. Monosaccharides are often referred to by the number of carbon atoms they contain; for example, a sugar with the molecular formula $(CH_2O)_5$ is a five-carbon sugar. Note that the parentheses work the same in this notation as in multiplication. A more common way to express the molecular formula for this sugar is $C_5H_{10}O_5$. The one monosaccharide that is found in almost all cells is **glucose** $(C_6H_{12}O_6)$. Glucose has a key role as an energy source within the cell, and nearly all the chemical reactions that produce energy for living organisms involve the manufacture or breakdown of this sugar.

Monosaccharides can combine to form larger, more complex molecules. When two monosaccharides combine they form a **disaccharide** (*di,* "two"). For example, two covalently bonded molecules of glucose form maltose, a disaccharide (Figure 4.8*a*). Other common disaccharides are sucrose (our familiar table sugar) and lactose (found in milk). Similarly, up to thousands of monosaccharides can be linked together to form a polymer called a **polysaccharide**. Monosaccharides, disaccharides, and polysaccharides are all carbohydrates.

Carbohydrates perform several different functions in living organisms. One function is to provide structural support. **Cellulose**, for example, is a polysaccharide that is bundled into strong parallel fibers that help support the plant body (Figure 4.8*b*). Carbohydrates also provide fuel (energy), as we have already seen in the case of glucose. Starch—which is found in many plant-based foods and is familiar to most of us as a dish of mashed potatoes or steamed rice—is a polysaccharide that serves as an energy store inside plant cells (Figure 4.8*c*). Animal digestive systems degrade starch to the constituent monomer, glucose, which is then delivered to individual cells by the bloodstream. Glycogen is the main storage polysaccharide in animal cells, although, as we shall see later, most of the surplus energy ingested by animals is stockpiled in the form of storage lipids ("fat") rather than carbohydrate. Glycogen is a polymer of glucose and very

Table 4.1

Some Important Functional Groups Found in Organic Molecules

Functional group	Formula	Ball-and-stick model
Amino group	—NH₂ —N(H)(H)	Bond to carbon atom
Carboxyl group	—COOH —C(=O)OH	
Hydroxyl group	—OH	
Phosphate group	—PO₄ O—P(=O)(O⁻)—O⁻ with O⁻	

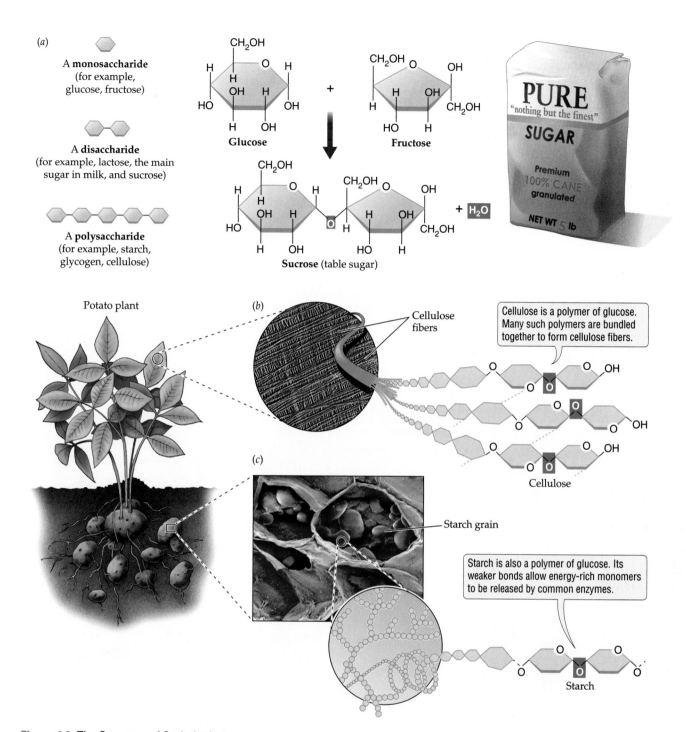

Figure 4.8 The Structure of Carbohydrates

Monosaccharides can bond with one another to form disaccharides and polysaccharides. (*a*) One molecule of glucose and one molecule of fructose can be linked by a covalent bond to form the disaccharide sucrose, or table sugar, releasing one water molecule. (*b*) Parallel strands of the polysaccharide cellulose are important components of plant cell walls. They help maintain the rigidity of cell walls, which is necessary for structural support of leaves and stems. Cellulose cannot be broken down in our digestive system, but it adds insoluble fiber to our diet, which is good for intestinal health. (*c*) Highly branched polysaccharides, such as starch, are used to store energy in plants. That is why starch-rich foods such as potatoes are good sources of energy for us as well.

similar to starch in its overall structure. The majority of the glycogen reserves in our bodies are stored within liver cells and skeletal muscle cells.

Nucleotides can deliver energy and they are critical components of information-storing molecules such as DNA

Nucleotides are important monomers in all organisms because they are the building blocks of the hereditary material. Some types of nucleotides also serve as energy delivering molecules. Nucleotides have three compo-

nents: a **nitrogenous base** (nitrogen-containing base) that is covalently bonded to a five-carbon sugar, which in turn is covalently bonded to a **phosphate group**, a functional group consisting of a phosphate atom and four oxygen atoms (Figure 4.9).

Five different nucleotides serve as the components for a class of polymers called **nucleic acids**. Nucleic acids in living cells are of two kinds: deoxyribonucleic acid (DNA) and ribonucleic acid (RNA). DNA is distinguished from RNA by the type of sugar in its nucleotides and by two of the nitrogenous bases that bond with that sugar. Ribose, the sugar in RNA, differs from

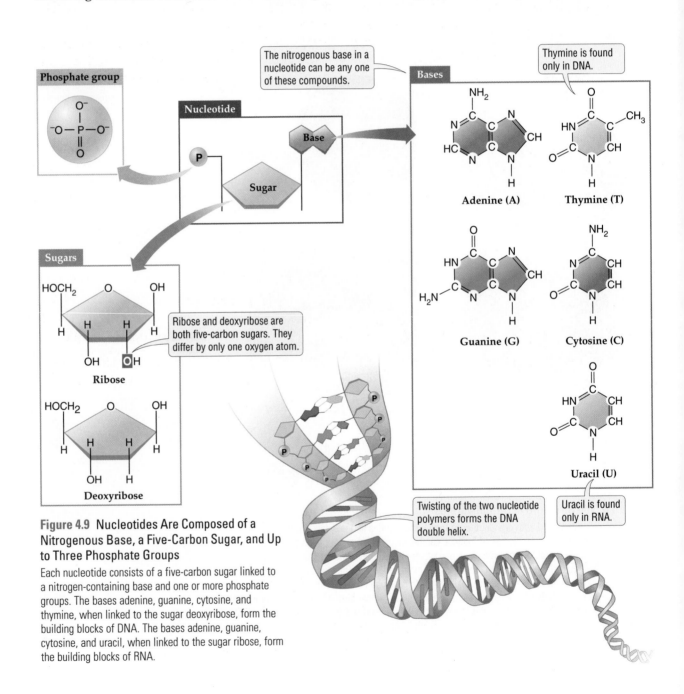

Figure 4.9 Nucleotides Are Composed of a Nitrogenous Base, a Five-Carbon Sugar, and Up to Three Phosphate Groups

Each nucleotide consists of a five-carbon sugar linked to a nitrogen-containing base and one or more phosphate groups. The bases adenine, guanine, cytosine, and thymine, when linked to the sugar deoxyribose, form the building blocks of DNA. The bases adenine, guanine, cytosine, and uracil, when linked to the sugar ribose, form the building blocks of RNA.

deoxyribose, the sugar in DNA in that it has one more oxygen atom (see Figure 4.9). Five different kinds of nitrogenous bases are found in nucleic acids: adenine, cytosine, guanine, thymine, and uracil (Figure 4.9). Thymine is found only in DNA, and uracil is found only in RNA.

Nucleotides play two essential functions in the cell: information storage and energy transfer. Every organism stores information in the form of a nucleic acid "blueprint" that dictates how that organism will live, grow, reproduce, and respond to the external world around it. The information is coded by the precise order in which nucleotides are joined together in the nucleic acid polymer. DNA is the nucleic acid that serves as hereditary material in all cells. It can be duplicated in a way that preserves its sequence of nucleotides and therefore the coded information in it. The transmission of duplicated DNA from one generation to the next is the basis of heredity, also known as genetics.

In addition to storing and transmitting genetic information, some types of nucleotides can also function as energy-delivery molecules, or energy carriers. The most universal of these energy carriers is the nucleotide known as **adenosine triphosphate**, or **ATP** (Figure 4.10). In ATP the base adenine is joined to a ribose molecule, which in turn is joined to a triphosphate, a chain of three phosphate groups. The ATP molecule is similar to the adenine-containing nucleotide that is one of the building blocks of RNA. ATP is the universal energy carrier for living organisms, and there are many chemical reactions in every cell that could not proceed without energy delivered by ATP. The energy of ATP is stored in the covalent bonds that link the three phosphate groups. The breaking of the bond between two phosphate groups releases

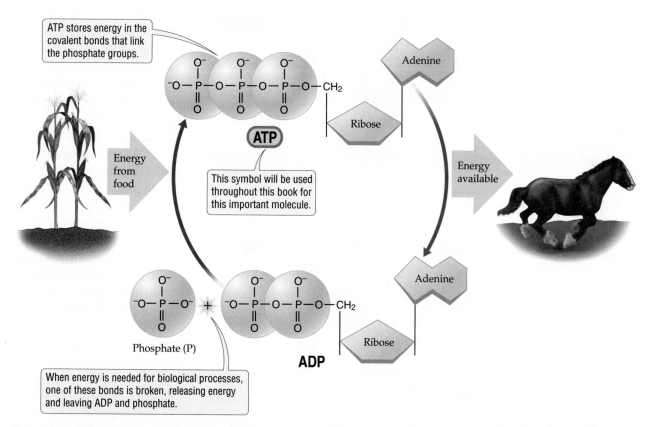

Figure 4.10 ATP Is a Universal Energy Carrier
ATP is an energy-laden molecule that serves as an energy carrier in every living cell. The phosphate groups are held together by energy-rich covalent bonds. Energy is released when these bonds are broken, and the released energy powers a great variety of chemical reactions in the cell. ATP is created by adding a phosphate group to ADP (adenosine diphosphate). In our bodies, the energy required for converting ADP into ATP comes from the breakdown of food molecules. ADP and a free phosphate group (P) are released whenever the terminal phosphate in ATP is broken to tap the chemical energy stored in the covalent bond.

energy that is used to power other chemical reactions. ATP is made by loading a phosphate group onto the lower-energy nucleotide ADP (adenosine diphosphate).

Animal cells must extract energy from food molecules and use that energy to convert ADP into ATP. Plants, algae, and certain bacteria, have the additional ability to make ATP using light energy.

Amino acids are the building blocks of proteins

Of the many different kinds of chemical compounds found in living organisms, **proteins** are among the most familiar, since we often hear and read about proteins and carbohydrates in connection with diet and nutrition. These polymers make up more than half the dry weight of living things. The steaks we throw on the grill in the summer are rich in proteins. Our own bodies contain thousands of different types of proteins. Some of these proteins form physical structures, such as keratin in hair and collagen in skin. Some carry oxygen throughout the body and others help us move by making muscle contraction possible. Proteins known as **enzymes** speed up the chemical reactions that are vital for life processes.

Amino acids are the monomers that build proteins. There are twenty different amino acids that can be found in proteins. All organisms on earth build their proteins from this same pool of 20 amino acids. The "anatomy" of these amino acids is similar. A carbon atom, called the alpha carbon, forms a central attachment site for four other components, as shown in Figure 4.11a. The alpha carbon is attached to a hydrogen atom, a chemical side chain called the R group, and two functional groups: an amino group (—NH₂), and a carboxyl group (—COOH).

In this group of 20 amino acids, one amino acid differs from another only with respect to the type of R group present in each. Each of the 20 possible R groups is unique to one amino acid. R groups vary in terms of size, acidic or basic properties, and whether they

Figure 4.11 The Structure and Diversity of Amino Acids
(a) Amino acids are the building blocks of proteins. Twenty different amino acids can be found in proteins, each differing from the other only in the nature of its R group. (b) Each of the 20 amino acids has a different R group bonded to the alpha carbon. The R group is responsible for the distinctive properties of each amino acid. The R groups found in amino acids can be broadly classified as hydrophobic or hydrophilic.

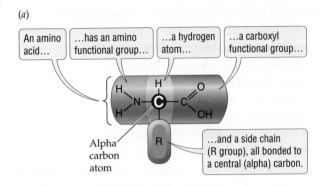

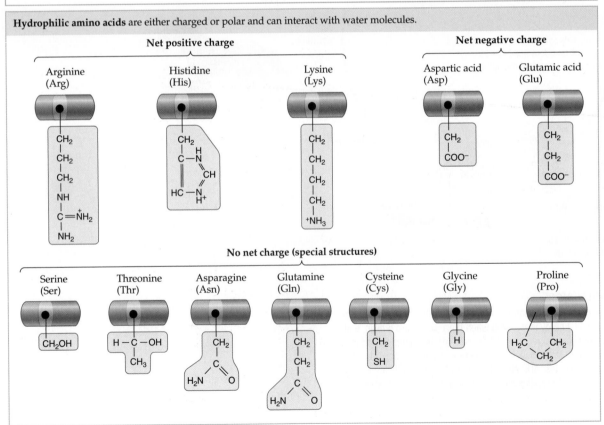

are hydrophobic or hydrophilic. The R groups found in amino acids range from just four atoms (in alanine), to complex arrangements of carbon chains (in arginine, for example) and ring structures (in tryptophan, for example), as depicted in Figure 4.11*b*. With an "alphabet" of 20 different amino acids, organisms have a diverse pool of building blocks from which they can make proteins with many different properties.

Linear chains of amino acids are covalently linked to create a polymer known as a **polypeptide**. Every protein consists of one or more polypeptides. In a polypeptide chain, the amino group of one amino acid is covalently linked to the carboxyl group of another via a covalent linkage called a **peptide bond** (Figure 4.12). A polypeptide may contain hundreds to thousands of amino acids held together by peptide bonds. These polypeptides are built from the same pool of 20 possible amino acids, so the crucial difference between one polypeptide and another is the sequence in which the amino acids are linked. Two polypeptides may differ in the amounts of various amino acids found in the chain; for instance, lysine may be absent in one, and extremely abundant in another. The many different polypeptides found in the average cell also vary enor-

mously in overall length, that is, in the total number of amino acids in each.

How can just 20 amino acids generate the thousands of different proteins found in nature? Well, consider this: the number of different sentences that can be written using the 26 letters of the English alphabet seems to have no bounds. If we think of the protein alphabet as having 20 amino acid letters (only 6 fewer than in the English alphabet), we can see that an enormous number of different protein sentences is possible. The complexity and diversity of life is fundamentally dependent on this variety in protein structure and function.

A polypeptide must be correctly folded to become a functional protein

The sequence of amino acids in a polypeptide is known as the **primary structure** of that polypeptide (Figure 4.13*a*). A polypeptide must acquire a higher level of organization, beyond its primary structure, before it can function as a protein or part of a protein. The next level of organization in a polypeptide is the regional folding of the amino acid chain into specific patterns that constitute the **secondary structure** (Figure 4.13*b*). Alpha (α) helices and beta (β) sheets are two of the most common patterns seen in the secondary structure of polypeptides. An alpha helix is a spiral pattern, resembling a phone cord, in which the spirals are created and maintained by hydrogen bonds. A beta sheet is created when the polypeptide backbone is bent into "ridges" and "valleys," such as we would see if a strip of paper were folded backward and forward and then released. In many polypeptides that have beta sheets, the polypeptide backbone is folded over in such a way that beta sheets are placed adjacent to each other and the parallel arrangement stabilized by hydrogen bonding between the side-by-side beta sheets.

In addition to having a secondary structure, most polypeptides must undergo an additional level of folding, to create a **tertiary structure**, before they can function as a protein (Figure 4.13*c*). The tertiary structure of a polypeptide is a very specific overall three-dimensional shape that is achieved not merely through local patterns of folding, as in the secondary structure, but requires long-distance interactions between different segments of the polypeptide chain. Tertiary structure is stabilized by noncovalent associations such as hydrogen bonds, but may also be stabilized by covalent links between distant amino acids (for example, between the sulfur atoms of distantly spaced cysteines).

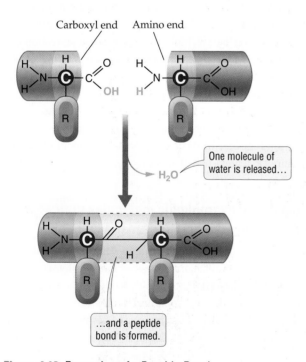

Figure 4.12 Formation of a Peptide Bond
A peptide bond is formed when the carboxyl group of one amino acid bonds covalently with the amino group of another amino acid. In the process, an OH group is eliminated from the carboxyl end and a hydrogen atom is released from the amino end; the OH and H come together to form one molecule of water.

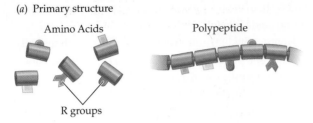

(a) Primary structure

Amino Acids   Polypeptide

R groups

(b) Secondary structure

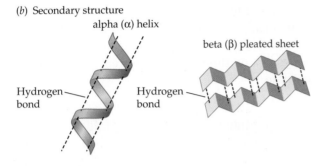

alpha (α) helix

beta (β) pleated sheet

Hydrogen
bond

Hydrogen
bond

(c) Tertiary structure

(d) Quaternary structure

Figure 4.13 The Four Levels of Protein Structure

(a) A chain of covalently bonded amino acids is known as a polypeptide. The polypeptide represents the primary structure, or most elemental organizational unit, of a protein. (b) The polypeptide may be locally organized in secondary structures. (c) A polypeptide must be folded into a stable three-dimensional shape, or tertiary structure, before it can function as a protein. (d) Some proteins are complexes of two or more polypeptides. This illustration is a schematic representation of the quaternary structure of hemoglobin, a protein comprised of four polypeptides. Attached to each polypeptide of hemoglobin is a special iron-containing functional group, known as heme (represented by the gray disc).

See how heat changes a protein, by frying an egg.　▶❚❚　4.5

Some proteins are composed of more than one polypeptide, in which case they have yet another level of organization, called the **quaternary structure** [KWAH-ter-nar-ee], that must be achieved before the protein can become biologically active. Hemoglobin, which transports oxygen in our blood, is an example of a protein that is only functional in its quaternary form. This protein consists of four separate polypeptides that must be precisely assembled to create the final three-dimensional shape that is characteristic of the quaternary structure of hemoglobin (Figure 4.13d).

The activity of most proteins is critically dependent on their three-dimensional structure. Extreme temperature, pH, and salt concentration can change or destroy the structure of a protein, and consequently change or destroy its activity. How? Most properly folded proteins tend to have hydrophilic R groups exposed on the surface and hydrophobic R groups buried deep inside the folded structure. This precise arrangement allows the protein to form hydrogen bonds with surrounding water molecules and remain dissolved. When a protein is heated beyond a certain temperature, its weak noncovalent bonds break, and the protein unfolds, losing its orderly three-dimensional shape. The destruction of a protein's three-dimensional structure, resulting in loss of protein activity, is known as **denaturation**. Some proteins have a three-dimensional structure that is relatively stable, while other proteins are more easily denatured. Conditions that can denature proteins include low pH, high pH, high salt concentrations, or high temperature. We witness the denaturation of egg proteins when we cook eggs. At room temperature, albumin, the predominant protein in eggs, is dissolved in the watery environment of the egg white. Albumin is denatured by the heat as the egg is cooked, causing the polypeptides to clump together into the solid mass we recognize as the whites in a hard-boiled or fried egg.

Lipids are hydrophobic molecules with many essential functions in the cell

Lipids are hydrophobic molecules made by living cells and they are built from chains or rings of hydrocarbon (covalently linked carbon and hydrogen atoms). Fatty acids, glycerides, sterols, and waxes are examples of lipids that we will discuss in greater detail shortly. Lipids that are solid or semisolid at room temperature are commonly called **fats** and those that are liquid at room temperature are generally known as oils, but we shall see later that this everyday usage is inconsistent.

Most lipids are built from one or more **fatty acids**. A fatty acid has a long hydrocarbon chain that is strongly hydrophobic. At the other end of the hydrocarbon chain

(a)

Stearic acid is a saturated fatty acid. It contains no double bonds between its carbon atoms.

(b)

Oleic acid is an unsaturated fatty acid. It has one double bond between two of its carbon atoms.

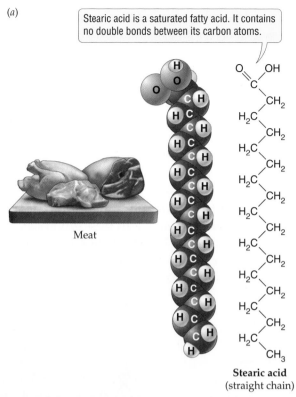

Meat

Stearic acid
(straight chain)

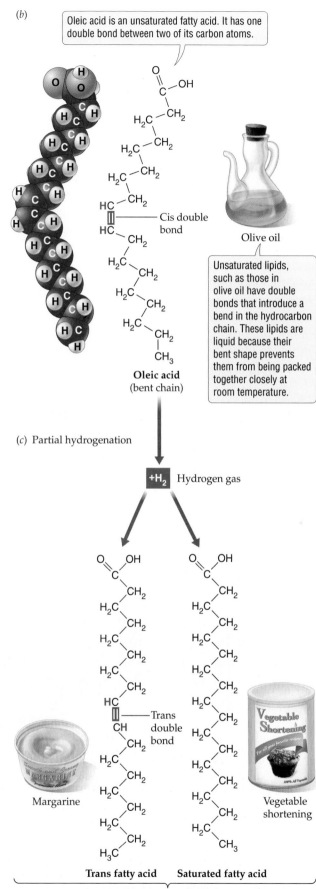

Cis double bond

Olive oil

Unsaturated lipids, such as those in olive oil have double bonds that introduce a bend in the hydrocarbon chain. These lipids are liquid because their bent shape prevents them from being packed together closely at room temperature.

Oleic acid
(bent chain)

(c) Partial hydrogenation

+H₂ Hydrogen gas

Margarine

Trans double bond

Vegetable Shortening

Vegetable shortening

Trans fatty acid   **Saturated fatty acid**

Straight chain molecules

Trans fatty acids have a semi-solid consistency (as in margarine) because their relatively straight hydrocarbon chains tend to pack together.

Figure 4.14 Saturated and Unsaturated Fatty Acids

(a) This space-filling model of stearic acid, with the chemical structure shown to the right of it, shows that the molecule is straight; as a result, it can pack tightly to form a solid at room temperature. (b) The cis double bond in oleic acid creates a kink in the molecule, preventing it from packing as tightly as stearic acid. Oleic acid therefore tends to be liquid at room temperature. It is commonly found in the seeds and fruits of such temperate species as the olive tree and canola plant (a type of mustard). (c) Partial hydrogenation of unsaturated lipids converts vegetable oils into semi-solid products. Hydrogen gas is used to convert some of the unsaturated fatty acids into trans fatty acids, which retain the double bond but have straight hydrocarbon chains. Some of the molecules are converted into saturated fatty acids through complete hydrogenation (and therefore loss) of the carbon–carbon double bond.

is a carboxyl group, a functional group that is polar and therefore hydrophilic.

The long hydrocarbon chains found in fatty acids contain many carbon atoms—16 to 22 in the types common in our foods—and these can vary in the way they are covalently bonded together. Fatty acids in which all the carbon atoms in the hydrocarbon chain are linked together by single covalent bonds are said to be **saturated** because each carbon in it is bonded to the maximum number of hydrogen atoms (Figure 4.14a). When one or more of these carbon atoms are linked by double bonds, the fatty acid is said to be **unsaturated**, because some of the carbon atoms are not bonded to a full complement of hydrogen atoms (Figure 4.14b).

The significance of the double bonds in unsaturated fatty acids goes beyond a mere difference in the number of hydrogen atoms linked to the hydrocarbon chain. Hydrocarbon

Figure: 04.14 CDLY, Discover Biology 4/e
100% of size

chains linked exclusively by single bonds tend to be straight, but the presence of double bonds can introduce kinks into the hydrocarbon chain. The consequence of these differences in shape is that the straight-chain saturated fatty acids can pack together very tightly, forming solids or semisolids at room temperature. Unsaturated fatty acids with kinks cannot pack tightly, so these lipids tend to be liquid at room temperature.

Familiar foods such as butter and olive oil are actually complex mixtures of different types of lipids, with small amounts of other substances such as milk protein in butter and vitamin E in most vegetable oils. The lipids in butter and olive oil include fatty acids of different types and fatty acid–containing molecules known as glycerides. A glyceride contains one to three fatty acids covalently bonded to a three-carbon molecule called glycerol. Glycerol has

three hydroxyl groups (—OH), each of which can form a covalent bond with the carboxyl group (—COOH) at the end of a fatty acid chain. As the bond forms, the OH on the carboxyl group of a fatty acid combines with an H in one of the hydroxyl groups on the glycerol, forming a molecule of water (Figure 4.15a). When all three hydroxyl groups in glycerol are bonded to a fatty acid, the resulting compound is called a **triglyceride** (Figure 4.15b), the most common glyceride in our diet.

Triglycerides built largely from saturated fatty acids tend to be solid at room temperature, and are familiar to us as fats. Because butter and lard are rich in triglycerides containing saturated fatty acids, these fatty foods are solid at room temperature. In contrast, most vegetable oils, such as canola oil and olive oil, are mixtures of unsaturated fatty acids and unsaturated triglycerides, which is why they are

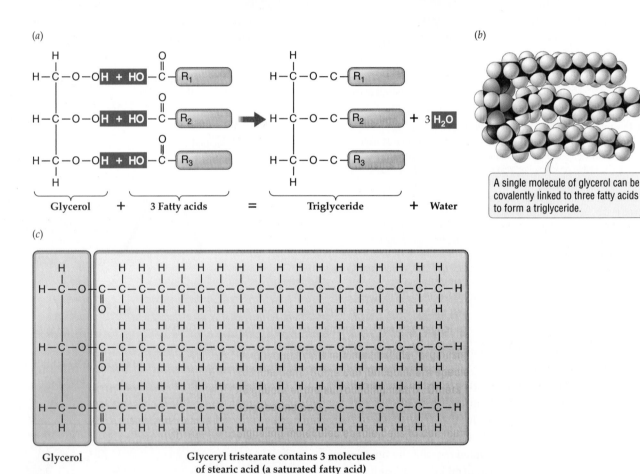

A single molecule of glycerol can be covalently linked to three fatty acids to form a triglyceride.

Glycerol + 3 Fatty acids = Triglyceride + Water

Glycerol

Glyceryl tristearate contains 3 molecules of stearic acid (a saturated fatty acid)

Figure 4.15 Triglycerides Contain Three Fatty Acids Bound to a Glycerol
Glycerides consist of a three-carbon sugar alcohol called glycerol, bound to one, two, or three fatty acids. Triglycerides have three fatty acids, one linked to each of the three carbons of glycerol (a). The bound fatty acids may be saturated, unsaturated, or a combination of the two (b). The triglyceride depicted in (c) is glyceryl tristearate, the most common storage lipid in animal cells. Glyceryl tristearate has three chains of stearic acid, a saturated fatty acid. Animal "fat" tends to be solid at room temperature because it is rich in triglycerides containing saturated fatty acids. Fish "oils" are an exception to this generalization; they contain long chains of unsaturated fatty acids known as omega-3 fatty acids.

liquids at room temperature. Some tropical "oils," such as those derived from coconut and palm kernels, are solid at room temperature because they are almost as rich in saturated lipids as butter and lard. This illustrates the inconsistency in how we use everyday words such as "fat" and "oil" and why special terminology, with a more precise meaning, is vital to the process of science.

A wide variety of organisms store surplus energy in the form of triglycerides, usually deposited in the cytoplasm as lipid droplets. Lipids are efficient as storage reserves because they pack slightly more than twice the energy found in an equal weight of carbohydrate or protein, while occupying only one-sixth the volume. Carbohydrates and proteins take up more space inside a cell because they are hydrophilic; these macromolecules are extensively associated with water molecules and all these extra molecules add bulk.

Another group of glycerides, the phospholipids, are major components of the plasma membrane, which is the outermost boundary of a cell. Most cells also have internal membranes, which make up the boundaries of internal compartments, and these internal membranes too are built mostly from phospholipids. A **phospholipid** is created when a phosphate group and two fatty acid chains form covalent bonds to glycerol. The negatively charged phosphate group on one end of the phospholipid is polar, which means that this "head" region of the molecule is hydrophilic and can interact with polar water molecules. On the other hand, the fatty acid chains ("tails") are nonpolar and therefore hydrophobic. These hydrophobic "tails" tend to move away from watery surroundings.

Because of their dual character, phospholipids exposed to water spontaneously arrange themselves in double-layered sheets, known as **phospholipid bilayers** (Figure 4.16). The double layers are arranged so that the hydrophilic head groups of the phospholipids are exposed to the watery world on either side, while the hydrophobic tails are tucked inward, away from

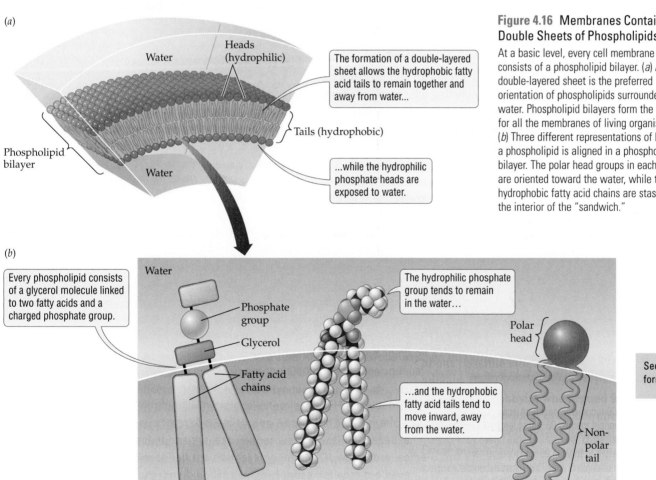

(a)

Water

Heads (hydrophilic)

The formation of a double-layered sheet allows the hydrophobic fatty acid tails to remain together and away from water...

Tails (hydrophobic)

Phospholipid bilayer

Water

...while the hydrophilic phosphate heads are exposed to water.

Figure 4.16 Membranes Contain Double Sheets of Phospholipids

At a basic level, every cell membrane consists of a phospholipid bilayer. (a) A double-layered sheet is the preferred orientation of phospholipids surrounded by water. Phospholipid bilayers form the basis for all the membranes of living organisms. (b) Three different representations of how a phospholipid is aligned in a phospholipid bilayer. The polar head groups in each layer are oriented toward the water, while the hydrophobic fatty acid chains are stashed in the interior of the "sandwich."

(b)

Every phospholipid consists of a glycerol molecule linked to two fatty acids and a charged phosphate group.

Water

Phosphate group

Glycerol

Fatty acid chains

The hydrophilic phosphate group tends to remain in the water...

...and the hydrophobic fatty acid tails tend to move inward, away from the water.

Polar head

Nonpolar tail

Phospholipid symbol

See how phospholipids form membranes.

4.6

the water. Nearly all cell membranes are organized as lipid bilayers. One phospholipid bilayer constitutes a single cell membrane. Membranes establish the boundaries of living cells and of compartments within cells. They control the exchange of ions and molecules between the cells and their external environment, and also between various compartments within a cell.

Sterols are lipids that play vital roles in a variety of life processes

Cholesterol, testosterone, estrogen, vitamin D—all of these are lipids with enough star quality, or medical notoriety, that they turn up on the evening news on a fairly regular basis. Although they are widely divergent in the functions they perform, all four compounds are classified in a group of lipids known as **sterols**. All sterols have the same fundamental structure: four hydrocarbon rings that are fused to each other. They differ in the number, type, and position of functional groups, and in the carbon side chains linked to the four hydrocarbon rings (Figure 4.17).

Cholesterol is a necessary component in the cell membranes of many animals. Cholesterol strengthens membranes and helps maintain their fluidity when temperature changes. Our livers can manufacture all the cholesterol we need. Ingesting a large excess from dietary sources can be harmful because surplus cholesterol tends to accumulate in the lining of our blood vessels, which can lead to cardiovascular disease.

Cholesterol is also the "starting" molecule in the manufacture of steroid hormones, which include sex hormones such as estrogen and testosterone. Vitamin D is important in the growth and maintenance of many tissues in the body, especially bone, and is known to have antitumor activity. It is partially manufactured by skin cells in response to ultraviolet radiation, but the process is completed in the liver. Bile salts aid in the digestion of fats. They are manufactured in the liver and secreted into the small intestine when food arrives there.

Liver enzymes also convert some cholesterol into steroid hormones, including sex hormones such as estrogen and testosterone. **Hormones** are signaling molecules that are active at very low levels and control a great variety of processes in plants and animals. The sex hormones, such as estrogen and testosterone, promote the development and maintenance of the reproductive system in animals. Testosterone, in its several natural forms and numerous synthetic forms, is an anabolic steroid (*anabolic*, "putting together"). Among its many effects is the promotion of muscle growth. The use of anabolic steroids by competitive athletes is seen as unfair advantage and the drugs are banned by all the major sports organizations. The regular use of anabolic steroids is associated with significant health risks, including higher odds of heart attack, stroke, liver damage, and liver and kidney cancer.

Concept Check

1. Which has a higher concentration of free hydrogen ions: vinegar, pH 2.8; or coffee, pH 5.5?
2. Which of the following is a polysaccharide and a key structural component of plant cell walls: glucose, sucrose, monosaccharide, cellulose, glycogen?
3. True or false: RNA and DNA are examples of nucleotides?
4. What is protein denaturation?
5. How is a saturated fatty acid different from an unsaturated one in terms of chemical structure?

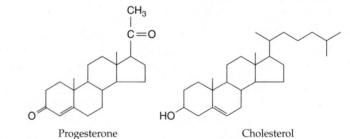

Four fused hydrocarbons rings, basic structure of all sterols

Progesterone

Cholesterol

Figure 4.17 Sterols Are Lipids Built from Four Fused Hydrocarbon Rings
All sterols share the same basic four-ring structure, but have different groups of atoms (functional groups) attached to these rings. Progesterone is a sex hormone important in the maintenance of fertility and pregnancy in female mammals. Cholesterol is an important constituent of the cell membranes of all birds and mammals, and of many other animals as well.

Applying What We Learned

The Chemistry of Life on the Edge

All serious attempts to find signs of extraterrestrial life are based on an understanding of life as we know it on earth. The assumption is that no matter where in the universe it exists, life must be carbon-based and water must be the primary solvent for life-sustaining chemical reactions. Is this too unimaginative, too earth-centric? A chemist contemplating the 92 naturally occurring elements will be hard-pressed to name an atom as versatile as carbon, or one that can be linked in anywhere near the variety of ways that carbon atoms are strung together in even the simplest form of earthly life. Silicon would be a distant, and rather pathetic, second to carbon. It is also difficult to imagine another solvent that could do for extraterrestrial life what water does for life on earth. Liquid ammonia has been proposed as a solvent for alien life forms, but it would make a poor substitute for water, and it is hard to imagine that such life forms could achieve the diversity and complexity displayed by terrestrial life. Furthermore, if we assume that extraterrestrial life exists but its chemistry is radically different from that of terrestrial life, we are left without a search strategy.

Because water has been detected on planets outside our solar system, and amino acids and simple carbohydrates (such as glycerol) have been identified in meteorites and asteroids, it seems reasonable that extraterrestrial life could be assembled from the same basic building blocks as life on earth. But to what sort of planet should we send our space probes, and on what kind of space rock or gas cloud should we train our earth-based instruments, in the search for alien life? The planetary habitability zone, or Goldilocks zone as it was quickly nicknamed, refers to a set of conditions that would be necessary for life on any planet. When the concept first emerged, about 40 years ago, the criteria for planetary habitability included temperatures from 0°C (32°F) to about 60°C (140°F) and an absence of intense radiation. Since then, scientists have discovered, right here on our seemingly mild blue planet, archaea that thrive in the boiling water of hot springs, algae that live in snow banks, shrimp that live in lakes ten times saltier than the ocean, and bacteria that grow best when exposed to intense nuclear radiation.

It has become abundantly clear in recent years that we don't really understand the full range of environmental extremes that can be tolerated by terrestrial life. Scientists at the NASA Astrobiology Institute are spearheading the exploration of extreme habitats on earth in an effort to hone the criteria for the planetary habitability zone. Some of the planets in our solar system are getting a closer look as potential hosts for life forms. Could there be life, perhaps in subsurface hydrothermal springs, on otherwise frigid Mars, with its polar ice caps, thin atmosphere, and intense UV radiation? What about Jupiter's moon, Europa, whose thick crust of ice probably hides an ocean of liquid water? The best way to start answering these questions would be to explore the extreme ecosystems of our own home planet.

The study of extremophiles is a young science, but investigation of these organisms that live at extreme temperature, pressure, pH, saltiness, dryness, or in the presence of intense radiation has already transformed our understanding of the limits of life. Learning how proteins, lipids, carbohydrates, and nucleic acids behave under severe conditions has improved our understanding of how more mundane organisms, like us, function or fail under stress. The new knowledge has brought far-reaching practical benefits, most notably a heat-stable enzyme, called *Taq* polymerase, that is now widely used to sequence DNA in everything from forensics to biodiversity conservation.

Pyrobolus fumarii holds the current world record for high-temperature living. It inhabits deep sea vents, cracks in the ocean floor that spew steam at 113°C, making them the hottest habitat on earth. What keeps this archaean's proteins from unfolding and becoming a useless tangle of polypeptides? Research on thermophilic (heat-loving) organisms has revealed that their heat-stable proteins sometimes differ by just a few amino acids from the heat-susceptible versions found in related species from milder environments. Based on those clues, scientists have engineered sturdier, heat-resistant versions of otherwise fragile proteins. Consider thermolysin, a protein-degrading enzyme found in bacteria that live at room temperature. Researchers changed 8 of the 319 amino acids in the protein to create an engineered thermolysin that is fully active at 100°C. The amino acid change produced a three-dimensional structure 340 times more stable than that of the native thermolysin. There are likely to be many uses for such engineered heat-stable proteins in fields ranging from medicine to crop production. Some vaccines, for example, are made up of heat-sensitive proteins, which makes it difficult to store and transport them in poor regions of the world. Heat-stable versions of these vaccines would be a tremendous boost for developing nations.

Trans Fat Banned In N.Y. Eateries; City Health Board Cites Heart Risks

By Annys Shin

The New York City Board of Health voted unanimously yesterday to require the city's roughly 20,000 restaurants to stop cooking with trans fats, making New York the first major U.S. city to adopt such a ban.

. . . Public health experts predicted that the city's ban could prevent up to 500 deaths a year from cardiovascular disease, while representatives of restaurant owners complained that it would probably have a negative impact, particularly on small ethnic establishments . . . Dariush Mozaffarian, a Harvard University cardiologist and epidemiologist, has estimated that up to 22 percent of heart attacks in the United States every year are caused by eating trans fat.

African Americans and Hispanics, who have among the highest rates of cardiovascular disease, are likely to benefit the most from the Board of Health's decision . . . "It's going to prevent a risk factor for cardiovascular disease most people don't see. You know if you smoke. You know if you're obese, but you don't know your cholesterol."

. . . [The ban] doesn't bode well for small and family-owned businesses, especially those that trade in affordable ethnic fare, said Richard Lipsky, a lobbyist for the Neighborhood Retail Alliance. He cited a survey of 1,000 restaurant owners conducted a month ago by the Latino Restaurant Association that found that 90 percent didn't know what trans fat was.

Trans fats are unsaturated lipids naturally present in trace amounts in animal foods such as meat and dairy products. They contain trans unsaturated fatty acids, whose hydrocarbon chains are straighter than the kinked shape of the more common cis unsaturated fatty acids. Because their straighter chains are readily compacted, trans fats are semisolid at room temperature, in contrast to the liquid consistency of cis unsaturated lipids.

The overwhelming majority of trans fat in the American diet comes not from natural sources, but from an artificial process known as partial hydrogenation. This industrial process converts liquid unsaturated plant lipids (vegetable oils) to a semisolid form familiar to us as margarine and vegetable shortening.

In the last 50 years, the use of artificial trans fats has exploded, especially in the processed food industry, because they are cheaper than the alternatives and less prone to becoming rancid. Foods prepared with trans fats last well on the shelf and don't need expensive refrigeration. Cis unsaturated fatty acids, in contrast, are very susceptible to attack by oxygen atoms. Some lipid mixtures that are rich in cis unsaturated fatty acids, such as olive oil, have natural substances, known as phenols, that inhibit rancidity. The darker the color of olive oil, the more phenolic compounds it contains, and the longer it will last in a kitchen cupboard. Vitamin E, an antioxidant, is often added to vegetable oils to ward off lipid oxidation (reaction with oxygen).

By the 1990s, alarm bells were being sounded about the health risks associated with the consumption of trans fats. Since then, many studies have confirmed that eating a significant amount of trans fat leads to higher rates of heart disease than consuming the same amount of other lipids and total calories but little or no trans fat. Since January 2007, the Nutrition Facts label on packaged foods must disclose the amount of trans fats in each serving. Health experts say there is no safe dose for artificial trans fats, and that consumers should minimize their intake and preferably eliminate these lipids from their diet. The ban imposed on New York City restaurants has galvanized other U.S. cities into passing similar measures. Many fast food giants and major food processors plan to replace trans fats with available substitutes, which include novel formulations of cis unsaturated lipids that are stable when used in high-temperature cooking such as frying and grilling.

Evaluating the News

1. How does a trans fatty acid resemble a saturated fatty acid? How do they differ? Why have trans fats been favored by the processed food industry?

2. Critics have described the ban imposed by the New York City Board of Health as an example of "Big Brother" telling people what to do. Supporters say it is government's responsibility to protect people's health. Do you favor the ban? Is there room for middle ground? Will the ban be beneficial for minorities overall? Explain your answers.

3. New York City's ban on trans fats in restaurant foods does not extend to manufactured foods—for example, cookies, crackers, and cake mix. Should it be extended to all types of commercial and packaged food products? Should it include added saturated fats in meals served at restaurants and in all types of foods? Explain your viewpoint.

Source: *Washington Post,* December 6, 2006

Chapter Review

Summary

4.1 The Physical World Is Made Up of Atoms

- The physical world is made up of chemical elements, each with unique properties. There are 92 different kinds of naturally occurring elements.
- An atom is the smallest unit of an element that has the chemical properties of that element.
- Individual atoms are made up of positively charged protons and uncharged neutrons in the nucleus, and negatively charged electrons moving around the nucleus. The number and arrangement of electrons in the atom of an element determines the chemical properties of that element.
- Isotopes of an element have different numbers of neutrons, but the same number of protons. Radioisotopes are isotopes that give off radiation.
- The atomic number of an element is the number of protons in its nucleus, and its atomic mass number is the sum of the number of protons and neutrons.
- The chemical attractions that cause atoms to associate with each other are known as chemical bonds. The unique chemical characteristics of the atoms dictate which type of chemical bonds will arise among them.
- When an atom loses or gains electrons, it becomes a positively or negatively charged ion, respectively. Ions of opposite charge are held together by ionic bonds, and atoms that are bound exclusively through such bonds are known as salts.
- A molecule contains at least two atoms that are held together by covalent bonds.
- Chemical compounds contain atoms from at least two different elements. All salts are compounds, but a molecule is a compound only if it contains atoms from at least two different elements.

4.2 Covalent Bonds Are the Strongest Chemical Bonds in Nature

- The nucleus of an atom is surrounded by a specific number of electrons that move in defined layers, or shells. Atoms share electrons with other atoms to fill their outermost electron shells to capacity. The bonding properties of an atom are determined by the number of electrons in its outermost shell.
- Covalent bonds, formed by the sharing of electrons between atoms, are one type of linkage that holds atoms within a molecule. They are the strongest chemical bonds in nature.

4.3 Hydrogen Bonds and Ionic Bonds Are Individually Weak, but Collectively Potent

- Noncovalent bonds are interactions among atoms that are not based on the sharing of electrons. Hydrogen bonds and ionic bonds are examples of noncovalent bonds that are important in biological systems.
- Hydrogen bonds are weak attractions between two molecules such that a partially positive hydrogen atom within one molecule is attracted to a partially negative region on the other compound.

- Partial electrical charges result from the unequal sharing of electrons between atoms, giving rise to a polar covalent bond. Molecules in which polar covalent bonds predominate are described as polar molecules.
- Water is a polar molecule. Hydrogen bonding between water molecules accounts for the physical properties of water and plays an important role in the structure and function of biologically important molecules.
- Ions are formed when electrons from one atom are transferred to another atom. The electrical attraction between a positively charged ion and negatively charged ion is known as an ionic bond. The ions in a salt are held together by ionic bonds.
- Ions and polar molecules are hydrophilic because they can readily interact with water molecules. Because they lack charged or partially charged atoms, nonpolar molecules cannot associate with water and are therefore hydrophobic. Nonpolar molecules are excluded by water, causing them to clump with each other. Polar molecules tend to be soluble in water, while nonpolar compounds tend to be insoluble because they do not interact readily with water.
- Chemical reactions involve the formation or rearrangement of bonds between atoms. Although the participants in a chemical reaction (reactants) are modified to give rise to new ions or molecules (products), atoms are neither created nor destroyed in the process.

4.4 The pH Scale Expresses the Acidity or Basicity of a Solution

- The life-supporting chemical reactions of a cell are conducted in a watery medium. Polar compounds known as acids and bases affect the amount of free hydrogen ions in a water solution. Acids donate hydrogen ions in a solution, forming hydronium ions (H_3O^+); bases accept hydrogen ions, producing a surplus of hydroxide ions (OH^-) in a water-based solution.
- The concentration of free hydrogen ions in water is expressed by the pH scale. A pH of zero, at the low end of the pH range, represents an extremely acidic solution, with a hydrogen ion concentration 10 million times greater than that of pure water. A pH of 14, at the upper end of the scale, represents an extremely basic solution, with a concentration of hydrogen ions 10 million times lower than that of pure water. Most living systems function best near the neutral pH of 7.
- Buffers are solutions containing a mixture of hydrogen ion donors and acceptors. They help maintain the constant internal pH that is necessary for the chemical reactions of life.

4.5 The Chemical Building Blocks of Living Systems

- Carbon atoms can link with each other, and with other atoms, to generate a great diversity of compounds. The four main building blocks of biologically important molecules are sugars, amino acids, nucleotides, and fatty acids. All four are built on a basic skeleton of carbon and hydrogen atoms, usually with atoms of a handful of other elements as well.
- Carbohydrates include simple sugars (monosaccharides) as well as disaccharides and more complex polymers (polysaccharides). Carbohydrates provide energy and physical support for living organisms.

- Each nucleotide consists of a five-carbon sugar, a nitrogen-containing base, and a phosphate group. Nucleotides are the building blocks of the nucleic acids DNA and RNA. DNA polymers, made up of four types of nucleotides, form the blueprint for life and govern the physical features and chemical reactions of a living organism. ATP is an energy-rich molecule that delivers energy for a great variety of cellular processes.
- Amino acids are the building blocks of proteins. The chemical properties of the 20 different amino acids depend on their different R groups.
- A chain of amino acids linked together by peptide bonds is a polypeptide, and this basic unit of organization constitutes the primary structure of a protein.
- A polypeptide can function as a protein only when it is folded into a distinctive three-dimensional shape. The function and three-dimensional shape of a protein (its secondary and tertiary structure) are determined by the chemical properties of the amino acids it contains.
- Some proteins contain more than one polypeptide, and the constituent polypeptides must be correctly assembled into the quaternary structure for such a protein to function properly.
- Lipids are hydrophobic substances containing one or more rings or chains of hydrocarbons. Fatty acids are the building blocks of all lipids except sterols; they are described as saturated or unsaturated, depending on the absence or presence of double covalent bonds in their hydrocarbon chains.
- Triglycerides are built from glycerol and three fatty acid units, and they are an important means of long-term energy storage.
- Phospholipids are the basic components of biological membranes.
- Sterols are lipids built on a fused ring structure; they include cholesterol, a component of animal cell membranes, and sex hormones such as testosterone and estrogen.

◉ Review and Application of Key Concepts

1. What is a monomer and what is its relationship to a polymer? Should lipids be regarded as polymers? Why or why not?

2. A sample of pure water contains no added acids or bases. Predict the pH of the water and explain your reasoning.

3. What are hydrogen bonds? Explain how the polarity of water molecules contributes to their tendency to form hydrogen bonds.

4. Describe the chemical properties of carbon atoms that make them especially suitable for forming so many different molecules of life.

5. Describe one function that is relevant to biological processes for each of the following compounds: carbohydrates, nucleic acids, amino acids, lipids.

Key Terms

acid (p. 81)
amino acid (p. 88)
atom (p. 74)
atomic mass number (p. 74)
atomic number (p. 74)
ATP (adenosine triphosphate) (p. 87)
base (p. 81)
buffer (p. 82)
carbohydrate (p. 84)
cellulose (p. 84)
chemical bond (p. 75)
chemical compound (p. 76)
chemical reaction (p. 80)
covalent bond (p. 75)
denaturation (p. 90)
disaccharide (p. 84)
electron (p. 74)
element (p. 74)
enzyme (p. 88)
fat (p. 90)
fatty acid (p. 90)
functional group (p. 84)
glucose (p. 84)
hormone (p. 94)
hydrogen bond (p. 78)
hydrophilic (p. 80)
hydrophobic (p. 80)
ion (p. 75)
ionic bond (p. 75)
isotope (p. 75)
lipid (p. 90)
macromolecule (p. 83)
molecule (p. 75)
monomer (p. 83)
monosaccharide (p. 84)

neutron (p. 74)
nitrogenous base (p. 86)
noncovalent bond (p. 77)
nonpolar molecules (p. 79)
nucleic acid (p. 86)
nucleotide (p. 86)
nucleus (p. 74)
organic molecule (p. 83)
peptide bond (p. 89)
pH (p. 81)
phosphate group (p. 86)
phospholipid (p. 93)
phospholipid bilayer (p. 93)
polar molecules (p. 78)
polymer (p. 83)
polypeptide (p. 89)
polysaccharide (p. 84)
primary structure (p. 89)
product (p. 80)
protein (p. 88)
proton (p. 74)
quaternary structure (p. 90)
radioisotope (p. 75)
reactant (p. 80)
salt (p. 80)
saturated (p. 91)
secondary structure (p. 89)
soluble (p. 78)
solute (p. 79)
solution (p. 79)
solvent (p. 79)
sterols (p. 94)
sugar (p. 84)
tertiary structure (p. 89)
triglyceride (p. 92)
unsaturated (p. 91)

Self-Quiz

1. The atoms of a single element
   a. have the same number of electrons.
   b. can form linkages only with atoms of the same element.
   c. can have different numbers of electrons.
   d. can never be part of a chemical compound.

2. Two atoms can form a covalent bond
   a. by sharing protons.
   b. by swapping nuclei.
   c. by sharing electrons.
   d. by sticking together on the basis of opposite electrical charges.

3. Which of the following statements about molecules is true?
   a. A single molecule contains atoms from only one element.
   b. Atoms in a molecule are linked only via ionic bonds.
   c. Molecules are found only in living organisms.
   d. Molecules can contain as few as two atoms.

4. Which of the following statements about ionic bonds is *not* true?
   a. They cannot exist without water molecules.
   b. They are not the same as hydrogen bonds.
   c. They involve electrical attraction between atoms with opposite charge.
   d. They are known to exist in crystals of table salt (NaCl).

5. Hydrogen bonds are especially important for living organisms because
   a. they occur only inside of organisms.
   b. they are stronger than covalent bonds and maintain the physical stability of molecules.
   c. they enable polar molecules to dissolve in water, which is the universal medium for life processes.
   d. once formed, they never break.

6. Glucose is an important example of a
   a. protein.
   b. carbohydrate.
   c. fatty acid.
   d. nucleotide.

7. Peptide bonds in proteins
   a. connect amino acids to sugar monomers.
   b. bind phosphate groups to adenine.
   c. connect amino acids together.
   d. connect nitrogen bases to ribose monomers.

8. An alpha helix is an example of
   a. primary protein structure.
   b. secondary protein structure.
   c. tertiary protein structure.
   d. quaternary protein structure.

9. Sterols are classified as
   a. sugars.
   b. amino acids.
   c. nucleotides.
   d. lipids.

10. Unlike saturated fatty acids, unsaturated fatty acids
    a. are solid at room temperature.
    b. pack more tightly because they have straight chains.
    c. have one or more double bonds in their hydrocarbon chain.
    d. have the full complement (maximum number) of hydrogen atoms covalently bonded to each carbon atom in the hydrocarbon chain.

CHAPTER 5

Cell Structure and Internal Compartments

Key Concepts

◉ All living organisms are made up of one or more basic units called cells.

◉ The plasma membrane forms the boundary of a cell. It controls the movement of substances into and out of a cell and determines how the cell communicates with the external world.

◉ Prokaryotes are single-celled organisms that lack a nucleus.

◉ Eukaryotes are single-celled or multicellular organisms whose cells have a nucleus and several other internal compartments specialized for distinctive functions.

◉ The specialized internal compartments of the eukaryotic cell have diverse functions, including the manufacture of lipids and certain types of proteins, the sorting and targeting of membrane proteins and secreted proteins, digestion and recycling of macromolecules, and the generation of energy to fuel cellular activities.

◉ The cytoskeleton is a network of protein cables and cylinders. There are three main types of cytoskeletal structures and they play an important role in giving shape and mechanical strength to a cell and enabling it to move.

◉ Organelles such as mitochondria and chloroplasts are thought to be descendants of primitive prokaryotes that were engulfed by ancestral eukaryotes and evolved a mutually beneficial relationship with the host.

The Hitchhiker's Guide to the Human Body

There's more to you than meets the eye. Of the roughly 100 trillion cells in your body, no more than 10 percent are genuinely yours. The rest are microscopic organisms (microbes) belonging to the Bacteria, Archaea, and Fungi, and even to benign members of the Arthropoda, an invertebrate group. Several hundred species of microbes, mostly bacteria and archaeans, inhabit our skin, moist external cavities, and all segments of the digestive tract.

The human body is host to benign organisms and to parasitic microbes that grow and multiply at our expense. We have defensive cells that engulf invading microbes and attempt to digest them. However, microbes such as the tuberculosis bacterium can escape such defenses and lie dormant inside the host cell. About one-third of the world's population harbors dormant tuberculosis bacteria, a condition known as latent TB. The bacterium will activate in about 10 percent of those people, often many years after the initial invasion, producing the devastating symptoms of tuberculosis. Some disease-causing microbes have evolved ways of subverting the cell's own internal machinery to get about. Consider *Listeria monocytogenes* [liss-TEER-ee-uh MON-oh-sye-TOJ-uh-neez], which can lurk in raw milk, meat, and fish, and contaminate foods prepared under unhygienic conditions. Each year in the United States, *Listeria* is responsible for nearly 2,000 cases of severe food poisoning, resulting in about 400 deaths. The bacterium invades a cell, hitches a ride on internal structures, and buries itself deep inside, where it can go undetected by the body's defense system.

This business of one cell invading another is widespread and very ancient. All kinds of bacteria, both nasty and nice, infect or inhabit the cells of many animals, plants, insects, amoebas, and even other bacteria! Aphids, common insect pests of houseplants, are known to harbor a bacterium that in turn has a resident bacterium within it. **Could our very cells be constructed from spare parts borrowed from ancient prokaryotic residents? What is the evolutionary link between prokaryotes and eukaryotes anyway?** We will address that primordial relationship in this chapter, after we examine the general structure of prokaryotic and eukaryotic cells.

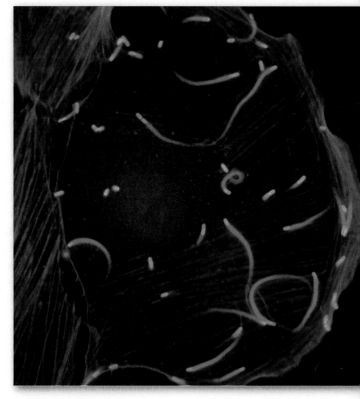

Invaders in Inner Space
The bacteria *Listeria monocytogenes* (green) are pushed along by cables of protein (red strands) inside an infected human cell.

In Chapter 4 we discussed atoms, molecules, and the large polymers and lipids that are common to all life forms. By themselves, these building blocks of life are inanimate. Only when they come together to constitute the highly organized, energy-dependent, self-replicating structure that is the **cell** can we recognize a living entity. From bacteria to blue whales, the cell is the smallest and simplest unit of life.

This chapter explores life at the level of the cell. After a broad overview of the unity and diversity of living cells, and a comparison of cellular organization in prokaryotes and eukaryotes, we examine the internal structures and compartments that allow a cell to function as an efficient and well-coordinated whole. We will begin the tour at the physical outer boundary of the cell, then work our way inward.

5.1 Cells Are the Smallest Units of Life

Every living organism consists of one to many billions of membrane-enclosed units called cells. Bacteria are single cells, while complex multicellular organisms such as humans contain trillions of cells.

There are about 220 different types of cells in the human body. Each cell performs many general functions as well as some highly specialized tasks unique to that cell type, or shared with just a few cell types. Compare, for example, the main cell type in our skeletal muscles with those in the clear lens of the eye. Skeletal muscles, which enable us to move body parts such as arms and legs, contain elongated cells that can generate physical force by shrinking or stretching. These cells contain bundles of specific proteins that ratchet against each other to produce a force, which then powers the movement of joints and bones. The energy-producing structures known as mitochondria are especially abundant in muscle cells. In contrast, the major cell type in the lens of the eye is so packed with a transparent protein, called crystalline, that many of the usual cell structures are absent or poorly developed. The crystalline-containing cells cannot move themselves, but they can be flexed by tiny muscles attached to the lens; the change in curvature changes the light-focusing properties of these cells, which is how our eyes are able to focus on a book, or the far horizon, as needed.

With millions of different species on Earth, the diversity of cell types found throughout the biosphere is enormous. None of the cell types found in our bodies, for example, is found in any plant. Yet even with so many different kinds of cells, certain basic components and structures are shared by all cells of all organisms on earth. Every cell, prokaryotic or eukaryotic, has a lipid-based outer boundary, called the **plasma membrane**. The plasma membrane is studded with proteins that serve as gatekeepers controlling what goes in and out of the cell.

All the contents of a cell internal to the plasma membrane, but excluding the nucleus, are collectively called the **cytoplasm**. The cytoplasm contains a water-based fluid called the **cytosol** that is composed of a multitude of free ions and molecules mixed in water. The chemical components of the cytosol include small organic molecules such as sugars, amino acids, and fatty acids; and macromolecules such as proteins, polysaccharides, and nucleic acids. There are so many small and large molecules crowded into the cytosol that it behaves more like a thick gel than a free-flowing liquid.

In addition to the cytosol, the cytoplasm contains structures called **organelles** that are part of the machinery of the cell. The different types of organelles vary in size but they are much larger than a single molecule of protein or other macromolecule. **Ribosomes** are small organelles found in large numbers in the cytoplasm of both prokaryotic and eukaryotic cells. The exact composition of ribosomes differs between prokaryotes and eukaryotes, although in both types of organisms each fully assembled ribosome is composed of several types of proteins combined with certain types of RNA molecules. These organelles are key components of the cell's protein-manufacturing machinery.

Many of the largest organelles, such as the **nucleus**, are wrapped in lipid membranes similar to the plasma membrane but with different types of proteins embedded in them. DNA, the nucleic acid that stores genetic information, is packed inside a nucleus in a eukaryotic cell, but lies in the cytoplasm in prokaryotes. As we shall see, eukaryotes have, in addition to the nucleus, a variety of other membrane-enclosed organelles.

5.2 The Plasma Membrane Separates a Cell from Its Surroundings

A key characteristic of every cell is the existence of a lipid-based boundary that separates that cell from its surrounding environment. The many chemical reactions required for sustaining life happen within the cytoplasm, the main compartment created by this boundary. The lipid boundary has the effect of enclosing and concentrating necessary raw materials in a limited space, which facilitates chemical processes.

The lipid-based outer boundary of a cell is known as the plasma membrane. As we saw in Chapter 4, biological membranes are composed mainly of a phospholipid bilayer. The phospholipids in each layer are oriented so that their hydrophilic heads are exposed to the watery environments inside and outside the cell, and their hydrophobic fatty acid tails are grouped together in the interior of the membrane (Figure 5.1).

If the plasma membrane had no function other than to define the boundary of the cell and to confine its contents, a simple phospholipid bilayer would suffice. However, the plasma membrane must also allow the cell to capture essential molecules, while shutting out unwanted ones; it must release waste products, but prevent needed molecules from leaving the cell; and it must interact with the outside world by receiving and sending signals as necessary. In other words, the plasma membrane must function as a selectively permeable barrier and also as a communications center.

The selective permeability of plasma membranes depends on various proteins that are embedded in the phospholipid bilayer. As Figure 5.1 shows, some of these proteins extend all the way through the phospholipid bilayer, forming tunnels that allow the passage of selected ions and molecules. Other membrane-spanning proteins, known as receptor proteins, are part of the communications system and act as sites for signal perception and transmission. Certain types of proteins are associated with, or embedded in, only one of the two lipid layers of the membrane. Some of the proteins associated with the membrane's inner layer link the plasma membrane to internal structures found within the cytoplasm.

Unless they are anchored to structures inside or outside the cell, most plasma membrane proteins are free to drift within the plane of the phospholipid bilayer. The concept of the plasma membrane as a highly mobile mixture of phospholipids, other types of lipids, and many

Test your knowledge of proteins in the plasma membrane. ▶❚❚ 5.1

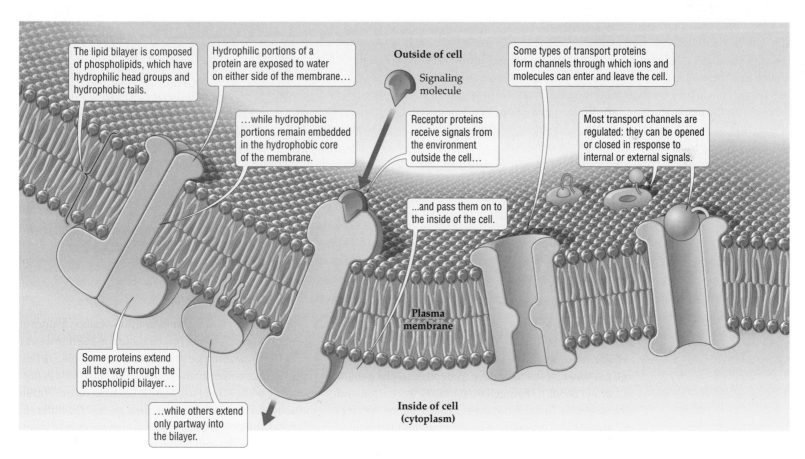

The lipid bilayer is composed of phospholipids, which have hydrophilic head groups and hydrophobic tails.

Hydrophilic portions of a protein are exposed to water on either side of the membrane…

…while hydrophobic portions remain embedded in the hydrophobic core of the membrane.

Outside of cell

Signaling molecule

Receptor proteins receive signals from the environment outside the cell…

…and pass them on to the inside of the cell.

Some types of transport proteins form channels through which ions and molecules can enter and leave the cell.

Most transport channels are regulated: they can be opened or closed in response to internal or external signals.

Some proteins extend all the way through the phospholipid bilayer…

…while others extend only partway into the bilayer.

Plasma membrane

Inside of cell (cytoplasm)

Figure 5.1 Proteins Are Embedded in the Plasma Membrane
Membrane proteins may span the entire phospholipid bilayer or extend only part way into it. These proteins serve a variety of functions, from transporting substances across the membrane to receiving external signals (red) and then relaying them into the interior of the cell.

Science Toolkit

Exploring Cells under the Microscope

A picture is worth a thousand words. This simple statement is as true in biology as it is in fine art. Our awareness of cells as the basic units of life is based largely on our ability to see them. The instrument that opened the eyes of the scientific world to the existence of cells—the light microscope—was invented in the last quarter of the sixteenth century. The key components of early light microscopes were ground-glass lenses that bent incoming rays of light to produce magnified images of tiny specimens.

The study of cells began in the seventeenth century when Robert Hooke examined a piece of cork under a microscope and noticed that it was made up of little compartments. Hooke described these structures as small rooms, or cells, originating the term we use today. Ironically, the compartments Hooke saw under the early microscope were not living cells, since cork is dead plant tissue. However, the discovery of previously invisible things proceeded rapidly, opening up a new world to scientific exploration.

While the light microscope has a place in the early history of biology, similar instruments are just as important in ongoing research today. The quality of lenses, however, has improved significantly since that time: the 200- to 300-fold magnification achieved in the seventeenth century has been improved to well over a 1,000-fold achieved by today's standard light microscopes. This degree of magnification allows us to distinguish structures as small as 1/2,000,000 of a meter, or 0.5 micrometer (μm). Light microscopes can therefore reveal not just animal and plant cells (5–100 μm), but also organelles such as mitochondria and chloroplasts (2–10 μm) as well as bacteria (1 μm).

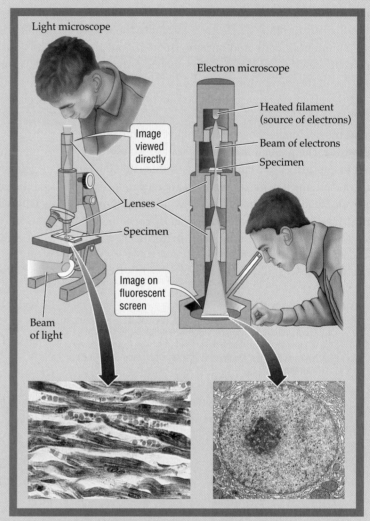

Light and Electron Microscopy

membrane proteins, is referred to as the **fluid mosaic model** of cell membranes. This mobility is critical for many cellular functions. For example, it is necessary for whole cell movement: cells that move from place to place could not crawl about if their cell membranes were rigid and unchangeable. In many instances, the perception of external signals also depends on membrane fluidity because some types of signaling molecules work by causing free-floating receptor proteins to clump together into active complexes, and it is these complexes that send the message into the interior of the cell to provoke a response.

Although the plasma membrane is a common feature of all cells, the specific set of proteins found in the membrane varies from one cell type to another. For example, certain types of plasma membrane proteins are found only in prokaryotes, while others are unique to eukaryotes. Within the body of a multicellular organism, each of the different cell types is likely to have a unique combination of membrane proteins. The particular combination of proteins in a certain cell type contributes to the distinctive properties of that cell and determines, for example, how that cell will interact with its external environment.

Size Range of Biological Structures

Microscopy enables us to visualize cells and cell structures. Light microscopy can reveal organelles as small as mitochondria, but scientists must resort to electron microscopy to capture images of structures as small as viruses or ribosomes.

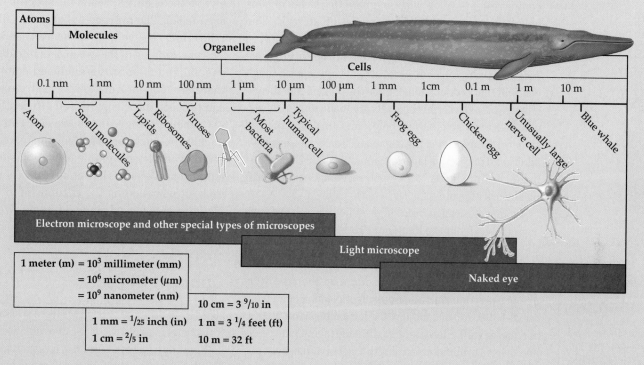

1 meter (m) = 10^3 millimeter (mm)
  = 10^6 micrometer (μm)
  = 10^9 nanometer (nm)

	10 cm = 3 $^9/_{10}$ in
1 mm = $^1/_{25}$ inch (in)	1 m = 3 $^1/_4$ feet (ft)
1 cm = $^2/_5$ in	10 m = 32 ft

Since the 1930s, an even more dramatic increase in magnification has been achieved by the replacement of visible light with streams of electrons that are focused by powerful magnets instead of glass lenses. Called electron microscopes, these instruments can magnify a specimen more than 100,000 times, revealing detail about the internal structure of cells, and even individual molecules such as proteins and nucleic acids. Both types of microscopy—electron and light—give us insights into how cells are organized and how each type of cell is physically adapted to a specific function in the body of a multicellular organism.

5.3 The Structural Organization of Prokaryotic and Eukaryotic Cells

As we learned in Chapter 3, living organisms can be informally classified into two broad categories, the prokaryotes and eukaryotes. Organisms whose DNA is not confined within a membrane-enclosed nucleus are known as **prokaryotes**. Those that have a nucleus, and an elaborate system of other membrane-enclosed compartments, are known as **eukaryotes**.

The first life forms to evolve on earth probably resembled present-day prokaryotes. Present-day prokaryotes are members of the domains Bacteria and Archaea. The great majority of these prokaryotes are single-celled, although some filament-forming species may be regarded as tending toward multicellularity. Prokaryotic cells, on the whole, are smaller than eukaryotic cells. For example, the well-studied bacterium *Escherichia coli*, a common resident of the human intestine, is only two-millionths of a meter or 2 micrometers long. About 125 *E. coli* would fit end to end across the period at the end of this sentence.

Prokaryotes have a simpler internal organization than eukaryotes and most lack membrane-enclosed organelles, although in recent years small, spherical compartments filled with magnetic particles or ions have been detected in some species of bacteria. The small size of bacteria and archaeans may account, in part, for their ability to get by with relatively little structuring of the cytoplasm. Most prokaryotes have a tough cell wall that is deposited outside the plasma membrane (Figure 5.2). Bacterial cell walls are made up of polysaccharides and proteins, and help maintain the shape and structural integrity of the organism. Some bacteria have additional protective layers, commonly called a capsule, made of slimy lipids, sticky polysaccharides, or a stiff meshwork of proteins.

Eukaryotes range from single-celled organisms, such as amoebas from the kingdom Protista or yeasts from the kingdom Fungi, to large multicellular organisms, such as redwood trees or blue whales. Eukaryotic cells are generally 10 times larger in diameter than the average prokaryotic cell. Most eukaryotic cells range from about 1/100,000 of a meter (10 micrometers) to 1/10,000 of a meter (100 micrometers), although some plant cells are several millimeters in length and certain nerve cells are nearly 1 meter long! Plant cells tend to be larger than animal cells. They contain one or more water-filled sacs, called vacuoles, which enable them to achieve a larger overall size without having to spend the extra energy that would be needed to make and maintain a larger volume of cytosol. Plants, fungi, and some protista (mostly algae) have cell walls made of polysaccharide, but the precise chemistry of the wall polysaccharides differs among these groups of eukaryotes and differs again from that of prokaryotic cell walls. Animal cells lack a polysaccharide cell wall but many of them deposit a protein-based network, known as the extracellular matrix, outside the cell on the surface of the plasma membrane. This matrix helps anchor an animal cell in place, and has other major functions, some of them specific to particular cell types.

The internal organization of eukaryotes is more complex than that found in prokaryotes, especially in the variety and complexity of membrane-enclosed organelles found in a typical eukaryotic cell (see Figure 5.2). As mentioned earlier, a major distinction between prokaryotic and eukaryotic cells is that eukaryotic cells have a membrane-enclosed organelle, the nucleus, which houses DNA and functions as the

storehouse of genetic information. Eukaryotes also have several other types of organelles, each with a distinct function, and commonly enclosed in one or two lipid bilayers. The membranes allow such organelles to accumulate high concentrations of the needed raw materials and to maintain an environment conducive to the chemical reactions specific to the various organelles. The average eukaryotic cell has roughly 1,000 times the volume of the average bacterial cell. The partitioning of the cytoplasm into discrete, highly specialized membrane-enclosed compartments is therefore vital for a eukaryotic cell. Cellular activities can proceed more rapidly and efficiently if all necessary chemical ingredients are concentrated within a membrane-enclosed organelle, instead of being scattered over a large volume of cytoplasm. Simply put, eukaryotic cells could not sustain their larger size without partitioning the cytoplasm into specialized membrane-enclosed compartments. We explore this concept in greater detail in the next section.

Concept Check

1. Which of these cell structures is found in eukaryotes but not in prokaryotes: plasma membrane, cytoplasm, ribosome, nucleus?
2. Name two functions performed by plasma membrane proteins.
3. What is meant by the fluid mosaic nature of the plasma membrane?
4. Eukaryotes have a variety of membrane-enclosed organelles. How is this type of internal organization beneficial for such organisms?

5.4 Eukaryotic Cells Have Many Specialized Internal Compartments

Imagine a large factory with many rooms housing different departments. Each department has a specific function and an internal organization that contributes to the overall mission of manufacturing certain products. Workers assembling a particular item are arranged in a specific order in a centralized assembly department, with packers and shipping agents taking over from them in the next department. Raw materials and finished products can be warehoused, there is a site for waste disposal and recycling, a power station that supplies energy, and the operations are conducted on the basis of directions issued by the administrative office. A eukaryotic cell is just such

Learn more about eukaryotic and prokaryotic cell structure; animal, plant, and bacterium.

▶❙❙ 5.2, 5.3, 5.4

Figure 5.2 Prokaryotic and Eukaryotic Cells Compared

The prokaryotic cell shown is a typical bacterium; the eukaryotic cells depicted here are a typical animal cell and a typical plant cell. The bacterium is easily distinguished from the eukaryotic cells by its much smaller size and the absence of a nucleus and a complex system of membrane-enclosed organelles. Ribosomes, however, are abundant in the cytoplasm of both eukaryotic and prokaryotic cells. The plant cell is distinguished from the animal cell by the presence of chloroplasts, a large water-filled vacuole, and a rigid cell wall. Both types of eukaryotic cells contain nuclei and other organelles.

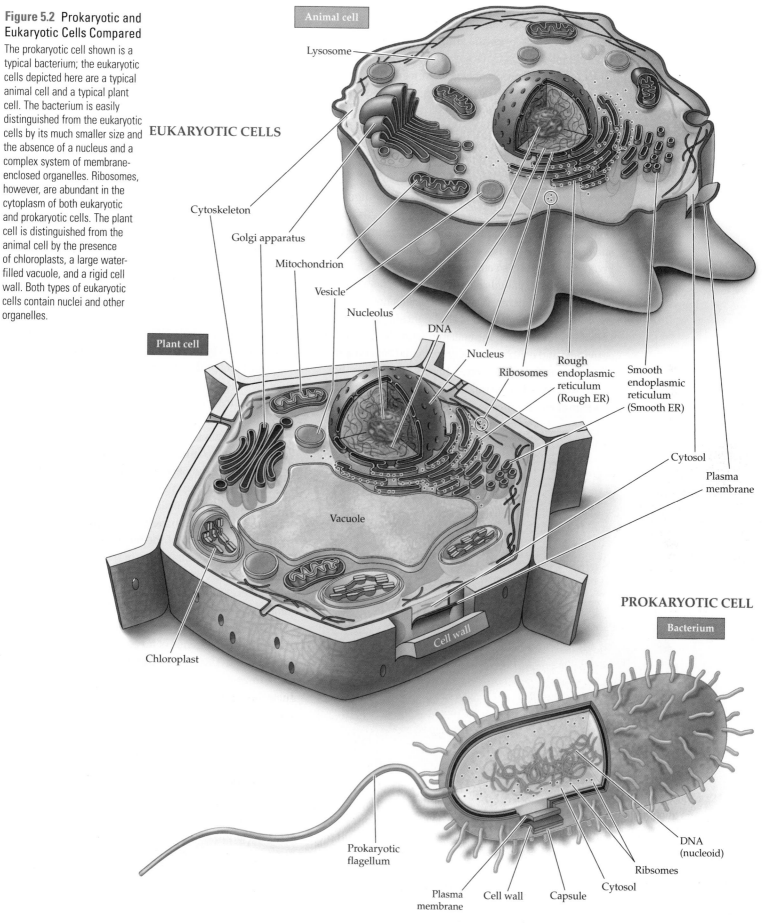

EUKARYOTIC CELLS

Animal cell

Lysosome

Plant cell

Cytoskeleton

Golgi apparatus

Mitochondrion

Vesicle

Nucleolus

DNA

Nucleus

Ribosomes

Rough endoplasmic reticulum (Rough ER)

Smooth endoplasmic reticulum (Smooth ER)

Cytosol

Plasma membrane

Vacuole

Chloroplast

Cell wall

PROKARYOTIC CELL

Bacterium

Prokaryotic flagellum

DNA (nucleoid)

Ribsomes

Cytosol

Capsule

Cell wall

Plasma membrane

an energy-dependent, highly efficient, highly structured factory. However, even the simplest amoeba is vastly more complex than any factory built by humans and it has the capacity for reproducing itself, a quality unique to living systems. We will begin our tour of the eukaryotic cell at the nucleus, the "administrative office."

The nucleus is the repository for genetic material

The presence of a nucleus fundamentally distinguishes eukaryotes from prokaryotes. Eukaryotic cells have a clearly delineated membrane-enclosed compartment—a nucleus—that contains most of their genetic material. Prokaryotic cells also concentrate their genetic material in one area, called the nucleoid, but this region is not wrapped in membranes. The nucleus houses DNA, the code-bearing molecules that contain information necessary for building the cell, managing its day-to-day activities, and controlling its growth and reproduction. The nucleus functions like a highly responsive head office, well-tuned to the talk on the shop floor. Hence, the readout of the DNA code can be modulated by signals received from other parts of the cell and even the world outside the cell.

The boundary of the nucleus, called the **nuclear envelope**, is made up of two concentric lipid bilayers. Inside the nuclear envelope, long strands of DNA are packaged with proteins into a remarkably small space. If all the 46 separate DNA molecules in one of your cells were laid end to end they would have a combined length of 1.8 meters, or almost 6 feet. The only way to fit that amount of DNA into a space only about 5 micrometers (0.005 millimeters) in diameter is to compact the DNA by winding it tightly with special proteins. Each DNA molecule, wound around spools of the compacting proteins, constitutes one **chromosome**. With the exception of certain reproductive cells, every cell in the average human contains 46 chromosomes.

The nuclear envelope contains many small openings called **nuclear pores** (see Figure 5.3). The pores are channels that are continuous through both membranes of the nuclear envelope. They allow free passage to ions and small molecules, but regulate the entry of larger molecules such as proteins, admitting some and shutting out others. Information stored in DNA is conveyed to the protein-manufacturing machinery in the cytoplasm by RNA molecules. These information-bearing nucleic acids must pass through nuclear pores to reach the cytoplasm, and they are important components of the outbound traffic from the nucleus.

The nuclei of most cells contain one or more distinct regions, known as nucleoli (singular "nucleolus"). A **nucleolus** is a region of the nucleus that specializes in churning out large quantities of a special type of RNA, called rRNA. In the nucleolus this RNA gets bundled with special proteins into partially assembled ribosomes, which exit through the nuclear pores to reach the cytoplasm, where they help in the manufacture of proteins. The presence of large amounts of the fibrillar ("threadlike") RNA and particulate ("particle-like") ribosomes creates the distinctive appearance of the nucleolus. Nucleoli are most readily seen in cells that are making large amounts of proteins and therefore have a high demand for ribosomes.

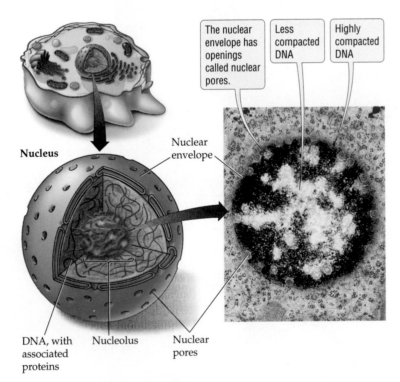

The nuclear envelope has openings called nuclear pores.

Less compacted DNA

Highly compacted DNA

Nucleus

Nuclear envelope

DNA, with associated proteins

Nucleolus

Nuclear pores

Figure 5.3 The Nucleus

The double-membrane nuclear envelope encloses DNA molecules compacted with proteins. The dark regions of the nucleus in this false color electron micrograph contain DNA that is more tightly compacted than the DNA in the lighter-appearing regions.

The endoplasmic reticulum is the manufacturing center for lipids and certain types of proteins

If the nucleus functions as the administrative office of the cell, the cytoplasm is the main factory floor manufacturing the great majority of proteins and other chemical components of the cell. Lipids and certain categories of proteins are made in the endoplasmic reticulum, which functions like a specialized department preparing items for export to the outside of the cell or to ship to other parts of the cell such as the plasma membrane. The **endoplasmic reticulum** [reh-TICK-yoo-lum] (**ER**) is an extensive and complex network of tubes and flattened sacs, all connected to one another (Figure 5.4). The boundary of the endoplasmic reticulum is formed by a single membrane, which is usually joined with the outer membrane of the nuclear envelope. The space inside is called the **lumen**, a general term for the space inside any closed structure inside the cell or inside the body.

The membranes of the ER can be classified into two types based on their appearance: smooth and rough (see Figure 5.4). These manufacture lipids and proteins, respectively. Enzymes associated with the surface of the **smooth ER** manufacture various types of lipids destined for other cellular compartments, including the plasma membrane. In most cells, the majority of ER membranes have ribosomes attached to them, which appear as small rounded particles attached to the cytosolic surface of the ER membrane. Such ER is referred to as **rough ER**, and it specializes in manufacturing proteins that are

destined for specific compartments within the cell or for export to the outside of the cell. These proteins may remain in the ER, they may be targeted to other compartments, or they may be exported to the cell surface for incorporation into the plasma membrane or release into the world outside the cell.

In certain regions of the ER, portions of the membrane bud off into rounded packages called transport vesicles. **Vesicles** are small, membrane-enclosed sacs with functions ranging from storage to the disassembly of macromolecules. **Transport vesicles** specialize in moving substances from one location to another within the cytoplasm and to and from the exterior of the cell. Transport vesicles are like carts used to move goods between different departments of a factory. They bud off from a membrane, such as the ER membrane, carrying their load of lipids and proteins, and deliver the cargo to their destination simply by fusing with its membrane. Because transport vesicles that bud off from the ER enclose a small portion of the ER lumen, and are bounded by a patch of ER membrane, they carry with them a cargo of lipids and proteins that were manufactured by the ER (Figure 5.5).

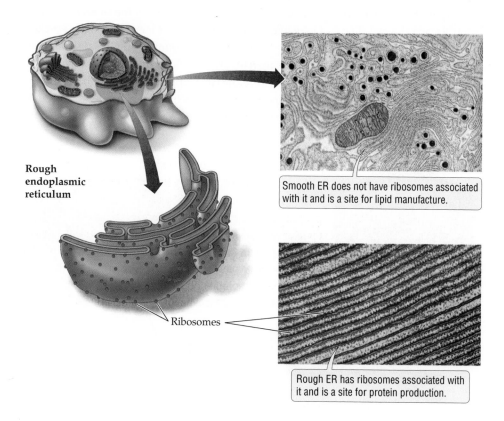

Rough endoplasmic reticulum

Ribosomes

Smooth ER does not have ribosomes associated with it and is a site for lipid manufacture.

Rough ER has ribosomes associated with it and is a site for protein production.

Figure 5.4 The Endoplasmic Reticulum
The ER forms an interconnected network of flattened membrane-enclosed sacs and branching tubes that are important sites for the synthesis of lipids and some types of proteins.

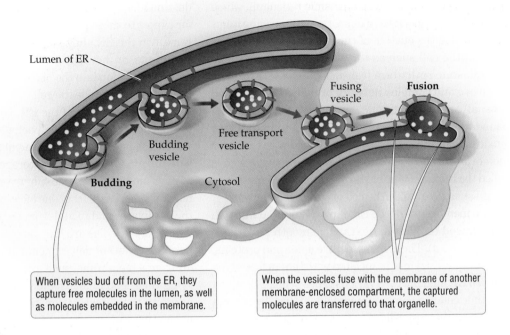

Explore vesicle budding and fusing. ►❚❚ 5.5

Lumen of ER

Fusing vesicle

Fusion

Budding vesicle

Free transport vesicle

Budding

Cytosol

When vesicles bud off from the ER, they capture free molecules in the lumen, as well as molecules embedded in the membrane.

When the vesicles fuse with the membrane of another membrane-enclosed compartment, the captured molecules are transferred to that organelle.

Figure 5.5 How Vesicles Move Proteins and Lipids from One Compartment to Another

The Golgi apparatus sorts proteins and lipids and ships them to their final destinations

Another membranous organelle, the **Golgi apparatus**, directs proteins and lipids produced by the ER to their final destinations, either inside or outside the cell. The Golgi apparatus functions as a sorting station, much like the shipping department in a factory. In a shipping department, goods destined for different locations get address tags that indicate where they should be sent. Similarly, in the Golgi apparatus, the addition of specific chemical groups to proteins and lipids helps target them to other destinations in the cell. These chemical address tags include carbohydrate molecules and phosphate groups.

Under the electron microscope, the Golgi apparatus looks like a series of flattened membrane sacs stacked together and surrounded by many small transport vesicles (Figure 5.6). These vesicles bring lipids and proteins from the ER to the Golgi apparatus and also carry them between the various sacs of the Golgi. Vesicles are the primary means by which proteins and lipids move from

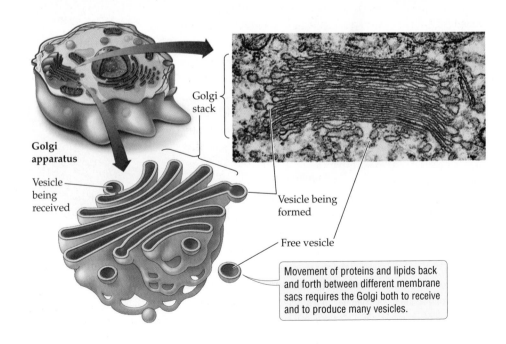

Golgi stack

Golgi apparatus

Vesicle being received

Vesicle being formed

Free vesicle

Movement of proteins and lipids back and forth between different membrane sacs requires the Golgi both to receive and to produce many vesicles.

Figure 5.6 The Golgi Apparatus
The Golgi apparatus consists of stacks of flattened membrane sacs in which proteins are chemically modified, sorted, and then directed to their final destinations inside or outside the cell. Proteins move between the various membrane sacs of the ER and Golgi apparatus in vesicles.

one sac to another in the Golgi apparatus, and from the Golgi apparatus to their final destinations, both inside and outside the cell.

Lysosomes and vacuoles are specialized compartments for recycling, storage, and structural support

The proteins produced in the ER and sorted by the Golgi apparatus are destined either for the cell surface or for other organelles. In animal cells, large molecules destined to be broken down are addressed to organelles called lysosomes. These macromolecules are delivered to lysosomes in vesicles. **Lysosomes** are like the junkyard and recycling center of the cell. A single membrane forms the boundary of each lysosome. Inside it are a variety of enzymes, each specializing in degrading specific macromolecules, such as a particular class of lipid or protein. Many of the breakdown products—among them, fatty acids, amino acids, and sugars—are transported across the lysosomal membrane and returned to the cytoplasm for reuse. Lysosomes can adopt a variety of irregular shapes (Figure 5.7), but all are characterized by an acidic interior with a pH of about 5. This acidic pH is the optimum environment for the lysosomal enzymes to do their work.

The vital importance of lysosomal enzymes is underscored by the existence of more than 40 different types of lysosomal storage disorders in humans. These inherited conditions are caused by the malfunction of one or more of the many lysosomal enzymes whose job it is to degrade a specific type of macromolecule. When a lysosomal enzyme is absent or fails to work properly, the macromolecule it would normally degrade accumulates inside the lysosomes instead. The consequences are devastating, and most of these disorders are fatal in childhood. Tay–Sachs disease is one such metabolic disorder. The lysosomal enzyme responsible for taking apart a membrane lipid found in brain tissue fails to do its job, with the result that large amounts of this lipid pile up in nerve cells, compromising the function of these cells and eventually destroying them. Tay–Sachs disease is rare in the population as a whole, but was unusually common in some populations that have tended to marry among themselves, such as certain French-Canadian communities and Orthodox Jews from Eastern Europe. Through a combination of genetic testing and genetic counseling for individuals contemplating marriage (Interlude C), Tay–Sachs disease has been virtually eradicated among Jewish communities in North America and Israel.

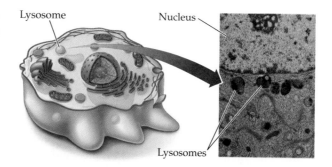

Figure 5.7 Lysosomes in an Animal Cell
Lysosomes are roughly spherical membrane-enclosed sacs containing enzymes that break down macromolecules. The cell shown on the right is from the stomach lining; it uses lysosomes to break down macromolecules such as proteins and lipids.

The plant organelles called **vacuoles** perform many of the same functions as the lysosomes of animal cells, as well as some additional ones. Most mature plant cells have a central vacuole that can occupy more than a third of a plant cell's total volume (Figure 5.8).

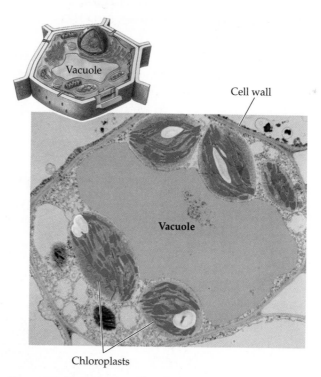

Figure 5.8 Vacuoles in a Plant Cell
Most mature plant cells have a single large vacuole. Each vacuole is a balloonlike organelle, enclosed in a single membrane. Plant vacuoles are similar to the lysosomes of animal cells in that they too contain enzymes for degrading large macromolecules. In addition, vacuoles store water, ions, sugars and other nutrients, pigments, and antiherbivore compounds.

Besides containing enzymes that break down macromolecules, plant vacuoles also store ions, sugars, colored pigments, and noxious compounds that could deter feeding by herbivores. For example, the vacuoles of tobacco leaves accumulate nicotine, a nervous system toxin that is released when the cells are damaged. Large vacuoles filled with water also contribute to the overall rigidity of the nonwoody parts of a plant. The vacuolar contents exert a physical pressure, known as turgor pressure, against the cytoplasm, the plasma membrane, and the cell walls. The fluid pressure keeps cells in nonwoody structures plumped up, and its absence results in the droopy appearance of leaves and young stems in a dehydrated plant.

Mitochondria are the power stations of the eukaryotic cell

So far, we have explored the administrative offices, factory floors, and shipping department within the cellular factory. However, none of the offices or specialized departments could function without a source of energy to run the machines that produce the goods and services. In most eukaryotic cells, the sole source of this energy is an organelle called the **mitochondrion** (plural "mitochondria"). We shall see shortly that photosynthetic organisms, such as plants, have an additional energy-generating unit, called the chloroplast, which gets its energy from sunlight. However, all eukaryotes, photosynthetic or not, need mitochondria to convert the chemical energy in food molecules into a form useful for powering cellular activities.

Mitochondria are pod-shaped and bound by double membranes, that is, two distinctly different lipid bilayers. The space between the two membranes is called the intermembrane space. The inner mitochondrial membrane is thrown into large folds, called **cristae** [KRIS-tee] (singular "crista"). The space interior to the cristae is called the matrix (Figure 5.9). Mitochondria use chemical reactions to transform the energy of food molecules into ATP, the universal cellular fuel. The energy stored in the covalent bonds of ATP is used, in turn, to power

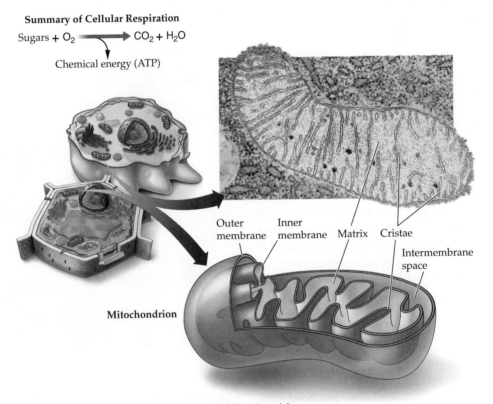

Summary of Cellular Respiration

Sugars + O_2 ⟶ CO_2 + H_2O

Chemical energy (ATP)

Outer membrane Inner membrane Matrix Cristae

Intermembrane space

Mitochondrion

Figure 5.9 Energy-Producing Organelles: Mitochondria
Mitochondria are energy-producing organelles that are needed by almost all eukaryotic cells. Each mitochondrion has a double membrane. The infoldings (cristae) of the inner membrane are studded with enzymes (not shown) that participate in energy production.

the many chemical reactions of the cell. The production of ATP by mitochondria is critically dependent on the activities of proteins embedded in the cristae. The availability of the intermembrane space and the matrix as two separate and distinct compartments is also crucial for the process. With this unique setup, mitochondria are able to trap some of the chemical energy released when food molecules are broken down and to use some of that energy to synthesize energy-rich ATP, through a process that requires oxygen. In other words, oxygen gas (O_2) and molecules derived from food are the raw materials that fuel the mitochondrial power station. As

in most man-made power stations, carbon dioxide (CO_2) and water (H_2O) are released as by-products. We will explore this process, known as **cellular respiration**, in some detail in Chapter 8.

Chloroplasts capture energy from sunlight

All eukaryotes have mitochondria that provide them with life-sustaining ATP, but the cells of plants and the protists known as algae have additional organelles, called **chloroplasts** (Figure 5.10), that capture energy from sunlight and convert it to chemical energy. The

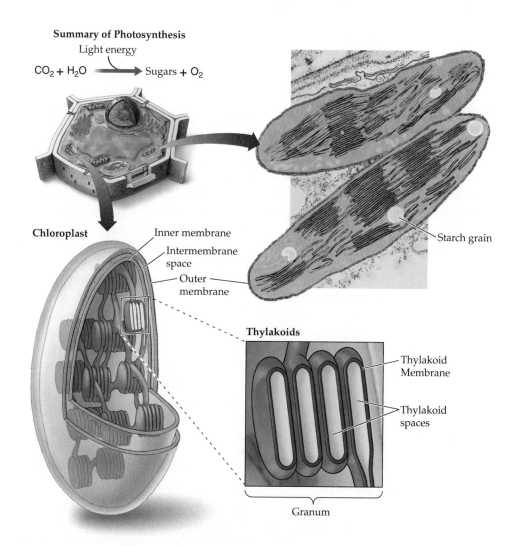

Summary of Photosynthesis

Light energy

$$CO_2 + H_2O \longrightarrow Sugars + O_2$$

Figure 5.10 Energy-Producing Organelles: Chloroplasts
Found only in green plant parts and protists known as algae, chloroplasts capture energy from sunlight and use it to generate chemical energy in the form of sugars. Each chloroplast has two concentric outer membranes and an internal network of stacked discs known as thylakoids. Each stack of thylakoid discs, called a granum, contains the proteins and pigments used to harness energy from light. The thylakoids are derived from the inner of the two concentric membranes during the development of chloroplasts.

Test your knowledge about chloroplasts.
5.7

light energy is first trapped in the form of energy-carriers such as ATP, which are then used to assemble sugar molecules from carbon dioxide (CO_2) and water (H_2O) in a process called **photosynthesis**. The energy in these sugars is used directly by plant cells and indirectly by all organisms that eat plants. At this very moment, as you read this page, your brain and the muscles that move your eyes are using chemical energy that was originally produced in chloroplasts through photosynthesis.

During photosynthesis, water molecules are broken down, releasing oxygen gas. The oxygen produced in photosynthesis is life sustaining for us and many other life forms. Mitochondria depend on a continuous supply of that oxygen to produce ATP in a process that is essentially the reverse of photosynthesis. Photosynthesis *generates* oxygen as a by-product when *creating* organic molecules (sugars) from carbon dioxide and water, while cellular respiration in the mitochondrion *consumes* oxygen while *breaking down* organic molecules and releasing carbon dioxide and water as by-products.

Chloroplasts are enclosed by double membranes, within which lies an internal system of membranes that are arranged like stacked pancakes (see Figure 5.10). Each "stack" is called a **granum** [*GRAH*-num] (plural "grana"), and every "pancake" in the stack is called a **thylakoid** [*THYE*-luh-koid]. Embedded in the thylakoid membranes are special light-absorbing pigments, notably chlorophyll, that enable chloroplasts to capture energy from sunlight. The green color of chlorophyll accounts for the green color of most plants. Enzymes present in the space surrounding the thylakoids use the captured energy to produce carbohydrates from water and from carbon dioxide the plant absorbs from the air around it.

5.5 The Cytoskeleton Provides Shape and Transport

The eukaryotic cell is not simply a formless bag of membrane with cytosol and organelles sloshing around inside. A network of protein filaments and tubules, collectively known as the **cytoskeleton**, creates a scaffold that supports the plasma membrane, positions organelles, and creates tracks on which transport vesicles and other cellular particles can move. The cytoskeleton gives shape and mechanical strength to cells that lack cell walls, such as those of animals and many protista. Because some components of the cytoskeleton are dynamic, many such cells can rapidly change their shape. Unlike the bony skeleton of an adult human, which has fixed connections between bones, the cytoskeleton of a cell has many noncovalent bonds between proteins that can break, re-form, and reshape the overall structure of the cell. Cells that can creep along on a solid surface are also crucially dependent on the changeable nature of cytoskeletal components because cell crawling involves dramatic changes in cell shape, particularly the extension and retraction of the plasma membrane at the leading and trailing edges of the cell, respectively. Cells that propel themselves in a liquid, or have plasma membrane extensions that they can lash about, depend on the ability of certain cytoskeletal elements to generate the necessary force.

The cytoskeleton consists of three basic components

Microtubules, intermediate filaments, and microfilaments are the components of the cytoskeleton. **Microtubules** are relatively rigid, hollow cylinders of protein that help position organelles, move transport vesicles and other organelles, and generate propulsive force in cell projections such as the cilia or flagella found in some eukaryotic cells. **Intermediate filaments** are ropelike cables of protein, the nature of which can vary from one cell type to another. These filaments provide mechanical reinforcement for the cell. **Microfilaments** are the thinnest and most flexible of the three types of cytoskeletal structures. They consist of strands of a protein called actin and are involved in creating cell shape and generating the crawling movements displayed by some eukaryotic cells.

Microtubules support movement inside the cell

Microtubules are the thickest of the cytoskeleton filaments, with diameters of about 25 nanometers (nm). Each microtubule is a helical polymer of the protein monomer **tubulin** [*TOOB*-yoo-lin] (Figure 5.11a) and has two distinct ends. Microtubules grow and shrink in length by adding or losing tubulin monomers at either of the two ends. This capacity allows microtubules to form dynamic structures capable of rapidly altering the internal organization

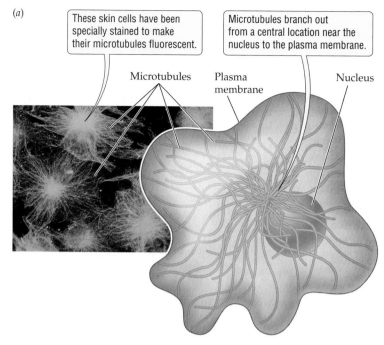

(a)

These skin cells have been specially stained to make their microtubules fluorescent.

Microtubules branch out from a central location near the nucleus to the plasma membrane.

Microtubules   Plasma membrane   Nucleus

(b)

This high-magnification microscopic image shows vesicles moving along microtubules in a nerve cell.

Microtubules   Vesicle

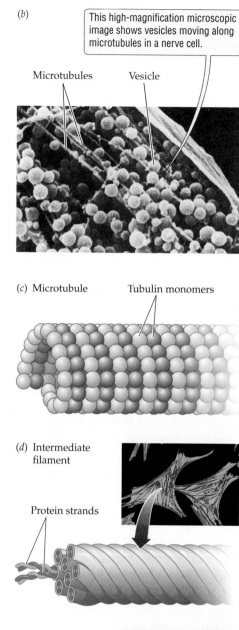

(c) Microtubule   Tubulin monomers

(d) Intermediate filament

Protein strands

(e) Microfilament

Actin monomers

Figure 5.11 Microtubules, Intermediate Filaments, and Microfilaments Are Three Main Types of Cytoskeletal Elements

(a) Microtubules are arranged in a radial pattern in most cells, extending from the center of the cell to the plasma membrane. (b) Microtubules function as tracks along which vesicles and other organelles are shuttled within the cell. (c) Microtubules are composed of tubulin monomers. (d) Intermediate filaments are multistranded, like a rope. (e) Microfilaments are the thinnest and most flexible of the three cytoskeletal elements. They are composed of actin monomers.

of the cell, or capturing organelles and hauling them through the cytosol. Most eukaryotic cells have a system of microtubules that radiate out from the center of the cell and terminate at the inner face of the plasma membrane (Figure 5.11b). This radial pattern of microtubules serves as an internal scaffold that helps position organelles such as the ER and the Golgi apparatus.

Microtubules also define the paths along which vesicles are guided in their travels from one organelle to another, or from an organelle to the cell surface. The ability of microtubules to act as "railroad tracks" for vesicles depends on the action of **motor proteins**, which attach to the vesicle by the "tail" end and associate with microtubules through the "head" end. Motor proteins convert the energy of ATP into mechanical movement, which allows them to move along a microtubule in a specific direction, carrying an attached cargo such as a vesicle (Figure 5.11c).

Intermediate filaments provide mechanical reinforcement

Intermediate filaments are a diverse class of ropelike filaments about 8–12 nm in diameter; they are thinner than microtubules but thicker than microfilaments (Figure 5.11d). Intermediate filaments serve as structural supports, in the way beams and girders support a building. For example, intermediate filaments consisting of the protein keratin strengthen the living cells in our skin. Skin cells lacking functional keratin cannot withstand even mild physical pressure and their rupture results in severe blistering and other types of skin lesions. Intermediate filaments also provide mechanical reinforcement for internal cell membranes. The nuclear envelope, for example, is supported by an underlying meshwork of intermediate filaments.

Microfilaments are involved in cell movement

Of the three filament types, microfilaments have the smallest diameter, about 7 nm (Figure 5.11e). Each microfilament is a cable formed by two polymeric strands that twist around each other. Each strand is built from monomers of the protein **actin**. Like microtubules, microfilaments are dynamic structures that can shorten or lengthen in either direction through rapid disassembly, or rapid assembly, at one or both ends.

Perhaps the best example of the rapid changes in microfilaments is illustrated by a cell creeping along on a solid surface. Skin cells called fibroblasts begin crawling by throwing forward a bump of plasma membrane known as a **pseudopodium** (*pseudo,* "false"; *podia,* "feet") (plural "pseudopodia"). In quick succession, the opposite end of the cell retracts by lifting the trailing portion of the plasma membrane off the substratum and pulling it forward (Figure 5.12). The movements are made possible by the rapid rearrangements of microfilaments, which support the extension of the plasma membrane at the leading edge of the cell and the contraction of the cytoplasm at the trailing edge.

In protruding pseudopodia, microfilaments are aligned in parallel arrays that grow in the forward direction. As these filaments lengthen, they may push against the plasma membrane or otherwise stabilize newly inserted membrane material, thereby extending the pseudopodia farther in the direction the cell will move. In contrast, microfilaments at the trailing

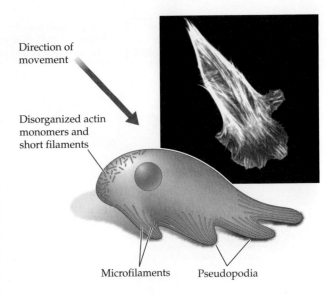

Figure 5.12 Microfilaments Drive Some Types of Cell Movement
Microfilaments help cells crawl on flat surfaces by allowing them to extend flattened projections called pseudopodia.

end of the cell display random organization, angling in all directions. During the retraction phase of cell crawling, the randomly oriented microfilaments disassemble altogether, detaching the plasma membrane from the solid surface, while protein "motors" generate forces that pull the entire rear of the cell forward.

Movement initiated by the cytoskeleton is essential to cell function. For example, cell crawling enables protists such as amoebas and slime molds to find food and mating partners. Fibroblasts, which play an important role in the healing of skin wounds, migrate into the site of injury using the microfilament-based system of cell crawling. Cell migration is also crucial in the embryonic development of many animals. However, cell migration is a devastating step in the development of cancer because it enables cancer cells to spread through the body.

Cilia and flagella enable whole cells to move

Many protists and animals have cells covered in a large number of hairlike projections, called **cilia** (singular "cilium"), that can be moved back and forth, like the oars of a rowboat, to move the whole cell through a liquid (Figure 5.13a) or to move a liquid over the cell surface. Many bacteria, archaea,

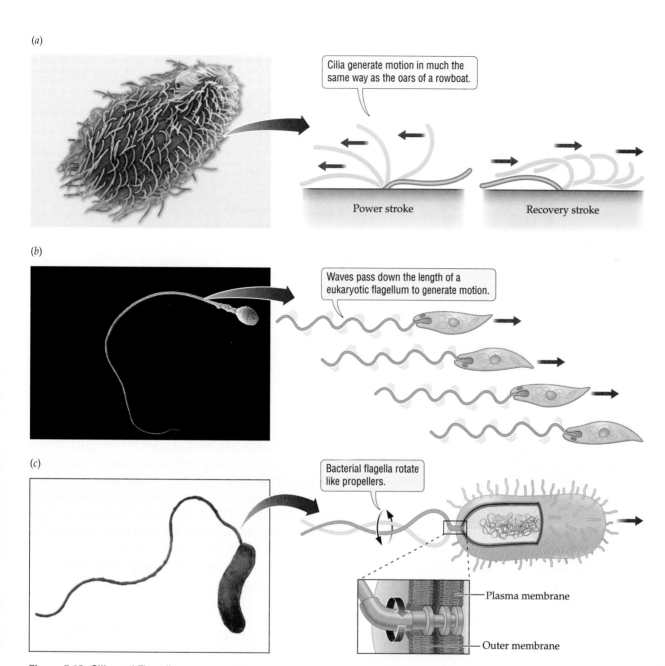

(a)

Cilia generate motion in much the same way as the oars of a rowboat.

Power stroke   Recovery stroke

(b)

Waves pass down the length of a eukaryotic flagellum to generate motion.

(c)

Bacterial flagella rotate like propellers.

Plasma membrane

Outer membrane

Figure 5.13 Cilia and Flagella

Many organisms, especially single-celled ones, use cilia or flagella to generate movement. (*a*) Cilia cover the single-celled protist known as *Paramecium*. The stroke action of a cilium (arrows) resembles the movements of a boat's oars. (*b*) Eukaryotic flagella, such as the one seen in these sperm cells, are much longer than cilia. Their movement resembles the lashing of a whip. (*c*) Bacterial flagella (such as the one seen in this colorized photo of *Bdellovibrio bacteriovorus*) differ in form and action from eukaryotic flagella. Each bacterial flagellum consists of ropelike proteins attached to protein complexes anchored in the cell membranes. Unlike eukaryotic flagella, the bacterial flagellum is not surrounded by an extension of the plasma membrane. It is a stiff, corkscrew-like structure that rotates like a propeller.

Explore flagellar movement.

▶ ‖

5.8

protists, and the sperm cells of some plants and all animals, can propel themselves through a fluid using one or more whiplike structures called **flagella** [fluh-*JEL*-uh] (singular "flagellum"). **Eukaryotic flagella** are lashed in a whiplike pattern (Figure 5.13*b*). They are much longer than cilia, but the internal structure of the two is very similar and both are covered by a lipid bilayer that is an extension of the plasma membrane. Prokaryotic flagella, on the other hand, lack a membrane covering, have a very different internal

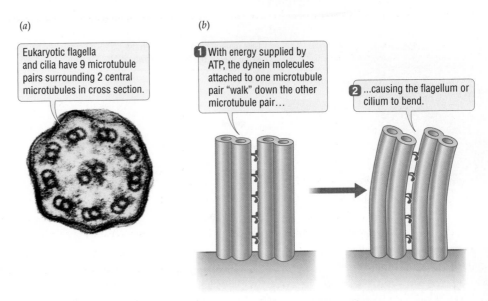

(a) Eukaryotic flagella and cilia have 9 microtubule pairs surrounding 2 central microtubules in cross section.

(b)

1 With energy supplied by ATP, the dynein molecules attached to one microtubule pair "walk" down the other microtubule pair...

2 ...causing the flagellum or cilium to bend.

Figure 5.14 How Eukaryotic Cilia and Flagella Operate
(*a*) The cilia and flagella of eukaryotes both share a characteristic arrangement of a central pair of microtubules attached to a ring with nine microtubule pairs. (*b*) Adjacent microtubule pairs are bowed as the dynein arms of one microtubule pair "walk" along the other.

structure, and are believed to have evolved separately from eukaryotic flagella (Figure 5.13). Instead of the whiplike motion displayed by eukaryotic flagella, **bacterial flagella** spin in a rotary motion, rather like a boat's propellers.

Many aquatic protists, like the one shown in Figure 5.13*a*, use cilia to move about in the waters they inhabit. On the other hand, some cells use cilia not to move themselves around, but to move an overlying fluid layer. This is true of the cells that line portions of our respiratory passages. The ciliated cells are embedded in a tissue layer and use the cilia on their surface to propel unwanted material, caught in a layer of mucus, out of the lungs and into the throat for elimination by coughing or swallowing. Eukaryotic cilia and flagella have nine pairs of microtubules arranged in a ring around a central pair of microtubules (Figure 5.14*a*). The bending of cilia and flagella in eukaryotes depends on the movement of one pair of tubules against an adjacent pair in the ring. Each tubule pair has an attached motor protein, called dynein [*DIE*-neen] that extends an "arm" to make contact with the adjacent tubule pair. Through

a series of attachments and detachments, the dynein arms "walk" along an adjacent microtubule pair using energy derived from ATP (Figure 5.14*b*). When all the tubule pairs on one side of the ring engage their neighbors through dynein "walking," a force is generated that bends the flagellum toward that side. If the tubule pairs on the other side of the ring flex against each other, the flagellum will bend in the opposite direction.

Concept Check

1. Which type of macromolecule is manufactured in (a) smooth ER; (b) rough ER?

2. What is the role of the Golgi apparatus?

3. List one important function performed by the lysosomes. What organelle performs a similar function in plant cells?

4. Chloroplasts and mitochondria both make ATP. What is the crucial difference in the function of these organelles?

5. List three important functions performed by the cytoskeleton.

Applying What We Learned

The Prokaryotic Heritage of Eukaryotes

The *Listeria* bacterium we described at the beginning of this chapter disrupts a major component of a cell's cytoskeleton. *Listeria* is able to use the actin proteins of the host cell for its own purposes. In fact, biologists have found proteins on the surface of *Listeria* that capture actin monomers and start the process of polymerizing them to form filaments. By propelling themselves on these actin filaments, the bacteria are able to burrow deep inside the host cell's cytoplasm, where they have a better chance of escaping the body's defense system. This hijacking of one organism's cellular machinery and energy resources by another organism is known as parasitism. It is one form of symbiosis, the living together of different species.

The entry of one cell into another does not always amount to a hostile takeover. Sometimes it results in a stable relationship that benefits both partners. This form of symbiosis is known as a mutualism. There is compelling evidence that in the distant evolutionary past, certain prokaryotic cells developed a mutualistic arrangement with the ancestors of eukaryotic cells, ultimately evolving into the organelles we know today as mitochondria and chloroplasts.

In the early 1900s, scientists studying eukaryotic cell structure first suggested that organelles such as mitochondria and chloroplasts may be descended from prokaryotes harbored within the cytoplasm of an ancestral eukaryotic cell. Lynn Margulis and other scientists elaborated on this idea in the 1980s to develop the endosymbiont theory of the origin of these organelles. According to this theory, a predatory ancestral eukaryote captured an oxygen-respiring prokaryotic neighbor in a "mitt" created from an indentation of the plasma membrane (Figure 5.15). The prokaryote was internalized in the form of a vesicle, but instead of being digested for food energy, the prokaryote survived and evolved into an endosymbiont, that is, an intracellular symbiotic partner.

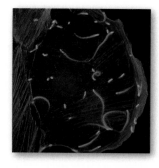

Primitive eukaryotes probably arose around 2 billion years ago, as large cells with internal membranes derived from the invagination of the plasma membrane. At a time when competition for food was becoming fierce, large size might have been advantageous because it would support a predatory lifestyle more readily; it is easier for a larger cell to eat cells smaller than itself. But larger cell size would have driven the evolution of internal compartments for more efficient cell function and the digestion of prey in a vacuole-like structure. At this turning point in evolution, some ancestral eukaryotes are thought to have engulfed oxygen-dependent prokaryotes that were somehow spared from digestion. In taking up residence inside the host cell, the prokaryote gained shelter from predators and a reliable food supply, since the larger ancestral eukaryote was presumably well equipped to gather food by engulfing less fortunate prokaryotes. In return, the host cell

Explore the evolution of organelles. ▶ ‖ 5.9

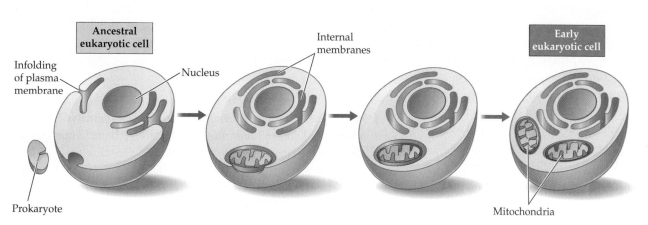

Figure 5.15 How Primitive Eukaryotes May Have Acquired Organelles
Some organelles, such as mitochondria and chloroplasts, are likely descendants of engulfed prokaryotes. Other organelles, such as the endoplasmic reticulum, probably arose through an infolding of the plasma membrane.

Scientists Manipulate "Fear Gene" in Mice

Scientists have identified a "fear gene" in mice that when removed turns them into daredevils, seemingly heedless to both inborn fears and risky situations that normal mice have learned to avoid through experience.

The gene, known as stathmin, controls the production of a protein linked to the creation of long-term fear patterns, said Rutgers University geneticist Gleb Shumyatsky, who led the study, published Friday in the journal Cell. Because the brain system that regulates fear is similar in all mammals, the research could help develop drugs that reduce anxiety by inhibiting the gene's expression.

The gene functions primarily in the amygdala, a part of the lower brain long recognized as a nimble "first responder" to danger in all mammals. Analyzing brain tissue, Shumyatsky found that stathmin works primarily by encoding a protein that . . . [associates with] microtubules . . . Mice lacking the gene were unable to disassemble and rebuild microtubules quickly enough to update their "fear maps," neural networks that relate the latest sensory experiences to memories of danger.

The dynamic nature of cytoplasmic microtubules is essential for cytoskeletal function because it allows the polymers to be quickly dismantled and reassembled in new ways in response to the needs of the cell. Cytoskeletal architecture is tightly controlled in most cells to allow the right configuration of microtubules to be organized at the right time in the right place. The agents responsible for organizing or overhauling the microtubule system are regulatory proteins known as microtubule-associate proteins (MAPs). Many different types of MAPs have been identified. Some are common to most eukaryotic cells, but most are unique to specific cell types. Certain well-known MAPs are responsible for bundling microtubules with each other or cross-linking them with other proteins or organelles. Some will sever microtubules into smaller fragments and yet others will trigger the sudden, rapid, and complete destruction of a microtubule array. The activity of a MAP is carefully controlled by signaling molecules, such as hormones or small proteins called growth factors. If these control systems malfunction, the consequences can be severe, leading to diseases such as cancer or Alzheimer's disease.

Microtubules are crucial to the function of neurons, nerve cells that enable us to process information, act on it, and record a memory of it. Neurotubules, a special class of microtubules, ferry vesicles and other organelles within a neuron and provide a scaffold of cell projections through which neurons communicate with each other and with other cell types. The strengthening of connections between neurons is a fundamental process in memory formation and stathmin appears to act by modulating the formation of microtubules in the amygdala, a brain region that specializes in risk recognition and orchestrates the fight-or-flight response when a danger is perceived. Stathmin is a MAP produced only in the amygdala and it destabilizes microtubules by binding tightly to tubulin monomers, keeping them from supporting polymer growth.

The Rutgers University researchers have demonstrated that mice deficient in stathmin lack the instinctive, as well as learned, capacity to recognize a dangerous situation and therefore they fail to respond to it. It is clear that the mice do not experience a wholesale failure of learning and memory, because they retain the capacity to learn how to run through a maze. What is the link between risk recognition and stathmin? Memory formation and recall requires strengthening of neuronal connections, which, in turn requires a remodeling of the microtubule arrays. The findings suggest that the reconfiguration of microtubule arrays in certain amygdala cells is crucially dependent on the microtubule-associated protein called stathmin. No stathmin means no microtubule remodeling, no memories of danger, and a mouse ready to tackle skydiving.

Evaluating the News

1. How is the dynamic behavior of microtubules regulated? In other words, what controls how, when, and where in the cell a microtubule array will be organized or reorganized?
2. Could the research on stathmins result in any practical benefits? Explain. Does this type of research represent a good use of taxpayer-funded research? Why or why not?
3. Could this type of research be abused or misapplied? Is that reason enough to discourage this type of investigation?

SOURCE: *Vancouver Sun*, Vancouver, British Columbia, November 19, 2005.

benefited from the unique and highly productive energy-releasing pathways that had evolved in this particular line of prokaryotes. With their outstanding capacity for converting food molecules into large amounts of ATP through a series of oxygen-utilizing reactions, these prokaryotic endosymbionts evolved over time into mitochondria.

Various lines of cells descended from the ancient mitochondrial eukaryote may have captured additional cells, this time internalizing a variety of photosynthetic prokaryotes, to give rise to the different groups of photosynthetic protists. Green algae and plants appear to have descended from a mitochondrial eukaryote that became host to a primitive blue-green bacterium (cyanobacterium). The captured cyanobacteria, acting as primordial chloroplasts, would have conferred a great advantage on their host cells by providing them with their own food factories fueled by sunlight.

The endosymbiont theory is based on the observation that mitochondria and chloroplasts are strikingly similar to prokaryotes in many ways. These organelles have their own DNA, which resembles prokaryotic DNA in its small size and its arrangement in a closed loop instead of the linear strands characteristic of nuclear DNA. Mitochondria and chloroplasts divide independently of the cell and they do so through binary fission, a type of cell division unique to prokaryotes (see Chapter 9). These organelles have two or more membrane layers, with the outermost membrane resembling eukaryotic plasma membranes in terms of structure and chemical composition, and the innermost resembling prokaryotic plasma membranes. The ribosomes found in mitochondria and chloroplasts are also more like bacterial ribosomes than eukaryotic ribosomes. These striking resemblances imply that mitochondria and chloroplasts were once free-living prokaryotes that were engulfed by primitive eukaryotes and went on to evolve a mutually beneficial relationship with their hosts.

Chapter Review

Summary

5.1 Cells Are the Smallest Units of Life

- Cells are the basic units that make up all living organisms.
- Multicellular organisms are made up of different types of specialized cells.

5.2 The Plasma Membrane Separates a Cell from Its Surroundings

- Every cell is surrounded by a plasma membrane that separates the chemical reactions of life from the surrounding environment.
- According to the fluid mosaic model, the plasma membrane is a highly mobile assemblage of lipids and proteins, many of which can move within the plane of the membrane.
- Proteins in the plasma membrane allow the cell to communicate with and respond to the environment outside the cell.

5.3 The Structural Organization of Prokaryotic and Eukaryotic Cells

- Living organisms can be classified as either prokaryotes or eukaryotes.
- Prokaryotes are single-celled organisms that lack complex internal compartments. Eukaryotes may be single-celled or multicellular, and their cells typically possess many membrane-enclosed compartments, such as the nucleus.

- The cytoplasm refers to all the cell contents enclosed by the plasma membrane, excluding the nucleus. It consists of the cytosol and organelles.
- The cytosol is a gel-like, water-based medium that contains many ions and molecules.
- Organelles are internal cell structures with a discrete function. Eukaryotic cells have many membrane-enclosed organelles.
- Eukaryotic cells can be 1,000 times larger than prokaryotic cells in volume. They require internal compartments that concentrate and organize cellular chemical reactions for optimal function.

5.4 Eukaryotic Cells Have Many Specialized Internal Compartments

- The nucleus is the "administrator" organelle of the cell. It houses the DNA-encoded instructions that influence every activity and every structural feature of the cell. The nucleus is bounded by the nuclear envelope, a double membrane studded with pores. Information stored in DNA is conveyed by RNA molecules that pass through the nuclear pores on their way to the cytoplasm.
- The endoplasmic reticulum (ER) is the site where lipids and certain proteins, such as membrane proteins and secreted proteins, are made. Lipids are made in the smooth ER. Ribosomes in the rough ER manufacture proteins.

- The Golgi apparatus receives proteins and lipids from the ER, sorts them, and directs them to their final destinations inside the cell, in the plasma membrane, or outside the cell.
- Lysosomes break down large organic molecules such as proteins into simpler compounds that can be used by the cell. Vacuoles are similar to lysosomes but also store ions and molecules and lend physical support to plant cells.
- Mitochondria produce energy for eukaryotic cells through cellular respiration. Their unique structure includes a highly folded inner membrane that isolates the matrix within from both the intermembrane space and the cytosol.
- Chloroplasts harness the energy of sunlight and convert it to chemical energy in a process called photosynthesis. The specialized inner membrane system of chloroplasts contains a pigment called chlorophyll.

5.5 The Cytoskeleton Provides Shape and Transport

- Eukaryotic cells depend on the cytoskeleton for structural support, and for the ability to move and change shape.
- The cytoskeleton consists of three types of filaments: microtubules, intermediate filaments, and microfilaments.
- Microtubules (polymers of tubulin) and microfilaments (polymers of actin) can change their length rapidly. Microtubules are essential for the movement of organelles inside the cell. Microfilaments enable movement of the entire cell during locomotion using pseudopodia. Intermediate filaments provide mechanical strength to cells and reinforce membranes such as the nuclear envelope.
- Some protists, sperm cells, archaea, and bacteria move using cilia or flagella. Eukaryotic flagella are different in structure and action from bacterial flagella.

◉ Review and Application of Key Concepts

1. What fundamental features are common to both prokaryotic and eukaryotic cells, and why are these features advantageous for living organisms?

2. Describe the major components of a plasma membrane and explain why we say that the membrane has a fluid mosaic nature.

3. Compare mitochondria and chloroplasts in terms of their occurrence (in what type of cells, in what type of organisms), structure, and function.

4. Describe the role of each major membrane-enclosed compartment in a eukaryotic cell.

5. Compare the mechanisms underlying the movement of a fibroblast cell and of a bacterium moving with the help of flagella.

6. Cite evidence in support of the theory that mitochondria and chloroplasts are descendants of primitive prokaryotes that were engulfed by ancestors of eukaryotic cells.

Key Terms

actin (p. 116)
bacterial flagellum (p. 118)
cell (p. 102)
cellular respiration (p. 113)
chloroplast (p. 113)
chromosome (p. 108)
cilium (p. 116)
crista (p. 112)
cytoplasm (p. 102)
cytoskeleton (p. 114)
cytosol (p. 102)
endoplasmic reticulum (ER) (p. 109)
eukaryote (p. 105)
eukaryotic flagellum (p. 117)
flagellum (p. 117)
fluid mosaic model (p. 104)
Golgi apparatus (p. 110)
granum (p. 114)
intermediate filament (p. 114)
lumen (p. 109)
lysosome (p. 111)
microfilament (p. 114)
microtubule (p. 114)
mitochondrion (p. 112)
motor protein (p. 115)
nuclear envelope (p. 108)
nuclear pore (p. 108)
nucleolus (p. 108)
nucleus (p. 102)
organelle (p. 102)
photosynthesis (p. 114)
plasma membrane (p. 102)
prokaryote (p. 105)
pseudopodium (p. 116)
ribosome (p. 102)
rough ER (p. 109)
smooth ER (p. 109)
thylakoid (p. 114)
transport vesicle (p. 109)
tubulin (p. 114)
vacuole (p. 111)
vesicle (p. 109)

Self-Quiz

1. Unlike prokaryotic cells, eukaryotic cells
   a. have no nucleus.
   b. have many different types of internal compartments.
   c. have ribosomes in their plasma membranes.
   d. lack a plasma membrane.

2. Which of the following would be found in a plasma membrane?
   a. proteins
   b. DNA
   c. mitochondria
   d. endoplasmic reticulum

3. Which of the following organelles have ribosomes attached to them?
   a. the Golgi apparatus
   b. smooth endoplasmic reticulum
   c. rough endoplasmic reticulum
   d. microtubules

4. Which organelle captures energy from sunlight?
   a. mitochondria
   b. cell nuclei
   c. the Golgi apparatus
   d. chloroplasts

5. Which organelle uses oxygen to extract energy from sugars?
   a. the chloroplast
   b. the mitochondrion
   c. the nucleus
   d. the plasma membrane

6. Which organelle contains both thylakoids and cristae?
   a. the chloroplast
   b. the mitochondrion
   c. the nucleus
   d. none of the above

7. The internal system of protein cables and cylinders that makes whole cell movement possible is called
   a. the endoplasmic reticulum.
   b. the cytoskeleton.
   c. the lysosomal system.
   d. the mitochondrial matrix.

8. Which of the following is *not* part of the cytoskeleton?
   a. pseudopodium
   b. intermediate filament
   c. microtubule
   d. microfilament

9. How is a bacterial flagellum different from a eukaryotic flagellum?
   a. It moves in a whiplike manner.
   b. It is not covered by plasma membrane.
   c. It evolved from eukaryotic flagella.
   d. It is composed of many cilia.

10. Which of the following organelles are thought to have arisen from primitive prokaryotes?
    a. the endoplasmic reticulum and the nucleus
    b. the Golgi apparatus and lysosomes
    c. chloroplasts and mitochondria
    d. vacuoles and transport vesicles

CHAPTER 6

Cell Membranes, Transport, and Communication

Key Concepts

◉ The movement of materials into and out of a cell across the plasma membrane is highly selective.

◉ Passive transport mechanisms move substances across the plasma membrane without energy input, whereas active processes require energy.

◉ Osmosis is the passive transport of water molecules across a selectively permeable membrane.

◉ Some materials can move within a cell, and into or out of a cell, in small membrane packages called transport vesicles.

◉ Because a multicellular organism is an assemblage of specialized cells, the various cell types must communicate with each other through signaling molecules and membrane-localized signal receptors.

◉ Neighboring cells in a multicellular organism are often connected through cell junctions. Some cell junctions are specialized for attaching cells to their neighbors or the surrounding matrix. Others facilitate communication between neighboring cells.

◉ Signaling molecules enable cells to communicate with each other over short and long distances. The cellular response to a signaling molecule can be rapid (less than a second) or slow (over an hour).

◉ Hydrophilic signal molecules bind to receptor proteins localized to the plasma membrane. Hydrophobic signal molecules can cross cell membranes and bind to an intracellular receptor.

Food, Family, and Membrane Receptors

Mimi La Fontaine had blood cholesterol levels five times as high as the values considered healthy and suffered her first heart attack at age 16. As the advertisements for anticholesterol drugs proclaim, cholesterol comes from two sources: food and family. Knowing that heart disease was rife on both sides of her family, Mimi stayed away from foods rich in cholesterol and animal fats. Cholesterol-lowering drugs, which work by obstructing cholesterol-manufacturing enzymes in the liver, did almost nothing for Mimi. So, cholesterol overproduction could not account for her high cholesterol levels, any more than dietary cholesterol could. Lab tests finally revealed the problem. The membrane proteins, known as LDL receptors, that enable normal cells to take up cholesterol-containing particles from the bloodstream were entirely absent from Mimi's plasma membranes. Cholesterol, an essential lipid, is made in small amounts by all cells, but the liver produces the great majority of the cholesterol in the body and pours much of it into the bloodstream enclosed in protein-lipid packets called LDL particles. Lacking LDL receptors, Mimi's cells could not pick up cholesterol from the bloodstream. As a result, cholesterol and the other lipids found in LDL particles were dumped on the lining of her arteries, in muscle tendons, in lumps under the skin and at other unusual sites in Mimi's body.

Mimi was diagnosed with familial hypercholesterolemia (FH) and learned that she had the more dangerous form of the disorder. About 1 in 500 Americans of northern European descent has the less dangerous form of FH. Because these individuals have at least half the normal amount of LDL receptors, some cholesterol is removed from their bloodstream, but these patients too have injuriously high blood cholesterol levels and an increased risk of arterial blockage. Severe narrowing of arteries can cut off blood flow, precipitating a heart

Cholesterol from Food and Family

The average American gets about 300 mg of cholesterol each day from animal foods, 60 percent higher than the intake recommended by health experts. However, cholesterol manufactured by the liver accounts for 80 percent or more of the levels found in the bloodstream. How much cholesterol our livers make, and how cells in our bodies deal with that output, has a substantial effect on our total blood cholesterol levels, which is why the genes we inherit ("family") are important players in this story.

attack, but it takes longer for the lipid deposit to build up in people with the less severe form of FH, which is why they don't usually experience serious ill effects until they reach middle age. Can anything be done to help Mimi and other FH patients? **Why is there such a fuss about LDL anyway? Why does it turn up on the evening news and why is it commonly called the "bad cholesterol?"** Before we can tackle these issues, we must examine how the plasma membrane performs its critical functions as the cell's gatekeeper and communications center.

Most of the chemical reactions that sustain life cannot take place outside of cells. Cells manage to maintain suitable conditions for the chemistry of life only by carefully controlling the uptake of materials from and loss of materials to their environment. All cells must have a way of moving materials into and out of themselves, as well as a way of controlling which materials can enter or leave at any given time.

In this chapter we consider how cells manage their relationship with their surroundings. We begin by examining the role of the plasma membrane as gate and gatekeeper for substances entering and leaving the cell. Then we consider the cell's ability to regulate the amount of water entering and leaving it, a key component of cellular health. We show how cell membranes serve as manufacturing and packaging centers and also as "luggage" transporters. We discuss the ways cells are physically connected to one another, and how these connections facilitate communication between cells. We conclude with a look at the role of signaling molecules in cell communication.

6.1 The Plasma Membrane Is Both Gate and Gatekeeper

As we saw in Chapter 5, the plasma membrane separates the inside of a cell from the environment outside. The plasma membrane is as universal a feature of life on earth as the DNA-based genetic code. A double layer of lipids, the phospholipid bilayer, provides the structural framework for the plasma membrane. Although some biologically important materials can pass directly through the phospholipid bilayer, most cannot. Embedded in the phospholipid bilayer are many proteins, which typically make up more than half the weight of the plasma membrane. Many of these pro-

teins completely span the plasma membrane, providing pathways by which materials can enter or leave cells. The phospholipid bilayer and its associated proteins together act as a sophisticated filter, a **selectively permeable membrane** that controls which materials enter and leave the cell.

Cells can move materials across the plasma membrane with or without the expenditure of energy

Two general rules can help us understand how materials move into and out of cells:

1. Molecules can move down a **concentration gradient**—that is, from an area of abundance to an area of scarcity—without any input of energy. This type of movement is called **passive transport**.
2. Molecules can move up a concentration gradient—that is, from an area of scarcity to an area of relative abundance—but they must do so *actively* (with an input of energy). This type of movement is called **active transport**.

Passive and active transport can be understood by using the physical example of a ball moving down or up a hill, in which the ball represents a chemical and the hill represents a concentration gradient (Figure 6.1). The ball rolls downhill on its own, but it cannot roll uphill unless energy is used to push it.

Imagine emptying a packet of drink mix at one spot in a pitcher of water. You can watch the coloring agents in the drink mix gradually move from the area of high concentration—the spot where the powder was poured—until they are spread throughout the water. This will happen even if you do not expend any energy to stir the mixture. The powder will dissolve and the color will continue to **diffuse**, or spread passively, until it is evenly distributed throughout the water

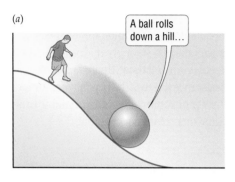

(a)

A ball rolls down a hill…

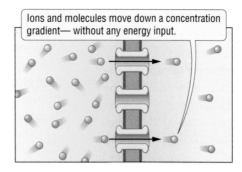

Ions and molecules move down a concentration gradient— without any energy input.

(b)

We must work to move the ball back uphill…

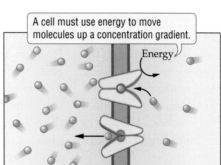

A cell must use energy to move molecules up a concentration gradient.

Energy

Figure 6.1 Active versus Passive Movement of Molecules

Materials can move into and out of organisms either passively (without an input of energy) or actively (with an input of energy). (*a*) Molecules can move passively through a membrane from areas of high concentration to areas of low concentration. (*b*) Energy is required to move materials from areas of low concentration to areas of high concentration.

(Figure 6.2). Once the food dye is distributed evenly, we say equilibrium is reached; the concentration differences driving the diffusion have disappeared and overall movement (net diffusion) of the colored chemicals has ceased. Although the chemicals in the drink mix will continue to move about in the water, the averages of these movements are equal in all directions and will therefore cancel themselves out, so the dye will remain evenly dispersed throughout the pitcher. In other words, at any given time, as many chemical particles are moving into the original release spot, or any other region of the pitcher, as are leaving that spot or region, so there is no overall (net) change in the distribution of these substances and therefore equilibrium is reached.

Organisms rely heavily on both passive and active transport to take up or get rid of nutrients, gases, and wastes. We shall see throughout this unit that the same passive process, diffusion, that allows the concentrated drink mix to spread through a pitcher of water plays a key role in the transfer of some molecules, such as water, oxygen, and carbon dioxide, into and out of cells. However, many of the ions and larger molecules the cell needs for its life-sustaining chemical reactions are present in fairly low concentration in a cell's surroundings, but at relatively high concentrations within the cell, where they are accumulated for ready availability. The continued uptake of these substances requires the expenditure of energy because

they must be brought in against a concentration gradient. It is safe to say that without active transport, or the energy to fuel it, no organism would survive for very long.

Test your knowledge of active versus passive movement of molecules. ▶❚❚ 6.1

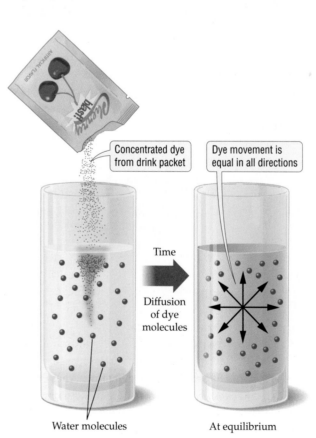

Concentrated dye from drink packet

Dye movement is equal in all directions

Time

Diffusion of dye molecules

Water molecules

At equilibrium

Figure 6.2 Diffusion Is a Passive Process

Diffusion refers to the movement of a substance (such as the dye in the drink mix) from a region of high concentration to a region of low concentration, without an input of energy. Net diffusion ceases, and equilibrium is reached, when the substance becomes uniformly distributed (same dye concentration throughout the water).

Small molecules can diffuse through the phospholipid bilayer

Materials that can readily cross the phospholipid bilayer of the plasma membrane do so simply by moving down their concentration gradients in a passive manner. Water, oxygen, and carbon dioxide usually enter and leave cells in this way (see Figure 6.3a). All these molecules are small and simple, consisting of just a few atoms each, so they can slip by the large molecules in the phospholipid bilayer without much hindrance. Most hydrophobic molecules, even fairly large ones, can pass through cell membranes because they mix readily in the hydrophobic core of the phospholipid bilayer. Many of the early pesticides, such as DDT, were effective in killing insect pests precisely because they could easily get into their cells this way. These pesticides tend to accumulate in animal fat, not only in the target species but also in the predators that eat them, precisely because they are hydrophobic ("water-hating") substances.

For the most part, however, the phospholipid bilayer is a barrier to the movement of large molecules and to electrically charged particles, which do not mix in lipids. Even nutrients such as the simplest sugars and amino acids, which consist of about 20 atoms each, are too large and too hydrophilic ("water-loving") to diffuse through the hydrophobic core of the phospholipid bilayer. Despite being small,

ions such as H^+ (hydrogen ion) or Na^+ (sodium ions) cannot get past the hydrophobic tails in the middle of the phospholipid bilayer because ions are strongly hydrophilic. As a result, larger molecules, and all those that bear an electrical charge, need assistance to cross the plasma membrane. That assistance is rendered by specialized membrane proteins broadly known as transport proteins.

Channel proteins and passive carrier proteins make it possible for ions and molecules to cross the plasma membrane passively

Hydrophilic substances, such as ions, and larger molecules, such as sugars and amino acids, cannot cross the plasma membrane without assistance. Two types of transport proteins help these substances move across the plasma membrane in a passive manner: channel proteins and passive carrier proteins. **Channel proteins** enable ions of the right size and charge to move through the plasma membrane, as long as they are moving *down* a concentration gradient (see Figure 6.3b). These proteins form water-filled tunnels that span the thickness of the phospholipid bilayer.

Passive carrier proteins also transport substances in a passive manner, but they function more like a revolving door than an open tunnel. A passive carrier

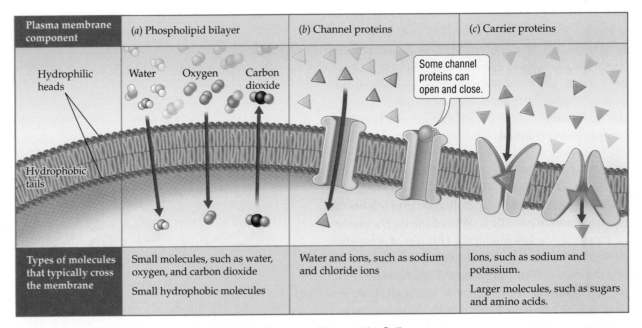

Plasma membrane component	(a) Phospholipid bilayer	(b) Channel proteins	(c) Carrier proteins
Hydrophilic heads / Hydrophobic tails	Water Oxygen Carbon dioxide	Some channel proteins can open and close.	
Types of molecules that typically cross the membrane	Small molecules, such as water, oxygen, and carbon dioxide Small hydrophobic molecules	Water and ions, such as sodium and chloride ions	Ions, such as sodium and potassium. Larger molecules, such as sugars and amino acids.

Figure 6.3 The Plasma Membrane Controls What Enters and Leaves the Cell

(a) The plasma membrane consists of a double layer of phospholipid molecules with hydrophilic "heads" and hydrophobic "tails." Proteins that span the plasma membrane (b, c) play an important role in moving materials into and out of cells. Different kinds of biologically important molecules cross the membrane in different ways.

protein recognizes, binds, and transports a specific molecule, such as a particular type of sugar or a certain amino acid. The selectivity comes from the fact that only a specific type of "cargo" molecule, such as a certain type of amino acid, can fit into the folds on the surface of a given carrier protein. Once the appropriate cargo is bound, the carrier protein changes shape in such a way that the bound cargo now becomes exposed on the other side of the membrane; the shape change also decreases the carrier protein's "clinginess" for the cargo, with the result that this molecule is now let go, having been picked up on one side and discharged on the other side of the membrane (see Figure 6.3c). Such carrier proteins can release bound molecules on the other side of the membrane only if the concentration of that molecule is relatively low on that side of the membrane. In other words, passive carrier proteins can only move their cargo down a concentration gradient.

All cells in the human body depend on glucose for energy. Therefore, every one of them needs special glucose carriers, known as GLUT proteins, to absorb this sugar from their surroundings. The majority of our cells can pick up glucose from the blood in a passive manner because blood normally contains about 10 times as much glucose as the cytoplasm of the average cell. Glucose can simply "roll" down its concentration gradient, aided by GLUT proteins acting as passive carrier proteins.

include sodium, potassium, calcium, and hydrogen ions, and a variety of sugars and amino acids. It takes a specialized active carrier protein to move any such substance from the low concentration side to the high concentration side of a cell membrane. Because the cell must spend energy anytime these substances have to be pumped "uphill," the amount of energy a cell spends on active transport is quite substantial. So much so, that 30 to 40 percent of the energy used by a resting human body goes into fueling active transport across plasma membranes.

The sodium–potassium pump is one of the most important active carrier proteins in our cells. It is present in the plasma membrane of virtually all the cells in our bodies, and it is so vital that most animal cells would die quickly if their sodium–potassium pump stopped working. The sodium–potassium pump creates and maintains the large, but opposite, concentration gradients of sodium and potassium ions across the plasma membrane of most animal cells. Blood and other body fluids have high concentrations of sodium ions (Na^+), but low concentrations of potassium ions (K^+). Within our cells, the situation is reversed: Na^+ is scarce in the cytoplasm but K^+ is plentiful. The sodium–potassium pump maintains these concentration differences by exporting sodium ions from the cell while importing potassium ions. It picks up Na^+ ions from the cytoplasm and moves them "uphill" to the outside of the cell, using energy from the breakdown

Only active carrier proteins can move materials against a concentration gradient

Molecules can cross a plasma membrane against a concentration gradient only by active transport. **Active carrier proteins** can move molecules across the plasma membrane with the aid of an energy-rich molecule such as ATP. Like passive carrier proteins, active carrier proteins bind only certain ions or molecules, those that can fit into specific folds in the protein (Figure 6.4). In this case, however, the addition of energy causes a shape change in the active carrier protein that forcibly releases the molecule being transferred, regardless of the concentration of that molecule near the site of release. This mechanism allows active carrier proteins to move ions or molecules from regions of low concentration to regions of high concentration.

Many ions and molecules are distributed across cell membranes in a lopsided manner, with very much more of the substance on one side and much less on the other side. These unevenly distributed substances

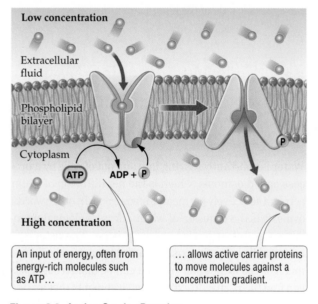

An input of energy, often from energy-rich molecules such as ATP…

… allows active carrier proteins to move molecules against a concentration gradient.

Figure 6.4 Active Carrier Proteins
Active carrier proteins use energy to move materials from areas of low concentration to areas of high concentration.

The sodium–potassium pump
is found in all human cells.
This active carrier protein uses
energy from ATP to move three
sodium ions out of the cell for
every two potassium ions that
enter. Both ions are transported
against their concentration
gradients.

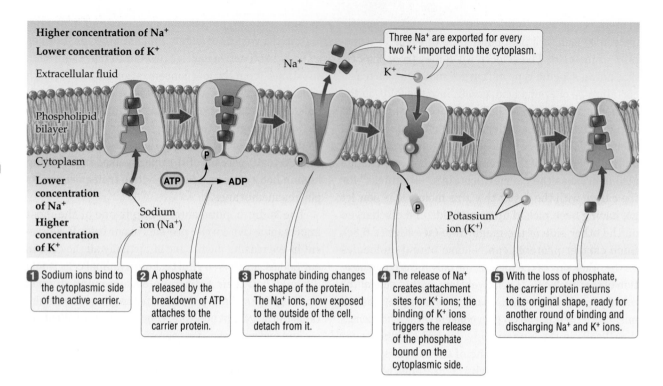

Higher concentration of Na⁺

Lower concentration of K⁺

Extracellular fluid

Three Na⁺ are exported for every
two K⁺ imported into the cytoplasm.

Na⁺

K⁺

Phospholipid
bilayer

Cytoplasm

ATP → ADP

Lower
concentration
of Na⁺

Sodium
ion (Na⁺)

Higher
concentration
of K⁺

Potassium
ion (K⁺)

1 Sodium ions bind to
the cytoplasmic side
of the active carrier.

2 A phosphate
released by the
breakdown of ATP
attaches to the
carrier protein.

3 Phosphate binding changes
the shape of the protein.
The Na⁺ ions, now exposed
to the outside of the cell,
detach from it.

4 The release of Na⁺
creates attachment
sites for K⁺ ions; the
binding of K⁺ ions
triggers the release
of the phosphate
bound on the
cytoplasmic side.

5 With the loss of phosphate,
the carrier protein returns
to its original shape, ready for
another round of binding and
discharging Na⁺ and K⁺ ions.

of ATP (Figure 6.5). The shape of the carrier protein changes when phosphate (P) released from ATP attaches to the protein, causing it to release the Na⁺ ions on the outside surface of the plasma membrane. The new shape is just right for picking up K⁺ ions from the exterior, but the binding of these ions "flips" the carrier protein to the original shape, causing the bound K⁺ ions to be ejected on the cytosolic side of the plasma membrane.

6.2 Cells Must Regulate Water Uptake and Loss to Maintain an Optimal Balance

Water is constantly moving into and out of cells by a process called **osmosis**, which is the net diffusion of water across a selectively permeable membrane. In other words, osmosis is the diffusion of water across biological membranes, and as such, it is a passive process. The water content of cells is continuously affected by osmosis, and too much or too little water in a cell can be disastrous. Cells can find themselves in external environments that are too watery, not watery enough, or just right (Figure 6.6).

A **hypotonic solution** is an external medium that is more watery (has fewer solutes, therefore a higher water concentration) than the cytosol of the cell, so more water flows into the cell than out of it; unchecked, this move-

ment can cause a cell to swell until it bursts. A **hypertonic solution** is an external medium that is less watery (has more solutes, less water) than the cytosol, so more water flows out of the cell than into it. This movement causes the cell to shrink. For example, if a dog were to drink a lot of seawater, which is hypertonic for this land mammal, the cells lining its digestive tract could lose enough water that they would shrink dangerously, possibly resulting in a life-threatening illness. Drinking an isotonic solution would present no such threat, however. An **isotonic solution** is "just right" in that its solute concentration is the same as that inside the cell under consideration. In this situation, the concentration of solutes is the same on both sides of the plasma membrane, so just as much water leaves the cell as enters it, with the result that there is no net diffusion of water across the membrane.

Because organisms inhabit such a diverse range of habitats, cells don't necessarily find themselves in a "just right" world of isotonic solutions at all times. Some organisms that live in a hypertonic world, such as ocean-dwelling fish, have special adaptations to help them counteract the tendency to lose water in a hypertonic environment. They actively transport salt out of their gills, while their kidneys help them retain more of the water they drink. Freshwater fish face the opposite challenge: a tendency to absorb too much water from their hypotonic surroundings. These fish have special transport proteins in the gills that allow them to absorb salts against a concentration gradient, while their kidneys help them

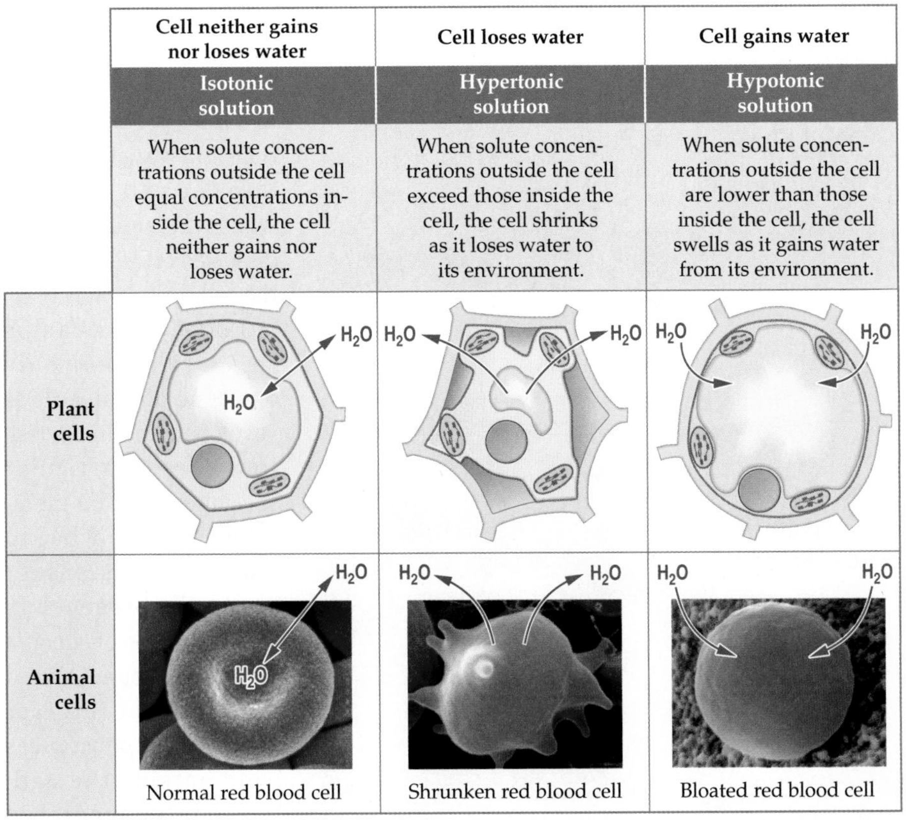

Cell neither gains nor loses water	Cell loses water	Cell gains water
Isotonic solution	**Hypertonic solution**	**Hypotonic solution**
When solute concentrations outside the cell equal concentrations inside the cell, the cell neither gains nor loses water.	When solute concentrations outside the cell exceed those inside the cell, the cell shrinks as it loses water to its environment.	When solute concentrations outside the cell are lower than those inside the cell, the cell swells as it gains water from its environment.

Plant cells

H_2O H_2O ← H_2O H_2O H_2O

Animal cells

H_2O H_2O ← H_2O H_2O H_2O

Normal red blood cell   Shrunken red blood cell   Bloated red blood cell

Figure 6.6 Water Moves into and out of Cells by Osmosis

Osmosis is the diffusion of water across a selectively permeable membrane. A hypertonic solution has a higher solute concentration than the cell's cytoplasm and therefore a lower water concentration than the cell. As a result, there is osmotic loss of water from any cell placed in a hypertonic solution. In contrast, a cell placed in a hypotonic solution (lower solute concentration, and hence higher water concentration, outside), gains water because of osmosis.

Investigate the effects of water gain and loss in plant and animal cells. 6.2

excrete excess water. This constant "balancing act" to maintain an appropriate amount of salt and water inside each cell is known as **osmoregulation**. As in the case of saltwater and freshwater fish, osmoregulation involves active transport and therefore it requires energy.

Concept Check

1. Explain why ions, such as Na$^+$, cannot move across a phospholipid bilayer unassisted.
2. What is the difference between passive and active transport across a biological membrane?
3. Are these statements true or false? (a) Freshwater fish live in an isotonic environment. (b) Osmoregulation in freshwater fish requires energy.

6.3 Membrane-Enclosed Organelles Mediate the Export and Import of Material by a Cell

In Chapter 5 we saw that many molecules are transported from place to place within a cell wrapped in small membrane packages called transport vesicles.

This packaging of chemical substances into molecular "ferries" also occurs in the process of importing or exporting substances from and to the plasma membrane.

Exocytosis [EX-oh-sye-TOE-siss] (*exo*, "outside"; *cyt*, "cell"; *osis*, "process") is the process by which cells release substances into their surroundings by fusing membrane-enclosed vesicles with the plasma membrane (Figure 6.7a). The substance to be exported is packaged into transport vesicles by the ER–Golgi network of membranes inside the cell. As the transport vesicle approaches the plasma membrane, a portion of the vesicular membrane makes contact with the plasma membrane and fuses with it. In the process, the inside of the vesicle (the lumen) is opened to the exterior of the cell, discharging the contents. Many of the chemical messages released into the bloodstream in humans, and many other animals, are discharged via exocytosis by the cells that produce them. For example, after we eat a sugary snack, specialized cells in the pancreas exocytose [EX-oh-sye-TOZE] the hormone insulin, which moves through the bloodstream to other cells and signals them to take up glucose derived from the snack.

The reverse of exocytosis is **endocytosis** (*endo*, "inside"). In this process, a section of plasma membrane

Figure 6.7 Cells Are Importers and Exporters

(a) Exocytosis exports materials from the cell. (b) Endocytosis imports materials into the cell. (c) Receptor-mediated endocytosis is a specialized importing process. (d) Phagocytosis is endocytosis on a large scale. (e) A mast cell releasing signaling molecules through exocytosis. The mast cell (large red sphere) seen in this colorized transmission electron micrograph has released a large number of particles (small red patches) containing histamines and other signaling molecules. The exocytosis is launched in response to the presence of foreign substances such as bacteria or allergens. (f) A macrophage (blue) engulfing an invading yeast cell (yellow). Like mast cells, macrophages are part of the body's defense system.

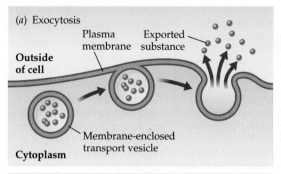

(a) Exocytosis

Outside of cell

Plasma membrane

Exported substance

Membrane-enclosed transport vesicle

Cytoplasm

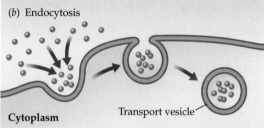

(b) Endocytosis

Cytoplasm

Transport vesicle

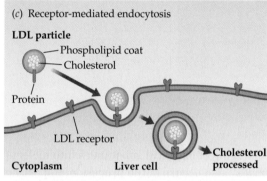

(c) Receptor-mediated endocytosis

LDL particle

Phospholipid coat

Cholesterol

Protein

LDL receptor

Cytoplasm   Liver cell

Cholesterol processed

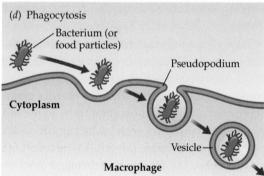

(d) Phagocytosis

Bacterium (or food particles)

Pseudopodium

Cytoplasm

Vesicle

Macrophage

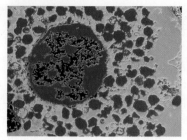

(e) Exocytosis

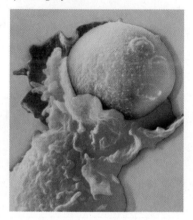

(f) Phagocytosis

as "cell drinking" because cells take in fluid this way. The cell does not attempt to collect particular solutions; the vesicle budding into the cell contains whatever solutes were dissolved in the fluid when the cell "drank."

Endocytosis can be so specific that only one type of molecule is enveloped and imported. How does a particular section of plasma membrane "know" what to endocytose? The answer lies in the presence in the membrane of specific **receptors**, specialized proteins that interact with specific substances from outside the cell. In **receptor-mediated endocytosis**, specialized receptor proteins embedded in the plasma membrane determine what substances will be selected for incorporation into the vesicles that arise from that membrane region (Figure 6.7c). The receptors select the cargo by recognizing specific surface characteristics of the material they bring in. For example, most cells in the human body take up cholesterol-containing particles (low-density lipoprotein, or LDL, particles) from the blood using receptor-mediated endocytosis. LDL receptors in the plasma membrane of these cells recognize specific proteins (lipoproteins) that are exposed on the surface of LDL particles.

Phagocytosis, or "cell eating," is a large-scale version of endocytosis in that it involves the ingestion of particles considerably larger than macromolecules, such as an entire bacterium or virus (Figure 6.7d). This remarkable process occurs in specialized cells, such as the white blood cells that defend us from infection. A single white blood cell can engulf a whole bacterium or yeast cell; this would be roughly equivalent to a person swallowing a big Thanksgiving Day turkey whole! As is the case with cholesterol uptake through internalization of LDL particles, receptors in the membrane of the white blood cell enable it to recognize and phagocytose harmful microorganisms.

bulges inward to form a pocket around extracellular fluid, selected molecules, or whole particles. The pocket deepens until the membrane it is made of breaks free and becomes a closed vesicle, now wholly inside the cytoplasm and enclosing extracellular contents. Endocytosis can be nonspecific or specific. In the nonspecific case all of the material in the immediate area is surrounded and taken in (Figure 6.7b). One form of nonspecific endocytosis is **pinocytosis** [PINN-oh ...], often described

Biology Matters

Water, Water Everywhere—But Not a Drop to Drink

After devastating tsunamis hit East Asia in December 2004, the availability of fresh water became a major problem in the stricken areas. Local water supplies became contaminated with seawater, threatening the spread of disease and dehydration. Seawater is not suitable for drinking, because it is hypertonic to our cells. When we drink seawater, the cells lining our gut become dehydrated, because water flows from them into the hypertonic solution. Drinking too much seawater can be fatal.

Certain ships in the U.S. Navy can extract fresh water from seawater through a process called reverse osmosis. Desalinization plants on these ships use high pressure and special membranes to force water molecules out of seawater and into storage tanks. For example, the USS *Bonhomme Richard* can produce over 30,000 gallons of fresh water a day from seawater by reverse osmosis. Such ships have the capacity to save many lives by supplying fresh water when desperately needed, as in the case of the tsunami disaster.

Why is fresh water so important? The body of an average person is about 70 percent water, meaning it contains about 10 gallons of life's most precious liquid. On an average day we lose around 10 cups of water, and if we don't replace what we lose, our bodies can't function properly. A child in a hot car can dehydrate, overheat, and die within a few hours. In normal surroundings we would not survive more than a few days with little or no water. The importance of the mission of the *Bonhomme Richard* becomes apparent when we consider how vital it is to maintain a proper water balance within our bodies.

Experts recommend taking in the equivalent of about 11 cups of water a day for women and 16 cups a day for men. Beverages obviously consist largely of water, but alcoholic beverages are an exception. Drinking them can lead to excessive water loss, because alcohol reduces the level of a hormone that helps our bodies retain water. Urine volume increases and we end up losing water. Many foods, such as lettuce and watermelon, also contain a lot of water.

USS *Bonhomme Richard* with Other Navy Vessels

6.4 Neighboring Cells Can Be Connected in a Variety of Ways

Multicellular organisms benefit from having a variety of cells and tissues, each of which performs a narrow range of specialized tasks. *Specialized cells* have specific chemical properties, and often a distinctive shape and structure as well, that enable them to conduct their functions more efficiently than any multitasking generalist cell could. By maximizing the efficiency of critical activities, an organism becomes better adapted to the challenges presented by its surroundings. But all the many cells in a multicellular body must be woven together properly, with appropriate means of **cell communication** among them, for the body to function as a well-knit, well-coordinated community of cells. In this section, we examine the physical connections that are commonly found among the cells of a multicellular organism.

Multicellular organisms, from seaweed to walruses, have at least some cells that are attached in special ways. Some of their cells may be intimately connected to allow direct and rapid communication among them. The structures, usually consisting of protein complexes, that anchor cells or hold them together or interconnect them, are known as **cell junctions**. Most animal cells secrete a viscous coating, called an **extracellular matrix** (Figure 6.8a), that helps anchor them or bind them to their neighbors. Vertebrate animals also possess three main types of cell junctions that hold cells in place or connect them to each other: anchoring junctions, tight junctions, and gap junctions.

Anchoring junctions act as protein "hooks" between cells, or between a cell and the extracellular matrix. They allow materials to pass in between cells while still holding them together. Anchoring junctions are especially abundant in tissues that experience heavy structural stress, such as heart muscle (see Figure 6.8a).

Tight junctions (Figure 6.8a) are areas where the plasma membranes of two adjacent cells are held together so tightly that they block the migration of membrane proteins, and the movement of extracellular substances, from one side of the junction to the other. A tight junction consists of a continuous band of proteins just beneath the

Review animal cell junctions.
▶II
6.3

plasma membrane of each cell. Each cell extends portions of these protein complexes through the plasma membrane to enmesh tightly with the adjacent cell's protein extensions. Membrane proteins are prevented from diffusing from one side of the junction to the other. Cells "stitched together" by tight junctions form leak-proof sheets that prevent large molecules from slipping between the cells that make up the sheet.

Gap junctions are the most widespread type of cellular connections in animals. They are found in most cell types in both vertebrates and invertebrates. Gap junctions are direct cytoplasmic connections between two cells. They consist of protein-lined tunnels that span the small intercellular space separating the adjacent cells (see Figure 6.8a). Gap junctions allow the rapid and direct passage of ions and small molecules, including signaling molecules. Electrical signals can be transmitted extremely quickly through gap junctions, which is critical for such activities as the coordinated function of heart tissues and our ability to think.

Despite being surrounded by thick cell walls, many plant cells are connected through tunnels, called plasmodesmata, which are similar to the gap junctions of animals. **Plasmodesmata** [plaz-moh-*DEZ*-muh-tuh] (singular "plasmodesma") are tunnels that breach the cell walls between two cells and connect their cytoplasm (Figure 6.8b). They are lined by the merged plasma membranes of the two cells and provide a pathway for the direct and rapid flow of ions, water, and molecules such as small proteins.

(a) Animal cell junctions

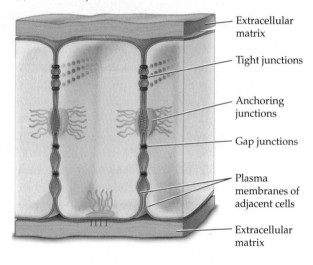

Extracellular matrix

Tight junctions

Anchoring junctions

Gap junctions

Plasma membranes of adjacent cells

Extracellular matrix

(b) Plasmodesmata between plant cells — Walls of two adjacent plant cells

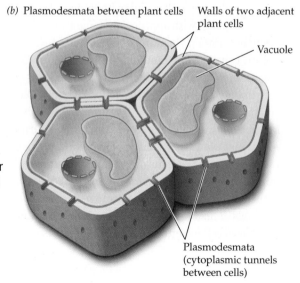

Vacuole

Plasmodesmata (cytoplasmic tunnels between cells)

Figure 6.8 Cells in Multicellular Organisms Are Interconnected in Various Ways
(a) Many animal cells secrete an extracellular matrix and are joined by different types of junctions.
(b) Plants cells are interconnected by plasmodesmata.

6.5 Signaling Molecules and Receptor Complexes Enable Cells to Communicate

The use of molecules to transmit signals between cells is widespread among multicellular organisms. In general, communication between cells is based on the release and transmission of **signaling molecules**, which could be ions, small molecules the size of amino acids, and larger molecules such as proteins. The signaling molecule is received by another cell, the **target cell**, usually through the means of **receptor proteins** (or simply, receptors) localized in the plasma membrane or somewhere in the cytoplasm. A signaling molecule and a target cell are therefore the key components of any signaling system in the living world. Most signaling molecules are short-lived, being destroyed or removed from the vicinity of the target cell within seconds. Some, however, are long-lived, lasting in the body for many days.

A signaling molecule can activate several different types of target cells, and a single cell can be the target of a variety of signaling molecules. The specificity in signal perception and response comes from the receptor protein. Receptor proteins are located either on the plasma membrane of a target cell or somewhere in the cytoplasm. Within the cytoplasm, the receptor proteins may lie in the cytosol or inside an organelle such as the nucleus. Plasma membrane receptors bind their signaling molecules at the cell surface, and must relay receipt of the signal to the cytoplasm through a series of cellular events collectively known as **signal transduction pathways**.

Signaling molecules that bind to receptor proteins inside the cell, that is, in the cytoplasm, must cross the plasma membrane and may have to traverse internal membranes such as the nuclear envelope (Figure 6.9).

Because they must diffuse across the hydrophobic phospholipid bilayer of cell membranes, signaling molecules that bind to intracellular (within the cell) receptors tend to be hydrophobic lipids themselves. Identifying where a receptor protein resides on or in a target cell is a necessary step in understanding how external signals affect the behavior of cells. Recent breakthroughs that allow scientists to tag and track various proteins in the cell have revealed many important features of cell signaling (see the box on page 136). These studies have confirmed that signaling molecules are important coordinators of a broad range of cellular processes (see the table below).

Multicellular organisms use a wide array of signaling systems. Some signaling systems act rapidly, so that the signal molecule elicits a response from the target cell within seconds. Other signaling systems involve signal molecules and specific target cells that together produce a slower response, on the order of hours in most cases. Some signals are narrowly

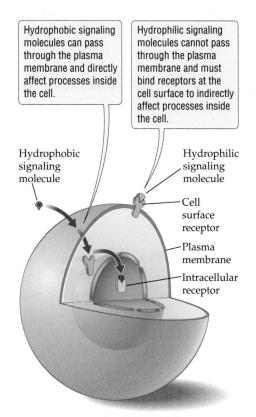

Figure 6.9 Receptors for Signaling Molecules
Two major classes of receptors, hydrophobic and hydrophilic, bind to signaling molecules. Intracellular receptors reside in the cytosol or the nucleus and bind to hydrophobic signaling molecules that can cross the plasma membrane. Cell surface receptors are embedded in the plasma membrane and bind to hydrophilic signaling molecules that cannot cross the membrane.

Hydrophobic signaling molecules can pass through the plasma membrane and directly affect processes inside the cell.

Hydrophilic signaling molecules cannot pass through the plasma membrane and must bind receptors at the cell surface to indirectly affect processes inside the cell.

Hydrophobic signaling molecule

Hydrophilic signaling molecule

Cell surface receptor

Plasma membrane

Intracellular receptor

Examples of Signaling Molecules			
Type of molecule	**Name of molecule**	**Site(s) of synthesis**	**Function(s)**
Animals			
Amino acid derivatives	Adrenaline	Adrenal glands	Releases stored fuels Increases heart rate
	Thyroxine	Thyroid gland	Increased metabolic rate
Choline derivative	Acetylcholine	Nerve cells	Assists signal transmission from nerves to muscles
Gas	Nitric oxide	Endothelial cells in blood vessel walls	Relaxes blood vessel walls
Proteins	Insulin	Beta cells of the pancreas	Promotes the uptake of glucose by cells
	Nerve growth factor (NGF)	Tissues with nerves	Promotes nerve growth and survival
	Platelet-derived growth factor (PDGF)	Many cell types	Promotes cell division
Steroids	Progesterone	Ovaries	Prepares the uterus for implantation of an embryo Promotes mammary gland development
	Testosterone	Testes	Promotes the development of secondary male sexual characteristics
Plants			
Amino acid derivative	Auxin	Young leaves and tips of shoots.	Promotes root formation Promotes stem elongation
Gas	Ethylene	Most plant cells	Promotes fruit ripening Inhibits stem elongation Promotes aging, leaf fall

Test your knowledge of signaling molecules. 6.4

Science Toolkit

Tagging Proteins

A living cell is a constant buzz of activity, in which proteins play a central role. Proteins are constantly interacting with one another and with other components of the cell. Many proteins move from place to place within the cell. Some proteins are being assembled or broken down. Thousands of different types of enzymes (most of them proteins) are speeding up reactions just about everywhere. Researchers hope that by learning about the movements, interplay, and fates of proteins, they will get closer to solving the secrets of cellular life and developing treatments for diseases such as cancer.

It is no easy task keeping track of proteins in the heavy traffic throughout the intricate compartments within a cell. Yet the past decade has seen the emergence of a powerful tool that eases the difficulty considerably. It is a molecular tag that can be attached to proteins with very little disturbance of their function. The tag is also a protein, one that glows for a while after external light shines on it.

Proteins that are "carrying lanterns" are a lot easier to track than those that are not.

This tagging tool provides great flexibility in studying proteins. Researchers can choose to tag and observe only particular proteins. They can then observe where these proteins are and where they go, measure how long they take to get to their destinations, and determine what other molecules they interact with along the way. This tool has revealed many previously unknown properties of proteins in cells.

One technique for using this tool works as follows (Figure A): After particular proteins in the cell have been tagged with the lantern protein molecule, the lantern tags are "turned on" by shining light on them. Next, an area of the cell is "bleached" with intense light, depriving the lanterns in the bleached region of the capacity to glow. Now the movement of other lantern-tagged protein molecules into the bleached area can be tracked visually because their glow will show up against the bleached background.

Figure A Tracking the Movement of Proteins Tagged with the Lantern Protein

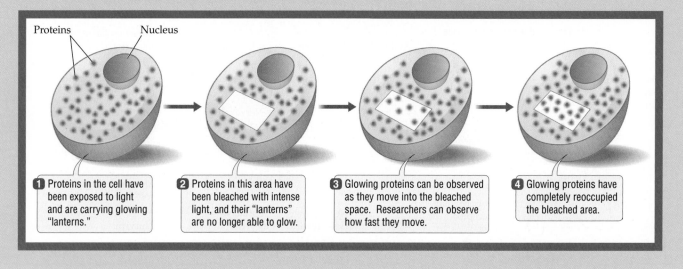

Proteins   Nucleus

1 Proteins in the cell have been exposed to light and are carrying glowing "lanterns."

2 Proteins in this area have been bleached with intense light, and their "lanterns" are no longer able to glow.

3 Glowing proteins can be observed as they move into the bleached space. Researchers can observe how fast they move.

4 Glowing proteins have completely reoccupied the bleached area.

targeted, released only near their target tissues, the way a cable TV show is sent only to subscribers. Other signals are widely dispersed throughout the body, like shows broadcast by satellite TV. Some signaling systems operate over long distances, while other signaling molecules do not have to travel far because they are generated and released in the vicinity of their target cells. If you've ever jumped in response to a sudden noise, you have experienced the almost instantaneous work of the fast-acting signaling molecules, called neurotransmitters, released by nerve cells. If these signaling molecules were slow acting, a "jump" would take hours or even days to occur. Nerve signals are also narrowly targeted, with a certain nerve releasing its neurotransmitters inside specific tissues such as the heart and skeletal muscle cells.

Hormones are long-range signaling molecules

All multicellular organisms use hormones to coordinate the activities of different cells and tissues. **Hormones** are long-lasting signaling molecules that can act over long distances. In contrast to nerve cell signals, most hormones are broadly disseminated through the body. Most hormones are slower-acting than neurotransmitters, but some, such as the hormone adrenaline, can act quite rapidly, triggering a response within seconds. Human growth hormone (hGH) is an example of a slower-acting signaling molecule. This hormone is at work, stimulating the growth of bones and other tissues, as we grow in size throughout childhood. If hGH acted as fast as a neurotransmitter, the term "growth spurt" would have a whole new meaning! In vertebrate animals, most hormones are produced by cells in one part of the body and transported to target cells in another part of the body through the bloodstream, ensuring rapid and widespread distribution. In plants, many of the hormones are dissolved in sap, the thick fluid that circulates throughout the plant body in "pipes" created by living cells, or in the more watery solution that is transported in hollow tubes formed by dead cells.

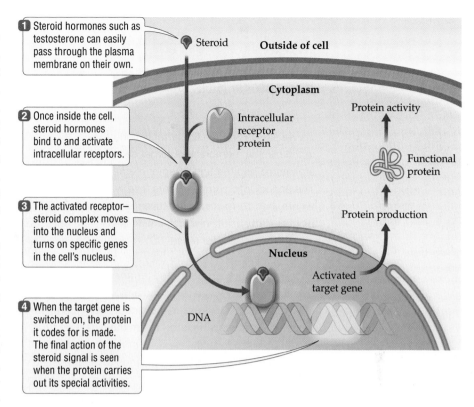

1 Steroid hormones such as testosterone can easily pass through the plasma membrane on their own.

2 Once inside the cell, steroid hormones bind to and activate intracellular receptors.

3 The activated receptor–steroid complex moves into the nucleus and turns on specific genes in the cell's nucleus.

4 When the target gene is switched on, the protein it codes for is made. The final action of the steroid signal is seen when the protein carries out its special activities.

Figure 6.10 A Cell's Response to a Steroid Hormone

Steroid hormones can cross cell membranes

Steroid hormones are an important class of signaling molecules that are essential for many growth processes, including the normal development of reproductive tissues in mammals. All steroid hormones are derived from cholesterol (see Figure 4.17). Because steroid hormones are hydrophobic, they can pass easily through the hydrophobic core of the target cell's plasma membrane and enter the cytosol. But being hydrophobic also means that they cannot move unaided through the bloodstream. They must be packaged with proteins that help them dissolve in the watery environment of the body. These associated proteins also extend the life span of steroids. The ability of steroids to remain in the bloodstream for up to several days improves the likelihood that they will reach target cells at a distance.

When a steroid molecule arrives at its target cell, it crosses the plasma membrane and alters the production of specific proteins inside the cell (Figure 6.10). To do this, the hormone must bind to an intracellular receptor in the cytosol. Together, the steroid and its receptor protein form an active molecular complex that can enter the nucleus and interact with the target cell's DNA. Recall that the DNA in the nucleus carries the instructions for making all of the proteins needed by the cell, each gene encoding a specific protein (as we will see in Chapter 12). The steroid–receptor complex acts on specific genes, activating the production of the proteins they encode. For example, in humans the steroid progesterone targets cells in the uterus (womb) and activates the genes encoding proteins needed to prepare the uterine lining for pregnancy. In general, the action of steroid hormones initiates and coordinates the production of specific proteins necessary for changes in the cell.

Not all signaling molecules enter the cell in the same way as steroids. Certain hormones send their signals into the cell via cell surface receptors. Some of the hormones that act in this way are small proteins; others are chemical derivatives of amino acids or fatty acids.

Concept Check

1. Which process is more selective in terms of the cargo transported: pinocytosis or receptor-mediated endocytosis?

2. What is the role of tight junctions?

3. What are some key differences between nerve cell signaling and hormonal signaling?

Applying What We Learned

Receptors 'R Us

As we saw in Section 6.3, after binding to LDL receptors in the plasma membrane of target cells, LDL particles are internalized through receptor-mediated endocytosis. The endocytotic compartments fuse with lysosomes, where their components are taken apart and the building blocks—fatty acids, glycerol, amino acids—are released into the cytoplasm for reuse. LDL receptors are generally spared degradation because they are retrieved from the endocytotic compartments and returned to the plasma membrane ready to intercept and endocytose another LDL particle. If LDL levels in the bloodstream are high—because too much is absorbed from the digestive tract, too much is produced by the liver, or not enough is removed by cells in the body—it tends to be deposited in the lining of arteries, leading to narrowing and stiffening of these blood vessels. Because high levels of total blood cholesterol are strongly linked to heart disease, resulting from reduced or blocked blood flow, LDL-associated cholesterol is popularly known as "bad cholesterol."

Food choices, and other lifestyle factors, can have a large impact on blood cholesterol levels and heart disease. Poor lifestyle choices are probably responsible, at least in part, for the heart disease risk faced by the roughly 12 million Americans who have unhealthy blood cholesterol levels. Recent research has shown, however, that the genes we inherit can also have a strong influence on how efficiently cholesterol and other lipids are absorbed by the digestive system, how much cholesterol is manufactured by the liver, and, as we saw in the case of FH, how efficiently cholesterol is cleared from the bloodstream.

Familial hypercholesterolemia is a relatively rare condition: about 1 in a million Americans of northern European descent has the homozygous form and 1 in 500 has the heterozygous form. The millions of others who have high blood cholesterol can be treated with fibrates, which reduce the absorption of cholesterol in the gut, or with statins and other drugs that inhibit cholesterol manufactured by the liver. Those who are minimally helped by these treatments may have a reduced density of LDL receptors. Researchers studying cell surface LDL receptors find wide variations in the density of these receptors among individuals and ethnic groups. In sharp contrast to FH patients, and to others with receptor-based causes of high blood cholesterol, about 1 to 2 percent of the American population appears to have an extremely large number of these receptors on their cell surface. This may allow them to internalize and "burn up" cholesterol much more efficiently than the rest of us, which would explain why they appear to be resistant to the ill effects of dietary cholesterol no matter how many supersized bacon triple cheeseburgers they down.

While lifestyle modifications and drugs that reduce cholesterol absorption or production save many lives, these interventions do little for those with FH. A number of patients with the severest form of FH have been successfully treated with gene therapy. Physicians removed a small number of liver cells from the patient and infected them with a harmless virus carrying a functional copy of the gene that codes for the LDL receptor. The virus "injected" the working version of the LDL gene into the liver cell DNA. The genetically modified cells were returned to the patient's liver, where they successfully endocytosed LDL particles, leading to a 20 to 30 percent drop in total blood cholesterol.

The LDL receptor is just one of hundreds of receptors and other cell surface components in each of the 220 cell types in the human body. These proteins and protein complexes help us see, hear, touch, taste, smell, experience emotions, think, and remember. Receptor proteins allow our immune systems to distinguish between our cells and those of foreign invaders and they are involved in the ensnaring and subsequent destruction of the latter. Each of us has about the same suite of receptor types in most of our cells, but subtle variations in the gene code can make substantial differences in our receptors, which contributes to our uniqueness as individuals.

A Spicy Sidekick That May Transform Surgery

By Will Dunham

A new approach to anesthesia using the chemical that gives chili peppers their kick promises an improved way to treat pain in surgery, dentistry and childbirth, researchers reported yesterday.

Current local anesthetics deaden all nerve cells and not just the pain-sensing ones, causing temporary paralysis and numbness . . .

Now, researchers have found a way to target only the pain-sensing nerve cells while avoiding the neurons responsible for muscle movement and sensations such as touch.

They demonstrated the approach in rats and feel confident it will also work in people.

They gave the rats injections containing capsaicin, the active ingredient in hot peppers, and a derivative of the common local anesthetic lidocaine. Working in concert, these chemicals targeted pain-sensing neurons, stopping them from transmitting "ouch" signals to the brain.

The rats were placed on an uncomfortable heat source and had their paws pricked, but showed no signs of feeling pain and moved and behaved normally. The pain relief lasted for several hours . . .

The researchers think this approach could be useful in dental procedures such as tooth extractions, knee surgery and other joint operations, pain relief for women during childbirth, and potentially for chronic pain.

Most nerve cells are highly specialized in the types of stimuli they can sense and how they react to them. The repertoire of receptors in their plasma membranes determines what types of signaling molecules can activate a nerve cell. Receptor proteins that do double duty as regulated ion channels are especially common in nerve cells.

The pain-sensing nerve cells described in the news report have TRP receptors ("trip receptors") in their plasma membranes. "Trip receptors" are noted for the broad range of stimuli they can detect. They bind not just capsaicin, but also compounds found in garlic, mustard, and wasabi, the pungent spice used in green sushi; they even bind the active ingredients of tear gas and tarantula venom! Besides responding to chemical cues, "trip receptors" sense noxiously hot temperatures and some types of injury to skin or mucous membranes. In all cases, the docking of a signaling molecule or the activation of the receptor by heat or injury triggers an onward signal that the brain interprets as acute discomfort or pain.

Capsaicin is a compound found only in hot peppers (*Capsicum anuum*). It activates pain receptors in such a wide variety of animals that it is used in insecticides and personal protection sprays, and even in marine paints designed to discourage colonization by barnacles. Some humans have learned to not only tolerate, but even relish, the sensations produced when capsaicin binds to "trip receptors" on the tongue and the mucous membranes of the mouth. The binding of capsaicin opens calcium-selective ion channels that are part of the receptor protein complex. The inflow of calcium ions triggers signal transduction pathways that ultimately generate an electrical pulse. The electrical pulse is received by certain regions of the brain that interpret it as an unpleasant burning sensation or a pleasantly piquant flavor, depending on one's experiences and preferences.

As the article notes, most painkillers are indiscriminate in their action because they interfere with the electrical activity of all types of nerve cells, not just the ones that communicate pain. The news report describes a new approach to pain relief in which researchers opened the trip receptors with small amounts of capsaicin in order to "sneak in" an anesthetic called QX-314. QX-314 is an effective painkiller because it blocks all onward communication by nerve cells, including pain perception in the brain, but QX-314 cannot cross cell membranes on its own. Capsaicin is the passport it needs to get inside a nerve cell, but capsaicin lets it only into pain-sensing nerve cells. The rats in this study were spared the temporary side effects of conventional anesthetics—numbness, muscle paralysis, memory impairment—because nerve cells not involved in pain perception are spared the electrical blackout QX-314 imposes *if* it gets into the cytoplasm.

Evaluating the News

1. Are "trip receptors" an example of intracellular receptors? Is QX-314 likely to be a hydrophobic molecule? Explain your answers.
2. Nearly 20 percent of all Americans are afflicted with chronic pain caused by conditions ranging from back injuries to migraines. How would you explain the significance of this research to a person suffering from chronic pain?

Source: *The Globe and Mail*, October 4, 2007.

Chapter Review

Summary

6.1 The Plasma Membrane Is Both Gate and Gatekeeper

- Everything that enters or leaves a cell crosses a selectively permeable plasma membrane composed of a phospholipid bilayer with embedded proteins.
- In passive transport, cells carry substances across the plasma membrane down concentration gradients without the expenditure of energy, whereas in active transport cells use energy to move substances up concentration gradients.
- Passive transport takes several forms. Certain small or hydrophobic molecules simply diffuse through the phospholipid bilayer unassisted. Proteins spanning the plasma membrane (channel proteins and passive carrier proteins) enable certain other large or hydrophilic molecules to cross the plasma membrane.
- Active transport requires the aid of active carrier proteins, which use energy to move materials across the membrane against a concentration gradient.

6.2 Cells Must Regulate Water Uptake and Loss to Maintain an Optimal Balance

- Osmosis is the passive movement of water across a selectively permeable membrane. The direction of movement depends on the relative concentrations of solute in the cell and its surrounding. In a hypotonic solution, more water moves into cells. In a hypertonic solution, more water moves out of cells. In an isotonic solution, water flows into and out of cells at the same rate.
- Cells can actively balance their water content by osmoregulation.

6.3 Membrane-Enclosed Organelles Mediate the Export and Import of Material by a Cell

- Cells can export materials by exocytosis: a vesicle carrying export-ready molecules merges with the plasma membrane to release its contents outside the cell.
- Cells can import materials by endocytosis: part of the plasma membrane buds inward, trapping extracellular material within it and pinching off to form a transport vesicle.
- Endocytosis can take several forms. In pinocytosis, cells ingest fluids. In receptor-mediated endocytosis, receptor proteins in the plasma membrane recognize and bind the substance to be brought into the cell. In phagocytosis, the plasma membrane surrounds a large particle, such as a bacterium, and engulfs it.

6.4 Neighboring Cells Can Be Connected in a Variety of Ways

- Cell junctions hold cell communities together, increasing the stability of the multicellular body and allowing neighboring cells to communicate.
- Most animal cells produce an extracellular matrix, made largely of proteins, which serves to bind adjacent cells.
- There are several kinds of cell junctions in animal cells. Anchoring junctions are protein "hooks" that attach cells to each other or to the extracellular matrix. Tight junctions bind cells together to form leak-proof sheets of cells. Gap junctions are cytoplasmic tunnels that connect animal cells and allow the passage of small molecules.
- Plasmodesmata are cytoplasmic tunnels that connect neighboring plant cells.

6.5 Signaling Molecules and Receptor Complexes Enable Cells to Communicate

- Target cells have receptors, specialized proteins that respond to signaling molecules dispatched by other cells.
- Signal-sending cells may be close to target cells or distant from them; responses to signals may be rapid or relatively slow.
- Hormones are long-distance signaling molecules that are broadly distributed in the body. Steroid hormones can cross cell membranes. Other kinds of hormones transmit signals using cell surface receptors.

◉ Review and Application of Key Concepts

1. Describe the ways in which a cell can "decide" what molecules to allow in or keep out.

2. What is osmosis and why is it critical for a cell to maintain an optimum osmotic balance? How would an animal cell, such as a red blood cell, respond to a hypertonic environment and why?

3. Compare exocytosis and endocytosis, and describe the role of receptors in receptor-mediated endocytosis.

4. Describe the major types of cell junctions found in animals and plants.

5. Compare the pathways used by slow-acting and fast-acting cell signals.

Key Terms

active carrier protein (p. 129)
active transport (p. 126)
anchoring junction (p. 134)
cell communication (p. 133)
cell junction (p. 133)
channel protein (p. 128)
concentration gradient (p. 126)
diffusion (p. 126)
endocytosis (p. 131)
exocytosis (p. 131)
extracellular matrix (p. 134)
gap junction (p. 134)
hormone (p. 137)
hypertonic solution (p. 130)
hypotonic solution (p. 130)
isotonic solution (p. 130)
osmoregulation (p. 131)
osmosis (p. 130)
passive carrier protein (p. 128)

passive transport (p. 126)
phagocytosis (p. 132)
pinocytosis (p. 132)
plasmodesma (p. 134)
receptor (p. 132)
receptor-mediated endocytosis
   (p. 132)
receptor protein (p. 134)
 selectively permeable membrane
   (p. 126)
signal transduction pathway
   (p. 135)
signaling molecule (p. 134)
steroid hormone (p. 137)
target cell (p. 134)
tight junction (p. 134)

Self-Quiz

1. Which of the following is *not* part of the plasma membrane?
   a. proteins
   b. phospholipids
   c. receptors
   d. genes

2. Energy is needed for
   a. diffusion.
   b. active transport.
   c. osmosis.
   d. passive transport.

3. Water would move out of a cell in
   a. a hypotonic solution.
   b. an isotonic solution.
   c. a hypertonic solution.
   d. none of the above

4. Which of the following describes movement of material out of a cell?
   a. pinocytosis
   b. phagocytosis
   c. endocytosis
   d. exocytosis

5. Which of these cellular connections creates a leak-proof sheet of cells, such as is found in the cells lining the human gut?
   a. anchoring junction
   b. tight junction
   c. plasmodesmata
   d. gap junction

6. Animal cells can directly exchange water and other small molecules through
   a. gap junctions.
   b. microfilaments.
   c. anchoring junctions.
   d. tight junctions.

7. Cell signaling involves
   a. receptors.
   b. signaling molecules.
   c. target cells.
   d. all of the above

8. Signaling molecules
   a. are all derived from cholesterol.
   b. are derived from active carrier proteins.
   c. affect only a cell's DNA.
   d. none of the above

9. A nerve signal (neurotransmitter)
   a. must travel through the bloodstream to reach target cells.
   b. acts on a target cell that is nearby.
   c. must be long-lived.
   d. must be hydrophobic in nature.

10. Steroid hormones
    a. bind to receptors in the plasma membrane.
    b. must be actively transported.
    c. bind to intracellular receptors.
    d. are hydrophilic molecules.

CHAPTER **7** **Energy and Enzymes**

Key Concepts

◉ Living organisms obey the universal laws of energy conversion and chemical change.

◉ The sun is the ultimate source of energy for most living organisms. Photosynthetic organisms capture energy from the sun and use it to synthesize sugars from carbon dioxide and water. Most organisms can break down sugars to release energy.

◉ Metabolism refers to the chemical reactions that occur inside living cells.

◉ Enzymes greatly increase the rate of chemical reactions in cells.

◉ Metabolic pathways are sequences of enzyme-controlled chemical reactions.

"Take Two Aspirin and Call Me in the Morning"

What if someone told you there was a drug that could reduce pain, fever, and inflammation and also help combat heart disease and cancer? As amazing as it may sound, such a drug exists. It is called aspirin. The active ingredient in aspirin, salicylic acid, has been curing aches and pains and reducing fevers for thousands of years. The ancients got their "natural aspirin" from willow bark and other plant sources. Despite that long history, we have only just begun to understand how salicylic acid works and why it has such a variety of effects on diverse organ systems in the human body.

In 1899, the German company, Bayer, introduced a pure and stable form of salicylic acid and named it aspirin. The drug swiftly became a staple in every pharmacy and hospital around the world and our familiarity with it is reflected in the humorous notion that doctors would recommend two aspirins for just about any medical issue that cropped up after hours. That exaggeration not withstanding, new research has revealed a host of therapeutic benefits of aspirin that go well beyond its well-established role as a pain reliever and fever reducer. A large number of studies have consistently shown that low doses of aspirin taken daily will reduce the risk of heart disease and blood clotting disorders in older adults. Other studies have revealed that people who take aspirin regularly have lower rates of colon cancer. However, taking aspirin for prolonged periods does have negative side effects, including damage to the stomach and kidneys. Given all the possible health benefits of taking aspirin, some obvious questions arise: **How does this wonder drug work? Can we make it better by reducing the negative side effects?**

Before we can address these questions, we must explore how cells use energy and how enzymes facilitate chemical reactions within the cell. In this chapter we will see that a living cell is a highly organized, energy-dependent chemical factory whose many thousands of chemical reactions cannot proceed to any appreciable degree without a boost from one or more enzymes. Disabling enzymes can yield an avalanche of effects. In a chemical system as complex as the human body, those effects may produce benefits, negative consequences, or some mixture of the two.

Aspirin: An Ancient Wonder Drug That Still Keeps Many Mysteries

Cells in willow bark and leaves use enzymes to manufacture salicylic acid, the key ingredient in aspirin. The compound protects the plant because it is toxic to many herbivores. For humans, however, small doses of the chemical can produce beneficial effects ranging from fever reduction to cancer prevention. Despite its popularity since ancient times, long-term use of the drug can produce serious side effects.

All living cells require energy, which they must extract from their environment. Organisms use this energy to manufacture and transform the many chemical compounds that make up living cells.

The capture and use of energy by living organisms involves thousands of chemical reactions, and is collectively known as metabolism. Most chemical reactions in a cell occur in chains of linked events known as metabolic pathways. The metabolic pathways that assemble or disassemble the key macromolecules of life, and their building blocks, are similar in all organisms.

In this chapter we examine the role played by energy in the chemical reactions that maintain living systems. We will discuss the role of specialized proteins, called enzymes, which speed up chemical reactions that would otherwise be too slow to sustain life. We will also explore the possible connections between the pace of metabolism in an organism and the length of its life span.

7.1 The Role of Energy in Living Systems

Any discussion about chemical processes in cells is at heart a discussion about the capture and use of energy. The idea that energy is behind every activity in the cell seems natural and unsurprising, since we have seen how this applies to vital processes such as active transport of materials across cell membranes. And in our daily lives we are accustomed to thinking of energy as a form of fuel. Living systems are shaped by the universal laws of energy, not just because they use energy as fuel, but also because these laws dictate which chemical reactions can occur and under what circumstances.

The laws of thermodynamics apply to living systems

The relationship between energy and the cell's activities is governed by the laws of thermodynamics, which also apply to everything else in the universe. When applied to living systems, the laws of thermodynamics explain why the highly organized state of a living cell cannot exist without an input of energy. They set the ground rules for energy conversions inside a cell, so that a rose bush can convert light energy into the chemical energy of sugars, then use the chemical energy of sugars for all its life processes, from growing bigger to manufactur-

ing the fragrances and brightly-colored pigments of its blossoms. It cannot, however, use the chemical energy in its fragrance molecules to make sugars. A firefly can convert the chemical energy of sugars into light energy and flying to a mate. It cannot, however, use its light flashes as a source of energy for duplicating its DNA. As we shall see in greater detail shortly, the laws of thermodynamics predict which chemical reactions in a cell will yield energy, but they also specify that a certain portion of that released energy will have to be "wasted" and will be unavailable for any cellular work.

The **first law of thermodynamics** states that energy can neither be created nor destroyed, only converted from one form to another. Consider what happens when we use an electric blender to frappe some strawberries and vanilla ice cream into a smoothie. The blender's motor converts electrical energy from the wall outlet into **kinetic energy**, the energy of motion, which we can see in the whirling of the blades. The electrical energy is not destroyed and the kinetic energy is not created out of nothing. Instead, the electrical energy is converted into the kinetic energy of moving blades, and also a fair amount of sound (depending on how noisy the blender is) and heat (which we can feel by touching the motor casing, if we run the blender long enough).

At a cellular level, the first law of thermodynamics is illustrated by the beating of a hummingbird's wings. The kinetic energy of wing movement is powered by the chemical energy in sugar molecules inside muscle cells. Chemical energy is a form of **potential energy**, which is a general term for any type of stored energy. Muscle cells break down sugar molecules and use some of the released energy to manufacture smaller energy-rich molecules, particularly ATP. Some of the chemical energy locked in ATP is released when this molecule is converted to ADP (see Figure 4.10, page 87). Part of the released energy is used to move millions of muscle proteins against each other in such a way that the shape of the whole muscle cell is changed, a kinetic process known as muscle contraction. When the millions of muscle cells in a bird's wing contract, the whole wing moves. So we see that the kinetic energy that moves a bird's wing does not come out of nowhere. It comes from the conversion of chemical energy (in molecules such as sugar and ATP) into the kinetic energy of muscle proteins pulling against each other, until whole muscles are moved and a hummingbird takes flight.

The **second law of thermodynamics** describes how energy use or transformation in any system affects the rest of the universe. This law states that the natural tendency of the universe is to become less organized, more disorderly. Any defined region of the universe, or system, will exhibit this tendency unless energy from elsewhere in the universe is used to organize that system. If any part of the universe, be it a cell or a tool shed, displays a high level of order, we can be certain that energy has been captured from elsewhere to create and maintain that order (Figure 7.1a). In using energy to create internal order, a system reduces the order of its surroundings, so on the whole the universe becomes more chaotic.

As we saw in Chapters 4 and 5, a cell is made up of many chemical compounds assembled into highly ordered structures, such as proteins built from amino acids. The tremendous structural and functional complexity of the cell exists in the midst of a general tendency towards chaos. To counteract the natural tendency toward disorganization, the cell must capture, store, and use energy. One of the many implications of the second law of thermodynamics is that the capture, storage, or use of energy is

never 100 percent efficient, so that at least some of the invested energy is dissipated as heat (metabolic heat). In other words, through the very act of creating order within, living systems add to the disorder of the universe by disbursing heat energy into the environment (Figure 7.1b). Consequently, only a portion of available energy, usually a relatively small portion, is available to fuel cellular processes, since a significant portion goes toward generating metabolic heat. No muscle cell can convert *all* the chemical energy in a sugar molecule into the chemical energy of ATP molecules. Only a portion of the chemical energy in each ATP actually moves the muscle proteins inside a muscle cell because a good deal is turned into heat. Some of the released heat may warm the hummingbird, or overheat the home improvement enthusiast working on a crumbling tool shed. Eventually, all of the energy released by ATP is radiated into the rest of the universe, rendering it a tiny bit more chaotic.

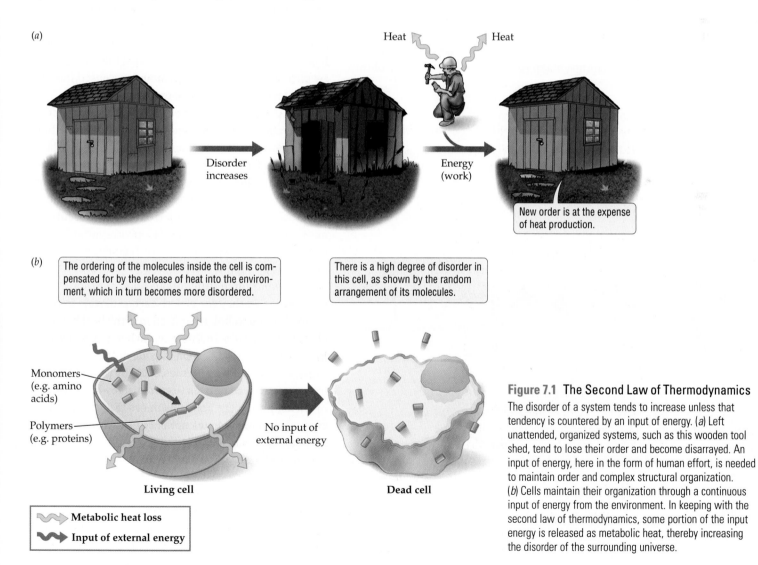

(a)

Heat   Heat

Disorder increases

Energy (work)

New order is at the expense of heat production.

(b)

The ordering of the molecules inside the cell is compensated for by the release of heat into the environment, which in turn becomes more disordered.

There is a high degree of disorder in this cell, as shown by the random arrangement of its molecules.

Monomers (e.g. amino acids)

Polymers (e.g. proteins)

No input of external energy

Living cell

Dead cell

⟿ Metabolic heat loss
⟿ Input of external energy

Figure 7.1 The Second Law of Thermodynamics
The disorder of a system tends to increase unless that tendency is countered by an input of energy. (a) Left unattended, organized systems, such as this wooden tool shed, tend to lose their order and become disarrayed. An input of energy, here in the form of human effort, is needed to maintain order and complex structural organization. (b) Cells maintain their organization through a continuous input of energy from the environment. In keeping with the second law of thermodynamics, some portion of the input energy is released as metabolic heat, thereby increasing the disorder of the surrounding universe.

The flow of energy and the cycling of carbon connect living things with the environment

What is the source of the energy that is used to generate order in the cell? We know from the first law of thermodynamics that the cell cannot create energy from nothing; the necessary energy must come from outside the cell. In other words, energy must be transferred into the cell in some fashion. In the case of photosynthetic organisms, the energy comes from sunlight. They trap light energy and use it to make sugars from carbon dioxide and water, thereby transforming light energy into chemical energy stored in the covalent bonds of sugar molecules. For most organisms that do not photosynthesize, energy comes from the chemical energy of food molecules, such as sugars and fats, gained from eating other organisms.

In the biosphere there is a tight relationship between energy and matter, and between producers and consumers. As we saw in Chapter 1, producers are organisms that can make their own food (molecules rich in energy and nutrients). Photosynthetic organisms, such as plants and certain bacteria, are producers that use sunlight to manufacture food. Consumers, such as animals and fungi, cannot make their own food and must obtain

energy, as well as the building blocks of macromolecules, by eating other organisms or absorbing their dead remains. Some bacteria and archaeans can make food from unusual sources of energy, such as minerals found in volcanic rock; however, photosynthetic organisms make up the bulk of producers found on land and in the sea. This means that, thanks to photosynthesis, the sun is the primary energy source in most ecosystems.

Energy streams through an ecosystem in a single direction, passing from producers to consumers and decomposers, with some escaping as metabolic heat during every biological process and at every step in the food chain, as dictated by the second law of thermodynamics. In contrast, the carbon atoms that constitute the backbone of biologically important molecules pass from producers to consumers, and are then recycled back to producers after passing through the nonliving part of the environment. How are carbon-containing molecules from living cells returned to the inanimate part of the ecosystem? Most organisms, producers and consumers alike, break down food molecules through an energy-releasing process known as cellular respiration. The carbon–carbon links in food molecules are taken apart during cellular respiration and each carbon atom is released into the environment as a molecule of carbon dioxide (CO_2).

Keep in mind that it is not just animals, and decomposers such as bacteria and fungi, that degrade food molecules and release carbon dioxide through cellular respiration. Producers such as plants also rely on cellular respiration. Photosynthetic cells manufacture energy-rich sugar molecules, but they must use cellular respiration to extract energy from these molecules to meet their daily energy needs. Nonphotosynthetic plant parts, such as roots, also depend on cellular respiration for an energy supply.

The remarkable thing about photosynthetic cells, however, is that they can absorb carbon dioxide from the environment and link its carbon into the backbone of a biological molecule. In this way, carbon atoms are continually cycled from carbon dioxide in the atmosphere to sugars and other molecules made by producers, and from these molecules back to carbon dioxide released by respiring producers and consumers (Figure 7.2). This kind of recycling occurs not only for carbon, but also for other elements found in living organisms, including oxygen, nitrogen, and phosphorus.

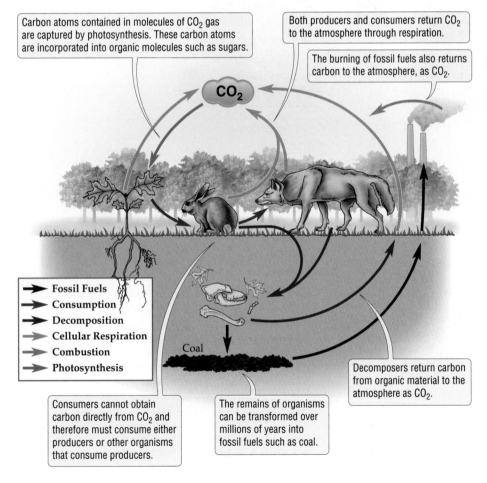

Carbon atoms contained in molecules of CO_2 gas are captured by photosynthesis. These carbon atoms are incorporated into organic molecules such as sugars.

Both producers and consumers return CO_2 to the atmosphere through respiration.

The burning of fossil fuels also returns carbon to the atmosphere, as CO_2.

CO_2

Legend:
→ Fossil Fuels
→ Consumption
→ Decomposition
→ Cellular Respiration
→ Combustion
→ Photosynthesis

Coal

Consumers cannot obtain carbon directly from CO_2 and therefore must consume either producers or other organisms that consume producers.

The remains of organisms can be transformed over millions of years into fossil fuels such as coal.

Decomposers return carbon from organic material to the atmosphere as CO_2.

Figure 7.2 Carbon Cycling
Carbon atoms cycle among producers, consumers, and the environment. Carbon atoms become parts of different kinds of molecules as they cycle between living organisms and the environment.

7.2 Using Energy from the Controlled Combustion of Food

Most living systems obtain energy from food by controlled "burning" (combustion) of food molecules to form carbon dioxide and water. If our cells were to convert food into carbon dioxide and water in a single chemical combustion reaction, however, we might burst into flame like a lit match. Here is the chemical equation that describes what happens when the wooden part of a match burns:

$$\text{Carbon compounds in wood} + O_2 \rightarrow CO_2 + H_2O + \text{energy (as heat and light)}$$

This combustion reaction is similar to what happens when our cells process the carbon compounds in food molecules, but fortunately for us, there are some important differences.

All of the energy released from the burning match is dispersed into the environment in a large burst of heat and light. The processing of food molecules inside a cell yields the same molecular products (carbon dioxide and water), but the energy is released in a stepwise manner through a series of chemical reactions. The small amount of energy made available at each step is just right for packaging into energy carriers such as ATP. The production of energy-rich molecules amounts to the capture of some of the released energy in chemical form. Cellular combustion of food molecules is inevitably accompanied by heat production, but because this heat is divvied up among the many steps in the metabolic pathway, a sudden and large increase in temperature is avoided, which is what saves the cell from going up in flames. The metabolic "burning" of food molecules can therefore be represented as

$$\text{Carbon compounds in food molecules} + O_2 \rightarrow CO_2 + H_2O + \text{energy (as heat and the chemical energy of ATP)}$$

Energy is extracted from food through a series of oxidation–reduction reactions

Many metabolic pathways, such as photosynthesis and cellular respiration, consist of a series of chemical reactions in which electrons are transferred from one molecule or atom to another. **Oxidation** is the loss of electrons from a molecule, atom, or ion, while **reduction** is just the opposite, the gain of electrons by a molecule,

atom, or ion. Because the two processes are complementary, oxidation and reduction go hand in hand, and the paired processes are called oxidation–reduction reactions, or simply **redox reactions**.

Although the modern definition of oxidation focuses on the loss of electrons, the term is derived from the nineteenth century description of certain chemical reactions in which oxygen is added to an atom or molecule. Oxygen is a powerful oxidizer in that it can pull electrons from many other atoms. Iron combines with O_2 to form the crumbly red material we know as rust (Fe_2O_3). Because of the electron-drawing power of oxygen atoms, electron pairs shared between the two types of atoms are closer to the oxygen atoms than to the iron atoms, so the iron atoms can be regarded as existing in an oxidized state. In contrast, a hydrogen atom covalently bonded to a different atom, such as carbon or oxygen, tends to become "electron-poor" because its one electron is pulled more strongly by the partner atom. An atom or molecule that gains one or more hydrogen atoms becomes "richer" in electrons and is therefore considered reduced (remember, reduction is the gain of electrons). Take, for instance, the combustion of an organic compound, such as the gas methane (CH_4). As shown in Figure 7.3, the products of this reaction are carbon dioxide (CO_2) and water (H_2O). The carbon atom is *oxidized* as it *gains oxygen atoms* and loses hydrogen atoms to become carbon dioxide. Each oxygen atom in O_2 strips two hydrogen atoms from a molecule of methane to become a molecule of water. The *gain of hydrogen atoms* by the oxygen atom signifies that it has been *reduced*.

Metabolism refers to all the chemical reactions within a living cell that capture, store, or use energy. Metabolic reactions that create complex molecules out of smaller compounds are called **biosynthetic**, or **anabolic**, **reactions**; those that break down complex molecules to release energy are called **catabolic reactions**.

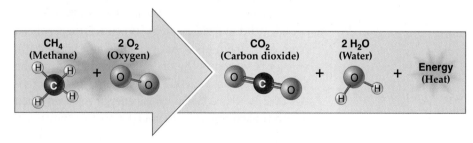

Figure 7.3 The Combustion of Methane Is a Redox Reaction

In some redox reactions, oxidation is marked by the gain of oxygen atoms and reduction by the gain of hydrogen atoms. Here, each carbon atom is oxidized to carbon dioxide, and each oxygen atom is reduced as it gains two hydrogen atoms to become a molecule of water. As in all combustion reactions, energy is released in the form of heat.

Science Toolkit

Counting Calories with a Bomb

Practically all labels on packaged food products include information on how many calories the food contains. The label on a bag of potato chips might read:

Nutrition Facts:
Serving Size: 1 oz (28 g/15 chips)
Amount per Serving: Calories—140, Calories from Fat—70

But just what are calories and how can anybody determine how many calories are in "15 chips"?

A calorie is one way of measuring heat energy. Chemists define 1 calorie as the amount of energy needed to raise the temperature of 1 gram of water by 1°C. That definition hints at the way food manufacturers and nutritionists determine how many calories are in a sample of food, such as 1 ounce of potato chips: they burn the chips and compare the amount of heat given off to the amount needed to heat a known amount of water.

To accomplish this, the food sample is burned in a device called a bomb calorimeter. The calorimeter is composed of a sealed container (the "bomb") surrounded by another sealed container with a known amount of water in it. A thermometer shows the temperature of the water before and after the sample is burned. Once the researchers know how much the temperature of the water was increased by burning the food, they can calculate the number of calories that were in the sample. The number of calories from fat can be determined by extracting the fats from the sample and burning them separately.

Bomb calorimeters measure 100 percent of the energy in a food sample, but our bodies are not 100 percent efficient, so the number of calories our bodies actually use is less than the calorimeter's number. Beer, and many other foods and beverages with a high water content, will not burn readily. To measure the caloric content of such foods, the water is evaporated, and the calorimeter burns only the dry remains. Based on such analysis, 12 fluid ounces of regular beer is reported to have 146 kilocalories (reported on Nutrition Fact labels as 146 Calories, with an uppercase C); a medium banana (118 grams) contains 109 Calories; and a small head of iceberg lettuce (55 grams) has only 7 Calories.

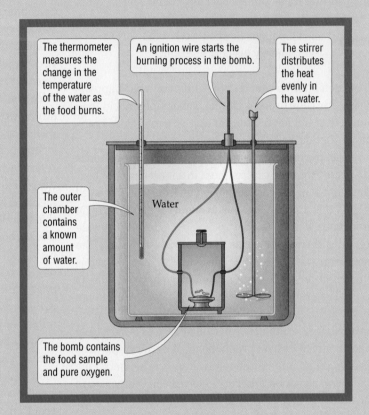

The thermometer measures the change in the temperature of the water as the food burns.

An ignition wire starts the burning process in the bomb.

The stirrer distributes the heat evenly in the water.

The outer chamber contains a known amount of water.

Water

The bomb contains the food sample and pure oxygen.

Inside a Bomb Calorimeter

Every living cell makes use of the universal energy carrier, ATP, to deliver energy. When ATP is produced from ADP and a phosphate group (see Figure 4.10), energy is stored in the chemical bonds holding the phosphate groups together. The majority of the activities of the cell are fueled by the energy released when ATP is broken down to ADP and phosphate (Figure 7.4). Continuous ATP production is an urgent priority for the human body; if it were halted, each cell would consume its entire supply of ATP in about a minute.

Energy from ATP is used for a variety of activities in the cell, such as moving molecules and ions between various cellular compartments and generating mechanical force during the contraction of a muscle cell. It is also required for biosynthetic reactions. The catabolic reactions that extract energy from food and harness it to generate ATP are tightly linked to the biosynthetic reactions that manufacture complex macromolecules from simpler building blocks. The two kinds of reactions in metabolism, catabolic (energy-releasing) pathways and anabolic (energy-consuming) pathways, are closely intertwined and tightly regulated within the organism. Paired reactions in which one provides the energy to make the other happen, are called **coupled reactions**.

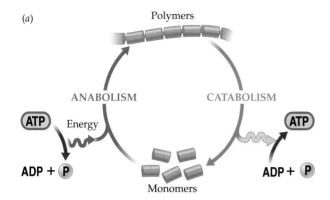

(a) Polymers

ANABOLISM   CATABOLISM

ATP

Energy

ADP + P   ADP + P

Monomers

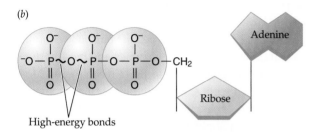

(b)

Adenine

Ribose

High-energy bonds

Figure 7.4 Energy Use and Release During Metabolism

(a) There are two main kinds of metabolic processes: anabolism and catabolism. Anabolism refers to metabolic processes that build biologically important molecules such as various polymers. Anabolic pathways require an input of energy, most commonly in the form of energy carriers like ATP. Catabolism refers to the breakdown of organic molecules, including polymers. Catabolic reactions release energy, some of which may be captured to make ATP. *(b)* Adenosine triphosphate (ATP) is called a universal energy carrier because it functions as an energy-storing molecule in all cells. ATP is a nucleotide, consisting of a nitrogenous base (adenine), a five-carbon sugar (ribose) and three phosphate groups (shown in yellow). When the terminal high-energy bond breaks, energy is released, and ATP is turned into ADP (adenosine diphosphate).

Chemical reactions are governed by the laws of thermodynamics

How does the cell control such a powerful event as combustion and break it down into smaller, more manageable and useful steps? The answer lies in the stepwise nature of these metabolic reactions. Let's review some of the fundamental principles that govern chemical reactions, as represented by the following general example:

$$A + B \rightarrow C + D$$

A and B are the starting materials, called **reactants**; C and D are the **products** formed by the reaction. Recall that a chemical reaction involves the creation or rearrangement of chemical bonds involving one or more substances. The second law of thermodynamics dictates that chemical reactions *can* occur spontaneously, that is, without an investment of energy, provided they generate products that are less organized, and therefore at a lower energy state than the reactants. Put another way, the reaction can occur spontaneously (on its own) provided some energy is lost. The other side of the coin is that reactions that create products with higher total energy than the reactants simply will not occur without an input of energy to "push" the reactants in this nonspontaneous direction. In other words, chemical reactions that are "downhill," because they go hand in hand with a loss of organization, can occur

without any "outside help"; but, those that are "uphill," because they generate a higher level of organization, won't occur unless energy is expended to make them happen.

Assuming that A + B have more total energy than C + D, the second law says that A and B *can* be converted to C and D spontaneously, that is without any outside input of energy. But that does not mean A and B *will* react with each other to any noticeable extent. The second law makes the reaction possible, but says nothing about the speed with which a reaction will take place or whether it will take place at all. What does it take for atoms or molecules (such as A and B in our example) to react with each other? They must bump into each other often enough, fast enough, and in the correct orientation, to allow their chemical bonds to be rearranged. This set of conditions constitutes an energy "hump," or energy barrier, that atoms and molecules must overcome *before* they can react with each other. One way to get reactants over the energy barrier is to invest a small amount of energy to get more of them moving faster and colliding more frequently. The minimum energy input that will allow atoms or molecules to overcome the energy barrier, thereby allowing them to react, is called the **activation energy**. The larger the fraction of atoms or molecules that overcome the energy barrier (because they possess enough activation energy), the faster a reaction will proceed.

Heat is one type of energy that can push atoms or molecules over the energy barrier, allowing them to react. Atoms and molecules move faster at higher temperatures, so raising the temperature will cause more of them to collide with each other forcefully enough to react. When the energy of these atoms or molecules equals the activation energy, the reaction will proceed. A safety match does not ignite spontaneously at room temperature because its chemical ingredients react very slowly with molecular oxygen in the air. But a small rise in temperature, from friction against a rough surface for example, can provide the activation energy necessary for the immediate combustion of the chemicals in the match head.

Concept Check

1. What is metabolism?
2. How is the combustion of carbon compounds in a match similar to, and different from, the metabolic "burning" of carbon-containing food molecules?
3. What is activation energy?

7.3 Cells Use Enzymes to Speed Up Chemical Reactions

How do chemical reactions inside a cell overcome the activation barrier? Relying on heat as a source of activation energy is not a workable solution for most cellular processes because heat acts indiscriminately. Whereas high temperatures will speed up almost all chemical reactions, the cell must exercise great selectivity in which chemical reactions are allowed to proceed at any given time. The great majority of chemical reactions in the cell take place when the activation barrier for the reaction is lowered by a vital class of macromolecules known as enzymes.

Enzymes speed up chemical reactions

A **catalyst** is a chemical that speeds up chemical reactions without itself being changed in the course of the reaction. An **enzyme** is a biological catalyst. Rather than *pushing* more of the reactants over the activation barrier, the way heat does, a catalyst *lowers* the activation barrier so more of the reactants can get across it. As we have seen, the larger the proportion of reactants that make it over the energy barrier, the faster a reaction proceeds.

An important characteristic of catalysts is that, unlike reactants, they remain chemically unaltered after the reaction is over. Because enzyme molecules are used over and over, relatively small amounts of it are needed to catalyze a reaction. Each cell in your body contains several thousand different enzymes and most of them are small or large proteins or protein complexes. Most enzymes are highly specific in their action, catalyzing only one type of chemical reaction or, at most, a small number of very similar reactions.

An enzyme lowers the activation barrier of a chemical reaction by binding tightly to the reactants and straining their chemical bonds in a way that promotes product formation. The specific reactants that bind to a particular enzyme are the **substrates** of that enzyme. Substrates bind to an enzyme in an orientation that favors the making and breaking of chemical bonds. Most enzymes have one or more pockets—known as active sites—that

(a)

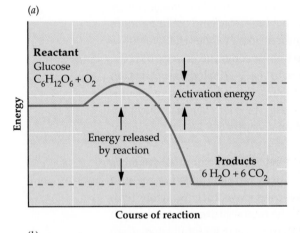

Figure 7.5 Enzymes Increase the Speed of Chemical Reactions by Reducing the Activation Energy Needed to Initiate a Reaction

(*a*) The oxidation of glucose is a spontaneous process because the reactants (glucose and oxygen) are at a higher energy level than the products (carbon dioxide and water), as represented by the solid red lines. However, on its own, this reaction would proceed very slowly, because under normal conditions (for example, moderate temperatures) not enough of the glucose and oxygen molecules collide with each other in the correct orientation for the reaction to take place at an appreciable rate. The minimum energy that is required for a reaction to get started is known as the activation energy for that reaction. The activation energy is recovered once products are formed. (*b*) An enzyme increases the speed of the specific reaction it catalyzes by reducing the activation energy needed to initiate the reaction. In the analogy shown here, the reactants are represented by water in the reservoir of a dam. On the left, the dam (representing the energy barrier) is so high that most of the waves (reactants) cannot spill over it. However, if the dam is lowered (as shown on the right), a large proportion of the waves (reactants) will overcome the barrier and run downhill (be turned into products). Enzymes lower the activation energy barrier, with the result that a much larger proportion of the reactants now have the energy to interact in the correct orientation and become converted into products.

(b)

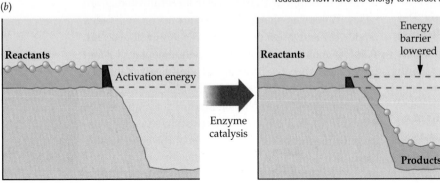

serve as docking points for one or more substrates. The size, geometry, and chemistry of an active site determine which reactant can bind to it, and this selectivity is the source of the substrate specificity of enzymes. One enzyme can bind only a particular reactant in the precise alignment necessary for product formation, which is why enzymes are specialists in terms of the reactions they catalyze. One enzyme will catalyze only one, or at most a few, of the thousands of possible reactions in a cell.

It is important to understand that an enzyme simply increases the rate, or speed, of a reaction that *could* occur on its own. As we have seen, an enzyme enhances the reaction speed by lowering the energy barrier that would otherwise make a spontaneous reaction extremely slow (Figure 7.5). An enzyme *does not* provide energy for a reaction that is thermodynamically uphill, that is, one that would create products at a higher energy state than the reactants. Because the enzyme is not altered by the chemical reaction in any way, an enzyme molecule can be used over and over again, which is why a small amount of enzyme can turn thousands of substrate molecules into product in less than a second.

Enzymes help us remove carbon dioxide from the body

Carbon dioxide is a by-product of cellular respiration, the process that helps us extract energy from food molecules. Excess carbon dioxide is a deadly poison for our cells, so it must be rapidly removed from body tissues. An enzyme called carbonic anhydrase helps us get rid of carbon dioxide. It speeds up the reaction of carbon dioxide with water molecules to make an ion, called bicarbonate (HCO_3^-), that dissolves readily in blood:

$$H_2O + CO_2 \xrightarrow{\text{carbonic anhydrase}} HCO_3^- + H^+$$

Bicarbonate circulating in the blood is carried to the lungs, where it is converted back to carbon dioxide, once again with the help of enzymes. The released carbon dioxide can no longer dissolve in the blood. It quickly diffuses into the air space in the lungs as CO_2 gas and is expelled when we breathe out.

The action of the enzyme carbonic anhydrase is a good example of the vital role played by enzymes in the proper functioning of our bodies. It speeds up the reaction of water and carbon dioxide by a factor of nearly 10 million. In fact, a single carbonic anhydrase molecule can process more than 10,000 molecules of carbon dioxide in a single second. Without it, carbon dioxide would

react with water so slowly that little of it would dissolve in the blood and we would not be able to rid our bodies of carbon dioxide fast enough to survive.

The shape of an enzyme determines its function

The binding of an enzyme to its particular substrate depends on the three-dimensional shapes of both the substrate and enzyme molecules. In the same way that a particular lock accepts only a key with just the right shape, each enzyme has an **active site** that fits only substrates with the correct three-dimensional shape and chemical characteristics (Figure 7.6a). The shape

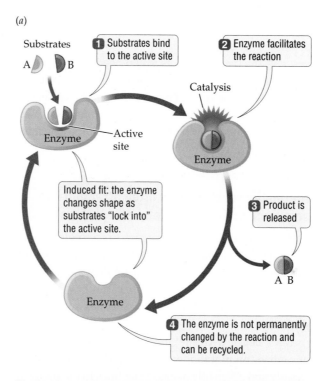

(a)

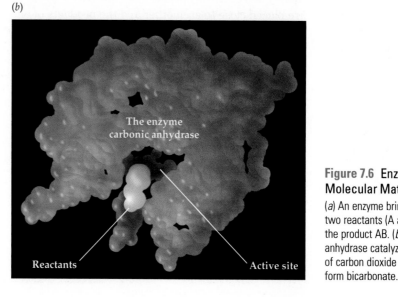

(b)

Figure 7.6 Enzymes as Molecular Matchmakers

(a) An enzyme brings together two reactants (A and B) to form the product AB. (b) Carbonic anhydrase catalyzes the reaction of carbon dioxide and water to form bicarbonate.

of an active site is relatively flexible and a substrate can "tweak" it further to create an even better match between it and the enzyme. According to the **induced fit model** of substrate–enzyme interaction, as a substrate enters the active site, the parts of the enzyme shift about slightly to allow the active site to mold itself around the substrate. This is similar to the way a limp glove (enzyme) takes on the shape of your hand (substrate) as you put it on. The ability of a substrate to induce a tighter and more accurate fit for itself in the active site of an enzyme stabilizes the interaction between the two and enables catalysis to proceed.

Learn more about enzyme catalysis.

▶❙❙

7.1

Because of its shape, carbonic anhydrase is able to bind both carbon dioxide and water in its active site. By bringing these two substrates together in just the right positions, the active site of carbonic anhydrase promotes the reaction (Figure 7.6b). If no enzyme were present, the two substrates would need to collide with each other in just the right way before the reaction could take place. These sorts of molecular collisions do occur, but not nearly as frequently as would be required for the continuous and rapid transfer of carbon dioxide from cells into the blood.

Metabolic reactions are often arranged in a sequence known as a metabolic pathway

So far, we have discussed the activity of a single enzyme acting alone to promote a single chemical reaction, but this state of affairs is not the most common in the cell. Typically, enzymes are involved in catalyzing steps in a sequence of chemical reactions known as a **metabolic pathway**.

Let's begin with the most noteworthy advantage of multi-step metabolic pathways: they can proceed rapidly and efficiently because enzymes in the pathway are placed close to each other and the products of one enzyme-catalyzed step serve as reactants for the next reaction in the series. In other words, the products of the first reaction are instantly available in large amounts, and in close proximity, to act as substrate for the next enzyme and are therefore rapidly processed by the second enzyme-catalyzed reaction. The net result is that a multi-step pathway of enzyme-catalyzed steps is "pushed" toward one specific outcome. Such a sequential arrangement of chemical reactions can be represented as follows:

$$A \xrightarrow{E1} B \xrightarrow{E2} C \xrightarrow{E3} D$$

The enzyme E1 catalyzes the conversion of A to B, enzyme E2 catalyzes the conversion of B to C, and so on, ensuring that D will be produced in the end. Metabolic pathways of this type are widely used in the production of key biological molecules, including those that generate important chemical building blocks of the cell, such as amino acids and nucleotides. The pathways that harness energy from food or sunlight are also organized in a multi-step sequence.

The challenge faced by all enzyme-catalyzed reactions is the need for the enzyme and its specific substrates to encounter each other often enough. Because enzymes in multi-step pathways are located close to one another in cells, the products of one reaction are in effect "aimed" at the next enzyme in the pathway. Another way of increasing the odds of encounters between an enzyme and its substrate is to contain and concentrate both of them inside a membrane-enclosed compartment, such as a mitochondrion (Figure 7.7). As we saw in Chapter 5, organelles concentrate the proteins and chemical compounds required for specific biological processes. Mitochondria, for example, are the sites where food molecules are oxidized, generating most of the cell's ATP in the process. The efficient production of ATP requires that the necessary enzymes and substrates be concentrated inside a small compartment. Enzymes present in the mitochondrial matrix participate in a sequence of reactions that produce large amounts of ATP. Other enzymes involved in the production of ATP are embedded in the inner mitochondrial membrane in a precise order.

At the molecular level, several enzymes can be physically connected in a single giant multienzyme complex (see Figure 7.7). This occurs in the biosynthesis of cellular building blocks, such as fatty acids and proteins, where many enzymes function as part of large assemblies of multiple enzymes.

Concept Check

1. What is an enzyme? Describe its characteristics.
2. How does an enzyme speed up chemical reactions within the cell?
3. What organizational features increase the speed and efficiency of enzyme-driven reactions in the cell?

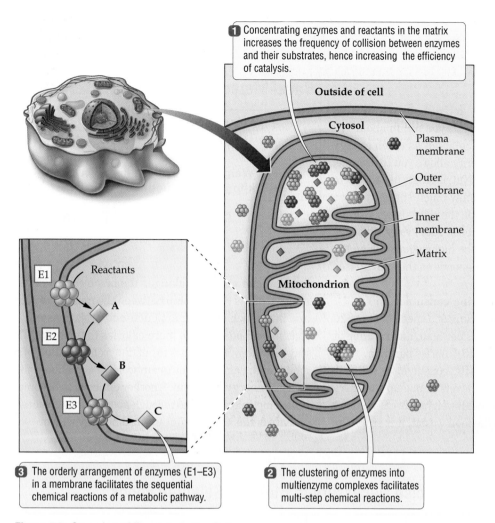

1 Concentrating enzymes and reactants in the matrix increases the frequency of collision between enzymes and their substrates, hence increasing the efficiency of catalysis.

Outside of cell

Cytosol

Plasma membrane

Outer membrane

Inner membrane

Matrix

Mitochondrion

Reactants

E1

A

E2

B

E3

C

3 The orderly arrangement of enzymes (E1–E3) in a membrane facilitates the sequential chemical reactions of a metabolic pathway.

2 The clustering of enzymes into multienzyme complexes facilitates multi-step chemical reactions.

Figure 7.7 Grouping of Enzymes in the Cell
Enzymes are often arranged in the cell in ways that facilitate the multiple chemical reactions of a metabolic pathway. These arrangements include the concentration of enzymes in organelles (in this case, a mitochondrion), their localization in membranes, and their clustering in multienzyme complexes.

7.4 Metabolic Rates and Life Span

The idea of immortality has fascinated human beings throughout recorded history. Today, researchers are seeking a scientific understanding of aging and the biological factors that limit the maximum life span attainable by an individual. Aging refers to the decline in cell and organ-level function over time, while life span is the maximum length of time an individual remains alive. Barring accidents or diseases caused by external factors, a slower rate of aging will increase life span. Researchers don't understand all the mechanisms underlying aging but it is clear that genes and lifestyle factors both have an impact on the rate at which we age. Numerous studies have identified metabolism as a key player influencing the rate of aging and therefore the maximum life span of an individual or species. In general, small animals have faster metabolic rates and shorter average life spans than larger animals. Furthermore, slowing down the metabolism of an animal can prolong its life span. For

Biology Matters

Energy Is a Weighty Subject

Energy lies at the center of all biological activity, even sustaining the seemingly simple act of staying alive. If you weigh about 140 pounds and are sitting still, you are using about as much energy per hour as a 75-watt light bulb. This energy allows you to breathe, to keep your heart beating, and to think. Nutritionists use the term *basal metabolic rate* (BMR) to describe how much energy we use each hour when we are at rest. The table on the right shows average BMRs for males and females. Our metabolic rates vary with how active we are. During any activity, it will be higher than during the resting rate, although how much higher depends on the activity (see the table below).

For a 150-pound person, brisk walking consumes about 350 Calories per hour. The same person bicycling at a speed of about 20 miles per hour burns around 720 Calories per hour, and running at a moderate pace burns approximately 846 Calories per hour.

If we take in more calories than our BMRs consume, we will gain weight, unless we increase our activity levels to use up the extra calories. Energy also lies at the center of all weight gain. If we don't use up all of the energy contained in the food we eat, it is stored as

Basal Metabolic Rates for Men and Women

	Weight in kilograms	Weight in pounds	Basal metabolic rate in watts per hour	Basal metabolic rate in Calories per hour
Woman	60	132	68	58
Man	70	154	87	75

fat. How can we avoid taking in too much energy? It is worth being mindful of the relationship between the number of calories in foods and how much energy particular activities consume. We all know that some foods contain more calories than others. Therefore, one way to avoid weight gain is to cut down on the intake of calorie-dense foods (high calorie count/gram weight) and choose foods that contain less energy per gram. Another option is to become more active. Balancing calorie intake (through good dietary choices) and physical activity is the key to maintaining a healthy weight.

Number of Minutes of Exercise Necessary to Burn Off Selected Foods

Exercise	Cheeseburger and fries	Large hamburger and fries	Fish and chips	Chicken and fries	Three pancakes with butter and syrup
Aerobics Active	88	114	74	102	67
Golf With cart	224	290	188	260	171
Dancing Energetic	102	132	85	118	78
Jogging 5 miles per hour	81	105	68	95	62
Swimming Steadily	81	105	68	95	62
Walking 3 miles per hour	136	176	114	158	104

NOTE: All figures are approximate, and are based on a 150-pound woman. A person who weighs more will burn more calories; a person who weighs less, will burn fewer.

Figure 7.8 Metabolism in the Lab
To measure the metabolism of lab rats, the animals are put in glass containers that measure the volume of air exhaled while they are fed sugar.

example, laboratory tests have shown that putting mice on a calorie-restricted diet slows down their metabolism (Figure 7.8) and that these mice live about 30 percent longer than mice that are allowed to eat as much as they want.

The link between a slower metabolism and a longer life holds true for another well-studied organism, the nematode worm *Caenorhabditis elegans* [KYE-noh-rab-DYE-tiss EL-eh-ganz]. Worms that lack proteins responsible for maintaining a normal metabolic rate, and that have an abnormally low metabolic rate as a result, mature more slowly and live up to five times longer than "normal" worms. The same phenomena has been observed in fruit flies, yeast, and monkeys as well, implying that the rate of metabolism is inversely related to overall life span in a wide variety of organisms. Does the relationship hold for humans? Women have slower metabolic rates than men, and they have a higher life expectancy (in the United States, 80.1 years versus 74.8 years for men). Even more striking is the fact that nine out of ten individuals 100 years old or older are women. Gender differences in susceptibility to chronic diseases and tendency toward risky behaviors account for some of the differences in the average life spans of men and women; for example, men have higher rates of heart disease and most cancers, and are more likely to smoke, abuse alcohol, commit suicide, or be victims of homicide. The innate rate of aging may also be lower in women, however, and may stem, at least in part, from their lower metabolic rates.

The idea that a higher metabolic rate can shorten one's life span may seem paradoxical, since metabolism encompasses all of the chemical reactions that *maintain* living organisms. How might metabolism shorten the life span of an organism? Free radicals generated by metabolic reactions appear to be partly responsible. Free radicals are toxic by-products that are inevitably generated in certain types of metabolic pathways—most notably, and in the largest quantity, during ATP production in the mitochondrion. Free radicals are highly reactive chemicals that damage critical cellular components, such as DNA and lipids, by oxidizing them. The gradual accumulation of this type of cellular damage is thought to contribute substantially not only to the rate of aging, but also to chronic diseases such as diabetes, heart and blood vessel disorders, and cancer. The net effect is a reduction in an individual's actual life span relative to the potential maximum life span.

Applying What We Learned

Making a Wonder Drug Even Better

Enzymes control the rates of specific chemical reactions in the body, so altering enzyme activity can have profound effects. Usually such effects cause dysfunction, but sometimes they promote healing or confer other health benefits on the complex metabolic machine that is the human body. The effects of aspirin, it turns out, are explained by its inhibitory action against two important enzymes, called COX-1 and COX-2. COX-1 is continuously produced in the body and catalyzes the biosynthesis of signaling molecules that help maintain the lining of the stomach. In contrast, COX-2 is produced in response to injury or stress and catalyzes the synthesis of a variety of signaling molecules that promote inflammation and fever, and activate pain-sensing nerves in the body. Although both enzymes are inhibited by aspirin, they participate in two very different pathways that have distinctly different functions. The therapeutic benefits of aspirin (reduction of pain, inflammation, and fever) are due to its inhibition of COX-2, while the negative side effects (damage to the stomach lining) come from the unavoidable blocking of COX-1 activity.

The inhibition of COX-2 activity may also explain aspirin's beneficial effects in reducing the risk of colon cancer. Some cancerous cells have an abnormally high level of COX-2, which appears to encourage the growth of blood vessels into the tumor. An enhanced blood supply nourishes a tumor and facilitates its spread through the bloodstream. By blocking COX-2 activity, aspirin may limit blood supply to tumors, restraining their growth and spread, and thereby limiting their destructive potential.

Is it possible to improve on a wonder drug, so its health benefits are retained or enhanced but its potential for negative side effects is reduced or eliminated? A number of drugs that selectively block COX-2 activity, with little or no effect on COX-1, have been developed. These "superaspirins" came about as researchers developed an understanding of the three-dimensional shape of the COX-2 enzyme, based on aspirin's interaction with this enzyme. Recall that the catalytic activity of an enzyme rests crucially on the shape of its active site. Knowing the shape of the COX-2 active site allowed researchers to design inhibitor molecules that would fit into the COX-2 active site but not into the COX-1 active site. Over a dozen such COX-2 selective anti-inflammatory drugs were developed and marketed as strong painkillers that were gentler on the stomach than aspirin. In recent years, however, these drugs have been shown to increase blood pressure and damage blood vessels in long-term users, thereby increasing the risk of heart attack and stroke, especially among those already at risk for these diseases. Vioxx was the first of the COX-2 inhibitors to be banned and today only one, Celebrex, remains available by prescription. The unanticipated side effects underscore the complexity of enzyme-catalyzed pathways in diverse tissues and organs and the value of improving our understanding of such pathways.

Biology in the News

Doctors Warned about Common Drugs for Pain; NSAIDs Tied to Risk of Heart Attack, Stroke

By Shankar Vedantam

Doctors treating people for chronic pain should avoid using all medications—at least at first—the American Heart Association advised yesterday in guidelines designed to have a significant impact on the use of medications known as nonsteroidal anti-inflammatory drugs, or NSAIDs [which includes aspirin, acetaminophen in Tylenol®, and ibuprofen in Advil®] . . .

The statement expressed particular concern over a subgroup of these drugs known as Cox-2 inhibitors. The only drug in this group currently on the market in the United States is Celebrex.

The professional association laid out a step-by-step approach that is very different from the way physicians typically have approached treating chronic pain and inflammation . . .

Patients should be treated first with nonmedicinal measures such as physical therapy, hot or cold packs, exercise, weight loss, and orthotics [supportive devices, such as knee braces] before doctors even consider medication, said the AHA scientific statement published in the journal *Circulation*.

"In general, the least risky medication should be tried first, with escalation only if the first medication is ineffective. In practice, this usually means starting with acetaminophen or aspirin at the lowest efficacious dose, especially for short-term needs."

Sometimes an exciting medical advance is later found to be associated with serious side effects. This appears to be true of some of the COX-2 inhibitors, painkillers initially heralded as "superaspirins." Aspirin blocks both COX-1 and COX-2, and its inactivation of the beneficial action of COX-1 has always been a concern for long-term users. At first glance, therefore, the development of COX-2 selective inhibitors seemed a major step forward in pain relief. The hope was that people suffering chronic pain due to diseases such as arthritis would get pain relief without being exposed to the negative side effects of aspirin, such as bleeding in the stomach. However, long-term studies with large numbers of participants have revealed that regular use of certain COX-2 inhibitors substantially raises the risk of serious heart problems, and some deaths have been linked to the use of the drugs. Pharmacologists have returned to the drawing board and are attempting to tweak the shape and chemistry of COX-2 inhibitors to reduce the risk of side effects. In 2007, Merck introduced the first of this new generation of COX-2 inhibitors (dubbed "son of Vioxx" by the press, Vioxx being the first COX-2 inhibitor to be banned). However, safety concerns kept the drug from getting approval from the FDA (Food and Drug Administration), the federal agency charged with ensuring that new medications are safe and effective.

In the United States, it is the responsibility of the FDA, pharmaceutical companies, pharmacies, and physicians to inform us to the best of their ability about the risks and benefits of prescription drugs and over-the-counter medications. Although regulatory mechanisms are in place to protect us from, or warn us about, potential harmful effects of medication, we strengthen those defenses by becoming better-informed consumers of medicines and other health-related products. This is especially true of health supplements, such as herbal remedies, produced by a billion-dollar industry that is largely unregulated, and whose health claims are usually not based on rigorous scientific evidence.

Evaluating the News

1. In what way are COX-2 inhibitors similar to aspirin and in what way are they different?
2. Pharmaceutical companies are required by law to inform users about possible side effects of approved drugs, and to include announcements about them in TV commercials. How would you respond to a lobbyist who argued for the removal of mandatory announcements about drug side effects because they are distracting, expensive, and "no one listens anyway"?
3. Some people insist that herbal remedies and health supplements cannot have harmful effects because these substances are "natural." Is that a reasonable argument? Explain your viewpoint, citing real examples. (*Hint*: There's one in the opening vignette of this chapter.)

Source: *Washington Post*, February 27, 2007.

Chapter Review

Summary

7.1 The Role of Energy in Living Systems

- Living organisms obey the laws of thermodynamics that apply to the rest of the physical world.
- Consistent with the first law of thermodynamics, energy used by organisms is converted from one form to another, but is never created or destroyed.
- As predicted by the second law of thermodynamics, the creation of biological order is always accompanied by the transfer of disorder to the environment, generally in the form of heat.
- The sun is the ultimate energy source for most living organisms.
- Atomic building blocks such as carbon are cycled between living organisms and the environment.

7.2 Using Energy from the Controlled Combustion of Food

- In oxidation, electrons are lost from a molecule, atom or ion. In reduction, electrons are gained by a molecule, atom, or ion.
- Metabolism consists of two kinds of reactions: catabolic reactions, which break down macromolecules and release energy, and anabolic reactions, which build macromolecules and require energy.
- Most chemical reactions must overcome an activation energy barrier to get started.

7.3 Cells Use Enzymes to Speed Up Chemical Reactions

- Catalysts speed up chemical reactions by lowering the amount of activation energy required. They are not consumed in the reaction.
- Most chemical reactions that support life are catalyzed by enzymes. Most enzymes are proteins.
- The activity of enzymes is highly specific. Each enzyme binds to a specific substrate or substrates and catalyzes a specific chemical reaction.
- The specificity of an enzyme is based on the three-dimensional shape and chemical characteristics of its active site.
- Each of the sequential steps in a metabolic pathway is catalyzed by a different enzyme.

7.4 Metabolic Rates and Life Span

- Life span may be determined by the overall metabolic rate of an organism, in combination with other genetic factors.
- Higher metabolic rates lead to greater production of free radicals, chemically reactive by-products that can damage cells and shorten life span.

◉ Review and Application of Key Concepts

1. Describe the role of the second law of thermodynamics in living systems.

2. Describe carbon cycling in relation to photosynthesis, the cellular respiration of sugars, and the burning of fossil fuels.

3. Compare anabolism and catabolism. Is photosynthesis an anabolic or catabolic process? Explain.

4. Cells use several methods to increase the efficiency of enzyme catalysis in metabolic pathways. Describe two of these methods and how they apply to mitochondria.

5. Explain the induced fit model of interaction between an enzyme and its substrate.

Key Terms

activation energy (p. 149)
active site (p. 151)
anabolic reaction (p. 147)
biosynthetic reaction (p. 147)
catabolic reaction (p. 147)
catalyst (p. 150)
coupled reactions (p. 148)
enzyme (p. 150)
first law of thermodynamics (p. 144)
induced fit model (p. 152)
kinetic energy (p. 144)

metabolic pathway (p. 152)
metabolism (p. 147)
oxidation (p. 147)
potential energy (p. 144)
product (p. 149)
reactant (p. 149)
redox reaction (p. 147)
reduction (p. 147)
second law of thermodynamics (p. 144)
substrate (p. 150)

Self-Quiz

1. Which of the following statements is true?
   a. Cells are able to produce their own energy from nothing.
   b. Cells use energy only to generate heat and move molecules around.
   c. Cells obey the same physical laws of energy as the nonliving environment.
   d. Most animals obtain energy from minerals to fuel their metabolic needs.

2. Living organisms need energy to
   a. organize chemical compounds into complex biological structures.
   b. decrease the disorder of the surrounding environment.
   c. cancel the laws of thermodynamics.
   d. keep themselves separate from the nonliving environment.

3. The carbon atoms contained in organic compounds such as proteins
   a. are manufactured by cells for use in the organism.
   b. are recycled from the nonliving environment.
   c. differ from those found in CO_2 gas.
   d. cannot be oxidized under any circumstances.

4. Oxidation is the
   a. removal of oxygen atoms from a molecule.
   b. gain of electrons by an atom.
   c. loss of electrons by an atom.
   d. synthesis of complex molecules.

5. Which of these molecules is in a reduced state?
   a. CO_2
   b. N_2
   c. O_2
   d. CH_4

6. The minimum input of energy that initiates a chemical reaction
   a. is called activation energy.
   b. is independent of the laws of thermodynamics.
   c. is known as the activation barrier.
   d. always takes the form of heat.

7. Activation energy is most like
   a. the energy released by a ball rolling down a hill.
   b. the energy required to push a ball from the bottom of a hill to the top.
   c. the energy required to get a ball over a hump and onto a downward slope.
   d. the energy that keeps a ball from moving.

8. Enzymes
   a. provide energy for anabolic but not catabolic pathways.
   b. are consumed during the reactions that they speed up.
   c. catalyze reactions that would otherwise never occur.
   d. catalyze reactions that would otherwise occur much more slowly.

9. The active site of an enzyme
   a. has the same shape for all known enzymes.
   b. can bind both its substrate and other kinds of molecules.
   c. does not play a direct role in catalyzing the reaction.
   d. can bring molecules together in a way that promotes a reaction between them.

10. Metabolic pathways
    a. always break down large molecules into smaller units.
    b. only link smaller molecules together to create polymers.
    c. are often organized as a multi-step sequence of reactions.
    d. occur only in mitochondria.

CHAPTER **8** # Photosynthesis and Cellular Respiration

Key Concepts

◉ The storage and transfer of usable energy in the cell requires the production of energy-carrier molecules, such as ATP.

◉ Photosynthesis is a series of chemical reactions that use sunlight and water to produce energy-carrier molecules that in turn are used to make sugars from carbon dioxide. The process releases oxygen gas into the environment.

◉ Photosynthesis occurs in two main steps: the first step captures light to produce energy carriers, generating oxygen as a by-product; the second step produces sugars from carbon dioxide and water, using energy carriers produced in the first step.

◉ Most eukaryotes rely on cellular respiration, a process that requires oxygen (O_2) to extract energy from sugars and other food molecules.

◉ Cellular respiration has three main stages: glycolysis occurs in the cytoplasm; the Krebs cycle and oxidative phosphorylation take place in the mitochondrion.

◉ Glycolysis generates a few energy carriers for each sugar molecule degraded.

◉ Carbon dioxide is released when sugars are further broken apart in the Krebs cycle. Many molecules of ATP are generated through oxidative phosphorylation, an O_2-dependent process.

◉ Fermentation allows certain organisms, and certain cell types, to make ATP when oxygen is absent or in low supply.

Food for Thought

The next time you feel hungry after skipping a meal, keep in mind that your brain consumes nearly 25 percent of the energy needed by your body when you are doing nothing but sitting and thinking. If you ignore those hunger pangs, and especially if you are very thin, you might feel dizzy or light-headed, which is a warning signal from the brain that your blood sugar levels are too low. Glucose, a six-carbon sugar, is the most common form in which energy is transported within the animal body, and maintaining an optimal glucose level in the bloodstream is a vital balancing act for your body.

A large brain is one of the defining features of the human species, *Homo sapiens*. Brain weight increases roughly with increasing body weight in vertebrates. Whales, the largest of the animals, also have the largest brain, ranging from about 4,000 to 7,000 grams (g). Elephants (4,000–5,000 g) and dolphins (about 1700 g) are next in brain size. The human brain commonly ranges from 1300 to 1700 g.

Although other animals, such as whales, certainly have larger brains by weight, the human brain is the largest when compared with the size of the human body. In other words, humans have the highest ratio of brain to body weight, which contributes to our status as the most intelligent species.

A daily challenge of having a large brain is the urgent need to supply it with energy. Our brains consume a large amount of energy while receiving, processing, and sending nerve impulses. The energy demands of a developing brain are even higher: about 60 percent of the nourishment consumed by an infant is used by its brain. **Given the tremendous energy demands of a large brain, how did the human brain evolve? Could a dietary switch have helped that process along?** We will explore these questions later in the chapter, but first we must consider how organisms capture and use energy. Animals like us eat plants, or animals that live on plants, to gain energy. Plants and other photosynthetic organisms use sunlight to manufacture sugars, and they use the chemical energy in those sugars to manufacture all the other substances that they need to survive. The chemical reactions that allow plants to make food molecules, and all types of organisms to extract energy from food, are the main focus of this chapter.

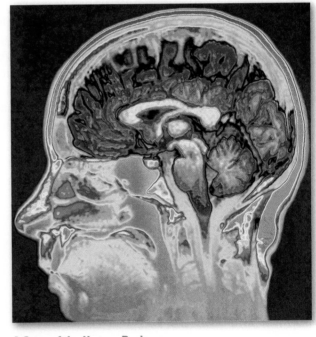

A Scan of the Human Brain

Given all the complex structures and chemical reactions in every living cell, obviously some kind of fuel must power these processes. Metabolism refers to the capture, storage, and utilization of energy in a cell, and it is a fundamental necessity for every living being. In most ecosystems on earth, the sun is the ultimate source of energy for living cells. Solar energy is used by plants and other photosynthetic organisms to make sugars from carbon dioxide and water. Photosynthetic organisms are producers, organisms that trap energy from their external environment to manufacture their own food. Consumers, in contrast, are organisms that acquire energy by eating producers or other consumers. Producers and consumers must both be able to extract energy stored in food molecules.

In this chapter we will begin our exploration of metabolism by looking at the critical role played by energy carriers, which function as an energy delivery service inside the cell. In the remainder of the chapter we will discuss two of the most important and widespread metabolic pathways on Earth: photosynthesis, which captures light energy to make sugars and other food molecules from carbon dioxide and water; and cellular respiration, which releases energy from food molecules to fuel cellular activities. We will see that photosynthesis releases oxygen, a by-product that is crucial for the survival of consumers like us because it is necessary for cellular respiration (Figure 8.1). While photosynthesizers *use up* carbon dioxide and water to make food consumers *release* these molecules as they extract energy from food.

8.1 Energy Carriers Power Cellular Activities

Energy carriers are molecules specialized for receiving, storing, and delivering energy within the cell. They receive energy from pathways in which energy is released, and they deliver energy to the many thousands of chemical reactions and cellular activities that could not proceed without an investment of energy. A striking feature of these energy carriers is that they are nearly universal throughout the biosphere: the energy carriers we will discuss in some detail in this chapter—ATP, NADH, NADPH—are used in every organism we know of. Each of these energy carriers is a specialist in terms of the amount of energy it carries and the types of chemical reactions it delivers to and takes delivery from.

Of the common energy carriers in a cell, **ATP** (adenosine triphosphate) carries the smallest load of energy but it is the most versatile: it delivers energy to the largest number and greatest diversity of chemical processes in the cell. ATP stores energy in the form of covalent bonds between its three phosphate groups. The addition of a phosphate group to a molecule is called **phosphorylation** [*FOSS*-fore-uh-*LAY*-shun]. ATP (adenosine <u>tri</u>phosphate) is made by the phosphorylation of ADP (adenosine <u>di</u>phosphate). ATP releases its stored energy when it loses one, or two, of its phosphate groups, to become ADP (adenosine <u>di</u>phosphate) or AMP (adenosine <u>mono</u>phosphate). In some cases, ATP actu-

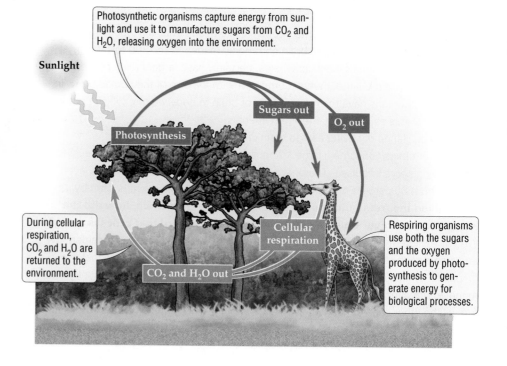

Photosynthetic organisms capture energy from sunlight and use it to manufacture sugars from CO_2 and H_2O, releasing oxygen into the environment.

Sunlight

Sugars out

O_2 out

Photosynthesis

During cellular respiration, CO_2 and H_2O are returned to the environment.

CO_2 and H_2O out

Cellular respiration

Respiring organisms use both the sugars and the oxygen produced by photosynthesis to generate energy for biological processes.

Figure 8.1 The Flow of Energy between Organisms and Their Environment

Sunlight is the ultimate source of energy in most ecosystems on earth. Photosynthetic organisms (the most common type of producers) convert inorganic carbon dioxide into organic molecules (sugars), using energy absorbed from the sun. They also release oxygen as a by-product of photosynthesis. All types of organisms, both producers and consumers, extract energy from organic molecules using an oxygen-dependent metabolic pathway known as cellular respiration.

ally transfers its terminal phosphate group to another molecule (phosphorylates it), energizing the recipient molecule and enabling it to participate in some energy-dependent cellular activity.

The energy carriers **NADPH** and **NADH** hold energy in the form of electrons and hydrogen atoms that they can deliver to various chemical reactions. NADPH is involved in metabolic pathways that build macromolecules (**anabolism**), while NADH picks up and delivers energy in metabolic pathways that take macromolecules apart (**catabolism**). NADPH and NADH are derived from low-energy precursor molecules. NADP$^+$ (nicotinamide adenine dinucleotide phosphate) is the precursor to NADPH, and NAD$^+$ (nicotinamide adenine dinucleotide) is the precursor to NADH. Each of these precursors can pick up two high-energy electrons along with a hydrogen ion (H$^+$) and be transformed into the high-energy form, NADPH and NADH, respectively.

In chemical terminology, the acceptance of electrons, or a hydrogen atom, is a reduction; therefore, NADPH and NADH are the reduced forms of NADP$^+$ and NAD$^+$. NADPH and NADH are *reducing agents* because they can donate two electrons and one hydrogen ion to another compound, reducing it in the process. The loss of electrons or hydrogen atoms is known as an oxidation. (Loss of a hydrogen atom is equivalent to loss of one electron plus a hydrogen ion.) NADP$^+$ and NAD$^+$, are *oxidizing agents* because they remove electrons and hydrogen ions from other substances, oxidizing them in the process. NADPH and NADH are oxidized to NADP$^+$ and NAD$^+$ when they lose electrons and hydrogen atoms. (For a review of oxidation and reduction, see Chapter 7.)

8.2 An Overview of Photosynthesis and Cellular Respiration

Photosynthesis and cellular respiration are two of the most important metabolic pathways in living organisms. Producers carry out photosynthesis, but their cells also respire. Consumers and decomposers, on the other hand, only respire. **Photosynthesis** is an anabolic pathway and in eukaryotes it is conducted in organelles called **chloroplasts**, which are bounded by a double membrane. Pigments, especially the green pigment known as chlorophyll, are embedded in an internal system of membrane sacs inside the two outer membranes of the chloroplast. When sunlight strikes chlorophyll, electrons are energized and knocked out of the pigment molecules. The electrons are transferred, through a series of steps, to NADP$^+$, which picks up hydrogen ions from its surrounding medium to become NADPH. The electron ejected from chlorophyll loses some of its energy as it is handed down a series of proteins and pigments, and some of that energy is harnessed in the phosphorylation of ADP to ATP. The energy carriers generated by these "light reactions" then fuel the reduction of carbon dioxide through a series of enzyme-catalyzed reactions, known as the Calvin cycle, to produce sugars (Figure 8.2). What about the chlorophyll molecules that had electrons ripped out of them when they were struck by light? Enzymes associated with the membrane sacs help catalyze the breakdown of water molecules (H$_2$O), which releases electrons, hydrogen ions, and molecular oxygen

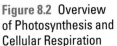
See an overview of photosynthesis. 8.1

Figure 8.2 Overview of Photosynthesis and Cellular Respiration

This diagram is a broad overview of the main inputs and outputs of photosynthesis and cellular respiration and the relationship between them. The two metabolic pathways are complementary: chloroplasts capture energy from sunlight and use it to synthesize sugar through the chemical reactions of photosynthesis; in cellular respiration, reactions that take place in the cytoplasm and mitochondria extract energy from sugars to synthesize ATP, an energy carrier.

PHOTOSYNTHESIS
(In chloroplasts)

CELLULAR RESPIRATION
(In mitochondria)

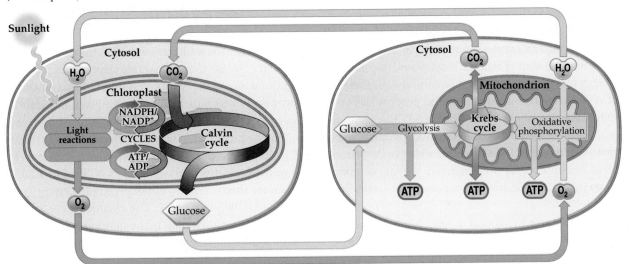

(O_2). The electrons "repair" the "holes" in the chlorophyll molecules, the hydrogen ions are used in the generation of energy carriers, such as NADPH, and as for the molecular oxygen . . . we're inhaling it right now.

Cellular respiration is an oxygen-dependent catabolic pathway through which sugars, lipids, and other organic molecules are broken down, and the released energy used to generate energy carriers, mainly ATP. Oxygen is necessary for the complete breakdown of sugars and other compounds with a carbon backbone. The carbon from the broken macromolecules is released as carbon dioxide. We refer to this process as *cellular* respiration to distinguish it from *whole body respiration,* which, in the case of animals with lungs, involves inhaling and exhaling of air (breathing). Breathing in animals, including us, is directly related to cellular respiration. The air we breathe out is rich in carbon dioxide, a by-product of cellular respiration. We must inhale air rich in oxygen because the gas is essential for the complete oxidation of carbon backbones through cellular respiration.

Cellular respiration, the three-phase oxidation of carbon backbones, begins in the cytoplasm and is completed in the mitochondrion, an organelle bounded by a double membrane. In the cytoplasmic phase, known as **glycolysis** [glye-*KOLL*-uh-siss], sugars (mainly glucose) are broken into a three-carbon compound (pyruvate), releasing ATP and NADH as energy carriers. Pyruvate enters the mitochondrion and is completely oxidized through a sequence of enzyme-driven reactions known as the Krebs cycle. The carbon backbone of pyruvate is taken apart, releasing carbon atoms that combine with oxygen atoms to make carbon dioxide. The carbon dioxide you are exhaling into the air right now originated in the carbon backbones of the food you ate, following the complete dismantling of those food molecules by the Krebs cycle reactions in the many hundreds of mitochondria inside the many trillions of cells in your body. The degradation of carbon backbones by the Krebs cycle produces energy carriers, particularly NADH. In the last and final step of cellular respiration, the chemical energy of NADH is converted into the chemical energy of ATP through a membrane-based process known as **oxidative phosphorylation** (Figure 8.1). Electrons and hydrogen ions removed from NADH are handed over to molecular oxygen (O_2 gas), creating water. In the process, large amounts of ATP are generated. We cannot survive more than a few minutes without oxygen because, in the absence of this gas, our cells cannot make enough ATP to run the many activities that rely on energy delivered by this extraordinary energy carrier.

Most cells in multicellular organisms rely on cellular respiration for their ATP needs. However, many single-celled organisms, and some cell types in multicellular organisms, can make do with a glycolysis-based catabolic pathway known as **fermentation**. This pathway consists of glycolysis, followed by a series of special reactions that process pyruvate, the end-product of glycolysis. These post-glycolysis reactions do not generate any additional energy beyond the ATP and NADH made by glycolysis. The post-glycolysis reactions are essential, however, because they help sustain glycolysis when oxygen is absent. Fermentation yields less than a tenth of the energy that cellular respiration releases from each sugar molecule.

In purely chemical terms, photosynthesis is the reverse of cellular respiration, and vice versa. Both pathways generate energy carriers, but in photosynthesis much of that output is used up in the manufacture of sugars. Photosynthesis can be summarized in general terms as shown below.

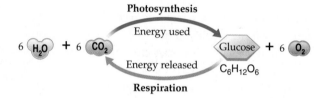

Concept Check

1. Name the three small energy carriers that are found in all organisms.
2. Compare photosynthesis and cellular respiration.
3. How is fermentation different from cellular respiration?

8.3 Photosynthesis: Capturing Energy from Sunlight

The next time you walk outside, look at the plants around you and consider the critical role they play in supporting the web of life. Plants and other photosynthetic organisms, which include some protists and bacteria, are able to capture energy from sunlight and store it in the chemical bonds of food molecules. The process of photosynthesis, as we have seen, uses solar energy to generate energy carriers such as ATP and NADPH, which are then used to synthesize complex, energy-rich molecules such as sugars. The chemical reactions of photosynthesis also result in the splitting of water (H_2O) and the release of oxygen gas (O_2) into the environment

(Figure 8.1). Plants and other photosynthesizers support humans and virtually all other organisms, which depend on them for both food and oxygen.

Chloroplasts are photosynthetic organelles

As mentioned earlier, photosynthesis in plants and protists takes place inside chloroplasts. The distinctive arrangement of membranes within the chloroplast is crucial for the special reactions of photosynthesis, especially light capture and the production of energy carriers. Like mitochondria, chloroplasts are enclosed by double membranes, which divide the organelle into compartments (Figure 8.3). The outermost compartment, between the two membranes, is called the **intermembrane space**; the second compartment, enclosed by the inner membrane, is called the **stroma**. Inside the stroma is a network of membrane-enclosed sacs that are derived from the inner membrane. It consists of groups of flattened, interconnected membrane sacs, called **thylakoids**, which lie one on top of another in stacks called grana. Each thylakoid is formed by a **thylakoid membrane** and encloses a **thylakoid space**.

Photosynthesis consists of two sets of reactions. The first set of reactions takes place in the thylakoid membrane and directly captures energy from sunlight; because these reactions have an absolute requirement for light, they are collectively known as the **light reactions**. The thylakoid membrane is studded with light-absorbing pigments, including chlorophyll, and a variety of proteins and other compounds needed to transform solar energy into the chemical energy of ATP and NADPH. The second set of reactions takes place in the stroma, a fluid medium containing many enzymes, ions and molecules, which collectively serve as the machinery for synthesizing sugars from carbon dioxide and water. The sugar-manufacturing reactions are know collectively as the **Calvin cycle**, after Melvin

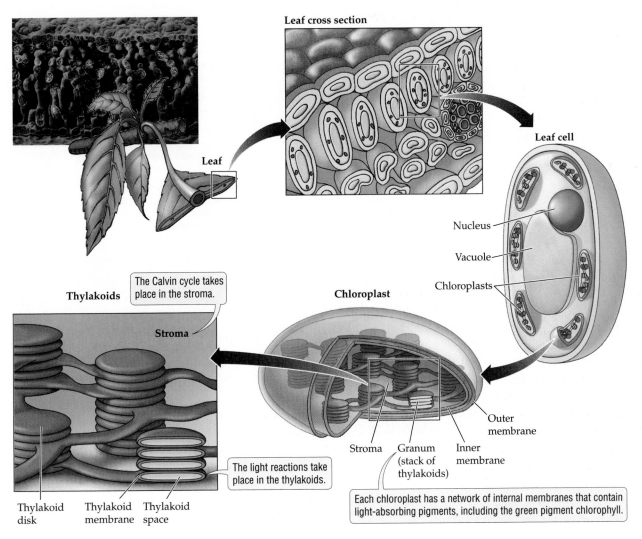

Leaf cross section

Leaf

Leaf cell

Nucleus

Vacuole

Chloroplasts

Thylakoids

The Calvin cycle takes place in the stroma.

Stroma

Chloroplast

The light reactions take place in the thylakoids.

Thylakoid disk   Thylakoid membrane   Thylakoid space

Stroma   Granum (stack of thylakoids)   Inner membrane   Outer membrane

Each chloroplast has a network of internal membranes that contain light-absorbing pigments, including the green pigment chlorophyll.

Figure 8.3 Chloroplasts in a Leaf Cell

Chloroplasts, the highly organized organelles in which photosynthesis occurs, are abundant in the cells that make up the bulk of the leaf.

Calvin, who won the Nobel Prize in 1961 for describing the main steps in this pathway. Together light reactions and the Calvin cycle form the foundation of life, not only for humans but also for virtually all other life forms on Earth, including plants themselves.

The light reactions use light to generate energy carriers

The capture of energy from sunlight is essential to photosynthesis and to life throughout the biosphere. In this process, solar energy is converted to the energy contained in the electrons and chemical bonds of organic molecules. Photosynthesis is carried out by specialized pigments, the most important being **chlorophyll** (from *chloro*, "greenish-yellow"; *phyll,* "leaf"), which accounts for the green color of most plant foliage. As rainbows reveal so dramatically, sunlight is composed of many colors. Plants absorb and use mostly red-orange and blue-violet light. Because these pigments do not absorb green light to any appreciable extent, green light bounces off photosynthetic tissues so they look green to us.

The thylakoid membrane is densely packed with disclike arrangements of pigment–protein complexes. Each disclike grouping is known as an **antenna complex** (Figure 8.4) and contains different types of pigments, including the yellow-orange pigments called carotenoids. Carotenoid pigments are readily seen in plant parts when chlorophyll is absent, as it is in vegetables such as carrots, yellow peppers, and corn, and in the bright yellow leaves we see in the fall (other fall colors come from water-soluble pigments stored in the vacuole). The antenna complex captures solar energy and funnels it to an enzyme–chlorophyll complex known as the **reaction center**, where light reactions are initiated. Electrons associated with certain chemical bonds in a chlorophyll molecule become more energized when they absorb light. Each energized electron, said to be in an "excited" state, is then handed over to an **electron transport chain** (**ETC**), a series of electron-accepting

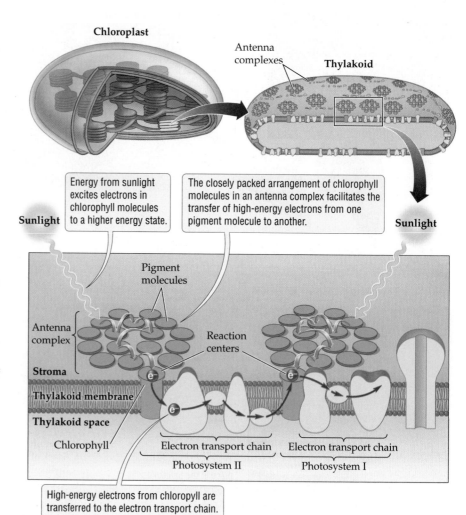

Figure 8.4 Each Photosystem Consists of an Antenna Complex and an Electron Transport Chain

The precise arrangement of molecules in the thylakoid membrane facilitates the transfer of energy from the antenna complex, which contains pigments such as chlorophyll (light green), to the proteins and other molecules that make up the electron transport chains (purple). There are two types of photosystems in plant chloroplasts, photosystems II and I, which function in the sequence shown here.

Chloroplast

Antenna complexes

Thylakoid

Energy from sunlight excites electrons in chlorophyll molecules to a higher energy state.

The closely packed arrangement of chlorophyll molecules in an antenna complex facilitates the transfer of high-energy electrons from one pigment molecule to another.

Sunlight

Sunlight

Pigment molecules

Antenna complex

Reaction centers

Stroma

Thylakoid membrane

Thylakoid space

Chlorophyll

Electron transport chain

Electron transport chain

Photosystem II

Photosystem I

High-energy electrons from chloropyll are transferred to the electron transport chain.

proteins located next to one another and embedded in the thylakoid membrane. (ETCs are found in a number of metabolic processes that involve the transfer of energy, as we shall see later in this chapter.) As electrons are passed down the ETC from one protein to another, small amounts of energy are released and used to generate the energy carriers ATP and NADPH.

The combination of an antenna complex with a neighboring ETC is called a **photosystem**. Each thylakoid has many photosystem units; indeed, they make up more than half the thylakoid membrane by weight. Photosystems come in two distinct types, photosystem I and II (Figure 8.4). The numbering of the two photosystems reflects the order in which each system was discovered by researchers, not the order of steps during photosynthesis. **Photosystem II** is associated with the splitting of water (*photolysis*) and generates electrons, O_2, and hydrogen ions (Figure 8.5a). **Photosystem I** is primarily responsible for the production of the powerful reducing agent NADPH (Figure 8.5b).

In most plant cells, the two photosystems operate in tandem to produce the main outputs of the light reactions: ATP, NADPH, and O_2. Electrons ejected from the photosystem II reaction center pass down the ETC. The electrons are eventually handed over to photosystem I, which then donates them to NADP$^+$ to generate NADPH. Each reaction center in photosystem II makes up its electron deficit by removing electrons from water, thereby catalyzing the splitting of water molecules to produce hydrogen ions and O_2 (Figure 8.5a). We will examine this overall scheme in greater detail next.

The transfer of high-energy electrons is at the heart of the light reactions. The journey begins at the reaction center of photosystem II. Absorbed light energy ejects excited state electrons from a photosystem II reaction center chlorophyll (Figure 8.5a). The high-energy electrons are picked up by the first component of the ETC, which transfers it to the next component in the chain, and so on. As the electrons travel down the ETC, they lose energy and some of that energy is used to drive the transport of hydrogen ions (protons) across the thylakoid membrane. A channel protein uses the energy released during electron transfer to move protons (H$^+$) from the stroma into the thylakoid space (Figure 8.5c). As protons accumulate inside the thylakoid space, their concentration builds up relative to the hydrogen ion concentration in the stroma, creating a **proton gradient** (an imbalance in the proton concentration), across the thylakoid membrane.

Pumping of ions to create a concentration gradient is a common means of harnessing energy for cellular processes. In this case, the gradient is used to manufacture

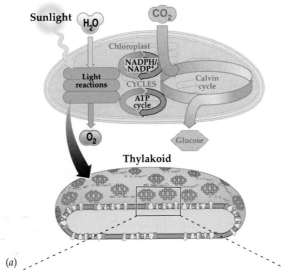

Figure 8.5 Production of Energy Carriers by the Light Reactions
Photosystems embedded in the thylakoid membrane capture light energy and use it to generate NADPH and ATP. Light absorbed by photosystem II strips electrons from H_2O, generating O_2 as a by-product (*a*). As electrons travel down the ETC, some of their energy is harnessed to create a proton gradient (*b*). The electrons are eventually transferred to NADP$^+$ to make NADPH. The H$^+$ gradient drives ATP production (*c*).

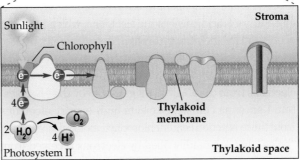

(*a*)

(*b*)

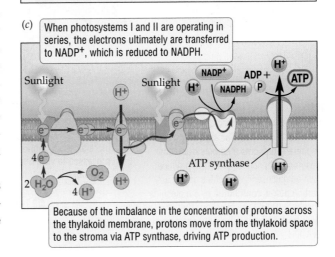

(*c*)

ATP. As we saw in Chapter 6, all dissolved substances, including protons, tend to move from a region of higher concentration to one of lower concentration. So the protons in the thylakoid space have a spontaneous tendency to move back down the proton gradient to the stroma. Since the thylakoid membrane will not allow protons to pass through it, the only way for them to do so is through a large channel-containing protein complex, called ATP synthase, that spans the thylakoid membrane. As protons rush through the ATP synthase channel, the energy of their concentration gradient is converted into chemical energy as enzymes associated with the complex catalyze the phosphorylation of ADP to ATP (Figure 8.5c).

With ATP production accounted for, let us return to tracking the excited electrons in their journey down the ETC. The last component in the ETC of photosystem II (see Figure 8.5c) transfers the electrons, which have lost energy during their travel through photosystem II, to chlorophyll molecules in the reaction center of photosystem I. Why are these chlorophyll molecules so quick to accept electrons from the ETC? Light gathered by the photosystem I antenna complexes ejects high-energy electrons from these reaction center chlorophyll molecules, creating "holes" that are readily filled by electrons received from the ETC. But what becomes of the electrons knocked out

of the photosystem I reaction center? They pass through the ETC associated with photosystem I and are then donated to NADP⁺. The transfer of two electrons to NADP⁺ gives it a net negative charge, so it takes up H⁺ from the stroma to make NADPH (see Figure 8.5b). Realize that these electrons and protons that are picked up by NADP⁺ come ultimately from water molecules: the electrons traveling down the ETC of photosystem II were initially ejected from the reaction centers of this photosystem, which replaced them by extracting electrons from water molecules to form hydrogen ions and O_2 (see Figure 8.5a). In sum, the two photosystems are arranged so that they must synchronize perfectly to generate ATP and NADPH, making oxygen as a by-product.

The Calvin cycle reactions manufacture sugars

The energy carriers produced by the light reactions—ATP and NADPH—are used in the Calvin cycle reactions. The Calvin cycle is a series of enzymatic reactions that take place in the stroma of the chloroplast and synthesize sugars from carbon dioxide and water (Figure 8.6). This process, also known as **carbon fixation**, illustrates the interconnectedness between life forms and their environment. By capturing inorganic

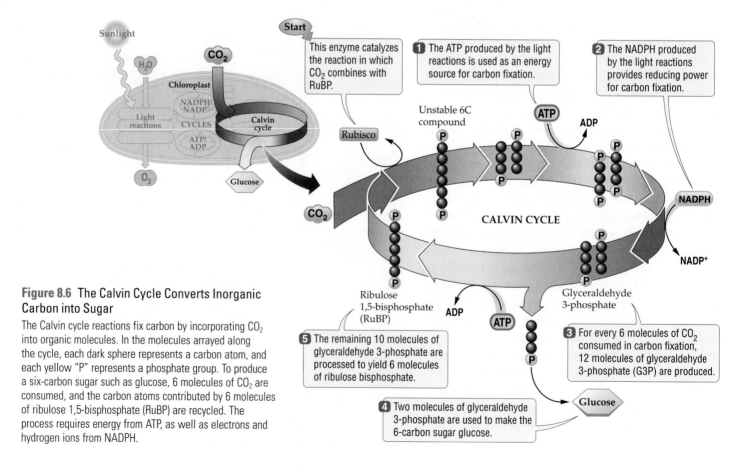

Figure 8.6 The Calvin Cycle Converts Inorganic Carbon into Sugar

The Calvin cycle reactions fix carbon by incorporating CO_2 into organic molecules. In the molecules arrayed along the cycle, each dark sphere represents a carbon atom, and each yellow "P" represents a phosphate group. To produce a six-carbon sugar such as glucose, 6 molecules of CO_2 are consumed, and the carbon atoms contributed by 6 molecules of ribulose 1,5-bisphosphate (RuBP) are recycled. The process requires energy from ATP, as well as electrons and hydrogen ions from NADPH.

This enzyme catalyzes the reaction in which CO_2 combines with RuBP.

1 The ATP produced by the light reactions is used as an energy source for carbon fixation.

2 The NADPH produced by the light reactions provides reducing power for carbon fixation.

3 For every 6 molecules of CO_2 consumed in carbon fixation, 12 molecules of glyceraldehyde 3-phosphate (G3P) are produced.

4 Two molecules of glyceraldehyde 3-phosphate are used to make the 6-carbon sugar glucose.

5 The remaining 10 molecules of glyceraldehyde 3-phosphate are processed to yield 6 molecules of ribulose bisphosphate.

carbon atoms from CO_2 gas and fixing (incorporating) them into organic compounds such as sugars, the Calvin cycle reactions make carbon from the nonliving world available to the photosynthetic producer and eventually to other living organisms as well.

The Calvin cycle reactions are catalyzed by enzymes present in the stroma. The most abundant of these enzymes is **rubisco**. Rubisco catalyzes the first reaction of the Calvin cycle, in which a molecule of the one-carbon compound CO_2 combines with a five-carbon compound called ribulose 1,5-bisphosphate [RYE-byoo-lohss . . . biss-FOSS-fayt], or RuBP for short, to eventually produce two three-carbon compounds. This reaction can be expressed as an equation displaying just the carbon atoms in the compounds involved: $1C + 5C = 2 \times 3C$. This first reaction in carbon fixation is followed by a multi-step cycle catalyzed by many different enzymes. These reactions manufacture sugars for use by the cell and also regenerate RuBP. This molecule is absolutely necessary to keep the Calvin cycle running because RuBP is the acceptor molecule for CO_2. Rubisco links RuBP to CO_2 through a covalent bond. The reduction of the fixed carbon in subsequent steps requires the input of energy from ATP together with electrons and hydrogen ions delivered by NADPH (Figure 8.6).

Three turns of the Calvin cycle bring in the three carbon atoms needed to make 1 molecule of a three-carbon sugar called glyceraldehyde 3-phosphate [gliss-er-AL-duh-hide . . .]. We can follow this process by tracking the number of carbon atoms as they get rearranged into different compounds at each step of the cycle (see Figure 8.6). For every 3 molecules of CO_2 ($3 \times 1C = 3C$) that combine with 3 molecules of ribulose 1,5-bisphosphate ($3 \times 5C = 15C$), 6 molecules of the three-carbon compound are produced ($6 \times 3C = 18C$). These molecules eventually produce 3 RuBP molecules ($3 \times 5C = 15C$) and 1 molecule of glyceraldehyde 3-phosphate (3C). As the arithmetic indicates, it takes three turns of the cycle to produce 1 molecule of the three-carbon sugar, with the other carbon atoms constantly recycled to maintain the pool of RuBP. The formation of 1 molecule of glyceraldehyde 3-phosphate is fueled by the input of 9 molecules of ATP and 6 molecules of NADPH.

Glyceraldehyde 3-phosphate (G3P, for short) is the chemical building block used to manufacture glucose, from which the cell can make the other carbohydrates it needs. Most of the glyceraldehyde 3-phosphate made in the chloroplasts is exported from these organelles and eventually consumed in various chemical reactions in the same or other cells. Some of the exported molecules of glyceraldehyde 3-phosphate immediately are used to manufacture sucrose (table sugar) in the cytoplasm. Sucrose is an important food source for all the cells in a plant and is transported from the leaves, where photosynthesis takes place, to other parts of the plant. Significant amounts of sucrose are stored in the vacuoles of sugarcane stems and sugar beet roots (Figure 8.7*a*), which is why these two crops are the mainstay of the sugar industry worldwide.

Not all the glyceraldehyde 3-phosphate made in the chloroplasts is shipped out. Some of it is converted into starch by enzymes in the stroma. Starch, a polymer of glucose, is an important form of stored energy in plants. It accumulates in chloroplasts during the day and is then broken down to simple sugars at night. The sugars are taken apart by catabolic pathways, mainly cellular respiration, to generate ATP for the cell's nighttime energy needs. Fruits, seeds, roots, and tubers such as potatoes, are also rich in stored starch, which provides for the energy needs of these nonphotosynthetic tissues (Figure 8.7*b*). The energy-rich nature of these plant parts explains why they are such an important food source for animals.

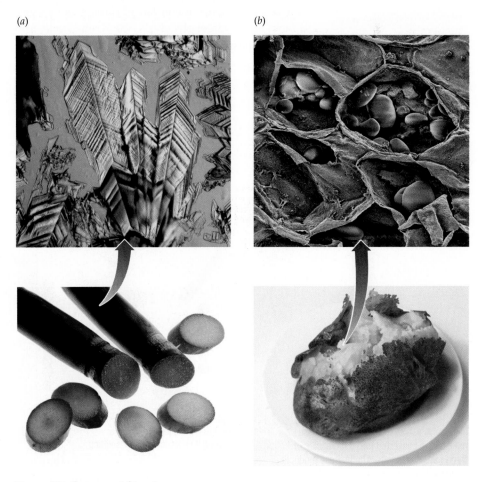

Figure 8.7 Sugar and Starch
(*a*) This polarized light micrograph displays the crystalline form of sucrose, or table sugar. Sucrose is made up of two simpler sugars, glucose and fructose, held together by a covalent bond. Some plants store large amounts of sucrose in the vacuoles of stem cells (as in the sugarcane shown here), root cells (as in sugar beets), and fruit cells (as in grapes). (*b*) This scanning electron micrograph of a section through a potato tuber shows many starch-storing organelles inside the cells. Starch is a polymer of glucose and it is the main form in which most plants store extra energy.

8.4 Cellular Respiration: Breaking Down Molecules for Energy

The catabolic reactions constantly occurring in our bodies are ultimately dependent on the products of photosynthesis. In complex multicellular animals, the first stage of catabolism is the digestion of large macromolecules in the stomach and intestines. Digestion involves the breakdown of macromolecules—such as carbohydrates, proteins, and fats—into smaller building blocks, such as simple sugars, amino acids, and fatty acids. These compounds are then absorbed by the intestine and passed on via the bloodstream to other cells in the body. Each cell then converts the simple sugars supplied by digestion into fuel for its own use. This final, cellular stage of catabolism is known as cellular respiration, and it consists of three major phases in most eukaryotes: glycolysis, the Krebs cycle, and oxidative phosphorylation.

Glycolysis is the first stage in the cellular breakdown of sugars

Glycolysis [glye-*KOLL*-uh-siss] literally means "sugar splitting." From an evolutionary standpoint, it was probably the earliest means of producing ATP from food molecules, and it is still the primary means of energy production in many prokaryotes. However, the energy yield from glycolysis is small because sugar is only partially oxidized through this process, resulting in just 2 molecules of ATP and 2 molecules of NADH. In most eukaryotic organisms, glycolysis is only the first step in energy extraction from sugars, and additional reactions in the mitochondrion help achieve the complete oxidation of carbon backbones, typically yielding at least 15 times as much ATP as does glycolysis.

Glycolysis consists of a series of chemical reactions that take place in the cytosol (Figure 8.8). These reactions break down glucose, a six-carbon sugar, and provide the mitochondria with the simpler substrate molecules they need in order to generate ATP. Through a series of enzyme-catalyzed reactions, glucose is converted to a six-carbon sugar intermediate, which is then split into 2 molecules of three-carbon sugars; these molecules are then converted into 2 molecules of a three-carbon organic acid called **pyruvate** [pye-*ROO*-vayt], which is transported into the mitochondria for further processing.

The net energy yield for glycolysis is calculated as follows: For each molecule of glucose consumed during glycolysis, 4 molecules of ADP are phosphorylated to produce

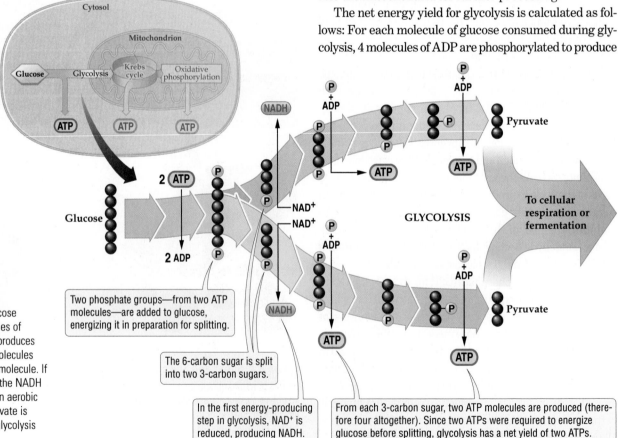

Figure 8.8 Glycolysis

In glycolysis, a single six-carbon glucose molecule is converted into 2 molecules of three-carbon pyruvate. The process produces a relatively low net yield of 2 ATP molecules and 2 NADH molecules per glucose molecule. If oxygen is present, the pyruvate and the NADH move into the mitochondria for use in aerobic respiration. If oxygen is absent, pyruvate is processed further through the post-glycolysis fermentation reactions.

Two phosphate groups—from two ATP molecules—are added to glucose, energizing it in preparation for splitting.

The 6-carbon sugar is split into two 3-carbon sugars.

In the first energy-producing step in glycolysis, NAD$^+$ is reduced, producing NADH.

From each 3-carbon sugar, two ATP molecules are produced (therefore four altogether). Since two ATPs were required to energize glucose before splitting, glycolysis has a net yield of two ATPs.

4 molecules of ATP, and electrons are donated to 2 molecules of NAD$^+$, generating 2 molecules of NADH. Since the early steps of glycolysis consume 2 molecules of ATP per glucose molecule, a single glucose molecule produces a net yield of 2 ATP molecules and 2 NADH molecules (see Figure 8.8). Glycolysis does not require oxygen (O_2), and it only partially oxidizes glucose to produce pyruvate. For complete oxidation, pyruvate molecules must enter the mitochondrion to be taken apart through an extremely efficient, oxygen-dependent ATP-generating system.

Fermentation facilitates ATP production through glycolysis when oxygen is absent

Glycolysis does not require O_2, therefore it is an **anaerobic** [AN-air-OH-bick] process. Glycolysis was probably the main source of energy for early life forms in the oxygen-poor atmosphere of primitive Earth. It is still the only means of generating ATP for some anaerobic organisms. Some anaerobic bacteria that live in oxygen-deficient swamps, in sewage, or in deep layers of soil, are actually poisoned by oxygen. Most anaerobic organisms use the catabolic pathway known as fermentation to extract energy from organic molecules. Fermentation consists of glycolysis and a special set of reactions (post-glycolysis reactions) whose only role is to help sustain glycolysis.

During fermentation, the pyruvate and NADH produced by glycolysis remain in the cytosol, instead of being imported by mitochondria. The post-glycolysis reactions convert pyruvate into other molecules, such as alcohol or lactic acid, depending on the type of fermentation pathways used by the cell type. In the process, NADH is converted back to NAD$^+$, an adequate supply of which is necessary to keep glycolysis running (look at Figure 8.8 to see the point at which NAD$^+$ is used in glycolysis). There is a finite pool of NAD$^+$ in the cell, and if all of it were converted into NADH, glycolysis itself would cease for lack of more NAD$^+$. The post-glycolysis reactions are a clever way of averting this problem: these reactions remove electrons and hydrogen ions from NADH, resulting in NAD$^+$ formation, which restores the cell's limited pool of this vital metabolic precursor. In regenerating NAD$^+$ in post-glycolysis reactions, the electrons and hydrogen ions from NADH create two- or three-carbon compounds, with alcohol and lactic acid being two of the most common by-products of fermentation.

Yeasts are single-celled fungi, some species of which are used in the production of beer, wine, and other alcoholic products. Fermentation by anaerobic yeasts converts pyruvate into an alcohol called ethanol, along with CO_2 gas, which gives beer its foamy effervescence (Figure 8.9a). The gas has an important role in bread

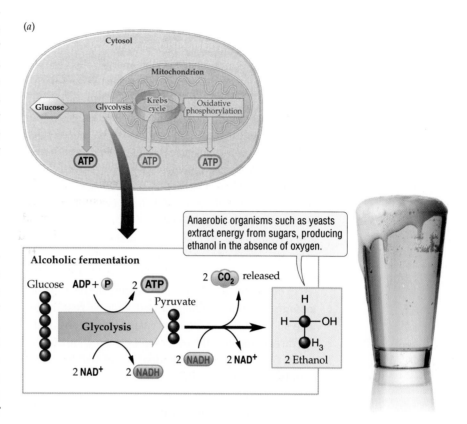

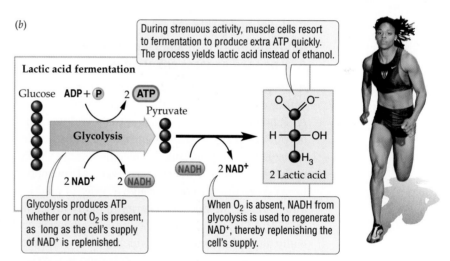

Figure 8.9 Ethanol and Lactic Acid are Two Examples of the By-products of Fermentation

When oxygen is absent, fermentation produces ATP through glycolysis alone. (*a*) Strains of yeast, which are single-celled fungi, are used in the brewing of alcoholic beverages such as beer. When oxygen is excluded from the fermentation tanks, the yeast resort to fermentation of sugars, producing ethanol and CO_2 as by-products of the post-glycolysis steps. (*b*) A similar process occurs in our muscles during short bursts of strenuous exercise, except that the post-glycolysis reactions turn pyruvate into a three-carbon organic acid known as lactic acid, instead of the two-carbon ethanol made during alcoholic fermentation.

making with baker's yeast. The CO_2 released by fermentation expands the dough and creates small holes that contribute a light texture to the baked product.

Fermentation is not limited to single-celled anaerobic organisms. It also takes place in the human body and the bodies of other **aerobic** [air-*OH*-bick] organisms (those that require O_2). When we exercise hard and push our muscles to the point of exhaustion, the burning pain we sometimes experience comes from fermentation: muscle cells convert pyruvate to lactic acid, which irritates nerve endings, resulting in the burning sensation felt in overtaxed muscles (Figure 8.9*b*). A rapid burst of strenuous exercise can exhaust the ATP stores in skeletal muscles in a matter of seconds; under these circumstances, the muscle cells resort to fermentation to produce extra ATP because blood vessels cannot deliver O_2 fast enough to support higher rates of cellular respiration. The short-duration, high-intensity effort required of sprinters and weight lifters gets a very significant assist from oxygen-independent lactic acid fermentation in muscle cells. In contrast, it takes sustained aerobic respiration to support the effort of marathon runners and other endurance athletes.

Cellular respiration in the mitochondrion furnishes much of the ATP needed by most eukaryotes

As long as oxygen is available, most eukaryotes use cellular respiration to satisfy their relatively large ATP needs. Glycolysis in the cytosol yields a small output of energy carriers, but much of the chemical energy of the original sugar molecule remains locked in the three-carbon molecule pyruvate. Mitochondria break down the pyruvate through a series of reactions and package the released energy in many molecules of ATP. ATP production in the mitochondrion is crucially dependent on oxygen, which is to say, the mitochondrial portion of cellular respiration is a strictly aerobic process.

Highly aerobic tissues, such as muscles, tend to have high concentrations of mitochondria and a rich blood supply that delivers the large amounts of oxygen needed to support their activity. Muscle cells in the human heart, for example, have an exceptionally large number of mitochondria to produce the enormous amounts of ATP needed to keep the heart beating. Blind mole-rats, which live underground in

Blind mole-rat

oxygen-poor burrows and must dig a lot every day, also display cellular and physiological adaptations that enhance the supply of oxygen to support ATP production by muscle cells. Compared with the muscles of laboratory rats, the muscles of blind mole-rats have 50 percent more mitochondria and 30 percent more blood capillaries.

The Krebs cycle produces NADH and carbon dioxide

The end-product of glycolysis, pyruvate, is transported into the mitochondria and enters the second major phase of cellular respiration, the **Krebs cycle**, a series of enzyme-driven reactions that take place in the mitochondrial matrix (Figure 8.10). Before the cycle begins, however, pyruvate entering the mitochondrion must be processed through several preparatory reactions. A large enzyme complex in the matrix breaks one of the two carbon–carbon covalent bonds in pyruvate, releasing a molecule of CO_2, and leaving behind a two-carbon unit known as an acetyl group. The same enzyme complex attaches this acetyl group to a "carbon carrier" known as coenzyme A (CoA), producing a molecule called acetyl CoA. Acetyl CoA enters the Krebs cycle when it donates the two-carbon acetyl group to a four-carbon acceptor molecule.

The Krebs cycle is sometimes called the **citric acid cycle** because citric acid is the first product in this enzyme-driven pathway. Citric acid, a six-carbon compound, is produced when the two-carbon acetyl group from acetyl CoA is added to a four-carbon acceptor. CoA is liberated in this process to recruit yet more acetyl groups for the cycle. The two-carbon unit that enters the cycle as part of citric acid is broken apart in a series of enzyme-driven steps. Each of the carbon atoms released in this process combines with oxygen atoms to become carbon dioxide. As the covalent bonds are broken, the energy stored in them is used to drive the formation of energy carriers, especially NADH. As we will see shortly, the energy in these NADH molecules will be used to make many molecules of ATP during the third and final stage of cellular respiration.

The Krebs cycle is like the "Grand Central Station" of metabolism because it is the meeting point of several different types of catabolic and anabolic pathways. For example, carbon skeletons derived from other

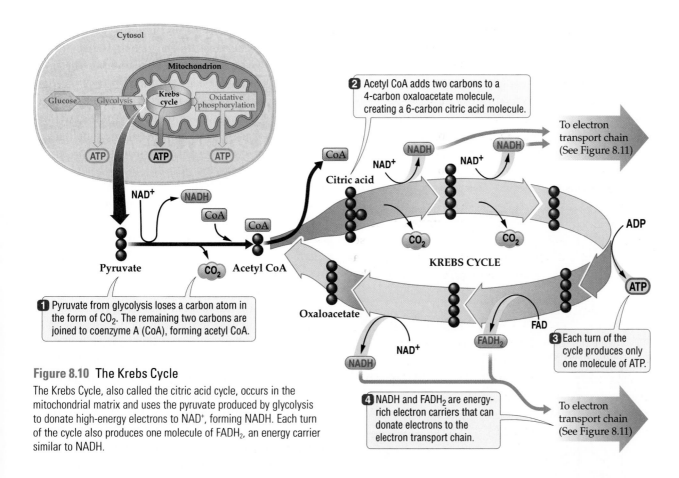

Figure 8.10 The Krebs Cycle

The Krebs Cycle, also called the citric acid cycle, occurs in the mitochondrial matrix and uses the pyruvate produced by glycolysis to donate high-energy electrons to NAD⁺, forming NADH. Each turn of the cycle also produces one molecule of $FADH_2$, an energy carrier similar to NADH.

Figure labels:

Cytosol

Mitochondrion

Glucose — Glycolysis — Krebs cycle — Oxidative phosphorylation

ATP — ATP — ATP

NAD⁺ → NADH

CoA

CoA

CoA

Pyruvate — CO_2 — Acetyl CoA

Citric acid

CoA

NAD⁺ → NADH

NAD⁺ → NADH

To electron transport chain (See Figure 8.11)

CO_2 — CO_2

KREBS CYCLE

ADP

ATP

Oxaloacetate

FAD

$FADH_2$

NAD⁺

NADH

1 Pyruvate from glycolysis loses a carbon atom in the form of CO_2. The remaining two carbons are joined to coenzyme A (CoA), forming acetyl CoA.

2 Acetyl CoA adds two carbons to a 4-carbon oxaloacetate molecule, creating a 6-carbon citric acid molecule.

3 Each turn of the cycle produces only one molecule of ATP.

4 NADH and $FADH_2$ are energy-rich electron carriers that can donate electrons to the electron transport chain.

To electron transport chain (See Figure 8.11)

types of food molecules, such as lipids, can also be converted to carbon dioxide in the citric acid cycle. Lipids such as triglycerides (see Chapter 4) are broken into fatty acids and glycerol in the cytosol; upon entering the mitochondrion these molecules are converted to acetyl CoA by special enzymes. The lipid-derived acetyl CoA is indistinguishable from that made from pyruvate during the preparatory reactions, and its carbon atoms are oxidized in the same manner by the Krebs cycle.

Oxidative phosphorylation uses oxygen to produce ATP in quantity

The largest output of ATP is generated in mitochondria during the third and last stage of cellular respiration. As we mentioned earlier, the mitochondrion has a double membrane that forms two separate compartments, the intermembrane space and the matrix. The

enzymatic reactions associated with Krebs cycle take place in the matrix. The final stage of cellular respiration takes place in the many folds (cristae) of the inner mitochondrial membrane. These folds create a large surface area on which are embedded many electron transport chains (ETC) and many units of an elaborate ATP-manufacturing machine known as ATP synthase. NADH molecules made by the Krebs cycle, and also those generated during glycolysis (see Figure 8.8), diffuse to the inner membrane and donate their high-energy electrons to the first component of the ETC (as does the lone $FADH_2$). Electron transfer along the ETC ultimately results in the phosphorylation of ADP to ATP in an oxygen-dependent process called oxidative phosphorylation (Figure 8.11). Let's examine the link between electron transfer through the ETC and phosphorylation of ATP.

The electron transport chains found in the inner mitochondrial membrane are similar in function to

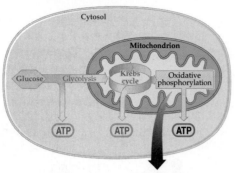

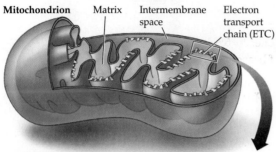

High-energy electrons in NADH are donated to the electron transport chain embedded in the inner mitochondrial membrane.

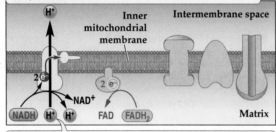

As electrons are passed down the electron transport chain, protons are pumped from the matrix into the intermembrane space, creating a proton imbalance across the inner mitochondrial membrane.

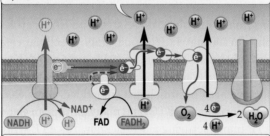

Oxygen is required as the final electron acceptor. In this final electron transfer, O_2 picks up 4 electrons and 4 protons and forms water.

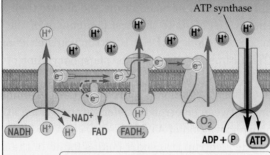

The passage of accumulated protons from the intermembrane space to the matrix through ATP synthase drives this enzyme complex to produce ATP.

those found in the thylakoid membranes of chloroplasts. In mitochondria, electrons donated by NADH are passed along a series of ETC components and the energy released is used to pump protons through channel proteins from the matrix into the intermembrane space. The resulting buildup of protons in the intermembrane space of the mitochondrion has the same effect as the buildup of protons in the thylakoid space: it creates a proton gradient across the membrane. As in chloroplasts, the proton gradient is depleted periodically as protons gush through the ATP synthases. Each ATP synthase is a multiprotein complex consisting of a membrane channel and associated motor proteins and enzymes. The movement of protons through the ATP synthase channel activates associated enzymes, which then catalyze the phosphorylation of ADP to form ATP (Figure 8.11). The striking similarities in the way ATP is produced in chloroplasts and mitochondria illustrate that diverse metabolic pathways can evolve through modifications of the same basic machinery. Both these organelles use electron transfer chains built from at least some similar components, both use energy released during electron transfer to move protons across an inner membrane, and the two use very similar ATP synthases to harness the energy of the proton gradient to make ATP.

Note the fate of the electrons, off-loaded by NADH, that travel down the ETCs of the inner mitochondrial membrane. They are accepted by oxygen (O_2), which combines them with H^+ picked up from the mitochondrial matrix, to form water (H_2O). In other words, electron transfer along the ETC terminates with O_2, which serves as the final electron acceptor (see Figure 8.11). Without oxygen to whisk the elec-

Figure 8.11 Oxygen Is Needed to Form ATP through Oxidative Phosphorylation

Oxidative phosphorylation is the last stage in cellular respiration and it produces the most ATP of any process in the cell. Electrons from NADH and $FADH_2$ produced by the Krebs cycle are donated to the electron transport chain (ETC) in the inner mitochondrial membrane. The electrons release energy as they travel down the ETC, and some of this energy is used to drive protons from the matrix into the intermembrane space. The energy of the resulting proton gradient in turn drives the phosphorylation of ADP to ATP, a reaction catalyzed by ATP synthase. Note that the electron traveling down the ETC is eventually accepted by oxygen gas, producing water.

Biology Matters

Our Fondness for Fermentation

"Fermented food" doesn't sound particularly appetizing, yet many of the most popular foods, drinks, and snacks depend on fermentation somewhere along the way to change unappealing starting ingredients into highly sought-after finished products. Consider that as a yearly average Americans consume approximately 11+ pounds of chocolate, 9+ pounds of coffee, 31 pounds of cheese, 5½ pounds of yogurt, 9 pounds of pickles, and 60 pounds of bread, all fermented foods.

Clearly, we love fermentation! Chocolate is a good example. Chocolate is a $60 billion-a-year industry. Worldwide, 5.5 billion pounds of chocolate are produced annually, mostly from Africa and South America. The Mayans and Aztecs of South America were making chocolate as long as 2,600 years ago, and the basic process is the same now as it was then. A natural fermentation process has always been central to the changes in cocoa beans that lead to chocolate.

To start the process of fermentation, beans are heaped into piles on plantain leaves, where microorganisms naturally present in the environment begin to ferment the pulp surrounding the beans. Some of the products of fermentation penetrate the beans and disrupt cell membranes. Enzymes released from the disrupted cells produce the main flavors of chocolate by breaking down certain proteins. Without fermentation no flavor would arise. Aerobic bacteria then take over and, through respiration, consume the undesirable products of fermentation, such as acids and ethanol. Stirring and mixing the beans exposes them to more oxygen, which generates a brown color when it combines with compounds in the beans. And with more

oxygen present, other flavors develop. Roasting completes the process. The bitter but richly flavored product is more or less identical to its millennia-old counterpart. The Spanish originally brought chocolate to Europe, and Europeans contributed to our modern-day version by adding sugar to this enticing product of fermentation.

Nearly everyone enjoys chocolate, and some chocolate consumers describe themselves jokingly as "chocoholics." But, joking aside, among the products arising from the enzymes released from membranes of the beans' disrupted cells, we find a stimulant similar to caffeine and another substance associated with pleasure responses in our nervous systems. Caffeine is known to have a variety of physiological effects. Like caffeine, substances in chocolate activate pleasure centers in the brain that are also activated by addictive drugs such as cocaine.

---

trons away, the electron traffic on the ETC "highway" would come to a standstill. Without electron transfer along the ETC, protons cannot be pumped into the intermembrane space, and without the energy of the proton gradient, ADP cannot be phosphorylated to make ATP. Like all aerobic organisms, we need oxygen because we cannot make enough ATP without it, so anything that interferes with that process is a danger to life. Hydrogen cyanide is a deadly poison because it binds to the last component in the electron transport chain, preventing it from handing electrons to oxygen, and thereby arresting energy production in mitochondria.

Cellular respiration (glycolysis, the Krebs cycle, and oxidative phosphorylation) has a net yield of about 30 ATP molecules per molecule of glucose. Mitochondrial respiration is therefore more productive than glycolysis alone, which yields only 2 ATP molecules per molecule of glucose consumed.

Concept Check

1. Compare the functions of photosystems I and II.
2. What is rubisco and what role does it play in photosynthesis?
3. Why do we need oxygen (O_2) to live?

Applying What We Learned

Fueling the Brain

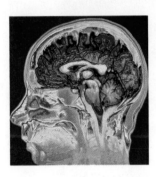

Having explored catabolic pathways that yield energy, let us consider the great energy demands of our large brains and how those needs may have been met at various stages in our evolutionary history. The high ATP demand of brain cells can only be met through oxidative phosphorylation, which is why brain cells start to die if deprived of oxygen for more than a few minutes.

The genus *Homo* appeared approximately 2.5 million years ago. The earliest members of this genus had a brain nearly twice as large as that of the apelike Australopithecines, who are thought to be the direct ancestors of our genus. "Lucy," the most famous Australopithecine fossil, had a brain that weighed about 400 grams, similar to the brain size of modern day chimpanzees.

What kind of nutritional good fortune, or behavioral or physiological change, could account for this leap in brain size among early humans? Paleoanthropologists have suggested that the earliest members of the genus *Homo* switched from a strictly vegetarian diet to a more varied diet that included calorie-dense foods, especially meat. Vegetable matter tends to be low in calories and nutrients, so animals that are largely vegetarians (herbivores) need longer digestive tracts so they can move a large bulk of vegetation slowly through the system in order to extract the most calories and nutrients from it.

Humans are typically omnivores (*omne,* "all"; *vorare,* "to devour") and the prominence of meat in our ancestral diet is reflected in the fact that our digestive tract is nearly 40 percent shorter than that of other primates. Meat-eating, according to this hypothesis, may have been a turning point in the evolution of the human brain because it would have provided the fuel necessary to support a large brain.

Energy supply in adulthood is clearly not the only challenge in the evolution of a larger brain. An early human mother would have faced enormous problems in giving birth to a large-brained infant and nurturing and protecting it over the span of a long childhood. A large part of our brain development occurs before birth, and because the fetus pulls all of its nutrients from the mother, pregnancy is metabolically demanding for the mother. Giving birth to a large-brained infant is a hazard human mothers have had to face until modern times. The transfer of metabolic energy from mother to infant continues after birth through lactation. A baby's brain grows most rapidly in the first year of life and reaches 85 percent of its adult size by age 4 or 5. Large brains bestowed on us complex behaviors, such as tool use, culture, and language, that clearly had enormous survival value. What it took to support those big brains was the gift of metabolic input from better nourished mothers, who were aided in turn by a protective family and tribal group.

Biology in the News

Biofuels Powering Town's Vehicles; Vegetable Oils Reduce Emissions

By Linda Bock

UXBRIDGE – The Department of Public Works and the Fire Department have begun using virgin soy and vegetable oils in about 25 vehicles in an effort to reduce reliance on oil products and reduce exhaust emissions. Irving Priest, assistant DPW superintendent, spearheaded the change from conventional diesel to B20 biodiesel fuel. "There are many benefits to using biofuels," according to Mr. Priest. Biodiesel fuel is a renewable energy source made in the United States, it can be used in most diesel engines and reduces tailpipe emissions. The biodiesel fuel currently costs a bit more per gallon, and may cost up to $2,000 to $3,000 more a year. Lawrence E. Bombara, superintendent of DPW, thinks over time there will be a savings to the town on reduced maintenance costs for the engines. . . .

Biodiesel is produced from agricultural resources such as vegetable oils. In the United States, most biodiesel is made from soybean oil; however canola oil, sunflower oil, recycled cooking oils and animal fats are also used, according to the U.S. Department of Environmental Protection.

To make biodiesel, the base oil is put through a process called esterification. This refining method uses an industrial alcohol (ethanol or methanol) and a catalyst (a substance that enables a chemical reaction) to convert the oil into a fatty-acid methyl-ester fuel (biodiesel), according to the EPA.

Biodiesel in its pure form is known as "neat biodiesel" or B100, but it can also be blended with conventional diesel, most commonly as B5 (5 percent biodiesel and 95 percent diesel) and B20. Biodiesel is registered with the EPA, and is legal for use at any blend level in both highway and nonroad diesel vehicles. Most diesel engines can run on biodiesel without needing special equipment.

Photosynthesis transforms the sun's energy into the usable chemical energy in plants. Fossil fuels, which consist of "aged" plant matter, are rich storehouses of energy that we received from the sun millions of years ago. We are consuming fossil fuels rapidly, and the supply is limited, so the search is on for new sources of energy, and new ways are emerging to connect photosynthesis to the immediate production of usable fuel. The production of biodiesel means that the "millions of years" are no longer needed to produce carbon-based fuels.

Biodiesel can be produced from many sources. Restaurants routinely discard huge quantities of waste cooking fats, plants like mustard and soy produce large amounts of oils, and some types of algae that grow extremely rapidly in shallow salt ponds are composed of 50 percent oil. Consider the last source: salt pond algae on algae "farms" fed by wastewater from animal farms and sewage treatment plants could theoretically produce enough oil to supply 100 percent of the diesel fuel needs of the United States. Moreover, a valuable by-product of the conversion process is glycerine, which is used in soaps and many other products.

It's true that the burning of biodiesel—like the burning of petroleum diesel—produces carbon dioxide, a greenhouse gas that contributes to global warming. However, when we burn fossil fuels we are discharging into the atmosphere carbon that has remained sequestered, or "locked up," in the dead remains of ancient plants for hundreds of millions of years. When we cultivate photosynthetic organisms for biodiesel, they remove carbon dioxide from the atmosphere, which is later returned to the air when we convert their carbon reserves to biodiesel and burn it for energy. Therefore, the use of biodiesel could be carbon neutral overall, neither adding carbon to the atmosphere nor removing it. Much depends on the source of our biodiesel (waste oil would be carbon neutral, but a fertilizer-intensive soybean crop may not be) and how we extract, process, and transport it. Biodiesel has other advantages. It is less polluting than petroleum-based diesel; it burns cleaner than petroleum diesel, so it produces fewer emissions, and it contains almost no sulfur, so it cannot contribute to the formation of acid rain. In addition, biodiesel is as biodegradable as sugar. We have seen the effects of petroleum oil spills. In contrast, any biodiesel "spills" would quickly disappear as organisms break them down into the products of catabolism: water and carbon dioxide. As public awareness grows, demand for biodiesel is likely to grow, and with it production.

Evaluating the News

1. What is biodiesel and how is it different from fossil diesel or gasoline?
2. Investing in the biodiesel industry carries great risk. Should federal and state governments offer tax subsidies to make biodiesel competitive with conventional fuels? Defend your position.
3. We are consuming fossil fuels at enormous and ever-increasing rates, and there is much controversy over whether we should stress conservation or drilling for more oil. Research the issues to decide what the U.S. energy strategy should be with regard to conventional fuels. Alone or with your classmates, write a letter that you might send to your representatives in Congress expressing your views.

Source: *Telegram & Gazette*, Worcester, MA, July 13, 2007.

Chapter Review

Summary

8.1 Energy Carriers Power Cellular Activities

- Energy is transferred within living organisms via special molecules called energy carriers.
- ATP, the most commonly used energy carrier, donates energy stored in chemical bonds to chemical reactions. ATP is created through phosphorylation of ADP.
- The energy carriers NADPH and NADH donate electrons and hydrogen ions (protons) in redox reactions. NADPH is used in anabolic pathways such as photosynthesis, while NADH participates in catabolic reactions such as cellular respiration.

8.2 An Overview of Photosynthesis and Cellular Respiration

- In chemical terms, photosynthesis is the reverse of cellular respiration.
- Photosynthesis is found in producers only. They use light energy to make ATP and NADPH, splitting water molecules and making oxygen gas in the process. The energy carriers are then used to convert carbon dioxide into sugar molecules during the Calvin cycle reactions.
- Cellular respiration occurs in producers and consumers. It begins in the cytoplasm and is completed in the mitochondria. A small amount of ATP and NADH are made during the first phase, glycolysis, during which sugar molecules are degraded to pyruvate. The next two steps occur in the mitochondria. Carbon dioxide is released during the degradation of pyruvate via the preparatory reactions and the Krebs cycle; the latter also yields many NADH. The final stage of cellular respiration is oxidative phosphorylation, during which many ATPs are made using an oxygen-dependent process.
- When oxygen supply is low or absent, many single-celled organisms, and some cell types in multicellular organisms, can make do with a glycolysis-based catabolic pathway known as fermentation.

8.3 Photosynthesis: Capturing Energy from Sunlight

- Photosynthesis takes place in chloroplasts. Each chloroplast has a network of membrane sacs, called thylakoids, surrounded by a large space known as the stroma. Light reactions occur in the thylakoid membrane, and the Calvin cycle reactions occur in the stroma.
- Light reactions capture energy from sunlight using chlorophyll molecules in the antenna complex. ATP and NADPH are produced as electrons flow in electron transport chains from photosystem II to photosystem I. The oxygen we breathe is a by-product of these reactions.
- Calvin cycle reactions use the ATP and NADPH produced by light reactions to fix atmospheric CO_2 and synthesize sugars such as glucose.

8.4 Cellular Respiration: Breaking Down Molecules for Energy

- Cellular respiration is a crucial catabolic pathway that is found in almost all eukaryotes. It requires oxygen and has three stages: glycolysis, the Krebs cycle, and oxidative phosphorylation.
- Glycolysis occurs in the cytosol and splits glucose molecules to produce pyruvate and a small amount of ATP and NADH.
- Fermentation is a catabolic pathway that occurs in the absence or near-absence of oxygen. In fermentation, the pyruvate from glycolysis remains in the cytosol and is converted into 1-, 2-, or 3-carbon compounds, such as CO_2 and alcohol (as in fermentation by yeasts) or lactic acid (as in skeletal muscles when the oxygen supply is low). Fermentation produces much less ATP than cellular respiration. The purpose of the post-glycolysis fermentation reactions is to regenerate NAD^+, which is essential for the continued operation of glycolysis.
- In the presence of oxygen, the pyruvate from glycolysis is used by mitochondria to generate many additional molecules of ATP.
- The Krebs cycle occurs inside the mitochondrial matrix and uses pyruvate to produce NADH and CO_2.
- Oxidative phosphorylation occurs at the inner membrane and uses O_2 and NADH to produce most of the cell's ATP.

⊙ Review and Application of Key Concepts

1. Compare photosynthesis and cellular respiration. Which of the two is an anabolic process and why? Which atmospheric gases are released and which are consumed by each of these processes? Which of the two occurs in producers but not in consumers?

2. In both chloroplasts and mitochondria, the transfer of electrons down an ETC involves hydrogen ions and leads to a similar outcome. Describe that outcome and explain how it contributes to the production of ATP in each of these organelles.

3. Describe the sequence of events in the light reactions and the Calvin cycle reactions of photosynthesis.

4. Certain drugs allow protons to pass through the inner mitochondrial membrane on their own, without the involvement of channel proteins. How would such drugs affect ATP synthesis?

5. Describe the flow of electrons from photosystem II to $NADP^+$ in the context of the chloroplast.

Key Terms

aerobic (p. 172)
anabolism (p. 163)
anaerobic (p. 171)
antenna complex (p. 166)
ATP (p. 162)
Calvin cycle (p. 165)
carbon fixation (p. 168)
catabolism (p. 163)
cellular respiration (p. 164)
chlorophyll (p. 166)
chloroplast (p. 163)
citric acid cycle (p. 172)
electron transport chain (ETC)
  (p. 166)
energy carrier (p. 162)
fermentation (p. 164)
glycolysis (p. 164)
intermembrane space (p. 165)

Krebs cycle (p. 172)
light reactions (p. 165)
NADH (p. 163)
NADPH (p. 163)
oxidative phosphorylation (p. 164)
phosphorylation (p. 162)
photosynthesis (p. 163)
photosystem (p. 167)
photosystem I (p. 167)
photosystem II (p. 167)
proton gradient (p. 167)
pyruvate (p. 170)
reaction center (p. 166)
rubisco (p. 169)
stroma (p. 165)
thylakoid (p. 165)
thylakoid membrane (p. 165)
thylakoid space (p. 165)

Self-Quiz

1. The chemical that functions as an energy-carrying molecule in all
   organisms is
   a. carbon dioxide.
   b. water.
   c. RuBP.
   d. ATP.

2. The element carbon cycles
   a. between the sun and Earth.
   b. only from producers to consumers.
   c. only from consumers to producers.
   d. between producers, consumers, and the environment.

3. The oxygen produced in photosynthesis comes from
   a. carbon dioxide.
   b. sugars.
   c. pyruvate.
   d. water.

4. Photosynthesis occurs in
   a. chloroplasts.
   b. mitochondria.
   c. the cytoplasm.
   d. glycolysis.

5. The light reactions in photosynthesis require
   a. oxygen.
   b. chlorophyll.
   c. rubisco.
   d. carbon fixation.

6. Glycolysis occurs in
   a. mitochondria.
   b. the cytosol.
   c. chloroplasts.
   d. thylakoids.

7. The electrons needed to replace those lost from chlorophyll in the light
   reactions of photosynthesis ultimately come from
   a. sugars.
   b. channel proteins.
   c. water.
   d. electron transport chains.

8. Which of the following statements is *not* true?
   a. Glycolysis produces most of the ATP required by aerobic organisms.
   b. Glycolysis produces pyruvate, which is consumed by the Krebs cycle.
   c. Glycolysis occurs in the cytosol of the cell.
   d. Glycolysis is the first stage of cellular respiration.

9. Which of the following is essential for oxidative phosphorylation?
   a. rubisco
   b. NADH
   c. carbon dioxide
   d. chlorophyll

10. Oxidative phosphorylation
    a. produces less ATP than glycolysis.
    b. produces simple sugars.
    c. is dependent on the activity of ATP synthase.
    d. is part of the photosystem I electron transport chain.

CHAPTER 9 Cell Division

Key Concepts

◉ In cell division, a cell splits to produce two daughter cells. This process is necessary for an organism to grow and develop, to replace old or worn out cells, and to pass on genetic information to the next generation.

◉ The lifespan of a cell consists of two main stages, interphase and cell division, which collectively make up the cell cycle. Each stage is marked by distinctive cell activities. DNA is replicated during interphase.

◉ Cell division is the final stage in the cell cycle.

◉ There are two types of cell division: mitotic division and meiosis.

◉ Mitosis leads to daughter cells that are identical to the parent cell, with both daughter cells receiving identical copies of all the parental chromosomes.

◉ Meiosis occurs exclusively in reproductive cells and involves two rounds of nuclear and cytoplasmic divisions. It produces four daughter cells, each with half of the chromosome set found in the parent cell. Meiosis gives rise to gametes (egg and sperm), which fuse during the process of fertilization to produce offspring with a double set of chromosomes.

◉ Meiosis and fertilization have the evolutionarily important role of generating genetic diversity in a population.

A Ray of Hope

"We have a disease for which we can't hope that we are going to outlive the odds, that we're going to beat the odds, because there aren't any odds to beat." Joan Samuelson, president of the Parkinson's Action Network, addressed these words to a U.S. Senate subcommittee in 2002. Actor Michael J. Fox, who was diagnosed with Parkinson's disease at the age of 30, also testified at the hearing, along with Lonnie Ali, speaking on behalf of her husband, boxer Mohammed Ali. Other activists, policy makers, and scientists joined them in urging increased federal funding for research into Parkinson's disease and stem cells. Parkinson's disease is a progressively disabling condition marked by uncontrolled body movements and speech impairments. The symptoms are caused by the death of dopamine-producing cells in a region of the brain that regulates movement. Dopamine is a neurotransmitter, a signaling molecule that allows brain cells to communicate with each other. Medications help to manage the symptoms of Parkinson's disease, but have significant side effects and become less effective over time. As underscored by Samuelson's poignant words, there is still no cure for the disease.

Yet scientists are investigating a number of innovative strategies for fighting Parkinson's, and the possibility of cures based on stem cells has generated considerable excitement. Researchers have cured Parkinson's disease in rats by implanting dopamine-producing cells in the brains of the affected animals. We cannot hope to replicate this success in humans any time soon because there is a great deal more we need to learn about manipulating human stem cells. But there are instances where the knowledge gap has narrowed, and the successes have been spectacular. Stem cells are widely used to treat burn victims and people with certain blood disorders, including blood cancers. Clinical trials of stem cell therapy in patients newly diagnosed with Type I diabetes have been encouraging. In a recent trial, children who suffered from the disease, which damages insulin-producing cells in the pancreas, were treated with stem cells. Most of the children were then able to produce enough insulin

Still Fighting for a Cure

About half a million Americans suffer from Parkinson's disease. They include former heavyweight boxing champion Mohammed Ali (left) and actor Michael J. Fox (right). Genetics and environmental influences both play a role in the development of this disease. Exposure to certain pesticides and severe head trauma are known risk factors. Most victims don't have obvious symptoms until they are past 40, but about 20 percent of sufferers are diagnosed in their 20s and 30s.

on their own and do not need injections of this vital hormone.

But what are stem cells and how are they different from all other cell types in the human body? Why is stem cell research considered controversial, when many victims of degenerative disease and **tissue injury see it as a ray of hope?** In this chapter we will see that stem cells are a special type of cells whose only job is to divide. Before we examine the special attributes of stem cells, we will explore the mechanisms that enable prokaryotic and eukaryotic cells to make more of their own kind through cell division.

Cell division is a distinctive property of all life forms. It is the means by which organisms produce more of themselves. In multicellular organisms, cell divisions grow the individual, help repair damaged tissues, and combat invading parasites such as bacteria and viruses. Skin is an example of an organ system that is maintained through continuous cell division. The inner portion of the skin consists of multiple layers of cells that gradually move to the surface as old surface layers are lost to wear and tear. In the process, the cells undergo dramatic physical changes, such that by the time they reach the surface layer, they form dead, flattened scales of protein (Figure 9.1). As cells move up through the layers and are lost from the skin surface, they are replaced by new cells produced by the division of special cells, called stem cells, in the deepest layer of the skin.

The stem cell undergoing division—the parent cell—must duplicate its genetic material before it divides. The products of cell division, the daughter cells, each receive one complete copy of genetic information from the parent cell they replace.

Single-celled organisms split themselves in two to give rise to the next generation, a process known as **asexual reproduction**. Daughter cells produced by asexual reproduction are genetically identical to the parent cell.

Prokaryotes, which are single-celled organisms, undergo asexual reproduction through a relatively simple process called **binary fission** (literally, "splitting in two"). The genetic material of most Bacteria and Archaea takes the form of one single loop of DNA. The first step in binary fission is the duplication of this DNA, giving rise to two molecules of DNA (Figure 9.2). The cell now expands and a partition appears roughly at the center of the cell. The partition, consisting of plasma membrane, and usually cell wall material as well, segregates the two DNA molecules into separate cytoplasmic compartments. Each compartment expands until it breaks loose of the other, so that two daughter cells now replace the parent cell. Binary fission is an asexual form of reproduction because it creates daughter cells that are genetically identical to each other and the parent cell. Binary fission is the only means by which prokaryotes can propagate themselves.

Cell division in eukaryotes is considerably more complex. Because eukaryotic cells have a nucleus, containing many molecules of DNA, it takes an elaborate cellular machinery, working in a precise sequence, to distribute the parent cell's genetic material evenly

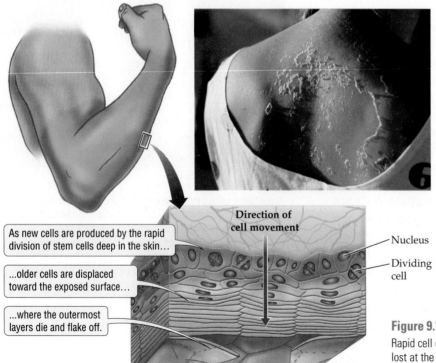

As new cells are produced by the rapid division of stem cells deep in the skin...

...older cells are displaced toward the exposed surface...

...where the outermost layers die and flake off.

Direction of cell movement

Nucleus

Dividing cell

Figure 9.1 Cell Division Replenishes the Skin
Rapid cell division in the deepest layer of the skin is necessary to replace dead cells lost at the surface of the skin. This loss can be due to normal wear or to severe damage, such as the aftereffects of the peeling sunburn shown in the photo.

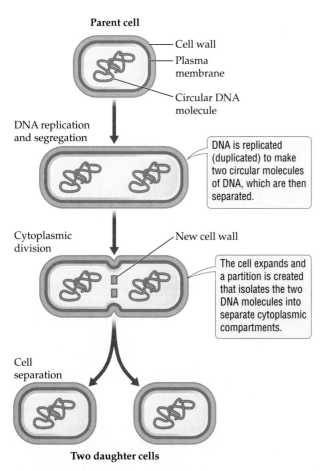

Parent cell

— Cell wall
— Plasma membrane
— Circular DNA molecule

DNA replication and segregation

DNA is replicated (duplicated) to make two circular molecules of DNA, which are then separated.

Cytoplasmic division

New cell wall

The cell expands and a partition is created that isolates the two DNA molecules into separate cytoplasmic compartments.

Cell separation

Two daughter cells

Figure 9.2 Cell Division in a Prokaryote

Prokaryotes, such as bacteria, propagate themselves asexually through a simple type of cell division known as binary fission.

into the daughter cells. In addition, most eukaryotes have two different types of cell division: mitotic divisions, which generate daughter cells identical to each other and the parent cell; and meiosis, which produces genetically distinct daughter cells that give rise to egg or sperm in most eukaryotes.

In multicellular organisms, mitotic divisions are responsible for the growth of tissues and organs and the body as a whole. They are necessary for repairing injured tissue and replacing worn out cells. In vertebrate animals, mitotic divisions enlarge the army of defensive cells that battle potentially harmful invaders. Single-celled eukaryotes, such as algae and amoeba, use mitotic divisions for asexual reproduction, much in the way prokaryotes reproduce asexually through binary fission. Many multicellular eukaryotes, including most seaweed, fungi, and plants, and some animals such as sponges and flatworms, reproduce asexually by creating new individuals, or offspring, through many rounds of mitotic divisions.

Meiosis is a specialized type of cell division that, in animals and many other eukaryotes, serves the function of **sexual reproduction**. It generates daughter cells that mature into sperm or egg (collectively known as gametes). The merging of egg and sperm in the process of fertilization creates a single cell, the zygote. In plants and animals, the zygote divides mitotically to create a mass of developing cells known as the embryo.

Cells in the very young embryo are not noticeably different from each other, but as the embryo develops, many of the cells in it stop dividing and acquire unique properties and highly specialized functions. The process through which a daughter cell becomes different from the parent cell is known as **cell differentiation**. The heart muscle is an example of a differentiated, or highly specialized, cell type. There are about 220 types of differentiated cells in the human body. However, a small number of cells in a variety of tissues and organs remain unspecialized. These **stem cells** retain the capacity to divide indefinitely without becoming committed to specialized functions. We will return to the subject of stem cells toward the close of this chapter, but first we must examine the mechanisms underlying cell division in eukaryotes and understand how and why mitotic divisions create genetic clones, while cells undergoing meiosis produce genetically diverse daughter cells.

9.1 The Cell Cycle Consists of Two Main Stages: Interphase and Cell Division

The **cell cycle** refers to the series of events that make up the life cycle of a eukaryotic cell that is capable of dividing. The cell cycle extends over the lifespan of a cell, from the moment of its origin to the time it divides to produce two daughter cells. The time it takes to complete the cell cycle depends on the organism, the type of cell, and the life stage of the organism. Dividing cells in tissues that require frequent replenishing, such as the skin or the lining of the intestine, require about 12 hours to complete the cell cycle. Cells in most other actively dividing tissues in the human body require about 24 hours to complete the cycle. By contrast, a single-celled eukaryote such as yeast can complete the cell cycle in just 90 minutes. There are two main stages in the cell cycle, interphase and cell division, each marked by distinctive cell activities. **Interphase** is the longest stage of the cell cycle in that most cells spend 90% or more of their lifespan in this state. The cell takes in nutrients, manufactures proteins and other substances, expands in size, and conducts its special functions during interphase. In cells that are destined to divide, preparations for cell division also begin during interphase. A critical event in these preparations is the copying of all the DNA molecules, which contain the organism's genetic information in the form of genes.

Cell division is the last stage in the life of an individual cell. Cell division is not only the most rapid stage of the cell cycle, it is also the most dramatic in visual terms. In large cells such as those of lily pollen or fertilized fish eggs, the cellular events of cell division can be readily seen with an ordinary light microscope. Mitotic cell divisions take place in two main steps: the genetic material is separated and symmetrically distributed to opposite sides of the cell during the first step, known as **mitosis**; then the cytoplasm is split to create two separate daughter cells during the second step, known as **cytokinesis** [sye-toh-kih-NEE-siss]. Cell division through mitosis and cytokinesis generates two genetically identical daughter cells from each parent cell. A special type of cell divi-

sion, called **meiosis** [mye-OH-siss] (*meio,* "less"), occurs in reproductive structures to make sperm or egg cells, which are collectively known as **gametes**. We will take a closer look at interphase, and the main events that take place at each of its three phases, before exploring cell multiplication through mitotic cell division and gamete production through meiosis.

9.2 Interphase Is the Longest Stage in the Life of Most Cells

In cells capable of dividing, interphase can be divided into three main segments: G_1, S, and G_2. These are defined by distinctive cellular events that take place. The duplication of DNA, requiring the manufacture, or synthesis, of new DNA, occurs in the **S phase** (S stands for "synthesis"). The **G_1 phase** is the first phase in the life of a "newborn" cell. The **G_2 phase** begins after S phase and before the start of division (see Figure 9.3). The G in G_1 and G_2 stands for "gap," the term early cell biologists bestowed on these phases, believing them to be minor

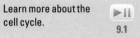
Learn more about the cell cycle.
▶❚❚
9.1

Figure 9.3 The Cell Cycle
The cell cycle consists of two major stages: interphase and cell division (shown by the central gray and black circle). During a mitotic cell division, the parent cell divides into two daughter cells. Interphase can be subdivided into three phases, as shown by the outer circle. The cell prepares itself for division by increasing in size and producing proteins needed for division during G_1 and G_2 phases, and by replicating its DNA during S phase.

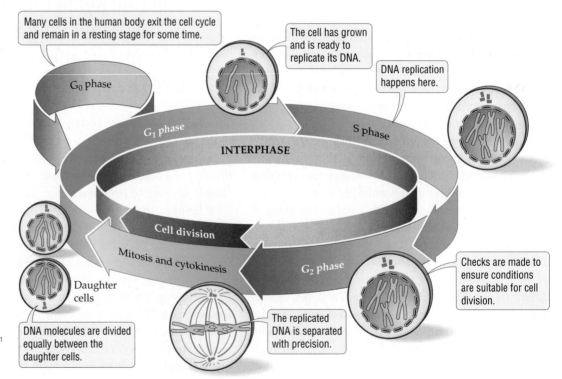

Many cells in the human body exit the cell cycle and remain in a resting stage for some time.

The cell has grown and is ready to replicate its DNA.

DNA replication happens here.

G_0 phase

G_1 phase

S phase

INTERPHASE

Cell division

Mitosis and cytokinesis

G_2 phase

Checks are made to ensure conditions are suitable for cell division.

The replicated DNA is separated with precision.

Daughter cells

DNA molecules are divided equally between the daughter cells.

and less interesting aspects of a cell's life compared to the S phase or cell division itself. We now know that many crucial events occur during the two "gap" phases and some of these events have a profound effect on the precision and stability of the cell division stage.

G_1 and G_2 are important phases for two reasons. They are periods of growth, during which both the size of the cell and its protein content increase. Furthermore, each phase prepares the cell for the phase immediately following it, serving as a checkpoint that ensures that the cell cycle will not progress to the next phase unless all conditions are suitable. The decision to divide the cell is made during the G_1 phase of the cell cycle. When the cell receives specific signals (such as hormones, or proteins called growth factors), and provided the conditions are right (the cell is large enough and there is an adequate nutrient supply, for instance), the cell makes a commitment to divide. Upon receiving the "go ahead" signal, special proteins trigger DNA replication and other processes associated with the next phase, the S phase.

Similarly, during the G_2 phase, another set of cell cycle regulatory proteins triggers the cellular events that prepare the cell for the upcoming cell division. As we saw for the G_1 phase, the G_2 phase serves as a checkpoint that ensures that "all systems are go," and the signal to launch cell division is sent out only if conditions are conducive. For example, if DNA duplication is not complete or if the cell has suffered DNA damage, the cell cycle will halt in the G_2 phase until the DNA is fully copied or DNA damage is repaired.

Not all of the cells in our bodies go through the cell cycle. Many tissues and organs do not normally increase in size in the adult body, nor do their constituent cells wear out or die under normal conditions. Such a cell proceeds to differentiate immediately after its "birth" through a prior cell division, and then it exits the cell cycle to enter a nondividing stage that cell biologists have named the **G_0 phase**. This phase can last for periods ranging from days to years. A cell in the G_0 phase and a nondividing cell in the G_1 phase may appear to behave similarly. The key difference is the complete absence of cell cycle regulatory proteins in G_0 cells. In contrast, these proteins are always present inside cells that are in the G_1 phase, though they might be in an inactive state.

Some cells, such as those that form the lens of the eye, remain in G_0 for life, as part of a nondividing tissue. Most of the nerve cells that make up the brain have also exited the cell cycle, which explains why brain cells lost as a result of physical trauma or chemical damage are usually not replaced. Most liver cells stay in the G_0 phase much of the time, but reenter the cell cycle about once a year on average, to make up for cells that have died because of normal wear and tear. The liver is a highly active organ that is also at the frontline in dealing with toxins, from antibiotics to alcohol, that we commonly ingest. The unusual regenerative capacity of the liver, compared to most other organs in the adult body, compensates for the cell damage the organ suffers in conducting its routine detoxification activities. The liver's regenerative capacity rests heavily on a large pool of G_0 cells that can re-enter the cell cycle when needed.

9.3 Mitosis and Cytokinesis: From One Cell to Two Identical Cells

The climax of the cell cycle is cell division, which, in the case of mitotic divisions, consists of two steps: mitosis and cytokinesis. These steps are not discrete in time; cytokinesis overlaps with the last phases of mitosis.

The central event of mitosis is the equal distribution of the parent cell's DNA into two daughter nuclei. This process, called **DNA segregation**, requires the coordinated actions of many different types of structural proteins. But before discussing the details of DNA segregation during mitosis, we must describe the packing of DNA into chromosomes, because it is these DNA-containing structures that are divided up between daughter nuclei during mitosis.

Each DNA molecule is packaged with proteins to create a chromosome

The DNA in the nucleus is not a disorganized tangle of naked nucleotide polymers. Instead, each long, double-stranded DNA molecule in a eukaryotic cell is attached to proteins that help pack it into a more compact physical structure called a **chromosome**. The packing is necessary because the DNA molecules are enormously long, even in the simplest eukaryotic cells. If all of the 46 different DNA molecules in one of your skin cells were lined up end-to-end, they would make a double helix nearly 2 meters (about 6 feet) in length. How can that much DNA be stuffed into a nucleus, which, in a human cell, has a diameter slightly under 5 micrometers (0.005 millimeters)? Extreme packaging is the answer. Each DNA double helix winds around special DNA

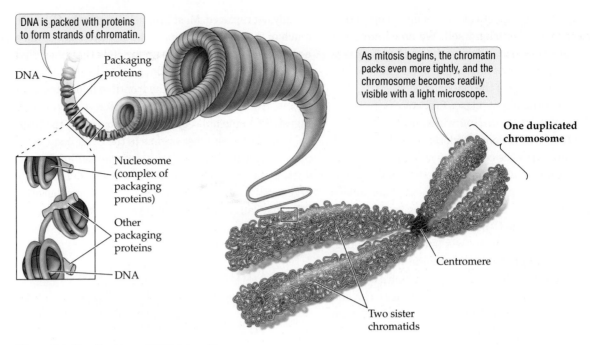

DNA is packed with proteins to form strands of chromatin.

DNA

Packaging proteins

Nucleosome (complex of packaging proteins)

Other packaging proteins

DNA

As mitosis begins, the chromatin packs even more tightly, and the chromosome becomes readily visible with a light microscope.

One duplicated chromosome

Centromere

Two sister chromatids

Figure 9.4 The Packing of DNA into a Chromosome

packaging proteins to create a DNA–protein complex known as **chromatin**. Chromatin is further looped and compressed to form an even more compact structure called a chromosome (Figure 9.4). We will learn much more about genes, chromosomes, and DNA packing and replication in Unit 3. For now, it is enough to know that each chromosome is a compacted DNA–protein complex, and within it is a single long molecule of DNA that bears many genes strung along its length. At the beginning of mitosis, the chromatin becomes packed and condensed even more densely than during interphase, which is why chromosomes are most easily seen at the cell division stage.

Every species has a characteristic number of chromosomes in the nucleus. **Somatic cells** (*soma,* "body") are all those cells in a multicellular organism that are not gametes (egg or sperm) or part of gamete-making tissues. We can think of them as the "generic" cells in the body, those not involved in sexual reproduction. Somatic cells of plants and animals may have anywhere from four to a few hundred chromosomes, depending on the species. During mitosis, when chromosomes are compacted to the maximum extent, the different types of chromosomes in a somatic cell can often be identified under a microscope by their size and distinctive shape. A display of all the mitotic chromosomes found in a somatic cell of any species is known as the **karyotype**

of that species (Figure 9.5*a*). For example, with the exception of gametes and their precursors, all cells in the human body contain a total of 46 chromosomes. Each somatic cell in a horse has 64 chromosomes, and in a corn plant, 20 chromosomes. The total number of chromosomes per cell has no particular significance other than being an identifying characteristic of the species. It does not reflect the number of genes found in that species, nor does it say anything about its structural or behavioral complexity.

A distinctive feature of nearly all eukaryotes, compared to prokaryotes, is that their somatic cells contain two matched copies, or a pair, of every type of chromosome. Each matched pair of the same type of chromosome is called a **homologous** [hoh-*MOLL*-uh-guss] **pair**. Returning to the example of the human karyotype, our 46 chromosomes are actually a double set consisting of 23 pairs of homologous chromosomes. You inherited one set (23 chromosomes) from your mother and the other set (the remaining 23) from your father, to create a double set of 46, or 23 pairs. In 22 of these homologous pairs (numbered 1 to 22), the two **homologues** (individual members of the pair) are alike in length, shape, and the location and type of genes they carry. But the twenty-third pair can be an odd couple, consisting of two very dissimilar homologues, an X chromosome and a Y chromosome. The X and Y chromosomes are called

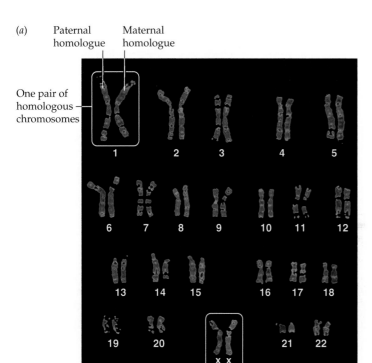

(a) Paternal homologue   Maternal homologue

One pair of homologous chromosomes

1 2 3 4 5

6 7 8 9 10 11 12

13 14 15 16 17 18

19 20 X X 21 22

Sex chromosomes

(b)

Sister chromatids in one duplicated homologue   One pair of duplicated homologues

Figure 9.5 Human Chromosomes

(a) The 46 chromosomes in this micrograph represent the karyotype of a human female. With the help of computer graphics, photos of the chromosomes have been aligned so that the two partners in a pair of homologues are placed next to each other. The non-sex chromosomes (also known as autosomes) are numbered. The sex chromosomes are represented by letters (XX, in the case of this individual). One chromosome in each homologous pair has been artificially shaded pink and the other blue, as a reminder that one homologue came from the individual's mother and the other one (paternal homologue) from her father. (b) By the beginning of mitosis, the chromosomes have been duplicated. Each duplicated chromosome consists of two sister chromatids held together at a constriction point, the centromere.

the **sex chromosomes** because, in mammals and some other vertebrates, these chromosomes determine the sex of the individual animal. In humans and other mammals, individuals with two X chromosomes in their cells are female and those with one X and one Y chromosome are male. The X chromosome is considerably longer than the Y chromosome and carries many more genes. Most, if not all, of the few genes that are found on the Y chromosome appear to be involved in controlling the development of male characteristics. Realize that all of the many genes that are unique to the X chromosome are also present in normal males because their cells contain one X chromosome; however, males (with their XY combination of sex chromosomes) have just the one copy of the X-specific genes; normal females, with their two copies of the X chromosome (XX) have two copies of all the X-specific genes.

Before cell division can proceed, the DNA of the parent cell must be copied (replicated) so that each daughter cell can receive a double set of chromosomes. The replication of DNA occurs during S phase and produces two identical double helices, known as **sister chromatids**, that remain linked to each other until the later stages of mitosis. Therefore, as mitosis begins, the nucleus of a human cell contains twice the usual amount of DNA, since each of the 46 chromosomes now consists of two identical sister chromatids, held together especially firmly at a constriction point called the **centromere** (Figure 9.5b).

Chromosomes become visible during prophase

Mitosis is divided into four main phases, each of them defined by easily identifiable events that are visible under the light microscope (Figure 9.6). Chromosomes undergo a high level of compaction during the first stage of mitosis, called **prophase** (*pro*, "before"; *phase*, "appearance"), and by the end of this phase they are 10 times more tightly wound than interphase chromosomes. In an interphase cell, each chromosome is dispersed through the nucleus in the form of a less compact string of chromatin that is too fine to be visualized by an ordinary light microscope. As the cell moves from G_2 phase into prophase, the chromatin is furled tighter, so that each chromosome becomes shorter and stouter, becoming readily visible in the nucleus.

During prophase, important changes occur in the cytoplasm as well as the nucelus. Two cytoskeletal structures, called **centrosomes** (*centro*, "center"; *some*, "body"), begin to move through the cytosol,

Helpful to know

One way to avoid confusing the similar-sounding terms *centromere* and *centrosome* is to focus on the end of each word. Centromeres (*mere*, "part") are the part of the chromosome where two chromatids are joined firmly at the beginning of mitosis. Centrosomes (*some*, "body") are not part of the chromosome. Like most bodies, they can move: very early in mitosis, they migrate to opposite sides of the cell.

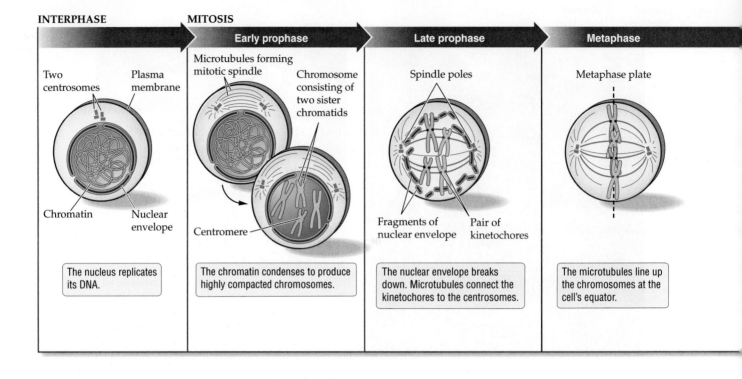

Early prophase   **Late prophase**   **Metaphase**

Two centrosomes

Plasma membrane

Microtubules forming mitotic spindle

Chromosome consisting of two sister chromatids

Spindle poles

Metaphase plate

Chromatin

Nuclear envelope

Centromere

Fragments of nuclear envelope

Pair of kinetochores

The nucleus replicates its DNA.

The chromatin condenses to produce highly compacted chromosomes.

The nuclear envelope breaks down. Microtubules connect the kinetochores to the centrosomes.

The microtubules line up the chromosomes at the cell's equator.

finally halting at opposite sides in the cell. As we shall see, this arrangement of centrosomes defines the opposite ends, or poles, of the cell; most cells split along an imaginary line roughly halfway between the two centrosomes, so that each of the two daughter cells that results from cytokinesis inherits one of the centrosomes.

At the same time that the centrosomes are moving toward the poles of the cell, microtubules are growing outward from each centrosome. Microtubules are long cylinders of special proteins (see Chapter 5), and during prophase some microtubules assemble themselves into an elaborate apparatus, called the **mitotic spindle**. The microtubules of the mitotic spindle will later attach to the chromosomes and help move them through the cytosol to sort them into the two daughter cells.

Chromosomes are attached to the spindle in late prophase

The nuclear envelope breaks down late in prophase, during a step cell biologists call prometaphase (Figure 9.6). With the nuclear envelope out of the way, the microtubules of the mitotic spindle, radiating out from the centrosome at each pole, seek out and attach to the now highly condensed chromosomes. The overall result is that each duplicated chromosome (now consisting of two sister chromatids) is "captured" by the mitotic

spindle and becomes linked to the two centrosomes through microtubules.

The physical structure of the centromere dictates how each chromosome will be attached to the spindle microtubules. Each centromere has two plaques of protein, called **kinetochores** [kih-*NET*-oh-korz], that are oriented on opposite sides of the centromere. Each kinetochore forms a site of attachment for at least one microtubule, so that the two chromatids making up a duplicated chromosome end up being linked to the centrosomes at the opposite poles of the cell. The successful "capture" of each pair of sister chromatids by the mitotic spindle sets the stage for the proper positioning of these duplicated chromosomes in the next phase of mitosis.

Chromosomes line up in the middle of the cell during metaphase

Once each duplicated chromosome becomes linked to both poles of the spindle, its microtubule attachments adjust their lengths to move the attached chromatids toward the middle of the cell. There the chromosomes are lined up in a single plane that, in most types of cell division, is equally distant from the two spindle poles. This stage of mitosis is called **metaphase** (*meta,* "after"), and the plane in which the chromosomes are arranged is called the metaphase plate (see Figure 9.6). The function

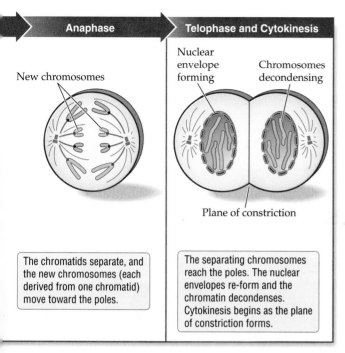

| Anaphase | Telophase and Cytokinesis |

New chromosomes

Nuclear envelope forming

Chromosomes decondensing

Plane of constriction

The chromatids separate, and the new chromosomes (each derived from one chromatid) move toward the poles.

The separating chromosomes reach the poles. The nuclear envelopes re-form and the chromatin decondenses. Cytokinesis begins as the plane of constriction forms.

Figure 9.6 The Stages of Mitotic Cell Division
The stages of mitosis and cytokinesis are shown by diagrams of a dividing cell.

See mitosis and cell division in action.
▶❚❚
9.2

of the elaborate mitotic machinery—spindle microtubules, chromosomes, and centrosomes—is to align the sister chromatids in the proper position and facilitate the equal and balanced segregation of chromatids to the opposite poles of the cell.

Chromatids separate during anaphase

During the next phase of mitosis, called **anaphase** (*ana,* "up"), the two chromatids in each pair of sister chromatids break free from each other and are dragged to opposite sides of the parent cell, achieving the equal and orderly partitioning (segregation) of the duplicated genetic information present in the parent cell. At the beginning of anaphase, the sister chromatids separate as the special proteins holding them together are broken down. Once separated, each chromatid is considered a separate daughter chromosome. The progressive shortening of the microtubules pulls the newly separated daughter chromosomes to opposite poles of the cell. This remarkable event results in the equal segregation of chromosomes into the two daughter cells. Each of the resulting daughter cells now contains the same, complete genetic blueprint as the parent cell. In a human cell, for example, each of the 46 chromosomes is duplicated during S phase, yielding 46 pairs of identical sister chromatids, representing 92 separate molecules of DNA. When the sister chro-

matids separate from each other during anaphase, an identical set of 46 new chromosomes arrives at each of the two poles of the dividing cell.

New nuclei form during telophase

The next phase of mitosis, **telophase** (*telo,* "end"), begins when a complete set of daughter chromosomes arrives at a spindle pole. Major changes also occur in the cytoplasm, changes that set the stage for division of the cytoplasm, and the cell as a whole, to create two new cells. The spindle microtubules break down, and the nuclear envelope begins to form around the chromosomes that have arrived at each pole (see Figure 9.6). As the two new nuclei become increasingly distinct in the cell, the chromosomes within them start to unfold, becoming less distinct under the microscope. Telophase is the last stage of mitosis, and the cell is now ready for physical separation of the cytoplasm to create two whole new daughter cells.

Animal cells, we shall see in the next section, divide by drawing the plasma membrane inward until it meets in the center of the cell, separating the cytoplasm into two compartments. A plant cell, however, has a relatively stiff **cell wall** around it that cannot be pulled in like the mouth of a drawstring bag. Instead, a plant cell is partitioned in a manner similar to the splitting of prokaryotic cells, which often possess cell walls. Guided by cytoskeletal structures, a partition, known as a **cell plate**, appears where the metaphase plate had been. The cell plate, consisting mostly of membrane vesicles, starts forming in telophase (Figure 9.7). Vesicles filled with cell wall components start to accumulate in the region that was previously the metaphase plate. These vesicles fuse with one another, mingling their membranes and their contents to create two new plasma membranes separated by cell wall material.

Cytoplasmic division occurs during cytokinesis

Cytokinesis (*cyto,* "cell"; *kinesis,* "movement") is the process of dividing the parent cell cytoplasm into two daughter cells (see Figure 9.6). In animal cells, the physical act of separation is performed by a ring of protein cables made of actin microfilaments. These

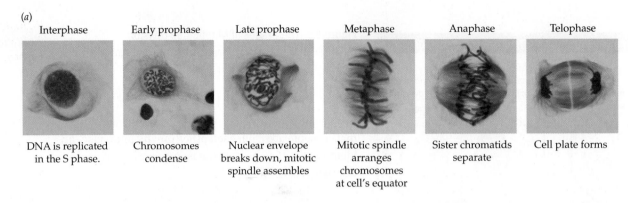

(a)

Interphase	Early prophase	Late prophase	Metaphase	Anaphase	Telophase
DNA is replicated in the S phase.	Chromosomes condense	Nuclear envelope breaks down, mitotic spindle assembles	Mitotic spindle arranges chromosomes at cell's equator	Sister chromatids separate	Cell plate forms

(b)

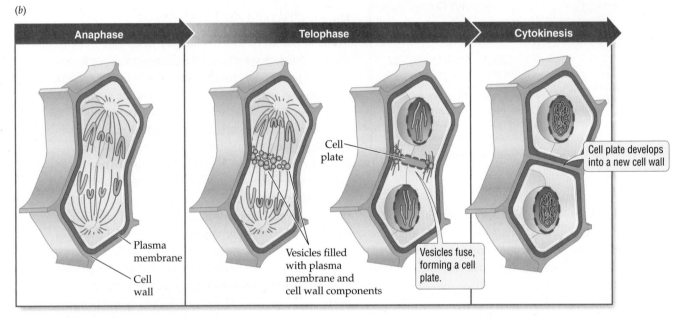

Figure 9.7 Cell Division in Plants

(*a*) A microscopic view of mitosis and cytokinesis in lily pollen. The relatively rigid cell wall that surrounds most plant cells necessitates a different kind of cytokinesis than is found in animal cells. Instead of pinching in two, as animal cells do, plant cells divide by depositing a cell plate down the middle. The cell plate appears as a pale line in the center of the cell, in the last photograph in the series (telophase). The cell plate contains vesicles filled with cell wall components, and its construction begins in telophase as shown in the schematic diagrams (*b*). As the vesicles fuse, they create a plasma membrane for each of the daughter cells (brown), which come to be separated by their newly formed cell walls (blue).

form against the inner face of the plasma membrane like a belt at the equator of the cell. When the actin ring contracts, it draws in the plasma membrane, pinching the cytoplasm and dividing it in two. Since the plane of constriction lies in the plane of the metaphase plate between the two newly formed nuclei, successful cytokinesis results in two daughter cells, each with its own nucleus.

During cytokinesis in plant cells, the cell plate that began forming in telophase matures into two new plasma membranes and two new cell walls, one for each resulting daughter cell (Figure 9.7).

Cytokinesis marks the end of the cell cycle. Once it is completed, the resulting daughter cells are free to enter

the G_1 phase and start the process anew, or to differentiate into a specialized cell type and perhaps take a rest from cell division by entering the G_0 phase.

Concept Check

1. Why is cell division in eukaryotes more complicated than it is in prokaryotes?

2. The house cat karyotype displays a total of 38 chromosomes. How many separate DNA molecules are present in a cat skin cell toward the end of: (a) G_0 phase; (b) G_1 phase; (c) S phase; (d) G_2 phase?

3. What is the difference between mitosis and cytokinesis?

Concept Check Answers

1. Because they have multiple chromosomes that must be sorted evenly into daughter cells.
2. (a) 38; (b) 38; (c) 76; (d) 76.
3. Mitosis refers to nuclear division, whereas cytokinesis is the partitioning of the cytoplasm, and ultimately the whole parent cell, into two daughter cells.

9.4 Meiosis: Halving the Chromosome Set to Make Gametes

Meiosis, we have seen, is a special type of cell division that produces daughter cells with half the chromosome set found in the parent cell. For example, when a human cell with 46 chromosomes undergoes meiosis, it produces daughter cells with only 23 chromosomes, half the set. These daughter cells are broadly known as gametes, and more specifically, *sperm* in male animals and *eggs* in female animals. Gametes are the only cells in the animal body that are created through meiosis. Gametes must have half the chromosome number of nonreproductive cells because of their role in sexual reproduction, a topic we examine next.

Gametes contain half the chromosome set found in somatic cells

Sexual reproduction requires the fusion of two gametes in a process known as **fertilization**. The successful union of an egg and sperm creates a single cell called a **zygote** (Figure 9.8). The zygote then undergoes multiple mitotic divisions to form an embryo, which develops into a new organism.

If the sperm and the egg both contained a double set of chromosomes (46 for humans), the resulting zygote would have twice that chromosome number (92 for humans), and this karyotype would be passed down to all the cells of the embryo through mitotic divisions. Such an embryo would have twice as many chromosomes as either of its parents, and therefore twice the number of genes characteristic of the species. The outcome of this genetic excess would be developmental chaos, resulting in death of the embryo. Therefore, for offspring to have the same karyotype as their parents, fertilization must yield the normal number of chromosomes in the zygote (46 chromosomes in the case of humans).

The simple solution to this problem is for the gametes to contain half of the full set of chromosomes found in somatic cells. This goal is readily achievable in eukaryotes because they possess two copies of every kind of chromosome found in their somatic cells. If *only one copy* from every homologous pair is inherited by a gamete, the chromosome number is halved, but the gamete now possesses one copy of *all* the genetic information (that is, one set of all the genetic information). In humans, for example, all somatic cells contain 23 homologous pairs,

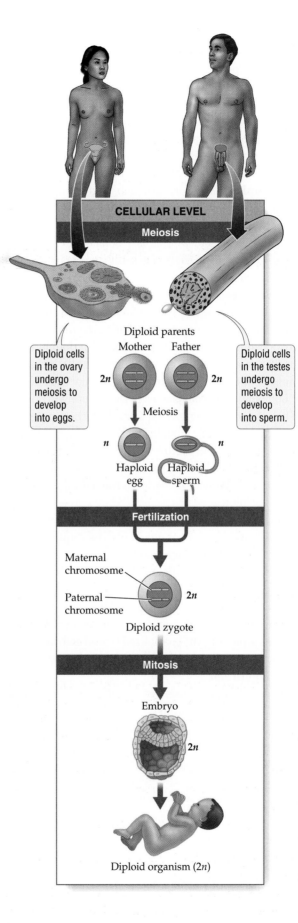

Figure 9.8 Sexual Reproduction Requires a Reduction in Chromosome Number

The fusion of a sperm and an egg at fertilization produces a zygote with the complete diploid (2*n*) chromosome set. This means that each gamete must be haploid (*n*), with half the chromosome set found in most cells in the body. Therefore, each gamete receives only one member of each homologous pair of chromosomes, so that when fertilization occurs, a full homologous pair is reestablished in the zygote. Human somatic cells have 23 pairs of chromosomes. For clarity, only one homologous pair (consisting of a maternal and paternal homologue) is shown in the diploid (2*n*) cells here.

See meiosis in action. ►❚❚
9.3

for a total of 46 chromosomes. Each gamete a person produces, however, contains only one chromosome from each homologous pair, for a total of 23 chromosomes per gamete. Where the sex chromosomes are concerned, all eggs produced by a woman normally contain a single X chromosome, while 50 percent of the sperm produced by a man contain an X chromosome and the rest carry a Y chromosome. Because they contain only one set of chromosomes, not the double set, gametes are said to be **haploid** (*haploos,* "single"), and we assign the symbol n to the number of chromosomes in haploid cells. Somatic cells are **diploid** (*di,* "double") because they have $2n$ chromosomes—that is, a double set of chromosomes compared to the single set haploid cells possess (see Figure 9.8).

Because each gamete contains the haploid number of chromosomes, the zygote formed by fertilization will contain $2n$ chromosomes, that is, the diploid set of chromosomes. In humans, that means a double set of the 23 homologues (46 chromosomes in all), including one pair of sex chromosomes. Furthermore, each pair of homologous chromosomes in the zygote will consist of one chromosome received from the father (**paternal homologue**) and one from the mother (**maternal homologue**), as shown in Figure 9.8. The equal contribution of chromosomes by each parent is the basis for genetic inheritance. We will investigate the details of inheritance in Unit 3.

The day-to-day processes that maintain the life of an individual organism need only the duplication of the diploid state, through mitotic cell divisions. But, as we have seen, the continued propagation of a species through sexual reproduction depends on the ability to generate the haploid state. So there must be a way of generating haploid gametes from diploid cells. That process is meiosis, the specialized type of cell division in which a single diploid cell ultimately yields four haploid cells (Figure 9.8).

The stages of meiosis are broadly similar to those of mitosis. However, unlike mitosis, in which a single nuclear division is sufficient, meiosis involves two rounds of nuclear divisions, each followed by cytokinesis. The two rounds of nuclear and cytoplasmic divisions are named meiosis I and meiosis II, and each has a distinct role in producing haploid cells from a diploid parent cell.

Meiosis I reduces the chromosome number

Let us begin our tour of meiosis by turning our attention to the diploid cells in reproductive tissues that are responsible for the production of gametes. **Meiosis I**, the first step in making haploid gametes from these diploid cells, reduces the chromosome number from $2n$ to n. The first unique aspect of meiosis I, not seen at any stage of mitosis, is the coming together of each duplicated homologous pair. In other words, early in meiosis I, each maternal homologue pairs off with its matching paternal homologue (see Figure 9.9). Furthermore, the sister chromatids in each duplicated chromosome (whether paternal or maternal) remain attached to each other throughout meiosis I, instead of coming apart and going to opposite poles as they do in mitosis. It is the paternal and maternal homologues, which were part of a homologous pair in the parent cell, that are sorted into two separate daughter cells at the end of meiosis I. Put another way, mitosis brings about the separation and sorting of sister chromatids into different daughter cells, but Meiosis I leads to the separation and sorting of each homologous chromosome pair so that daughter cells receive only one member of the pair.

The pairing off and orderly sorting of homologous chromosomes during meiosis I is what makes it possible for the resulting daughter cells to inherit exactly half the chromosome set of the parent cell. It is important to realize that each of the two daughter cells produced by meiosis I receives *one* of the two chromosomes that make up a pair of homologous chromosomes in the parent cell. That is, each daughter cell gets one copy of chromosome 1 (either the maternal or paternal one), one copy of chromosome 2 (either the paternal or maternal one), and so on, including, in the case of mammals, one or the other copy of the two sex chromosomes. Normally, if meiosis proceeds smoothly, neither daughter cell is "shorted" in this division of the parent cell's chromosomal legacy; for example, one daughter cell is not likely to be missing chromosome 3 altogether, with the other daughter cell inheriting both the maternal and paternal homologues of chromosome 3. The basis of this even-handed distribution of homologous pairs will become clearer as we examine the mechanics of meiosis I in greater detail.

The paternal and maternal partners of each homologous chromosome pair align themselves next to each other during prophase I of meiosis (Figure 9.9). Realize that each of these chromosomes was replicated during the S phase and therefore consists of two chromatids. Like a pair of "Siamese" twins, each paternal and maternal chromosome is made up of two chromatids that will remain bound to each other throughout meiosis I. When these duplicated paternal and maternal chromosomes pair off with each other, they are

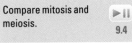
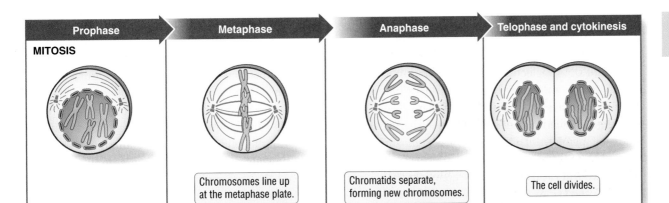

Prophase	Metaphase	Anaphase	Telophase and cytokinesis
MITOSIS	Chromosomes line up at the metaphase plate.	Chromatids separate, forming new chromosomes.	The cell divides.

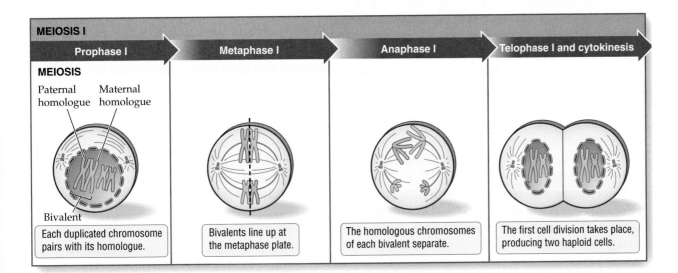

MEIOSIS I

Prophase I	Metaphase I	Anaphase I	Telophase I and cytokinesis
MEIOSIS Paternal homologue Maternal homologue Bivalent Each duplicated chromosome pairs with its homologue.	Bivalents line up at the metaphase plate.	The homologous chromosomes of each bivalent separate.	The first cell division takes place, producing two haploid cells.

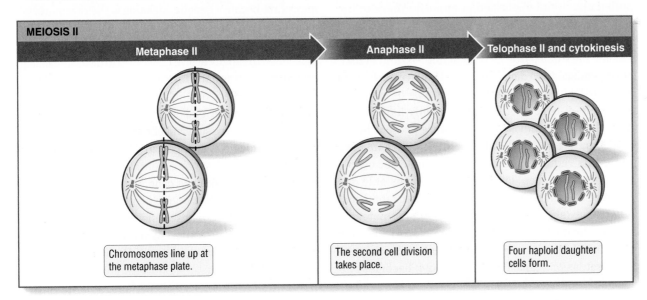

MEIOSIS II

Metaphase II	Anaphase II	Telophase II and cytokinesis
Chromosomes line up at the metaphase plate.	The second cell division takes place.	Four haploid daughter cells form.

Figure 9.9 Similarities and Differences between Meiosis and Mitosis

The major difference between meiosis and mitosis can be seen in meiosis I. During meiosis the homologous chromosomes are paired during prophase I through metaphase I, resulting in a separation of the homologues at the end of meiosis I (the reduction division). In contrast, meiosis II is more similar to mitosis. For the sake of simplicity and greater clarity, not all the steps described in the text are illustrated here.

collectively known as a **bivalent**. This means each bivalent, consisting of a pair of duplicated homologous chromosomes, contains a total of *four* chromatids (four individual DNA molecules). At this point, an extraordinary process unfolds: the maternal and paternal partners in a bivalent swap pieces of themselves! This exchange of genetic material between the paternal and maternal members of a homologous chromosome pair is brought about by a process called **crossing-over**, a subject we will return to after we complete the tour of meiosis I and II.

Late in prophase I, spindle microtubules extend from the two centrosomes to capture each bivalent. They do so in such a manner that the microtubules from one centrosome attach themselves to only one member of the bivalent, either the maternal homologue or the paternal homologue, as shown in Figure 9.9. As metaphase I begins, the shortening and lengthening of the attached microtubules positions each bivalent at the metaphase plate. Anaphase I begins after all the bivalents have been captured and positioned at the equator of the cell. As the spindle microtubules begin to shorten during anaphase I, the paternal and maternal partners in each bivalent are pulled to opposite poles of the cell. Although this process looks superficially similar to the anaphase of mitosis, realize that during anaphase I it is homologous chromosome pairs, not chromatids, that are pulled apart and deposited at opposite poles of the cell (Figure 9.9).

After anaphase I of meiosis, the events of telophase I follow the same patterns seen in mitosis, and cytokinesis produces two daughter cells. Unlike cells formed by mitosis, however, the daughter cells of meiosis I are haploid, because they contain only half the set of duplicated chromosomes that were present in the diploid parent cell at the start of meiosis. As we have seen, this reduction in the chromosome set comes about because each daughter cell receives only one member of each homologous pair, either the maternal homologue (shown in pink in Figure 9.9) or paternal homologue (shown in blue in Figure 9.9). Meiosis I is called a *reduction division* because it halves the chromosome set, as one diploid parent cell (2*n*) becomes two haploid daughter cells (*n*).

Test your knowledge of the meiotic stages.

9.5

Meiosis II achieves the segregation of sister chromatids

The two haploid cells formed after meiosis I go through a second division, called **meiosis II**. This time

the stages of the division cycle are more like those of mitosis. In particular, sister chromatids separate at anaphase II, leading to an equal segregation of chromosomes into the daughter cells (see Figure 9.8). In this manner, the two haploid cells produced by meiosis I give rise to a total of four haploid cells. The haploid cells differentiate into gametes containing half the number of chromosomes that are found in a somatic cell. The reduction in chromosome number that is achieved through meiosis I offsets the combining of chromosomes when gametes fuse during fertilization. It is nature's way of maintaining the constant chromosome number of a species during sexual reproduction.

Meiosis and fertilization contribute to genetic variation in a population

Meiosis and fertilization are the means of sexual reproduction in eukaryotes. Individuals in a population tend to be genetically different from each other because sexual reproduction leads to offspring that are not only genetically different from their parents but also from their siblings. You may resemble one or the other, or both, of your parents, but you cannot be genetically identical to either of them. Similarly, you may resemble a brother or sister, but you are not a clone of anyone, unless you have an "identical twin." Genetic diversity in a population is important because genetic variation is the raw material evolution acts on. For example, the genetic diversity created by sexual reproduction may allow a eukaryotic population to rapidly evolve resistance to disease-causing organisms such as harmful strains of viruses or bacteria. A diverse population is likely to contain at least a few individuals who can fend off an otherwise fatal infection and thereby ensure survival of the species.

Where does the genetic variation in a population come from in the first place? Mutations, which are accidental changes in the DNA code, are the ultimate source of genetic variation in all types of organisms. The mutation of a given gene creates a different "flavor" or genetic variant of that gene. Different versions of a particular gene, created ultimately through DNA mutations, are known as **alleles**.

While mutations create alleles, meiosis is exceptionally effective at shuffling them. We will see shortly that meiosis in a single individual can generate a staggering diversity of gametes, and meiosis does so by shuffling alleles between homologous pairs and then sorting these scrambled homologues *randomly* into gametes. The randomness of fertilization adds to genetic diversity in sexually reproducing populations. Entirely new combinations of genetic information are created when an egg with a unique genetic makeup fuses with one of many genetically diverse sperm cells.

Meiosis generates genetic diversity in two ways: crossing-over between the paternal and maternal members of each homologous pair, and the independent assortment of paternal and maternal homologues during meiosis I. Let us consider crossing-over first. As mentioned earlier in our discussion of prophase I, crossing-over is the name given to the physical exchange of chromosomal segments between nonsister chromatids in paired off paternal and maternal homologues. Early in prophase I, every paternal and maternal homologue, now duplicated and therefore consisting of two chromatids, finds its homologous partner and the two align themselves closely (Figure 9.10). Crossing-over is initiated when a chromatid belonging to the paternal homologue makes contact with the maternal chromatid across from it. These nonsister chromatids contact each other at one or more random sites along their length. Special proteins at the crossover site break these nonsister chromatids and then connect the broken ends of one chromatid with the free ends of the other broken chromatid.

As Figure 9.10 shows, the overall result of this breaking and joining is that the nonsister chromatids exchange one or more segments along their length. The swapped segments contain the same genes positioned in the same order. But as we have seen, genes can exist in different versions, called alleles. Crossing-over exchanges alleles between the paternal and maternal chromatids. Therefore, chromatids produced by crossing-over are genetic mosaics, bearing new combinations of alleles compared to those originally carried by the paternal and maternal homologues in the diploid parent cell. The mosaic chromatid is said to be recombined, and the creation of new groupings of alleles through exchange of DNA segments is known as **genetic recombination**. Without crossing-over, every chromosome inherited by a gamete would be just the way it was in the parent cell. But because of crossing over, meiosis can produce at least four different types of gametes, if we consider crossing-over between *just one* pair of homologous chromosomes, as shown in the bottom panel of Figure 9.10.

The possibilities for creating genetically diverse gametes isn't restricted to crossing-over. The **independent assortment of chromosomes**, which refers to the random distribution of maternal and paternal chromosomes into daughter cells during meiosis I, also contributes to the genetic variety of the gametes produced. It comes about because each homologous

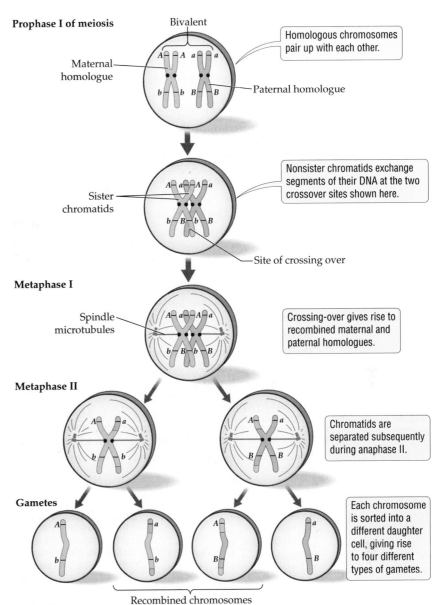

Figure 9.10 Crossing-Over Produces Recombinant Chromosomes
Crossing-over is the physical exchange of corresponding segments between the nonsister chromatids in a pair of homologous chromosomes that have paired off during prophase I. For clarity, only one maternal and one paternal chromatid are depicted as exchanging segments. In reality, each of the two maternal chromatids may be involved in crossovers with the paternal chromatid aligned closest to it. In human cells undergoing meiosis, most bivalents display one to three crossover sites, with longer chromosomes more likely to have multiple crossovers. A crossover site can be located at any point along the length of the paired homologues (bivalent), not just at the tips of the chromatids. The letters (*A/a* and *B/b*) represent alternative alleles of two genes, *A* and *B*. Note that the parental combinations of these alleles have been shuffled in the recombinant chromosomes.

chromosome pair in a given meiotic cell orients itself randomly, that is, without regard to the alignment of any other homologous pair, when it lines up at the metaphase plate during meiosis I.

To illustrate why the random alignment of homologous pairs produces varied patterns in the subsequent sorting of paternal and maternal homologues, and therefore genetically varied gametes, let us consider a cell with just two homologous chromosome pairs ($n = 2$, $2n = 4$). During metaphase I, there are two ways of arranging each homologous pair at the metaphase plate: option A places the maternal homologue "on the left" and paternal homologue "to the right," for *both* chromosome pairs; option B would keep the first homologous pair in the same orientation as in option A, but reverse the orientation of the second homologous pair, so this second couple is oriented paternal homologue "on the left" and maternal homologue "to the right." As we can see in Figure 9.11, option A would sort the maternal and paternal homologues of the two pairs in one particular pattern, creating two different types of gametes. Orienting the two homologous pairs according to option B produces a *different* pattern for the assortment of paternal and maternal homologues, leading to two types of gametes, both with a different combination of homologues than is seen in gametes created by option A.

What this means is that four types of gametes, each with a different combination of paternal and maternal homologues, can potentially be generated through meiosis in a diploid cell containing just 2 pairs of homologous chromosomes ($n = 2$). How many types of gametes can be made by a meiotic parent containing 3 pairs of homologous chromosomes ($n = 3$)? Since there are two ways to arrange each of the three pairs of homologues, 2^n patterns are possible; therefore, 2^3, or 8, different types of gametes could be created.

What about meiosis in human cells, with $n = 23$? Since each of the 23 pairs of homologous chromosomes can be oriented one of two ways, there are 2^{23}, or 8,388,608, different ways of combining homologues in the gametes. Of these 8,388,608 ways of mixing and matching homologues, only one is the combination found in the original diploid cell that underwent meiosis. As we saw in the case of crossing-over, the independent assortment of chromosomes creates gametes that are likely to be different from the parent, and also from each other, with respect to the mix of chromosomes they contain.

Finally, fertilization has the potential to add a tremendous amount of genetic variation to that already produced by crossing-over and the independent assortment of chromosomes. In the previous paragraph we saw that each sperm cell represents one of over 8 million possible outcomes of independent assortment during meiosis in humans; similarly, every egg represents one

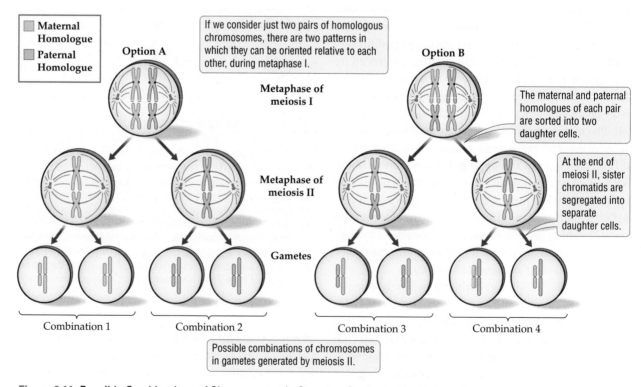

Figure 9.11 Possible Combinations of Chromosomes in Gametes Generated by Meiosis II

out of over 8 million egg types with respect to chromosomal combinations. If we set aside the variation caused by crossing-over, this means there are over 64 trillion (8 million possible sperm × 8 million possible eggs) genetically different offspring that could be formed each time a sperm fertilizes an egg. The total number of humans that have ever lived is estimated to be no more than 100 billion. Therefore, the chance that two siblings will be genetically identical is less than 1 in 64 trillion. (The exception is identical twins; because they arise from a single zygote, the chance that they will be genetically identical is 100 percent.) The random events in meiosis and the random fertilization of a certain egg by a particular sperm give each of us our genetic uniqueness.

Concept Check

1. If the diploid number for cats is 38, how many chromosomes are present in the egg cell of a house cat?
2. How many molecules of DNA are present in each daughter cell at the end of meiosis I, when a cat reproductive cell undergoes meiosis?
3. Meiosis I achieves the reduction of the diploid chromosome set. What role does meiosis II play in gamete formation?
4. How does meiosis contribute to genetic variation in a population?

Applying What We Learned

The Biology of Stem Cells

Stem cells are undifferentiated cells with the exceptional attribute of self-renewal: a stem cell can undergo mitotic divisions to produce more of itself and, theoretically, it can continue to do so forever. Some descendants of stem cells may go on to differentiate into specialized cell types—skin cells, or nerve cells, or heart muscle cells—but enough of the daughter cells must turn into stem cells to maintain the stem cell pool. There are two main categories of stem cells: embryonic stem cells and adult stem cells. Before we describe the properties of these two types of stem cells, their potential in advancing human health, and why the research on embryonic stem cells has stirred impassioned debate, let's take a closer look at the development of the human body.

Female meiosis produces haploid eggs, male meiosis creates haploid sperm. The fusion of egg and sperm during fertilization produces a diploid zygote that starts dividing through mitosis to make a ball of cells known as the **morula** (Latin for "mulberry"). All the cells in the morula are **totipotent**, which means they are capable of giving rise to any human cell type. The cells in the early stage morula are developmentally flexible because they are not committed to any specific developmental pathway.

About 5 days after fertilization, the morula is transformed into a hollow ball of cells known as a **blastocyst** (Figure 9.12). It contains about 150 cells. Inside the blastocyst is a group of about 30 cells, known as the **inner cell mass**, which gives rise to the embryo proper. The inner cell mass is **pluripotent**, which means these cells are capable of producing all of the 220 cell types in the adult body. The outer cell layer of the blastocyst turns into tissues that will surround the developing embryo, and later participate in forming the placenta. The cells of the outer layer are said to be **multipotent**, because the range of cells that their descendants can differentiate into is relatively narrow. They lack the *complete* developmental flexibility of the cells in the morula, and the *nearly* complete flexibility of the inner cell mass. About 5–7 days after fertilization, the blastocyst attaches to the lining of the uterus (womb) in a process known as implantation. Mitosis continues at a rapid pace but from this point on, many of the daughter cells begin differentiating into special cell types, as the three main tissue layers of the vertebrate embryo emerge.

Around 10 weeks after fertilization, most of the organs systems are established and the embryo is transformed into a fetus. The developmental flexibility of most fetal cells declines as development proceeds.

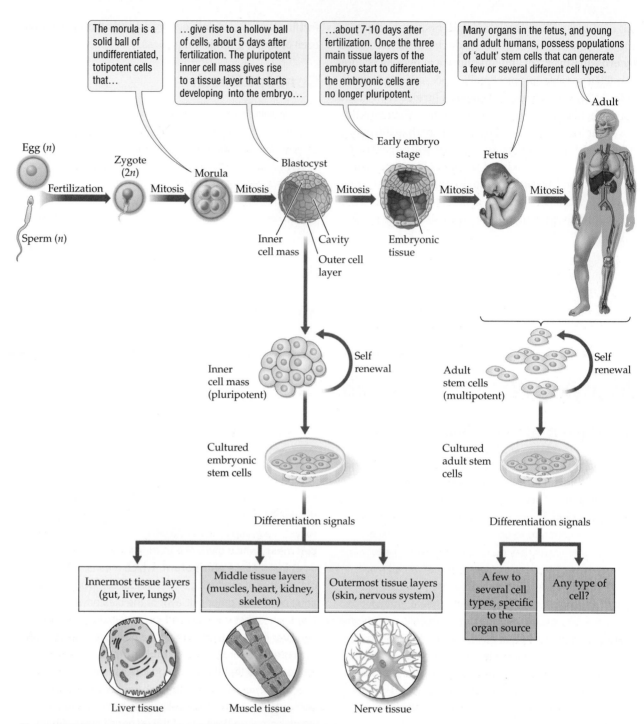

The morula is a solid ball of undifferentiated, totipotent cells that...

...give rise to a hollow ball of cells, about 5 days after fertilization. The pluripotent inner cell mass gives rise to a tissue layer that starts developing into the embryo...

...about 7-10 days after fertilization. Once the three main tissue layers of the embryo start to differentiate, the embryonic cells are no longer pluripotent.

Many organs in the fetus, and young and adult humans, possess populations of 'adult' stem cells that can generate a few or several different cell types.

Adult

Egg (n)

Sperm (n)

Zygote (2n)

Fertilization   Mitosis   Morula   Mitosis   Blastocyst   Mitosis   Early embryo stage   Mitosis   Fetus   Mitosis

Inner cell mass   Cavity   Outer cell layer   Embryonic tissue

Inner cell mass (pluripotent)   Self renewal

Adult stem cells (multipotent)   Self renewal

Cultured embryonic stem cells

Cultured adult stem cells

Differentiation signals   Differentiation signals

| Innermost tissue layers (gut, liver, lungs) | Middle tissue layers (muscles, heart, kidney, skeleton) | Outermost tissue layers (skin, nervous system) | A few to several cell types, specific to the organ source | Any type of cell? |

Liver tissue   Muscle tissue   Nerve tissue

Figure 9.12 Human Development and the Origins of Stem Cells
Embryonic stem cells are pluripotent and their descendants can therefore differentiate into a great variety of specialized cell types in response to the appropriate molecular signals. Adult stem cells, present in low numbers in well-developed tissues and organs, are multipotent and normally give rise to a more limited range of cell types.

At birth, only small populations of cells in various tissues and organs retain the capacity for self-renewal and such cells are called **adult stem cells** because they persist into adulthood (though they are present in newborns and children as well). Adult stem cells, also known as somatic stem cells, are multipotent. They retain the potential to generate a number of different cell types, but that potential is not as broad as that of the inner cell mass or morula. Adult stem cells have been identified in a number of organs and tissue types in the adult body, including skin, muscle, bone marrow, liver, brain, and the eye (Figure 9.13).

The identification of stem cells in the adult body immediately suggested a number of medical therapies and some of this promise has been substantially realized, for example, in the treatment of blood disorders such as leukemia and in treating severe burns. Skin stem cells can be isolated from any patch of healthy skin that survives on a burn victim, and when given the right conditions in the lab, they quickly generate sheets of skin tissue that can be grafted on the patient's body. Hopes were raised that the same methods could be used to repair and regenerate damaged tissue in conditions such as Alzheimer's disease, spinal cord injury, muscular dystrophy, multiple sclerosis, and of course, Parkinson's disease. But adult stem cells from organs other than skin and bone marrow are generally difficult to work with.

With the advent of assisted reproduction in the 1990s, scientists were able to grow the inner cell mass, derived from donated embryos, in a lab dish. The technique of in vitro fertilization (IVF) involves combining egg and sperm in a lab dish. The resulting zygote is supplied special nutrients and kept at controlled temperatures as it develops into a morula and then into a blastocyst. The blastocyst must be implanted in the woman seeking fertility treatment, or else frozen for later use, because it cannot be kept alive in a dish during the later stages of development. More than 400,000 human blastocysts are held in deep freezers at fertility clinics across the United States.

In the late 1990s, a number of "surplus" blastocysts were donated by couples to researchers who had developed methods for growing human cells derived from the inner cell mass in a nutrient medium, building on knowledge gained from manipulating mouse blastocysts. The pluripotent cells displayed their stem cell characteristics by dividing over and over without differentiating. Several lines of such human **embryonic stem cells** (hES cells) were developed by labs in the United States and elsewhere, using embryos donated by fertility clinics. (Cells descended from a common parent cell through an unbroken chain of mitotic divisions are said to belong to a line.) The announcement in 1998 that human embryonic stem cells had been multiplied in a dish, and been made to differentiate into certain types through the addition of specific signaling molecules, was met with elation as well as condemnation.

Researchers have been interested in hES cells because these cells are pluripotent and can therefore generate a much greater variety of cell types than multipotent adult stem cells. Adult stem cells from tissues

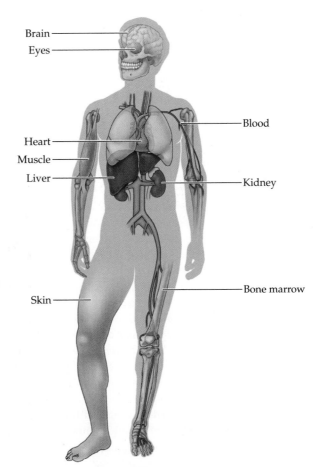

Figure 9.13 Adult or Somatic Stem Cells
Small numbers of these cells have been identified in certain tissues and organs in the human body. Adult stem cells are multipotent, rather than pluripotent like embryonic stem cells. Adult stem cells from most organs (except skin and bone marrow) are harder to identify, isolate, and culture (grow in a lab dish), compared to embryonic stem cells derived from the inner cell mass.

such as the brain, the heart, and the kidney are small, scarce, hard to identify, and difficult to grow in a lab dish.

Most opponents of embryonic stem cell research believe that the life of a human being begins at conception; therefore an embryo has moral status and it is unethical to use its cells for someone else's benefit. In 2001, the United States government cut off funding for any research involving human embryos and banned the creation of new hES lines using public funds. No new hES lines have become publicly available since then, although about a dozen cell lines developed before the moratorium are still in use. Researchers say those pre-2001 cell lines have deteriorated as a result of contamination and the inevitable accumulation of mutations. Advocates of stem cell research maintain that creating new hES lines would greatly increase the pace of basic research and the therapeutic applications of stem cell technology.

New Type of Stem Cells May Help Regenerate Heart Tissues

By Adrienne Law

When cold, sweaty skin and pressure, burning and tightness in the chest strikes, it could just be a sign of a heart attack. But heart transplants, pacemakers and multiple types of drugs may not be the only solution for fixing permanently damaged heart tissue post-heart attack.

Last week, the scientific journal *Stem Cells* published a UCLA study about stem cells that may regenerate different kinds of human cardiovascular cells. The article, written by Dr. Robb MacLellan and his colleagues at the David Geffen School of Medicine at UCLA, points to a new type of stem cell that may be used to regenerate heart tissue that dies during a heart attack because of a lack of blood flow.

Induced pluripotent stem cells function like the more commonly used embryonic stem cells, which are currently used in medical research to clone cells from various parts of the human body . . . [They] mimic signals similar to embryonic stem cells found in human embryos, said Dr. Martin Pera, the director of the Center for Stem Cell and Regenerative Medicine at the University of Southern California.

"This [research] is an important step forward showing that these cells can do what embryonic stem cells can do," he said. But unlike embryonic stem cells, induced pluripotent stem cells do not come from human embryos, and as such can help avoid many ethical issues, he said. Like embryonic stem cells, these stem cells can be used to regrow organ system-specific tissue following an adverse medical event.

One of the more severe effects of a heart attack is dead heart muscles that cannot be regrown during recovery. Instead, doctors can use induced pluripotent stem cells to replace what is missing by growing cardiac muscle cells and blood vessel cells, Pera said.

The creation of induced pluripotent stem cells (iPS) is one of the most exciting developments in cell biology and biomedicine in recent years. This extraordinary achievement was first described by Shinya Yamanaka, then at Kyoto University, Japan, and Rudolf Jaenisch and his colleagues at the Massachusetts Institute of Technology in Boston. An iPS is any cell, even a highly differentiated cell in the adult body, that has been genetically reprogrammed to mimic the pluripotent behavior of embryonic stem cells (hES). Scientists reached this milestone by sorting through the many genes that are uniquely active in embryonic stem cells. They identified four in particular that function as master switches, turning other genes on or off and thereby controlling the pluripotency of these cells. Amazingly, manipulating just these four genes in the nucleus of a differentiated cell turned back the developmental clock, making the differentiated cell behave as though it were still embedded in the inner cell mass inside a blastocyst. Researchers introduced the critical genes into mouse skin cells (fibroblasts) using harmless retroviruses as delivery vehicles.

The scientists whose work is described in this news report were the first to achieve the feat using human cells. One large benefit of tissue repair therapies based on iPS is that the adult cell subjected to reprogramming would be obtained from the patient. As a result, the patient's defense system would not reject the iPS cells if they were placed in injured or diseased tissue after undergoing reprogramming. The problem of tissue rejection has been a major obstacle in the use of hES in regenerative medicine because hES cell lines are genetically different from any potential recipient of stem cell therapy, and will therefore be perceived as "foreign" and attacked by the patient's defense system.

The discoverers of iPS are quick to point out that the technology is still in its infancy and it may take five to ten years before iPS-based therapies can be tested in human subjects. The public debate on the use of stem cells has often centered on whether adult stem cells are sufficient as a source of therapeutic stem cells and whether blastocysts, even the "unwanted" blastocysts of donated embryos, should be used to create new hES cell lines for therapeutic uses. Most stem cell researchers emphasize the vital importance of studying all three types of stem cells because each has added, and continues to add, to our understanding of how, when, and where cells divide or differentiate.

Evaluating the News

1. Compare iPS with embryonic stem cells and adult stem cells, both in terms of the developmental flexibility of these cells and their origins.
2. In 2004, Californians passed Proposition 71, which generated $3 billion in funding for embryonic stem cell research and turned the state into a magnet for stem cell researchers. Do you think controversial issues in science should be resolved by panels of experts, elected government officials, or put to a vote by the citizens, as was done in California? Explain your rationale.
3. It is possible to remove a cell from an early stage morula without harming the embryo. Should it be permissible to remove a cell in such a nondestructive manner and use it to make new hES lines? Explain your viewpoint.

SOURCE: *Daily Bruin,* Los Angeles, CA, May 5, 2008.

Chapter Review

Summary

- Cell division is necessary for the renewal and repair of tissues in multicellular organisms and for asexual and sexual reproduction.
- Stem cells have the capacity for self-renewal and can also give rise to descendants that develop into differentiated cell types with specialized cell functions.
- Prokaryotes divide through binary fission.
- Cell division in eukaryotes is more complicated than in prokaryotes.

9.1 The Cell Cycle Consists of Two Main Stages: Interphase and Cell Division

- The cell cycle refers to the series of events that extend over the lifespan of a eukaryotic cell, beginning with the origin of that cell and ending with its division into daughter cells.
- Interphase and cell division are the two main stages in the cell cycle.

9.2 Interphase Is the Longest Stage in the Life of Most Cells

- Interphase consists of three phases: G_1, S, and G_2.
- During S phase, the cell's DNA is replicated. During phases G_1 and G_2, which straddle the S phase, the cell increases in size and produces specific proteins needed for division.
- Cells in the G_0 phase are in a resting state and do not divide.

9.3 Mitosis and Cytokinesis: From One Cell to Two Identical Cells

- During mitotic cell divisions, the parent cell's replicated DNA is distributed in such a way that each daughter cell receives all the genetic information that was present in the parent cell.
- Each DNA molecule in the nucleus winds around bundles of packaging proteins to create a DNA-protein complex known as chromatin. Chromatin is further compacted into a chromosome. Each chromosome contains a single DNA molecule, bearing many genes along its length.
- The karyotype of a species refers to a display of all the chromosomes that are characteristic of a somatic cell of that species. In eukaryotes, all the chromosomes in a somatic cell are part of a pair, known as a homologous pair, matched on the basis of their size, shape, and other characteristics.
- One homologue in each pair is inherited from the maternal parent, the other from the paternal parent. In mammals, the sex chromosomes (X and Y) determine gender. Females have two copies of the X chromosome, males have one X and one Y chromosome.
- In the S phase (prior to mitosis), DNA replication produces chromosomes made up of two identical side-by-side chromatin strands, called sister chromatids, that are held together firmly at the centromere.
- The four main stages of mitosis are prophase, metaphase, anaphase, and telophase.
- In early prophase, chromosomes shorten and the nuclear envelope breaks down. The two centrosomes move to opposite poles of the cell, and the mitotic spindle forms between them.

- In late prophase, the duplicated chromosomes become attached to the mitotic spindle at their kinetochores.
- In metaphase, the chromosomes are positioned at the center of the cell, at the metaphase plate.
- In anaphase, the two chromatids of each chromosome move to opposite ends of the cell. Once separated, each former chromatid becomes a new chromosome.
- In telophase, new nuclei form as the nuclear envelope assembles around each set of daughter chromosomes at the opposite poles of the cell.
- During cytokinesis, the cytoplasm of the parent cell is physically divided to create two daughter cells. In animal cells, a ring of actin microfilaments pinches the cell in two. In plant cells, a new cell wall forms to separate the two daughter cells.

9.4 Meiosis: Halving the Chromosome Set to Make Gametes

- Meiosis is the special division process that produces gametes, which are haploid (n), containing only one chromosome from each homologous pair. Other cells in the body (somatic cells) are produced by mitosis and are diploid ($2n$), containing a double set of each type of chromosome.
- Meiosis consists of two rounds of nuclear and cytoplasmic divisions.
- Meiosis I is the reduction division because it produces haploid daughter cells from a diploid parent cell.
- During meiosis I, the maternal and paternal members of each homologous pair align themselves parallel to each other to form bivalents; each bivalent contains four chromatids in all.
- The two members of each homologous pair (the maternal and paternal partners) are sorted into different daughter cells at the end of meiosis I. Because each such daughter cell has only one member of each homologous pair, it has only one set (the haploid set, n) of chromosomes.
- Meiosis II is similar to mitosis in that sister chromatids are segregated into separate daughter cells at the end of cytokinesis.
- The fusion of two gametes is called fertilization and results in a diploid zygote ($2n$).
- Meiosis and fertilization introduce genetic variation in a population.
- Meiosis generates diverse gametes through two means: crossing-over and the independent assortment of homologous chromosomes.

◉ Review and Application of Key Concepts

1. Describe what happens in each stage of the cell cycle.

2. Horses have a karyotype of 64 chromosomes. How many separate DNA molecules are present in a horse cell that is in the G_2 phase just before mitosis? How many separate DNA molecules are present in each cell produced at the end of meiosis I in a horse?

3. Compare mitotic cell division with meiosis by completing the table below. Consider whether the statement is correct as it applies to the type of division. Write "True" or "False" in each column, as in the first example.

	Mitotic cell division	Meiosis
1. In humans, the cell undergoing this type of division is diploid.	True	True
2. The daughter cells have half as many chromosomes as the parent cell.		
3. A total of four daughter cells are produced when one parent cell undergoes this type of division.		
4. In male animals, this type of division directly gives rise to sperm.		
5. Stem cell self-renewal involves this type of division.		
6. The daughter cells are genetically identical to the parent cell.		
7. This type of division involves two nuclear divisions.		
8. This type of division involves two rounds of cytokinesis.		
9. Maternal and paternal homologues pair up to form bivalents at some point during this type of division.		
10. Sister chromatids separate from each other at some point during this type of cell division.		

4. What difference would you find in the karyotypes of male and female humans? How would this be reflected in the gametes produced by males and females?

5. For sexually reproducing organisms, how would the chromosome numbers of offspring be affected if gametes were produced by mitosis instead of meiosis?

Key Terms

adult stem cell (p. 198)
allele (p. 194)
anaphase (p. 189)
asexual reproduction (p. 182)
binary fission (p. 182)
bivalent (p. 194)
blastocyst (p. 197)
cell cycle (p. 183)
cell differentiation (p. 183)
cell division (p. 184)
cell plate (p. 189)
cell wall (p. 189)
centromere (p. 187)
centrosome (p. 187)
chromatin (p. 186)
chromosome (p. 185)
crossing-over (p. 194)
cytokinesis (p. 184)
diploid (p. 192)
DNA segregation (p. 185)
embryonic stem cell (p. 199)
fertilization (p. 191)
G_0 phase (p. 185)
G_1 phase (p. 184)
G_2 phase (p. 184)
gamete (p. 184)
genetic recombination (p. 195)
haploid (p. 192)
homologous pair (p. 186)

homologue (p. 186)
independent assortment of chromosomes (p. 195)
inner cell mass (p. 197)
interphase (p. 184)
karyotype (p. 186)
kinetochore (p. 188)
maternal homologue (p. 192)
meiosis (p. 184)
meiosis I (p. 192)
meiosis II (p. 194)
metaphase (p. 188)
mitosis (p. 184)
mitotic spindle (p. 188)
morula (p. 197)
multipotent (p. 197)
paternal homologue (p. 192)
pluripotent (p. 197)
prophase (p. 187)
S phase (p. 184)
sex chromosome (p. 187)
sexual reproduction (p. 183)
sister chromatids (p. 187)
somatic cell (p. 186)
stem cell (p. 183)
telophase (p. 189)
totipotent (p. 197)
zygote (p. 191)

Self-Quiz

1. In the cell cycle, duplication of DNA occurs in the
   a. G_1 phase.
   b. S phase.
   c. G_2 phase.
   d. division stage.

2. A karyotype is
   a. a display of all the chromosomes in a diploid cell derived from an individual of a particular species.
   b. necessary for the physical separation of the daughter cells.
   c. a pair of identical chromosomes.
   d. the same in all species.

3. Which of the following statements is true?
   a. Chromatin is more highly compacted in prophase than it is during the G_2 phase.
   b. The key event of S phase is segregation of sister chromatids.
   c. The mitotic spindle first appears during anaphase.
   d. The cell increases in size during metaphase.

4. Which of the following statements is *not* true?
   a. DNA is packed into chromatin with the help of proteins.
   b. All chromosomes in the somatic cell of a particular species have the same shape and size.
   c. Each chromosome contains a single DNA molecule.
   d. Somatic cells are diploid.

5. Which of the following correctly represents the order of the phases in the cell cycle?
   a. mitosis, S phase, G_1 phase, G_2 phase
   b. G_0 phase, G_1 phase, mitosis, S phase
   c. S phase, mitosis, G_2 phase, G_1 phase
   d. G_1 phase, S phase, G_2 phase, mitosis

6. Cytokinesis occurs
   a. at the end of prophase.
   b. just before telophase.
   c. at the end of mitosis.
   d. at the end of G_1 phase.

7. In fertilization, gametes fuse to form
   a. a bivalent zygote.
   b. a haploid zygote.
   c. a diploid zygote.
   d. a triploid zygote.

8. Human gametes contain
   a. twice the number of chromosomes than our skin cells.
   b. only sex chromosomes.
   c. half the number of chromosomes than our skin cells.
   d. only X chromosomes.

9. The reduction division is
   a. prophase of mitosis.
   b. anaphase II of meiosis.
   c. metaphase II of mitosis.
   d. meiosis I.

10. Meiosis results in
    a. four haploid cells.
    b. two diploid cells.
    c. four diploid cells.
    d. two haploid cells.

Cancer: Cell Division Out of Control

The Battle against Rogue Cells

Cancer represents a breakdown in the cooperative functioning of the cells in a multicellular organism. Every cancer begins with a single rogue cell that starts dividing with wild abandon, giving rise to a mass of abnormal cells. With the passage of time, the descendents of these abnormal cells adopt yet more malicious behaviors. They change shape, increase in size, quit their normal job, and sprawl over their neighbors (see Figure B.1). At their very worst, these highly abnormal cells break loose from their neighborhood, invade the bloodstream, and disperse to set up satellite colonies in other parts of the body. Wherever they establish themselves, they multiply furiously, overrunning their neighbors and monopolizing oxygen and nutrients to the extent that they starve the normal cells in the vicinity. The ability to spread through the body identifies these rogue cells as cancer cells, also known as malignant cells. Without restraints on their growth and migration, cancer cells take over, steadily destroying tissues, organs, and organ systems, until the body can no longer survive.

Cancer accounts for more than 500,000 deaths in the United States each year. While the past decade has seen improvements in treatment and prevention, more than a million Americans are diagnosed with some form of cancer every year. The National Cancer Institute estimates that the collective price tag for the various forms of cancer is more than $100 billion per year.

About 30 years ago, President Richard Nixon declared a war on cancer in the United States by making anticancer research a high priority. Since then, some major victories have been won, thanks to improvements in radiation and drug therapies. Whereas in the early twentieth century very few individuals survived cancer, today roughly 40 percent of patients are alive 5 years after treatment is begun. Nevertheless, the war against cancer is far from over, and the need for powerful new treatments to stop or kill the malignant cells is as urgent as ever. The greatest challenge in battling cancer is the selective destruction of rogue cells, while sparing healthy cells in the process. The standard war plan today relies on high-energy radiation, high doses of chemical poisons (chemotherapy), or both in sequence, to disable any and all rapidly dividing cells. The devastating side effects of radiation therapy and chemotherapy are terrible precisely because this all-out assault also kills many innocent bystanders, cells necessary for the normal functions of the human body.

The good news is that discoveries in basic cell biology, and the large investment in cancer research, has produced a variety of innovative strategies for destroying malignant cells selectively. One inventive method uses genetically engineered viruses that infect and destroy specific types of cancer cells, while leaving normal cells alone. Another line of attack seeks to disable proteins that give cancer cells their unusual immortality, the capacity to divide indefinitely as long as oxygen and nutrients are available. A particularly promising approach is to choke off the supply of oxygen and nutrients to a cancerous mass by blocking the growth of blood vessels in its immediate vicinity. These and several other novel anticancer strategies are now in clinical trials. The early results are encouraging, although a lot of groundwork will have to be done before any of them become a routine treatment for any type of cancer.

The biggest lesson learned from the past three decades of cancer research is a surprisingly simple one: instead of dealing with them only when they become monsters, we should try to reduce the odds that our cells will start to progress toward cancer in the first place. It is now abundantly clear that environmental factors, including lifestyle choices, have a very large impact on a person's risk of developing cancer. The link between cancer and tobacco use is the most dramatic illustration of how chemical exposure can transform healthy cells into dangerous ones. Lung

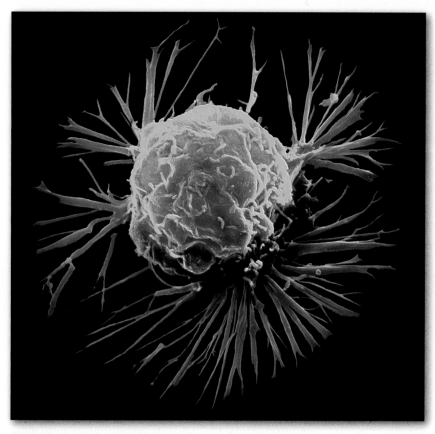

Figure B.1 A Home-Grown Monster
This color-enhanced photograph, captured with a scanning electron microscope, shows a breast cancer cell. A lab technician can recognize it as a cancer cell because of its large size, abnormally rounded shape, and altered cell surface. The spikes are membrane projections that help the cell creep about within the body. The ability to spread from one part of the body to another through the bloodstream is a hallmark of cancer cells.

cancer was a rare type of cancer just before the turn of the twentieth century, when few people smoked tobacco. Now, with nearly a third of the world's population lighting up, lung cancer is the most common and the deadliest cancer worldwide, killing more than 1.2 million people annually. The overall incidence of cancer has started to decline in the United States over the last 10 years, not because of some billion dollar technological innovation, but mostly because fewer people are smoking and therefore fewer people are developing tobacco-related cancers. Cancer prevention, through increased public awareness of the risk factors and the adoption of cancer-protective behaviors, is now a high priority in the war against cancer.

In this essay we will explore the biology of cancer by considering how a cancer is launched when normal restraints on cell division and cell movement are lost. The restraint system fails when information carried within certain genes is corrupted, or mutated. We will see that the vast majority of cancers are caused by the gradual accumulation of mutations in the cells of the adult body, often as a result of DNA damage caused by environmental factors. We will examine some of the environmental factors known to contribute to cancerous development and we will remind ourselves that, as with any disease, prevention is better than having to find cures.

Cancer Develops When Cells Lose Normal Restraints on Cell Division and Migration

To achieve a high level of organization, a multicellular organism must have a means of controlling and coordinating the behavior of the cells within it. Any large community that does not have rules of restraint quickly falls into chaos, and the same is true of a multicellular organism. Therefore, all aspects of cell behavior, especially the frequency with which a cell divides, are closely regulated in a healthy body. Controlling which cell divides, where and when, is quite literally a matter of life and death in something as complex as the human body.

As we saw in Chapter 9, the cell cycle is controlled by a variety of external and internal signals and regulatory proteins. Some of these agents are **positive growth regulators** in that they stimulate cell division, while others act as **negative growth regulators** because they put the brakes on cell division. Under normal circumstances these agents are released only when and where they are needed. A paper cut on your finger will trigger the release of positive growth regulators that will cause skin cells to divide until the wound is closed. How do cells at the wound edge "know" when they have divided enough and it is time to quit? Negative growth regulators become active when the newly made cells find themselves pushing up against other cells. This phenomenon, known as contact inhibition, halts the cell cycle, preventing overgrowth at the wound site.

The life of every cell is managed by a delicate interplay between a variety of positive and negative growth regulators. A multicellular individual, such as a human, is a cooperative community of cells, so the failure of just one cell to respond appropriately to the balance of positive and negative growth regulators could have serious consequences for the organism as a whole. Runaway growth is the consequence if the cell cycle is excessively stimulated through the pathways controlled by positive growth regulators. Unbridled growth can also be triggered in a cell that ignores the antiproliferation message of negative growth regulators. In either case, excessive cell proliferation sets the stage for the development of cancer.

The solid cell mass formed by the inappropriate proliferation of cells is known as a **tumor** (Figure B.2). Tumors that remain confined to one site are **benign tumors**. They are rarely dangerous nowadays because they can be surgically removed in nearly all cases. However, a benign tumor that is growing actively is like a cancer-in-training. It becomes a full-fledged **malignant tumor**, synonymous with a cancerous tumor, when its cells "learn" how to break away and invade the bloodstream. The majority of cells in the adult animal body are firmly anchored in one place. Even if released from their moorings, most such cells cannot crawl over their neighbors or enter blood vessels by squeezing through their walls. When benign cells acquire the extraordinary capacity to migrate and invade other tissues, they are transformed into a **cancer** (Figure B.2). But what causes a cell to behave in such reckless ways that endanger the whole organism? We shall see in the next section that malfunctioning genes are responsible for the wayward behavior of tumor cells: a benign tumor cell harbors some damaged genes, leading to some bad behavior; a cancerous cell has yet more gene malfunctions, producing the most malicious of cell activities.

Learn more about the development of cancer.

B.1

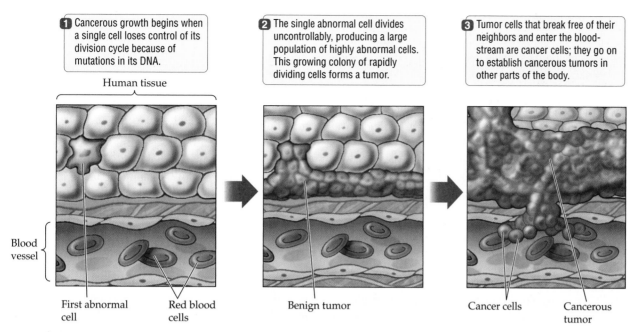

① Cancerous growth begins when a single cell loses control of its division cycle because of mutations in its DNA.

② The single abnormal cell divides uncontrollably, producing a large population of highly abnormal cells. This growing colony of rapidly dividing cells forms a tumor.

③ Tumor cells that break free of their neighbors and enter the bloodstream are cancer cells; they go on to establish cancerous tumors in other parts of the body.

Human tissue

Blood vessel

First abnormal cell

Red blood cells

Benign tumor

Cancer cells

Cancerous tumor

Figure B.2 Cancers Start with a Single Cell That Starts Dividing in an Uncontrolled Manner

Gene Mutations Are the Root Cause of All Cancers

A **mutation** is a change in the DNA sequence of a gene. A gene with mutations is like a corrupted computer file: the information in it is garbled. Genes commonly code for proteins. A gene with a mutation is likely to produce an altered protein; if the mutation is severe enough, protein production might fail altogether. Most commonly, gene mutations reduce or eliminate the activity of the protein encoded by that gene. In some instances, however, a gene mutation can alter the encoded protein in such a way that its activity is actually *increased*. DNA mutations can also cause certain proteins to be made in larger than normal amounts, or to be manufactured in cells that do not normally make them.

Genes that have been implicated in tumor development fall into two main groups: proto-oncogenes and tumor suppressor genes. Genes that code for positive growth regulators, that is, genes that promote cell division, are broadly classified as **proto-oncogenes**. These genes can mutate in such a way that they become hyperactive, in which case they tend to trigger runaway cell proliferation. Realize that proto-oncogenes are perfectly normal genes with essential roles in the body. It is only when they become overactive as a consequence of muta-

tions that they lead to excessive cell proliferation and tumor development. Proto-oncogenes that have become hyperactive as a result of DNA mutations are known as **oncogenes** (*onkos,* "tumor").

Genes that code for negative growth regulators are known as **tumor suppressor genes**. The normal activity of such genes is to inhibit the cell cycle and prevent cell proliferation. Mutations in tumor suppressor genes can reduce or eliminate the activity of the encoded protein, in which case the cell bearing the mutations becomes capable of uncontrolled proliferation and tumor development.

Most human cancers develop only after several critical genes have been hit by mutations. If any of these mutations occur in gametes (egg or sperm), or the gamete-producing cells, they can be passed on to offspring. A child receiving a mutated proto-oncogene or a mutated tumor suppressor gene from parents has an inherited risk of cancer. However, only 1 to 5 percent of all human cancers can be traced exclusively to inherited genetic defects. Most people who have an inherited cancer risk will not be victimized by cancer unless they have the bad luck to develop additional, noninherited mutations in critical proto-oncogenes or tumor suppressor genes. About 90 percent of the people diagnosed with cancer seem to have no inherited risk of the disease. That means their cancers were caused by a series of unfortunate changes in their DNA, either due to environmental agents such as viruses or toxic chemicals, or because of unavoidable cellular accidents such as cell division errors, or some combination of these two sources of noninherited mutations. The requirement for multiple mutations within a single cell explains why cancer is rare among the young, becomes more common as we get older, and increases steeply past middle age. As we shall see in the next section, a normal cell is transformed into a rogue cell in multiple steps, each step enabled by at least one hyperactive proto-oncogene or one crippled tumor suppressor gene.

Cancer is a multi-step process

Cancer is a group of diseases that is likely to affect many lives (Table B.1) over the decades to come. Over the course of a lifetime, an American male has a 45 percent chance of developing an invasive cancer. American women fare slightly better, with a 38 percent chance of developing cancer. In the United States, one in four deaths is due to cancer, and more than 8 million Americans alive today have been diagnosed with cancer and are either cured or undergoing treatment. More than 1,500 Americans die from cancer each day.

Given such a high incidence of cancer, you might think that the human body is exceptionally prone to cancer. Actually, we have robust defenses against cancer. We have a number of safeguards that reduce the likelihood of runaway cell proliferation and tumor development, at least during the reproductive years. As we age, however, mutations start to accumulate in the genes that orchestrate our anticancer defenses, bringing us closer to the unlucky string of failures that yield a cancerous tumor.

Consider cancer of the colon (the large intestine), which afflicts more than 100,000 individuals each year in the United States. In many cases of colon cancer, the tumor cells contain at least one overactive proto-oncogene and several completely inactive tumor suppressor genes. The mutations in different genes that eventually lead to colon cancer usually occur over a period of years, and the gradual accumulation of these mutations often goes hand-in-hand with stepwise changes in cell behavior that mark the progression toward cancer.

We can illustrate the step-by-step sequence of chance mutations, and the accompanying changes in cell activities, by following the disease progression characteristic of colon cancer. In most cases, the first cancer-

Table B.1

Selected Human Cancers in the United States

Types of cancer	Observation	Estimated new cases per year	Estimated deaths per year
Breast cancer	The second leading cause of cancer deaths in women (after lung cancer)	211,200	40,400
Colon and rectal cancer	The number of new cases is leveling off as a result of early detection and polyp removal	145,300	56,300
Leukemia	Often thought of as a childhood disease, this cancer of white blood cells affects more than 10 times as many adults as children every year	38,800	22,600
Lung cancer	Accounts for 28 percent of all cancer deaths and kills more women than breast cancer does	172,600	163,500
Ovarian cancer	Accounts for 3 percent of all cancers in women	22,200	16,200
Prostate cancer	The second leading cause of cancer deaths in men (after lung cancer)	232,100	30,400
Malignant melanoma	The most serious and rapidly increasing form of skin cancer in the United States	59,600	10,600

promoting mutation results in a relatively harmless, or benign, growth described as a polyp (Figure B.3). The cells that make up the polyp divide at an inappropriately rapid rate. These cells are the descendants of a single cell in the lining of the colon that has suffered one or more mutations.

As a polyp grows larger, the odds increase that one or more cells within this larger population of cells will acquire additional mutations. The complete loss of one tumor suppressor's activity, combined with the presence of an overactive cell proliferation protein, is enough to accelerate cell division. Even so, most such polyps do not spread to other tissues and can be safely removed surgically at this stage.

The progression from a benign polyp to a malignant tumor depends on the inactivation of additional tumor suppressor genes. In many colon tumors, the start of true malignancy coincides with the loss of a part of chromosome 18 (see Figure 9.5 for a photo of the human karyotype) that contains at least two important tumor suppressor genes. This complete loss of two additional tumor suppressors results in a far more aggressive and rapid multiplication of the cancerous cells, greatly increasing the chance that these cells will spread to other tissues.

One of the last key events in the path to full malignancy is the complete inactivation of an especially critical tumor suppressor gene, named *p53*. For reasons that are not entirely clear, loss of the p53 protein seems to remove all controls on cell division, allowing the cells to break free of the original tumor and travel through the bloodstream to other parts of the body as cancer cells. At this point the cancer cells are entirely resistant to signals from the body's regulatory and immune systems, and the worst possible scenario—a malignancy—has emerged.

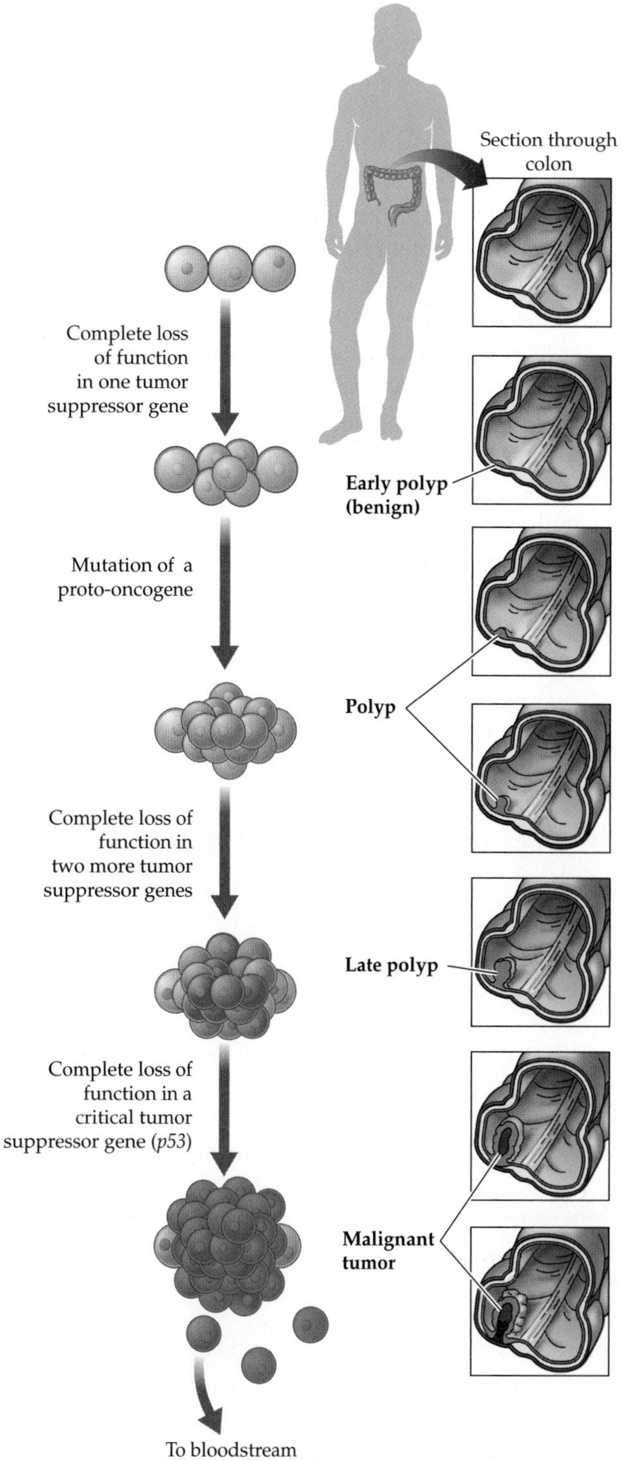

Environmental factors can cause mutations that lead to cancer

The relative contributions of inherited and environmental factors to an individual's cancer risk have been debated for decades. In recent years, large-scale studies have settled this issue by tracking cancer incidence in thousands of pairs of "identical" twins, who share the same genetic makeup. If inherited genetic defects are more important than environmental factors in causing cancer, then one would expect to see a very similar incidence of cancer in both twins. On the other hand, if environmental factors play a greater role, one would expect to see significant differences in cancer incidence due to differences in the twins' adult environments or habits. In one Scandinavian study that tracked over

Figure B.3 Development of Colon Cancer Is a Multistep Process
The sequential mutation of several genes that code for positive and negative growth regulators coincides with the progression from a benign polyp in the colon to a malignant tumor. The order in which the various proto-oncogenes and tumor suppressor genes mutate is not set, and varies from individual to individual.

Helpful to know

By convention, the names of genes are always given in italic type, while the names of their protein products are in roman type. Often this font difference is the only thing distinguishing the protein name from the gene name; for example, the product of the *Rb* gene is the Rb protein.

44,000 pairs of identical twins, environmental influences were by far the most important determinant of individual cancer risk.

Environmental influences that affect cancer risk include infectious agents such as viruses and bacteria, chemical toxins such as the nitrosamines that form when fatty meat is grilled at high temperatures, and lifestyle choices such as our consumption

Biology Matters

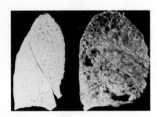

Before and After
Lung tissue from a nonsmoker (left) and a smoker (right).

The Truth about Cigarette Use among Young People

About one-third of American college students smoke cigarettes regularly, although only about 13 percent admit to being daily smokers. And, in the 18- to 25-year-old age group, about half have tried marijuana at least once. Both tobacco and marijuana cigarettes contain a compound called benzopyrene, a powerful cancer-causing agent. This substance suppresses a gene that controls the cell cycle. If that gene is not working as it should, cells can begin to multiply without restraint. An average marijuana cigarette contains about 50 percent more benzopyrene than a cigarette made of tobacco. And, marijuana smoke is typically inhaled more deeply and held in the lungs longer than cigarette smoke. But smoking tobacco is much more addictive than smoking marijuana, and that means that people who start smoking cigarettes are far more likely to be trapped in the habit for a long time. And a regular marijuana smoker will consume far fewer cigarettes per day than even a moderate cigarette smoker. The bottom line, however, is that both types of smoke are dangerous.

There's a lot that we can do to shape our own health destiny. Being informed and thinking ahead can be the road to a longer and healthier life.

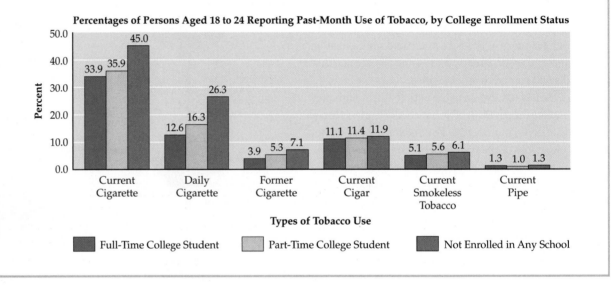

Percentages of Persons Aged 18 to 24 Reporting Past-Month Use of Tobacco, by College Enrollment Status

of vegetables and whole grains and how much we exercise. The strong link between environmental factors and the vast majority of cancers suggests that changes in personal behavior can substantially reduce our risk. Let us consider one cause of cancer that is particularly amenable to behavioral change: tobacco use.

Since 1982, cigarette smoking has been recognized as the single leading cause of cancer mortality in the United States. This acquired behavior is a major cause of lung and oral cancer, and it contributes to a wide range of other cancers, including those of the kidney, stomach, and bladder. There are close to 50 million American smokers, so it is not surprising that tobacco use accounts for one in five deaths in the United States. Among the thousands of chemical compounds that have been identified in tobacco smoke, over 40 have been confirmed to be carcinogens. A **carcinogen** [kar-*SIN*-uh-jin] is any physical, chemical, or biological agent that causes cancer.

Polycyclic [polly-*SYKE*-lick] aromatic hydrocarbons, or PAHs, are an important class of carcinogens found in tobacco smoke. These organic compounds can bind to DNA, forming a physical complex known as an adduct. Adducts cause mistakes in DNA synthesis (see Chapter 12), which introduce mutations into the DNA sequence. PAHs can form adducts at several sites on the tumor suppressor gene, *p53*. The resulting mutations in the

Figure B.4 The Rous Sarcoma Virus Causes Cancer in Chickens

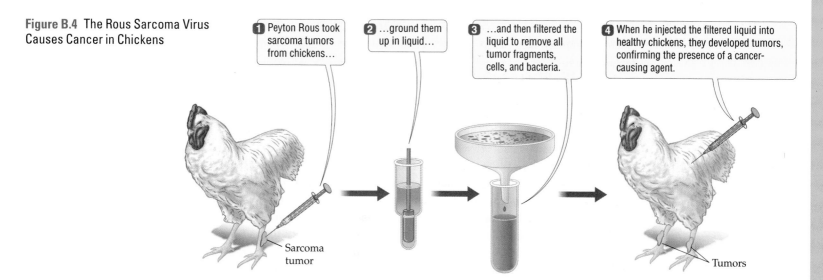

1 Peyton Rous took sarcoma tumors from chickens…

2 …ground them up in liquid…

3 …and then filtered the liquid to remove all tumor fragments, cells, and bacteria.

4 When he injected the filtered liquid into healthy chickens, they developed tumors, confirming the presence of a cancer-causing agent.

Sarcoma tumor

Tumors

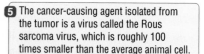

5 The cancer-causing agent isolated from the tumor is a virus called the Rous sarcoma virus, which is roughly 100 times smaller than the average animal cell.

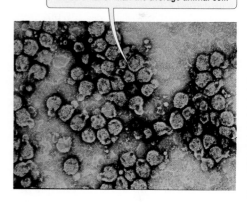

p53 gene prevent the production of functional p53 tumor suppressor protein in the affected cells. In the same way that the inactivation of *p53* contributes to colon cancer, its inactivation in lung cells allows them to divide uncontrollably, leading to lung cancer. Furthermore, the formation of adducts due to PAHs is not restricted to lung cells. The white blood cells of smokers also show PAH-related genetic damage, which can contribute to other forms of cancer.

The good news is that stopping smoking can dramatically reduce an individual's cancer risk. People who quit smoking before the age of 50 reduce their risk of dying in the subsequent 15 years by half. Regardless of age, people who quit smoking live longer than those who continue to smoke. While nicotine, which is the addictive drug in tobacco, makes quitting smoking difficult, all should find inspiration in the fact that one in five Americans is a former smoker.

Oncogenes from Animal Viruses Revealed the Cellular Basis of Cancer

Mutations, as we have seen, can cause proto-oncogenes to become hyperactive, turning them into oncogenes. Oncogenes responsible for human cancers most often come from mutations caused by chemical pollutants and other environmental factors, such as overexposure to sunlight. In nonhuman animals, however, oncogenes are frequently brought into a cell by an infecting

virus, and viral infections are a leading cause of cancer in such animals. Oncogenes, and their role in cancer development, were first described in poultry viruses.

Our understanding of how cells respond to signals that promote cell division began with observations of cancer in animals. Biologist Peyton Rous studied cancerous tumors called sarcomas in chickens in the first decade of the twentieth century. Rous discovered that he could grind up sarcomas and extract an unidentified substance that, when injected into healthy chickens, caused cancer. He knew that the extract contained no bacteria because it had been carefully filtered, so the cause of the cancer had to be something much smaller—something that could pass through his filters. His work led to the discovery of the first animal tumor virus, which was named the Rous sarcoma virus in honor of its discoverer and the type of tumor from which it was obtained (Figure B.4).

Viruses bearing oncogenes promote cancer by stimulating cell division

As we saw in Chapter 1, viruses are tiny assemblages of either RNA or DNA surrounded by a protein coat. The nucleic acids found inside a virus contain genes that are necessary for the viral reproductive cycle. Viruses are more than a hundred times smaller than the average animal cell. They are invaders that can multiply only by infecting the cells of other organisms and using the biochemical machinery of the infected cell for their own replication.

The discovery that a virus could cause cancer in chickens was a major breakthrough in our understanding of cancer, but it took many more decades before scientists discovered how the Rous sarcoma virus derails the normal controls that regulate cell division. The solution to this mystery came with the discovery of a particular strain of Rous sarcoma virus that could multiply in cells without causing the cells to divide rapidly. Biologists could then compare the viral genes found in this virus with those found in cancer-causing strains. The virus that did not cause cancerous cell division was missing a single gene. Protein-coding genes direct the manufacture of a specific protein, so the absence of this gene meant that this strain of the Rous sarcoma virus was not able to produce one particular protein. Further research showed that this protein is an exceptionally strong positive growth regulator, belonging to a class of enzymes called protein kinases.

Protein kinases activate their target proteins by adding phosphate groups to them. Under normal conditions, these activation events are counterbalanced by the action of other enzymes that remove the phosphates, effectively turning the target proteins off. When an overactive viral protein kinase acts on its target proteins, however, there is no way to turn the signal cascade off because the phosphates cannot be removed fast enough. The avalanche of enzymatic reactions that drive the cell toward mitosis roars out of control, producing a cell that just keeps dividing. When the Rous sarcoma virus infects a cell, it inserts all of its genes into the cell's DNA, so that all the daughter cells receive the cancer-promoting protein kinase gene. Eventually, the growing colony of rapidly dividing cells forms a tumor.

Viral oncogenes may have descended from host cell proto-oncogenes

The protein kinase gene in the Rous sarcoma virus is named *Src* [SARK]. It is just one of several oncogenes that have been identified in viruses of different types. Comparison of DNA samples has revealed that the *Src* oncogene is simply a slightly altered version of a gene normally found in the genetic material of the host organism's cells.

As we saw previously, viruses like the Rous sarcoma virus multiply by "stitching" themselves into the infected cell's DNA. At some point in evolutionary history, a mutant protein kinase gene may have been accidentally "cut" from an abnormal host cell's DNA and "pasted" into the genetic material of the virus. The version of *Src* acquired by the virus codes for an overactive kinase that cannot be controlled like its normal counterpart in the cell.

The realization that the *Src* oncogene has a normal, controllable counterpart in host cells was an important breakthrough in identifying the cellular genes that regulate cell division, including proto-oncogenes. Today scores of proto-oncogenes are known, based, in most cases, on the prior discovery of their oncogene cousins in tumor viruses infecting nonhuman animals such as chickens, mice, and cats. Most proto-oncogenes identified in these animals have also been found in human cells, underscoring the similarity in the organization and control of the cell division machinery of animals.

Tumor Suppressor Genes Normally Restrain Cell Division

As dangerous as they may seem, oncogenes are usually not the sole villains in the rampant cell growth that leads to cancer. This is because normal cells have internal safeguards that ordinarily prevent uncontrolled cell division. These barriers to runaway cell proliferation take the form of a family of proteins that are collectively known as tumor suppressors because their normal activities were found to inhibit tumor growth. Tumor suppressors are therefore negative growth regulators that stop cells from dividing by opposing the action of the proteins encoded by proto-oncogenes.

Whether a normal cell divides depends on the activities of both proto-oncogenes and tumor suppressor genes. For a cell to divide, proto-oncogenes must be activated, and tumor suppressor genes must be inactivated. Because controlling the timing and extent of cell division is so important, cells use proto-oncogenes and tumor suppressor genes as opposing forces in a system of checks and balances that normally controls the cell cycle with great precision. Under the right circumstances, signaled by proto-oncogenes and unopposed by tumor suppres-

Explore how to suppress tumors by controlling cell division.
▶ ❚❚
B.2

sor genes, a cell will be allowed to divide a certain number of times and no more.

How do tumor suppressor genes oppose the activity of proto-oncogenes when the situation calls for restraint in cell division? Before we tackle that question, we must consider how positive growth regulators stimulate cell division. External signals that promote cell division commonly do so by triggering a stepwise sequence of protein activations inside the cell that are collectively known as a **signal cascade**. One category of positive growth regulators, known as **growth factors**, plays an especially important role in initiating and maintaining cell proliferation in the human body. Scores of these proteins have been identified, most of them acting on a narrow range of cell types. Growth factors activate signal cascades that typically turn on cell proliferation genes, such as those that advance the cell cycle or initiate the copying of DNA in preparation for mitosis.

Tumor suppressors block specific steps in growth factor signal cascades

In the same way that proto-oncogene proteins induce cell division by activating the components of a growth factor signal cascade, tumor suppressors block cell division by *inactivating* some of the same components. A well-known example of tumor suppressor activity was discovered during a study of a rare childhood cancer known as retinoblastoma [RET-ih-noh-blass-TOH-muh]. As the name indicates, retinoblastoma (*retino*, "net"; *blastoma,* "bud") is a cancer that forms in the retina of the eye, and generally leads to blindness (Figure B.5). Retinoblastoma strikes one in every 15,000 children born in the United States and accounts for about 4 percent of childhood cancers.

What causes this kind of cancer to develop? As we saw in Chapter 9, each species has a characteristic set of chromosomes, termed the karyotype, which can be seen under the microscope. In cancer cells from some children with retinoblastoma, a portion of chromosome 13 appeared to be missing, hinting that the cancer might be caused by the absence of a particular gene. Today we know that the gene in question normally produces a protein called Rb. The missing *Rb* gene, and the resulting lack of Rb protein, results in retinoblastoma as retinal cells start dividing without restraint.

This is not the effect we would expect an oncogene to have. An oncogene causes cancer by producing

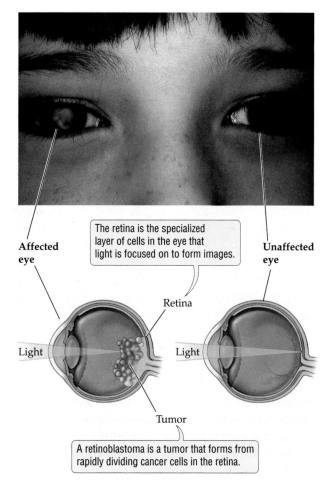

The retina is the specialized layer of cells in the eye that light is focused on to form images.

Affected eye

Unaffected eye

Retina

Light

Light

Tumor

A retinoblastoma is a tumor that forms from rapidly dividing cancer cells in the retina.

Figure B.5 Retinoblastoma

This child has a visible retinoblastoma in her right eye. A retinoblastoma both blocks light from reaching retinal cells and destroys the ability of retinal cells to respond to light, usually resulting in blindness.

an overactive protein that pushes the cell to divide. The *Rb* gene must have the opposite effect, since its *absence* promotes cell division. A simple explanation would be that the Rb protein normally *inhibits* a process required for cell division; when it is missing, the brakes on cell division no longer work, and cells divide uncontrollably.

The Rb protein inhibits a key process in the cell's preparations for division. It binds to and inactivates a protein that is required for the cell's response to growth factor signals. When cells are stimulated to divide by growth factors under normal conditions, the resulting signal cascade involves not just the activation of proto-oncogenes, but also the inactivation of tumor suppressors such as Rb.

The Rb protein is inhibited by a protein kinase that is activated by the growth factor signal cascade. The

Figure B.6 How the Rb Protein Inhibits Cell Division

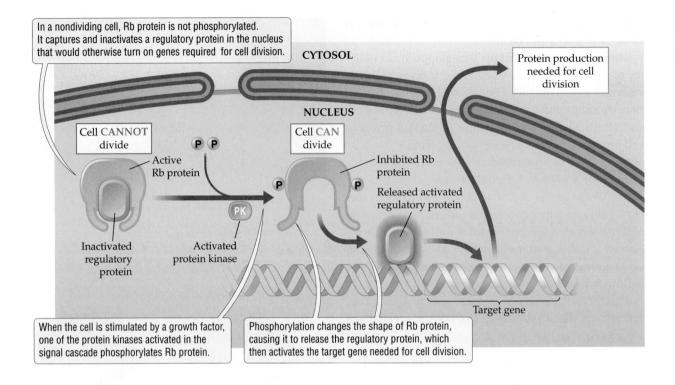

In a nondividing cell, Rb protein is not phosphorylated. It captures and inactivates a regulatory protein in the nucleus that would otherwise turn on genes required for cell division.

CYTOSOL

Protein production needed for cell division

NUCLEUS

Cell **CANNOT** divide

Active Rb protein

Cell **CAN** divide

Inhibited Rb protein

Released activated regulatory protein

Inactivated regulatory protein

Activated protein kinase

PK

Target gene

When the cell is stimulated by a growth factor, one of the protein kinases activated in the signal cascade phosphorylates Rb protein.

Phosphorylation changes the shape of Rb protein, causing it to release the regulatory protein, which then activates the target gene needed for cell division.

Explore making the most of losing p53.

▶ ❚❚

B.3

activated kinase phosphorylates the Rb protein, causing it to change its shape and release its target protein, which can then activate the genes needed for cell division (Figure B.6). This example shows that the phosphorylation resulting from growth factor signals acts to turn on some proteins and turn off others. This supports the hypothesis that a balance of positive and negative regulatory controls must come into play before a cell can divide.

Both copies of a tumor suppressor gene must be mutated to cause cancer

The differences in how oncogenes and tumor suppressor genes function highlight differences in the kinds of genetic mutations that can lead to cancer. Because chromosomes exist in pairs, and because the two chromosomes in each pair have the same set of genes, there are two copies of each gene in the cell, one contributed by each parent. Just one copy of a mutated proto-oncogene is enough to launch cancerous change. For example, one mutated copy of the gene might produce an overactive protein that can push the cell to divide.

In contrast, for a tumor suppressor gene to promote cancer, both copies of the gene must be mutated to an inactive form. In other words, if only one copy were inactivated, the other copy might still produce enough tumor suppressor protein to inhibit cell division. Complete breakdown of this negative control mechanism—meaning that no tumor suppressor protein is being made—requires that both copies of a tumor suppressor gene be inactivated (Figure B.7).

The p53 protein helps to maintain normal cell division

The critical importance of the p53 tumor suppressor protein is illustrated by the dramatic acceleration of malignancy in colon polyps and lung tumors in which the *p53* gene is nonfunctional (Figure B.3). The p53 protein is an especially distinctive member of the family of tumor suppressors because it has multiple roles in guarding the integrity of cellular processes. It not only prevents the cell from dividing at inappropriate times, but it also halts cell division when there is DNA damage that could result in harmful mutations. By halting the cell cycle, the p53 protein gives the cell an opportunity to repair the damage to its DNA. If the repair process fails, for instance because the damage is so extensive that it is beyond repair, the p53 protein goes so far as to induce a cascade of enzymatic reactions that kills the cell. In other words, if the cell's DNA is too badly damaged to repair, the cell commits suicide, rather than passing on mutations that could potentially harm the entire organism.

Given the important guardian functions of the p53 protein, it is not surprising that more than half of all cancers involve a complete loss of p53 activity in tumor cells. The number goes as high as 80 percent in some types of cancer, such as colon cancer.

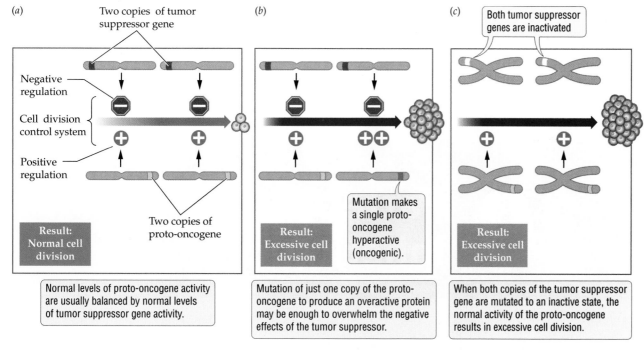

(a) Two copies of tumor suppressor gene

Negative regulation

Cell division control system

Positive regulation

Two copies of proto-oncogene

Result: Normal cell division

Normal levels of proto-oncogene activity are usually balanced by normal levels of tumor suppressor gene activity.

(b)

Mutation makes a single proto-oncogene hyperactive (oncogenic).

Result: Excessive cell division

Mutation of just one copy of the proto-oncogene to produce an overactive protein may be enough to overwhelm the negative effects of the tumor suppressor.

(c) Both tumor suppressor genes are inactivated

Result: Excessive cell division

When both copies of the tumor suppressor gene are mutated to an inactive state, the normal activity of the proto-oncogene results in excessive cell division.

Figure B.7 The Control of Cell Division by Proto-oncogenes and Tumor Suppressor Genes
Whether a cell divides or not depends on the balance between proto-oncogene and tumor suppressor gene activity. (a) The activities of proto-oncogenes, which promote cell division, are counterbalanced by the activities of tumor suppressor genes, which inhibit cell division, so just the normal, regulated, amount of cell division takes place. (b) The mutation of one copy of a proto-oncogene to an oncogene results in excessive cell division. (c) When both copies of a tumor suppressor gene are inactivated, the result is excessive cell division.

◉ Review and Discussion

1. Describe the possible consequences when a proto-oncogene becomes an oncogene.

2. Describe the interplay between cellular signals that promote cell division and those that inhibit cell division.

3. Colon cancer develops in a series of stages. Outline the stages and what happens in each stage.

4. Polycyclic aromatic hydrocarbons (PAHs) are found in tobacco. Discuss the possibility that commonly used products also contain carcinogens, some of which may not have been identified.

5. In light of the clear link between tobacco usage and cancer, many have questioned the right of tobacco companies to continue selling such a deadly substance. Consider the issues of personal freedom versus public health policy and explain what restraints, if any, you think should be placed on the sale of tobacco.

6. As environmental causes of cancer receive increasing attention, the warning labels on food have become lengthier and more ominous in tone. Since many factors contribute to cancer, do you think that expanding food warning labels is an effective approach to reducing cancer risk? If so, how might one combat the public's tendency to ignore long and complex warning labels?

Key Terms

benign tumor (p. B3)
cancer (p. B3)
carcinogen (p. B7)
growth factor (p. B10)
malignant tumor (p. B3)
mutation (p. B4)
negative growth regulator (p. B3)

oncogene (p. B4)
positive growth regulator (p. B3)
proto-oncogene (p. B4)
signal cascade (p. B10)
tumor (p. B3)
tumor suppressor gene (p. B4)

10 Patterns of Inheritance

Key Concepts

- Genetics is the study of inherited characteristics (genetic traits) and the genes that affect those traits.

- A gene is the basic unit of information affecting a genetic trait. It consists of a stretch of DNA.

- A phenotype is the specific version of a genetic trait that is displayed by a particular individual.

- Alternative versions of a gene (alleles) arise by mutation. The alleles responsible for a given phenotype in a particular individual constitute the genotype of that individual.

- Sexually reproducing organisms typically have two copies of every gene, one copy from the male parent, one from the female. Homozygotes have identical copies (alleles) of a gene, whereas heterozygotes have two different alleles.

- A dominant allele controls the phenotype in a heterozygote. The masked allele is described as recessive.

- Mendel's laws of inheritance help us to predict the phenotype of offspring from the genotype of the parents.

- Mendel's laws apply broadly to most sexually reproducing organisms, but geneticists have extended those basic laws to describe more complex patterns of inheritance.

- Many aspects of an organism's phenotype are determined by groups of genes that interact with one another and with the environment, so offspring with identical genotypes can have very different phenotypes.

The Lost Princess

In the early hours of July 17, 1918, the Russian royal family was awakened and taken to the basement of a house in the industrial city of Yekaterinburg. Told they were to be photographed, the Tsarina Alexandra, and her young son, Alexis, who suffered from hemophilia, were seated in chairs. The rest of the family—Tsar Nicholas II and his four daughters, Olga, Tatiana, Maria, and Anastasia—stood behind Alexandra and Alexis, as did the family physician, cook, maid, and valet. Suddenly, eleven men burst into the room, each with a different intended victim, and began firing with revolvers. In a brutal act that brought to an end the Romanov dynasty of pre-Communist Russia, all seven members of the royal family, and their four servants, were killed.

Or were they? In 1920, a young woman was pulled freezing from a Berlin canal. At first known simply as "Fraulein Unbekannt," or "Miss Unknown," and later as Anna Anderson, she claimed that she was the Princess Anastasia. Her knowledge of minute details of life at the Russian imperial court convinced many that she was indeed Anastasia, the youngest daughter of Nicholas and Alexandra. Others, troubled by her inability to speak Russian and her bouts of erratic behavior, thought she was a pretender. Anna Anderson herself never doubted that she was Anastasia, a conviction she held to her death in 1984.

Over the years, the legend of Princess Anastasia grew to become the subject of books, movies, and magazine articles. While the escape of a beautiful princess from execution made a wonderful story, was Anna Anderson really Anastasia? Ultimately, the mystery was solved with a combination of careful detective work and genetic analysis. Investigators used the basic principles of genetics to determine whether Anna Anderson could have been the lost princess.

What are the rules that govern how characteristics are inherited? How could those rules be used to determine whether Anna Anderson was a member of the Russian royal family? To answer these and many other questions about inherited characteristics, we must understand the principles of genetics, the main focus of this chapter.

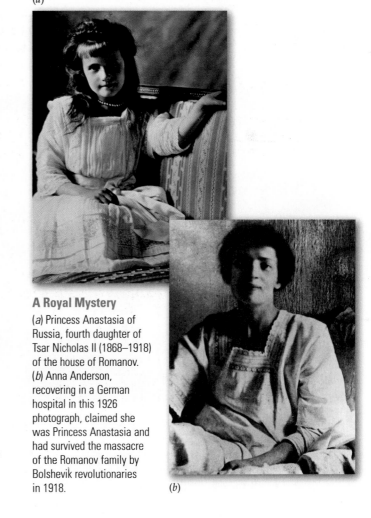

(a)

A Royal Mystery
(a) Princess Anastasia of Russia, fourth daughter of Tsar Nicholas II (1868–1918) of the house of Romanov. (b) Anna Anderson, recovering in a German hospital in this 1926 photograph, claimed she was Princess Anastasia and had survived the massacre of the Romanov family by Bolshevik revolutionaries in 1918.

(b)

Humans have used the principles of inheritance for thousands of years. Noticing that offspring tend to be similar to their parents, people raised animals and plants by mating individuals with desirable characteristics and selecting the "pick of the litter" for further breeding. Our ancestors used such methods to domesticate wild animals and to develop agricultural crops from wild plants (see Figure 1.10). As a field of science, however, genetics did not begin until 1866, the year that an Augustinian monk named Gregor Mendel (Figure 10.1) published a landmark paper on inheritance in pea plants. Prior to Mendel's work, some aspects of inheritance were understood, but no one had conducted extensive and systematic experiments to explain the patterns in which inherited characteristics, or **genetic traits**, are passed from parent to offspring.

Mendel's extraordinary insights were made possible by his exceptional training. As a young monk, Mendel attended the University of Vienna, where he took courses that ranged from mathematics to botany. Upon assuming his duties at the monastery of St. Thomas, Mendel put his training in probability statistics and plant breeding to good use. He used mathematics to analyze the inheritance of seven different genetic traits in garden peas. From the patterns he observed, Mendel was able to deduce the basic principles that govern how genetic information is passed from one generation to the next.

Based on his experiments with peas, Mendel arrived at some basic generalizations, now known as the laws of inheritance, that have stood the test of time. Even though Mendel's laws have been modified considerably by modern genetics, his predictions about the outcomes of certain types of mating experiments still hold true and apply broadly to most organisms that multiply themselves through sexual reproduction. Mendel's experiments led him to propose that the inherited characteristics of organisms are controlled by hereditary factors—now known as genes—and that one factor for each trait is inherited from each parent. Although he did not use the word "gene," Mendel was the first to propose the concept of the gene as the basic unit of inheritance. The emphasis that Mendel placed on genes continues today.

In the more than 100 years since Mendel's work, we have learned a great deal about genes, especially about their physical and chemical properties. We now know that genes are located on chromosomes, which consist of DNA molecules complexed with packaging proteins (Chapter 9). A **gene** is a stretch of DNA that governs one or more genetic traits. A single chromosome typically contains many hundreds of genes. Genes commonly contain instructions for the manufacture of proteins, and to a very large extent it is these proteins that shape the chemical, structural, and behavioral characteristics of the individual organism. In other words, the genetic traits we display are mostly brought about by proteins, which in turn are encoded by genes.

Before we describe Mendel's remarkable insights into the principles of inheritance, and the further elaborations of those principles in modern times, we will go over some basic terminology used in genetics. Some of this vocabulary, and its associated concepts, did not exist in Mendel's time, which probably made it difficult for his contemporaries to appreciate the significance of his largely mathematical analysis. The lack of an understanding of meiosis, especially the random assortment of chromosomes during meiosis I (Chapter 9), was perhaps the biggest obstacle for Mendel's contemporaries, who included Charles Darwin. Mendel's work was largely ignored for about 30 years after it was published. Upon its "rediscovery" in the early 1900s, his principles became the foundation for the modern discipline of **genetics**, the study of genes. Today, Mendel's principles have been extended to reveal, in much greater complexity, how genes, especially as influenced by the environment, shape the observable characteristics of organisms.

Figure 10.1 Mendel and the Monastery Where He Performed His Experiments
Gregor Mendel (inset) was a monk at the monastery of St. Thomas in Brno, now in the Czech Republic. For many years it was believed that Mendel had performed his experiments behind the fence visible here immediately in front of the monastery. Staff members at a museum devoted to Mendel recently discovered that Mendel's garden, no longer evident, was located in the foreground of this photograph.

10.1 Essential Terms in Genetics

If we examine a large natural population, there are likely to be at least some characteristics that differ from one individual to another. Individuals in a herd of zebras differ in their stripe patterns, one carnation plant produces red flowers while another makes white flowers, and one canary sings a slightly different tune from another. A genetic trait is any inherited feature of an organism that can be measured or observed. Stripe patterns, flower color, and song patterns are examples of genetic traits. Genetic traits are determined at least in part by genes. At the molecular level, a gene consists of a stretch of DNA. Geneticists commonly use one to four letters, written uppercase and italicized, to symbolize a particular gene. For example, the *Orange* gene, which leads to orange fur in cats, can be designated by an italicized *O*. The shorthand for the gene that codes for the alpha polypeptide of hemoglobin, an oxygen-carrying protein, is *HBA*. The display of a particular version of a genetic trait in a specific individual is the **pheno-type** of that individual. For example, black, bay, and sorrel are different types of coat color in horses and each of them is therefore a specific phenotype of the coat color trait.

As we described in Chapter 9, somatic cells in plants and animals are **diploid** in that they possess two copies of each type of chromosome. The two copies make up a **homologous pair** for each type of chromosome. One member of each homologous pair is inherited from the paternal parent of the individual, and is therefore known as the **paternal homologue**; the other member of the pair is inherited from the female parent, and is therefore known as the **maternal homologue**. Human somatic cells, for example, have a double (diploid) set of 23 different types of chromosomes. This diploid set consists of 23 pairs of homologous chromosomes for a total of 46 chromosomes.

Because there are two copies of each type of chromosome (with one exception), a diploid cell has two copies of every gene located on these chromosomes: one copy is located on the paternal homologue and the other copy is on the maternal homologue of each homologous pair (Figure 10.2). The genes located on

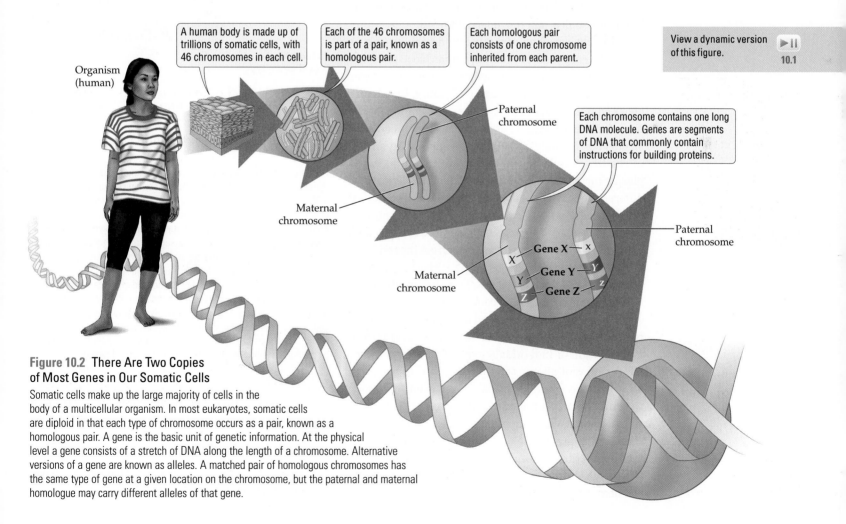

A human body is made up of trillions of somatic cells, with 46 chromosomes in each cell.

Each of the 46 chromosomes is part of a pair, known as a homologous pair.

Each homologous pair consists of one chromosome inherited from each parent.

View a dynamic version of this figure. ▶❙❙ 10.1

Organism (human)

Paternal chromosome

Maternal chromosome

Each chromosome contains one long DNA molecule. Genes are segments of DNA that commonly contain instructions for building proteins.

Maternal chromosome

Paternal chromosome

X — Gene X — x
Y — Gene Y — Y
Z — Gene Z — z

Figure 10.2 There Are Two Copies of Most Genes in Our Somatic Cells
Somatic cells make up the large majority of cells in the body of a multicellular organism. In most eukaryotes, somatic cells are diploid in that each type of chromosome occurs as a pair, known as a homologous pair. A gene is the basic unit of genetic information. At the physical level a gene consists of a stretch of DNA along the length of a chromosome. Alternative versions of a gene are known as alleles. A matched pair of homologous chromosomes has the same type of gene at a given location on the chromosome, but the paternal and maternal homologue may carry different alleles of that gene.

the X chromosomes of male mammals are an exception to this two-copies-of-every-gene rule for diploid cells. Recall from Chapter 9 that, in contrast to the two matched X chromosome pairs found in the diploid cells of female mammals, male mammals have only one X chromosome. This means that in all of their diploid cells male mammals have *only one copy* of the genes that are located on the X chromosome, while females of the species have two copies, one on each of their two X chromosomes. This difference between males and females has consequences for the inheritance of certain disorders in humans, a pattern of inheritance we will explore in greater depth in Chapter 11.

We mentioned in Chapter 9 that gametes—sperm and egg cells—are **haploid** because these sex cells have only one set of chromosomes, that is, half the total number of chromosomes found in a diploid cell. Because a gamete has only one copy of each pair of homologous chromosomes, either the paternal one or the maternal one, it possesses *only one copy* of every gene. Gametes are created through meiosis. Meiosis generates haploid daughter cells by halving the diploid chromosome number of the parent cell. A human gamete, for example, has only 23 chromosomes, half the diploid set of 46.

Different versions of a given gene are known as **alleles** of that gene. For example, human blood groups—A, B, AB, and O—are controlled by at least three alleles of the I gene. These alleles are I^A, I^B, and i. Recall that genes commonly affect a genetic trait by coding for the manufacture of a specific protein. The I^A allele codes for an enzyme that puts an "A type" of sugar on the surface of red blood cells (on certain proteins found in the plasma membrane of these cells). The I^B allele makes a different version of the enzyme, one that inserts the "B type" of sugar on the blood cells. The i allele simply fails to make a sugar-inserting enzyme that works, so both the A and B type of sugars are absent from the red blood cells of a person with the ii genotype; such a blood type is represented by the letter O ('Oh') (see Figure 10.3).

Notice that a diploid cell contains at most two different alleles for a given gene, and a haploid cell can have only one of all the possible alleles of a gene. However, a population of individuals may collectively harbor many different alleles, the way every one of the three alleles of the I gene is probably represented among the students in your classroom. The genetic diversity we see in natural populations comes about because there are many different alleles of the many genes in that population. The main reason just about everyone in your classroom looks recognizably different from the next person is that each of you carries different

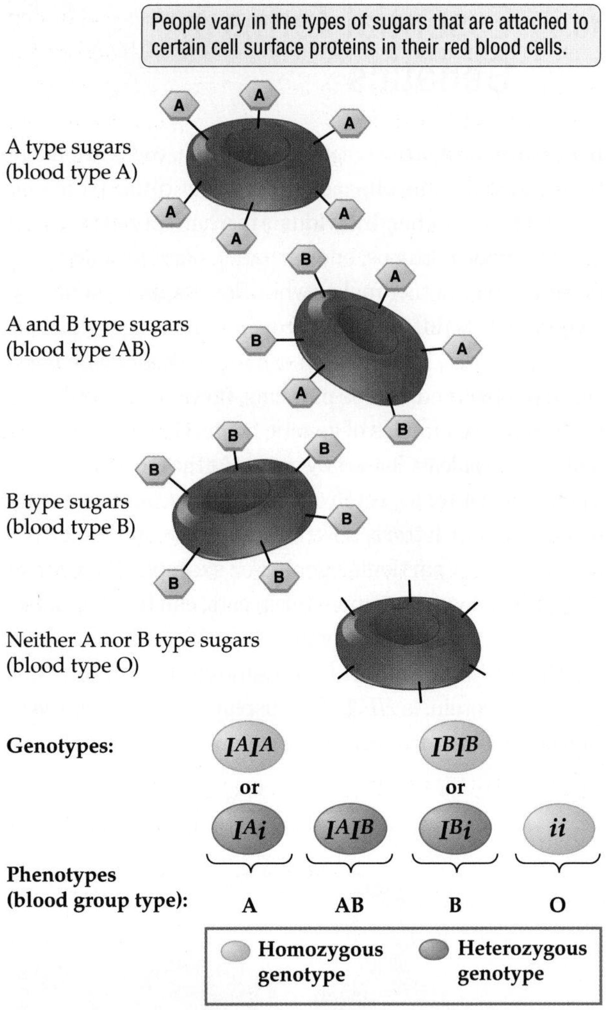

People vary in the types of sugars that are attached to certain cell surface proteins in their red blood cells.

A type sugars (blood type A)

A and B type sugars (blood type AB)

B type sugars (blood type B)

Neither A nor B type sugars (blood type O)

Genotypes: I^AI^A or I^Ai I^AI^B I^BI^B or I^Bi ii

Phenotypes (blood group type): A AB B O

Homozygous genotype Heterozygous genotype

Figure 10.3 Genetic Basis of the ABO Blood Types in Humans
People can be classified into distinct blood group types based on the chain of sugars that are attached to certain cell surface proteins in their red blood cells. Those with blood type A have an "A type" of sugar (chemically, *N*-acetyl galactosamine) as a side branch in their sugar chains, those who are type B have a "B type" of sugar (galactose) as a side branch in their sugar chains. In those with the AB blood type, about half the sugar chains have an A type of branch, half have the B type of pattern. The sugar side branch is added to the chain by variant forms of a single enzyme, coded by the *I* gene. What blood group type you have (your phenotype) is determined by the particular alleles of the *I* gene you carry (your genotype). The *i* allele codes for a nonfunctional version of the enzyme that cannot tack on either A or B type of sugars to the chain. Those who have two copies of the *i* allele (homozygotes) are said to have blood type O.

alleles for many of the roughly 25,000 genes that all humans possess.

The **genotype** of an individual is the allelic makeup of that individual leading to a particular phenotype. In other words, a genotype refers to the pair of alleles an individual has for a given phenotype. It is the genotype that partly, or wholly, creates the phenotype. For example, if an individual's phenotype for the blood group trait is type O, then the genotype responsible for that phenotype is *ii*. An individual who carries two copies of the *same* allele (such as an $I^A I^A$, $I^B I^B$, or *ii* individual) is said to be a **homozygote** or to have a homozygous genotype. An individual whose genotype consists of two *different* alleles for a given phenotype (as in an $I^A I^B$, $I^A i$, or $I^B i$ individual) is a **heterozygote**. Some genes have alleles that show dominance interactions such that one allele prevents a second allele from affecting the phenotype when the two alleles are paired together in a heterozygote. The allele that exerts a controlling influence on the phenotype, to the point of masking the effect of a second allele it is paired with, is said to be **dominant**. Alleles that have a dominant effect on the phenotype are denoted by uppercase letters, such as I^A or I^B. An allele that has no effect on the phenotype when it is paired

with a dominant allele is considered **recessive** to that dominant allele. Alleles that act recessively are denoted by lowercase letters, for example *i*.

A **genetic cross**, or cross for short, is a controlled mating experiment performed to examine the inheritance of a particular trait. "Cross" can also be used as a verb, as in "individuals of genotype *AA* were crossed with individuals of genotype *aa*." The parent generation in a genetic cross is called the **P generation**. The first generation of offspring in a genetic cross is called the **F_1 generation** ("F" is for "filial," a word that refers to a son or daughter). When the individuals of the F_1 generation are crossed with each other, the resulting offspring are said to belong to the **F_2 generation**.

Definitions of these important genetics terms are collected in Table 10.1. Study these terms carefully, and refer to them as needed as you study the rest of this chapter.

Table 10.1

Essential Terms in Genetics

Term	Definition
Allele	One of two or more alternative versions of a gene.
Dominant allele	Allele that has exclusive control over the phenotype of an organism when paired with a different allele.
F_1 generation	The first generation of offspring in a genetic cross.
F_2 generation	The second generation of offspring in a genetic cross.
Gene	The basic unit of genetic information for a specific trait. Genes consist of a stretch of DNA that is part of a chromosome.
Genetic cross	A controlled mating experiment, usually performed to analyze the inheritance of a particular trait.
Genotype	The genetic makeup of an organism; more specifically, the two alleles of a given gene that affect a specific phenotype in a given individual.
Heterozygote	An individual that carries one copy of each of two different alleles (for example, an *Aa* individual or a $C^W C^R$ individual).
Homozygote	An individual that carries two copies of the same allele (for example, an *AA*, *aa*, or $C^W C^W$ individual).
P generation	The parent generation in a genetic cross.
Phenotype	The specific version of a genetic trait that is displayed by a given individual; for example, black, brown, red, and blond are phenotypes of the hair color trait in humans.
Recessive allele	An allele that does not have a phenotypic effect when paired with a dominant allele.
Trait	Any inherited feature of an organism that can be measured or observed; height, flower color, or the chemical structure of a protein, are all examples of genetic traits.

Review basic genetics terminology.

► ||

10.2

Biology Matters

Know Your Type

One of the most important inherited characteristics to know about yourself and your family members is blood type, since the majority of Americans will receive donated blood at some point in their lives. Knowing your blood type and the blood types of your close relatives could save valuable time in an emergency, since the type of blood that you can receive is determined by the type of blood that you have. Type O– blood is the only type of blood that can be transfused to a person of any blood type, but only 7 percent of people have type O– blood. Blood centers often run short of type O and type B blood, but shortages of all types of blood occur during the summer and winter holidays.

The following table summarizes the ability of each blood type to either be given to or received from other blood types:

If your blood type is:	You can give blood to:	You can receive blood from:
A+	A+, AB+	A+, A–, O+, O–
O+	O+, A+, B+, AB+	O+, O–
B+	B+, AB+	B+, B–, O+, O–
AB+	AB+	Everyone
A–	A+, A–, AB+, AB–	A–, O–
O–	Everyone	O–
B–	B+, B–, AB+, AB–	B–, O–
AB–	AB+, AB–	AB–, A–, B–, O–

According to current data, 4.5 million Americans would die annually without life-saving blood transfusions. Approximately 32,000 pints of blood are used each day in the United States, with someone receiving blood about every 3 seconds. Whole blood or its components are used for many surgical procedures as well as for ongoing treatment of chronic diseases. For example, people who have been in car accidents and suffered massive blood loss may need as many as 100 pints of blood. Some patients with complications from severe sickle-cell disease, which affects 80,000 people in the United States, need as much as 4 pints of blood every month.

We all expect blood to be there for us, but only a small fraction of those who can give, do so. Yet sooner or later, virtually all of us will face a situation in which we will need blood. And that time is all

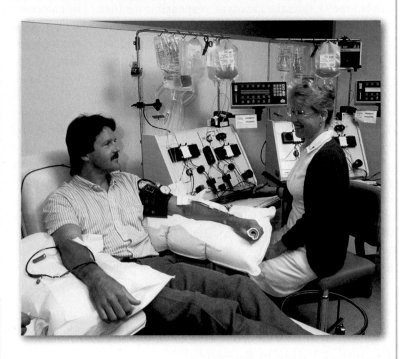

too often unexpected. To find out where you can donate, visit www. givelife.org or call 1-800-GIVE-LIFE (1-800-448-3543).

Some Facts about Blood

- An adult body contains 10 to 12 pints of blood.
- It takes about 6 to 10 minutes to donate a pint of blood and 24 hours for the body to replace the blood fluid volume. The red cells may take up to 2 months for full restoration.
- Just one unit of blood can be used to treat up to three patients.
- An average healthy person will be eligible to give blood every 56 days, or more than 330 times in a lifetime.
- Blood donation takes four steps: medical history, quick physical, donation, and snacks. The actual blood donation usually takes less than 10 minutes. The entire process, from signing in to leaving, takes about 45 minutes.
- Giving blood will not decrease your strength.
- You cannot get AIDS or any other infectious disease by donating blood.
- Fourteen tests, eleven of which are for infectious diseases, are performed on each unit of donated blood.
- 60 percent of the U.S. population is eligible to donate, but only 5 percent do so on a yearly basis.

NOTE: Facts and figures included here are courtesy of the American Red Cross, Northern Ohio Blood Services Region, and Blood Centers of the Pacific.

10.2 Gene Mutations: The Source of New Alleles

Different alleles of a gene arise by **mutation**, which we can define briefly here as any change in the DNA that makes up a gene (see Chapter 13 for a more detailed discussion). When a mutation occurs, the new allele may contain instructions for a protein whose form differs from the version specified by the original allele. By specifying different versions of a protein, the different alleles of a gene produce genetic differences among the individuals in a population.

Mutations are sometimes harmful. For example, a mutation may lead to the production of a protein that performs a vital function poorly or not at all. Consider, for example, the genes that control the manufacture of the pigment responsible for black scales in the black rat snake. When one of these genes mutates into a nonfunctional version (known as a null allele), it fails to make an enzyme that is essential for pigment production, so no pigment is made in a snake that is homozygous for the null allele (Figure 10.4). Snakes with the white-scale phenotype are rare in nature, probably because predators can see them more easily against dark-colored rocks or because they suffer serious damage when basking in the sun's rays.

Harmful or nonfunctional alleles tend to be recessive. Such recessive alleles can be fairly common in a population because heterozygotes can harbor such alleles and pass them on to future generations, without suffering any ill effects themselves. The one "good copy" of the normal allele masks the effect of the harmful recessive allele, so the heterozygote does not display the negative phenotype.

The most common mutations are those that are neither harmful nor beneficial to the individual. Such neutral mutations arise, for instance, when a new allele specifies a protein that is nearly identical to the protein specified by the original allele. In some cases a cell can tolerate a wide range in the activity of a protein with no harm to the organism, as is the case for the sugar-inserting enzyme coded by the *I* gene. Occasionally, mutations produce alleles that improve on the original protein or carry out new, useful functions. Such beneficial mutations are the rarest of the three types of mutations.

Mutations have two other important characteristics. First, mutations occur at random with respect to their usefulness. There is no evidence, for example, that beneficial mutations occur because they are "needed" by the organism. Second, mutations can happen at any time and in any cell of the body. In multicellular organisms, however, only mutations that are present in gametes, or in the cells that ultimately produce gametes, can be passed on to offspring.

Figure 10.4 Not True to Its Name
This is a white phenotype of the black rat snake! This snake has a mutation in a gene for scale color, causing it to be white, not black.

10.3 Basic Patterns of Inheritance

Now that we have defined some key genetic concepts and discussed how mutations produce new alleles, we are ready to explore how genes are transmitted from parents to offspring. Prior to Mendel, many people argued that the traits of both parents were blended in their offspring, much as paint colors blend when they are mixed together. According to this notion, which was known as the theory of blending inheritance, offspring should be intermediate in phenotype to their two parents, and it should not be possible to see "lost" traits reappear in the generations that come later. For example, if a white-flowered plant were mated with a red-flowered plant, the offspring should have pink flowers, and the original flower colors of white and red should not be seen in later generations.

Many observations, including Mendel's, do not match these predictions, however. The features of offspring often are not intermediate between those of their parents, and it is common for traits to skip a generation (for example, a child may have blue eyes like one of its grandparents, but unlike its brown-eyed parents). How can such observations be explained? Gregor Mendel answered this question with a series of experiments on plants.

Mendel conducted genetic experiments on pea plants

During 8 years of investigation, Mendel conducted many experiments to analyze inheritance in pea plants. His results led him to reject the idea of blending inheritance. Mendel proposed instead that for each trait, offspring inherit two separate units of genetic information (genes), one from each parent.

Figure 10.5 Genotype and Phenotype

Flower color in peas is controlled by a gene with two alleles (*P* and *p*). Although there are three genotypes (*PP*, *Pp*, and *pp*), there are only two phenotypes (purple flowers and white flowers). This happens because genotypes *PP* and *Pp* both produce purple flowers, while *pp* produces white flowers.

Genotype		
PP (homozygous)	*Pp* (heterozygous)	*pp* (homozygous)
Phenotype		
Purple	Purple	White

Peas are excellent organisms for studying inheritance. Ordinarily, peas self-fertilize; that is, an individual pea plant contains both male and female reproductive organs, and it fertilizes itself. But because peas can also be mated experimentally, Mendel was able to perform carefully controlled genetic crosses. In addition, peas have true-breeding varieties, which means that when these plants self-fertilize, all of their offspring have the same phenotype as the parent. For example, one variety has yellow seed and it produces only offspring with yellow seeds when it is bred with itself. We now know this happens because the parent plants are homozygous for the allele that causes seeds to be yellow. Mendel based all of his experiments on homozygous varieties that were true-breeding for traits such as plant height, flower color, or the color or shape of the seeds. Based on what we know today about how genotypes affect phenotypes, we can say individuals that breed true for a given phenotype have a homozygous genotype, such as *PP* or *pp* for the flower color trait shown in Figure 10.5.

In his experiments, Mendel observed inherited traits in each of three generations of plants: a set of original, true-breeding parents (P generation), and two generations of hybrid (non-true-breeding) offspring. For example, he crossed plants that bred true for purple flowers with plants that bred true for white flowers (Figure 10.6). Mendel then allowed the F_1 plants (the first generation of offspring) to self-fertilize, thereby producing the F_2 generation.

Inherited traits are determined by genes

According to the theory of blending inheritance, the cross shown in Figure 10.6 should have yielded F_1 generation plants bearing flowers of intermediate color. Instead, all the F_1 plants had purple flowers. Furthermore, when the F_1 plants self-fertilized, about 25 percent of the F_2 offspring had white flowers. The occur-

rence of white flowers skipped a generation, something that should not happen with blending inheritance.

Mendel studied seven traits in peas, and his results for each of those traits were similar to those shown in Figure 10.6. These results led him to propose a new theory of inheritance, in which genes behave like separate units, or particles, not like colors of paints that blend together. Using modern terminology, Mendel's concepts can be summarized as follows:

1. *Alternative versions of genes cause variation in inherited traits.* For example, peas have one version (allele) of a certain gene that causes flowers to be purple, and another version (a different allele) of the same gene that causes flowers to be white. One individual carries at most two different alleles for a particular gene.

2. *Offspring inherit one copy of a gene from each parent.* In his analysis of crosses like that in Figure 10.6, Mendel reasoned that for white flowers to reappear in the F_2 generation, the F_1 plants must have had two copies of the flower color gene (one copy that caused white flowers and one copy that caused purple flowers). Mendel was right: with the exception of our gametes, somatic cells in the adult organism typically have one

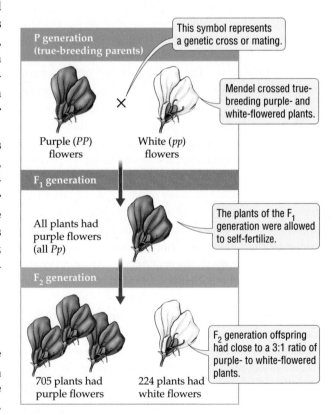

P generation (true-breeding parents)

This symbol represents a genetic cross or mating.

Purple (*PP*) flowers ✕ White (*pp*) flowers

Mendel crossed true-breeding purple- and white-flowered plants.

F_1 generation

All plants had purple flowers (all *Pp*)

The plants of the F_1 generation were allowed to self-fertilize.

F_2 generation

705 plants had purple flowers

224 plants had white flowers

F_2 generation offspring had close to a 3:1 ratio of purple- to white-flowered plants.

Figure 10.6 Three Generations in One of Mendel's Experiments

maternal and one paternal copy of each of their many genes (see Figure 10.2).

3. *An allele is dominant if it has exclusive control over the phenotype of an organism when paired with a different allele.* For example, let's call the allele for purple flower color *P* and the allele for white flower color *p*. Plants that breed true for purple flowers must have two copies of the *P* allele (that is, they are of genotype *PP*), since otherwise they would occasionally produce white flowers. Similarly, plants that breed true for white flowers have two copies of the *p* allele (genotype *pp*). Working back from the phenotypes of both the F_1 and F_2 generations, Mendel correctly deduced that the F_1 plants in Figure 10.6 must have genotype *Pp*. He realized that the genotype of the F_1 plants was best explained by assuming that each such plant received a *P* allele from the *PP* parent with purple flowers and a *p* allele from the *pp* parent with white flowers. Since all the F_1 plants had purple flowers, one must also assume that the *P* allele is dominant over the *p* allele. The recessive allele, *p*, has no effect on the phenotype because its effects are masked by the *P* allele's impact on flower color.

4. *The two copies (alleles) of a gene separate during meiosis and end up in different gametes.* Each gamete receives only one copy of each gene. If an organism has two copies of the same allele for a particular trait, as in the homozygous varieties used by Mendel, all of its gametes will contain that allele. However, if the organism has two different alleles, like an individual of genotype *Pp*, then 50 percent of the gametes will receive the one allele and 50 percent of the gametes will receive the other allele.

5. *Gametes fuse without regard to which alleles they carry.* When gametes fuse to form a zygote, they do so randomly with respect to the alleles they carry for a particular gene. As we'll see, this element of randomness allows us to use a simple method to determine the chance that offspring will have a particular genotype.

10.4 Mendel's Laws of Inheritance

Mendel summarized the results of his experiments in two laws: the law of segregation and the law of independent assortment. Let's take a look at each of Mendel's laws and how he developed them.

Mendel's first law: segregation

The **law of segregation** states that the two copies of a gene separate during meiosis and end up in different gametes. This law can be used to predict how a single trait will be inherited. As an illustration, let's revisit the experiment shown in Figure 10.6. In that experiment, Mendel crossed plants that bred true for purple flowers (genotype *PP*) with individuals that bred true for white flowers (genotype *pp*). This cross produced an F_1 generation composed entirely of heterozygotes (individuals with genotype *Pp*). According to the law of segregation, when the F_1 plants reproduced, 50 percent of the pollen (sperm) should have contained the *P* allele, and the other 50 percent the *p* allele. The same is true for the eggs.

We can represent the separation of the two copies of a gene during meiosis, and their random recombining through fertilization, using a diagram called a **Punnett square** (Figure 10.7). This visual method was first used in 1905 by the British geneticist Reginald Punnett. It charts how alleles are distributed into gametes by meiosis and combined in all possible ways during fertilization. To create a Punnett square, we list all the possible genotypes of the male gametes across the top of a grid (the "Punnett square"), writing each unique genotype once. We also list all possible genotypes of the female

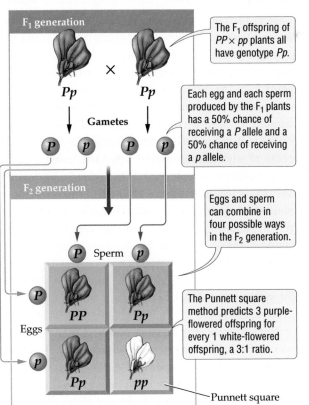

F₁ generation

The F₁ offspring of *PP* × *pp* plants all have genotype *Pp*.

Pp × *Pp*

Gametes

P p   P p

Each egg and each sperm produced by the F₁ plants has a 50% chance of receiving a *P* allele and a 50% chance of receiving a *p* allele.

F₂ generation

Eggs and sperm can combine in four possible ways in the F₂ generation.

P Sperm p

Eggs

	P	p
P	PP	Pp
p	Pp	pp

The Punnett square method predicts 3 purple-flowered offspring for every 1 white-flowered offspring, a 3:1 ratio.

Punnett square

Investigate the law of independent assortment.

▶❚❚ 10.3

Figure 10.7 The Punnett Square Method
The Punnett square method can be used to predict the outcome of a genetic cross. It diagrams the separation of alleles into gametes and all the possible ways in which each type of male and female gamete can be combined to produce an offspring. A single-trait cross in which both parents are heterozygotes is known as a monohybrid cross.

gametes, writing them along the left edge of the grid, again writing each unique genotype only once. Then we fill out the grid by combining in each box in the grid the male genotype at the top of each column with the female genotype shown at the beginning of each row (follow the blue and pink arrows in Figure 10.7). As shown in Figure 10.7, regardless of whether it has a P or a p allele, each sperm has an equal chance of fusing with an egg that has a P allele or an egg that has a p allele. The Punnett square shows all four ways in which the two types of sperm can combine with the two types of egg. The four genotypes shown within the Punnett square are all equally likely outcomes of this cross.

Using the Punnett square method, we can predict that ¼ of the F_2 generation is likely to have genotype PP, ½ to have genotype Pp, and ¼ to have genotype pp. Because the allele for purple flowers (P) is dominant, plants with PP or Pp genotypes have purple flowers, while pp genotypes have white flowers. Therefore, we predict that ¾ (75 percent) of the F_2 generation will have purple flowers and ¼ (25 percent) will have white flowers. This prediction is very close to Mendel's actual results for the F_2 generation: 705 (76 percent) had purple flowers and 224 (24 percent) had white flowers.

Mendel's second law: independent assortment

Mendel also performed experiments in which he simultaneously tracked the inheritance of *two* genetic traits. For example, pea seeds can have a round or wrinkled shape, and they can be yellow or green. Two different genes control the two different traits: the R gene controls seed shape, the Y gene controls the color of the seed. Mendel wanted to know what would happen if true-breeding round, yellow-seeded individuals (genotype $RRYY$) were crossed with true-breeding wrinkled, green-seeded individuals (genotype $rryy$). When Mendel performed the two-trait cross (Figure 10.8), all of the resulting F_1 plants had round, yellow seeds. From the phenotypes of this F_1 generation, Mendel could see that the allele for round seeds (denoted R) was dominant over the allele for wrinkled seeds (denoted r). Similarly, with respect to the seed color trait, Mendel deduced that the allele for yellow seeds (Y) was dominant over the allele for green seeds (y).

Next, Mendel crossed large numbers of the F_1 plants (genotype $RrYy$) to raise a generation of F_2 plants. He knew that the phenotypes of the F_2 generation would answer the question that most interested him: Is the inheritance of seed shape independent of the inheritance of seed color? If his hunch was right, the distribution of the alleles of one gene into gametes would be independent of the distribution of alleles of the other gene, so that all possible combinations of the alleles would be found in the gametes (illustrated in Figure 10.8b). If this were not the case, and a particular seed color always went with a particular seed shape, we would *not* find novel combinations of phenotypes among the F_2 offspring. Instead, the F_2 offspring would show one of the two parental phenotypes: round, yellow seeds or wrinkled, green seeds.

Mendel tested the two possibilities by crossing $RrYy$ plants with each other (Figure 10.8). He obtained the following results in the F_2 generation: approximately ⁹⁄₁₆ of the seeds were round and yellow, ³⁄₁₆ were *round and green,* ³⁄₁₆ were *wrinkled and yellow,* and ¹⁄₁₆ were wrinkled and green (a 9:3:3:1 ratio). As shown in Figure 10.8, Mendel's results were what we would expect if the genes for these two traits were inherited independently of each other. If the alleles of the R gene segregate independently from the alleles of the Y gene during gamete formation (as depicted in Figure 10.8b), we would predict four phenotypes among the F_2 offspring: ⁹⁄₁₆ displaying the dominant phenotypes for both traits, ¹⁄₁₆ with the recessive phenotypes of both traits, and two novel combinations of phenotypes not seen in either parent. So, the two nonparental combinations of phenotypes Mendel was looking for did turn up among the F_2 plants: ³⁄₁₆ of the plants had round, green seeds, and another ³⁄₁₆ had wrinkled, yellow seeds.

Mendel made similar crosses for various combinations of the seven traits he studied. His results led him to propose the **law of independent assortment**, which states that when gametes form, the separation of the two copies (alleles) of one gene during meiosis is independent of the separation of the copies of other genes. Most of the seven genes Mendel was tracking in his breeding experiments happen to be on a different pair of homologous chromosomes; for example, the seed shape gene is located on chromosome pair 7, while the seed color gene sits on chromosome pair 1. As explained in Chapter 9, the maternal and paternal members of a pair of homologous chromosomes are randomly distributed into the daughter cells during gamete formation through meiosis. This means that if we consider two pairs of homologous chromosomes, say pair 7 and pair 1, there is no telling whether a particular gamete will receive the paternal copies of both 7 and 1, or whether the paternal copy of 7 will be combined with the maternal copy of chromosome 1 in a gamete. Because the homologous pairs are randomly shuffled into gametes

(a)

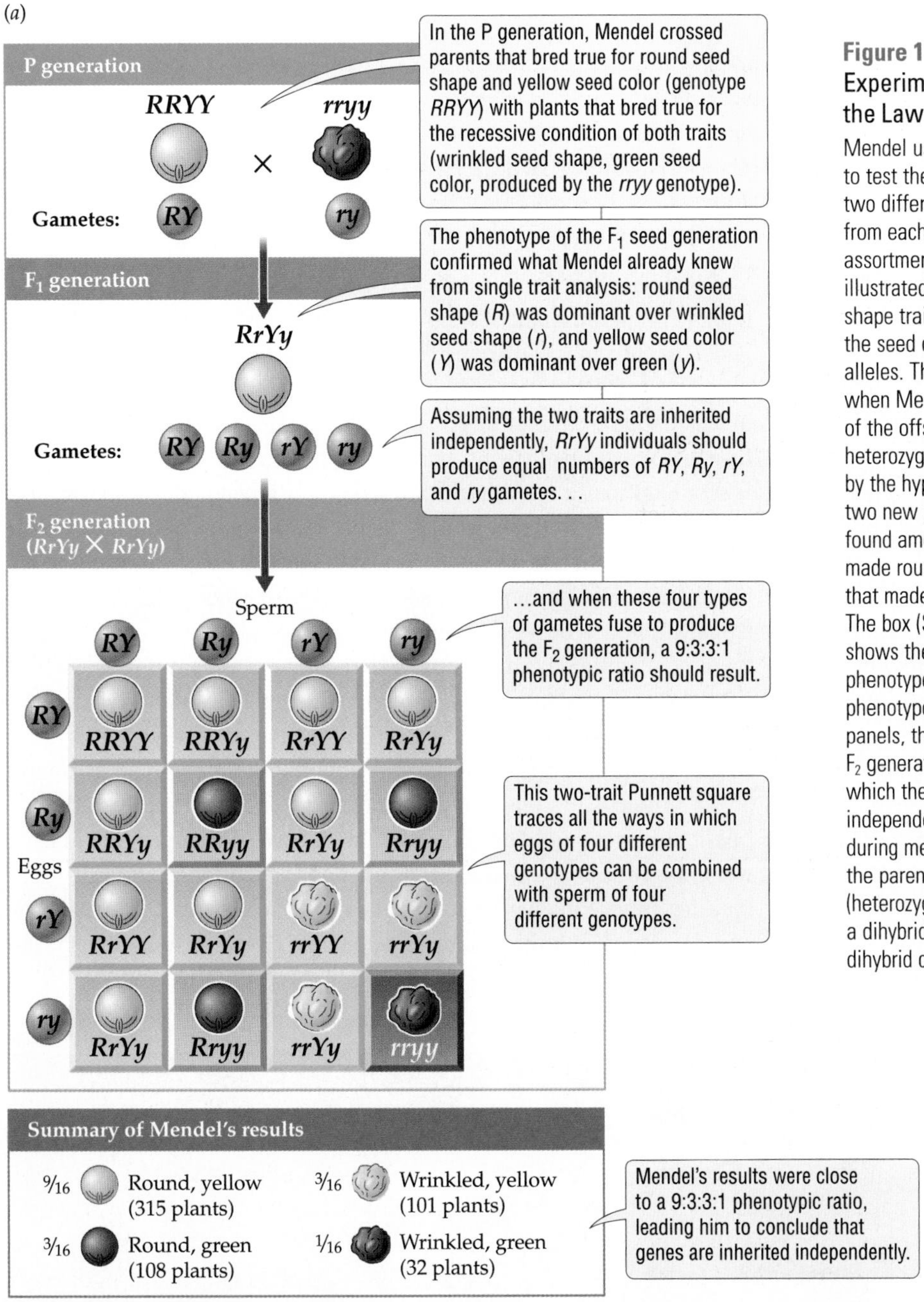

P generation

RRYY   ×   rryy

Gametes: RY   ry

In the P generation, Mendel crossed parents that bred true for round seed shape and yellow seed color (genotype *RRYY*) with plants that bred true for the recessive condition of both traits (wrinkled seed shape, green seed color, produced by the *rryy* genotype).

F₁ generation

RrYy

Gametes: RY   Ry   rY   ry

The phenotype of the F₁ seed generation confirmed what Mendel already knew from single trait analysis: round seed shape (*R*) was dominant over wrinkled seed shape (*r*), and yellow seed color (*Y*) was dominant over green (*y*).

Assuming the two traits are inherited independently, *RrYy* individuals should produce equal numbers of *RY*, *Ry*, *rY*, and *ry* gametes. . .

F₂ generation
(RrYy × RrYy)

Sperm

	RY	Ry	rY	ry
RY	RRYY	RRYy	RrYY	RrYy
Ry	RRYy	RRyy	RrYy	Rryy
rY	RrYY	RrYy	rrYY	rrYy
ry	RrYy	Rryy	rrYy	rryy

Eggs

...and when these four types of gametes fuse to produce the F₂ generation, a 9:3:3:1 phenotypic ratio should result.

This two-trait Punnett square traces all the ways in which eggs of four different genotypes can be combined with sperm of four different genotypes.

Summary of Mendel's results

9/16 ◯ Round, yellow (315 plants)

3/16 ◉ Round, green (108 plants)

3/16 ⬡ Wrinkled, yellow (101 plants)

1/16 ⬤ Wrinkled, green (32 plants)

Mendel's results were close to a 9:3:3:1 phenotypic ratio, leading him to conclude that genes are inherited independently.

(b)

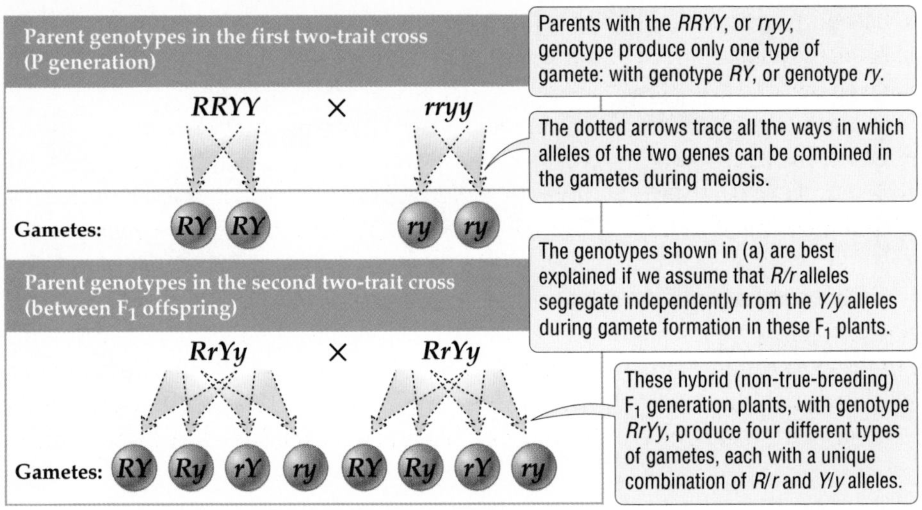

Parent genotypes in the first two-trait cross (P generation)

RRYY   ×   rryy

Gametes: RY   RY   ry   ry

Parents with the *RRYY*, or *rryy*, genotype produce only one type of gamete: with genotype *RY*, or genotype *ry*.

The dotted arrows trace all the ways in which alleles of the two genes can be combined in the gametes during meiosis.

Parent genotypes in the second two-trait cross (between F₁ offspring)

RrYy   ×   RrYy

Gametes: RY Ry rY ry   RY Ry rY ry

The genotypes shown in (a) are best explained if we assume that *R/r* alleles segregate independently from the *Y/y* alleles during gamete formation in these F₁ plants.

These hybrid (non-true-breeding) F₁ generation plants, with genotype *RrYy*, produce four different types of gametes, each with a unique combination of *R/r* and *Y/y* alleles.

Figure 10.8 Mendel's Two-Trait Experiments Provide Evidence for the Law of Independent Assortment

Mendel used two-trait breeding experiments to test the hypothesis that the alleles of two different genes are inherited separately from each other (law of independent assortment). In one set of experiments, illustrated here, Mendel tracked the seed shape trait controlled by the *R/r* alleles and the seed color trait controlled by the *Y/y* alleles. The real test of the hypothesis came when Mendel examined the phenotypes of the offspring produced by crossing the heterozygous F₁ plants (*RrYy*). As predicted by the hypothesis, and illustrated in (a), two new phenotypic combinations were found among the F₂ offspring: plants that made round, green seeds (*R-yy*) and those that made wrinkled, yellow seeds (*rrY-*). The box (Summary of Mendel's results) shows the ratio of the two parental phenotypes and the two novel, nonparental phenotypes. (b) As depicted in these panels, the 9:3:3:1 phenotypic ratios in the F₂ generation are explained by a model in which the alleles of the *R* gene segregate independently of the alleles of the *Y* gene during meiosis. A two-trait cross in which the parents are double heterozygotes (heterozygous for both traits) is called a dihybrid cross. Here, *RrYy × RrYy* is a dihybrid cross.

Test your understanding of linked inheritance versus independent assortment.   ▶❚❚   10.4

Science Toolkit

Tossing Coins and Crossing Plants: Probability and the Design of Experiments

The probability of an event is the chance that the event will occur. For example, there is a probability of 0.5 that a fair coin will turn up "heads" when it is tossed. A probability of 0.5 is the same thing as a 50 percent chance. As an illustration, consider a coin-tossing experiment. If you toss a penny only a few times, the observed percentage of heads may differ greatly from 50 percent. For example, if you tossed a coin only 10 times, you might get 70 percent (7) heads. However, if you tossed a coin 10,000 times, it would be very unusual to get 70 percent (7,000) heads.

Each toss of a coin is an independent event, in the sense that the outcome of one toss does not affect the outcome of the next toss. For a series of independent events, we can estimate the probability of each of the possible outcomes. Suppose we toss a coin 10,000 times and get 5,046 heads. From these results, it would be reasonable to estimate the chance of getting heads on the next toss as 50 percent, the percentage we expect from that coin. When only a small number of events are observed, our estimates of the underlying probabilities are less likely to be accurate, as when we toss a coin only a few times and get 7 of 10 tosses coming up heads.

How does all this relate to the design of a scientific experiment? Consider Mendel's work on peas. When Mendel crossed heterozygous plants (for example, Pp individuals) with each other, the offspring always had a phenotypic ratio close to 3:1. The reason for this consistent ratio is that the chances of getting offspring with PP, Pp, and pp genotypes are 25, 50, and 25 percent, respectively (see Figure 10.6). Because the 25 percent PP individuals and the 50 percent Pp individuals have the same phenotype, 75 percent of the individuals should look alike, giving a 3:1 phenotypic ratio of 3 purple to 1 white.

The Punnett square method predicts the percentages of PP, Pp, and pp offspring that Mendel should have observed. The method assumes that all sperm cells and all egg cells have an equal chance of achieving fertilization. When large numbers of offspring are considered, this assumption is not too far off, because the successes or failures of the different types of gametes tend to balance one another. But if Mendel had used only small numbers of offspring in his experiments, his results might have differed greatly from a 3:1 ratio. If that had been the case, he might not have discovered his two fundamental laws: the law of segregation and the law of independent assortment.

What was true for Mendel's experiments is true for scientific experiments in general: chance events not under the control of the scientist can affect the outcome of an experiment, so large amounts of data are needed to compensate for the effect of chance events. This is true of any genetic cross in which the researcher has no control over which sperm and which egg fuse to produce an offspring, but it is also true in many other settings.

For example, a scientist wanting to know if pesticides were producing deformities in frogs would need to collect a lot of data to isolate the effects of chance events. One way scientists have investigated the impact of agricultural pesticides on frogs is to record the frequency of frogs with extra or missing legs in farm ponds that receive pesticides from nearby fields and compare that with data from ponds without pesticides. The researchers would have to collect data from as many ponds as possible to minimize the chance that some other factor, such as another pollutant or another organism found only in some ponds, affects the results one way or another. In one such study, although pesticides were not the direct cause of the deformities (a parasite was), the addition of pesticides weakened the frogs' immune systems, making it more likely that individuals would succumb to the parasite and develop deformities. (For an expanded discussion on deformities in frog populations, see Section 10.6.)

A Deformed Frog

during meiosis, and can be mixed in all possible combinations during fertilization, the alleles on these chromosomes can also be mixed and matched in all possible allelic combinations. That is how offspring can turn out with genotypes and phenotypes (such as *RRyy* and *rrYY* in Figure 10.8*a*) that were not present in either parent. To this day, Mendel's law of independent assortment applies *as long* as we are following the inheritance of two genes that are each physically separated from each other because they lie on different chromosomes. For reasons we will explore in Chapter 11, this law may not apply to a pair of genes that are located relatively close to each other on the *same* chromosome.

Mendel was able to deduce these patterns of inheritance because he conducted a large number of genetic crosses, which gave him data for a large number of offspring. We explore some general reasons for using a large number of "experimental units" in the "Science Toolkit." When performing genetic crosses in particular, it is a good idea to obtain as many offspring as possible because the chance of obtaining an offspring with a particular genotype is just that—a chance.

It is important to understand that the ratios predicted by Mendel (for example, the 3:1 ratio illustrated in Figure 10.7) simply place the odds that a particular offspring will have a certain phenotype or genotype. We cannot know with certainty what the *actual* phenotype or genotype of a particular offspring is going to be (except when true-breeding individuals are crossed). Moreover, the odds that a particular offspring will display a specific phenotype are completely unaffected by how many offspring there are. It is just that the likelihood that we will see a 3:1 outcome increases when we analyze a greater number of offspring. For example, if there is a ¼ chance that each offspring will be a homozygous recessive (*pp*) individual, that means that if many offspring are produced, it is likely that 25 percent of them will have genotype *pp*. But there is no guarantee that if there are four offspring, one of them will always have genotype *pp*. That may happen, but it is also possible for none of the offspring to have genotype *pp*, or for more than one to have genotype *pp*. Such outcomes can occur because the 25 percent probability that an offspring will have genotype *pp* applies not only to the first offspring, but to the second, third, and fourth offspring as well. Hence it is even possible (but not likely, since the chance is 0.25 × 0.25 × 0.25 × 0.25, or less than half of 1 percent) that all four offspring will have genotype *pp*. When many offspring are examined, the chance of obtaining unusual results (such as all of them having genotype *pp*) becomes very small.

Concept Check

1. What is the ultimate source of genetic variation in a population of organisms?
2. For the offspring of a cross between an *Rr* plant and an *rr* plant, where *R* is dominant, predict the number and ratio of genotypes and phenotypes.
3. What central concept of genetics is outlined by Mendel's law of segregation?
4. Mendel found that wrinkled shape (controlled by dominant allele, *R*) is always inherited with green seed color (controlled by dominant allele, *Y*). Explain what is wrong with this statement.

10.5 Extensions of Mendel's Laws

Try an interactive coin-flipping exercise.

10.5

Mendel's laws describe how genes are passed from parents to offspring. In some cases—such as the seven traits of pea plants that Mendel studied—these laws allow offspring phenotypes to be predicted accurately from parental genotypes. In particular, Mendel's laws allow us to make accurate predictions whenever a genetic trait is controlled by a single gene with two alleles, one dominant, the other recessive. But many traits are not under such simple genetic control. Geneticists refined and extended Mendel's laws through much of the twentieth century to explain more complex patterns of inheritance. They have discovered, for example, that sometimes a single allele can produce a number of different phenotypes. The most complex patterns of inheritance are created when a phenotype is controlled by more than one gene, each with multiple alleles, and especially if the phenotype is also affected by environmental factors rather than by the genotype alone.

Many alleles do not show complete dominance

For dominance to be complete, a single copy of the dominant allele must be enough to produce its phenotypic effect in a heterozygote; for example, one *P* allele ensures that even a *Pp* pea plant has purple flowers, since *P* is dominant over *p*. But in the case of some allele combinations, no one allele completely dominates over the other when the two are paired together in a heterozygote. Because neither allele is able to exert its full effect, the heterozygote displays an "in-between" phenotype. When neither allele in a heterozygote is completely dominant over the other, a third phenotype may be observed that is intermediate between the phenotypes of the two

H^cH^c

H^cH^{cr}

$H^{cr}H^{cr}$

Figure 10.9 Incomplete Dominance in Horses

Palominos (genotype H^cH^{cr}) are intermediate in color to chestnuts and cremellos. In a horse with the appropriate genetic background (the "red horse" type), the H^c allele yields a chestnut coat in the homozygous state. The presence of the H^{cr} allele "dilutes" the chestnut color to create the palomino phenotype in the heterozygote (H^cH^{cr}). Two copies of the H^{cr} allele lead to the cream (cremello) coat color. H^c is incompletely dominant over H^{cr} because the two alleles produce an intermediate phenotype when they are combined in a heterozygote.

C^RC^R
Red flowers

C^WC^W
White flowers

C^RC^W
Pink flowers

homozygous alleles. In this case the two alleles are said to display **incomplete dominance**. Two of the alleles that control flower color in snapdragons are incompletely dominant. When a homozygote with red flowers (C^RC^R) is crossed with a homozygote with white flowers (C^WC^W), the heterozygous offspring (C^RC^W) produce pink flowers. Some of the genes controlling coat color in animals also display incomplete dominance. Figure 10.9 shows how two incompletely dominant alleles, H^C and H^{Cr}, produce an intermediate phenotype in a heterozygote. The chestnut horse (which has a reddish-brown coat) is homozygous for H^C (H^CH^C). A cream-colored horse (known as *cremello*) is produced when the H^{Cr} allele is present in a homozygous state ($H^{Cr}H^{Cr}$). In a heterozygote, the effect of the H^C allele is "diluted" by the presence of the cream-color allele, H^{Cr}, resulting in the palomino horse, beloved in parades and shows for its golden coat and flaxen mane.

Although incomplete dominance superficially resembles the old idea of blending inheritance, it is really just an extension of Mendelian inheritance. We can predict the genotypes and phenotypes of F_1 and F_2 offspring using Mendelian laws of inheritance and the Punnett square methods described earlier, except that we must assign an intermediate phenotype to heterozygotes. For example, if two heterozygous snapdragons (C^RC^W) are crossed, the odds are that ¼ of the offspring will have red flowers (genotype C^RC^R), ½ are likely to have pink flowers (genotype C^RC^W), and ¼ are likely to have white flowers (genotype C^WC^W). Work this out for yourself using the Punnett square method. You will see that Mendel's laws apply to alleles that show incomplete dominance, just as they apply to alleles that display a dominant–recessive relationship; the main difference

is that the heterozygotes (C^RC^W) look different from C^RC^R individuals when there is incomplete dominance, whereas they would look identical if C^R were dominant to C^W. Also, when pink-flowered plants are bred with each other, we are likely to see among these F_2 offspring the flower colors—red or white—that were seen in the purebred parents of the F_1 generation. This could not happen if blending inheritance were true because, according to that notion, the two colors would have become inseparably mixed in the pink-flowered F_1 offspring.

A pair of alleles can also show **codominance**, in which the effect of both alleles is equally visible in the phenotype of the heterozygote. In other words, the influence of each codominant allele is fully displayed in the heterozygote, without being diminished or diluted by the presence of the other allele (as in incomplete dominance) or being suppressed by a dominant allele (as in the case of dominant–recessive alleles). The ABO blood typing system we mentioned earlier provides an example (Figure 10.10). Three alleles determine a person's blood type: the I^A, I^B, and i alleles. The first two of these alleles, I^A and I^B, are codominant to each other, producing an AB blood type in the heterozygote. The AB blood type is not halfway between an A and B blood type, as would be the case if these alleles were incompletely dominant. Instead, people with the AB blood type (I^AI^B genotype) have *both* A and B types of sugars deposited on the surface of their red blood cells. In contrast, the I^AI^A homozygote has only the A type sugars, and therefore an A blood type. Similarly, an I^BI^B individual has just the B type of sugars, and therefore blood type B. (The i allele is recessive to the other two alleles, so the I^Ai individual has blood type A, the I^Bi individual has blood type B, and the *ii* individual

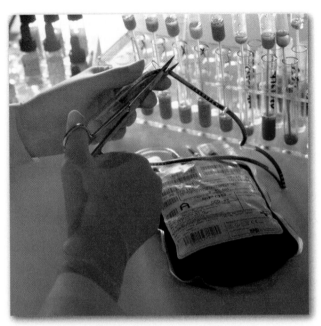

Figure 10.10 Codominance in the Blood Typing System
People who have an AB blood type have codominant I^A and I^B alleles. The i allele is recessive to both I^A and I^B. Those with type A blood are either homozygous for the I^A allele or have the $I^A i$ genotype. Similarly, people with type B blood are either homozygous for the I^B allele or have the $I^B i$ genotype.

has blood type O.) It is easy to get confused between the terms incomplete dominance and codominance because it might sound like they mean similar things. To help keep these two types of inheritance straight in your mind, remember that _in_complete dominance produces an _in_termediate phenotype.

A gene displaying pleiotropy can influence multiple phenotypes

Each of the seven genes that Mendel studied controls a single clear-cut trait, such as flower color, seed color, seed shape, or plant height. We now know, however, that some genes control functions that are of such central importance that multiple biological processes are impacted if the gene in question fails to act normally. The situation when a single gene influences a variety of different traits is called **pleiotropy** (*pleio*: many; *trepein*, turnings). A pleiotropic gene is one that can influence two or more distinctly different traits.

Such seemingly different traits as skin color and vision problems are affected by the action of a single gene in the genetic condition known as albinism (Figure 10.11). There are various forms of albinism, all of them marked by the absence, or reduced production, of a brown-black pigment called melanin. About 1 in 17,000 Americans has albinism. Most affected individuals produce very little

melanin in the skin, hair, and eyes; contrary to popular belief, most of them have blue eyes, although a minority have red or violet eyes. Everyone diagnosed with albinism has eye problems, which can range from being "cross-eyed" to being legally blind. The gene involved in the most common type of albinism controls a step in the pathway for melanin production. It is easy to imagine how a malfunction in this gene would affect pigment deposition and therefore skin color, but what does that have to do with the eyes? For reasons that are not fully understood, melanin production is necessary for the proper development of the eye, including the nerves that help the eye communicate with the brain.

Marfan's syndrome is a pleiotropic disorder in which many organ systems are affected because a single gene, coding for a protein called fibrillin 1, does not work properly. The protein is crucial for the normal function of connective tissues, which act as a gluing and scaffolding system for all types of organs, from bones to the walls of blood vessels. The 1 in 5,000 Americans diagnosed with Marfan's syndrome show a wide range of phenotypes, depending on which allele of the fibrillin 1 gene they possess and also on their other genetic characteristics. Most have problems with vision and also the skeleton. Many are tall and gangly, with long arms, legs, fingers, and toes. The nervous system, lungs, and skin are commonly affected. The weakening of the largest blood vessel carrying blood away from the heart is the most serious phenotype associated with the disorder. The rupture of this blood vessel is believed to have killed volleyball star Flo Hyman, who won the Olympic silver medal with

Figure 10.11 Person with Albinism
A mother and her son, who has albinism, sitting in their living room in Douala, Cameroon, Africa. Albinism affects eyesight in addition to pigmentation, and is an example of a pleiotropic condition.

Figure 10.12 Flo Hyman
U.S. Olympics volleyball silver medalist Flo Hyman shown during practice in 1984. Complications from Marfan's syndrome contributed to her death in 1986. Marfan's syndrome is an example of a pleitropic genetic disorder.

her team in 1984 (Figure 10.12). Flo was nearly six and a half feet tall. Her condition was diagnosed only after her death in 1986, at the age of 31, during an exhibition game in Japan.

Alleles for one gene can alter the effects of another gene

Another twist in the inheritance of genetic traits comes from the interplay of different genes. Genes can interact in such a way that the phenotypic effect of gene A depends not only on the genotype of an individual with respect to gene A, but also the particular alleles of another gene, gene B, that this individual has. Such interactions among the alleles of different genes are common in all types of organisms. For example, in yeast (a single-celled organism used in making bread and beer), each gene tested was found to interact with at least 34 other genes.

When the phenotypic effect of the alleles of one gene depends on which alleles are present for another, independently inherited gene, the phenomenon is known as **epistasis**. Epistasis is seen among the many genes that control coat color in mice and other animals. Many of these genes code for enzymes that are involved in the multi-step pathway that converts the amino acid tyrosine into melanin. One such gene, acting further down in the pathway, has a dominant allele (*B*) that leads to black fur and a recessive allele (*b*) that produces brown fur. But the effects of these alleles (*B* and *b*) can be eliminated completely, depending on which alleles of the *C* gene are present. The *C* gene codes for an enzyme (called tyrosinase) that acts at the very first step in this pathway. In mice with the *cc* genotype, the enzyme fails to do its work, and the entire "assembly line" grinds to a halt. No melanin is made in the fur or eyes, resulting in an albino phenotype (Figure 10.13). It doesn't matter whether the genotype with respect to the *B* gene is *BB* or *Bb* or *bb*; as long as the genotype

is *BBcc*, *Bbcc*, and *bbcc*, the albino phenotype prevails. The *C* gene is therefore said to be epistatic to the *B* gene, meaning that alleles of the *C* gene can completely obscure any potential effect of any allele of the *B* gene. This does *not* mean that the alleles of the *C* gene are dominant over those of the *B* gene. The protein encoded by the *C* gene simply acts earlier in the pathway, so how the alleles of the *C* gene function (or don't) affects whether the phenotype controlled by any allele of *B* is displayed or not.

The environment can alter the effects of a gene

The effects of many genes depend on internal and external environmental conditions, such as body temperature, carbon dioxide levels in the blood, external temperature, and amount of sunlight. For example, in Siamese cats, the C^t allele of the tyrosinase gene is sensitive to temperature, which means that the production of melanin is dependent on the temperature of the surroundings (Figure 10.14). This allele codes for a tyrosinase that works well at colder temperatures, but not at all at warmer temperatures. Because a cat's extremities tend to be colder than the rest of its body, melanin can be produced there, and hence the paws, nose, ears, and tail of a Siamese cat tend to be dark. If a patch of light fur is shaved from the body of a Siamese cat and the skin is covered with an ice pack, when the fur grows back, it will be dark. Similarly, if dark fur is shaved from the tail and allowed to grow back under warm conditions, it will be light-colored.

Chemicals, nutrition, sunlight, and many other environmental factors can also alter the effects of genes. In plants, genetically identical individuals (clones) grown in different environments often differ in many aspects of their phenotype, including height and the number of flowers they produce. For example, plants on a wind-swept mountainside may be short and have few flowers, while clones of the same plants grown in a warm, protected valley are tall with many flowers. Similar effects are found in people. For example, a person who was malnourished as a child is likely to be shorter as an adult than if he or she had received plenty of food.

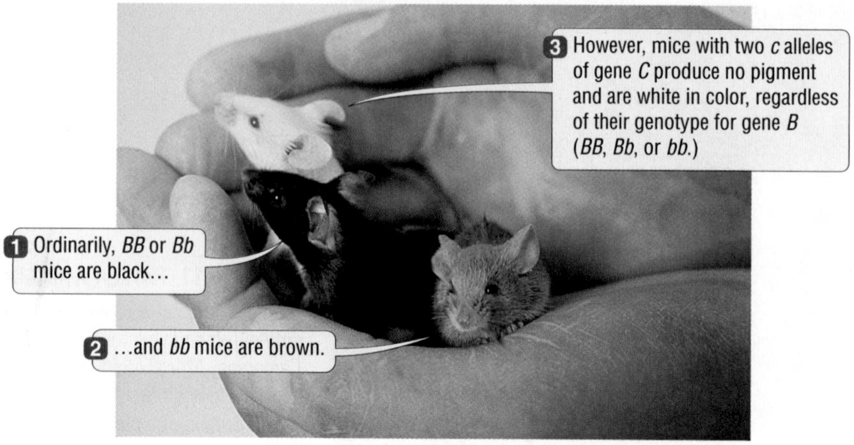

3 However, mice with two *c* alleles of gene *C* produce no pigment and are white in color, regardless of their genotype for gene *B* (*BB*, *Bb*, or *bb*.)

1 Ordinarily, *BB* or *Bb* mice are black...

2 ...and *bb* mice are brown.

Figure 10.13 Gene Interactions
Gene interactions are very common. The example illustrated here shows how the *c* allele of gene *C* masks the alleles of another gene, *B*. In mice with the *CC* or *Cc* genotype, the dominant *B* allele directs the production of melanin resulting in black fur, while the recessive *bb* genotype alters melanin production in such a way that brown fur results. Mice with the *cc* genotype are albinos because melanin production gets blocked at an early point in the pathway, before the point at which the *B* gene exerts its influence. The *cc* genotype always results in albinism, regardless of whether the genotype with respect to the *B* gene is *BB*, *Bb*, or *bb*.

Figure 10.14 The Environment Can Alter the Effects of Genes Coat color in Siamese cats is controlled by an allele that produces dark pigment (as on the nose, tail, paws, and ears) only at low temperatures.

Most traits are determined by two or more genes

Mendel studied characteristics that were under simple genetic control: a single gene determined the phenotype for each of the traits he studied. Most traits, however, are **polygenic**—that is, they are determined by the action of more than one gene. Examples of polygenic traits include skin color, running speed, and body size in humans; and height, flowering time, and seed number in plants. Let's look in more detail at one of these examples, the inheritance of skin color in humans.

As we mentioned earlier, the pigment melanin is the main source of skin color in humans and most other mammals. At least three genes are involved in controlling how much melanin gets deposited in the specialized pigment-bearing cells in our skin. There are probably more than three such genes, but let us assume for simplicity that there are only three (A, B, and C) and that each affects skin color equally. Assume also that there are only two incompletely dominant versions of each gene: one version (A^1, B^1, C^1) that causes melanin production and another (A^0, B^0, C^0) that prevents melanin production. In this model, the phenotype of a person's skin color is controlled by how many "units" of the melanin-producing allele he or she has; the more such units a person has, the darker the skin. If two individuals heterozygous for each of the three genes were to marry, we would predict seven different shades of skin color among their children, as shown in Figure 10.15.

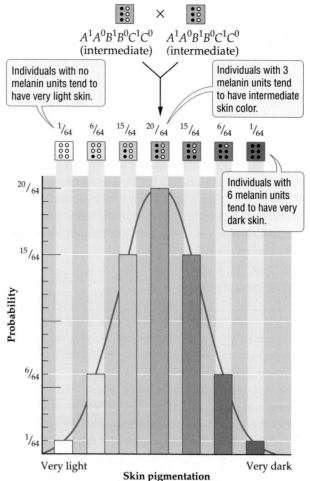

$A^1A^0B^1B^0C^1C^0$ (intermediate) × $A^1A^0B^1B^0C^1C^0$ (intermediate)

Individuals with no melanin units tend to have very light skin.

Individuals with 3 melanin units tend to have intermediate skin color.

Individuals with 6 melanin units tend to have very dark skin.

$1/64$ $6/64$ $15/64$ $20/64$ $15/64$ $6/64$ $1/64$

Probability

Very light — **Skin pigmentation** — Very dark

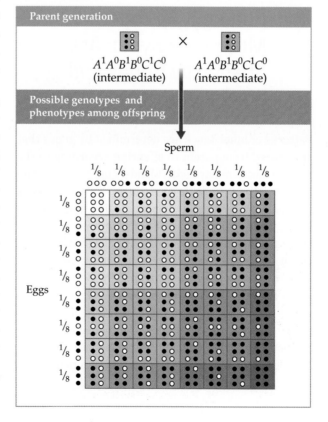

Parent generation

$A^1A^0B^1B^0C^1C^0$ (intermediate) × $A^1A^0B^1B^0C^1C^0$ (intermediate)

Possible genotypes and phenotypes among offspring

Sperm

Eggs

Figure 10.15 Three Genes Can Produce a Range of Skin Color in Humans

For each of the three hypothetical genes shown here (A, B, and C), the superscript 1 (for example, A^1) identifies an allele that leads to the production of one "unit" of melanin, while the superscript 0 (for example, A^0) identifies an allele that produces no melanin. The two alleles of each of these genes are incompletely dominant. Alleles that do not contribute to melanin production (A^0, B^0, and C^0) are represented by open circles, while alleles that contribute to melanin production (A^1, B^1, and C^1) are represented by solid circles. In this model, a person's skin color is influenced by the total number of melanin "units" specified by their genotype. The Punnett square shows the proportion of gametes with a specific genotype that are produced by two heterozygote ($A^1 A^0 B^1 B^0 C^1 C^0$) parents and all the possible ways in which the 8 different genotypes can come together during fertilization. The bar graph depicts the phenotypic outcome. Bar heights indicate the relative proportions of children of each phenotype. Each of the seven different phenotypes (that is, children with 0, 1, 2, 3, 4, 5, or 6 melanin units) can be specified by one to several different genotypes. For example, children producing 2 melanin units can have any one of six different genotypes. Additional variation in skin color would result from different levels of sun exposure.

A suntan is another example of an environmental influence on the phenotype. In most people, melanin production increases in response to the ultraviolet radiation in sunlight. If we add different degrees of sun tanning to the seven shades of skin color predicted on the basis of genotype alone, we get an even greater variety of skin colors, enough to fit a smooth curve to the bar graph shown in Figure 10.15. In summary, human skin comes in just about every shade, from very light to very dark, because human skin color is a polygenic trait, and such a trait yields a greater range of phenotypes than a trait influenced by a single gene. In addition, this trait is affected by environmental factors, which adds yet more variety to the palette of skin color phenotypes.

Concept Check

1. Allele *H* produces straight hair, while allele *H'* produces curly hair. Individuals with the *HH'* genotype have wavy hair, somewhere between straight and curly. Are alleles *H* and *H'* codominant?
2. What is pleiotropy?
3. The ABO blood groups are determined by several alleles of the *I* gene. Is this a polygenic trait?
4. Explain why this statement is incorrect: Polygenic traits are affected by environmental influences, but single-gene traits are not.

10.6 Putting It All Together: Most Phenotypes Are Shaped by Genes and the Environment

Patterns of inheritance are determined by genes that are passed from parent to offspring according to the simple rules summarized in Mendel's laws. Some traits are controlled by a single gene and are little affected by environmental conditions. For such traits—for example, seed shape and flower color in pea plants—it is possible to predict the phenotypes of offspring just from knowing which alleles the parents have for a single gene.

Many phenotypes are controlled by more than one gene. The alleles of some genes can interact with the alleles of another gene so that the action of one gene can be altered by the alleles of another gene. The relationship between genotype and phenotype is more complex for traits that show such non-Mendelian pat-

terns of inheritance. In such cases, a given gene does not act in isolation; rather, its effect depends not only on its own function, but also on the function of other genes with which it interacts and on the environment (Figure 10.16). For example, scientists investigating deformities that were turning up in frog populations in certain parts of the country found that a fungal parasite was the culprit. The fungus is widespread, so why were frogs in some areas affected, when genetically similar frogs in other areas were free of the disease? Further analysis revealed that pesticide pollution weakened the animals' immune system, making them more vulnerable to the infection. Resistance to the fungus is thought to be regulated by multiple genes coding for immune system function. The phenotype (disease resistance) encoded by one or more of these genes is influenced by environmental

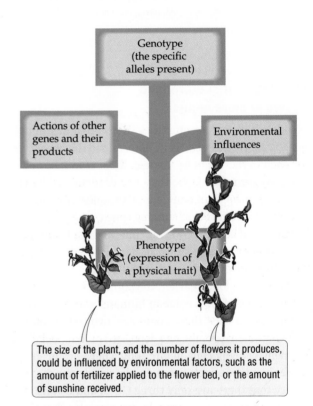

The size of the plant, and the number of flowers it produces, could be influenced by environmental factors, such as the amount of fertilizer applied to the flower bed, or the amount of sunshine received.

Figure 10.16 From Genotype to Phenotype: The Big Picture

The effect of a gene on an organism's phenotype can depend on a combination of the gene's own function, the function of other genes with which it interacts, and the environment. The environmental effects could be natural influences such as difference in the amount of sunshine received by a plant, or they could be human interventions such as the amount of fertilizer we apply. As a result, two individuals with the same genotype for a gene may have show very different phenotypes, as illustrated by the differences in growth between the two pea plants.

conditions, and in the presence of pesticide pollution, the normal phenotype (disease resistance) appears to be absent.

Many human diseases, including heart disease, cancer, alcoholism, and diabetes, are strongly influenced by multiple genes and by many different environmental factors, such as smoking, diet, and overall mental and physical health. Predicting the phenotypes of offspring for such traits requires a detailed understanding of how genes and the environment influence the final product of the genes—the phenotype. Such prediction is a challenging and important task. We could reduce the death rate from heart disease and cancer, for example, if we knew how specific genes interacted with the environment to cause these diseases. Recent developments in genetics (as we shall see in Chapters 14 through 16) hold the promise of future improvements in our ability to identify those at risk for these and other human disease phenotypes and to devise better methods of prevention and cure.

Applying What We Learned

Solving the Mystery of the Lost Princess

Tsar Nicholas II abdicated in 1917, ending over three centuries of rule by the House of Romanov. While alive, Tsar Nicholas and his wife, the Tsarina Alexandra, served as a rallying point for people opposed to the new Communist revolutionary government. Even after death, the royal family was viewed as a threat, and so Communist officials spread misleading information about them. Tsar Nicholas, they said, had been shot, but his family had been moved to a place where they were safe from the turmoil of civil war. The resulting uncertainty over the fate and location of Alexandra and her children set the stage for the legend of Anastasia, the lost princess.

The mystery of Anna Anderson and the lost princess was finally solved in 1994. The secret grave of Tsar Nicholas and his family was unearthed in 1989, and the remains were carefully analyzed. Investigators electronically superimposed photographs of the skulls on archive photographs of the family, and they compared skeletal measurements with clothing known to have belonged to the Tsar and his family. They also matched the platinum dental work on one skull with the Tsarina's dental records. All these and other tests yielded a match: the skeletons seemed to be those of the Russian royal family. Finally, to make their case airtight, the investigators turned to the ultimate arbitrator, genetic analysis.

When DNA obtained from the skeletons was compared with DNA obtained from relatives of the Russian royal family (including the Tsar's brother, who died in 1899), the results showed conclusively that the skeletal remains were those of the Russian royal family. However, two sets of bones were missing, those of Prince Alexis and one of the two princesses, either Maria or Anastasia. Could it be that Anastasia had escaped and that Anna Anderson was who she claimed to be?

Here, too, the answer came from genetic analysis: Did Anna Anderson have the alleles the Romanov children would be predicted to have, based on Mendelian principles? DNA was obtained from small samples of Anna Anderson's intestinal tissue that had been preserved after a 1979 surgical operation. Next, her DNA was compared with DNA obtained from the skeletons of the Tsar and Tsarina. In one region of human DNA, at least five codominant alleles (A^1, A^2, A^3, A^4, and A^5) are commonly found among European populations. The Tsar had genotype A^1A^2, and the Tsarina had genotype A^2A^3. According to Mendel's laws, for Anna Anderson to have been the daughter of the Tsar and Tsarina, she should have had one of the following genotypes: A^1A^2, A^1A^3, A^2A^2, or A^2A^3. But her actual genotype was A^4A^5, which was not consistent with the Tsar and Tsarina being her parents. Three other regions of DNA yielded similar results, indicating that Anna Anderson was not the lost princess.

Rodents May Teach Us How to Stay Married

By Ken-Yon Hardy

Prairie voles . . . may shed light on how to have a long, happy marriage. "Prairie voles are socially monogamous, which is unusual among mammals. They typically mate for life," said Brian Keane, Miami University Hamilton Zoology professor . . . "Field and lab studies show that males display numerous behaviors associated with monogamy including paternal care," said Keane . . .

Previous lab studies suggest that variations in social attachment among prairie voles might be due in part to genetic differences in a single gene that codes for a receptor in the brain. This receptor binds a specific chemical messenger called vasopressin. Scientists know that vasopressin is involved in regulating complex social behaviors such as recognition, aggression and affiliation in mammals including humans,

according to Keane. "In the lab, male prairie voles with longer versions of this gene spent much more time in contact with their female partner, compared to an unfamiliar female, relative to males with shorter forms of the gene," Keane said. "These data are intriguing because they suggest that size differences in the gene coding for the vasopressin receptor affect social behavior and possible mate fidelity among male prairie voles."

Can a single gene alter the social behavior of an animal species? A spate of recent investigations suggests that the answer is "Yes." One study sought to find out whether changing one gene could change the sexual behavior of male meadow voles. In captivity, male meadow voles are solitary and promiscuous—after they mate with a female, they do not spend extra time with or show preference for that female. Males of a closely related species, the prairie vole, are more monogamous than the meadow vole in that they bond with the female they mate with, preferring to spend time with her and keeping other males away from her.

Scientists suspected that the hormone vasopressin, which is released in the brain after mating, might influence the tendency for males to be monogamous. Copies of a single gene were inserted into the brains of male meadow voles to find out whether that gene would change the way they behaved. The gene contains instructions for building the receptor of vasopressin. Although male meadow voles already have such receptors in their brains, they do not have as many as the brains of monogamous prairie voles. By inserting extra copies of the vasopressin gene into the brains of male meadow voles, the researchers experimentally boosted the number of vasopressin receptors that these voles have in their brains. The results confirmed the scientists' hypothesis: increasing the number of vasopressin receptors in the

brain made the meadow vole males more sensitive to the hormone, and that, apparently, was enough to change their behavior from promiscuous to monogamous.

Evaluating the News

1. A male vole with one copy of the long vasopressin receptor (long allele) and one copy of the short vasopressin receptor (short allele) showed an intermediate level of monogamy compared to the low fidelity displayed by homozygotes for the short allele and the high fidelity displayed by homozygotes for the long allele. Based on this information, would you infer that the vasopressin gene shows incomplete dominance or codominance? Can the inheritance of the vasopressin receptor gene be predicted based on Mendel's laws of inheritance?

2. If we knew that a human behavior could be altered by a single gene, would it be ethical to insert that gene into the brain of a person in order to change his or her behavior?

3. Should research on the genetics of human behavior receive funding from national governments?

Source: *Journal News,* July 16, 2006.

Chapter Review

Summary

10.1 Essential Terms in Genetics

- Genes—the basic units of inheritance—are segments of DNA that help determine an organism's inherited characteristics, or genetic traits.
- Diploid individuals generally have two copies of each gene, one inherited from the male parent, the other from the female parent.
- Alternative versions of a gene are called alleles.
- The genotype is an individual's allelic makeup, while the phenotype is the specific version of an observable trait that it displays.
- A dominant allele is one that determines the phenotype of an individual even when it is paired with a different allele. A recessive allele has no phenotypic effect when it is paired with a dominant allele.

10.2 Gene Mutations: The Source of New Alleles

- In a population of many individuals, a particular gene may have one, a few, or many alleles.
- The different alleles of a gene arise by mutation.
- A new allele (produced by mutation) may cause the organism to produce a protein that differs in form and function from the versions of the protein specified by other alleles of the gene.
- By specifying different proteins, the different alleles of a gene result in hereditary differences among organisms.
- Some mutations are harmful, many have little effect, and a few are beneficial. The effect of a mutation depends on the functionality of the protein specified by the new allele.
- Mutations can happen at any time and in any cell in an organism, but only mutations in gametes or their precursor cells can be passed on to the next generation.
- Mutations occur at random with respect to their usefulness; the fact that a mutation may be useful does not make it more likely to occur.

10.3 Basic Patterns of Inheritance

- Over an 8-year period, Gregor Mendel performed a pioneering series of experiments designed to elucidate how inheritance works.
- Mendel's experiments with pea plants led him to reject the old notion of blending inheritance. Instead, they suggested that the inherited characteristics of organisms are controlled by specific units of inheritance, which we now know as genes.
- Using modern terminology, Mendel's view of inheritance can be summarized as follows: (1) Alleles of genes account for the variation in genetic traits. (2) Offspring inherit one allele of a gene from each parent. (3) Alleles can be dominant or recessive. (4) The two copies of a gene separate into different (haploid) gametes. (5) Fertilization combines gametes in a random manner to give rise to diploid offspring.

10.4 Mendel's Laws of Inheritance

- Mendel summarized the results of his experiments in two laws: the law of segregation and the law of independent assortment.
- The law of segregation states that the two copies of a gene (the two alleles) end up in different gametes during meiosis.
- The law of independent assortment states that when gametes form during meiosis, the distribution of the copies of one gene is independent of the distribution of the copies of other genes.
- The Punnett square method considers all possible combinations of gametes to predict the outcome of a genetic cross.

10.5 Extensions of Mendel's Laws

- For some traits, Mendel's laws do not predict the phenotype of the offspring. There are several reasons for this: (1) Many alleles do not show complete dominance; instead, they may show incomplete dominance (as in the coat color of horses) or codominance (as in the ABO blood types in humans). (2) One gene may affect more than one genetic trait (pleiotropy). (3) Alleles for one gene can suppress the alleles of another gene (epistasis). (4) The environment can alter the effect of a gene. (5) Most traits are polygenic; that is, they are governed by two or more genes.
- Even when Mendel's laws do not accurately predict offspring phenotype, the genes in question are inherited according to Mendel's laws. In these cases, what differs is how the genes affect the phenotype, not how the genes are passed from parent to offspring.

10.6 Putting It All Together: Most Phenotypes Are Shaped by Genes and the Environment

- Some traits are controlled by a single gene that is little affected by other genes or by environmental conditions.
- Most traits are influenced by sets of genes that interact with one another and with the environment.
- The effect of a gene on an organism's phenotype can depend on the gene's own function, the function of other genes with which it interacts, and the environment. As a result, two individuals with the same genotype may have different phenotypes.

Review and Application of Key Concepts

1. Describe what genes are and how they work. Your explanation should include a description (using modern terminology) of Mendel's theory of inheritance, as well as a short summary of what we now know about (a) the chemical and physical structure of genes and (b) the information that they encode.

2. How many copies of each gene does a (sexually reproducing) organism have? Why? If all copies of a gene in an individual are identical (that is, the individual is homozygous for that gene), what can you infer about the genotypes of its parents?

3. Explain how new alleles arise, and how different alleles cause hereditary differences among organisms.

4. Draw a diagram that shows meiosis for an organism with four chromosomes (two pairs of homologous chromosomes). Label two genes, one on each chromosome, and use the diagram to explain why the distribution of copies of one of these genes during meiosis is independent of the distribution of copies of the other gene.

5. The allele for purple flowers (P) is dominant over the allele for white flowers (p). A purple-flowered plant, therefore, could be of genotype PP or Pp. What genetic cross could you make to determine the genotype of a purple-flowered plant? Explain how such a test cross enables you to deduce whether the purple-flowered plant is a homozygote or heterozygote.

6. Although we are accustomed to thinking of "identical" twins as looking exactly alike, they can have different phenotypes. Why? Provide specific reasons why identical twins might not look exactly alike.

7. Many lethal human genetic disorders are caused by a recessive allele, whereas relatively few are caused by a dominant allele. Why might dominant alleles for lethal human diseases be uncommon? (*Hint*: Solve problems 5 and 6 in the Sample Genetics Problems on page 227 and use the results to guide your answer to this question.)

Key Terms

allele (p. 208)
codominance (p. 218)
diploid (p. 207)
dominant (p. 209)
epistasis (p. 220)
F_1 generation (p. 209)
F_2 generation (p. 209)
gene (p. 206)
genetic cross (p. 209)
genetic trait (p. 206)
genetics (p. 206)
genotype (p. 208)
haploid (p. 208)
heterozygote (p. 209)
homologous pair (p. 207)

homozygote (p. 209)
incomplete dominance (p. 218)
law of independent assortment (p. 214)
law of segregation (p. 213)
maternal homologue (p. 207)
mutation (p. 211)
P generation (p. 209)
paternal homologue (p. 207)
phenotype (p. 207)
pleiotropy (p. 219)
polygenic (p. 221)
Punnett square (p. 213)
recessive (p. 209)

Self-Quiz

1. Alternative versions of a gene for a given trait are called
   a. alleles.
   b. heterozygotes.
   c. genotypes.
   d. copies of a gene.

2. If an allele for long hair (L) is dominant to an allele for short hair (l), then a cross of $Ll \times ll$ should yield
   a. ¼ short-haired offspring.
   b. ¾ short-haired offspring.
   c. ½ short-haired offspring.
   d. all offspring with intermediate hair length.

3. If A and a are two alleles of the same gene, then individuals of genotype Aa are
   a. homozygous.
   b. heterozygous.
   c. dominant.
   d. recessive.

4. Genes
   a. are the basic units of inheritance.
   b. are located on chromosomes and composed of DNA.
   c. commonly contain instructions for building a single protein.
   d. all of the above

5. Coat color in horses shows incomplete dominance. $H^C H^C$ individuals have a chestnut color, $H^C H^{cr}$ individuals have a palomino color, and $H^{cr} H^{cr}$ individuals have a cremello (white) color (see Figure 10.9). What is the predicted phenotypic ratio of chestnut to palomino to cremello if $H^C H^{cr}$ individuals are crossed with other $H^C H^{cr}$ individuals?
   a. 3:1
   b. 2:1:1
   c. 9:3:1
   d. 1:2:1

6. When the phenotypes controlled by two alleles are equally displayed in the heterozygote, the two alleles are said to show
   a. codominance.
   b. complete dominance.
   c. incomplete dominance.
   d. epistasis.

7. Traits that are determined by the action of more than one gene are
   a. recessive.
   b. not common.
   c. common in some organisms, but not in people.
   d. polygenic.

8. Select the term indicating that the phenotypic effects of alleles for one gene can be suppressed by the alleles of another, independently inherited gene.
   a. phenotypic variation
   b. codominance
   c. gene–environment interaction
   d. epistasis

Sample Genetics Problems

1. One gene has alleles A and a, a second gene has alleles B and b, and a third gene has alleles C and c. List the possible gametes that can be formed from the following genotypes:
   a. *Aa*
   b. *BbCc*
   c. *AAcc*
   d. *AaBbCc*
   e. *aaBBCc*

2. For the same three genes described in problem 1, what are the predicted genotype and phenotype ratios of the following genetic crosses? (Following our standard notation, alleles written in uppercase letters are dominant to alleles written in lowercase letters; the phenotype produced by allele A is therefore dominant over the phenotype controlled by allele a.)
   a. Aa × aa
   b. *BB* × *bb*
   c. *AABb* × *aabb*
   d. *BbCc* × *BbCC*
   e. *AaBbCc* × *AAbbCc*

3. Sickle-cell anemia is inherited as a recessive genetic disorder in humans. That means that in terms of disease onset, the normal hemoglobin allele (S) is dominant to the sickle-cell allele (s). For two parents of genotype Ss, construct a Punnett square to predict the possible genotypes and phenotypes (does or does not have the disease) of their children. Also list the genotype and phenotype ratios. Each time two Ss individuals have a child together, what is the chance that the child will have sickle-cell anemia?

4. Alleles for a gene (C) that determines the color of Labrador retrievers show incomplete dominance. Black labs have genotype $C^B C^B$, chocolate labs have genotype $C^B C^Y$, and yellow labs have genotype $C^Y C^Y$. If a black lab and yellow lab mate, what proportions of black, chocolate, and yellow coat colors would you expect to find in a litter of their puppies?

5. For any human genetic disorder caused by a recessive allele, let n be the allele that causes the disease and N be the normal allele (the capital "N" is for "normal" individuals).
   a. What are the phenotypes of NN, Nn, and nn individuals?
   b. Predict the outcome of a genetic cross between two Nn individuals. List the genotype and phenotype ratios that would result from such a cross.
   c. Predict the outcome of a genetic cross between an Nn and an NN individual. List the genotype and phenotype ratios that would result from such a cross.

6. For any human genetic disorder caused by a dominant allele, let D be the allele that causes the disorder and d be the normal allele (where the capital "D" stands for "disorder").
   a. What are the phenotypes of DD, Dd, and dd individuals?
   b. Predict the outcome of a genetic cross between two Dd individuals. List the genotype and phenotype ratios that would result from such a cross.
   c. Predict the outcome of a genetic cross between a Dd and a DD individual. List the genotype and phenotype ratios that would result from such a cross.

7. If blue flower color (B) is dominant to white flower color (b), what are the genotypes of the parents in the following genetic cross: blue flower × white flower yields only blue-flowered offspring?

8. The fruit pods of peas can be yellow or green. In one of his experiments, Mendel crossed plants that were homozygous for the allele for yellow fruit pods with plants that were homozygous for the allele for green fruit pods. All fruit pods in the F_1 generation were green. Which allele is dominant, the one for yellow or the one for green? Explain why.

Key Concepts

- A gene is a stretch of DNA that affects a genetic trait. Each gene has a specific location on a chromosome.

- Human males have one X and one Y chromosome, and human females have two X chromosomes. A specific gene on the Y chromosome is required for human embryos to develop as males.

- Genes that are located near one another on the same chromosome tend to be inherited together, and are said to be linked. Genes that are located far from one another on the same chromosome are often not linked.

- Many inherited genetic disorders in humans are caused by mutations of single genes.

- Dominant disorders whose harmful effects prevent a person from bearing children are generally rare in large populations, compared to recessive disorders that similarly keep a person from reproducing.

- Recessive alleles located on the X chromosome often produce gender-specific phenotypes known as X-linked traits. Recessive X-linked disorders tend to be more common among males than females.

- Some human genetic disorders result from abnormalities in chromosome number or structure.

A Horrible Dance

As a child in the mid-1800s, George Huntington went with his father, a medical doctor, as he visited patients in rural Long Island, New York. On one such trip, the boy and his father saw two women by the roadside, a mother and daughter, who were bowing and twisting uncontrollably, their faces contorted in a series of strange grimaces. Huntington's father paused to speak with them, but was unable to help them or even understand why they were so ill.

For any child, such an encounter would be unforgettable. For the young Huntington, it was that and more. Like his father and grandfather before him, George Huntington became a doctor, and in 1872 he described and named the disorder that had plagued the two women. He called it "hereditary chorea" (*chorea* [koh-*REE*-uh] comes from the Greek word for "dance").

Huntington described hereditary chorea as an inherited disorder, one that destroyed the nervous system and caused jerky, involuntary movements of the body and face. Hereditary chorea had no cure. Eventually, it killed its victims, but first it reduced them to a shell of their former selves. In addition to causing extreme motor impairments, it led to memory loss, severe depression, mood shifts, personality changes, and intellectual deterioration.

If a parent had hereditary chorea, it did not necessarily strike all the children of that parent. The children who did get it usually showed no symptoms until they were in their thirties, forties, or fifties. Hereditary chorea was like a genetic time bomb. In Huntington's words, the combination of the terrible symptoms of the disorder and its late and uncertain onset caused "those in whose veins the seeds of the disease are known to exist [to speak of it] with a kind of horror."

Hereditary chorea is now known as Huntington disease, in honor of George Huntington. **What is the genetic basis of Huntington disease? Why does it run in families and why does it affect some children but not their siblings? Can a person in whose "veins the seeds of the disease are known to exist" be tested to find out his or her risk of actually developing this condition? And if the test is positive, is there a cure?** Before we can answer these questions,

The Inheritance of Huntington Disease
Misty Oto, shown here with her mother and baby daughter, works to educate people about Huntington disease. Her mother, Rosie Shaw, has the disease, and Misty has 50 percent odds of developing the condition in middle age. In recent years, scientists have made significant strides in understanding the genetic basis of this disease.

we must explore the chromosomal basis of inheritance in greater detail. We will see why some genes tend to be inherited together, in apparent violation of Mendel's law of independent assortment, and why some genetic conditions are more common among males than in females. We will examine the patterns of inheritance that explain some common and rare human genetic disorders, including Huntington disease.

In Chapter 10, we described Mendel's discovery that inherited traits are governed by discrete hereditary units, or genes. We begin this chapter with a second cornerstone of modern genetics, the chromosome theory of inheritance. We then explain how an individual's sex is determined in humans and other organisms, and how new combinations of alleles, different from those of either parent, can occur in offspring. This information about chromosomes, sex determination, and new allele combinations sets the stage for the discussion of human genetic disorders in the remainder of the chapter.

11.1 The Role of Chromosomes in Inheritance

When Mendel published his theory of inheritance in 1866, he had no idea what genes were made of or where they were located within a cell. By 1882, studies using microscopes had revealed that threadlike structures—the chromosomes—exist inside dividing cells. The German biologist August Weismann (Figure 11.1) hypothesized that the number of chromosomes was first reduced by half during the formation of sperm and egg cells, then restored to its full number during fertilization. In 1887, meiosis was discovered, and this supported Weismann's hypothesis. Weismann also suggested that the hereditary material was located on chromosomes, but at that time there was no experimental evidence for or against that idea.

Genes are located on chromosomes

Early in the twentieth century, geneticists gathered a great deal of experimental evidence in support of Weisman's hypothesis. The concept that genes are located on chromosomes came to be known as the **chromosome theory of inheritance**. Modern genetic techniques allow us to pinpoint which chromosome contains a particular gene and where on the chromosome that gene is located.

How are chromosomes, DNA, and genes related? As we described in Chapters 9 and 10, chromosomes that

Figure 11.1
August Weismann

pair during meiosis are called **homologous chromosomes**. In each homologous [hoh-*MOLL*-uh-guss] pair, one member is inherited from the mother, the other from the father. Recall that each chromosome consists of a single long DNA molecule attached to many bundles of packaging proteins. Each gene is a small region of the DNA molecule and there are many genes on each chromosome. For example, we humans are estimated to have about 25,000 genes located on one set of our chromosomes, which consists of 23 different types of chromosomes. So, on average, we have 25,000/23, or 1,087, genes per chromosome.

The physical location of a gene on a chromosome is called a **locus** (plural "loci"). With some exceptions (to be discussed shortly), every diploid cell has two copies of every gene, one on each of the chromosomes in a homologous pair (see Figure 11.2). Because a gene can come in different versions, or alleles, a diploid cell can have two *different* alleles at a given genetic locus, in which case it has a heterozygous genotype for the gene at that locus. But if the two alleles at a genetic locus are identical, then the diploid cell has a homozygous genotype for the gene at that locus. The allelic makeup, or genotype, at a particular genetic locus

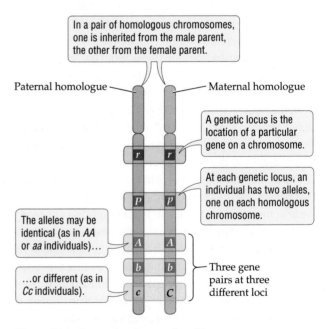

In a pair of homologous chromosomes, one is inherited from the male parent, the other from the female parent.

Paternal homologue — — Maternal homologue

A genetic locus is the location of a particular gene on a chromosome.

At each genetic locus, an individual has two alleles, one on each homologous chromosome.

The alleles may be identical (as in *AA* or *aa* individuals)…

…or different (as in *Cc* individuals).

Three gene pairs at three different loci

Figure 11.2 Genes Are Located on Chromosomes
The genes shown here take up a larger portion of the chromosome than they would if they were drawn to scale.

Karyotype of human male

Homologous pair

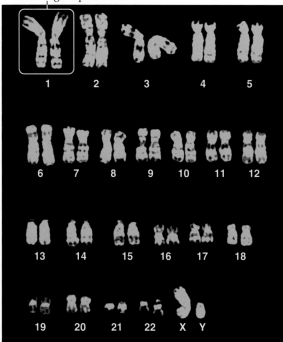

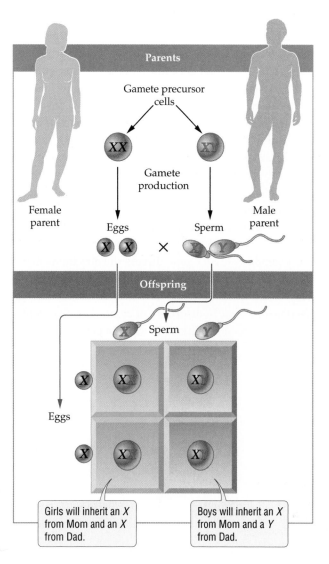

Parents

Gamete precursor cells

XX

XY

Gamete production

Female parent

Eggs

X *X*

×

Sperm

Male parent

Offspring

Sperm

X

Y

Eggs

X

XX

X

X

XX

X

Girls will inherit an *X* from Mom and an *X* from Dad.

Boys will inherit an *X* from Mom and a *Y* from Dad.

influences the phenotype, or outward display of a genetic trait. We will see shortly that the inheritance of various genes can be affected by how close or far apart they are on a chromosome and whether they are located on a sex chromosome or an autosome (non–sex chromosome).

Autosomes differ from sex chromosomes

As we saw in Chapter 9, most pairs of homologous chromosomes are exactly alike in terms of length, shape, and the genetic loci they carry. But in humans and many other organisms, this is not true of the chromosomes that determine the sex of the organism. These sex chromosomes are assigned different letter names. In humans, for example, males have one X chromosome and one Y chromosome (Figure 11.3), whereas females have two X chromosomes. The Y chromosome in humans is much smaller than the X chromosome, and few of its genes have a counterpart on the X chromosome. Since human males have one X and one Y chromosome, they have only one copy (instead of the usual two) of each gene that is unique to either the X or the Y chromosome. In some other organisms, such as birds, butterflies, and some fish, males have two identical chromosomes, denoted ZZ, whereas females have one Z chromosome and one W chromosome.

Chromosomes that determine sex are called **sex chromosomes**; all other chromosomes are called **autosomes**.

Human autosomes are labeled not with letters, but with the numbers 1 through 22 (for example, chromosome 4).

In humans, maleness is specified by the Y chromosome

Because human females have two copies of the X chromosome, all the gametes (eggs) they produce contain one X chromosome. Males, however, have one X chromosome and one Y chromosome, so the odds are that half their gametes (sperm) will contain an X chromosome and half will contain a Y chromosome. The sex chromosome carried by the sperm therefore determines the sex of a child. If a sperm carrying an X chromosome fertilizes an egg, the resulting child will be a girl; if a sperm carrying a Y chromosome fertilizes the egg, the child will be a boy (see Figure 11.3).

Helpful to know

Biologists often name particles found inside cells using the suffix *some* [sohm] ("body"). For example, autosomes ("same bodies") are chromosomes whose homologues look similar in both sexes. Chromosomes (literally, "colored bodies") were named for their appearance under the microscope when first discovered, before it was known what they were. Other examples include *lysosomes* ("splitting bodies") and *centrosomes* ("central bodies").

Compared with the X chromosome, the Y chromosome has few genes. It does, however, carry one very important gene: the **SRY gene** (short for "sex-determining region of Y"). The *SRY* gene functions as a master switch, committing the sex of the developing embryo to "male." In the absence of this gene, a human embryo develops as a female. The *SRY* gene does not act alone: in both males and females, other genes on the autosomes and sex chromosomes directly influence the development of the many sexual characteristics that distinguish men and women. The *SRY* gene plays a crucial role because when it is present, it causes the other genes to produce male sexual characteristics, whereas when it is absent, the other genes produce female sexual characteristics. For example, the *SRY* gene directs the development of testes, the male reproductive organs, in place of ovaries. The role of the non-*SRY* genes in gender determination becomes evident in disorders of sexual development, a broad range of conditions in which a person appears to be, or identifies with, a gender different from the one predicted by his or her sex chromosomes. For example, individuals with androgen insensitivity syndrome (AIS) are female in appearance and say they "feel female," even though they are XY in their chromosomal makeup, and their ovaries fail to develop normally. AIS is most commonly due to a genetic mutation that makes these individuals unable to respond normally to male hormones such as testosterone.

11.2 Genetic Linkage and Crossing-Over

As we saw in Chapter 10, the results of Mendel's experiments indicated that genes are inherited independently of one another. These results led him to propose his law of independent assortment, which states that the two copies of one gene are separated into gametes independently of the two copies of other genes. Early in the twentieth century, however, results from several laboratories indicated that certain genes were often inherited together, contradicting the law of independent assortment. Much of this work was done on fruit flies, a species that reproduces rapidly and is easy to study genetically.

Linked genes are those that tend to be inherited together

In his research on fruit flies, which began in 1909 at Columbia University in New York City, Thomas Hunt Morgan discovered genes that were inherited together. In one experiment, Morgan crossed a fruit fly that was homozygous for both a gray body (*G*) and wings of normal length (*W*) with another that was homozygous for both a black body (*g*) and wings that were greatly reduced in length (*w*). That is, he crossed *GGWW* flies with *ggww* flies to obtain flies of genotype *GgWw* in the F₁ generation. He then mated those *GgWw* flies with *ggww* flies. As shown in Figure 11.4, Morgan's

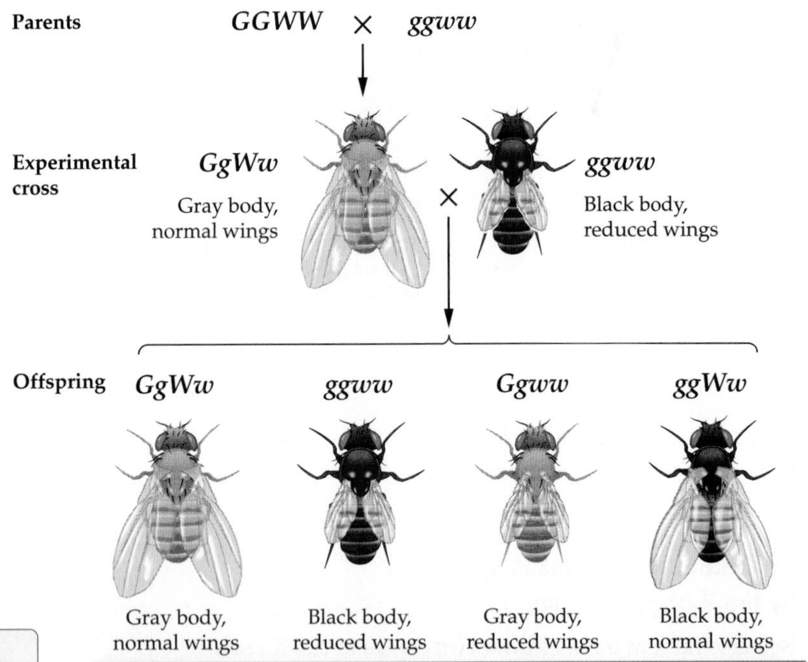

Parents	GGWW	×	ggww	

Experimental cross	GgWw	ggww
	Gray body, normal wings	Black body, reduced wings

Offspring	GgWw	ggww	Ggww	ggWw
	Gray body, normal wings	Black body, reduced wings	Gray body, reduced wings	Black body, normal wings

The law of independent assortment predicts these results…

Expected results			
575	575	575	575

…but these results were observed.

Actual results			
965	944	206	185

Parental phenotypes | Nonparental phenotypes

Conclusion: These two genes do not assort independently. They are linked together on the same chromosome.

Figure 11.4 Some Alleles Do Not Assort Independently
By crossing flies of genotype *GgWw* with flies of genotype *ggww*, Thomas Hunt Morgan found that the gene for body color (dominant allele *G* for gray, recessive allele *g* for black) was linked to the gene for wing length (dominant allele *W* for normal length, recessive allele *w* for reduced length). This linkage occurred because the two genes are located relatively close to each other on the same chromosome.

results were very different from the results he expected based on the law of independent assortment. What had happened?

Morgan concluded that the genes for body color and wing length must be located on the same chromosome. Because they were on the same chromosome, the genes were physically connected to each other and therefore the body color and wing length phenotypes were more likely to be inherited "in one lump" than genes that were on completely separate chromosomes, such as the seed color and seed shape genes studied by Mendel (Chapter 10). As a result, the law of independent assortment did not always hold in the case of the body color and wing length traits in Morgan's fruit flies. Genetic loci that are neighbors or positioned close to each other on the same chromosome tend to be inherited together and are said to be **genetically linked**. As we shall see shortly, some genes that are located far from one another on the same chromosome are not linked. Genes located on different chromosomes also are not genetically linked.

Crossing-over reduces genetic linkage

If the linkage between two genes on a chromosome were complete, a gamete could never have a chromosome type that was not originally present in one of the parents of the individual that produced that gamete. Consider, for example, the offspring of the $GGWW \times ggww$ cross shown in Figure 11.4. Recall that the two genes (one with alleles G and g, the other with alleles W and w) are on the same chromosome. As a result, the $GgWw$ flies would have inherited a GW chromosome from the $GGWW$ parent and a gw chromosome from the $ggww$ parent. Therefore, if linkage were complete, the $GgWw$ flies would have been able to make gametes only with chromosomes like those in one of their parents; namely, GW gametes or gw gametes (Figure 11.5). In that case, half of the offspring from the $GgWw \times ggww$ cross shown in Figure 11.4 would have had genotype $GgWw$, and the other half would have had genotype $ggww$. Since the majority of the offspring did have those two genotypes, Morgan realized that the two genes were linked. But how can we explain the appearance of the $Ggww$ and $ggWw$ offspring,

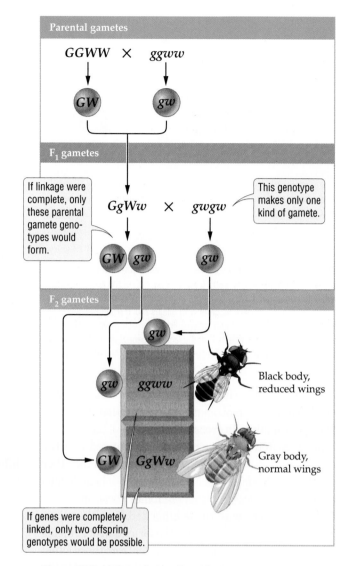

Explore some of Morgan's initial experiments.

▶II

11.1

Figure 11.5 Linkage Is Not Complete
If genes were completely linked, the F₂ progeny in Morgan's experiment would all have had genotype $GgWw$ or genotype $ggww$, as shown in this diagram. The results of Morgan's experiment (Figure 11.4) showed that this was not the case.

which have chromosomes (such as a Gw chromosome or a gW chromosome) that differ from those found in either parent?

To explain the appearance of these offspring genotypes, Morgan suggested that genes are physically exchanged between homologous chromosomes during meiosis. As we explained in Chapter 9, random segments of the chromosome are "swapped" between pairs of homologous chromosomes in a process known as **crossing-over** (or **recombination**), which takes place during meiosis I. To visualize this concept in the context of alleles, imagine that the two chromosomes illustrated

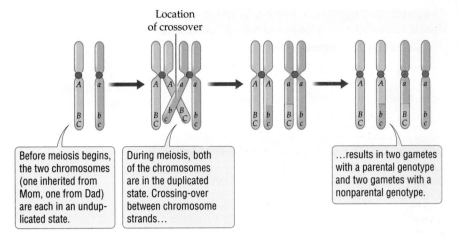

Location of crossover

Before meiosis begins, the two chromosomes (one inherited from Mom, one from Dad) are each in an undup-licated state.

During meiosis, both of the chromosomes are in the duplicated state. Crossing-over between chromosome strands…

…results in two gametes with a parental genotype and two gametes with a nonparental genotype.

Figure 11.6 Crossing-Over Disrupts the Linkage between Genes

In the case shown here, a crossing-over event occurs at a point on the chromosome between two linked genes, one gene with alleles *A* and *a*, the other with alleles *B* and *b*. As a result, half of the gametes have a parental genotype (*ABC* or *abc*), while the other half have a nonparental genotype (*Abc* or *aBC*). Although less likely, crossing-over also could have occurred between the gene with alleles *B* and *b* and the gene with alleles *C* and *c*, again producing nonparental genotypes (not shown). Finally, crossing-over could have occurred outside of the region bounded by the gene with alleles *A* and *a*, and the gene with alleles *C* and *c*, in this case (also not shown), all of the gametes would have had a parental genotype (either *ABC* or *abc*).

Review the process of crossing-over.

▶❙❙ 11.2

in Figure 11.6 come from one of your cells. You inherited one of these chromosomes from your father, the other from your mother. In crossing-over, part of the chromosome inherited from one parent is exchanged with the corresponding region of DNA inherited from the other parent. By physically exchanging pieces of homologous chromosomes, crossing-over combines alleles inherited from one parent with those inherited from the other. This exchange makes possible the formation of gametes with combinations of alleles that differ from those found in either parent, such as the gametes that resulted in the *Ggww* and *ggWw* offspring shown in Figure 11.4.

Crossing-over can be compared to the cutting of a string at a few random locations stretching from one end of the string to the other. Two points that are far apart on the string will be separated from each other in most cuts, whereas points that are close to each other will rarely be separated. Similarly, genetic loci that are far from each other on a chromosome are more likely to be separated by crossing-over than are genes that are close to each other. In fact, two genes on the same chromosome that are very far from each other may be separated by crossing-over so often that they are not genetically linked. Such genes are inherited independently of one another even though they are located on the same chromosome. Among the traits that Mendel studied in pea plants, we now know that the genes for flower color and seed color are both located on chromosome 1 (of the pea plant's seven pairs of chromosomes), but are so far apart that they are not linked. The law of independent assortment holds for these genes.

11.3 Origins of Genetic Differences between Individuals

Inheritance is both a stable and a variable process. It is stable in that most of the time genetic information is transmitted accurately from one generation to the next. Despite this stability, offspring are not exact genetic copies, or clones, of their parents, and hence the individuals of a sexually reproducing species differ genetically. These genetic differences are important, for they provide the genetic variation on which evolution can act. Genetic differences explain why one person has a genetic disorder, such as cystic fibrosis, and another does not. Genetic differences also explain why some individuals suffer severely from a disease, such as asthma, while others have a milder form. How do genetic differences among individuals arise? As we saw in Chapter 9, new alleles arise by mutation. Once formed, those alleles are shuffled or arranged in new ways by crossing-over, independent assortment of chromosomes, and fertilization.

Crossing over is part of the reason that one set of parents typically produces offspring with a range of different genotypes; in other words, why siblings don't always look exactly like each other. Every time meiosis occurs, crossing-over produces chromosomes with new combinations of alleles. These "recombined" chromosomes contain some alleles inherited from one parent and other alleles inherited from the other parent. By exchanging alleles between chromosomes, crossing-over causes some offspring to have a genotype that differs from the genotype of either parent (see Figures 11.4 and 11.6).

New combinations of alleles can also be produced by the **independent assortment of chromosomes**, which refers to the random distribution of maternal and paternal chromosomes into gametes during meiosis. Put another way, the paternal and maternal members of one homologous pair are sorted into gametes independently of the way any other homologous pair is sorted into the same gametes. The independent assortment of homologous pairs comes about because the maternal and paternal homologues of each homologous pair are randomly oriented at the center of the cell (metaphase plate), just before they are sorted into separate daughter cells during meiosis; as a result the daughter cells, which mature into gametes, acquire a random mix of paternal and maternal homologues. Because the members of a homologous pair usually carry different alleles for at least some of the genes, a gamete may acquire a combination of alleles that was not present in the parental cell undergoing meiosis.

Recall that maternal and paternal homologues pair up with each other very early in meiosis (during prophase I). The way each homologous pair is positioned at the metaphase plate by the spindle microtubules is random. The maternal and paternal homologues can be arranged in one of two ways: maternal homologue "to the left" of the metaphase plate; paternal homologue "to the right" of the metaphase plate, or the other way around. To picture this, imagine a cell with a total of 4 chromosomes (haploid number = 2), so it has just 2 pairs of homologous chromosomes, pair 1 and pair 2. The panels below illustrate the two ways in which the maternal and paternal members of pair 1 and pair 2 can be arranged at the metaphase plate. In pattern A, the maternal chromosomes of *both* pairs happen to be on the same side of the metaphase plate. In the lower panel (pattern B) we see the other possibility: the placement of the maternal and paternal homologues of pair 2 is switched, so that this time the *paternal* member is "to the left" of the metaphase plate and the maternal member is "to the right." Whether the maternal and paternal members of the two homologous pairs are positioned according to pattern A or pattern B during meiosis in a particular cell is purely a matter of chance.

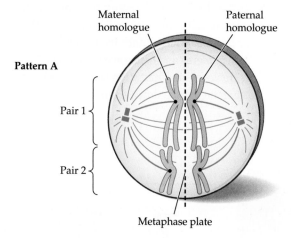

Maternal homologue   Paternal homologue

Pattern A

Pair 1

Pair 2

Metaphase plate

Or the other way around:

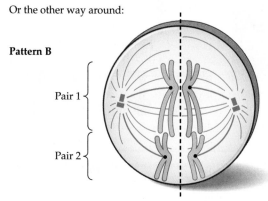

Pattern B

Pair 1

Pair 2

Note that patterns A and B will generate gametes containing a very different mix of homologues. Pattern A produces two types of gametes: one type contains the maternal members of pair 1 and pair 2, the other type contains the paternal members of pair 1 and pair 2. What combination of homologues is found in gametes generated by pattern B? Half the gametes will have the maternal member of pair 1 and the paternal member of pair 2; the other half will have the paternal member of pair 1 and the maternal member of pair 2.

The larger the number of homologous pairs in an organism, the greater the variety of patterns in which the homologues can be arranged during metaphase, and therefore the greater the variety of gametes that can be generated through meiosis. In the example illustrated above, we saw that just 2 homologous pairs can be arranged in two different ways, and can therefore generate 2^2 different types of gametes.

The diploid cells that undergo meiosis in the human body have 23 pairs of homologous chromosomes (diploid number = 46). Since each of the 23 pairs of homologous chromosomes lines up at random in one of two ways, there are 2^{23}, or 8,388,608, different ways that our chromosomes can be arranged during meiosis I. This means a person can produce at least 8,388,608 different types of gametes—either egg or sperm—in terms of the mix of homologues that are found in each type of gamete. Like crossing-over, the independent assortment of chromosomes generates gametes with unique combinations of the genetic information found in the parental chromosomes. In Chapter 9, we explained how fertilization adds to the genetic variation produced by crossing-over and the independent assortment of chromosomes.

Develop your understanding of how independent assortment produces gamete diversity through a dynamic animated tutorial. ▶II 11.3

Concept Check

1. What are genes made of and where are they located?
2. How are sex chromosomes different from autosomes?
3. Will two genes that are close together on a chromosome show stronger genetic linkage than a pair that are far apart? Explain.

11.4 Human Genetic Disorders

Many of us know someone with a genetic disorder, such as cystic fibrosis, sickle cell anemia, a hereditary form of cancer, or one of the many other disorders

Figure 11.7 Living with a Genetic Disorder
A physiotherapist is testing lung function in a young patient with cystic fibrosis. Cystic fibrosis is a recessive genetic disorder in which mucus builds up in the lungs, digestive tract, and pancreas, causing chronic bronchitis, poor absorption of nutrients, and recurrent bacterial infections, often leading to death before the age of 35.

caused by inherited gene mutations (Figure 11.7). It is important to study genetic disorders, since such studies could lead to the prevention or cure of much human suffering. But the study of human genetic disorders is beset by daunting problems. We humans have a long generation time, we select our own mates, and we decide when and whether to have children. Geneticists cannot perform experiments directly involving humans to help them figure out how human genetic disorders are inherited. In addition, our families are much smaller than would be ideal in a scientific study (see the "Science Toolkit" in Chapter 10). How can we get around these problems?

bers over two or more generations of a family's history. Pedigrees provide geneticists with a way to analyze information from many families in order to learn about the inheritance of a particular disorder. The pedigree shown in Figure 11.8, for example, shows the inheritance of cystic fibrosis, the most common lethal genetic disorder in the United States. Individuals 2 and 3 in generation III have cystic fibrosis, but their parents do not. The pedigree in Figure 11.8 indicates that the allele that causes cystic fibrosis is not dominant, for if it were, one of the parents of the affected individuals would have had the disorder.

Genetic disorders may or may not be inherited

As mentioned earlier, humans can be afflicted by a variety of genetic disorders. Some of these disorders—including most cancers (see Interlude B)—result from new mutations that occur in the cells of an individual sometime during his or her life. Such mutations usually occur in cells other than gametes and hence are not passed down to offspring. However, any mutation that is present in the gametes can be passed down from parent to child.

Test your understanding of human pedigree analysis with an interactive quiz.

▶❚❚

11.4

Pedigrees are a useful way to study human genetic disorders

One way to study human genetic disorders is to analyze pedigrees. A **pedigree** is a chart similar to a family tree that shows genetic relationships among family mem-

Figure 11.8 Human Pedigree Analysis
The pedigree shown here—in this case for cystic fibrosis—illustrates symbols commonly used by geneticists. The Roman numerals at left identify different generations. Numbers listed below the symbols identify individuals of a given generation. Individuals 1 and 2 in generation II each had genotype *Aa*. Neither suffered from cystic fibrosis because the *a* allele is recessive to *A*, but they were both genetic carriers for the disorder.

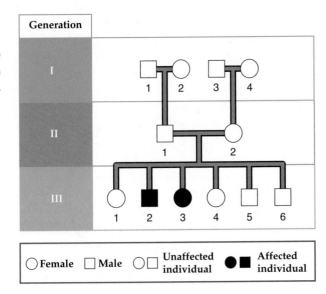

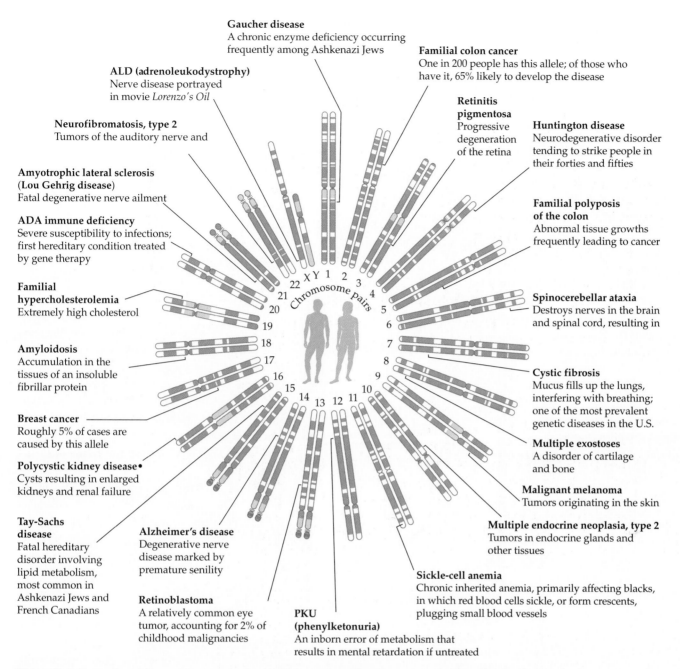

Gaucher disease
A chronic enzyme deficiency occurring frequently among Ashkenazi Jews

Familial colon cancer
One in 200 people has this allele; of those who have it, 65% likely to develop the disease

ALD (adrenoleukodystrophy)
Nerve disease portrayed in movie *Lorenzo's Oil*

Retinitis pigmentosa
Progressive degeneration of the retina

Huntington disease
Neurodegenerative disorder tending to strike people in their forties and fifties

Neurofibromatosis, type 2
Tumors of the auditory nerve and

Amyotrophic lateral sclerosis (Lou Gehrig disease)
Fatal degenerative nerve ailment

Familial polyposis of the colon
Abnormal tissue growths frequently leading to cancer

ADA immune deficiency
Severe susceptibility to infections; first hereditary condition treated by gene therapy

Spinocerebellar ataxia
Destroys nerves in the brain and spinal cord, resulting in

Familial hypercholesterolemia
Extremely high cholesterol

Amyloidosis
Accumulation in the tissues of an insoluble fibrillar protein

Cystic fibrosis
Mucus fills up the lungs, interfering with breathing; one of the most prevalent genetic diseases in the U.S.

Breast cancer
Roughly 5% of cases are caused by this allele

Multiple exostoses
A disorder of cartilage and bone

Polycystic kidney disease•
Cysts resulting in enlarged kidneys and renal failure

Malignant melanoma
Tumors originating in the skin

Tay-Sachs disease
Fatal hereditary disorder involving lipid metabolism, most common in Ashkenazi Jews and French Canadians

Multiple endocrine neoplasia, type 2
Tumors in endocrine glands and other tissues

Alzheimer's disease
Degenerative nerve disease marked by premature senility

Sickle-cell anemia
Chronic inherited anemia, primarily affecting blacks, in which red blood cells sickle, or form crescents, plugging small blood vessels

Retinoblastoma
A relatively common eye tumor, accounting for 2% of childhood malignancies

PKU (phenylketonuria)
An inborn error of metabolism that results in mental retardation if untreated

Chromosome pairs
X Y 1 2 3 4 5 6 7 8 9 10 11 12 13 14 15 16 17 18 19 20 21 22

Figure 11.9 Examples of Genes Known to be Associated with Inherited Genetic Disorders in Humans
Mutations of single genes that lead to genetic disorders are found on the X chromosome and on each of the 22 autosomes in humans. Thousands of single-gene genetic disorders are known; for clarity, we show only one such disorder per chromosome.

Inherited genetic disorders can be caused by mutations in individual genes (Figure 11.9) or by abnormalities in chromosome number or structure. Clinical genetic tests can be used to determine whether a prospective parent carries an allele for one of these disorders, and variations on these methods can be used to test for genetic disorders long before a baby is born (see the box on page 238).

In the remainder of this chapter we focus on inherited genetic disorders that have relatively simple causes: those caused by mutations of a single gene and those caused by chromosomal abnormalities. As you read this material, however, it is important to bear in mind that the tendency to develop some diseases, such as heart disease, diabetes, and some inherited forms of cancer, is caused by interactions among multiple genes and the environment. For most diseases caused by multiple genes, the identity of the genes involved and how they lead to disease is poorly understood.

Science Toolkit

Prenatal Genetic Screening

How is the baby? This is one of the first questions we ask after a child is born. Usually everything is fine, but sometimes the answer can be devastating. Today, some parents choose to have one of several prenatal genetic screening methods performed to check their baby's health long before it is born.

This practice has been around a surprisingly long time. In the 1870s, doctors occasionally withdrew some of the fluid in which the fetus is suspended to obtain information about its health. Modern versions of that practice have been standard medical procedure since the early 1960s. In **amniocentesis**, a needle is inserted through the abdomen into the uterus to extract a small amount of amniotic fluid from the pregnancy sac that surrounds the fetus. This fluid contains fetal cells (often sloughed-off skin cells) that can be tested for genetic disorders. Another method is **chorionic** [*KORE*-ee-ah-nick] **villus sampling** (**CVS**), in which a physician uses ultrasound to guide a narrow, flexible tube through a woman's vagina and into her uterus, where the tip of the tube is placed next to the villi, a cluster of cells that attaches the pregnancy sac to the wall of the uterus. Cells are removed from the villi by gentle suction, then tested for genetic disorders.

Risks associated with amniocentesis and CVS, including vaginal cramping, miscarriage, and premature birth, have declined quite dramatically in the past 10 years because of advances in technology and more extensive training. Recent studies suggest that the risk of miscarriage after CVS and amniocentesis is essentially the same, about 0.06 percent. The tests are widely used by parents who know they face an increased chance of giving birth to a baby with a genetic disorder. Older parents, for example, might want to test for Down syndrome, since the risk of that condition increases with the age of the mother. A couple in which one parent carries a dominant allele for a specific genetic disorder (such as Huntington disease), or both parents are carriers for a recessive genetic disorder (such as cystic fibrosis), might also choose prenatal genetic screening.

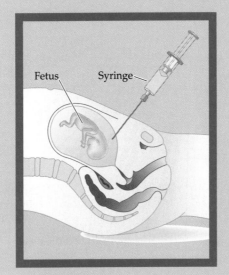

Fetus   Syringe

Amniocentesis

In amniocentesis, amniotic fluid, which contains fetal cells, is extracted from the uterus.

Until recently, couples who elected to have such tests performed had only two choices if their fears were confirmed: they could abort the baby, or they could give birth to a child who would have a genetic disorder. Since 1989, however, a third option has been available to couples who are willing, and can afford, to have a child by in vitro fertilization (in which fertilization occurs in a Petri dish, after which one or more embryos are implanted into the mother's uterus). In **preimplantation genetic diagnosis** (**PGD**), one or two cells are removed from the developing embryo, usually 3 days after fertilization occurs. (It is important to perform PGD at this time because then the embryo typically has 4 to 12 loosely connected cells; in another day or two, the cells will begin to fuse more tightly to one another, making PGD more difficult.) Next, the cell or cells removed from the embryo are tested for genetic disorders. Finally, one or more embryos that are free of disorders are implanted into the mother's uterus, and the rest of the embryos, including those with genetic disorders, are discarded.

PGD is typically used by parents who either have a serious genetic disorder, or carry alleles for one; for example, cystic fibrosis or Huntington disease. Like all genetic screening methods, the use of PGD raises ethical issues. People who support the use of PGD think that amniocentesis and CVS provide parents with a bleak set of moral choices: if the fetus has a serious genetic disorder, the parents can either abort the baby or allow it to live a life that may be short and full of suffering. In their view, it is morally preferable to discard an embryo at the 4- to 12-cell stage than it is to abort a well-developed fetus, or to give birth to a child that will suffer the devastating effects of a serious genetic disorder. Those opposed to the use of PGD argue that the moral choices are the same: their view is that once fertilization has occurred, a new life has formed, and it is immoral to end that life, even at the 4- to 12-cell stage. What do you think?

11.5 Autosomal Inheritance of Single-Gene Mutations

We organize our discussion of single-gene genetic disorders by whether the gene is located on an autosome or a sex chromosome. We further subdivide our discussion of autosomal disorders by whether the disease-causing allele is recessive or dominant. As we shall see, recessive genetic disorders are much more common than dominant ones.

There are many autosomal recessive genetic disorders

Several thousand human genetic disorders are inherited as recessive traits. Most of these disorders, such as cystic fibrosis, sickle-cell anemia (see Figures 13.10 and 15.5), and Tay–Sachs disease (see chromosome 15 in Figure 11.9), are caused by recessive mutations of genes located on autosomes.

Recessive genetic disorders vary in severity; some are lethal, whereas others have relatively mild effects. Tay–Sachs disease is a lethal recessive genetic disorder in which the disease-causing allele encodes a defective version of a crucial enzyme that does not work properly, so that lipids accumulate in brain cells. As a result, the brain begins to deteriorate during a child's first year of life, and death occurs within a few years. At the other end of the severity spectrum, albino skin color in humans can be caused by a variety of single-gene mutations, including one similar to that which produces a white coat color in mice and other mammals (see Figure 10.13).

The only individuals who actually get a disorder caused by an autosomal recessive allele (say, *a*) are those who have two copies of that allele (*aa*). Usually, when a child inherits a recessive genetic disorder, both parents are heterozygous; that is, they both have genotype *Aa*. (It is also possible for one or both parents to have genotype *aa*.) Because the *A* allele is dominant and does not cause the disorder, heterozygous individuals are said to be **carriers** of the disorder: they carry the disorder allele (*a*), but do not get the disorder.

If two carriers of a recessive genetic disorder have children, the patterns of inheritance are the same as for any recessive trait: ¼ of the children are likely to have genotype *AA*, ½ to have genotype *Aa*, and ¼ to have genotype *aa*. As shown in Figure 11.10, each child has a 25 percent chance of not carrying the disorder allele (genotype *AA*), a 50 percent chance of being a carrier

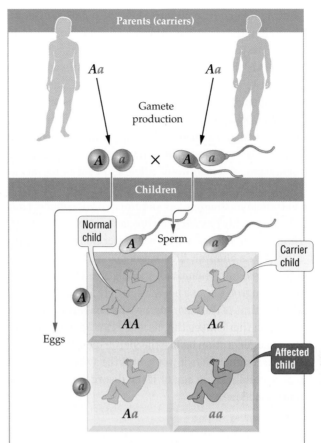

Figure 11.10 Inheritance of Autosomal Recessive Disorders
The patterns of inheritance for a human autosomal recessive genetic disorder are the same as for any recessive trait (compare this figure with the pattern shown by Mendel's pea plants in Figure 10.7). Recessive disorder alleles are colored red and denoted *a*. Dominant, normal alleles are black and denoted *A*. Here, the parents are a carrier male (genotype *Aa*) and a carrier female (genotype *Aa*).

(genotype *Aa*), and a 25 percent chance of actually getting the disorder (genotype *aa*).

These percentages reveal one way in which lethal recessive disorders such as Tay–Sachs disease can persist in the human population: although homozygous recessive individuals (with genotype *aa*) die long before they are old enough to have children, carriers (with genotype *Aa*) are not harmed by the disorder. In a sense, the *a* alleles can "hide" in heterozygous carriers, and those carriers are likely to pass the disorder allele to half of their children. Recessive genetic disorders also arise in the human population because new mutations produce new copies of the disorder alleles.

Dominant genetic disorders that produce serious effects are less common than recessive ones

A dominant allele (*A*) that causes a genetic disorder cannot "hide" in the same way that a recessive allele can. In this case, *AA* and *Aa* individuals get the disorder; only *aa* individuals are symptom-free. When a dominant genetic disorder produces serious negative effects, the individuals that have the *A* allele may not live long enough to reproduce; hence few of them pass the allele on to their children. As a result, those dominant alleles that prevent a sufferer from reproducing are uncommon in the population. Such lethal dominant alleles are rarely handed down through the generations in a family line, and instead, most such alleles appear in the population because of new mutations that arise during gamete formation. If a dominant allele expresses its lethal effects later in life, however, after the person carrying that allele has had an opportunity to reproduce, that allele is readily passed from one generation to the next. Huntington disease, which was described at the beginning of this chapter, illustrates how such a dominant lethal allele can remain in the population. The symptoms of Huntington disease begin relatively late in life, often after victims of the disorder have had children. Because the allele that causes the disorder can be passed on to the next generation before the victim dies, the disorder is more common than it would be if the symptoms began before childbearing age or if the disorder persisted by mutation alone.

Use an interactive Punnett square to explore the genetic mechanisms of several inherited diseases. ▶❚❚ 11.5

11.6 Sex-Linked Inheritance of Single-Gene Mutations

Roughly 1,200 of the estimated 25,000 human genes are found only on the X chromosome or only on the Y chromosome; such genes are said to be **sex-linked**. Because sex-linked genes are found on the X chromosome or the Y chromosome—but not both—males receive only one copy of each sex-linked gene (whether it is on the X chromosome or the Y chromosome), while females receive two copies of genes on the X chromosome and no copies of genes on the Y chromosome. About 15 genes are shared by the X and Y chromosomes. Therefore, males and females receive two copies of each of these genes, just as they do for all autosomal genes; as a result, these 15 genes are not sex-linked.

Approximately 1,100 of the 1,200 human sex-linked genes are located on the X chromosome, while 80 are located on the much smaller Y chromosome. Sex-linked genes on the X chromosome are said to be **X-linked**; similarly, all sex-linked genes on the Y chromosome are said to be **Y-linked**. Although there are no well-documented cases of disease-causing Y-linked genes, X chromosomes do contain genes known to be involved in human genetic disorders (see Figure 11.9, ALD disease).

Sex-linked genes have different patterns of inheritance than do genes on autosomes.

Consider how an X-linked recessive allele for a human genetic disorder is inherited (Figure 11.11). We label the recessive disorder allele a, and in the Punnett square we write this allele as X^a to emphasize the fact that it is on the X chromosome. Similarly, the dominant allele is labeled A and is written as X^A in the Punnett square. If a carrier female (with genotype $X^A X^a$) has children with a normal male (with genotype $X^A Y$), each of their sons will have a 50 percent chance of getting the disorder (see Figure 11.11). This result differs greatly from what would happen if the same disorder allele (*a*) were on an autosome: in that case, none of the children, male or female, would get the disorder because then males could be heterozygous (*Aa*) as well, in which case they would be shielded from the harmful effects of the a allele by the presence of the normal *A* allele. Instead, what happens in X-linked disorders is that males of

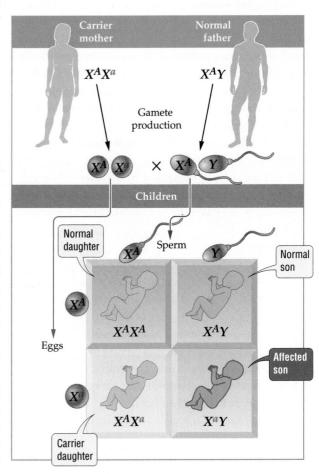

Figure 11.11 Inheritance of X-Linked Recessive Disorders
The recessive disorder allele (*a*) is located on the X chromosome and is denoted by X^a. The dominant normal allele (*A*) on the X chromosome is denoted by X^A.

genotype X^{a}Y suffer from the condition because the Y chromosome does not have a copy of that gene at all; in other words, because males cannot be heterozygous for any X-linked genes, the effects of an *a* allele cannot be masked. In general, males are more likely than females to get recessive X-linked disorders because they have to inherit only a single copy of the disorder allele to exhibit the disorder, whereas females must inherit two copies to be affected. In contrast, both sexes are equally likely to be affected by autosomal recessive disorders, since both have two copies of autosomal chromosomes and identical odds of being homozygous or heterozygous for a disorder allele.

X-linked genetic disorders in humans include hemophilia, a serious disorder in which minor cuts and bruises can cause a person to bleed to death, and Duchenne muscular dystrophy, a lethal disorder that causes the muscles to waste away, often leading to death at a young age. Both of these X-linked disorders are caused by recessive alleles. An example of a dominant X-linked disorder with a daunting name—congenital generalized hypertrichosis [*HYE*-per-trih-*KOH*-siss], or CGH—is shown in Figure 11.12.

The structure of chromosomes can change in several ways

When chromosomes are being aligned or separated during cell division, breaks can occur that alter the length of one or more chromosomes (Figure 11.13). Sometimes a piece breaks off and is lost from the chromosome (**deletion**). Other times the broken piece is reattached, but incorrectly. For example, in an **inversion**, the fragment returns to the correct place on the original chromosome, but with the genetic loci in reverse order. This would be similar to snapping a pencil in two, then reattaching the eraser end to the broken end of the other piece. In a **translocation**, a broken piece from one chromosome becomes attached to a different, nonhomologous chromosome. Translocations are frequently reciprocal, meaning two nonhomologous chromosomes are involved in a mutual exchange of fragments. This type of translocation is shown in Figure 11.13*c*; it is rather like breaking pieces off a blue pencil and an orange one, then swapping the broken pieces between the two pencils, so the blue one acquires a piece of the orange one and vice versa. **Duplication** is a type of chromosomal

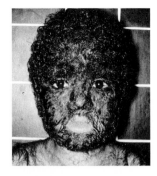

Figure 11.12 Congenital Generalized Hypertrichosis (CGH)

A 6-year-old boy with CGH, a rare genetic disorder that causes extreme hairiness of the face and upper body. CGH is caused by a dominant allele of a single gene on the X chromosome.

11.7 Inherited Chromosomal Abnormalities

Every species has a characteristic number of chromosomes, and each chromosome has a particular structure, with specific genetic loci arranged on it in a precise sequence. Any change in the chromosome number or structure, compared to what is typical for the species, is considered a chromosomal abnormality. Two main types of chromosomal changes are seen in humans and other organisms: changes in the overall number of chromosomes and changes in chromosome structure, such as a change in the length of an individual chromosome. A cell is especially vulnerable to developing chromosomal abnormalities when it is dividing: chromosomes can be misaligned, misdirected, or even ripped into pieces during the delicate business of lining them up in the center of the cell and then segregating them into daughter cells. For chromosomal abnormalities to be passed on from a parent to offspring, the chromosomal changes have to occur in the gametes or gamete-producing cells. However, relatively few human genetic disorders are caused by inherited chromosomal abnormalities, probably because most large changes in the chromosomes kill the developing embryo.

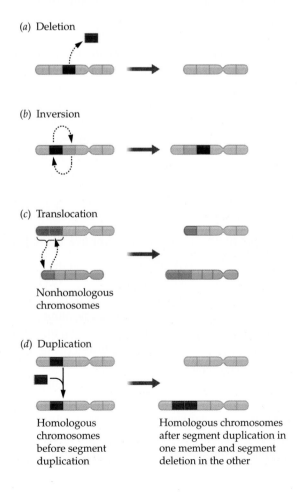

(*a*) Deletion

(*b*) Inversion

(*c*) Translocation

Nonhomologous chromosomes

(*d*) Duplication

Homologous chromosomes before segment duplication

Homologous chromosomes after segment duplication in one member and segment deletion in the other

Figure 11.13 Structural Changes to Chromosomes

(*a*) Deletion: A segment (black) breaks off and is lost from the chromosome. (*b*) Inversion: A segment (black and purple) breaks off and is reattached, but in reverse order. (*c*) Translocation: A segment (dark bluish-gray; or dark orange) breaks off one chromosome and becomes attached to a different, nonhomologous chromosome. (*d*) Duplication: A chromosome becomes longer after acquiring an extra copy of one of its chromosome segments (black), commonly due to unequal crossing-over between a pair of homologues during meiosis I.

abnormality in which a chromosome becomes longer because it ends up with two copies of a particular chromosome fragment. There are a number of ways in which a chromosome can acquire duplicates of a particular stretch of the chromosome. Errors in crossing-over are one source of duplications. When a pair of homologous chromosomes pair up to exchange chromosome segments during meiosis I (Chapter 9), sometimes the trade is uneven in that one homologue receives a chromosomal fragment from the other homologue, but fails to reciprocate by "giving back" an equivalent length of itself to its partner. The first homologue therefore ends up with duplicates of that stretch of the chromosome (one copy of its own plus one acquired from its partner homologue), while the other homologue suffers a deletion.

Changes in chromosome structure can have dramatic effects. If breakage occurs in the sex chromosomes, it can cause a change in the expected sex of the developing fetus. A deletion in which the portion of the Y chromosome that contains the *SRY* gene is lost produces an XY individual who develops as a female. A translocation that results in an X chromosome to which the Y chromosome's sex-determining region is attached produces an XX individual who develops as a male. XY females and XX males are always sterile (unable to produce offspring).

Changes to the structure of autosomes can have even more dramatic effects. Cri du chat [KREE-doo-SHAH] syndrome occurs when a child inherits a chromosome 5 that is missing a particular region (Figure 11.14). *Cri du chat* is French for "cry of the cat," which describes the characteristic mewing sound made by infants with this condition. Other characteristics of this condition are slow growth and a tendency toward severe mental retardation, a small head, and low-set ears.

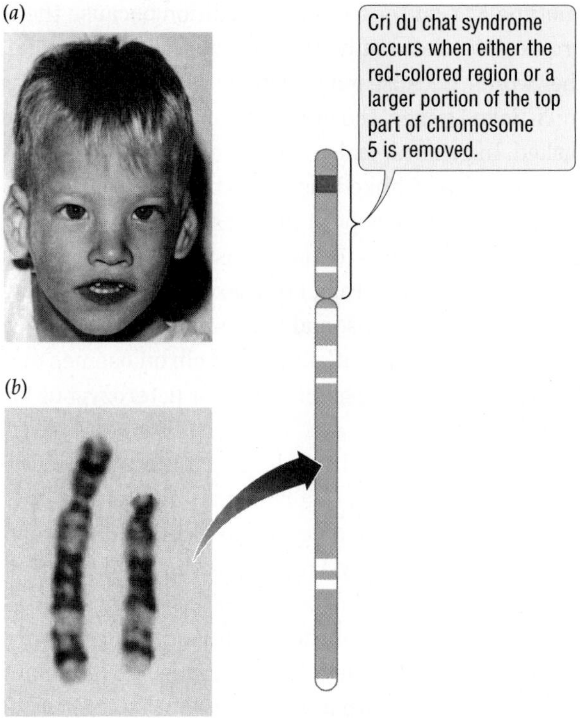

Changes in chromosome number are often fatal

Unusual numbers of chromosomes—such as one or three copies instead of the normal two—can be produced when chromosomes fail to separate properly during meiosis. In humans, such changes in chromosome number often result in fetal death. At least 20 percent of human pregnancies abort spontaneously, largely as a result of such changes in chromosome number.

There is only one genetic anomaly in which a person who inherits the wrong number of autosomes commonly reaches adulthood: Down syndrome. Individuals with this condition usually have three copies of chromosome 21, the smallest autosome in humans; for this

(a)

(b)

Figure 11.14 Cri du Chat Syndrome
Cri du chat syndrome is caused by the deletion of a portion of the top part of chromosome 5.

Cri du chat syndrome occurs when either the red-colored region or a larger portion of the top part of chromosome 5 is removed.

reason, Down syndrome is also known as trisomy 21, where **trisomy** [TRYE-suh-mee] refers to the condition of having three copies of a chromosome (instead of the usual two). A small minority (3–4 percent) of Down syndrome cases occur when an extra piece of chromosome 21 breaks off during cell division and attaches to another chromosome. These individuals have two copies of chromosome 21 plus part of another chromosome 21. People with Down syndrome tend to be short and mentally retarded, and they may have defects of the heart, kidneys, and digestive tract. With appropriate medical care, most people with this condition lead healthy lives, and many live to their sixties or seventies (their average life expectancy is 55). Live births can also result when an infant has three copies of chromosome 13, 15, or 18. However, such children have severe birth defects, and they rarely live beyond their first year.

Compared with having too few or too many autosomes, changes in the number of sex chromosomes have more minor effects. Klinefelter syndrome, for example, is a condition found in males that have an extra X chromosome (XXY males). Such men have a normal life span and normal intelligence, and they

Biology Matters

Many of the Most Common Diseases Have Complex Patterns of Inheritance

A disease is a condition that impairs health. It may be caused by external factors, such as infection by viruses, bacteria, and other parasites, or injury produced by harmful chemicals or high-energy radiation. Nutrient deficiency can also lead to disease, the way inadequate vitamin C consumption produces scurvy. Disease may be also caused by internal factors, controlled by one or more genes. Diseases that are entirely hereditary—caused exclusively by gene malfunction and present in the individual since conception—are described as genetic disorders, to distinguish them from infections and other types of diseases. Cystic fibrosis and sickle cell anemia are examples of genetic disorders caused by errors in the activity of a single gene. The inheritance of single-gene (monogenic) disorders can usually be explained by Mendel's laws. Myotonic dystrophy, in which muscle cells are affected, is an example of a genetic disorder created by malfunctions in more than one gene. Disorders in which multiple genes are involved (polygenic disorders) are often inherited in a non-Mendelian pattern, in part because the alleles of these genes interact in complicated ways that cannot be predicted based on Mendel's laws.

The diseases that are most common in industrialized countries—heart disease, cancer, diabetes, asthma, and arthritis, for example—are caused by multiple genes interacting in complex ways with each other and with various external factors. Malfunctions in key genes make a person susceptible to developing these diseases, but environmental factors affect whether the disease will actually appear or how severe the symptoms will be. As shown in the graph here, a large part of the estimated risk of developing colon cancer, stroke, coronary heart disease, and type 2 diabetes is avoidable. Such lifestyle choices as maintaining good nutrition, exercising regularly, and avoiding tobacco, have a significant impact on our risk of developing such chronic diseases. (The term chronic means unceasing, a reference to the fact that once we develop one of these diseases, we have it for the rest of our lives.)

To reduce our risk of these diseases, we should maintain a proper body weight, engage in physical activity equivalent to at least 30 minutes of brisk walking per day, avoid smoking, and consume fewer than three alcoholic beverages per day. It helps to limit our intake of saturated fat and trans fat (Chapter 4) and

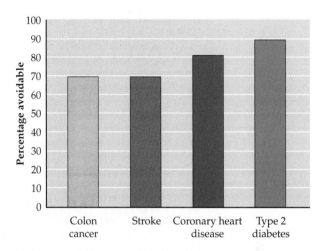

also sugar and refined carbohydrates. Consuming enough folic acid (at least 400 mg/day) and dietary fiber (21–38 g, depending on size and age) is linked to a lower risk of several chronic diseases, including heart disease and colon cancer. People who eat fewer than three servings of red meat per week lower their odds of getting colon cancer.

A major goal of modern genetics is to identify genes that contribute to human disease when they fail to function normally. Researchers have identified alleles associated with increased risk of a number of common ailments, including high blood pressure, heart disease, diabetes, Alzheimer's disease, several types of cancer, and schizophrenia. For example, a particular allele on chromosome 9 raises the risk of heart disease. Those who are homozygous for the risk allele are about 30 to 40 percent more likely to have coronary heart disease than individuals who are homozygous for the harmless allele. Heterozygotes are about 15 to 20 percent more likely to do so than individuals who lack the risk allele.

The hope is that genetic tests (see page 238) will tell us if we are predisposed to a disease before we become ill with it. A person carrying a risk allele could take preventive measures to reduce his or her risk of actually developing the condition, and treatment could be made more effective by customizing it based on the particular allele involved. This customized approach to treatment, or personalized medicine, is already being used to treat breast cancer and some other chronic diseases.

tend to be tall. Many XXY males also have small testicles (about one-third normal size) and reduced fertility, and some have feminine characteristics, such as enlarged breasts. Females with a single X chromosome (Turner syndrome) have normal intelligence and tend to be short (with adult heights under 150 centimeters (about 4 feet 11 inches), to be sterile, and to have a broad, webbed neck. Other changes in the number of sex chromosomes, as in XYY males and XXX females, also produce relatively mild effects. However, when there are two or more extra sex chromosomes, as in XXXY males or XXXX females, a wide range of problems can result, including severe mental retardation.

Concept Check

1. What is a genetic pedigree?
2. Why are dominant autosomal disorders that produce serious negative effects less common than recessive autosomal disorders?
3. Why are males more likely than females to have recessive X-linked genetic disorders?

Applying What We Learned

Uncovering the Genetics of Huntington Disease

In 1872, George Huntington wrote the classic paper describing the disorder named for him. For over a century, there was little progress in treating the disorder. It was known that the gene for the disorder is on an autosome, and that the Huntington allele is dominant (*A*). But there was no cure, and little helpful information could be given to potential victims of Huntington disease (HD). By constructing a pedigree, a geneticist might learn that a person's father (who had HD) was of genotype *Aa*. If the mother did not have the disorder, and hence was of genotype *aa*, all the geneticist could say was that the person had a 50 percent chance of developing the disorder (see problem 3 in the Sample Genetics Problems at the end of this chapter).

In 1983 this situation changed dramatically when pedigree analyses that looked for patterns of genes inherited together indicated that the *HD* gene was linked to other genes known to be located on chromosome 4. This discovery set off an intense effort to isolate the *HD* gene. Over the next 10 years, researchers identified the genes on chromosome 4 that were linked most closely to the *HD* gene, a process that eventually allowed them first to pinpoint the gene's location on chromosome 4, then in 1993 to isolate the gene itself.

By isolating the *HD* gene, scientists were able to learn the identity of an abnormal protein produced by the Huntington allele. Portions of this protein form clumps in the brains of people with the disorder; these clumps are correlated with, and may cause, the symptoms of Huntington disease.

Dramatic new results within the past few years suggest that knowledge of the *HD* gene and its associated protein may help in the design of effective treatments. For example, in mice genetically engineered to have the human *HD* gene, scientists have developed ways to slow the progression of the symptoms, or even reverse them. And human cells lacking the Huntington allele have been transplanted into an HD patient's brain, where they survived and remained free of HD protein clumps for 18 months. We return to some of the new discoveries about the HD gene at the end of Chapter 13. Collectively, these results offer hope that brain repair may eventually be possible.

In addition to providing clues to possible treatment, isolation of the *HD* gene had the immediate effect of allowing scientists to design a diagnostic genetic test for the disorder. With this test, a person at risk (because one parent had the disorder) can now learn with near certainty whether or not he or she also will get the disorder. In some cases, the genetic test offers hope. For example, people at risk who want to have a family without the fear of passing the disorder on to their children can take the test and use it to make an informed decision. But the test poses an agonizing choice for these and all other individuals at risk: If they take the test, they may experience tremendous relief if the results show that they will not get the disorder. Alternatively, they may experience a crushing loss of hope if they find out they have a lethal disease with horrible symptoms that, as of now, cannot be cured. Given these alternatives, would you take the test?

Genetic Tests Offer Promise, but Raise Questions, Too

BY DENISE CARUSO

A growing industry is hoping to spin gold from DNA's double helixes by using ultra-sensitive genetic tests to personalize medical treatment for cancer, lupus and other diseases.

These molecular diagnostic tests can give doctors more detailed information than ever about their patients. Genetic information can help them decide whether their lymphoma patients would respond better to surgery, chemotherapy or radiation treatment; make more accurate diagnoses of abnormal cells or tissues; and more readily detect serious autoimmune disorders.

What's more, the development of genetic tests has given academic researchers the tools to begin establishing causal links between common bacteria and viruses—streptococcus, say, or influenza—and diseases like autism, cervical cancer, type 1 diabetes, schizophrenia and even obsessive-compulsive disorder.

More than 1,000 genetic tests are available clinically, and hundreds more are available to researchers.

Most of the roughly 4 million babies born in the United States each year are tested for genetic conditions. Sickle cell anemia, cystic fibrosis, and phenylketonuria (PKU) are among the 29 screening tests mandated in all 50 states. Some states require testing for over 50 different inherited disorders. The value of these tests is illustrated by the case of phenylketonuria, an autosomal recessive condition that leads to severe brain damage if untreated, but whose ill effects can be prevented by prompt management of the child's diet.

Advances in DNA technology have increased the number of tests for genetic conditions. These can now be used to estimate an individual's risk of developing certain types of breast and colon cancer, some forms of type 1 diabetes, and Crohn disease. Because the chance of developing these conditions is influenced by environmental factors, timely intervention could delay or arrest development of disease. Genetic testing can also guide treatment or the choice of medication, since genetic makeup can affect an individual's response to drugs used for heart disease or as anti-cancer medication. Targeted medicine, or "personalized medicine," in which treatment strategies are matched to a person's genotype, may be the next great innovation in medicine.

The availability of over 1,000 genetic tests has stirred controversy and created ethical dilemmas. How would knowledge of a persons's genotype be used by potential employers or insurance companies? Activists for the disabled wonder if the quest for "perfect babies" has already gone too far. Studies show that 90 percent of women who learn that their fetuses have Down syndrome choose to abort them, so that the number of Americans who have Down syndrome is in sharp decline. Conversely, there are instances of people with genetic conditions, such as genetic dwarfism (achondroplasia) or congenital deafness, who have used genetic testing to *select* fetuses with the same genetic condition, in the belief that such a child would relate to them better than a child with an average phenotype.

Genetic testing is a growing commercial enterprise, and many companies advertise directly to consumers. Critics charge that this leads to misrepresentation of the science and overselling the benefits of expensive tests. For instance, some companies have been exhorting women with a family history of breast or ovarian cancer to get tested for a mutation in the *BRCA1* and *BRCA2* that confers a 35 to 85 percent chance of developing breast cancer. Women who test positive are faced with difficult choices, ranging from lifestyle changes, to taking preventive drugs with serious side effects, or opting for removal of all breast tissue. Mutant alleles of these two genes account for less than 10 percent of the diagnosed cases of breast cancer, so the test is of benefit to a very small segment of the U.S. population. Worse, some tests run by commercial groups miss about 12 percent of those actually at risk. In any case, genetic testing is most useful when undertaken with the guidance of medical professionals in the context of a person's entire medical history, instead of as a do-it-yourself project.

Evaluating the News

1. Name a genetic disorder that can be tested in each of the following: fetuses, newborns, adults. What is the value, if any, of each test? Should everyone in the population be routinely tested for each disorder? Explain.

2. Certain alleles make a person prone to getting a particular disorder, such as type 2 diabetes. Would you want to know whether you have such alleles? Should anyone other than you have access to that information, for example, insurance companies? Why or why not?

3. Should commercial companies be allowed to freely market any genetic test, including those for conditions that cannot be cured and for which no preventive measures exist? Explain your rationale.

SOURCE: *New York Times,* February 18, 2007.

Chapter Review

Summary

11.1 The Role of Chromosomes in Inheritance

- The chromosome theory of inheritance states that genes are located on chromosomes.
- Each chromosome is composed of a single DNA molecule and many associated proteins.
- A gene is made up of DNA and in a living cell it is located on a chromosome. The physical location of a gene is known as its genetic locus.
- Chromosomes that pair during meiosis are homologous chromosomes. A pair of homologous chromosomes has the same genetic loci.
- Chromosomes that determine the sex of the organism are called sex chromosomes; all other chromosomes are called autosomes.
- Humans have two types of sex chromosomes: X and Y. Males have one X and one Y chromosome. Females have two X chromosomes.
- A specific gene on the Y chromosome (the *SRY* gene) is required for human embryos to develop as males.

11.2 Genetic Linkage and Crossing-Over

- Genes that tend to be inherited together are said to be genetically linked.
- Crossing-over, the exchange of chromosomal segments between a pair of homologues, can reduce the linkage between two genes located on the same chromosome.
- Two genes that are far apart on a chromosome are more likely to be shuffled by crossing-over than are genes located near one another, and are therefore less likely to show genetic linkage.
- Mendel's law of independent assortment holds for genes that are located very far apart on a chromosome, just as it does for a pair of genes that are located on two different chromosomes.

11.3 Origins of Genetic Differences between Individuals

- Genetic differences among individuals provide the genetic variation on which evolution can act. Mutations are the source of new alleles.
- Offspring differ genetically from one another and from their parents because each pair of homologues exchanges random stretches of the chromosome during crossing-over; there is independent distribution of maternal and paternal homologues into gametes during meiosis; and, fertilization combines male and female genotypes randomly.

11.4 Human Genetic Disorders

- Pedigrees are useful for studying the inheritance of human genetic disorders.
- Humans suffer from a variety of genetic disorders, including those caused by mutations of a single gene and those caused by abnormalities in chromosome number or structure.

- Clinical genetic testing can be performed on both fetuses and adults. Three fetal testing procedures are amniocentesis, chorionic villus sampling (CVS), and preimplantation genetic diagnosis (PGD).

11.5 Autosomal Inheritance of Single-Gene Mutations

- Most genetic disorders are caused by a recessive allele (*a*) of a gene on an autosome. Only homozygous (*aa*) individuals get these disorders; heterozygous (*Aa*) individuals are merely carriers of the disorders.
- In dominant autosomal genetic disorders, both *AA* and *Aa* individuals are affected. If defects of the phenotype are so severe that the *AA* or *Aa* individual cannot reproduce, such a dominant genetic disorder is likely to be rare.
- Alleles that cause lethal dominant genetic disorders can remain in the population if symptoms begin late in life, as in Huntington disease, or if new disorder alleles are produced by mutation during each generation.

11.6 Sex-Linked Inheritance of Single-Gene Mutations

- Because males inherit only one X chromosome, genes on sex chromosomes have different patterns of inheritance than do genes on autosomes.
- Genes found on one of the sex chromosomes, but not on both, may show sex-linked patterns of inheritance. Genes found only on the X chromosome are said to be X-linked; those found only on the Y chromosome are Y-linked.
- Males are more likely than females to have recessive X-linked genetic disorders because males need to inherit only one copy of the disorder allele to be affected, whereas females must inherit two copies to be affected. In contrast, both sexes are equally likely to be affected by autosomal genetic disorders.

11.7 Inherited Chromosomal Abnormalities

- Chromosomal abnormalities include changes in chromosome structure and chromosome number.
- Changes in the structure of an individual chromosome can happen if breakage occurs during cell division, resulting in deletion, inversion, duplication, or translocation of a chromosome fragment. Such structural changes can have profound effects.
- Changes in the number of autosomes in humans are usually lethal. Down syndrome, a form of trisomy in which individuals receive three copies of chromosome 21, is an exception to this rule.
- People who have one too many or one too few sex chromosomes may experience relatively minor effects; however, if there are four or more sex chromosomes (instead of the usual two), serious problems can result, including severe mental retardation.

Review and Application of Key Concepts

1. In terms of its physical structure, describe what a gene is and where it is located.

2. Consider the XY chromosome system by which sex is determined in humans. Do patterns of inheritance for genes located on the X chromosome differ between males and females? Why or why not?

3. Look carefully at Figure 11.4. Explain in your own words why the results shown in that figure convinced Morgan that genes located near one another on a chromosome tend to be inherited together, or linked. What results would be expected if these genes were located on different chromosomes?

4. Explain how crossing-over occurs. Assume that genes *A*, *B*, and *C* are arranged in that order along a chromosome. Using your understanding of how crossing-over occurs, will it occur more often between genes *A* and *B* or between genes *A* and *C*?

5. Explain how nonparental genotypes are formed.

6. Are genetic disorders caused by single-gene mutations more common or less common in human populations than those caused by abnormalities in chromosome number or structure? Explain why.

Key Terms

amniocentesis (p. 238)
autosome (p. 231)
carrier (p. 239)
chorionic villus sampling (CVS) (p. 238)
chromosome theory of inheritance (p. 230)
crossing-over (p. 233)
deletion (p. 241)
duplication (p. 241)
genetic linkage (p. 233)
homologous chromosome (p. 230)
independent assortment of chromosomes (p. 234)
inversion (p. 241)
locus (p. 230)
pedigree (p. 236)
preimplantation genetic diagnosis (PGD) (p. 238)
recombination (p. 233)
sex chromosome (p. 231)
sex-linked (p. 240)
SRY gene (p. 232)
translocation (p. 241)
trisomy (p. 242)
X-linked (p. 240)
Y-linked (p. 240)

Self-Quiz

1. Genes commonly exert their effect on the phenotype by
   a. promoting DNA mutations.
   b. creating chromosomal structures such as centromeres.
   c. coding for a protein.
   d. all of the above

2. Which of the following is an autosomal dominant disorder in which the symptoms begin late in life and the nervous system is destroyed, resulting in death?
   a. Tay–Sachs disease
   b. Huntington disease
   c. Down syndrome
   d. cri du chat syndrome

3. Crossing-over is more likely to occur between genes that are
   a. close together on a chromosome.
   b. on different chromosomes.
   c. far apart on a chromosome.
   d. located on the Y chromosome.

4. Comparatively few human genetic disorders are caused by chromosomal abnormalities. One reason is that
   a. most chromosomal abnormalities have little effect.
   b. it is difficult to detect changes in the number or length of chromosomes.
   c. most chromosomal abnormalities result in spontaneous abortion of the embryo.
   d. it is not possible to change the length or number of chromosomes.

5. Nonparental genotypes can be produced by
   a. crossing-over and the independent assortment of chromosomes.
   b. linkage.
   c. autosomes.
   d. sex chromosomes.

6. Sometimes a segment of DNA breaks off from a chromosome, then returns to the correct place on the original chromosome, but in reverse order. This type of chromosomal structural change is called
   a. crossing-over.
   b. a translocation.
   c. an inversion.
   d. a deletion.

7. The prenatal genetic screening method that tests cells attaching the pregnancy sac to the wall of the uterus is called
   a. chorionic villus sampling (CVS).
   b. amniocentesis.
   c. preimplantation genetic diagnosis (PGD).
   d. in vitro fertilization.

8. Which of the following can most precisely be described as a master switch that commits the sex of the developing embryo to "male"?
   a. an X chromosome
   b. a Y chromosome
   c. an XY chromosome pair
   d. the *SRY* gene

Sample Genetics Problems

1. Recall that human females have two X chromosomes and human males have one X chromosome and one Y chromosome.
   a. Do males inherit their X chromosome from their mother or from their father?
   b. If a female has one copy of an X-linked recessive allele for a genetic disorder, does she have the disorder?
   c. If a male has one copy of an X-linked recessive allele for a genetic disorder, does he have the disorder?
   d. Assume that a female is a carrier of an X-linked recessive disorder. With respect to the disorder allele, how many types of gametes can she produce?
   e. Assume that a male with an X-linked recessive genetic disorder has children with a female who does not carry the disorder allele. Could any of their sons have the genetic disorder? How about their daughters? Could any of their children be carriers for the disorder? If so, which sex(es) could they be?

2. Cystic fibrosis is a recessive genetic disorder. The disorder allele, which we'll call *a*, is located on an autosome. What are the chances that parents with the following genotypes will have a child with the disorder?
   a. $aa \times Aa$
   b. $Aa \times AA$
   c. $Aa \times Aa$
   d. $aa \times AA$

3. Huntington disease (HD) is a genetic disorder caused by a dominant allele—we'll call it *A*—that is located on an autosome. What are the chances that parents with the following genotypes will have a child with HD?
   a. $aa \times Aa$
   b. $Aa \times AA$
   c. $Aa \times Aa$
   d. $aa \times AA$

4. Hemophilia is a recessive genetic disorder whose disorder allele, which we'll call *a*, is located on the X chromosome. What are the chances that parents with the following genotypes will have a child with hemophilia?
   a. $X^A X^A \times X^a Y$
   b. $X^A X^a \times X^a Y$
   c. $X^A X^a \times X^A Y$
   d. $X^a X^a \times X^A Y$
   e. Do male and female children have the same chance of getting the disorder?

5. Explain why the terms "homozygous" and "heterozygous" do not apply to X-linked traits in males.

6. Study the pedigree shown below. Is the disorder allele dominant (*D*) or recessive (*d*)? Is the disorder allele located on an autosome or on the X chromosome? What are the genotypes of individuals 1 and 2 in generation I?

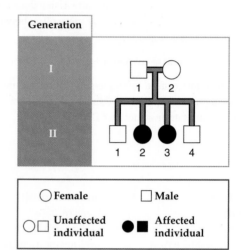

7. In the text, we state that males are more likely than females to inherit recessive X-linked genetic disorders. Are males also more likely than females to inherit dominant X-linked genetic disorders? Illustrate your answer by constructing Punnett squares in which
   a. an affected female has children with a normal male.
   b. an affected male has children with a normal female.

8. Study the pedigree shown below. Is the disorder allele dominant or recessive? Is the disorder allele located on an autosome or on the X chromosome? To answer this question, assume that individual 1 in generation I, and individuals 1 and 6 in generation II, do not carry the disorder allele.

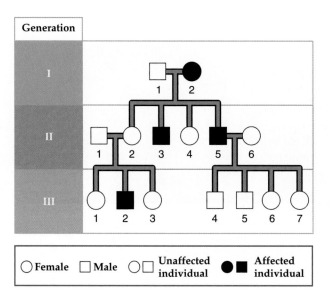

Generation	

○ Female   ☐ Male   ○☐ Unaffected individual   ●■ Affected individual

9. Imagine you are conducting an experiment on fruit flies, and you are tracking the inheritance of two genes, one with alleles A or a, the other with alleles B or b. $AABB$ individuals are crossed with $aabb$ individuals to produce F_1 offspring, all of which have genotype $AaBb$. These $AaBb$ F_1 offspring are then crossed with $aaBB$ individuals. Construct Punnett squares and list the possible offspring genotypes that you would expect in the F_2 generation

a. if the two genes were completely linked.

b. if the two genes were on different chromosomes.

CHAPTER 12 DNA

Key Concepts

- Genes are composed of DNA.

- Four nucleotides serve as the building blocks of DNA. Each of these nucleotides contains one of four nitrogen-containing bases: adenine, cytosine, guanine, or thymine.

- DNA consists of two strands of nucleotides twisted into a helix. The two strands are held together by hydrogen bonds that form between adenine in one strand and thymine in the other, and similarly, between cytosine in one strand and guanine in the other.

- The sequence of nitrogenous bases in DNA differs between species and between individuals within a species. These differences are the basis of inherited variation.

- Because adenine pairs only with thymine and cytosine pairs only with guanine, each strand of DNA can serve as a template from which to duplicate the other strand.

- DNA in cells is subject to damage by various physical, chemical, and biological agents. Up to a point, such damage can be repaired.

Main Message: DNA is the genetic material in all living cells. It consists of two strands of nucleotides twisted into a double helix.

The Library of Life

The book you are reading has more than a million letters in it. How long would it take you to copy those letters by hand, one by one? How many mistakes do you think you would make? Would you check your work for mistakes, and if so, how long would that take?

Difficult as the job of copying all the letters in this book would be, it pales in comparison to the job your cells do each time they divide. Before a cell divides, it must make a copy of all its genetic information. That information is stored in deoxyribonucleic acid, DNA, the material that genes are made of. The amount of information stored in your DNA is mind-boggling: whereas this book contains roughly a million letters, the DNA in your cells has the equivalent of 6,600,000,000 letters. If single-letter codes for the DNA bases were printed end-to-end in letters the same size as those on this page, the information that is in your DNA would fill thousands of books similar in length to this one.

Your cells copy the phenomenal amount of information in your DNA in a matter of hours. Despite the speed with which they work, on average, they make only one mistake for every billion "letters" that are copied. For comparison, a professional typist entering text at 120 words per minute is likely to make one mistake for every 250 words copied. **But what are the consequences of mistakes made in the copying of DNA? Do such errors contribute to the hundreds of inherited disorders that are seen in human populations? More broadly, what is the structure of the DNA molecule, and what implications does this structure have for the processes of life?** We examine these questions in this chapter, after we describe one of the most momentous discoveries in biology: the experimental evidence that established DNA as the storehouse of genetic information. Those findings spurred the race to decipher the structure of DNA. Knowing how DNA is put together enabled us to understand how DNA, indeed life itself, is propagated.

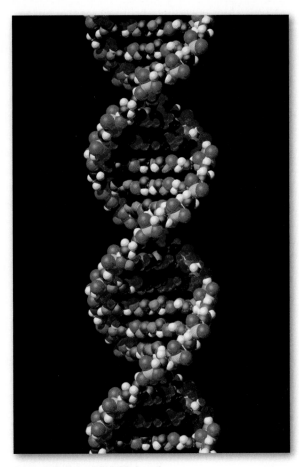

A Model of the DNA Molecule

In the previous two chapters, we learned that genes control the inheritance of traits and are located on chromosomes. However, this knowledge leaves unanswered several fundamental questions: What are genes made of? When a cell divides to form two daughter cells, how is the information in the genes copied? How are errors in copying corrected, and how is damage to the cell's genetic material repaired?

To answer such questions, geneticists had to discover the substance of which genes are made, and they had to learn the physical structure of this substance. As they began their search, they were guided by three basic biological facts about the nature of the genetic material. First, the genetic material had to contain the information necessary for life. It had to contain, for example, the information needed to form the organism and to control the complex metabolic reactions on which life depends. Second, the genetic material had to be composed of a substance that could be copied accurately. If this could not be done, reliable genetic information could not be passed from one generation to the next. Despite this overall need for accuracy, the genetic material had to be a changeable molecule, able to code for slightly different versions of the same information; otherwise, there would be no genetic differences within or among species.

Parallel to the search for the chemical composition and physical structure of genes was a search for the function of genes: how exactly did genes produce their effects? In this chapter we describe how scientists discovered that genes are composed of DNA. We also describe the physical structure of genes, how the genetic material is copied and repaired, and how genetic disorders can arise. In Chapter 13 we will see how genes produce their effects.

12.1 The Search for the Genetic Material

By the early 1900s, geneticists knew that genes control the inheritance of traits, that genes are located on chromosomes, and that chromosomes are composed of DNA and proteins. With this knowledge in hand, the first step in the quest to understand the physical structure of genes was to determine whether the genetic material was DNA or protein.

Initially, most geneticists thought that protein was the more likely candidate. Proteins are large, complex molecules, and it was not hard to imagine that they could store the tremendous amount of information needed to govern the lives of cells. Proteins also vary considerably within and among species; hence it was reasonable to assume that they caused the inherited variation observed within and among species.

DNA, on the other hand, was initially judged a poor candidate for the genetic material, mainly because DNA was thought to be a small, simple molecule whose composition varied little among species. Over time, through a number of key experiments, these ideas about DNA were shown to be wrong. In fact, DNA molecules are large and vary tremendously within and among species. Still, as we shall see, variations in the DNA are more subtle than the variation in shape, electrical charge, and function shown by proteins, so it is not surprising that most researchers initially favored proteins as the genetic material.

Over a period of roughly 25 years (1928–1952), geneticists became convinced that DNA, not protein, was the genetic material. Let's consider three important studies that helped cause this shift of opinion.

Harmless bacteria can be transformed into deadly bacteria

In 1928, a British medical officer named Frederick Griffith published an important paper on *Streptococcus pneumoniae* [noo-*MOH*-nee-eye], a bacterium that causes pneumonia in humans and other mammals. Griffith was studying two genetic varieties, or strains, of *Streptococcus* to find a cure for pneumonia, which was a common cause of death at that time. The two strains, called strain S and strain R, were named after differences in their appearance. When the bacteria were grown on a petri dish, strain S produced colonies that appeared smooth, while strain R produced colonies that appeared rough. While conducting experiments to develop a vaccine against the disease-causing S strain, Griffith made a startling discovery: some heat-resistant material from the S strain could transform the harmless R strain into the disease-causing strain.

Griffith conducted four experiments on these bacteria (Figure 12.1). First, when he injected bacteria of strain R into mice, the mice did not develop pneumonia and survived. Second, when he injected bacteria of strain S into mice, the mice developed pneumonia and died. In the third experiment, he injected heat-killed strain S bacteria into mice, and this time the mice survived. His original plan was to test mice from the third experiment to see if they were resistant to later exposure to live strain S bacteria, but his 1928 paper did not include results from such a test.

The results from the first three experiments were not particularly unusual: Griffith had simply shown that there were two strains of bacteria, one of which (strain S) killed mice and was itself killed and rendered harmless by heat. In the fourth experiment, however, something unexpected happened. Griffith mixed heat-killed bacteria of strain S with live bacteria of harmless strain R. On the basis of the results from the first three experiments, he expected the mice to survive. Instead, the mice died, and Griffith recovered large numbers of live strain S bacteria from the blood of the dead mice.

In Griffith's fourth experiment, something that survived heat treatment had caused harmless strain R bacteria to change into deadly strain S bacteria. Griffith showed that the change was genetic: when they reproduced, the altered strain R bacteria produced cells resembling strain S bacteria. Overall, the results of Griffith's fourth experiment suggested that genetic material from heat-killed strain S bacteria had somehow changed living strain R bacteria into strain S bacteria.

This remarkable result stimulated an intensive hunt for the material that caused the change. We now know that the strain R bacteria had absorbed a small piece of DNA from the heat-killed strain S bacteria, causing the genetic characteristics (genotype) of the strain R bacteria to change. Such a change in the genotype of a cell or organism after it has absorbed DNA of another genotype is called **transformation**.

DNA can transform bacteria

For 10 years, beginning in 1934, Oswald Avery, Colin MacLeod, and Maclyn McCarty (all at Rockefeller University in New York) struggled to identify the genetic material that had transformed the bacteria in Griffith's experiments. They isolated and tested different compounds from the bacteria. Only DNA was able to transform harmless strain R bacteria into deadly strain S bacteria. In 1944, Avery and his colleagues published a landmark paper that summarized their results. The paper created quite a stir.

In addition to showing that DNA transforms bacteria, Avery, MacLeod, and McCarty's paper led many biologists to a broader conclusion: that DNA, not protein, is the genetic material. As a leading DNA researcher later remarked, the paper stimulated an "avalanche" of new research on DNA. Some biologists remained skeptical, arguing, for example, that DNA was too simple a molecule to be the genetic material. However, the tide was turning in favor of DNA.

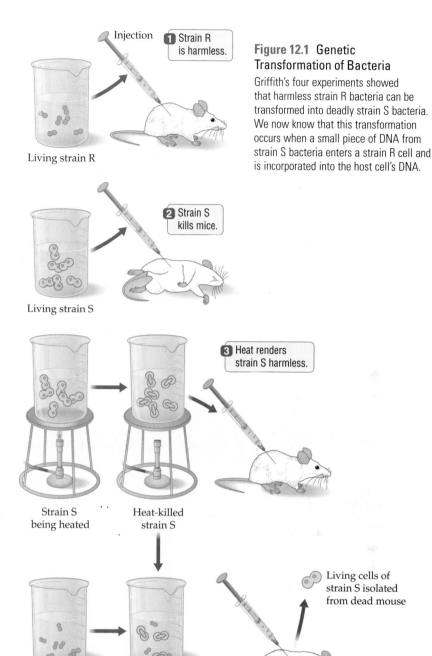

Figure 12.1 Genetic Transformation of Bacteria
Griffith's four experiments showed that harmless strain R bacteria can be transformed into deadly strain S bacteria. We now know that this transformation occurs when a small piece of DNA from strain S bacteria enters a strain R cell and is incorporated into the host cell's DNA.

1 Strain R is harmless.

Living strain R

2 Strain S kills mice.

Living strain S

3 Heat renders strain S harmless.

Strain S being heated

Heat-killed strain S

Living strain R

Mixture of heat-killed strain S and living strain R

Living cells of strain S isolated from dead mouse

4 A substance from the heat-killed strain S changes strain R from harmless to deadly.

The genetic instructions of a virus are contained in its DNA

In Griffith's experiments, heat killed the strain S bacteria, but did not destroy its genetic material. Since most proteins are destroyed by heat, this result suggested that protein was not the genetic material. DNA is not destroyed even at boiling temperatures, and the

Recreate Griffith's experiments.
 12.1

experiments of Avery and co-workers provided strong evidence that DNA extracted from strain S contained information that could transform the benign strain R into a disease-producing form. Additional proof came in 1952, when Alfred Hershey and Martha Chase published an elegant study on the genetic material of viruses.

Hershey and Chase studied a virus that consists only of a DNA molecule surrounded by a coat of proteins (Figure 12.2). To reproduce, this virus attaches to the cell wall of a bacterium and injects its genetic material into the bacterium. The genetic material of the virus then takes over the bacterial cell, eventually killing it, but first causing it to produce many new viruses. Because the virus is composed only of proteins and DNA, it provided an excellent experimental system in which to test which substance—DNA or protein—was the genetic material.

Hershey and Chase demonstrated that only the DNA portion of the virus was injected into the bacterium (see Figure 12.2). They knew that sulfur atoms are found in proteins but not in DNA, and that phosphorus atoms are found in DNA but are not a component of proteins. By growing the virus in solutions containing radioactive sulfur compounds or radioactive phosphorus compounds, they could selectively label the protein-based coat of the virus, or the DNA inside it, with radioactivity. When bacteria were infected with sulfur-labeled viruses, the radioactivity remained outside the cell, showing that viral proteins did not enter the host cell. However, when phosphorus-labeled viruses were used, the radioactivity turned up inside the bacterial cell. Since the researchers knew from previous experiments that the virus injects its genetic material into the cells it infects, they could deduce that DNA, not protein, was the genetic material of this virus. These experiments convinced most remaining skeptics that it is DNA, not protein, that holds genetic information.

See the Hershey-Chase experiment in motion. ▶❚❚ 12.2

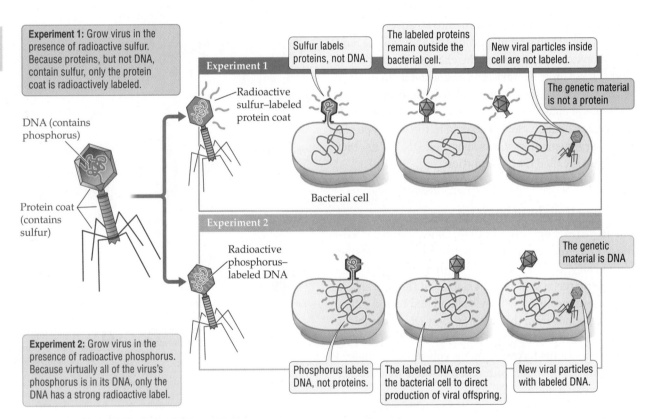

Experiment 1: Grow virus in the presence of radioactive sulfur. Because proteins, but not DNA, contain sulfur, only the protein coat is radioactively labeled.

DNA (contains phosphorus)

Protein coat (contains sulfur)

Radioactive sulfur–labeled protein coat

Sulfur labels proteins, not DNA.

The labeled proteins remain outside the bacterial cell.

New viral particles inside cell are not labeled.

The genetic material is not a protein

Experiment 1

Bacterial cell

Experiment 2

Radioactive phosphorus–labeled DNA

The genetic material is DNA

Phosphorus labels DNA, not proteins.

The labeled DNA enters the bacterial cell to direct production of viral offspring.

New viral particles with labeled DNA.

Experiment 2: Grow virus in the presence of radioactive phosphorus. Because virtually all of the virus's phosphorus is in its DNA, only the DNA has a strong radioactive label.

Figure 12.2 DNA Is the Genetic Material
Hershey and Chase conducted experiments using viruses that infect bacteria. They knew that the virus injected genetic material into the bacterial host, and this genetic material directed the production of new viruses within the cytoplasm of the host cell. They used a radioactive labeling technique to demonstrate that it is DNA, not protein, that enters the host cell when the virus infects it. Hershey and Chase grew viruses in two different radioactive solutions that labeled either the viruses' DNA or their proteins. They then exposed bacteria to these viruses. When Hershey and Chase found labeled DNA—but not labeled protein—inside bacterial cells and inside new viruses, they concluded that DNA was the genetic material of this virus.

12.2 The Three-Dimensional Structure of DNA

By the early 1950s, genes were known to be composed of DNA. The next step was to determine the three-dimensional structure of DNA. This structure needed to be determined down to the level of atoms before gene function could be understood in molecular terms.

In 1951, Linus Pauling and his colleagues published a series of papers on the three-dimensional structure of certain proteins and described how portions of these proteins could twist into a spiral, or helical, configuration. Pauling's work was a breakthrough in understanding the geometry of biologically important molecules, and it suggested that the three-dimensional structure of DNA could also be determined with similar methods. Over the next few years, major research laboratories from around the world, including Pauling's, devoted great effort to analyzing the three-dimensional structure of DNA.

The effort to discover the structure of DNA was a race to unlock some of the greatest mysteries of life: How is genetic information stored in DNA? And how is the cell's genetic material copied so that it can be passed from parent to offspring?

DNA is a double helix

Working at Cambridge University in England, the American James Watson and the Englishman Francis Crick won the race to determine the physical structure of DNA. In a two-page paper published in 1953, they proposed that DNA was a **double helix**, a structure that can be thought of as a ladder twisted into a spiral coil (Figure 12.3). Watson was 25 at the time, Crick 37. Nine years later Watson and Crick were awarded the Nobel Prize in physiology or medicine for their discovery. They shared the prize with Maurice Wilkins, who had also worked to discover the structure of DNA. Missing from the Nobel ceremony was Rosalind Franklin, a gifted young scientist whose research provided Watson and Crick with critical data. Rosalind Franklin died of cancer in 1958 at age 37. Nobel prizes cannot be awarded to a deceased person, so we will never know whether she might have shared the 1962 prize.

As Watson and Crick described it, DNA's untwisted (two-dimensional) form has two long strands of repeating units called **nucleotides**. Each nucleotide is composed of the sugar deoxyribose, a phosphate group, and one of four **nitrogenous bases**: adenine (A),

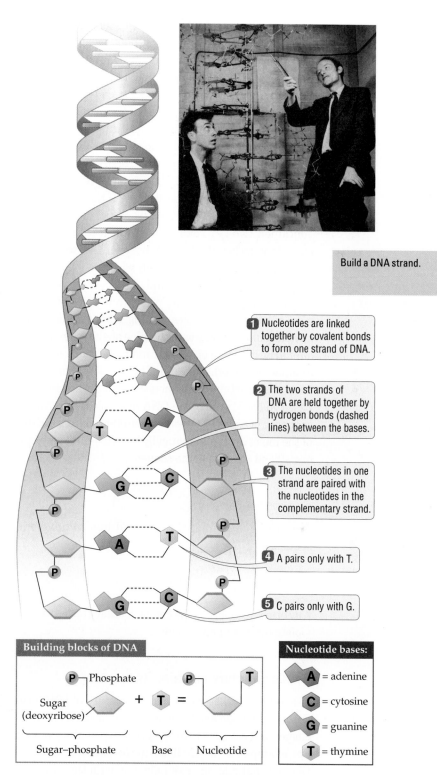

Build a DNA strand. ▶❚❚ 12.3

1. Nucleotides are linked together by covalent bonds to form one strand of DNA.

2. The two strands of DNA are held together by hydrogen bonds (dashed lines) between the bases.

3. The nucleotides in one strand are paired with the nucleotides in the complementary strand.

4. A pairs only with T.

5. C pairs only with G.

Building blocks of DNA

P — Phosphate
Sugar (deoxyribose)
+ T =
Sugar–phosphate   Base   Nucleotide

Nucleotide bases:
A = adenine
C = cytosine
G = guanine
T = thymine

Figure 12.3 The DNA Double Helix and Its Building Blocks
A nucleotide consists of three types of chemical groups: phosphate, sugar, and a nitrogen-containing base. Four types of nucleotides are found in DNA, varying only in the type of base found in them. DNA consists of two complementary strands of nucleotides that are twisted into a spiral around an imaginary axis, rather like the winding of a spiral staircase. The two strands are held together by hydrogen bonds between their complementary bases. *Inset*: James Watson and Francis Crick, with a model of the DNA double helix.

cytosine (C), guanine (G), or thymine (T). The way the nucleotides are held together explains why a ladder is a good image for DNA's two-dimensional structure (see Figure 12.3). Along each strand, the nucleotides are connected by covalent bonds between the phosphate of one nucleotide and the sugar of the next nucleotide. These sugar-phosphate links constitute the "sides" of the ladder. Hydrogen bonds connect the bases on one strand to the bases on the other strand, thereby holding the two strands together. These **base pairs** connected by hydrogen bonds are the "rungs" of the ladder.

Among other insights, Watson and Crick's paper contained two key realizations. First, there are two strands of nucleotides in DNA, not one. The two strands are twisted around each other, which is why DNA is called a double helix. Second, only certain pairings between bases are possible. Watson and Crick proposed a set of base-pairing rules, stating that adenine on one strand could pair only with thymine on the other strand; similarly, cytosine on one strand could pair only with guanine on the other strand. These base-pairing rules have an important consequence: when the sequence of bases on one strand of the DNA molecule is known, the sequence of bases on the other, **complementary strand** of the molecule is automatically known. For example, if one strand consists of the sequence

ACCTAGGG,

then the complementary strand has to have the sequence

TGGATCCC.

Any other sequence in the complementary strand would violate the base-pairing rules.

DNA's structure explains its function

We now know that the physical structure of DNA proposed by Watson and Crick is correct in all its essential elements. This structure has great explanatory power. For example, as we shall see in the following section, the fact that adenine can pair only with thymine and that cytosine can pair only with guanine suggested a simple way in which the DNA molecule could be copied: the original strands could serve as templates on which new strands could be built. This suggestion turned out to be correct.

Knowledge of the three-dimensional structure of DNA also suggested that the information stored in DNA

could be represented as a long string of the four bases: A, C, G, and T. Although A has to pair with T and C has to pair with G, the four bases can be arranged in any order along a single strand of DNA. The fact that each strand of DNA is composed of millions of these bases suggested that a tremendous amount of information could be contained in the order of the bases along the DNA molecule, or **DNA sequence**; this suggestion has also proved to be correct (see Chapter 13).

The sequence of bases in DNA differs between species and between individuals within a species (Figure 12.4). We now know that different alleles of a gene have different DNA sequences, and hence that differences in DNA sequence are the basis of inherited variation. For example, people with a genetic disorder such as Huntington disease or cystic fibrosis inherit particular alleles that cause the disorder, as we saw in Chapter 11. At the molecular level, one allele causes a disease and another allele does not

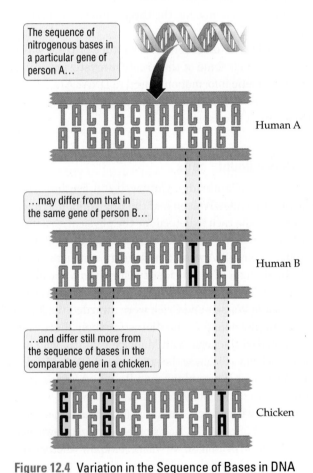

Figure 12.4 Variation in the Sequence of Bases in DNA
The sequence of bases in DNA differs among species and among individuals within a species. Here, the sequence of bases in a hypothetical gene is compared for two humans (A and B) and a chicken. Base pairs highlighted in yellow are variant, that is, they differ between the genes of persons A and B, or the comparable human and chicken genes.

because the two alleles have a different sequence of bases. Sometimes a difference of only one base pair in hundreds or thousands within the gene can make the difference between life and death. The severe consequences of some alleles that cause genetic disorders can lead people who are at risk of inheriting them to seek the guidance of a genetic counselor (see the box on page 258).

Concept Check

1. Describe one historic experiment that provided evidence that genetic material can be transferred from one cell to another.

2. What is the percentage of thymines (T) in a DNA double helix in which 20 percent of the nitrogenous bases are guanine (G)?

3. If all genes are composed of just four nucleotides, how can different genes carry different types of information?

12.3   How DNA Is Replicated

As Watson and Crick noted in their historic 1953 paper, the structure of the DNA molecule suggested a simple way that the genetic material could be copied. They elaborated on this suggestion in a second paper, also published in 1953. Because A pairs only with T, and C pairs only with G, each strand of DNA contains the information needed to duplicate the complementary strand. For this reason, Watson and Crick suggested that **DNA replication**—the duplication of a DNA molecule—might work in the following way (Figure 12.5):

1. The hydrogen bonds connecting the two strands of the DNA molecule are broken.

2. Breaking of the hydrogen bonds causes the two strands to unwind and separate.

3. Each strand is then used as a template for the construction of a new strand of DNA.

4. When this process is completed, there are two identical copies of the original DNA molecule, each with the same sequence of bases. Each copy is composed of one "old" strand of DNA (from the original DNA molecule) and one newly synthesized strand of DNA. This mode of replication is known as **semi-conservative replication**, because one "old" strand from the template DNA is retained, or conserved, in each new double helix.

Five years later, other researchers confirmed that DNA replication produces DNA molecules composed of one old strand and one new strand, as predicted by Watson and Crick. The main enzyme involved in the replication of DNA has now been identified and is called **DNA polymerase** [puh-*LIM*-er-ayss].

The Watson–Crick model of DNA replication is elegant and simple, but the mechanics of actually copying DNA are far from simple. More than a dozen enzymes and proteins are needed to unwind the DNA, to stabilize the separated strands, to start the replication process, to attach nucleotides to the correct positions on the template strand, to "proofread" the results, and to join partly replicated fragments of DNA to one another.

Although DNA replication is such a complex task, cells can copy DNA molecules containing billions of nucleotides in a matter of hours—about 8 hours in people (over 100,000 nucleotides per second). This speed is achieved in part by starting the replication of the

Explore how DNA replicates.

12.4

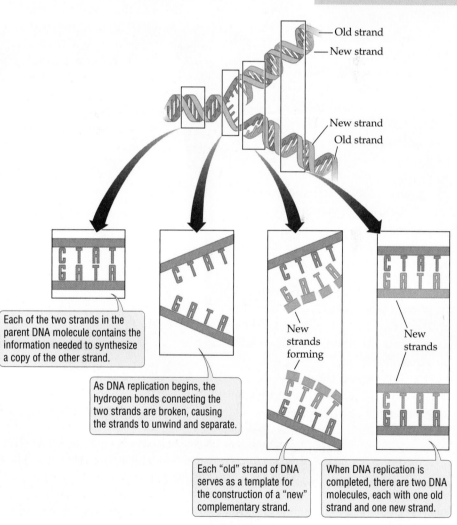

Figure 12.5 Semi-Conservative Replication of DNA
In this overview of DNA replication, the template DNA strands are gray, and the newly synthesized strands are orange. One strand (gray) from the parent double helix is conserved in each newly made daughter double helix (gray and orange).

Biology on the Job

Helping People Cope with Genetic Disorders

Each year, thousands of patients seek the guidance of genetic counselors like Robin L. Bennett, who is a senior genetic counselor and the associate director of the Medical Genetics Clinics at the University of Washington in Seattle.

How did you become interested in genetic counseling? Was there a key moment when you suddenly knew what you wanted to do? In tenth grade, I learned about the new field of genetic counseling. I was interested in working in an area like this because my mother's best friend had a son with profound mental and physical handicaps. I saw how this devastated the family, and I wanted to make a difference for families like this.

How does the type of genetic counseling that you do compare to the range of jobs that are open to a person trained as a genetic counselor? The majority work with either children or pregnant women with genetic conditions, or with adults and their families. Most genetic counselors are involved in education at some level—they give talks or help develop educational materials (pamphlets, CDs, online programs). More genetic counselors are becoming involved in setting policies related to genetic diseases, such as initiating newborn screening programs.

What are some of the most exciting new discoveries or advances in your field? Individuals can now be tested for a predisposition to a genetic disease. Often there is no treatment for the disease, but there is a test to see if a person has a high chance of developing the condition—although the test won't predict exactly when the condition will occur, or even how bad the disease will be. Also, in the last few years, there are more treatments for genetic disorders (particularly enzyme replacement therapy).

You just mentioned that genetic tests raise psychological issues for patients. How often do genetic counselors face ethical and psychological issues? Genetic counselors face ethical issues almost on a daily basis. Psychological issues are faced by each of our clients, so genetic counseling is important before a person has a genetic test—not just after. In most cases, genetic counseling can help individuals make

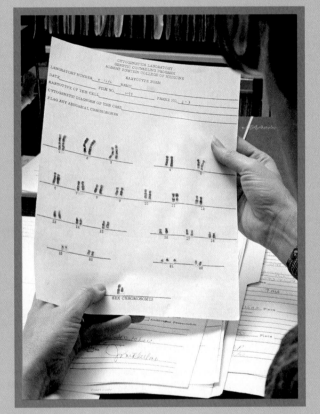

A genetic counselor reviewing the full set of chromosomes (the karotype) of a patient with Down syndrome.

informed life choices. A person who has a fetus with multiple birth defects identified by ultrasound may choose to deliver the baby at a hospital that has a pediatric intensive care unit or doctors that specialize in the problems (for example, cardiac surgery).

What do you enjoy most about your job? Being a genetic counselor helps me to put my own life into perspective. I try to live life to the fullest because I can appreciate my blessings. Even when I have to give bad news, I feel like I'm giving people information in a supportive way that can bring meaning to their lives, even options and hope. I love the challenge of being a genetic counselor—every day is different—never boring.

Is there anything else you'd like to tell our readers? There are wonderful career resources through the National Society of Genetic Counselors Web site (www.nsgc.org). Also, the American Board of Genetic Counselors has information about genetic counseling programs (www.abgc.net).

The Job in Context

As we've seen in this and the preceding chapters, there have been rapid advances in our ability to test whether an adult, fetus, or embryo has specific alleles that influence human genetic disorders. Information from such tests can benefit patients, relieving them of worry (if the test reveals they or their children don't have the disorder), or enabling them to begin treatment early or take other therapeutic measures (if the test reveals that they or their children have the disorder); but sometimes the results of these tests are hard to interpret. This uncertainty stems from a fundamental aspect of how genes work: often, the effect of a gene depends not only on its own function, but also on the function of other genes and on environmental factors.

DNA molecule at thousands of different places at once. Despite their speed, cells make remarkably few mistakes when they copy their DNA.

12.4 Repairing Replication Errors and Damaged DNA

When DNA is copied, there are many opportunities for mistakes to be made. In humans, for example, more than 6 billion base pairs of template DNA must be copied each time a diploid cell divides. Two new strands must be made from each of the two template strands, so there are over 12 billion opportunities for mistakes. In addition, the DNA in cells is constantly being damaged by various sources. Replication errors and damage to the DNA—especially to essential genes—disrupt normal cell functions. If not repaired, this damage would lead to the death of many cells and, potentially, to the death of the organism.

Few mistakes are made in DNA replication

The enzymes that copy DNA sometimes insert an incorrect base in the newly synthesized strand. For example, if DNA polymerase were to insert a cytosine (C) across from an adenine (A) located on the template strand, an incorrect C–A pair bond would form instead of the correct T–A pair bond (Figure 12.6). Such mistakes are made about once in every 10,000 bases. However, nearly all of these mistakes are corrected immediately by enzymes (including DNA polymerase itself) that check, or "proofread," the pair bonds as they form. This form of error correction is similar to what happens as you type a paper, realize you made a mistake, and correct it right away with the "delete" key.

When an incorrect base is added but escapes the mechanism for immediate proofreading, a **mismatch error** has occurred. Mismatch errors occur about once in every 10 million bases. Cells contain repair proteins that specialize in fixing mismatch errors; these proteins play a role similar to the error checking you perform when you complete the first draft of a paper, print it, and carefully review it for mistakes. Proteins that fix mismatch errors correct 99 percent of those errors, reducing the overall chance of an error to the incredibly low rate of one mistake in every billion bases.

On the rare occasions when a mismatch error is not corrected, the DNA sequence is changed, and the new sequence is reproduced the next time the DNA is replicated. A change to the sequence of bases in an organism's DNA is called a **mutation**. When a mistake in the copying process is not corrected, a mutation has occurred. Mutations can also occur when cells are exposed to **mutagens**, substances or energy sources that alter DNA.

Mutations result in the formation of new alleles. Some of the new alleles that result from mutation are beneficial, but most are either neutral or harmful. Among the harmful alleles are those that cause cancer and other human genetic disorders, such as sickle-cell anemia and Huntington disease. Note that our definition of mutation includes not only changes in the DNA sequence of a single gene, but also the larger scale changes in DNA sequence that are created by chromosomal abnormalities (Chapter 11). Changes in chromosome number or chromosome organization generally affect large numbers of genes, and such changes also amount to DNA mutations because they have the effect of adding, deleting, or rearranging nucleotide sequences.

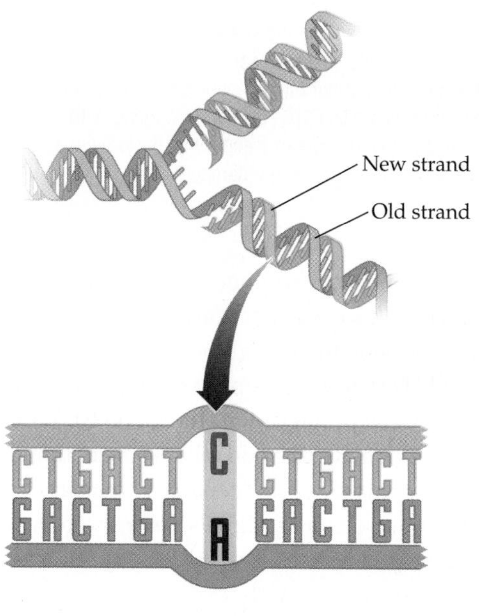

Explore how DNA is repaired.

▶ ❚❚
12.5

New strand
Old strand

Figure 12.6 Mistakes Can Be Made in DNA Replication
Here, a cytosine (C) has been incorrectly inserted opposite an adenine (A). DNA repair enzymes usually fix such mismatch errors before the cell's DNA is replicated again. If the mismatch is not repaired before the next round of DNA replication, half the daughter helices made by this DNA will have a C–G base pair in place of the original A–T base pair.

Normal gene function depends on DNA repair

Every day, the DNA in each of our cells is damaged thousands of times by chemical, physical, and biological agents. These agents include energy from radiation or heat, collisions with other molecules in the cell, attacks by viruses, and random chemical accidents (some of which are caused by environmental pollutants, but most of which result from normal metabolic processes). Our cells contain a complex set of repair proteins that fix the vast majority of this damage. Single-celled organisms such as yeasts have more than 50 different repair proteins, and humans probably have even more.

Although humans are very good at repairing damaged DNA, some damage far exceeds our ability to repair it. Humans exposed to 1,000 rads (radiation units) of energy die in a few weeks, in part because their DNA is damaged beyond repair. (A rad is a unit for measuring the amount of absorbed radiation; to give you a sense of scale, overall our tissues absorb less than 0.1 rads of energy during a dental X-ray.) Some of the people who initially survived the atomic blasts at Hiroshima and Nagasaki were exposed to about 1,000 rads. Over the next few weeks they died from acute radiation poisoning as cells in the bone marrow and digestive system died from the severe DNA damage. Although 1,000 rads kills a person, such a dose would barely faze the bacterium *Deinococcus radiodurans* [DYE-noh-KOCK-uss ray-dee-oh-DER-unz]. This species is so efficient at repairing damage to DNA that a dose of 1,000,000 rads merely slows its growth, but does not kill it. Even when the dose is raised to 3,000,000 rads—3,000 times greater than a lethal dose for a person—a small percentage of the bacteria survive.

In humans, *Deinococcus*, and all other organisms, there are three steps in **DNA repair**: the damaged DNA must be recognized, removed, and replaced (Figure 12.7). Different sets of repair proteins specialize in recognizing defects in DNA structure and removing the damaged segment of DNA by cutting it out with special enzymes. Once these first two steps have been accomplished, the final step is to add the correct sequence of nucleotides to fill the gap created by the removal of the damaged DNA strand. This third step of the repair process uses the intact strand of DNA as template for re-creating the missing segment in the complementary strand.

Mutations are the consequence of failure in DNA repair. When an animal cell that is capable of multiplying through cell divisions acquires many mutations in a vari-

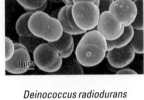

Deinococcus radiodurans
(bacterium)

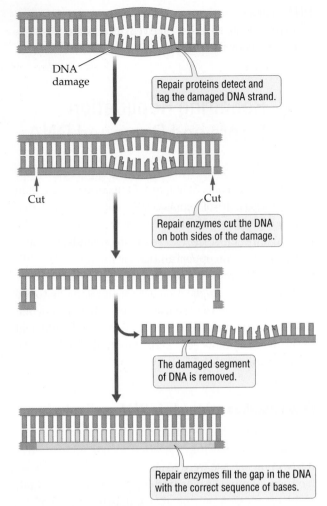

Figure 12.7 Repair Proteins Fix DNA Damage
Large complexes of DNA repair proteins work together to fix damaged DNA. They (1) recognize DNA damage, (2) remove the segment of DNA strand that is damaged, and (3) replace the missing segment through new DNA synthesis. Some of the repair proteins recognize damaged DNA and bind to it, flagging the site where the damage is located. Other proteins (enzymes), that are recruited to the flagged site, cut out the damaged DNA strand and remove it from the helix. DNA polymerases, along with other enzymes, then use the exposed template to build a new DNA segment to replace the damaged one.

ety of genes, it becomes dangerous because such a cell is more likely to have malfunctions in genes that normally keep cell division under tight control. Consequently, it is more likely to become cancerous, a condition marked by runaway cell proliferation (Interlude B). When DNA damage is extensive, the odds that some of that damage will go unrepaired, or be "repaired" incorrectly, rises dramatically. In complex animals, severely damaged cells are such a high cancer risk that the body often responds

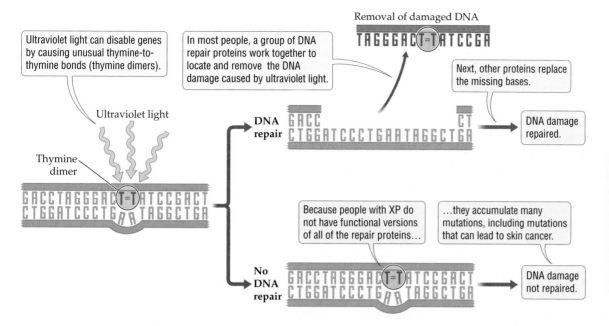

Ultraviolet light can disable genes by causing unusual thymine-to-thymine bonds (thymine dimers).

In most people, a group of DNA repair proteins work together to locate and remove the DNA damage caused by ultraviolet light.

Removal of damaged DNA

TAGGGACT=TATCCGA

Next, other proteins replace the missing bases.

Ultraviolet light

Thymine dimer

DNA repair

GACC   CT
CTGGATCCCTGAATAGGCTGA

DNA damage repaired.

GACCTAGGGACT=TATCCGACT
CTGGATCCCTGAA TAGGCTGA

Because people with XP do not have functional versions of all of the repair proteins...

...they accumulate many mutations, including mutations that can lead to skin cancer.

No DNA repair

GACCTAGGGACT=TATCCGACT
CTGGATCCCTGAA TAGGCTGA

DNA damage not repaired.

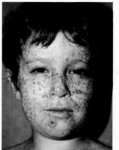

This child has XP. The large growth on his chin is a skin cancer.

Figure 12.8 The Importance of DNA Repair Mechanisms
When DNA repair mechanisms fail to work properly, the consequences can be severe. This is illustrated by the high frequency of skin cancers in people who have xeroderma pigmentosum (XP), a recessive genetic disorder in which cells are unable to make a protein used to repair DNA damage caused by ultraviolet light.

by triggering the death of damaged cells. The process by which a cell brings about its own death, through an orderly series of cellular events, is known as **apoptosis**.

A peeling sunburn is an uncomfortable, and very visual, display of apoptosis in skin cells. If you expose yourself to high levels of UV radiation for several hours, you may cause irreparable damage to the DNA in your skin cells, especially if you have light skin. The damaged skin cells shrink, and their nucleus and cytoplasm are progressively destroyed, as they undergo apoptosis. Like any other process, apoptosis itself has a certain failure rate, which is why we shouldn't count on apoptosis to protect us from the havoc that can be unleashed by DNA-damaged cells. If apoptosis fails in even a few of the millions of skin cells that are potentially damaged when large parts of the body are overexposed to ultraviolet (UV) light, those survivors can turn into especially aggressive cancer cells. Being exposed to more than two peeling sunburns before age 18 doubles a person's lifetime risk for developing malignant melanoma. While other forms of skin cancer are more common and usually easily treated, malignant melanoma tends to be deadly if not caught at an early stage, and it kills nearly 8,000 Americans every year.

The importance of DNA repair mechanisms is also highlighted by genetic disorders in which the repair system itself is inactive or seriously inefficient. The child in Figure 12.8 has xeroderma pigmentosum (XP), a recessive genetic disorder in which even brief exposure to sunlight causes painful blisters. The allele that causes XP produces a nonfunctional version of one of the many human DNA repair proteins. The job of the normal form of this protein is to repair the kind of damage to DNA caused by UV light. The lack of this DNA repair protein makes individuals with XP highly susceptible to skin cancer. Several inherited tendencies to develop cancer, including some types of breast and colon cancer, also stem from less effective versions of genes that participate in DNA repair.

Concept Check

1. Name the enzyme that creates covalent bonds between successive nucleotides to create a polymer complementary to a template DNA strand.

2. Describe the semi-conservative nature of DNA replication.

3. What types of mechanisms reduce the chance that the DNA in a cell will mutate? Are these mechanisms 100 percent effective?

Applying What We Learned

Errors in the Library of Life

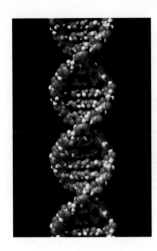

The nucleotide sequence of a gene may vary from one individual to another, producing a variety of phenotypes that are harmless to the individual. But all is not perfect in the library of life: errors in DNA replication can produce new, mutant alleles that cause genetic disorders, some of which are lethal. In Chapter 11 we saw that serious genetic disorders, including sickle-cell anemia, cystic fibrosis, and Huntington disease, can result from mutations in a single gene. In these and literally thousands of other cases, genetic disorders are caused by alleles that have a DNA sequence that differs from the sequence found in a normal allele.

An allele that causes a genetic disorder can differ by as little as one base from the normal allele, as does the sickle-cell anemia allele. In other genetic disorders, the responsible allele differs by several bases from the sequence found in the normal allele. Cystic fibrosis, for example, is a fatal disorder caused by a recessive allele located on chromosome 7 (see Figures 11.7 and 11.9). In the critical portion of the normal allele, the sequence of bases is TAGTAGAAA, whereas in the cystic fibrosis allele, the sequence is TAGTAA. Thus cystic fibrosis is caused by an allele that is missing three bases (the sequence GAA) found in the normal allele (Figure 12.9a). The cystic fibrosis allele alters a protein that regulates the movement of chloride ions into and out of cells; this change causes cells in the lungs, digestive tract, and other parts of the body to become covered with thick, sticky secretions.

Other alleles that cause genetic disorders differ from normal alleles by a larger number of bases. Huntington disease (HD) occurs when a person inherits a dominant allele that codes for an abnormal version of the protein usually produced by the *HD* gene. It turns out that many different alleles can cause Huntington disease, all sharing a common feature: each has extra copies of the sequence GTC inserted near the beginning of the *HD* gene (Figure 12.9b). Thus, instead of a single *HD* allele, there is a family of *HD* alleles that can cause the disorder.

Many other genetic disorders can be caused by more than one allele. In such cases, the particular allele a person inherits can be important because it can affect the severity of the disorder. For example, there are over 800 known alleles of the gene *BRCA1* that can cause inherited forms of breast cancer. If a woman with one of these alleles develops breast cancer, the severity of the disease—and hence the type of treatment a doctor might recommend—is influenced by which of the 800 *BRCA1* breast cancer alleles she possesses. Similarly, a person who inherits an *HD* allele with many extra copies of the sequence GTC is likely to get the disease at an earlier age than someone whose *HD* allele has relatively few extra copies of the sequence GTC.

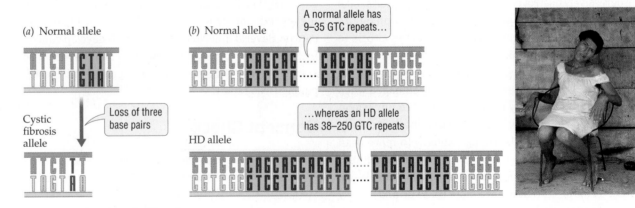

Figure 12.9 Two Deadly Disorders
(*a*) Cystic fibrosis is a genetic disorder caused by a mutation in which three bases (the sequence GAA) are missing. (*b*) Huntington disease (HD) is a lethal neurological disorder caused by a family of mutant alleles in which 3 to 215 extra copies of the sequence GTC are inserted near the beginning of the *HD* gene.

Chernobyl Wildlife Baffles Biologists

By Douglas Birch

Many assumed the 1986 meltdown of one reactor, and the release of hundreds of tonnes of radioactive material, would turn much of the 2,850 square-kilometre evacuated area around Chernobyl into a nuclear dead zone. It certainly doesn't look like one today. Wildlife has returned despite radiation levels in much of the evacuated zone that remain 10 to 100 times higher than background levels . . .

Biologist Robert Baker of Texas Tech University was one of the first Western scientists to report that Chernobyl had become a wildlife haven. He says the mice and other rodents he has studied at Chernobyl since the early 1990s have shown remarkable tolerance for elevated radiation levels. But Timothy Mousseau of the University of South Carolina, a biologist who studies barn swallows at Chernobyl, says a high proportion of the birds he and his colleagues have examined suffer from radiation-induced sickness and genetic damage. Survival rates are dramatically lower for those living in the most contaminated areas . . . Roughly one-third of 248 Chernobyl nestlings studied were found to have ill-formed beaks, albino feathers, bent tail feathers and other malformations . . . In other studies, Mousseau and his colleagues have found increased genetic damage, reduced reproductive rates and what he calls "dramatically" higher mortality rates for birds living near Chernobyl.

I n the early hours of April 26, 1986, reactor 4 at the Chernobyl Nuclear Power Plant in Ukraine exploded, releasing a plume of steam and smoke heavily laced with radioactive atoms. The accident was caused by a combination of poor reactor design, miscommunication between the day and night shifts, and a series of mistakes made by an inexperienced crew. Within minutes of the explosion, the crew had been exposed to more than 100 times the lethal dose of radiation. Radiation-measuring devices were not working, so the firefighters called in were unaware they were being exposed to deadly radiation as they tried to extinguish what they thought was an ordinary fire. Not until the next day, when radioactive particles began falling out of the air above Scandinavia, did the world learn that a terrible nuclear accident had taken place in the Ukraine, then part of the Soviet Union.

Forty-seven reactor workers and firefighters died from radiation poisoning within 3 weeks of the accident. Nine children who lived close by died of thyroid cancer in the months that followed. Released in the explosion were large quantities of radioactive iodine, which tends to build up in the thyroid gland, and strontium, which displaces calcium atoms and accumulates in bones.

Nearly 5 million people live in the contaminated zone, which extends from the Ukraine into the now-independent nations of Belarus and Russia. Scientists have been evaluating this population for decreased fertility, birth defects, and increased rates of cancer and other health conditions, but in more than two decades of monitoring, no direct links have been found between any such ill effects and exposure to radiation from the Chernobyl explosion. Some believe this is because these populations were exposed to relatively low levels of radiation, which the body's robust DNA repair system can handle. Others say certain consequences of DNA mutations, such as increased incidence of cancer, will not become apparent until many years have passed. In the near-absence of human activity in the Chernobyl exclusion zone, wildlife has made an unexpected resurgence. The dead pine forests around the power plant have re-seeded, and animals that are uncommon elsewhere, such as the European wolf, have increased dramatically.

Evaluating the News

1. What is radiation poisoning? Explain how DNA damage contributes to the symptoms of radiation poisoning. Explain how DNA damage contributed to the deaths of the nine children who were the only civilian victims of the Chernobyl nuclear plant explosion.
2. In an effort to reduce its dependence on Russian oil and coal, Ukraine is considering reactivating old nuclear reactors and building new ones to obtain electricity. Nuclear power plants have also been advocated as sources of energy that do not release carbon dioxide and other greenhouse gases. Based on the lessons learned from the Chernobyl accident, evaluate the claim that the threat to humans and the environment from a similar accident in the future would be limited, as long as nuclear reactors are located more than 10 km from human populations.

SOURCE: *Toronto Star*, Toronto, Ontario, June 8, 2007.

Chapter Review

Summary

12.1 The Search for the Genetic Material

- Geneticists initially thought that protein was the genetic material. A few landmark experiments showed that this initial view was wrong and that DNA, not protein, is the genetic material.
- Griffith's experiment showed that harmless strain R bacteria could be transformed into deadly strain S bacteria when exposed to heat-killed strain S bacteria.
- Avery and his colleagues showed that only the DNA from heat-killed strain S bacteria was able to transform strain R bacteria into strain S bacteria.
- Hershey and Chase demonstrated that the DNA of a virus, not its proteins, was responsible for taking over a bacterial cell and producing the next generation of viruses.

12.2 The Three-Dimensional Structure of DNA

- In 1953, James Watson and Francis Crick determined that DNA is a double helix formed by two long strands of covalently bonded nucleotides.
- The two strands of nucleotides are held together by hydrogen bonds between the nucleotides' nitrogen-containing bases: adenine, cytosine, guanine, and thymine.
- The hydrogen-bonded nucleotides follow these base-pairing rules: An adenine on one strand pairs only with a thymine on the other strand; a cytosine on one strand pairs only with a guanine on the other strand.
- The base-pairing rules allow each strand to serve as a template from which the other (complementary) strand can be duplicated.
- The sequence of bases in DNA, which differs among species and among individuals within a species, is the basis of inherited variation.

12.3 How DNA Is Replicated

- Because of the base-pairing rules (A pairs only with T, and C pairs only with G), each strand of DNA contains the information needed to duplicate the complementary strand.
- A complex set of enzymes and other proteins guides the replication of DNA; the primary enzyme involved is DNA polymerase. To replicate DNA, the protein complexes must first break the hydrogen bonds connecting the two nucleotide strands.
- Breaking of the hydrogen bonds causes the two strands to unwind and separate. Each of these strands is then used as a template from which to build a new strand of DNA.
- DNA replication is semiconservative: it produces two copies of the original DNA molecule, each composed of one old strand (from the parent DNA molecule) and one newly synthesized daughter strand.

12.4 Repairing Replication Errors and Damaged DNA

- On rare occasions, mistakes occur during DNA replication. Most mistakes in the copying process are corrected, either immediately by "proofreading" or by later correction of mismatch errors.
- Uncorrected mistakes in DNA replication are one source of mutations.
- The DNA in our cells is altered thousands of times every day by accidental chemical changes, and many of our cells are also exposed to DNA-damaging agents, such as chemical mutagens and radiation, that are common in our environment. If none of the DNA damage in an organism's cells is repaired, the accumulated mutations would likely kill the individual.
- Replication errors and damage to DNA are fixed by a complex set of DNA repair proteins.
- Several inherited genetic disorders result from the failure of DNA repair proteins to work properly.

◉ Review and Application of Key Concepts

1. Summarize the key findings of the landmark experiments that helped to convince geneticists that genes are composed of DNA.

2. Draw a diagram that shows the three main components of a nucleotide found in a DNA molecule. There are four different nucleotides in DNA, each consisting of three main chemical components. Of the three components, which two are identical in the four nucleotides, and which differs from one nucleotide to another?

3. What type of chemical bond holds the two strands of the DNA molecule together?

4. A gene has two codominant alleles, A^1 and A^2. Each allele produces a different, but related version of a protein found on the surface of a type of white blood cell. In physical terms, each allele is a segment of DNA. Explain how the DNA of one allele might differ from the DNA of the other allele. Describe the possible effects of these differences in the two DNA segments.

5. Explain why the structure of DNA proposed by Watson and Crick suggested a way DNA could be replicated.

6. Describe the relationship between the bases in DNA, mutations in a protein-coding gene, and alleles that cause human genetic disorders.

7. Summarize how DNA repair works and why the repair mechanisms are essential for cells and whole organisms to function normally.

Key Terms

apoptosis (p. 261)
base pair (p. 256)
complementary strand (p. 256)
DNA polymerase (p. 257)
DNA repair (p. 260)
DNA replication (p. 257)
DNA sequence (p. 256)
double helix (p. 255)

mismatch error (p. 259)
mutagen (p. 259)
mutation (p. 259)
nitrogenous base (p. 255)
semiconservative replication
   (p. 257)
transformation (p. 253)

Self-Quiz

1. The base-pairing rules for DNA state that
   a. any combination of bases is allowed.
   b. T pairs with C, A pairs with G.
   c. A pairs with T, C pairs with G.
   d. C pairs with A, T pairs with G.

2. DNA replication results in
   a. two DNA molecules, one with two old strands and one with two new strands.
   b. two DNA molecules, each of which has two new strands.
   c. two DNA molecules, each of which has one old strand and one new strand.
   d. none of the above

3. Experiments performed by Oswald Avery and colleagues, in which they transformed harmless strain R with material extracted from strain S, helped demonstrate that
   a. protein, not DNA, transformed bacteria.
   b. DNA, not protein, transformed bacteria.
   c. carbohydrates, not protein, transformed bacteria.
   d. either DNA by itself or protein by itself transformed bacteria.

4. The DNA of cells is damaged
   a. thousands of times per day.
   b. by collisions with other molecules, chemical accidents, and radiation.
   c. not very often and only by radiation.
   d. both a and b

5. The DNA of different species differs in
   a. the sequence of bases.
   b. the base-pairing rules.
   c. the number of nucleotide strands.
   d. the location of the sugar-phosphate portion of the DNA molecule.

6. If a strand of DNA has the sequence CGGTATATC then the complementary strand of DNA has the sequence
   a. ATTCGCGCA.
   b. GCCCGCGCTT.
   c. GCCATATAG.
   d. TAACGCGCT.

7. Hershey and Chase conducted experiments with viruses that attack bacteria. They found that
   a. sulfur atoms are found in DNA.
   b. phosphorus atoms are found in protein.
   c. radioactively labeled protein entered bacterial cells.
   d. radioactively labeled DNA entered bacterial cells.

8. Mutation
   a. can produce new alleles.
   b. can be harmful, beneficial, or neutral.
   c. is a change in an organism's DNA sequence.
   d. all of the above

CHAPTER **13** From Gene to Protein

Key Concepts

◉ Most genes contain instructions for building proteins. The DNA sequence of such a gene encodes the amino acid sequence of its protein product.

◉ Some genes encode ribonucleic acid (RNA) molecules as their final product.

◉ The flow of information from gene to protein requires two steps: transcription and translation.

◉ In eukaryotic cells, transcription occurs in the nucleus and produces a messenger RNA (mRNA) version of the information stored in the gene. The mRNA moves from the nucleus to the cytoplasm, where it is used to guide the manufacture of a protein.

◉ Translation occurs in the cytoplasm and converts the sequence of bases in an mRNA molecule to the sequence of amino acids in a protein.

◉ Gene mutations can alter the sequence of amino acids in a gene's protein product. Such changes, in turn, can alter the protein's function. Although changes in protein function are often harmful, occasionally they benefit the organism.

Finding the Messenger and Breaking the Code

We live in a global economy. The headquarters of a corporation may be located in one country—say, Germany—but the company's factories may be located elsewhere—say, the United States. Immediately a problem arises: decisions made in Germany need to be communicated to employees in the United States. This problem is easy to solve: A message must be sent from one location to the other. In addition, the message must be translated from German, the language in which the decision was made, to English, the language in which the decision must be implemented.

Eukaryotic cells face similar challenges. A gene works by providing the information needed to make a specific protein, and proteins produce phenotypes. Whereas genes are located in the nucleus of the cell, the proteins they specify are made on ribosomes, which are located outside of the nucleus, in the cytoplasm. So a gene must control the construction of a protein from a distance. Like our imaginary corporate headquarters, the gene does this by sending a message. **What is the chemical messenger that carries the gene's instructions from the nucleus to the ribosomes? And once the message reaches the ribosomes, how do the ribosomes "read" it? We know that poor communication within a business corporation can be disastrous, but what are the consequences for the cell when the information stored in a gene is garbled?**

In this chapter we will explore the means by which information contained in genes is converted from the "language" of DNA (composed of a molecular alphabet of four nitrogenous bases) to the "language" of proteins (composed of different types of amino acids). We will see that much of the time information flows from gene to protein with flawless precision, but serious genetic disorders may result if the DNA code is corrupted by mutations.

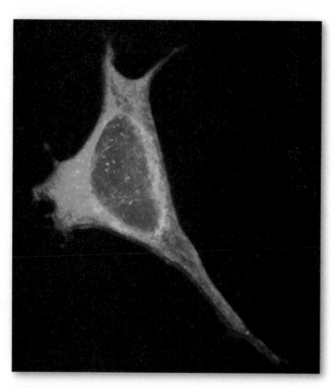

In Eukaryotes, Genetic Information Is Stored in the Nucleus

Most genes code for proteins, which are made in the cytoplasm. In eukaryotes, the information in a gene must be copied from the DNA and sent out from the nucleus to the cytoplasm. A special type of nucleic acid, called messenger RNA (mRNA), carries information to the protein-manufacturing machinery in the cytoplasm. The brain cell (neuron) seen here was prepared using a special technique to reveal the presence of a cytoskeletal protein called tubulin (green). The gene coding for this protein resides in the nucleus (blue).

Chapters 10 through 12 have described how genes are inherited, where they are located (on chromosomes), and what they are made of (DNA). But we have yet to describe how genes work. How do genes store the information needed to build their final products, RNA and proteins? How do RNA molecules direct the manufacture of proteins?

Knowing how genes work can help us understand how mutations produce new phenotypes, including disease phenotypes. We begin this chapter by describing how genetic information is encoded in genes and how the cell uses that information to build proteins. We then describe how a change in a gene can change an organism's phenotype. At the end of the chapter, we apply what we've learned by focusing on the effects of alleles that cause two human genetic disorders, sickle-cell anemia and Huntington disease. Our discussion of how cells use the information stored in genes to build proteins focuses mostly on eukaryotes; except where noted, events are similar in prokaryotes.

13.1 How Genes Work

Learn about the flow of genetic information in a eukaryotic cell. ▶‖ 13.1

Proteins are essential to life. They are used by cells and organisms in many ways: Some provide structural support, others transport materials through the body, still others defend against disease-causing organisms. In addition, the many chemical reactions on which life depends are controlled by a crucial group of proteins, the enzymes. Enzymes and other proteins influence so many features of the organism that they, along with the organism's internal and external environment, have the largest impact on an organism's phenotype.

How do genes affect the phenotype of an organism? Early clues came at the beginning of the twentieth century from the work of British physician Archibald Garrod, who studied several inherited disorders in which metabolism was disrupted. In 1902, he argued that these metabolic disorders were caused by an inability of the body to produce specific enzymes. Garrod was particularly interested in alkaptonuria, a condition in which the urine of otherwise healthy infants turns black when exposed to air. He proposed that infants with alkaptonuria have a defective version of an enzyme that, in its normal form, breaks down the substance that causes urine from these infants to turn black. But Garrod did not stop there; he and his collaborator, William Bateson, went on to suggest that in general, genes are responsible for the production of enzymes.

Genes contain information for the synthesis of RNA molecules

Garrod and Bateson were on the right track, but they were not entirely correct: genes control the production of all proteins, not just enzymes. Furthermore, although most genes contain instructions for building particular proteins, some genes do not directly specify proteins. Rather, these genes contain instructions for building any one of a variety of ribonucleic acid (RNA) molecules that do not code for proteins, although some of them help in the construction of proteins encoded by other genes. This forces us to modify the simpler definition of a gene we presented in Chapter 10. We now redefine a **gene** as a DNA sequence that contains information for the synthesis of one of several types of RNA molecules, most of which are used in the production of proteins.

All genes make an RNA molecule as their initial product. RNA and DNA share a number of structural similarities as well as several important differences. Both are nucleic acids consisting of nucleotides covalently bonded to one another. But whereas DNA molecules are double-stranded, the various types of RNA molecules are all single-stranded; overall, the structure of an RNA molecule is similar to the structure of a single strand of DNA. The two strands in a DNA molecule are twisted into a double helix, while a single RNA molecule may fold back on itself to assume a variety of three-dimensional shapes. As in DNA, each nucleotide in RNA is composed of a sugar, a phosphate group, and one of four nitrogen-containing bases (Figure 13.1). However, the nucleotides in RNA and DNA differ in two respects. First, RNA uses the sugar ribose, whereas DNA contains the sugar deoxyribose. Second, in RNA, the base uracil (U) replaces the base thymine (T), which is found only in DNA. The other three bases—adenine (A), cytosine (C), and guanine (G)—are the same in RNA and DNA. In general, RNA is chemically less stable than DNA, and most RNA molecules in the cell have a limited life in the cytoplasm. In Chapter 14 we describe how a cell regulates how much of which type of RNA is made and how long a particular RNA survives in the cytoplasm. As the permanent store of genetic information, the DNA in the nucleus of most cells is extremely stable, being destroyed only if the cell itself is destined to die soon.

Three types of RNA are involved in the production of proteins

The nucleic acids DNA and RNA play key roles in the construction of proteins. Several types of RNA, as well as many enzymes and other proteins, are required for the

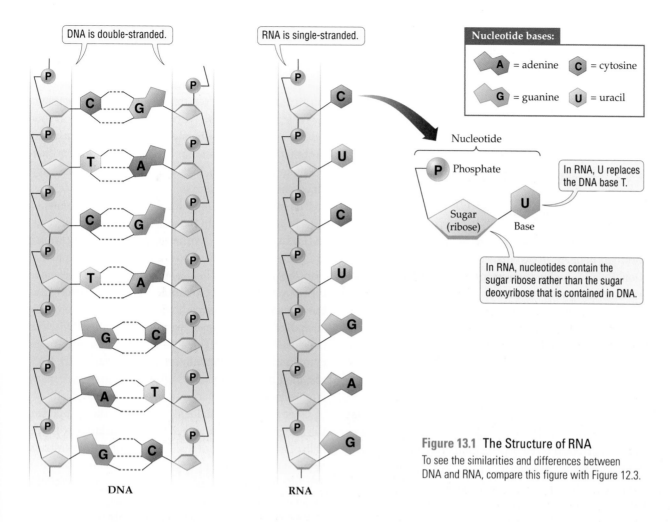

DNA is double-stranded.

RNA is single-stranded.

Nucleotide bases:

A = adenine   C = cytosine

G = guanine   U = uracil

Nucleotide

P Phosphate

In RNA, U replaces the DNA base T.

Sugar (ribose)   Base   U

In RNA, nucleotides contain the sugar ribose rather than the sugar deoxyribose that is contained in DNA.

DNA

RNA

Figure 13.1 The Structure of RNA
To see the similarities and differences between DNA and RNA, compare this figure with Figure 12.3.

cell to make proteins. As already described, DNA directs the production of all these essential molecules, so the information for manufacturing a specific protein comes ultimately from DNA.

Cells use three main types of RNA molecules to construct proteins: **messenger RNA (mRNA)**, **ribosomal RNA (rRNA)**, and **transfer RNA (tRNA)**. The function of each of these three kinds of RNA is defined in Table 13.1

Table 13.1

RNA Molecules and Their Functions

Type of RNA	Function	Shape
Messenger RNA (mRNA)	Specifies the order of amino acids in a protein using a series of three-base codons, where different amino acids are specified by particular codons	
Ribosomal RNA (rRNA)	As a major component of ribosomes, rRNA assists in making the covalent bonds that link amino acids together to make a protein	
Transfer RNA (tRNA)	Transports the correct amino acid to the ribosome, based on the information encoded in the mRNA; contains a three-base anticodon that pairs with a complementary codon revealed in the mRNA	

and discussed in more detail in the sections that follow. Cells also produce several other types of RNA that affect the production of proteins, but we will not discuss them in this chapter.

13.2 How Genes Guide the Manufacture of Proteins

In both prokaryotes and eukaryotes, the production of proteins happens in two steps: transcription and translation. Briefly, during **transcription**, an mRNA molecule is made using the information in the DNA sequence of a gene. The base sequence of that mRNA molecule specifies the amino acid sequence of a protein. During **translation**, the information in the mRNA molecule is used to synthesize the protein. This process requires ribosomes, which are composed of rRNA and proteins, and it also relies on many different types of tRNA molecules.

In the case of some genes, for example those that encode rRNAs or tRNAs, transcription of the gene produces an RNA molecule as its final product; there is no translation step. For most of this chapter, however, we shall focus on the transcription and translation of genes that encode proteins.

Test your knowledge of protein synthesis. ▶❚❚ 13.2

Before we discuss transcription and translation in detail, let's consider how genes work from the perspective of information flow. We will describe how eukaryotic cells use the information stored in genes to synthesize proteins. Events are similar in prokaryotes except that, because prokaryotes lack a nucleus, both genes and ribosomes are located in the cytoplasm.

To make a protein in eukaryotes, the information in a gene must be sent from the gene, located in the nucleus, to the site of protein synthesis, the cytoplasm (Figure 13.2). Messenger RNA is the intermediary molecule that transfers the information from the gene in the nucleus to the ribosomes in the cytoplasm. During gene transcription, the sequence of bases in one DNA strand (the template strand) of a gene is copied (transcribed) into mRNA. The information stored in the gene is not only transmitted to the cytoplasm, it is substantially amplified, through gene transcription, as each template DNA typically generates many hundreds of mRNA molecules when it is transcribed.

Once the mRNA molecule arrives in the cytoplasm, the information it contains must be translated, with the help of ribosomes, from the language of mRNA (nitrogenous bases) to the language of proteins (amino acids). The information is translated at the ribosomes by tRNA molecules. To do this, a three-base sequence on each tRNA molecule binds to its complementary sequence on the mRNA by forming hydrogen bonds with it; one end of the tRNA molecule carries the particular amino acid specified by the three-base sequence in the mRNA. We will examine the binding rules involved in this process shortly. For now, it is enough to know that the base sequence of the mRNA dictates which tRNAs are bound, and therefore which amino acids are delivered at the ribosomes. The ribosomes hold the mRNA, position the incoming tRNAs, and also join the amino acids delivered by the tRNAs to synthesize a protein with the exact sequence of amino acids called for by the gene.

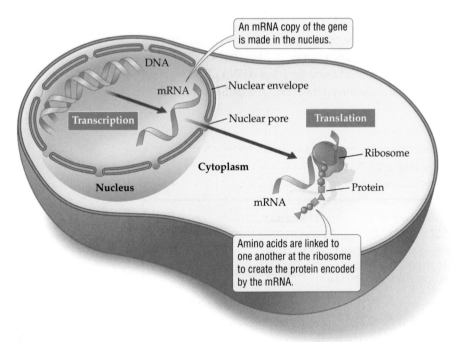

Figure 13.2 The Flow of Genetic Information in a Eukaryotic Cell
Genetic information flows from DNA to RNA to protein in two steps, transcription and translation. Transcription produces an mRNA molecule, which is then transported to the cytoplasm, where translation occurs, and the protein is made with the help of ribosomes. Different amino acids in the protein being constructed at the ribosome are represented by different colors and shapes.

13.3 Gene Transcription: Information Flow from DNA to RNA

Gene transcription is similar to DNA replication in that one strand of DNA is used as a template from which a new strand—in this case, a strand of mRNA—is formed. However, transcription differs from DNA replication in

Figure 13.3 An Overview of Transcription

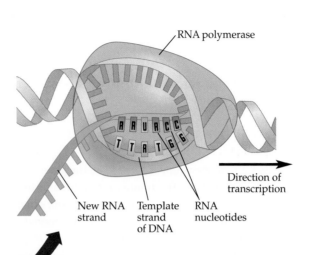

1 Transcription begins when RNA polymerase binds to the promoter.

RNA polymerase

DNA of gene

Promoter (in red)

Terminator (in red)

Direction of transcription

2 An mRNA molecule is produced as RNA polymerase moves down the template strand of DNA.

3 Transcription ends when RNA polymerase reaches the terminator.

RNA polymerase

Direction of transcription

New RNA strand   Template strand of DNA   RNA nucleotides

three important ways. First, a different enzyme guides the process: the key enzyme in DNA replication is DNA polymerase, while the key enzyme in transcription is **RNA polymerase**. Second, the entire DNA molecule is duplicated in DNA replication, but in the transcription of a particular gene, only a small portion of the chromosome is copied into RNA. Finally, whereas DNA replication produces a double-stranded DNA molecule, transcription produces a single-stranded RNA molecule

that is complementary to one strand (the template strand) of the DNA double helix.

Transcription of a gene begins when the enzyme RNA polymerase binds to a segment of DNA near the beginning of the gene, called a **promoter**. Although the promoters of different genes vary in size and sequence, all contain several specific sequences of 6 to 10 bases that enable the RNA polymerase to recognize and bind to them. Once bound to the promoter, the RNA polymerase unwinds the DNA double helix at the beginning of the gene. This separates a short portion of the two strands. Then the enzyme begins to construct an mRNA molecule (Figure 13.3). Only one of the two DNA strands is used as a template, and this strand is called the **template strand** (lower strand in Figure 13.3). If the opposite strand (upper strand in Figure 13.3) were used as template, it would result in a completely different sequence of amino acids, and hence a different protein. How does the RNA polymerase "choose" which strand to use as template? The RNA polymerase binds to the promoter in a specific orientation, and that determines which strand it will be able to "read." The location and orientation of the promoter sequence guides the binding of the RNA polymerase, so ultimately it is the positioning of the promoter that specifies which strand will serve as the template.

The four kinds of bases in RNA pair with the four kinds of bases in DNA according to specific rules: A in RNA pairs with T in DNA, C in RNA pairs with G in DNA, G in RNA pairs with C in DNA, and U in RNA pairs

Helpful to know

The names of proteins that are enzymes usually end with *ase,* while many names of proteins that are not enzymes end with *in.* For example, the protein RNA polymerase is an enzyme, but the protein huntingtin is not. (We encounter both of these molecules in this chapter.)

with A in DNA. These base-pairing rules determine the sequence of bases in the mRNA molecule that is made from a DNA template. For example, if the sequence of the DNA template is

TTATGGCACCG,

then an mRNA molecule synthesized from this template will have the sequence

AAUACCGUGGC.

As an RNA polymerase moves away from the gene promoter and travels down the template strand, another RNA polymerase can bind at the promoter and start synthesizing an mRNA "fast on the heels" of the previous RNA polymerase. At any given time, therefore, many RNA polymerases can be traveling down a DNA template, each synthesizing its own mRNA at the rate of 60 bases per second in human cells. The death cap mushroom produces an extremely toxic substance, called α-amanitin, that binds to RNA polymerase and reduces its travel speed to just 2 or 3 bases per second. Eating just one death cap mushroom can kill a person in about 10 days, primarily

because the liver and pancreas fail. The cells in these organs are highly active in gene transcription, pouring out a great variety of enzymes and other proteins that are vital for life. A slowdown of transcriptional activity in these organs has devastating consequences. The lethal effect of this toxin underscores the value of mRNA. These molecules not only relay the information stored in DNA, they represent an amplification of the information because hundreds of mRNA molecules can be made from a single gene template and these mRNAs in turn direct the manufacture of many thousands of protein molecules.

In prokaryotes, synthesis of an mRNA molecule from the DNA template continues until the RNA polymerase reaches a special sequence of bases called a **terminator**. When the terminator sequence is copied into mRNA, it generates three-dimensional shapes, known as hairpins, in the mRNA sequence. The hairpins destabilize the RNA polymerase and cause it to drop off the template. Transcription ends at this point and the newly formed mRNA molecule separates from its DNA template. Transcription termination in eukaryotes is more complicated than in prokaryotes and is linked to RNA processing, an elaborate sequence of steps that modifies RNA and gets underway even as the RNA is being transcribed. One of the last steps in this process destabilizes the RNA polymerase, causing it to fall off the template DNA.

RNA splicing is another key event in RNA processing. Most freshly transcribed mRNAs in eukaryotic cells have "extra" base sequences embedded in them that do not carry protein-building information. These internal sequences are called **introns** (Figure 13.4); the base sequences that do code for the amino acid sequence of the protein are called **exons**. Most newly made mRNA in a eukaryote is a patchwork of exons and introns, and is therefore like an uncut video film with "extra footage." The introns must be snipped out, and the remaining pieces of mRNA re-joined, in a process known as RNA splicing. Only properly spliced mRNA leaves the nucleus. On arriving in the cytoplasm, this mature mRNA carries the "final cut" of information, which is translated into a very specific protein.

After transcription and processing, each mature mRNA molecule is transported out of the nucleus through a nuclear pore. In the cytoplasm, mRNA become attached to ribosomes, where the protein speci-

Amanita phalloides,
the death cap mushroom

Explore the removal of introns from RNA. ▶❚❚ 13.5

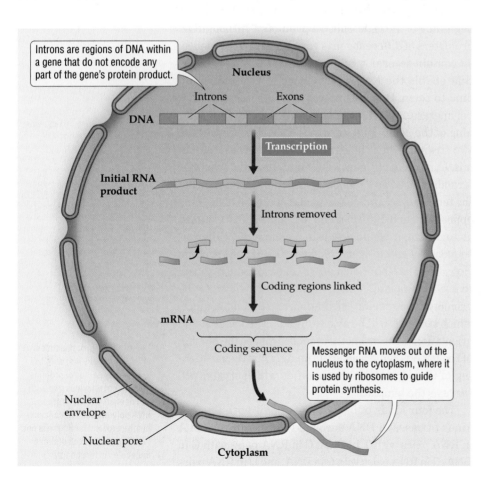

Introns are regions of DNA within a gene that do not encode any part of the gene's protein product.

Nucleus

Introns   Exons

DNA

Transcription

Initial RNA product

Introns removed

Coding regions linked

mRNA

Coding sequence

Messenger RNA moves out of the nucleus to the cytoplasm, where it is used by ribosomes to guide protein synthesis.

Nuclear envelope

Nuclear pore

Cytoplasm

Figure 13.4 Removal of Introns in Eukaryotic Cells
Most eukaryotic genes contain both coding sequences (exons) and noncoding sequences (introns). Before the mRNA transcribed from such genes can be exported to the cytoplasm, enzymes in the nucleus must remove the introns and link the remaining exons to one another.

fied by the gene will be built. In this way the information for making a protein (encoded in the DNA) will be carried from one part of the cell (the nucleus) to another (the cytoplasm) by a messenger molecule (mRNA).

Concept Check

1. Compare the chemical structures of RNA and DNA. Which is more stable chemically and how is that consistent with its function?
2. What is meant by gene transcription?
3. The template strand of a gene has the base sequence TGAGAAGACCAGGGTTGT. Write the sequence of RNA transcribed from this DNA, assuming RNA polymerase travels from left to right on this strand.

13.4 The Genetic Code

The information in a gene is encoded in its sequence of bases. As we learned in the previous section, the gene's DNA sequence is used as a template to produce an mRNA molecule. Recall that the final products of most genes are proteins, and that proteins consist of one or more folded strings of amino acids. How does mRNA encode, or specify, the sequence of amino acids in a protein?

The information in an mRNA molecule is "read" by ribosomes in sets of three bases, and each unique sequence of three bases is called a **codon**. Since there are 64 different ways in which we can arrange four bases to create a three-base sequence, there are 64 possible codons. If there were a language with only four letters (A, U, C, and G), there would be 64 different words in this language, if only three-letter words were allowed. The **genetic code** refers to the information specified by each of the 64 possible codons (the "meaning" of every word in the language). The genetic code is shown in its entirety in Figure 13.5. Most of the 64 codons specify a particular amino acid, but some act as signposts that communicate to the ribosomes where they should start or stop reading the mRNA. A few amino acids are specified by only one codon; for example, there is only one codon (UGG) that stands for the amino acid tryptophan. Other amino acids are specified by anywhere from two to six different codons.

When reading the code, the cell begins at a fixed starting point on an mRNA molecule, called a **start codon** (the codon AUG), and ends at one of several **stop codons** (such as UGA or UAA). By beginning at a fixed point, the cell ensures that the message from the gene is read precisely the same way every time. The start codon specifies the amino acid methionine (look it up in Figure 13.5), which is why most proteins inside a cell have this amino acid at their "starting point," the portion of the protein that was translated first.

To examine how the rest of the amino acid–specifying codons are read, let's consider the example shown in Figure 13.6, which shows a portion of an mRNA molecule with the base sequence UUCACUCAG. Because the mRNA code is read in sets of three bases, the first codon (UUC) specifies one amino acid (phenylalanine, abbreviated as phe), the next codon (ACU) specifies a second amino acid (threonine, abbreviated as thr), and the third codon (CAG) specifies a third amino acid (glutamine, abbreviated as gln). Next, let's consider the crucial role played by the start codon in establishing which trio of bases is interpreted as a codon by the ribosomal–tRNA machinery. Use Figure 13.5 to determine the amino acid sequence that would result if the sequence UUCACUCAG in Figure 13.6 were read in codons that began with the *second* U, not the first. If the code were read starting with the second U, (UUCACUCAG), we would get a very different protein chain: one containing serine, followed by a leucine (ser, leu . . .), instead of one containing phenylalanine, followed by a threonine and a glutamine (phe, thr, gln . . .). The bases that follow the start codon (AUG) are read consecutively, with each three-base sequence being read as one codon. The start codon therefore establishes the correct order, or reading frame, by which the three-letter language of mRNA is translated into protein.

Concept Check Answers

1. RNA is single-stranded. It contains ribose and the bases A, G, C, and U. DNA is double-stranded. It contains deoxyribose and A, G, C, and T. DNA is more stable, a property it must have to serve as the storehouse of genetic information.
2. Gene transcription is the copying of genetic information from DNA to RNA, by RNA polymerase, using one strand of the DNA as a template.
3. The corresponding RNA sequence, written from left to right, is ACUCUUCUGGUCCCAACA.

Build a protein from mRNA. 13.6

Figure 13.5 The Genetic Code

UAA, UAG, and UGA do not code for an amino acid. Translation stops when these codons are reached.

Like arginine, most amino acids are specified by more than one codon.

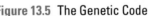

First letter of codon	Second letter: U	Second letter: C	Second letter: A	Second letter: G	Third letter of codon
U	UUU, UUC Phenylalanine	UCU Serine	UAU, UAC Tyrosine	UGU, UGC Cysteine	U, C
U	UUA, UUG Leucine	UCC, UCA, UCG Serine	UAA Stop codon; UAG Stop codon	UGA Stop codon; UGG Tryptophan	A, G
C	CUU, CUC, CUA, CUG Leucine	CCU, CCC, CCA, CCG Proline	CAU, CAC Histidine	CGU, CGC, CGA, CGG Arginine	U, C
C			CAA, CAG Glutamine		A, G
A	AUU, AUC, AUA Isoleucine	ACU, ACC, ACA, ACG Threonine	AAU, AAC Asparagine	AGU, AGC Serine	U, C
A	AUG Methionine; start codon		AAA, AAG Lysine	AGA, AGG Arginine	A, G
G	GUU, GUC, GUA, GUG Valine	GCU, GCC, GCA, GCG Alanine	GAU, GAC Aspartate	GGU, GGC, GGA, GGG Glycine	U, C
G			GAA, GAG Glutamate		A, G

Figure 13.6 How Cells Use the Genetic Code

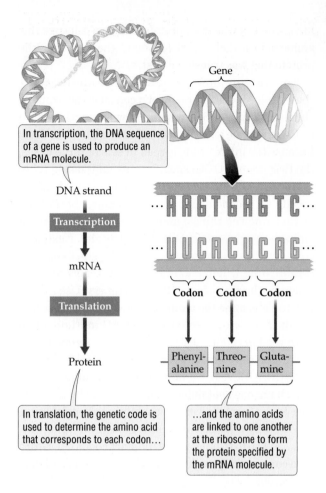

In transcription, the DNA sequence of a gene is used to produce an mRNA molecule.

Gene

DNA strand

Transcription

mRNA

Translation

Protein

...AAGTGAGTC...

...UUCACUCAG...

Codon Codon Codon

Phenyl-alanine Threo-nine Gluta-mine

In translation, the genetic code is used to determine the amino acid that corresponds to each codon...

...and the amino acids are linked to one another at the ribosome to form the protein specified by the mRNA molecule.

are only 20 amino acids, so some of these codons specify the same amino acid. For example, there are six different codons that specifiy the amino acid serine. Third, the code is virtually universal: nearly all organisms on Earth use the same code, a feature that illustrates the common descent of all organisms. The discovery of the genetic code and its near universality revolutionized our understanding of how genes work and helped pave the way for what is now a thriving biotech industry (see "Biology on the Job" on page 276). There are a few minor exceptions to the universality of the genetic code: in certain species some of the 64 codons are read differently than they are in most other species.

13.5 Translation: Information Flow from mRNA to Protein

The genetic code provides the cell with the equivalent of a dictionary with which to transform the language of genes into the language of proteins. The conversion of a sequence of bases in mRNA to a sequence of amino acids in a protein is called translation.

Translation is the second major step in the process by which genes specify the manufacture of specific proteins (see Figure 13.2). It occurs at ribosomes, which are composed of more than 50 different proteins and several different strands of rRNA. Ribosomes link amino acids in a precise order to make a particular protein.

Another type of RNA, known as a tRNA, also plays a crucial role in the manufacture of proteins (protein synthesis). There are many types of tRNA, but they all have a similar structure, as shown in Figure 13.7. Each type of tRNA specializes in binding to one specific amino acid, and it recognizes and pairs with specific codons in the

The genetic code has several significant characteristics. First, the code is unambiguous: each codon specifies only one amino acid. Second, the code is redundant: several different codons may have the same "meaning," that is, they call for the same amino acid. There are four possible bases at each of the three positions of a codon, so there are a total of 64 codons ($4 \times 4 \times 4 = 64$). However, there

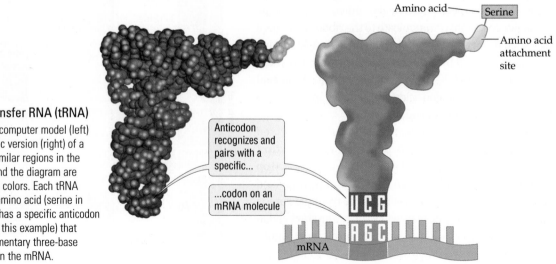

Figure 13.7 Transfer RNA (tRNA)

Shown here are a computer model (left) and a diagrammatic version (right) of a tRNA molecule. Similar regions in the computer model and the diagram are drawn in matching colors. Each tRNA carries a specific amino acid (serine in this example) and has a specific anticodon sequence (UCG, in this example) that binds to a complementary three-base sequence (codon) in the mRNA.

Amino acid — Serine

Amino acid attachment site

Anticodon recognizes and pairs with a specific...

...codon on an mRNA molecule

UCG

AGC

mRNA

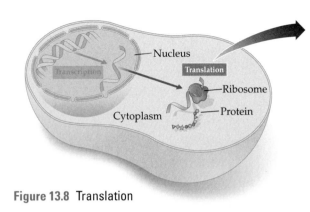

Figure 13.8 Translation

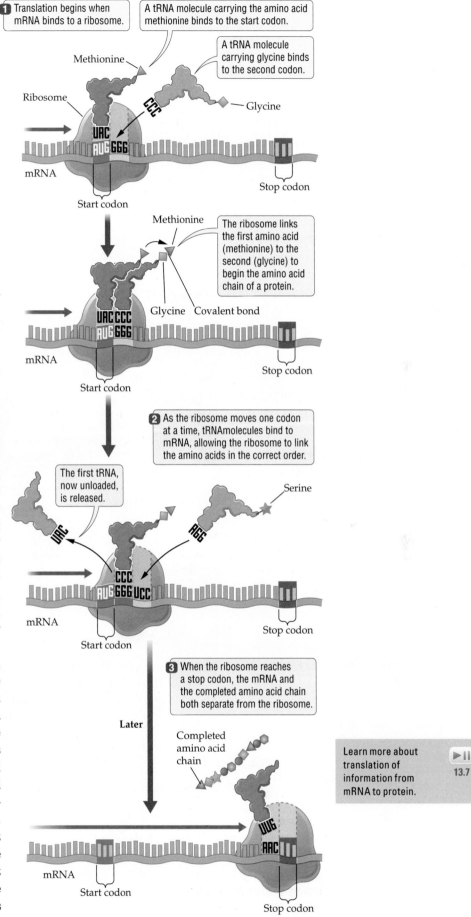

1 Translation begins when mRNA binds to a ribosome.

A tRNA molecule carrying the amino acid methionine binds to the start codon.

A tRNA molecule carrying glycine binds to the second codon.

Methionine

Glycine

Ribosome

mRNA

Start codon

Stop codon

Methionine

The ribosome links the first amino acid (methionine) to the second (glycine) to begin the amino acid chain of a protein.

Glycine   Covalent bond

mRNA

Start codon

Stop codon

2 As the ribosome moves one codon at a time, tRNAmolecules bind to mRNA, allowing the ribosome to link the amino acids in the correct order.

The first tRNA, now unloaded, is released.

Serine

mRNA

Start codon

Stop codon

3 When the ribosome reaches a stop codon, the mRNA and the completed amino acid chain both separate from the ribosome.

Later

Completed amino acid chain

mRNA

Start codon

Stop codon

mRNA. At one end, the tRNA molecule has a site where its special amino acid attaches. Each tRNA also has a sequence of three nitrogenous bases that collectively make up the **anticodon**. The three-base sequence of the anticodon determines which codons on the mRNA this tRNA can recognize through complementary base pairing. For example, the tRNA that carries the amino acid serine will recognize and pair with any AGC codon in an mRNA (Figure 13.7).

Some tRNAs can recognize more than one codon because the base at the third position of their anticodon can "wobble": it can pair with any one of two or three different bases in the codon. For example, one serine-bearing tRNA can base pair with either UC<u>U</u> or UC<u>C</u> in the mRNA, while another serine-bearing tRNA pairs with either UC<u>A</u> or UC<u>G</u>. A third tRNA recognizes the two other serine-specifying codons: AG<u>U</u> and AG<u>C</u> (consult the genetic code in Figure 13.5). This flexibility in the pairing between some anticodons and codons means that a cell doesn't need 61 different tRNAs, one for each of the 61 amino acid-specifying codons. In fact, most organisms have only about 40 different tRNAs, because many tRNA anticodons can recognize and pair with more than one codon in the mRNA.

For translation to occur, as Figure 13.8 illustrates, an mRNA molecule must first bind to a ribosome. The ribosomal machinery "scans" the mRNA until it finds a start codon, which is the first AUG codon in the mRNA sequence. Then the ribosomes recruit the appropriate tRNAs, as determined by the codons they encounter as they proceed to read the message in the mRNA. With all the necessary components held together in the required three-dimensional orientation, a special site on the ribosome facilitates the linking of one amino acid to another.

Translation begins when the tRNA molecule that specializes in carrying the amino acid methionine [meh-*THYE*-oh-neen] recognizes the AUG of the start codon and pairs with it. Next, another tRNA molecule that specializes in carrying another amino acid (in this

Learn more about translation of information from mRNA to protein.

▶ǁ
13.7

Biology on the Job

Investing in Biotech

Biotechnology or "biotech" refers to the use of biology to make many different types of products, including agricultural products, DNA fingerprinting kits, and medicinal drugs. As techniques for working with DNA and proteins improved in the 1960s and 1970s, new companies using these innovations sprang up around the world. Here we interview Russell T. Ray, a specialist in biotechnology companies and a managing partner for HLM Venture Partners, a venture capital firm.

What do you do during a typical day at work? I arrive at the office around 7:30 AM and spend the next several hours responding to e-mail messages from coworkers and from companies we might invest in. I often meet with the managers of companies seeking to raise investment capital. I spend a few hours each day reviewing the scientific literature and reading health care newsletters, many of which are published daily. I also call colleagues at other venture capital firms to compare notes on what they are seeing in the way of interesting investments, and to share information on projects that we might work on together.

How did you become interested in investing in biotechnology companies? Almost by chance—in my first year in Merrill Lynch's investment banking department, I was selected to work on a financing project for a plant biotechnology company. Prior to that, I had completed an MBA in finance, as well as a master of science in biology that focused on the territorial behavior in a species of Costa Rican hummingbird. My scientific training helped me to understand the newly emerging field of biotechnology. I have now been working in this area for 23 years and I still love it!

What do you look for in a successful business plan or idea? In evaluating a biotechnology or medical device company, we focus on several areas:

- *Intellectual property.* Does the company have patents that protect its technology and science from other companies? If not, this is usually a deal killer.
- *Stage of development.* For example, does a pharmaceutical company have drug candidates that have successfully passed through the early stages of development? If not, we are unlikely to invest due to the high risk of failure associated with products that have not progressed from animal testing to human clinical testing.

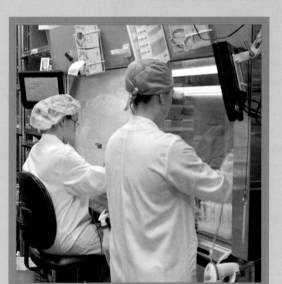

Technicians at a CBR System Cord blood storage facility in Tucson, Arizona.

- *Market analysis.* How big is the potential market? What are the competing products?
- *Quality of management.* Has management successfully done the job before? Who are the people behind the science?

Can you describe an example of an interesting company or a novel idea in the biotech industry? Take Cbr Systems, Inc. This company is the market leader in collecting, processing, and storing stem cells extracted from a newborn baby's umbilical cord. These cells are frozen for future use to provide the donor or a closely related family member with a stem cell infusion instead of a bone marrow transplant. The company has stored over 80,000 samples, and has used stored stem cells in 36 cases, with a success rate of 100 percent. The company receives a fee when the stem cells are collected from the umbilical cord, and an annual storage fee for 18 years. The company is profitable, and its revenues have been growing almost 120 percent per year for the past year or so.

How do you think the biotech industry is likely to change over the next five years? I hope that we will see the arrival of targeted medicines that match the patient's unique genotype.

What do you enjoy most about your job? I really enjoy working with bright entrepreneurs whose companies are on the cutting edge of science and medical technology. Each project I work on exposes me to new areas of science, so I'm constantly challenged to expand my horizons by learning more—something I love to do.

The Job in Context

Russell Ray and others like him straddle two worlds—they must understand both science and business. They invest in companies that take the information presented in this unit and apply it to solve real-world problems.

example, glycine), recognizes the second codon on the mRNA molecule (GGG) and pairs with it through its anticodon. The ribosome now forms a covalent bond between the first amino acid (methionine) and the second amino acid (glycine). At the same time that the bond between the first two amino acids is formed, the first tRNA (the one bound to AUG) releases its amino acid (methionine). This tRNA, now freed from the amino acid it had been carrying, is ejected from the mRNA–ribosome complex, and its place is taken by the next tRNA (the one carrying glycine). The ribosome is now ready to read the third codon in the mRNA, which is UCC in our example. The tRNA that carries serine specifically pairs with this codon, since it bears the complementary anticodon (AGG), and this serine-specific tRNA takes the place previously occupied by the glycine codon.

The ribosome now links the amino acid chain it has built so far (consisting of just methionine and glycine) to the newly delivered amino acid (serine), and the second tRNA (the glycine specialist) is now freed of its cargo and ejected from the ribosome–mRNA complex. This cycle continues: Each codon encountered by the ribosome is recognized by a specific tRNA, the one whose anticodon can pair with that codon, and the ribosome adds the amino acid delivered by this tRNA to the growing amino acid chain. Finally, a stop codon is reached. The amino acid chain cannot be extended any further because none of the tRNAs will recognize and pair with any of the three stop codons. At this point the mRNA molecule and the completed amino acid chain both separate from the ribosome. The new protein then folds into its compact, specific three-dimensional shape (discussed in Chapter 4).

13.6 The Effect of Mutations on Protein Synthesis

Recall that a mutation is a change in the sequence of an organism's DNA. As we've seen in the previous chapters of this unit, mutations range in extent from a change in the identity of a single base pair to the addition or deletion of one or more chromosomes.

How do mutations affect protein synthesis? In answering this question, we will focus on mutations that occur in the portions of a gene that code for parts of proteins (exons, that is), rather than on mutations that occur in introns or on the large-scale mutations that disrupt entire chromosomes.

Mutations can alter one or many bases in a gene's DNA sequence

There are three major types of mutations that can alter a gene's DNA sequence: substitutions, insertions, and deletions. For simplicity, we will describe **point mutations** first: those in which a single base is altered. We will then discuss mutations that involve changes in multiple bases.

In a **substitution** mutation, one base is substituted for another in the DNA sequence of the gene. In the substitution mutation shown in Figure 13.9, for example, the sequence of the gene is changed when a thymine (T) is replaced by a cytosine (C). As the figure shows, this particular change causes the substitution of one amino acid for another: When the TAA sequence in the DNA is changed to CAA, the mRNA codon is changed from AUU to GUU; GUU is recognized by the valine-carrying

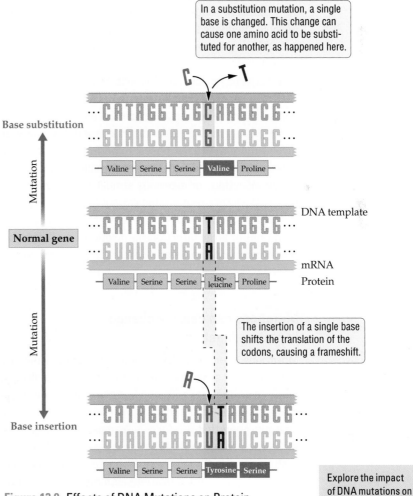

Figure 13.9 Effects of DNA Mutations on Protein Production

Two kinds of mutations are shown here: a substitution and an insertion. In each case, the mutation and its effects on transcription and translation are shown in red.

Explore the impact of DNA mutations on protein synthesis. ▶❚❚ 13.8

tRNA, so a valine is inserted at this position in the protein, instead of an isoleucine.

Insertion or **deletion** mutations occur when a base is inserted into, or deleted from, a DNA sequence. Single-base insertions and deletions cause a genetic **frameshift**. Consider what happens in a multiple-choice test if you accidentally record the answer to a question twice on a machine-gradable answer sheet: all your answers from that point forward are likely to be wrong, since each is an answer to the previous question. This is equivalent to a frameshift caused by the insertion of a single base. If you forget to answer a question but record the next question's answer in its space, all your entries from that point on will get scrambled. This is equivalent to a frameshift caused by the deletion of a base. The insertion or deletion of one or two bases shifts all "downstream" codons by one or two bases. This scrambles the whole message downstream of the insertion or deletion because the ribosomes assemble a very different sequence of amino acids from that point onward, compared to the protein encoded by the original DNA sequence and its corresponding mRNA (see Figure 13.9).

Insertions or deletions involving three bases do not shift the reading frame of an mRNA, and therefore do not change the resulting protein as extensively as frameshifting mutations do. Insertions and deletions can involve more than a few bases: sometimes thousands of bases may be added or deleted as a result of mutations. Large insertions or deletions almost always result in the synthesis of a protein that cannot function properly. Frameshift mutations, whether they are caused by point mutations or large insertions or deletions, also alter the resulting protein so severely that it fails to function in most cases.

Mutations can cause a change in protein function

Mutations alter the DNA sequence of a gene, which in turn alters the sequence of bases in any mRNA molecule made from that gene. Such changes can have a wide range of effects on the resulting protein.

A mutation that produces a frameshift usually prevents the protein from functioning properly because it alters the identity of *many* of the amino acids in the protein. Frameshift mutations can also stop protein synthesis before it is complete: if a frameshift converts a codon specifying an amino acid into a stop codon, a full-length version of the protein will not be made. Regardless of whether it causes a frameshift or not, a mutation that alters an active site in a protein (such as the substrate-binding site of an enzyme) is usually harmful. Such mutations change the way the protein interacts with other molecules, decreasing or destroying its function. Finally, a mutation that inserts or deletes a series of bases causes the protein to have extra or missing amino acids, which can change the protein's shape and hence its function.

Sometimes changing a few bases in a gene's DNA sequence has little or no effect. For example, if a single-base substitution mutation does not alter the amino acids specified by the gene, then the structure and function of the protein will not be changed. Although a change in the DNA sequence from GGG to GGA would alter the mRNA sequence from CCC to CCU, both CCC and CCU code for the same amino acid, proline (see Figure 13.5). In such cases, the substitution mutation is said to be "silent" because it produces no change in the structure (and hence the function) of the protein, and therefore no change in the phenotype of the organism.

A few mutations may even be beneficial. Changes to the binding region of a protein, for example, might improve its efficiency or allow it to take on a new and useful function, such as reacting with a new substrate.

13.7 Putting It All Together: From Gene to Phenotype

Based on current estimates, humans have approximately 25,000 genes, arranged linearly on 23 different chromosomes. A large number of these genes code for proteins; the rest code for RNA molecules such as tRNA and rRNA. Here we review the major steps in how cells go from gene to protein to phenotype, focusing on genes that encode proteins. However, it is important to remember that transcription—the first step in the process that leads from gene to protein—is similar in all genes, including the small percentage of genes that specify tRNA and rRNA molecules. Translation does not occur for these genes because the tRNA and rRNA molecules are their final product.

Each gene is composed of a segment of DNA on a chromosome and consists of a sequence of the four bases adenine (A), cytosine (C), guanine (G), and thymine (T). The particular sequence of bases in the DNA a protein-coding gene specifies the amino acid sequence of the gene's protein product.

Transcription and translation are the two major steps in the synthesis of a protein from the information in its corresponding gene (see Figure 13.2). In transcription, the sequence of bases in a gene is used as a template to produce mRNA molecules. The cell then transports the mRNA out of the nucleus to a ribosome in the cytoplasm, where translation occurs. In translation, the sequence of bases in each mRNA molecule is used as a template to synthesize the gene's protein product by stringing together the correct sequence of amino acids.

The proteins encoded by genes are essential to life. A mutation in a gene can alter the sequence of amino acids in the gene's protein product, and this change can disable or otherwise alter the function of the protein. When a critical protein is disabled, the entire organism may be harmed. In people with the genetic disorder sickle-cell anemia, for example, a single base in the gene that encodes hemoglobin is altered (Figure 13.10). Hemoglobin is a protein involved in the transport of oxygen by red blood cells. The red blood cells of people with sickle-cell anemia become curved and distorted under low-oxygen conditions, such as those found in narrow blood vessels like our capillaries. The distorted red blood cells clog these narrow blood vessels, thereby leading to a wide range of serious effects, including heart and kidney failure. With little or no medical care, people with sickle-cell anemia usually die before they reach childbearing age. However, with intensive medical care, they can now live to their mid-forties or beyond.

In sickle-cell anemia, a gene mutation alters the gene's protein product, which in turn produces a change in the organism's phenotype. A similar chain of events occurs for other genes. Overall, the phenotype of an organism is determined by the organism's proteins and by its internal and external environment. Because genes carry information that affects the production of proteins, they play a crucial role in determining the phenotype of the organism.

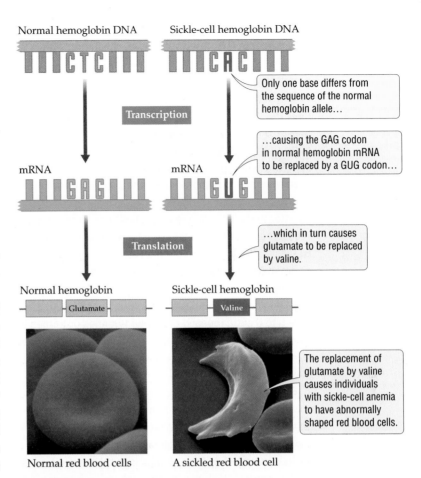

Figure 13.10 A Small Genetic Change Can Have a Large Effect
Sickle-cell anemia is caused by a change in a single base.

Concept Check

1. What is meant by translation of the genetic code?
2. A single-base addition or deletion in a gene is likely to alter the protein product more than the substitution of a single base, such as C for T. Why?

Applying What We Learned

From Gene to Protein, to New Hope for Huntington Disease

Huntington disease (HD), as we saw in Chapters 11 and 12, is a dominant genetic disorder that strikes the brain, causing involuntary movements, personality changes, intellectual deterioration, and eventually, death. There is no cure, but since the isolation of the *HD* gene in 1993, new research has improved our understanding of the

disease and provided hope that we may eventually be able to develop treatments that can control—or even reverse—its symptoms.

How did the isolation of the *HD* gene lead to these promising developments? In essence, scientists were able to use the genetic code to solve the

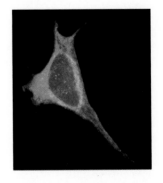

communication problem described in the opening pages of this chapter. Specifically, they compared the DNA sequences of normal and *HD* alleles. This comparison found that *HD* alleles have 3 to 215 extra copies (called repeats) of the sequence GTC inserted near the beginning of the gene (see Figure 12.9); the rest of the sequence is normal. By using the genetic code to determine the amino acids specified by the *HD* gene, scientists identified both the normal and mutant versions of huntingtin, the protein produced by the *HD* gene (Figure 13.11).

After that, our understanding of HD improved rapidly. We learned that the number of GTC repeats in a mutant *HD* allele affects the age of onset and the severity of the disorder. Most people with Huntington disease have 5 to 15 extra GTC repeats. On average, their disease symptoms begin when they are in their fifties or sixties (if they have 5 to 6 extra repeats) or in their twenties or thirties (if they have 10 to 15 extra repeats). People with more than 15 extra GTC repeats usually develop symptoms before the age of 20.

Researchers have also learned that certain enzymes in brain cells cut the mutant huntingtin into pieces, separating a long string of the amino acid glutamine from the rest of the protein. The glutamine strings enter the nucleus, where they form clumps with other such pieces.

Researchers think the clumps of glutamine contribute to the death of brain cells, which leads to the symptoms of HD. If these researchers are right, we could search for drug therapies that stop the clumping process and may possibly decrease, or even reverse, the symptoms of the disorder. Researchers were able to do this in mice. Here's what they did.

First, the researchers chose to work with a dye called Congo red because they knew it could prevent the formation of glutamine clumps such as those seen in people with HD. Next, they injected the dye into the brains of mice with the *HD* allele. Their findings were startling: the clumps were broken apart, and the mice lost less weight, walked in a more normal way, and survived significantly longer.

Eight other neurological disorders, including Machado–Joseph disease and Haw River syndrome, are also caused by alleles that have extra GTC repeats. Furthermore, clumps of glutamine form in the nerve cells of people with these disorders. These observations provide a tantalizing prospect: if the glutamine clumps contribute to disease symptoms, the development of a therapy that prevents or reverses the formation of these clumps could offer hope for people with HD and other terrible neurological disorders.

New research has revealed the precise function of huntingtin. The normal version of the protein acts as a regulator of about 50 other genes. In particular, it promotes the transcription of a growth factor called BDNF (brain-derived neurotrophic factor) that is essential for the growth of certain types of brain cells, the very ones damaged in those afflicted with HD. When the glutamine repeats destroy the normal function of huntingtin, the growth factor is not made and brain cells start to die, a process that may be accelerated by the presence of glutamine clumps. Recent studies with genetically engineered mice have shown promising results. When scientists restored BDNF function in mice with early stage HD, the treated animals showed much less brain cell death and better muscle coordination than the untreated animals, which showed the expected progression of the disease. Experiments of this type raise the hope that someday newly diagnosed HD patients will receive gene therapy to restore the transcription of genes necessary for the survival, growth, and normal function of brain cells. Combined with strategies to prevent the formation of glutamine clumps, such an approach offers the exciting possibility that this terrible disease may be curable in the not too distant future.

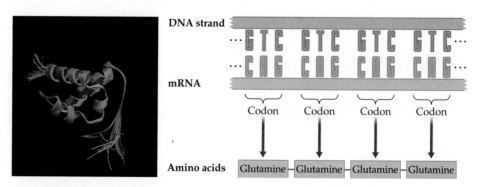

DNA strand

... GTC GTC GTC GTC ...

... CAG CAG CAG CAG ...

mRNA

Codon   Codon   Codon   Codon

Amino acids   Glutamine – Glutamine – Glutamine – Glutamine

Figure 13.11 A Deadly Mutant
The mutant version of the protein huntingtin contains longer strings of the amino acid glutamine (shown in purple) than the normal version of huntingtin (inset), and has a different overall shape.

Researchers Delve into "Gene for Speed"

PARIS – If you were a prehistoric human, would you prefer to be able to sprint fast for short distances or to jog comfortably for kilometres? That's one of the questions thrown up by the so-called "gene for speed," known as ACTN3 . . .

Around 18 percent of the world's population has a truncated variant of the gene that blocks [the corresponding] protein. The stubby variant, called *577X*, is common among successful endurance athletes. On the other hand, elite sprinters, who need explosive speed, are likelier to have the reverse—a functioning variant of ACTN3.

Keen to find out more, researchers led by Kathryn North, a professor at the Children's Hospital at Westmead, in Sydney, Australia, created a batch of mice that had been engineered to lack ACTN3. A study published Sunday in the journal *Nature* *Genetics* says the "knockout" mice and ordinary mice with a functioning ACTN3 gene were put on a treadmill . . .

The winners were the knockout mice, which were able to run three times as long. North's team also looked at genetic profiles from individuals of European and East Asian descent and found that the ability to run longer distances . . . became incorporated into many *Homo sapiens*.

Skeletal muscles contain bundles of muscle cells, or muscle fibers. There are two main types of fibers: fast twitch and slow twitch. Fast-twitch fibers specialize in producing large bursts of power, but they tire quickly. Slow-twitch fibers are more efficient in extracting energy from sugars, and their power output can be sustained much longer. Most of us have roughly equal numbers of the two types of fibers in our skeletal muscles. In contrast, the muscles of elite athletes in strength-based sports, such as sprinting or weight lifting, may be 80 percent fast-twitch fibers. Those excelling in endurance sports, such as long-distance running or cycling, can have muscles that are 80 percent slow-twitch fibers.

Alpha actinin 3, the protein encoded by the *ACTN3* gene, is made only in skeletal muscles. It anchors the contractile proteins so muscle fibers can generate power. Australian scientists have found two alleles of the *ACTN3* gene seemingly linked to athletic ability. The R allele codes for a functional alpha actinin 3, whereas the X allele leads to the production of a shortened, nonfunctional version. As shown in the table below, the XX genotype was rare in athletes in strength sports, but was found in about 24 percent of endurance athletes. The advantage of the X allele in endurance sports is supported by the findings of Professor North and her colleagues who used genetic engineering technology (Chapter 15) to "knockout" the activity of both copies of the *ACTN3* gene in lab mice, creating the "marathon mice."

Success in sports depends on many things, including such psychological attributes as personal drive; and top-level performance always requires extensive training and conditioning. The physical component of athletic success is likely influenced by many genes, not just one or two genes such as *ACTN3*. In fact, in one study as many as 92 different genes were potentially associated with athletic ability and health-related fitness. Note that the predictive power of the *ACTN3* alleles is limited, since as many as 31 percent of the elite distance runners lacked the X allele, and 45 percent had only one copy.

Association between Athletic Ability and Prevalence of *ACTN3* Alleles in Strength and Endurance Athletes

Genotype	Percentage of the group that has the genotype		
	Control (non-athlete)	Strength sport elite athletes (sprinters)	Endurance sport elite athletes (distance runners)
RR	30	50	31
RX	52	45	45
XX	18	6	24

Evaluating the News

1. The X allele of *ACTN3* is caused by a mutation that prematurely halts protein synthesis. Is this mutation likely to be: (i) a base substitution that changes one amino acid into another, (ii) a base substitution that changes an amino acid codon to a stop codon, (iii) an insertion or deletion producing a frameshift? Explain your answer.
2. Researchers justify studies of genes influencing physical performance by pointing out their relevance in conditions such as muscular dystrophy and other muscle diseases. Opponents say the knowledge gained will lead to abuse by overambitious athletes and pushy parents of would-be sports stars. Do you think taxpayers should fund investigation of the genetic basis of physical performance? Do you think the potential for medical benefits outweighs the risk for abuse?
3. An Australian company offers a commercial test for athletic potential, based on the R and X alleles of the *ACTN3* gene. Would you want to know your allelic makeup? Would you want your children tested for this gene, and if so, how would you use that information?

SOURCE: *Vancouver Sun*, September 10, 2007.

Chapter Review

Summary

13.1 How Genes Work

- Genes code for RNA molecules. Protein-coding genes code for messenger RNA (mRNA), which contains instructions for building a protein.
- An RNA molecule consists of a single strand of nucleotides. Each nucleotide is composed of the sugar ribose, a phosphate group, and one of four nitrogen-containing bases. The bases found in RNA—adenine (A), cytosine (C), guanine (G), and uracil (U)—are the same as those in DNA except that uracil replaces thymine (T).
- At least three types of RNA (mRNA, rRNA, and tRNA) and many enzymes and other proteins participate in the manufacture of proteins.

13.2 How Genes Guide the Manufacture of Proteins

- In both prokaryotes and eukaryotes, protein synthesis requires two steps: transcription and translation.
- In transcription, an RNA molecule is made using the DNA sequence of the gene.
- In translation, ribosomes, mRNA, and tRNA molecules together direct protein synthesis. Ribosomes consist of rRNA and certain proteins.
- In eukaryotes, information for the synthesis of a protein is transmitted from the gene, located in the nucleus, to the site of protein synthesis, the ribosomes, located in the cytoplasm.

13.3 Gene Transcription: Information Flow from DNA to RNA

- During transcription, one strand of the gene's DNA serves as a template for synthesizing many copies of mRNA.
- The key enzyme in transcription is RNA polymerase.
- Each gene has a sequence (a promoter) at which RNA polymerase begins transcription. In prokaryotic genes, transcription stops at a special sequence called a terminator. Transcription termination in eukaryotes is more complicated.
- The mRNA molecule is constructed according to specific base-pairing rules: A, U, C, and G in mRNA pair with T, A, G, and C, respectively, in the template strand of DNA.
- In eukaryotes, most genes contain internal noncoding sequences of DNA (introns) that must be removed from the initial mRNA product while it is still in the nucleus. The remaining segments of mRNA correspond to exons, the DNA sequences of the gene that code for the protein.

13.4 The Genetic Code

- The information in a gene is encoded in its sequence of bases.
- The information encoded by an mRNA is read in sets of three bases; each three-base sequence is a codon. Of the 64 possible codons, most specify a particular amino acid, but certain codons signal the start or stop of translation. The information specified by each of the codons is collectively called the genetic code.

- When reading the genetic code, ribosomes begin at a fixed starting point on the mRNA (the start codon) and stop reading the code when any one of the three stop codons is encountered.
- The genetic code is unambiguous (each codon specifies no more than one amino acid), redundant (several codons specify the same amino acid), and nearly universal (used in almost all organisms on Earth).

13.5 Translation: Information Flow from mRNA to Protein

- In translation, the information in the sequence of bases in mRNA determines the sequence of amino acids in a protein.
- Translation occurs at ribosomes, which are composed of rRNA and more than 50 different proteins.
- Each transfer RNA (tRNA) molecule specializes in carrying a particular amino acid and has a three-base sequence, called the anticodon, that recognizes and pairs with a specific codon in the mRNA. A specific tRNA attaches to the ribosome–mRNA complex when its anticodon is able to bind a codon in the mRNA by complementary base pairing. In this way, a specific tRNA molecule delivers a specific amino acid based on the codon message present in the mRNA.
- The ribosome makes the covalent bonds that link the amino acids together into a protein. It holds the mRNA and tRNA in an orientation that allows the amino acid carried by the tRNA to be covalently bonded to the growing polypeptide.

13.6 The Effect of Mutations on Protein Synthesis

- Many mutations are caused by the substitution, insertion, or deletion of a single base in a gene's DNA sequence.
- Insertion or deletion of a single base causes a genetic frameshift, resulting in a different sequence of amino acids in the gene's protein product.
- Mutations involving insertions or deletions of a series of bases are also common.
- Mutations can change the function of a gene's protein product, especially when a frameshift is involved or when the mutation affects a protein's binding site. Mutations causing frameshifts usually destroy the protein's function.
- Other mutations have neutral effects, and a few even have beneficial effects.

13.7 Putting It All Together: From Gene to Phenotype

- Some genes code for RNA only, but a large number encode proteins.
- Proteins are essential to life. In conjunction with the environment, they determine an organism's phenotype.
- Because genes guide the production of proteins, they play a key role in determining an organism's phenotype.

Review and Application of Key Concepts

1. What is a gene? In general terms, how does a gene store the information it contains?

2. Discuss the different products specified by genes. What are the function(s) of each of these products?

3. Describe the flow of genetic information from gene to phenotype.

4. How is the information contained in a gene transferred to another molecule? In eukaryotic cells, why must the molecule that "carries" the information stored in the gene be transported out of the nucleus?

5. What is RNA splicing? Does it occur in both eukaryotes and prokaryotes? Explain your answer.

6. Describe the roles played by rRNA, tRNA, and mRNA in translation.

7. Scientists have discovered a mutation in a gene encoding a tRNA molecule that appears to be responsible for a series of human metabolic disorders. The mutation occurred at a base located immediately next to the anticodon of the tRNA, a change that destabilized the ability of the tRNA anticodon to bind to the correct mRNA codon. Why might a single-base mutation of this nature result in a series of metabolic disorders?

8. Write a paragraph explaining to someone with little background in biology what a mutation is and why mutations can affect protein function.

Key Terms

anticodon (p. 275)
codon (p. 273)
deletion (p. 278)
exon (p. 272)
frameshift (p. 278)
gene (p. 268)
genetic code (p. 273)
insertion (p. 278)
intron (p. 272)
messenger RNA (mRNA) (p. 269)
point mutation (p. 277)
promoter (p. 271)

ribosomal RNA (rRNA) (p. 269)
RNA polymerase (p. 271)
RNA splicing (p. 272)
start codon (p. 273)
stop codon (p. 273)
substitution (p. 277)
template strand (p. 271)
terminator (p. 272)
transcription (p. 270)
transfer RNA (tRNA) (p. 269)
translation (p. 270)

Self-Quiz

1. For genes that specify a protein, what molecule carries information from the gene to the ribosome?
   a. DNA
   b. mRNA
   c. tRNA
   d. rRNA

2. During translation, each amino acid in the growing protein chain is specified by how many nitrogenous bases in mRNA?
   a. one
   b. two
   c. three
   d. four

3. Which molecule delivers the amino acid specified by a codon to the ribosome?
   a. rRNA
   b. tRNA
   c. anticodon
   d. DNA

4. During transcription, which molecule or molecules are produced?
   a. mRNA
   b. rRNA
   c. tRNA
   d. all of the above

5. A portion of the template strand of a gene has the base sequence CGGATAGGGTAT. What is the sequence of amino acids specified by this DNA sequence? (Use the information in Figure 13.5 and assume that the corresponding mRNA sequence will be read from left to right.)
   a. alanine–tyrosine–proline–isoleucine
   b. arginine–tyrosine–tryptophan–isoleucine
   c. arginine–isoleucine–glycine–tyrosine
   d. none of the above

6. Which of these is responsible for creating the covalent bonds that link the amino acids of a protein together in the order specified by the gene that encodes that protein?
   a. tRNA
   b. mRNA
   c. rRNA
   d. ribosome

7. A mutation occurs in which the fourth, fifth, and sixth bases are deleted from a gene that encodes a protein with 57 amino acids. Which of these would be expected to happen?
   a. The resulting frameshift will prevent protein synthesis.
   b. A protein with 56 amino acids will be constructed.
   c. A protein that differs from the original one—but still has 57 amino acids—will be constructed.
   d. A protein with 54 amino acids will be constructed.

8. Most eukaryotic genes contain one or more segments that are transcribed but not translated. Each such segment is known as
   a. a start codon.
   b. a promoter.
   c. an intron.
   d. an exon.

CHAPTER 14

Control of Gene Expression

Key Concepts

◉ Eukaryotic DNA is organized by a packing system that enables cells to store an enormous amount of information in a small space.

◉ Prokaryotes have relatively little DNA, most of which encodes proteins. Eukaryotes have more DNA and more genes than prokaryotes have. Unlike prokaryotes, eukaryotes have large amounts of DNA that does not encode proteins.

◉ Gene expression begins with the activation of a gene and ends with a detectable phenotype influenced by that gene. Transcription and translation are included in the gene expression pathway of protein-coding genes.

◉ Aspects of the environment, such as short-term changes in food availability, can alter gene expression. In multicellular organisms, different sets of genes are activated in different cell types, and patterns of gene expression change dramatically during development.

◉ Most gene expression pathways are regulated at the level of transcription. Cells can also control gene expression in other ways, such as by regulating the life span of a protein.

◉ Genetics has begun to move from the study of single genes to the study of interactions among many different genes and how these are influenced by signals from the surroundings.

Greek Myths and One-Eyed Cows

Among his many adventures, the Greek mythological hero Odysseus encountered (and outwitted) a Cyclops, a gigantic humanlike creature with great strength and a single large eye. The faces of the Cyclopes of legend had characteristics that resemble those caused by some rare genetic and developmental disorders. Cows, mice, and humans occasionally are born with a single large eye, along with other abnormalities of the brain and face. But these individuals die soon after birth.

What might cause an animal to be born with only one eye? Two possibilities are known, both of which relate to the function of genes. Certain master-switch genes guide the development of an organism by activating, or "turning on," a series of other crucial genes. Some one-eyed individuals have a defect in one of these master-switch genes that prevents it from turning on certain genes that control the normal development of the brain and face. Other one-eyed individuals were exposed as embryos to chemicals that prevented the protein product of the master-switch gene from having its normal effect. Whether due to a defective master-switch gene or exposure to chemicals, the formation of a single large eye results from the failure of cells to regulate (control) how a series of crucial genes are turned on.

Deciphering the causes of a single eye brings us to one of the most exciting areas of modern genetics: the control of gene expression. To develop and function normally, organisms must express the right genes at the right place and time, producing just the right amount of RNA and protein. This is a task of bewildering complexity, but each of us does it, many, many times, every day.

When the control of gene expression fails to work properly, disaster results: the organism may not develop properly (for example, it may form a single eye), or a group of cells may become cancerous, activating a set of genes very different from those

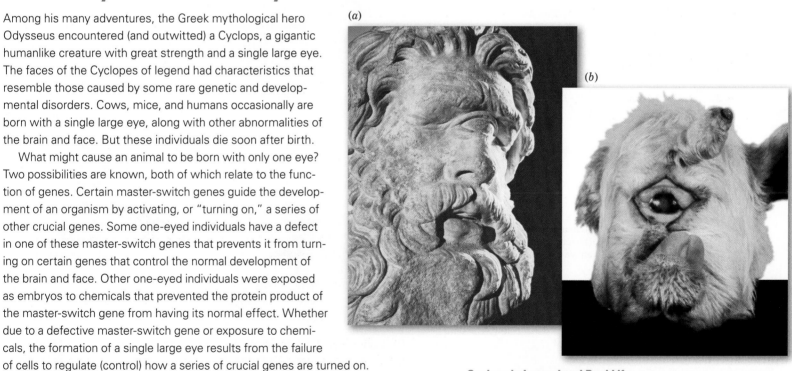

(a)

(b)

Cyclops in Legend and Real Life
(a) Polyphemus, a Cyclops mentioned in Homer's *Odyssey*. In Greek mythology, the Cyclopes are an ancient race of one-eyed giants. (b) Cyclopia, or prosencephaly, in a newborn calf. This rare birth defect is also recorded in other mammals, including humans. It is caused when parts of the head region fail to divide symmetrically during embryo development, resulting in a single, enlarged eye located in the center. Typically, the nose and mouth are also malformed (as in this example), and newborns rarely live past a few days.

activated in a normal cell of the same type. **Is it possible nowadays to monitor gene expression in our cells and detect if something is wrong? Could such monitoring lead to early detection and better treatment for genetic disorders such as cancer?** We will consider these issues toward the close of the chapter, but first we must examine how prokaryotic and eukaryotic cells regulate which genes are turned on and to what degree at any one time. We will see that although the risk of dangerous mistakes looms at every step in this process, most of the time our cells control the expression of our 25,000 genes with astonishing accuracy.

Helpful to know

Biologists use several words to describe a gene whose product is made: the gene is "activated," is "turned on," or is being "expressed."

Even though they all have the same genes, the various cell types in a multicellular organism differ greatly in their structure and function (Figure 14.1). If genes are the blueprint for life, how can cells with the same genes be so different? The answer lies in how the genes are used: one cell type is different from another because different sets of genes are turned on in the two. Gene expression also changes, sometimes dramatically, as an organism grows and develops, and as it senses and responds to its surroundings.

Recall from Chapter 13 that each gene contains instructions for the synthesis of a particular RNA. Often, these RNA molecules are mRNA, each directing the manufacture of a particular protein; however, some types of RNA, such as tRNA or rRNA, have other roles in the cell and do not themselves contain instructions for building proteins. **Gene expression** begins with the activation, or "turning on," of a gene and ends with a detectable phenotype controlled or influenced by that gene. Gene transcription, the synthesis of the RNA

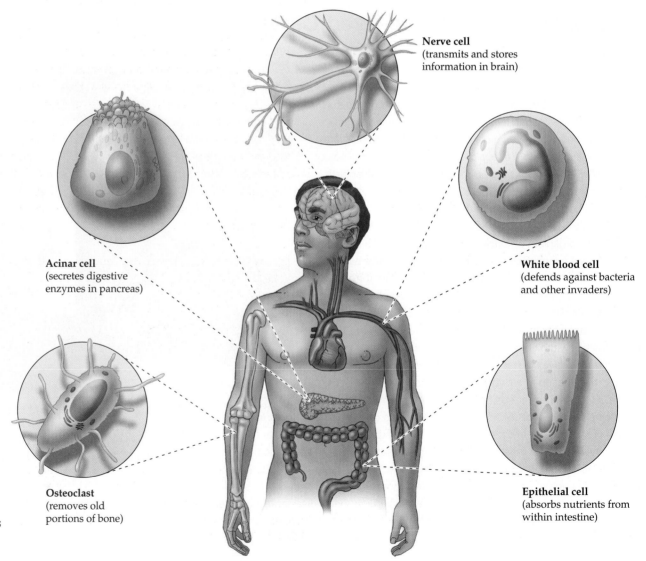

Figure 14.1 Different Cells Have the Same Genes
The nucleus of each cell type contains the same DNA. Although all cells within a multicellular organism have the same genes, these cells can differ greatly in structure and function because different genes are active in different types of cells.

Nerve cell
(transmits and stores information in brain)

Acinar cell
(secretes digestive enzymes in pancreas)

White blood cell
(defends against bacteria and other invaders)

Osteoclast
(removes old portions of bone)

Epithelial cell
(absorbs nutrients from within intestine)

encoded by that gene, is a critical step in the expression of any gene. A large number of our genes code for proteins, so the synthesis of proteins (translation) is also included in the concept of gene expression. Put another way, gene expression is the series of events that makes the action of a gene evident.

In this chapter we describe how environmental cues and internal signals regulate gene expression and how gene expression helps build the phenotype. We begin by examining the structural and functional organization of DNA, which is necessary for understanding patterns of gene expression in developing and adult organisms. Then we look at some of the ways in which cells fine-tune gene expression in response to environmental cues and how changes in gene expression shape development in multicellular organisms. We close with a look at how one DNA technology—the DNA chip—is changing our views about gene expression and promises to improve many aspects of medical diagnosis and treatment.

14.1 The Structural and Functional Organization of DNA

How much DNA does a cell contain? Does all of the DNA in an organism consist of genes? These questions call for a comparison of prokaryotes (Bacteria and Archaea) and eukaryotes (all other organisms) because the DNA of these two major groups is organized differently. First, a typical bacterium has several million base pairs of DNA, all in a single chromosome, whereas most eukaryotic cells have hundreds of millions to billions of base pairs distributed among several chromosomes. Second, most of the DNA in prokaryotes encodes proteins, and prokaryotic genes rarely contain noncoding segments of DNA. In contrast, eukaryotes have noncoding DNA *inside* most genes, as well as large amounts of noncoding DNA *between* genes. Finally, prokaryotic genes tend to be organized by function: the different genes needed for a given metabolic pathway are usually clustered together and are turned on or turned off as one unit. In contrast, although some eukaryotic genes with related functions are grouped near one another on a chromosome, most are not organized in this way; genes with related functions may even be found on entirely different chromosomes. Overall, the organization of eukaryotic DNA is more complex. Let's consider some of these differences in greater detail.

Compared to prokaryotes, eukaryotes have much more DNA per cell

The **genome** of an organism is one copy of all the DNA-based information contained in its chromosomes. In prokaryotes, this amounts to all the genetic information contained in the single chromosome that is typically present in these organisms.

The genome of a eukaryote is all the genetic information contained in a haploid set of chromosomes, such as that found in a sperm or egg. The genome size of prokaryotes varies from 0.6 million to 30 million base pairs. Eukaryotes show much greater variation in genome size: their genomes range in size from 12 million base pairs in yeast (a single-celled fungus) to over a trillion base pairs in a certain species of *Amoeba* (a single-celled protist). Most vertebrates have genomes that contain hundreds of millions to billions of base pairs. For example, puffer fish have 400 million base pairs of DNA in their genome (Figure 14.2). The range among mammals is 1.5 billion to 6.3 billion base pairs (with humans at 3.3 billion), and some salamanders have 90 billion base pairs.

As these examples illustrate, eukaryotes usually have far more DNA than prokaryotes have. Why? In part, the reason is that eukaryotes generally have greater complexity in structure and behavior than prokaryotes and hence need more genes to direct that complexity. A typical prokaryote has about 2,000 genes, although certain tiny bacteria have no more than about 500. Among the eukaryotes studied to date, the single-celled yeast *Saccharomyces cerevisiae* has 6,000 genes, the nematode worm *Caenorhabditis elegans* has 19,100 genes, several plant species each have an estimated 20,000 genes, and humans have about 25,000 genes.

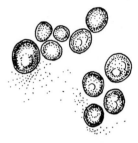

Saccharomyces cerevisiae

Caenorhabditis elegans
(three individuals)

Figure 14.2 Vertebrates with Millions to Billions of Base Pairs
This puffer fish has only 400 million base pairs of DNA in one set of its chromosomes, compared to the diver's 3.3 billion.

Genes constitute only a small percentage of the DNA in most eukaryotes

Although eukaryotes have roughly 3 to 15 times as many genes as a typical prokaryote has, this difference in gene number does not fully explain why eukaryotes often have hundreds to thousands of times more DNA than prokaryotes. Only a small percentage of the DNA in eukaryotic genomes consists of genes, defined as DNA sequences that code for RNA and affect some phenotype. Some of the remaining DNA has regulatory functions—for example, controlling gene expression. Some of it has architectural functions, such as giving structure to chromosomes or positioning them at precise locations within the interphase nucleus. Much of the genome, however—in fact, a majority of the genome in most eukaryotes—has no apparent function. At least some of this noncoding DNA appears to be nonessential and is popularly called "junk DNA" because researchers have found no impact on the phenotype if this DNA is lost from the cell. Although various hypotheses have been proposed, we don't yet understand why so many eukaryotes have large amounts of DNA with no clear-cut function.

Scientists estimate that genes that encode a protein make up less than 1.5 percent of the human genome. Other genes in our cells encode different types of nonprotein RNA molecules. The rest of our genome consists of various types of noncoding DNA, defined as DNA that does not code for any kind of functional RNA.

Learn more about the structural organization of the chromosome. 14.1

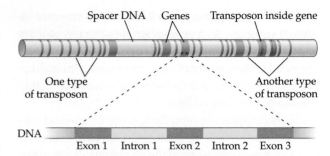

Figure 14.3 The Composition of Eukaryotic DNA
Eukaryotic genomes contain a small proportion of genes (purple) interspersed with a large amount of spacer DNA (light blue) and transposons (red and dark blue). Each of the two different types of transposons shown here is found in many copies throughout the human genome. Note that one transposon (dark blue) has inserted itself into one of the five genes. Most eukaryotic genes consist of coding regions (exons) interspersed with noncoding regions called introns.

Noncoding DNA includes introns and spacer DNA (Figure 14.3, Table 14.1). As we saw in Chapter 13, most eukaryotic genes are interrupted by stretches of DNA, called **introns**, that do not code for the amino acid sequence of the protein and must be spliced out of a newly made mRNA molecule before translation. **Spacer DNA** are noncoding DNA sequences that separate one gene from another; spacer sequences found in eukaryotic genomes are exceptionally long, compared to those found in the much more compact genomes of prokaryotes.

Table 14.1

Types of Eukaryotic DNA

Type	Description
Exons (of genes)	Transcribed portions of a gene that code for the amino acid sequence of a protein
Noncoding DNA:	
Introns (in a gene)	Transcribed portions of a gene that do not code for the amino acid sequence of a protein, and whose corresponding sequence in RNA is removed before the RNA leaves the nucleus.
Spacer DNA	DNA sequences that separate genes
Regulatory DNA	DNA sequences that control the expression of genes
Structural DNA	DNA sequences that have architectural function, creating structures such as the centromere of a chromosome
DNA of unknown function	DNA sequences that have no known function in the cell
Transposons ("jumping genes")	DNA sequences that can move from one position on a chromosome to another, or from one chromosome to another

Prokaryotic and eukaryotic genomes also contain "jumping genes," or **transposons**: sequences that can move from one position on a chromosome to another, or even from one chromosome to another. The activity of a gene can be disrupted if a transposon inserts itself somewhere within that gene. Although transposons can encode proteins, and these proteins are necessary for mobilizing a transposon from one location to another, the great majority of transposon sequences in the human genome are "DNA fossils" in that they do not make functional proteins and are usually incapable of moving. Transposons may constitute a large proportion of a eukaryote's DNA. For example, transposons make up an estimated 36 percent of the human genome and more than 50 percent of the 5.4 billion base pairs in the corn genome. Most transposons appear to be viral in origin, consisting of genetic material from ancient viruses that have become noninfective (not transmitted between genetically unrelated individuals). Other transposons appear to be fragments of "selfish DNA," segments of the cell's own DNA that have acquired the ability to make copies of themselves and insert them at random locations in the genome.

14.2 DNA Packing in Eukaryotes

To be expressed, the information in a gene must first be transcribed into an RNA molecule. A gene cannot be transcribed unless the enzymes that guide transcription are able to reach that gene. The task may sound simple, but it is complicated by what may be the ultimate storage problem: how to store an enormous amount of genetic information (the organism's DNA) in a small space (the nucleus), and still be able to retrieve each piece of that information precisely when it is needed.

In humans and other eukaryotes, every chromosome in each cell contains one DNA molecule. These chromosomes hold a vast amount of genetic information. As we learned in Chapter 9, the haploid number of chromosomes in humans is 23; these chromosomes together contain about 3.3 billion base pairs of DNA. Stretched to its full length, the DNA from all 46 chromosomes in a single human cell would be more than 2 meters long (taller than most of us). That is a huge amount of DNA, especially considering that it is packed into a nucleus only 0.000006 meter (0.0002 inch) in diameter. The combined length of DNA in our bodies is staggering: the human body has about 10^{13} cells, each of which contains roughly 2 meters of DNA. Therefore, each of us has about 2×10^{13} meters of DNA in the body—a length

more than 130 times the distance between Earth and the sun.

How can our cells stuff such an enormous amount of DNA into such a small space? They use a variety of packaging proteins to wind, fold, and compress the DNA double helix, going through several levels of packing to create the DNA–protein complex we call a chromosome. Let's examine the packing of DNA in a metaphase chromosome, beginning with the DNA double helix, which is about 2 nanometers wide (Figure 14.4, bottom panel). Short lengths of this double helix are wound around "spools" of proteins, known as **histone proteins**,

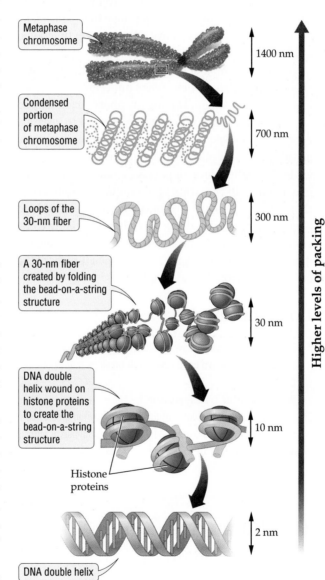

Test your knowledge of DNA packing.
14.2

Figure 14.4 DNA Packing in Eukaryotes
The DNA of eukaryotes is highly organized by a complex packing system. The arrow indicates increasingly higher levels of DNA packing. The chromosome is shown at metaphase, the phase of the cell cycle at which DNA is most tightly packed.

to create a bead-on-a-string structure that is about 10 nanometers wide and consists of many histone "beads" linked by a DNA molecule that wraps around each bead. The bead-on-a-string structure is compressed into a more compact form, known as the 30-nanometer fiber, by yet other types of packaging proteins. This fiber is thrown into loops and attached to a protein scaffold that organizes the interior of the interphase nucleus. During interphase, much of the genome exists as a looped 30-nanometer fiber, but any region of the chromosome that is being transcribed is "unpacked" into a bead-on-a-string state, which can be accessed by RNA polymerase and other components of the gene transcription machinery. Some portions of an interphase chromosome remain tightly condensed at all times because they contain no genes, although they may have other types of functional DNA, such as structural DNA.

At the beginning of cell division, whether by mitosis or meiosis, all chromosomes undergo yet another level of condensation. During prophase, each looped 30-nanometer fiber is further compressed to create a cable that is considerably shorter and more than twice as thick. This mitotic (or meiotic) chromosome represents the highest level of packing that a chromosome can acquire. In this state of maximum condensation, these shorter, stouter chromosomes are less liable to become entangled and are therefore less likely to be ripped apart when they are aligned at the cell center during metaphase or segregated to opposite sides of the cell during anaphase and telophase.

Concept Check

1. Each human cell contains over 1,000 times as much DNA as *E. coli* bacteria have. Do we have over 1,000 times as many genes as *E. coli*? Explain.
2. What is meant by "gene expression"?
3. What is the functional value of the roughly twofold greater condensation that chromosomes undergo during prophase of mitosis or meiosis?

14.3 Patterns of Gene Expression

Gene expression is the means by which genes influence the structure and function of a cell or organism. In other words, gene expression enables a gene to influence a particular phenotype, the actual display of an inherited characteristic. At any given time, less than a third of the genes in a typical human cell are actively expressed. The rest of the genes are not in use. Different cells express different sets of genes, and within a given cell the pattern of gene expression can change over time. But what determines which genes an organism expresses at a particular time in a particular cell?

Organisms can turn genes on or off in response to environmental cues

Single-celled organisms such as bacteria face a big challenge: they are directly exposed to their environment, and they have no specialized cells to help them deal with changes in that environment. One way they meet this challenge is to express different genes as conditions change.

Bacteria respond to changes in nutrient availability, for example, by turning genes on or off. If the *E. coli* bacteria in a petri dish are given lactose (a sugar found in milk) as the only source of energy, within a matter of minutes they turn on genes that code for lactose-digesting enzymes (Figure 14.5). When the lactose is used up, the bacteria stop producing those enzymes. In effect, these bacteria reorganize gene expression to match the food resources available to them. When a resource runs out, they switch on genes that will enable them to exploit the next resource that becomes available (in Figure 14.5, the sugar arabinose). By producing the enzymes to process a particular food only when that food is available, bacteria avoid wasting energy and cellular resources making enzymes that are not needed.

Like single-celled organisms, multicellular organisms change which genes they express in response to internal signals (arising inside the body) or external cues present in the environment. For example, we humans change the genes we express when our blood sugar or blood pH levels change, enabling us to keep those levels from becoming too high or too low. Similarly, humans, plants, and many other organisms, when exposed to high temperatures, turn on genes encoding proteins that protect cells against heat damage.

Different cells in eukaryotes express different genes

Throughout the life of a multicellular organism, different types of cells express different sets of genes. Whether a cell expresses a particular gene depends on whether that gene function is needed by that cell at that time under the prevailing conditions. Not surprisingly,

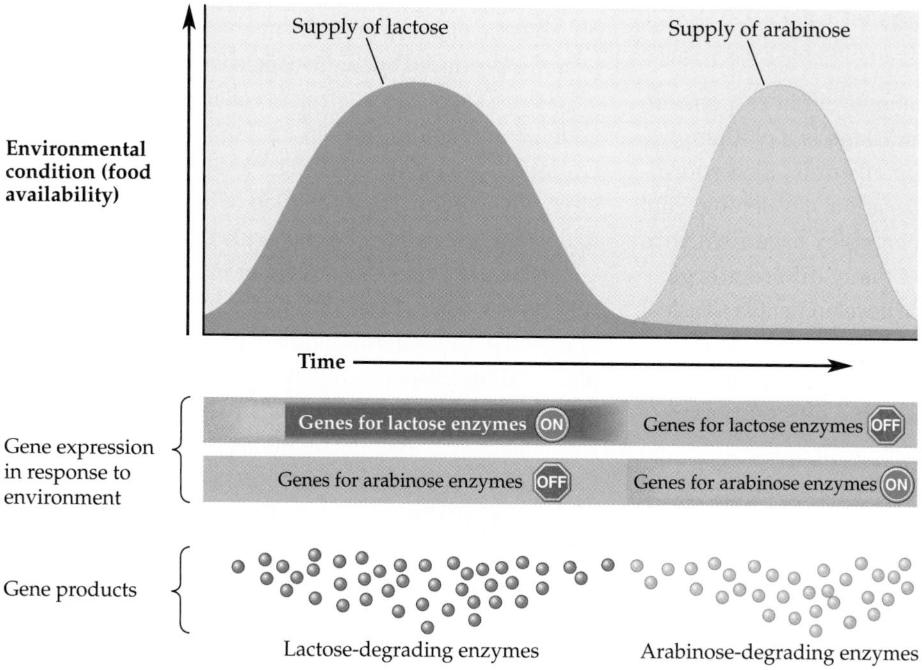

Figure 14.5 Bacteria Express Different Genes as Food Availability Changes

The sugars lactose and arabinose are not always available, but both can serve as food for the bacterium *E. coli*.

a gene that encodes a specialized protein is expressed only in cells that need either to use that protein within the cell or to produce it for transport to other cells that will use it. For example, among the 220 cell types in the human body, red blood cells are the only cells that use the oxygen transport protein hemoglobin; therefore, as they develop to maturity, red blood cells are the only cells that express the gene for this protein (Figure 14.6).

Similarly, the gene for crystallin, a protein that is the main component of the lens in our eyes, is expressed only in certain cells in the developing eye. The gene for insulin, a hormone produced in the pancreas and used elsewhere in the body, is expressed only in certain cells of the pancreas.

Some genes, known as **housekeeping genes**, play an essential role in the maintenance of cellular activities in all kinds of cells. These genes are therefore expressed most of the time by most cells in the body. Genes for rRNA, for example, are expressed by almost all cells (see Figure 14.6). This is not surprising, since virtually all cells need to make proteins, and rRNA is a key component of ribosomes, the structures on which proteins are built. Housekeeping genes tend to be highly conserved in evolutionary history, meaning that their base sequence and general function is similar in diverse groups of organisms. Genes that are critical for survival—because they help conduct a function as elemental as protein synthesis, for example—tend not to change much as new groups evolve from the ancestral forms.

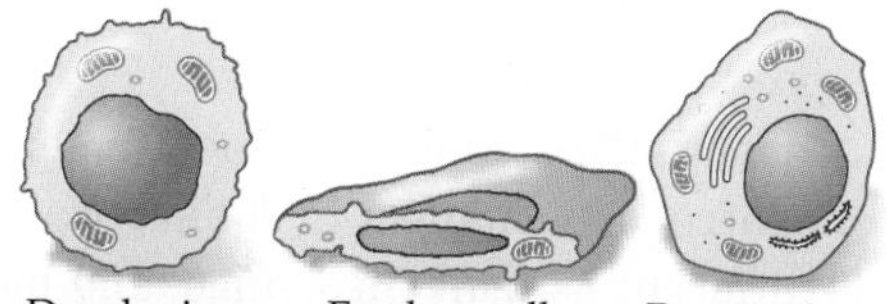

	Developing red blood cell	Eye lens cell (in embryo)	Pancreatic cell
Hemoglobin gene	ON	OFF	OFF
Crystallin gene	OFF	ON	OFF
Insulin gene	OFF	OFF	ON
rRNA gene	ON	ON	ON

Figure 14.6 Different Types of Cells Express Different Genes

Some genes, such as those encoding hemoglobin, crystallin, and insulin, are active ("on") only in the types of cells that use or produce the protein encoded by the gene. Housekeeping genes, such as the rRNA gene, are active in most types of cells.

Development in eukaryotes relies on gene cascades

Turning the appropriate gene on or off in response to changing environmental conditions is a challenging task. But multicellular organisms must also coordinate an even more difficult operation: developing from a single-celled zygote into a complex organism with many different cell types and many different organs and organ systems. Embryonic development is a task of great complexity, orchestrated by an orderly sequence of gene expression events that lead to the establishment of specific body plans and the differentiation of particular cell types, often in tune with signals received from the surrounding world. Errors in the expression of a single critical gene can scramble the process enough to result in deformity or death (Figure 14.7).

In many organisms, the identity of cell types or body parts is governed by specific cascades of gene expression. In a **gene cascade**, genes are turned on or off, one after another in sequence—like a series of falling dominoes, in which one domino knocks over the next, which in turn knocks over the next, and so on. In a typical gene cascade, one or more "master-switch genes" are activated inside a cell, usually by a signal received from the surrounding environment. The function of the protein product of the master gene is to turn on, or off, specific clusters of genes in specific cell types. The proteins produced by this second set of genes turn on, or off, a third set of genes, often in collaboration with other gene products and in response to additional signals. Eventually, genes are expressed whose protein products alter the structure and function of cells, enabling cells to become specialized for particular tasks.

Homeotic genes are a class of master-switch genes that have a central role in the development of the body plan and in the differentiation of organs in animals and plants. Gene cascades involving homeotic genes drive gene expression patterns along an ever narrower pathway, until a very specific outcome, such as the differentiation of an antenna, is achieved. Given this crucial role, it is not surprising that defective versions of homeotic genes can have striking phenotypic effects, such as those shown in Figure 14.7. In fact, the existence of homeotic genes was revealed through the analysis of developmental mutants, such as the *antennapedia* mutants of fruit flies. The *Antennapedia* gene (*Ant*) controls gene cascades that drive the formation of legs. If a mutation causes this gene to be expressed in the head segment of the fly body, legs appear where the antennae would normally be (Figure 14.7*b*).

At different times during an individual's development, different sets of homeotic genes are active in the different cell types in the body. For example, a homeotic gene that coordinates the development of the eye will be active in cells that give rise to the eyes. In other parts of the body, although this gene is present, it is not likely to be active; instead, other homeotic genes are expressed that specify the identity of other body parts. As the body changes during development, the homeotic genes expressed by cells also change.

In the 1990s, similar homeotic genes were found to control development in similar ways in organisms as different as fruit flies, mice, and humans (Figure 14.8). The homeotic genes of multicellular animals first evolved hundreds of millions of years ago, and since then they have been used in similar ways to organize the body plans of animals as different as sea anemones and orca whales. Homeotic genes are an example of highly **conserved genes**, those that have a similar base sequence and general function among diverse groups of organisms. Genes that control early stages of development—events such as specification of the head and tail ends of an embryo or the demarcation of body segments—are more likely to be conserved, compared to genes that act later in development (genes that control brain development, for example).

(*a*) Head of a normal fruit fly

(*b*) Head of a developmental mutant

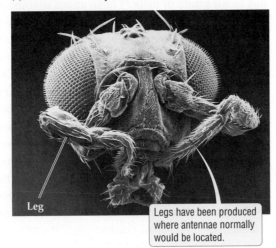

Legs have been produced where antennae normally would be located.

Figure 14.7 A Developmental Mutation with a Bizarre Phenotype

(*a*) The head of a normal (wild type) fly. Note the location of the antenna. (*b*) The head of a mutant fly, known as *antennapedia*. Note the presence of legs where the antenna should be. The *Antennapedia* gene (*Ant*) is a homeotic gene that normally guides the development of legs on the body segments immediately behind the head. When a mutation causes *Ant* to be expressed in the head segment, the gene represses antenna formation and activates leg development instead.

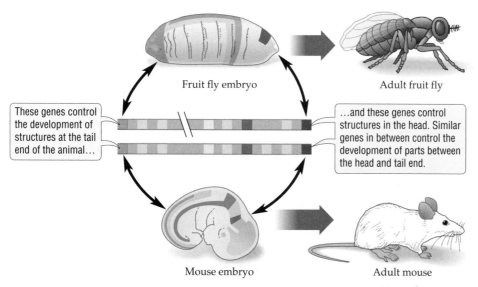

Figure 14.8 The Similarity of Homeotic Genes among Different Organisms Reflect Shared Evolutionary Origins
Development is controlled by similar homeotic genes in organisms as different as fruit flies and mice. The homeotic genes that control development in fruit flies and mice are arranged in a similar order on fruit fly and mouse chromosomes. As it happens, the body parts affected by these genes are positioned in the same order in which the genes are arranged on the chromosomes. Similar homeotic genes, and the structures whose development they control, are matched here by color. The segment of the fruit fly chromosome shown here contains additional DNA sequences that are not found in the matching region of the mouse chromosome. Those additional sequences have been omitted and the omission is represented by the "gap" in the chromosome (between the slanted lines).

14.4 How Cells Control Gene Expression

Cells receive signals that determine which of their genes are turned on or off. Some of these signals are sent from one cell to another, as when one cell releases a signaling molecule that alters gene expression in another cell. Cells also receive signals from the organism's internal environment (for example, blood sugar level in humans) and external environment (for example, sunlight in plants). Overall, cells process information from a variety of signals, and that information affects which genes are expressed.

The expression of most genes is controlled at the level of transcription

The most common way for cells to control gene expression is to regulate the transcription of the gene. In the absence of gene transcription, the RNA encoded by the gene is not made, the protein encoded by the RNA cannot be made, and any phenotype directly produced by the protein is therefore absent. The bacterium *E. coli*, for example, requires a supply of the amino acid tryptophan. If tryptophan is available in the external environment, the bacterium absorbs it rather than wasting cellular resources to make it. But if tryptophan is not readily available, the bacterium expresses a series of five genes that together encode the enzymes needed for making tryptophan "from scratch" inside the cytoplasm.

E. coli controls these five genes in the following way: When tryptophan is present in the environment, it binds to a repressor protein in the bacterial cell (**repressor proteins** are so named because they stop the expression of one or more genes). This tryptophan–repressor protein complex can then bind to the tryptophan operator. An **operator** is a sequence of DNA that controls the transcription of a gene cluster—in this case, the five genes needed to make tryptophan. When bound to the operator, the tryptophan–repressor protein complex blocks access to the promoter of the tryptophan genes (Figure 14.9a). As we saw in Chapter 13, the promoter is the segment of a gene that recruits the RNA polymerase and guides it to the transcription start site. In

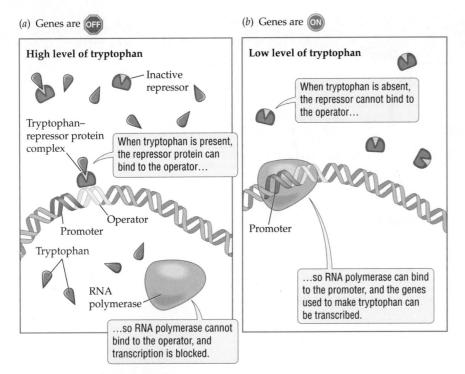

(a) Genes are OFF

High level of tryptophan

Inactive repressor

Tryptophan–repressor protein complex

When tryptophan is present, the repressor protein can bind to the operator…

Operator

Promoter

Tryptophan

RNA polymerase

…so RNA polymerase cannot bind to the operator, and transcription is blocked.

(b) Genes are ON

Low level of tryptophan

When tryptophan is absent, the repressor cannot bind to the operator…

Promoter

…so RNA polymerase can bind to the promoter, and the genes used to make tryptophan can be transcribed.

Figure 14.9 Repressor Proteins Turn Genes Off

In the bacterium *E. coli*, a repressor protein binds to an operator to control the transcription of a group of genes that encode the enzymes needed to make tryptophan. (*a*) When tryptophan is present, it binds to a repressor protein; this tryptophan–repressor protein complex then turns the genes off by binding to the operator. (*b*) On its own (that is, when tryptophan is absent), the repressor protein cannot bind to the operator, but RNA polymerase *can* bind to the promoter, which turns on the genes for tryptophan synthesis.

Explore the tryptophan operator. 14.3

the presence of tryptophan, the tryptophan–repressor protein complex prevents RNA polymerase from binding to the promoter. But in the absence of tryptophan, the repressor protein cannot bind to the operator, so RNA polymerase is free to bind to the promoter, and the genes are transcribed (Figure 14.9*b*). As a result, the bacterial cells do not make tryptophan when it is present in the surrounding medium, but they do manufacture it when tryptophan levels are low. This control of gene expression ensures two things: the cell always has an adequate supply of tryptophan, and the cell does not waste resources producing tryptophan when it is readily available in the environment.

A few genes, such as those in *E. coli* that encode the tryptophan–repressor protein complex, are always expressed at a low level; their transcription is not regulated, because they are needed at all times. The tryptophan–repressor protein complex must be readily available, for example, in case the need arises to prevent tryptophan synthesis because tryptophan becomes available from the surroundings. But most genes in pro-

karyotes and eukaryotes are regulated, and usually the control is exerted at the level of gene transcription.

In general, transcription is controlled through two essential elements: **regulatory DNA** that can activate or inactivate gene transcription; and gene **regulatory proteins** that interact with signals from the environment and also with the regulatory DNA to promote or repress gene transcription. Both of these elements of transcription control are illustrated by the regulation of tryptophan synthesis in *E. coli*. The tryptophan operator, which can switch gene transcription on or off, is an example of regulatory DNA. The repressor protein that binds to the tryptophan operator when tryptophan is present is an example of a gene regulatory protein. Together, regulatory DNA and gene regulatory proteins are responsible for the transcriptional activation or inactivation of genes in both prokaryotes and eukaryotes.

Gene expression can be regulated at several levels

In eukaryotes, gene expression can be controlled at a number of key steps in the pathway from gene to protein (Figure 14.10) to phenotype. The activity of a gene can be regulated before transcription, at transcription, during RNA processing, or at translation. Gene expression can also be regulated after translation by control of the activity and life span of the protein. For genes that code for proteins, the activity of the encoded protein is what ultimately enables that gene to influence a phenotype. Let's first examine how the pathway of gene expression from gene activation to protein function is controlled (see Figure 14.10). We will then round out the concept of gene expression by connecting this pathway with its final outcome: the manifestation of a specific phenotype.

1. *Tightly packed DNA is not expressed* (Figure 14.10, control point 1). During interphase, when most gene expression occurs, each chromosome exists as a looped 30-nanometer fiber, with regions that are "unpacked" down to the bead-on-a-string state. The unpacking allows the transcriptional proteins access to regulatory DNA and gene promoters, which makes it possible for these regions to be transcribed. In contrast, the tightly packed regions of the chromosome are transcriptionally inactive, in part because the proteins necessary for transcription, such as gene regulatory proteins and RNA polymerase, do not have access to their target DNA sequences.

DNA packing is itself a regulated process, and most eukaryotic cells have an elaborate complex of proteins that can selectively condense or decondense portions of a chromosome, depending on the cell type or specific signals received by that cell.

2. *Transcription can be regulated* (Figure 14.10, control point 2). Regulation of transcription, as described in the previous section, is the most common means of controlling gene expression. This control mechanism is efficient because it enables the cell to conserve resources when it does not need the gene product. The downside is that transcriptional activation of gene expression is relatively slow in eukaryotes: even in the best-case scenarios it takes 15 to 30 minutes to make a protein following the transcriptional activation of a gene.

3. *The breakdown of mRNA molecules can be regulated* (Figure 14.10, control point 3). Most mRNA molecules are broken down within a few minutes to hours after they are made; a few persist for days or weeks in the cytoplasm. The longer the life of an mRNA molecule, the more protein molecules can be made from it. How long an mRNA molecule will persist is often determined by its chemical properties: some mRNAs have chemical features that make them less stable. By limiting the life span of many types of mRNA, a cell prevents the wasteful synthesis of proteins that are needed only temporarily. But why continue making mRNA if the protein is not needed? If circumstances change and the protein is needed, its mRNA can be quickly stabilized, and it can then be translated and accumulated rapidly. Time is saved because transcription does not have to be activated, so protein levels often start rising just a few minutes after the need for the protein is sensed.

4. *Translation can be regulated* (Figure 14.10, control point 4). Specific RNA binding proteins can attach to their target mRNA molecules and block their translation into protein. This method of control is especially important in the regulation of some types of long-lived mRNA. It allows the cell to deactivate mRNA whose protein product cannot be used immediately by the cell, but that must be kept in readiness for rapid protein synthesis in case circumstances change. For example, some immune cells in the body make large amounts of mRNA for certain signaling proteins, called cytokines, but don't allow them to be translated. If these cells detect an invading bacterium, the translation block is immediately lifted. Cytokines are translated within minutes and poured into the bloodstream, where they act like an early

Test your understanding of the control of gene expression in eukaryotes. ▶⏸ 14.4

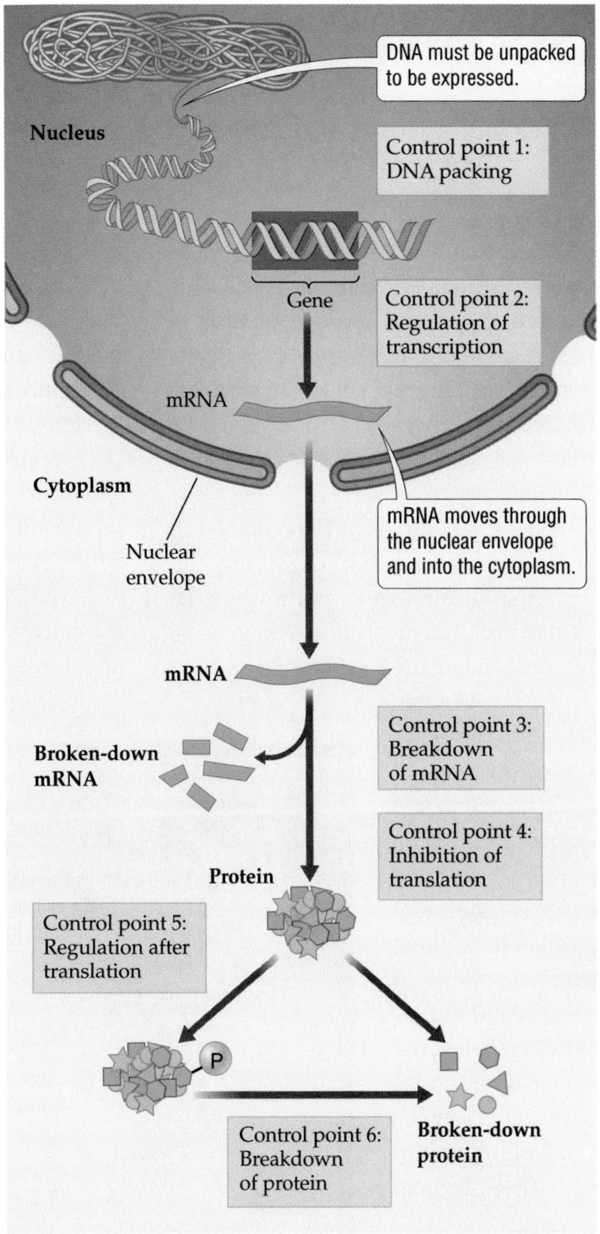

Figure 14.10 Control of Gene Expression in Eukaryotes
Eukaryotes can control gene expression in many ways. Each point in the pathway from gene to protein represents a point at which cells can regulate the production or activity of proteins.

warning system that prepares other components of the immune system to defend the body.

5. *Proteins can be regulated after translation* (Figure 14.10, control point 5). Many proteins must be trimmed, or chemically modified, before they can exert their effect on a phenotype. Some of the blood-clotting proteins, for example, are synthesized as inactive precursors; a segment of the protein must be cleaved off from

Science Toolkit

Using DNA Chips to Monitor Gene Expression

Geneticists have long realized that the metabolism and phenotype of an organism are influenced by many genes. But organisms have thousands of genes, and until recently it was not possible to study how the expression of large numbers of genes changes during particular stages of development or under a particular set of environmental conditions. With the advent of DNA chips, first developed in the late 1990s, it is now possible to monitor the expression of many genes at once.

A **DNA chip** consists of thousands of samples of DNA placed in a regimented order on a small glass surface, or "chip," roughly the size of a dime. DNA chips are constructed in two main ways. In one method, a robotic arm that has multiple printing tips is used to deliver minute droplets of single-stranded DNA (500–5,000 bases long) to the chip's glass surface. The DNA is then treated and dried so that it will bind to the glass. The DNA sequence in each droplet is known (it may, for example, correspond to the exons of a gene), and each droplet has a known location on the chip.

In the second method, shorter segments of DNA (20–80 bases long) are used. These short pieces of single-stranded DNA can be synthesized directly on the glass chip, via a technique similar to the way computer chips are made (hence the name "DNA chip"). It is also possible to synthesize the fragments of DNA elsewhere (using a machine designed for that purpose); the fragments are then placed on the chip in a specific order and immobilized.

With either of these methods, DNA representing some or all of an organism's genes can be placed on a single DNA chip and used to screen the expression of many genes simultaneously. How is this done? Although a considerable number of technical steps are necessary, the basic idea is simple. When a gene is expressed, an mRNA copy of the information in that gene is produced. To monitor which genes are being expressed, mRNA

is isolated from the organism or cells being tested, labeled (as with a dye that glows red or green), and then washed over the DNA chip. Both the mRNA and the DNA on the chip are single-stranded, so the labeled mRNA can bind to the DNA representing the gene from which it was originally produced.

Next, a scanner is used to detect the locations on the chip where DNA was able to bind to labeled mRNA. Because the gene that corresponds to each location on the chip is known, results from the scanner tell us which of the organism's genes produced mRNA—and hence which of the organism's many genes were expressed (and which were not). Finally, the genes expressed in different circumstances can be compared, providing valuable information about how organisms regulate development and cope with environmental variation.

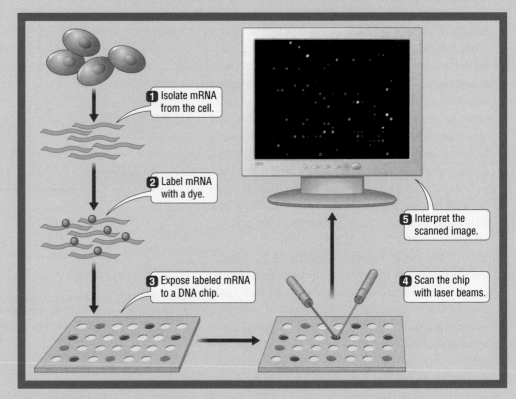

Using DNA Chips

such a protein before it can help plug a wound. A translated protein can also be rendered inactive by chemical modification or through the action of other molecules that bind to it. For example, liver cells respond to a hormone called glucagon by attaching a phosphate group on glycogen synthase, an enzyme that makes an energy-storing carbohydrate called glycogen. Glycogen synthase becomes inactive when it has a phosphate group on it. With the enzyme blocked, sugars cannot be stockpiled as glycogen and are instead released into the blood.

6. *Protein function can be regulated through control of a protein's life span* (Figure 14.10, control point 6). The activity of a protein provides one final opportunity to control the pathway from gene to protein. Most proteins in the cell have a limited life span, but a few, such as collagen and crystallin, last us through our lifetime. Proteins that are no longer needed, or are damaged, are taken apart and their amino acids recycled into new proteins. The recycling of unnecessary proteins enables the cell to invest resources where they are most needed. The excessive accumulation of proteins can even kill a cell. Brain cells, perhaps because of the complexity of their function, appear to be especially vulnerable to damage from protein deposits. Alzheimer's, Parkinson's, and Huntington diseases are marked by the gradual death of brain cells after large protein clumps appear inside the cells, probably because of a failure to dispose of damaged proteins.

A gene expression pathway can be influenced by other genes and by environmental signals

Having described how cells control when, where, and how much of a gene product is made, we turn now to an overarching question: how does the activation of a gene influence a particular phenotype? In many instances, the activation of a single gene has a powerful and direct effect on a phenotype. Consider, for example, the effect of homeotic genes on a developing organism. Recall that a homeotic gene controls the expression of a series of other genes that influence development. If the protein specified by a single homeotic gene fails to function properly, the expression of all the other genes in the cascade will be affected, and extreme phenotypic effects may result (see page 292 and Figure 14.7).

However, the genetic information encoded in a gene does not always exercise complete control over the phenotype. Signals from the environment can not only activate or inactivate a gene, but also alter any of the many steps between a gene and the function of the protein it encodes. Other gene products, whether produced by the same cell or by neighboring cells, can also tweak the gene expression pathway of a given gene. Any modification of the gene expression pathway can potentially affect the phenotype by changing when, where, and how much of a gene product is made and the extent to which it is active. For example, the gene coding for the enzyme that adds phosphate groups on glycogen synthase has the potential to alter any phenotype that the glycogen synthase gene influences. As we also discussed in Chapter 10 (see Figure 10.15), a phenotype is not necessarily the product of a single genotype (the alleles of a specific gene); a phenotype can also be altered by environmental influences and the activity of other genes. We have seen in this chapter how the many steps in a given gene expression pathway provide opportunities for environmental signals and other genes to exert their influence.

Concept Check

1. Why are most genes controlled at the level of transcription?

2. If transcriptional control is the most favored method of gene regulation, why aren't all genes controlled at the level of transcription?

3. How can gene expression be modified at the posttranslational level—that is, after the protein has been made?

4. Is the expression of a gene controlled exclusively by the information encoded in its DNA? Explain.

Applying What We Learned

From Gene Expression to Cancer Treatment

A revolution has begun in the field of genetics—one that may ultimately rival the computer revolution in its effect on society. Like the computer revolution, this genetic revolution is being pushed forward by technical advances. However, the revolution in genetics is not centered on a particular new type of machine or technology. Instead, it is due to a shift in perspective, from the study of single genes to the study of large numbers of genes simultaneously.

What can we learn by studying many genes at once? Scientists are using new technologies to discover what happens to patterns of gene expression when cells face changing environmental conditions. Do cells turn a small handful of genes on or off, or do they change the expression of large numbers of their genes? Early results suggest that environmental changes alter the expression of large numbers of genes.

In 1996, for example, scientists published the complete DNA sequence of *Saccharomyces cerevisiae*. They found that this yeast has about 6,000 genes. The researchers built multiple, identical copies of a DNA chip (see the "Science Toolkit" on page 296) that would hold all of these genes, and they used these chips to monitor the activity of all the yeast genes at once. They learned, for example, that as the yeast adjusted to changes in food availability, 710 genes that had been inactive were turned on, while 1,030 genes that had been active were turned off.

Scientists can now build DNA chips that have the potential to offer "personalized medicine" to patients. For example, by surveying how individuals with different genetic characteristics respond to different drugs, doctors may be able to use DNA chips to decide which of several possible drugs would be likely to work best in any given patient. In a family practice setting, mass-produced DNA chips could be used to identify the exact strain of bacteria causing a particular individual's sore throat, along with the antibiotics to which that bacterial strain is resistant (and which, therefore, should not be prescribed by the physician).

DNA chip results can even provide new hope for cancer patients. Consider breast cancer. In the United States, one in seven women will get breast cancer, and half of those women will die from it. When a breast cancer is detected and the tumor removed, doctors use characteristics of the tumor (such as its size and appear-

ance) to help them decide what treatment the patient should receive after surgery. The goal is to use the most aggressive treatments (such as various forms of chemotherapy, which can have toxic side effects) only when needed—that is, only for those cancers that are most likely to spread to other parts of the body.

Most methods currently used to predict whether a cancer will spread are not very accurate. As a result, many patients receive unnecessary, toxic treatments: roughly 80 percent of breast cancer patients who could be cured by surgery and radiotherapy alone are advised to have chemotherapy. To improve our ability to predict whether a patient's cancer will spread, a team of doctors in the Netherlands and the United States used DNA chips to study the expression of all 25,000 human genes in breast cancer tumors. Their findings revealed that different sets of genes are turned on or off in breast cancer tumors, depending on the severity of the cancer. Applying these results, the doctors correctly predicted whether the cancer would spread in 65 of 78 cases. This amounts to a successful-prediction rate of 83 percent, a big improvement over previous methods. In addition, the percentage of patients who could be cured by surgery and radiotherapy alone but who were advised incorrectly to have chemotherapy dropped from 80 percent to 25 percent.

In the examples just discussed, DNA chips could help doctors by removing the guesswork from the treatment of conditions ranging from sore throats to cancer. Keep in mind, though, that many human diseases are heavily influenced by factors other than our genes. Recall that about 70 percent of the cases of colon cancer, stroke, heart disease, and type 2 diabetes may be prevented simply by adopting a "low-risk" lifestyle (see the "Biology Matters" box in Chapter 11). For such disorders, results from DNA chips alone cannot provide doctors with the guidance they need. Instead, as we battle such disorders, the best approach may be to identify which of our many genes influence whether people with similar lifestyles will contract particular diseases. Although this is a daunting task, it is well worth the effort: by understanding the effects of both lifestyle and genetics, we may be able to dramatically improve our ability to treat and even prevent disease.

Biology in the News

A Race for Miracle Cures—and Billions

By John Lauerman

Two Nobel Prize winners . . . are racing to turn a 10-year-old genetic discovery into novel therapies that may help drug development at major pharmaceutical companies. Phillip Sharp . . . and Craig Mello . . . are seeking to be the first to use a technology called RNA interference, or RNAi, which blocks the actions of genes. RNAi's potential against almost any disease may lead to drugs with sales "in the billions" . . . [and] the technology may lead to therapies against infectious diseases, cancer and arthritis . . . Viruses, for example, are composed almost entirely of genetic material and cancer is principally a disease of flawed DNA. By turning off certain genes through the process of RNAi, scientists are attempting to squelch the biological machinery that drives illnesses as varied as bird flu and AIDS.

. . . Sharp said he had been stunned by the implications. "I understood that this was really a very general biological phenomenon," Sharp recalled. "It was ancient and broad and deep and powerful, and no one had ever recognized it. It had to be a fundamental principle." Even one-celled algae use RNAi to turn off genes when they need to stop making certain proteins or stop a biological process . . . That suggests that RNAi evolved as part of a molecular on-off switch millions of years sooner than multicelled plants and animals themselves.

RNA interference is a mechanism for controlling gene expression. Most eukaryotes have RNAi components, suggesting that RNAi appeared very early in their evolution. RNAi probably evolved as a defense against foreign genetic material, especially viruses and transposons. In many eukaryotes, however, RNAi is also used to regulate native genes—genes normally found in the organism's genome that have useful functions.

RNAi works by selectively destroying certain types of RNA or suppressing the translation of certain target mRNAs. The way RNAi fights against the takeover of a cell by an RNA virus illustrates the basic strategy. The genetic material of many viruses is a double-stranded RNA (dsRNA), and certain RNAi proteins can recognize these "strange" nucleic acids. Next, a protein called Dicer chops the strange dsRNA into small chunks known as small interfering RNAs (siRNAs). A complex of proteins, called the effector complex, picks up one of the two strands in siRNA and then goes on a "search and destroy" mission for any RNA that is complementary to the bound siRNA. When such RNA is found, the effector complex either degrades it or prevents it from being translated into protein. In essence, the RNAi machinery grabs a piece of the offending double-stranded RNA and uses it as means of recognizing, and shutting down, all RNA molecules that resemble it.

The discovery of RNAi has generated much excitement because of its potential as a tool for understanding basic biology and for diagnosing and treating infectious diseases and genetic disorders. That is because RNAi offers a relatively straightforward means for eliminating the expression of a specific gene. If we can get the cell to make "strange" double-stranded RNA that is complementary to the mRNA of the gene we want to knock out, RNAi will do the rest. It will block the target gene from being expressed in that cell. So far, RNAi has had the most dramatic impact on basic research. In hundreds of labs around the world, scientists have designed synthetic siRNAs and introduced them into living cells to silence the expression of a specific gene, in an effort to understand the relationship between that gene and its phenotype. Several clinical trials of RNAi-based therapy are under way in humans. Some are directed at knocking out the HIV virus or the hepatitis virus. Some are designed to destroy the expression of mutated genes that produce harmful effects, leading to such conditions as age-related macular degeneration (the most common cause of blindness in older people) and pneumonia caused by a virus called RSV.

Evaluating the News

1. What is RNA interference? How might it benefit the eukaryotic organisms that use it?
2. Some people object to gene therapy on the grounds that altering a person's genome is not ethical. Since RNAi does not alter genes but rather blocks their effects, do you think RNAi-based therapy is less objectionable? Explain your position.
3. Some mammals, including humans, have a rare mutation in a gene that codes for myostatin, a protein that blocks muscle growth. Lack of myostatin leads to exceptionally large muscles. In principle, human athletes could use RNAi for "gene doping": synthetic RNA molecules could be used to knock out the myostatin gene. Gene doping using RNAi may be difficult to detect. As a society, what stance should we take on undetectable gene doping? Should we ban RNAi research because it represents a "slippery slope"?

SOURCE: *International Herald Tribune*, June 8, 2007.

Chapter Review

Summary

14.1 The Structural and Functional Organization of DNA

- Compared with eukaryotes, prokaryotes have a small amount of DNA, and it is mostly on a single chromosome. Most prokaryotic DNA encodes proteins, and functionally related genes in prokaryotes are grouped together in the DNA.
- With regard to DNA, eukaryotes differ from prokaryotes in several ways: (1) In eukaryotic cells, the DNA is distributed among several chromosomes. (2) Eukaryotes have more DNA per cell than prokaryotes have, in part because they have more genes than prokaryotes have. In addition, genes constitute only a small portion of the genome in many eukaryotes; the rest consists of noncoding DNA (including introns and spacer DNA) and transposons. (3) Eukaryotic genes with related functions often are not located near one another.

14.2 DNA Packing in Eukaryotes

- Cells can pack an enormous amount of DNA into a very small space because their DNA is highly organized by a complex packing system. In eukaryotes, segments of DNA are wound around histone spools, packed together tightly into a narrow fiber, and further folded into loops.
- Chromosomes are most tightly packed during mitosis and meiosis. The packing is looser during interphase, when most gene expression occurs.
- In DNA regions that are tightly packed, genes cannot be expressed, because the proteins necessary for transcription cannot reach the genes.

14.3 Patterns of Gene Expression

- In both prokaryotes and eukaryotes, genes are turned on and off selectively in response to short-term changes in environmental conditions.
- Development in multicellular eukaryotes is controlled by changes in gene expression, which may be influenced by internal and external signals.
- The different cell types in a multicellular organism express different sets of genes.
- Housekeeping genes, which have an essential role in the maintenance of cellular activities, are expressed in most cells of the body.
- Eukaryotes control gene expression in development through gene cascades, many of which are regulated by homeotic (master-switch) genes.

14.4 How Cells Control Gene Expression

- Cells regulate most genes by controlling transcription.
- Transcription is controlled by regulatory DNA sequences that interact with regulatory proteins to switch genes on and off.
- An operator is a regulatory DNA sequence that controls the expression of a single gene, or a group of genes, in prokaryotes.
- Gene regulatory proteins link gene expression to internal and external signals.
- Some regulatory proteins (repressor proteins) inhibit transcription, others (activator proteins) promote it.
- Eukaryotic cells can control gene expression at several points in the pathway from gene to protein.
- The phenotype of an organism results from the combined effects of the organism's genotype (which may include multiple interacting genes) and the environment.

◉ Review and Application of Key Concepts

1. The total length of the DNA in a eukaryotic cell is hundreds of thousands of times greater than the diameter of the nucleus in which it is found. Explain how so much DNA can be packed inside the nucleus.

2. Summarize the major differences between DNA in prokaryotes and in eukaryotes, emphasizing differences in the amount and function of DNA.

3. Imagine you transferred a bacterium from an environment in which glucose was available as food to an environment in which the only source of food was the sugar arabinose. A specific enzyme is required to digest arabinose but not glucose. How do you think gene expression in the bacterium would change as a result of your action?

4. Cell types in multicellular organisms often differ considerably in structure and in the metabolic tasks they perform, yet each cell of a multicellular organism has the same set of genes. Explain how cells with the same genes can be so different (a) in structure and (b) in the metabolic tasks they perform.

5. What are homeotic genes? How would you explain the observation that homeotic genes are conserved in organisms as diverse as fruit flies and humans?

6. Summarize how gene expression can be controlled at various steps in the process of generating a protein that is fully active.

7. Describe the function and method of operation of the tryptophan operator in *E. coli*.

8. What is a DNA chip? How can DNA chips be used to study many genes at once?

Key Terms

conserved gene (p. 292)
DNA chip (p. 296)
gene cascade (p. 292)
gene expression (p. 286)
genome (p. 287)
histone protein (p. 289)
homeotic gene (p. 292)
housekeeping gene (p. 291)

intron (p. 288)
noncoding DNA (p. 288)
operator (p. 293)
regulatory DNA (p. 294)
regulatory protein (p. 294)
repressor protein (p. 293)
spacer DNA (p. 288)
transposon (p. 289)

Self-Quiz

1. A segment of noncoding DNA that separates two genes is called
   a. an intron.
   b. spacer DNA.
   c. a transposon.
   d. an exon.

2. In prokaryotes and eukaryotes, gene expression is most often controlled by regulation of which of the following?
   a. the destruction of a gene's protein product
   b. the length of time mRNA remains intact
   c. transcription
   d. translation

3. Which of the following is a regulatory DNA sequence?
   a. repressor protein
   b. operator
   c. intron
   d. housekeeping gene

4. During development, different cells express different sets of genes, resulting in
   a. the formation of different cell types.
   b. gene mutation.
   c. DNA packing.
   d. developmental abnormalities.

5. The DNA of eukaryotes is packed most loosely at which time?
   a. during mitosis
   b. during meiosis
   c. during interphase
   d. both b and c

6. Select the term that best describes one copy of all the DNA-based information of an organism.
   a. genes
   b. exon
   c. gene expression
   d. genome

7. Genes that have an essential role in cellular activities and that are expressed in most cell types are called
   a. regulatory genes.
   b. housekeeping genes.
   c. homeotic genes.
   d. enzymes.

8. A DNA sequence that can move from one position on a chromosome to another is called
   a. an operator.
   b. a transposon.
   c. an exon.
   d. an intron.

9. Assume that an organism's genome has 20,000 genes and a large quantity of DNA, most of which is noncoding DNA. What kind of organism is it most likely to be?
   a. an insect
   b. a plant
   c. a bacterium
   d. either a or b

CHAPTER 15 DNA Technology

Key Concepts

- DNA technology refers to the techniques used to modify and manipulate DNA and the use of such techniques for research, health, forensic, and commercial purposes.

- Restriction enzymes cut DNA molecules at specific target sequences. Gel electrophoresis is a technique for separating pieces of DNA on the basis of their size.

- We clone genes by isolating them, joining them with other pieces of DNA to make recombinant DNA, and introducing the recombinant DNA into a host cell, such as a bacterial cell, that can generate many copies of the genes.

- Automated sequencing machines can be used to determine the nucleotide sequence of a cloned gene.

- Cloning and sequencing a gene can provide vital clues about the role of that gene in the life of a cell or organism. It can also help us understand how allelic forms of that gene lead to various phenotypes, including inherited genetic disorders in humans.

- Genetic engineering is the introduction of foreign DNA into an organism such that the introduced DNA becomes incorporated into the genetic material of the recipient. Expression of the introduced gene usually changes one or more phenotypes in the recipient, which is considered a genetically modified organism (GMO).

- DNA technology provides many benefits, but its use also raises ethical concerns and poses potential risks to human society and natural ecosystems.

Main Message: DNA technology makes it possible to isolate genes and produce many copies of them, and to introduce them into whole organisms.

From Glowing Bunnies to Hopes for New Cures

Some jellyfish produce flashes of light that may serve to ward off attacks from predators. A few years ago, these same lights made the headlines: the gene that enables jellyfish to produce light was transferred to a rabbit named Alba, creating a piece of "living art" meant to confront people with a creature that was both lovable and alien. Alba was never shown in an art exhibit because the outcry over her creation caused her to be confined to the laboratory in which she was made.

Alba was a product of DNA technology, which enables scientists to isolate genes, analyze them, manipulate them, modify them, and introduce them into almost any organism. DNA technology is used for more than "stupid pet tricks" or controversial art. In fact, it is fair to say that DNA technology has transformed our world in far-reaching ways. Consider some recent news items. A team of researchers concluded that beaked whales don't just live on squid. They eat certain species of bony fishes as well, since DNA from the fishes could be identified in the well-digested meals recovered from the guts of these elusive mammals. Anthropologists suggest that some Neandertals may have been redheads, because DNA from the fossil bones of these human relatives contains an allele that is responsible for red hair and pale skin in modern humans. A geneticist in Spain extracted DNA from the bones of Christopher Columbus and is attempting to compare it to DNA volunteered by thousands of people, including many named Colom or Colombos, who hope that genetic analysis will show them to be related to the Great Navigator. To certify the authenticity of sports memorabilia, all the balls used in Super Bowl XXXIV were marked with a unique DNA fragment that can be "read" with a special laser-based detector.

In the past three decades, DNA technology has helped convict hundreds of killers and has exonerated over 200 people who were wrongfully imprisoned before DNA-based forensic technology became widely available. Those who like to explore

Alba, a White Rabbit Genetically Modified to Glow

A gene coding for a jellyfish protein called green fluorescent protein (GFP) was introduced into the fertilized egg from which this rabbit developed. Many other "glowing" organisms, from tobacco plants to zebra fish, have been raised through DNA technology. As flamboyant as these experiments may sound, the ease with which GFP can be tracked inside a cell has led to tremendous advances in our understanding of basic cell biology. The three scientists who contributed to the discovery and use of GFP in biology and biomedicine shared the 2008 Nobel Prize in Chemistry.

genealogies as a hobby can buy DNA kits that will let them compare their DNA with that of an obliging stranger in another city who happens to have the same last name. For a few thousand dollars, some companies will create a DNA profile of you: they will list your genotype for hundreds of genetic loci, including alleles associated with such traits as having a good verbal memory, or perfect pitch, or a tendency to take risks.

Almost 75 percent of the prepared foods sold in the United States contain at least one ingredient that has been modified by having DNA inserted into it. Nearly a million diabetics rely on human insulin, which is just one of the many lifesaving pharmaceutical products made through DNA technology. At the very forefront of DNA technology, and still in its infancy, is the strat-egy of curing human genetic disorders through gene therapy.

How can scientists isolate and analyze DNA, sometimes from just a few cells, and sometimes from cells that have been dead half a million years? What is gene cloning? Are foods that have been modified through DNA technology potentially harmful to humans or our ecosystems? Can gene therapy work? In this chapter we will consider some of the tools and techniques of DNA technology, and see how they have been applied to expand scientific knowledge or generate practical results such as medical advances and commercial products. We will also examine some of the ethical and social concerns that have been raised about the potential risks of this new technology.

I ndirectly, people have been manipulating the genetic material of other organisms for thousands of years. This fact is well illustrated by the many differences between domesticated species and their wild ancestors. For example, because of genetic changes brought about through selective breeding, dog breeds differ greatly from one another and from their wild ancestor, the wolf. Similarly, because of selective breeding, food plants such as corn bear little resemblance to the wild species from which they arose.

Although we have a long history of manipulating the genetic characteristics of other organisms through selective breeding (Figure 15.1a), the past 40 years have witnessed a huge increase in the power, precision, and speed with which we can alter the genes of organisms ranging from bacteria to mammals. We can now select a particular gene, produce many copies of it, and then transfer it into a living organism of our choice. In doing so, we can alter DNA directly and rapidly in ways that do not happen naturally (Figure 15.1b).

(a)

(b)

Figure 15.1 Traditional Breeding versus Direct DNA Manipulation
When it comes to strange appearances, conventional breeding methods and DNA technology can both generate some extraordinary phenotypes. (a) This British Belgian Blue bull was developed through conventional selective breeding practices for heavily muscled, low-fat meat. (b) These "naked" chickens were genetically altered using the tools of DNA technology. The goal of the research project was to develop low-fat poultry that would be cheap and environmentally friendly, in part because there would be no feathers to remove and dispose of.

We begin this chapter with a broad overview of DNA technology. We then describe the main procedures and special techniques that scientists use to isolate and analyze a gene and to introduce it into an organism so as to alter the phenotype of the recipient. We then turn to some of the many practical applications of this technology, including DNA fingerprinting and genetic engineering. We close by considering some of the ethical issues and social and environmental risks associated with DNA technology.

15.1 The Brave New World of DNA Technology

The strategies and techniques that scientists use to analyze and manipulate DNA are broadly known as **DNA technology**. DNA is a polymer of nucleotides and it serves as the genetic material in all organisms. Although the nucleotide (base) sequence of DNA can vary greatly between species, and among individuals of a species, the general chemical structure of the DNA molecule (see Figure 12.3) is the same in all species. This consistency means that similar laboratory techniques can be used to isolate and analyze DNA from organisms as different as bacteria and people.

DNA can be readily extracted from most cells and tissues. Extracted DNA can be cut into smaller fragments with certain enzymes (called restriction enzymes), separated in a gel-like material, stained with certain dyes, and viewed under special lights. A revolutionary technique known as the polymerase chain reaction (PCR) makes it possible to amplify a target DNA so that billions of copies can be generated from a single molecule of the desired DNA. The information in a piece of DNA—its nucleotide sequence—can be determined by robotlike machines. Such machines have made it possible to decipher *all* the information held in the chromosomes (**genomes**) of many species of bacteria, fungi, plants, and animals. In the year 2000, teams of scientists from across the globe published the nucleotide sequence of the human genome, as part of the Human Genome Project (described in greater detail in Interlude C).

Fragments of DNA can be joined together with the help of special enzymes, creating an artificial assembly of genetic material known as **recombinant DNA** (Figure 15.2a). Many bacteria, for example, have closed loops of DNA, called plasmids, in addition to the single large chromosome typical of prokaryotic cells; it is a relatively simple matter to extract these plasmids from the bacterium and insert a foreign gene into them to create recombinant DNA molecules. **DNA cloning**, or gene cloning, is the introduction of recombinant DNA into a host cell that can generate many copies of the introduced DNA (Figure 15.2b). The purpose of DNA cloning is to multiply a particular type of recombinant DNA so that a large amount of this DNA is made available for further analysis and manipulation. Bacteria are the most common host cells in DNA cloning. Introduced recombinant plasmids can multiply rapidly, creating hundreds of copies of the desired DNA within the cytoplasm of a host bacterium.

Genetic engineering is the permanent introduction of one or more genes into a cell, a certain tissue, or a whole organism, leading to a change in at least one genetic characteristic in the recipient. The organism receiving the DNA is said to be genetically modified (GM) or genetically engineered (GE). The gene introduced into a genetically modified organism is known as a transgene (*trans*, "across from"), so GM individuals are also known as "transgenic organisms." Genetic engineering is aimed at changing one or more inherited characteristics (phenotypes) in the cell or organism receiving the DNA. A great variety of genetically modified organisms have been created, including bacteria, single-celled protists such as algae, fungi such as yeasts, invertebrate animals such as the fruit fly, and vertebrate animals such as the mouse. **Gene therapy** is the use of genetic engineering techniques to alter the characteristics of specific tissues and organs in the human body, with the goal of treating serious genetic disorders or diseases.

DNA technology has produced dramatic advances in biology, and the wide-ranging applications of this technology touch many aspects of our daily life today, from the foods we eat to the diagnosis we might hear in a doctor's office. The ability to analyze DNA from different species, and from individuals in populations of the same species, has helped us understand biological processes at all levels in the biological hierarchy. DNA analysis has played a large role in helping us reconstruct the history of life on Earth (see Chapter 2) and it has revealed, perhaps more clearly than any other approach, the common ancestry and the genetic diversity of life.

Today, the classification of organisms into related groups (such as genus and family) relies heavily on a comparison of DNA sequences. Wildlife biologists use

Helpful to know

"Biotechnology" is a broad term that describes the use of biology to make a wide variety of products, including brewing, baking, and cheese making. Recently, the word has also been applied to the production of genetically modified organisms or the manufacture of products from genetically modified organisms. When used in this way, the terms "biotechnology" and "DNA technology" overlap considerably, which is why many people use them interchangeably.

Figure 15.2 Genetic Engineering: Introducing a Human Gene into a Bacterium

(a) This diagram illustrates the general scheme for genetic engineering, using the introduction of a human gene into a bacterium as an example. (1) Preparing the DNA. Human DNA containing the gene of interest (blue) is isolated with the help of restriction enzymes. Bacterial plasmid DNA (red) is extracted and cut open with the same restriction enzymes. (2) Creating recombinant DNA. The human DNA of interest is "pasted" into the cut plasmid to create the recombinant plasmid. (3) Creating a genetically modified bacterium. The recombinant plasmid can be inserted into the desired strain of bacterium in a number of different ways. The electroporation method uses electrical pulses to create short-lived holes in the bacterial plasma membrane through which DNA can enter. The bacterium makes many copies of the recombinant plasmid, which are inherited by the daughter cells when the cells divide. Human genes can be active inside the bacterial cell, directing the production of the proteins they code for. (b) Cloned DNA can be used in a variety of ways, including the production of pharmaceuticals and crops with desired qualities, as well as gene therapy in humans.

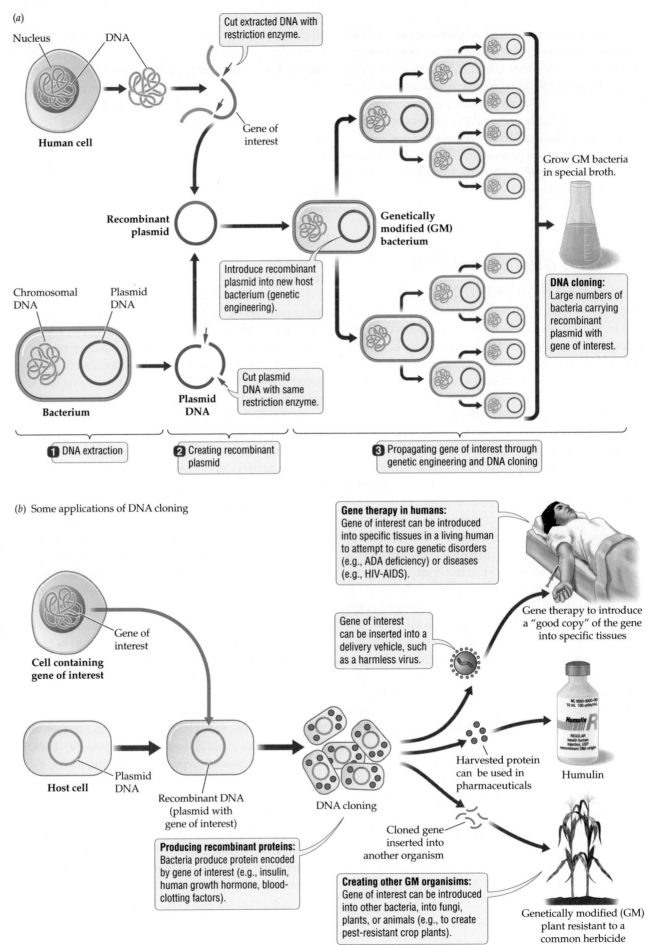

(a)

Nucleus   DNA

Cut extracted DNA with restriction enzyme.

Human cell

Gene of interest

Recombinant plasmid

Chromosomal DNA   Plasmid DNA

Bacterium

Plasmid DNA

Cut plasmid DNA with same restriction enzyme.

Introduce recombinant plasmid into new host bacterium (genetic engineering).

Genetically modified (GM) bacterium

Grow GM bacteria in special broth.

DNA cloning: Large numbers of bacteria carrying recombinant plasmid with gene of interest.

1 DNA extraction

2 Creating recombinant plasmid

3 Propagating gene of interest through genetic engineering and DNA cloning

(b) Some applications of DNA cloning

Gene of interest

Cell containing gene of interest

Host cell   Plasmid DNA

Recombinant DNA (plasmid with gene of interest)

DNA cloning

Producing recombinant proteins: Bacteria produce protein encoded by gene of interest (e.g., insulin, human growth hormone, blood-clotting factors).

Gene therapy in humans: Gene of interest can be introduced into specific tissues in a living human to attempt to cure genetic disorders (e.g., ADA deficiency) or diseases (e.g., HIV-AIDS).

Gene of interest can be inserted into a delivery vehicle, such as a harmless virus.

Gene therapy to introduce a "good copy" of the gene into specific tissues

Harvested protein can be used in pharmaceuticals

Humulin

Cloned gene inserted into another organism

Creating other GM organisims: Gene of interest can be introduced into other bacteria, into fungi, plants, or animals (e.g., to create pest-resistant crop plants).

Genetically modified (GM) plant resistant to a common herbicide

DNA analysis to measure genetic diversity in populations of endangered species, and they use that information to design conservation strategies. Anthropologists have used DNA from fossil material and from diverse groups of modern-day humans to conclude that anatomically modern humans evolved in southeastern Africa and spread out from this one area of Africa some 50,000 years ago, displacing other early humans (such as Neandertals) as they proceeded to colonize the world.

DNA technology has enabled us to identify many critical genes, and it has enhanced our understanding of gene function in diverse cell types, in a great range of organisms. Hundreds of genetic disorders and infectious diseases in humans can now be diagnosed through genetic tests (discussed in greater detail in Interlude C). DNA chips (see Chapter 14), a recent innovation in DNA technology, have made it possible to understand which sets of genes are expressed ("turned on") in specific cell types under certain conditions, and how the normal pattern might be altered when a person is sick. Rather than relying on a one-size-fits-all approach to treatment, such techniques make it possible to determine exactly which genes are behaving in destructive ways in a certain patient and which drug will be most effective in correcting the problem. Personalized medicine—consisting of treatments and therapies based on a patient's DNA profile—has been launched in a limited way and is likely to become quite common in the years ahead.

Despite its controversial aspects, genetic engineering is a fact of everyday life. Many lifesaving pharmaceuticals, such as human insulin, human growth hormone, human blood-clotting proteins, and anticancer drugs, are manufactured by GM bacteria or GM mammalian cells grown in laboratory dishes. Most of the corn, soybean, canola, and cottonseed products we consume are likely to be products of genetic engineering. The alteration of crop plants through DNA technology has the potential to improve the nutritional quality of our foods and possibly reduce the environmental impact of intensive agriculture. However, for reasons we examine later, this application of DNA technology has been more controversial than most others so far. Human gene therapy, in contrast, has been the most challenging on technical grounds, but the least controversial in the social arena. In the next section, we will examine the methods of DNA technology in greater detail, then return to a consideration of the applications of DNA technology and the ethical and social dimensions of this new frontier.

Concept Check

1. What important goal is achieved by the polymerase chain reaction (PCR)?

2. What is genetic engineering? Describe a pharmaceutical application of genetic engineering.

15.2 Working with DNA: Basic Techniques

Extracting DNA is a fairly straightforward process. Scientists first break open the cell membranes to release the cell contents. Then they strip away other macromolecules, such as proteins and lipids, with the help of specific chemicals, until just the DNA remains. The many enzymes that are used to cut and join DNA, and to amplify it through such methods as PCR, come from natural sources. Cells, and even viruses, deploy a diverse and versatile collection of enzymes in order to replicate DNA (see Chapter 12), swap chromosome segments during crossing-over (see Chapter 9), invade a cell, or fight off an invading virus or other parasite.

As researchers have come to understand these natural processes better, they have been able to "borrow" these same enzymes to manipulate DNA in a test tube. Some of these enzymes can function under extreme conditions, such as near-boiling temperatures, because the organisms from which they are derived are adapted to living in some of the most hostile habitats on the planet. As you read the details of some of the procedures that are widely used in DNA technology, keep in mind that most of these "tricks" are inspired by nature. Just about everything in the DNA technologist's tool chest comes from living organisms or viruses, and has therefore been honed by evolutionary processes.

Enzymes are used to cut and join DNA

Each of us has 3.3 billion base pairs of DNA on 23 unique chromosomes. The DNA molecule in each chromosome is so large (140 million base pairs, on average) that after DNA has been extracted from a cell, it must be broken into smaller pieces before it can be analyzed further. DNA can be split into more manageable pieces by **restriction enzymes**, which cut the DNA at highly specific sites. A restriction enzyme called *Alu*I, for example, cuts DNA everywhere the sequence AGCT occurs, but nowhere else

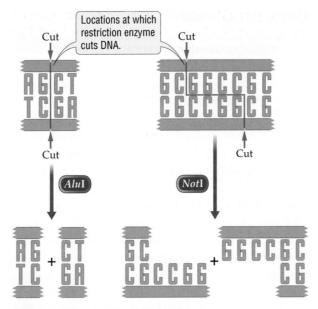

Figure 15.3 Restriction Enzymes Cut DNA at Specific Places

The restriction enzyme *Alu*I specifically binds to and then cuts the DNA molecule wherever the sequence AGCT occurs. Another restriction enzyme, *Not*I, specifically binds to and cuts the DNA sequence GCGGCCGC. Each enzyme binds to and cuts its own special target sequence, and no other.

(Figure 15.3). There are many different restriction enzymes, each of which recognizes and cuts DNA if its unique target sequence is present in that DNA, and it cuts the DNA only at those sites. Restriction enzymes were discovered in bacteria in the late 1960s. Bacteria appear to have evolved restriction enzymes to do battle against foreign DNA, such as viral DNA. Infecting viruses begin by injecting their DNA into a bacterium. The bacterium then deploys its restriction enzymes in an attempt to chop up the viral DNA, thereby "restricting" viral growth.

DNA **ligase** is another important enzyme for making recombinant DNA. DNA ligases join two DNA fragments. They are commonly used to insert one piece of DNA, such as a human gene, into another DNA molecule, such as a plasmid extracted from bacterial cells (as illustrated in Figure 15.2). A recombinant DNA molecule is created when scientists link two or more pieces of DNA to create an artificial, or synthetic, combination of DNA segments. For instance, the gene coding for the green fluorescent protein (GFP) can be cut out of jellyfish DNA with restriction enzymes, and "pasted" into a bacterial plasmid with DNA ligase, creating a recombinant plasmid. The recombinant plasmid can then be put into bacterial cells by means

of genetic engineering, and if the recombinant plasmid is constructed in the right manner, the resulting GM bacteria will glow when exposed to bluish light.

Gel electrophoresis sorts DNA fragments by size

Once a DNA sample has been cut into fragments by one or more restriction enzymes, researchers often use a process called gel electrophoresis to help them see and analyze the fragments. In **gel electrophoresis**, DNA is placed into a depression (a "well") in a gelatin-like slab (a "gel"). When an electrical current passes through the gel, it causes the DNA (which has a negative electrical charge) to move toward the positive end of the gel (Figure 15.4). (Recall that opposite

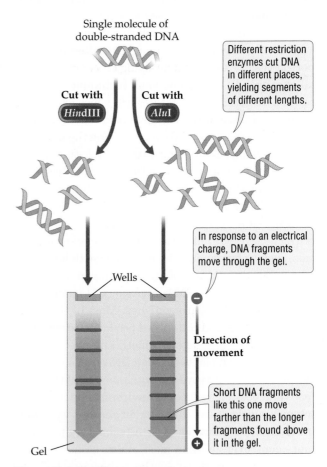

Figure 15.4 Gel Electrophoresis

When subjected to an electrical current for a given period of time, DNA fragments move through a gel at different rates, depending on their size. Fragments found toward the positive end of the gel are shorter than fragments found toward the negative end of the gel. DNA cut by different restriction enzymes produces different patterns on the gel—a characteristic that is exploited in analyzing DNA for the presence of certain alleles or sequences.

charges attract each other and like charges repel each other.) Long pieces of DNA pass through the gel with more difficulty, and therefore move more slowly, than short pieces. The distance a fragment travels through the gel is related to its speed of movement. After a fixed time period, the shorter, more rapidly moving fragments are found toward the positive end of the gel, and the longer, more slowly moving fragments are located closer to the negative end. Because the fragments are invisible to the human eye, they must be stained or labeled in special ways before they can be seen.

By using restriction enzymes and gel electrophoresis together, researchers can examine differences in DNA sequences. For example, the restriction enzyme *Dde*I cuts the normal hemoglobin allele into two pieces, but it cannot cut the sickle-cell allele. This difference provides a simple test for the disease allele (Figure 15.5).

DNA hybridization experiments can detect the presence of a gene in a sample of DNA

Another way of testing for the sickle-cell allele is to use a DNA probe. A **DNA probe** is a short, single-stranded segment of DNA with a known sequence. A probe can pair with another single-stranded segment of DNA if the sequence of nucleotides in the probe (for example, CTGAGGA) is complementary to the nucleotide sequence in the other DNA segment (GACTCCT in this case). DNA or RNA probes are commonly used in **DNA hybridization**

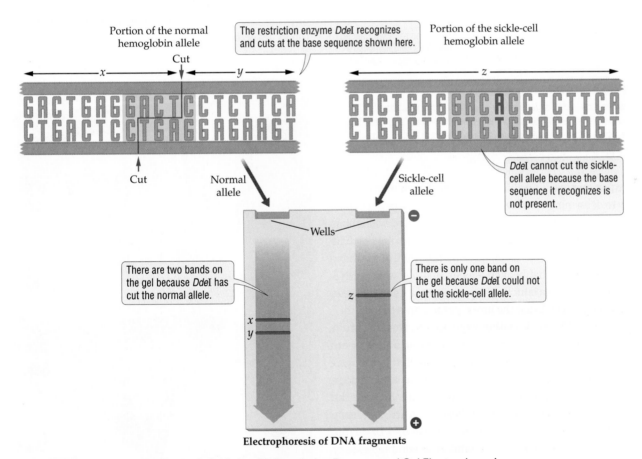

Figure 15.5 Identifying the Sickle-Cell Allele with Restriction Enzymes and Gel Electrophoresis
The restriction enzyme *Dde*I cuts DNA wherever it encounters the sequence GACTC (shown in yellow in the top panels). Only one such sequence occurs in the normal allele of the hemoglobin gene. A single base pair mutation (in which T–A becomes A–T) causes sickle-cell anemia; this mutation occurs in the GACTC sequence, and changes the sequence to GACAC. Since *Dde*I cannot recognize the mutant sequence GACAC, it does not cut the DNA at this location. As a result, the sickle-cell allele produces only one band on the gel, whereas the normal hemoglobin allele shows up as two bands on the gel, representing two fragments of different sizes.

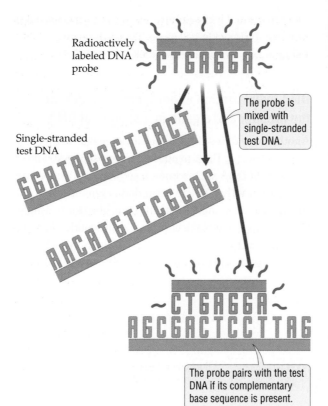

The probe is mixed with single-stranded test DNA.

Radioactively labeled DNA probe

Single-stranded test DNA

The probe pairs with the test DNA if its complementary base sequence is present.

Figure 15.6 DNA Hybridization
When exposed to several single-stranded segments of test DNA (bluish-gray), a DNA probe (orange) binds only to the segment that contains a base sequence that is complementary to the probe.

experiments (Figure 15.6). The goal of such experiments is to determine whether DNA extracted from a particular source (such as the gut contents of a beaked whale) has base pair complementarity to a specific probe (such as a gene unique to a deep-sea bony fish that might have been eaten by that beaked whale). In other words, DNA hybridization is a hunt for the presence of a known gene in a sample of DNA that the investigator wants to analyze.

In a DNA hybridization experiment, the DNA that will be exposed to a probe is first cut into fragments (by restriction enzymes) and then converted to a single-stranded form by the application of heat and certain chemicals that break the hydrogen bonds holding the two strands of DNA together. These steps are necessary to make the DNA easier to manipulate and to enable the single-stranded probe to bind to its complementary sequence. The probe is chemically labeled, for example with a radioactive tag, to make it easier to identify the DNA segments to which it binds. Finally, the DNA to be tested is exposed to the probe. If the probe binds (hybridizes) to a fragment of the test DNA, the two pieces of DNA have a complementary nucleotide sequence.

DNA sequencing and DNA synthesis are used in basic research and for practical applications

DNA sequencing enables researchers to identify the sequence of nucleotides in a DNA fragment, a gene, or even the entire genome of an organism. Sequences can be determined by several methods, the most efficient of which rely on automated sequencing machines (Figure 15.7). One of these machines can identify over a million bases per day, making it possible to determine the sequence of a single gene quickly. DNA can also be sequenced manually by slower, but still highly effective, methods.

Machines can also be used to create, or synthesize, a DNA fragment with a specific, made-to-order nucleotide sequence. In less than an hour, a DNA synthesis machine can produce single-stranded DNA segments hundreds of nucleotides long. Synthesized single-stranded DNA segments, usually less than 30 nucleotides long, are used as probes in DNA hybridization experiments and to prime the enzymatic reactions that are the basis of the PCR technique.

Figure 15.7 Machines Can Sequence DNA
Automated DNA sequencing machines can rapidly determine the nucleotide (base) sequence of a DNA fragment. Here a scientist examines a computer display showing part of an automated DNA sequencing gel. Each of the four chemical bases in DNA is represented by a different color (red, green blue, and yellow).

DNA cloning is a means of propagating recombinant DNA

A single copy of a gene is difficult to analyze and manipulate for such purposes as gene sequencing or genetic engineering. Biologists generally clone such a DNA fragment by isolating it, ligating it with other DNA fragments to create a recombinant molecule, and then introducing the recombinant molecule into a host cell that will make many identical copies of it. **Cloning** means to make a copy that is genetically identical to the original, so copying a whole organism is also a form of cloning. Whole-organism cloning includes such traditional methods as propagating houseplants by making cuttings of them or, as we shall see later, using DNA technology to make genetically identical copies of animals such as sheep.

DNA that is cloned can be sequenced, transferred to other cells or organisms, or used as a probe in DNA hybridization experiments. In addition, with cloning and gene sequencing as tools, researchers can use the genetic code to determine the amino acid sequence of the gene's protein product. Knowledge of a gene's product can provide vital clues to the gene's function, as it did in the case of the *HD* gene involved in Huntington disease (see Chapter 13). For this reason, DNA cloning is a key step in the study of genes that cause inherited genetic disorders and cancers.

A DNA library contains many pieces of cloned DNA derived from one source

A **DNA library** is a collection of cloned DNA fragments that, together, represents all of the information in the genetic material of one organism. The collection of cloned fragments is introduced into, and stored within, a suitable host organism such as a bacterium. A complete library of all the DNA in the human genome would contain tens to hundreds of thousands of DNA fragments.

The principles underlying the construction of a DNA library are similar to the DNA cloning procedures described earlier. First, scientists use restriction enzymes to cut the DNA from the source cells (jellyfish cells, for example). Next, they randomly insert each of the many resulting DNA fragments into at least one copy of a **DNA vector**. A DNA vector is a loop or chain of DNA designed to serve as a "DNA vehicle." It is useful in the cloning of DNA fragments and in the transfer of recombinant DNA from one cell to another.

Plasmids are commonly used as DNA vectors (Figure 15.8). The DNA of certain viruses, or the chromosomes of bacterial or yeast cells, are also used as DNA vectors, especially when large DNA fragments need to be cloned. Once a DNA fragment is ligated into the vector DNA, it becomes a recombinant vector. The DNA library is complete when the many molecules of recombinant vectors, each harboring a cloned DNA fragment from the source cells, are moved into host cells (in the case of viral vectors, the recombinant DNA is packaged into virus particles). Each of the many fragments created by cutting up the source DNA (jellyfish DNA, in our example) now lies within a vector DNA molecule that in turn resides in the cytoplasm of a cell or in the genetic material of a viral particle.

A DNA library is usually made with the goal of identifying, isolating, and cloning a particular gene from the genome of an organism (for example, the GFP gene from the jellyfish genome). The idea is analogous to searching for a book in a real library, checking it out, then making many photocopies of it. The process of searching the DNA library to identify a specific gene is known as library screening.

Watch how a DNA library is created. ▶❚❚ 15.2

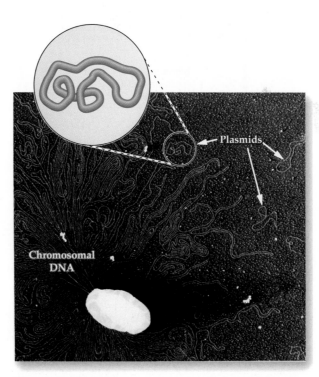

Figure 15.8 Plasmids
Plasmids are small, closed loops of DNA that are found naturally in many species of bacteria. Plasmids are separate from, and much smaller than, the single main chromosome of the bacterium. Here, a ruptured *E. coli* bacterium spills out its chromosome and several plasmids, three of which are indicated with arrows.

Plasmids

Chromosomal DNA

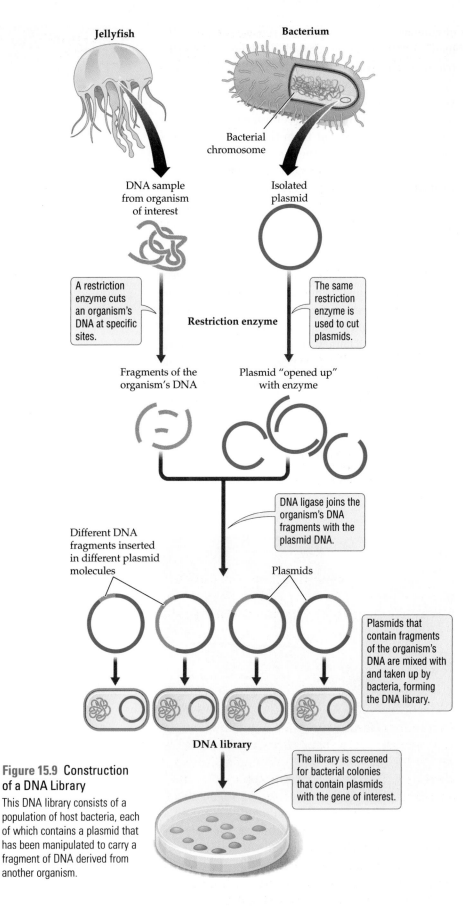

Figure 15.9 Construction of a DNA Library
This DNA library consists of a population of host bacteria, each of which contains a plasmid that has been manipulated to carry a fragment of DNA derived from another organism.

Image labels:
Jellyfish

Bacterium

Bacterial chromosome

DNA sample from organism of interest

Isolated plasmid

A restriction enzyme cuts an organism's DNA at specific sites.

The same restriction enzyme is used to cut plasmids.

Restriction enzyme

Fragments of the organism's DNA

Plasmid "opened up" with enzyme

DNA ligase joins the organism's DNA fragments with the plasmid DNA.

Different DNA fragments inserted in different plasmid molecules

Plasmids

Plasmids that contain fragments of the organism's DNA are mixed with and taken up by bacteria, forming the DNA library.

DNA library

The library is screened for bacterial colonies that contain plasmids with the gene of interest.

Scientists can make a DNA library in plasmid vectors using the steps outlined in Figure 15.9. The first step in screening such a library is to grow small batches of the bacterial cells on nutrient gel in a petri dish. Each bacterial cell gives rise to a mass of cells, called a colony, on the surface of the gel. Every cell in the colony contains the same DNA fragment, embedded in a recombinant plasmid. Bacterial colonies can be screened in a variety of ways to find the ones that contain the DNA fragment of interest (for example, the DNA fragment containing the GFP gene).

DNA hybridization is used for library screening when a DNA or RNA probe is available that will form complementary base pairs with a region of the target gene (the GFP gene, in our example). The probes are commonly labeled with radioactive tags or light-releasing tags, and after breaking the colonies open and exposing their DNA to the probe, researchers can home in on bacterial colonies "highlighted" by the probe. These colonies harbor plasmids containing the gene of interest, and they can be grown in a liquid broth to make billions of bacterial cells, each with many copies of the recombinant plasmid containing the target gene. Through this process, the gene of interest is effectively cloned forever, giving researchers many options for additional analysis and manipulation, including the possibility of introducing the gene into the cells of another organism (such as Alba, the rabbit featured in our opening story).

The polymerase chain reaction is used to amplify small quantities of target DNA

The **polymerase chain reaction** (**PCR**) uses a special type of DNA polymerase to make billions of copies of a targeted sequence of DNA in just a few hours. To amplify a piece of DNA by PCR, researchers must use two short segments of synthetic DNA, called **DNA primers**. Each primer is designed to bind to one of the two ends of the target DNA by complementary base pairing. By the series of steps shown in Figure 15.10, DNA polymerase then produces many copies of the DNA flanked by the two primers. To amplify a target DNA, scientists must have at least some information about its nucleotide sequence. Without this knowledge, they cannot synthesize the specific primers required in every PCR reaction.

The power of PCR technology lies in the fact that it can amplify extremely small amounts of DNA, amounts extracted from just a few cells or a single blood stain, for example. As a result, PCR has come to be widely used in basic research and in fields as diverse as medical diagnostics, forensics, paternity testing, paleoanthropology,

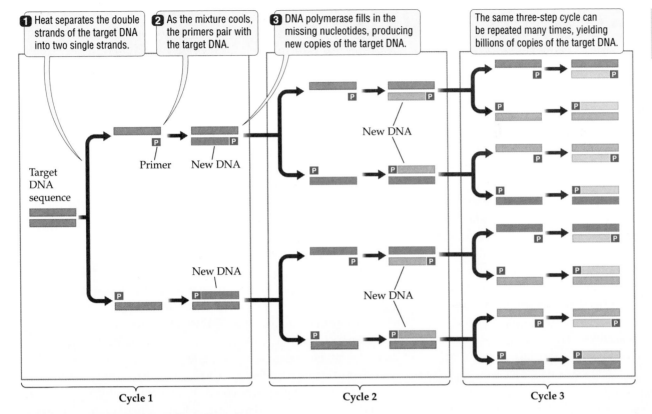

① Heat separates the double strands of the target DNA into two single strands.

② As the mixture cools, the primers pair with the target DNA.

③ DNA polymerase fills in the missing nucleotides, producing new copies of the target DNA.

The same three-step cycle can be repeated many times, yielding billions of copies of the target DNA.

See the polymerase chain reaction in motion. ►❚❚ 15.3

Target DNA sequence

Primer   New DNA

New DNA

New DNA

New DNA

Cycle 1   Cycle 2   Cycle 3

Figure 15.10 The Polymerase Chain Reaction
Short primers (red) that can pair with the two ends of a gene of interest (bluish-gray) are mixed in a test tube with a sample of the target DNA, the enzyme DNA polymerase, and all four nucleotides (A, C, G, and T). A machine then processes the mixture through the three steps shown, in which the temperature is first raised and then lowered, to double the number of double-stranded versions of the template sequence. The doubling process can be repeated many times (only three cycles are shown here). The template DNA sequence in the first cycle (bluish-gray), the DNA that is newly made in each cycle (shades of orange), and the primers (red) are color-coded here for clarity.

as well as the authentication of delicacies such as caviar and expensive vintage wine. The technique became so successful so quickly that in 1991, only 6 years after the first paper on PCR was published, the PCR patent was sold for $300 million. Two years later, in 1993, Kary Mullis won a Nobel Prize for the development of PCR.

15.3 Applications of DNA Technology

DNA technology has many important applications, recent examples of which include prenatal screening for genetic disorders (see Interlude C) and genetic profiling to determine which of several drugs will work best in a given patient (see the box in Chapter 13, page 276). Other examples include gene therapy (addressed later in this chapter) and the use of DNA technology in therapeutic

and reproductive cloning (see the box on page 314). In this section we will elaborate on two common uses of DNA technology: DNA fingerprinting and genetic engineering.

DNA fingerprinting can distinguish one individual from another

The process of identifying DNA unique to a species, or a specific individual within a species, is called **DNA fingerprinting**. It is a method for generating a DNA-based identity, or DNA profile, of an individual or species. DNA fingerprinting can be used to detect the contamination of food or water by certain harmful microorganisms, and to identify remains of endangered species found in the possession of alleged poachers. It is used to match organ donors with patients seeking organ transplants, and to identify victims of mass attacks such as those on the World Trade Center on September 11, 2001.

Science Toolkit

Reproductive Cloning of Animals

The goal of **reproductive cloning** is to produce an offspring (baby) that is a genetic copy of a selected individual. In 1996, Dolly the sheep, born on a Scottish farm, became the first mammal to be created through reproductive cloning.

There are three key steps in reproductive cloning. First, an egg cell is obtained from a cytoplasmic donor (a Scottish Blackface ewe, in Dolly's case) and its nucleus removed (enucleation). Next, an electrical current is used to fuse the enucleated egg with a somatic cell from a nuclear donor (for example, a Finn-Dorset ewe, Dolly's genetic parent). Chemicals are used to activate this product of cell fusion—that is, to trick it into dividing so that it begins to form an embryo. Finally, the embryo is transferred to the uterus of a surrogate mother where, if all goes well, it continues to develop, ultimately resulting in the birth of a healthy baby animal. Such a baby—referred to as a clone—is genetically identical to the ewe who provided the donor nucleus, not to the egg donor or the surrogate mother.

To date, reproductive clones have been developed in a variety of mammals, including sheep, pigs, mice, cows, horses, dogs, and cats.

Why would anyone want to clone a sheep, a pig, or a cow? Reproductive cloning can be used to produce multiple copies of an organism with useful characteristics. For example, a company in South Dakota has produced cloned calves genetically engineered to produce human disease-fighting proteins called immunoglobulins. The ultimate aim is to create herds of genetically identical cows, each of which would serve as a "biological factory," producing large quantities of commercially valuable immunoglobulin proteins.

Reproductive cloning is also being used to produce pigs that could save the lives of people in need of organ transplants. Each year, thousands of people die while waiting for an organ transplant. Pig organs are roughly the same size as human organs, and could work well in people except for one major problem: the human immune system rejects them as foreign. In recent studies, scientists have used reproductive cloning to produce pigs whose organs lack a key protein that stimulates the human immune system to attack. This work represents an important step toward the production of pig clones whose organs could be used to save the lives of people who would otherwise die for lack of a suitable transplant organ.

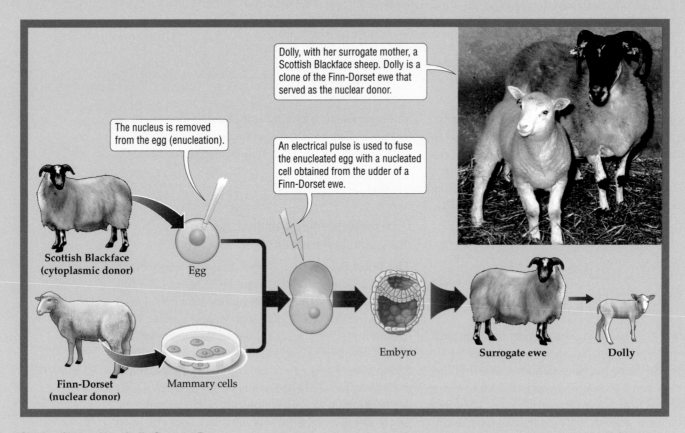

Dolly, with her surrogate mother, a Scottish Blackface sheep. Dolly is a clone of the Finn-Dorset ewe that served as the nuclear donor.

The nucleus is removed from the egg (enucleation).

An electrical pulse is used to fuse the enucleated egg with a nucleated cell obtained from the udder of a Finn-Dorset ewe.

Scottish Blackface (cytoplasmic donor)   Egg

Finn-Dorset (nuclear donor)   Mammary cells

Embryro   Surrogate ewe   Dolly

Cloning Sheep: How Dolly Came to Be

DNA fingerprinting is such a powerful tool in forensics today that it has traveled beyond the real world of law enforcement to be dramatized in crime novels, TV dramas, and movie thrillers. A laboratory technician can take a biological sample, such as blood, tissue, or semen, from a crime scene and develop a DNA fingerprint, or profile, of the person from whom the sample came. That profile can then be compared with another profile—for example, that of a crime victim or suspect—to see whether they match.

It is theoretically possible for two people to have the same DNA profile. Therefore, a match such as the one in Figure 15.11, between a victim's DNA profile and blood stains found on a suspect's clothes, does not provide definitive proof that the two samples are from the same person. In most cases, however, the probability that two people will have the same DNA profile by sheer accident is between one in 100,000 and one in a billion. (The actual odds, within this range, depend on the methods used to create the DNA profile, as we shall see shortly.)

DNA fingerprinting takes advantage of the fact that all individuals (except identical twins and other multiples) are genetically unique. To distinguish between different individuals, scientists examine certain regions of the human genome that are known to vary greatly from one person to the next. Such highly variable regions include noncoding portions of our DNA, such as introns and spacer DNA, which tend to be quite different in size and nucleotide sequence among different individuals.

DNA fingerprinting can be done by various methods, including RFLP analysis and PCR amplification. In **RFLP analysis**, restriction enzymes are used to cut a DNA sample into small pieces. Next, the fragments are separated by size on a gel. Differences (between individuals) in the lengths of these fragments are called restriction fragment length polymorphisms, or RFLPs. The RFLP patterns generated by a person's DNA can produce a DNA profile, or fingerprint, that is unlikely to be shared by any other person except an identical sibling from a multiple birth. The greater the number of DNA regions used in the RFLP analysis, the lower the chance that two individuals will have the same RFLP fingerprint just by accident.

We can also use PCR to amplify certain highly variable regions of the human genome. This type of PCR generates a pattern of amplified DNA fragments that is likely to vary from one person to another. As with RFLP, the more regions we compare by PCR analysis, the less likely it is that there will be a coincidental match between all of the PCR fragments generated from the DNA of two individuals who are not twins. PCR-based

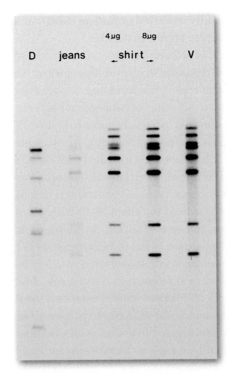

Figure 15.11 Using DNA Fingerprinting to Identify Criminals
The DNA profile on the far left is that of the defendant (D) in a murder trial. The profile on the far right is that of the victim (V). The defendant's jeans and shirt were splattered with blood. The fingerprint of the DNA extracted from the blood splatters is an exact match with the victim's DNA profile.

fingerprinting has almost completely replaced RFLP-based fingerprinting, because RFLP analysis requires larger amounts of DNA, takes longer, and is more expensive. The Federal Bureau of Investigation (FBI) and most state law enforcement agencies use PCR-based amplification from 13 different regions of the human genome to produce a near-unique DNA profile of an individual. DNA fingerprints of missing persons, victims of unsolved crimes, and those convicted of serious crimes are maintained in a DNA database named CODIS (Combined DNA Index System). The CODIS database contains more than 5 million DNA profiles, including PCR-based fingerprints of biological material collected at the site of unsolved crimes.

Genetic engineering is used to transfer genes from one species to another

If a gene can be transferred between species, it can often make a functional protein product in the new species. The gene for the light-producing protein (GFP), for example, has been transferred to and expressed in organisms as different as plants, mice, and bacteria. The deliberate transfer of a gene from one species to another is one example of genetic engineering. Genetic engineering is usually a three-step process in which a DNA sequence (usually a gene) is isolated, modified in some cases before being cloned, and then introduced into the same species or into a different species. As mentioned earlier, an individual receiving such a DNA sequence

Explore whether human cloning is possible.   ▶︎II   15.4

Learn more about genetic engineering.   ▶︎II   15.5

is called a **genetically modified organism (GMO)** or a genetically engineered organism (GEO).

Recombinant DNA or foreign genes can be introduced into a cell or whole organism in many different ways. We have already seen how plasmids can be used to transfer a gene from humans or other organisms to bacteria. Plasmids can also be used to transfer genes to plant cells or animal cells. In some species, including many plants and some mammals, genetically modified (GM) adults can be generated, or cloned, from these altered cells (see the box on page 314). Other means of gene transfer include viruses, which can be used to "infect" cells with genes from other species, and gene guns that fire microscopic pellets coated with the gene of interest into target cells.

Genetic engineering is commonly used to alter the genetic characteristics of the recipient organism, often focusing on a particular aspect of performance or productivity. Atlantic salmon, for example, have been given genes that cause them to grow up to six times faster than normal (Figure 15.12). Crop plants have been genetically engineered for a wide variety of traits, including increased yield, insect resistance, disease resistance, frost tolerance, drought tolerance, herbicide resistance (which enables crops to survive the application of weed-killing chemicals), increased shelf life, and improved nutritional value. GM crops now occupy more than 200 million acres worldwide—a coverage that includes over 20 percent of the world's soybeans, corn, cotton, and canola.

Genetic engineering is commonly used to churn out large amounts of a gene product, usually one with therapeutic or commercial value (Table 15.1). In 1978, the gene for human insulin was transferred into *E. coli* bacteria, and this hormone became the first genetically engineered product to be mass-produced. Before GM insulin, diabetics had to use insulin extracted from pigs and cows to control their blood sugar levels. The animal-derived hormone was often in short supply and could cause allergic reactions in some people. GM insulin is safer and less expensive, and every year more than 300,000 Americans who suffer from insulin-dependent diabetes use the GM product to control their disease.

Another application of genetic engineering is in the manufacture of certain types of vaccines. A vaccine is any substance that stimulates the immune system in such a way that it shields the body from future attack by a specific invading organism. Bacteria have been genetically engineered to produce large amounts of the distinctive proteins that are present on the surface of many disease-causing organisms, including a number of viruses, bacteria, and infectious protists. When injected into the human body, the GM versions of these cell surface proteins stimulate the immune system to recognize

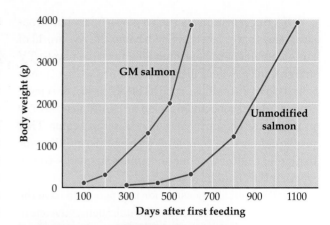

Figure 15.12 Genetically Modified Fish
In fish farms, genetically modified salmon eat more food and grow more rapidly than their unmodified relatives.

the disease organisms that normally carry such proteins. The GM proteins act like a vaccine because they prepare the immune system to quickly fend off a real invasion by the disease organism. Such a GM approach has been used to develop effective vaccines for flu and malaria, among other diseases. There are ongoing attempts to develop an AIDS vaccine using genetically engineered surface proteins from the AIDS virus.

15.4 Ethical and Social Dimensions of DNA Technology

As we have seen, DNA technology provides many benefits to human society. At the same time, the immense power and scope of genetic engineering raises ethical concerns in the minds of some and poses potential risks, especially to the genetic integrity of wild populations. At the most basic level, some people ask how we can assume we have the right to alter the DNA of other species. Some see no ethical conflict in altering the DNA of a bacterium or a virus, but object to changing the

Table 15.1

Methods of Production and Uses for Some Products of Genetic Engineering

Product	Method of production	Use
PROTEINS		
Human insulin	*E. coli*	Treatment of diabetes
Human growth hormone	*E. coli*	Treatment of growth disorders
Taxol biosynthesis enzyme	*E. coli*	Treatment of ovarian cancer
Luciferase (from firefly)	Bacterial cells	Testing for antibiotic resistance
Human clotting factor VIII	Mammalian cells	Treatment of hemophilia
Adenosine deaminase (ADA)	Human cells	Treatment of ADA deficiency
DNA SEQUENCES		
Sickle-cell probe	DNA synthesis machine	Testing for sickle-cell anemia
BRCA1 probe	DNA synthesis machine	Testing for breast cancer mutations
HD probe	*E. coli*	Testing for Huntington disease
Probe *M13,* among many others	*E. coli,* PCR	DNA fingerprinting in plants
Probe *33.6,* among many others	*E. coli,* PCR	DNA fingerprinting in humans

NOTE: For each product, the gene or DNA sequence that codes for the product is either inserted into host cells, such as E. coli or mammalian cells, or used in one of several automated procedures, such as a DNA synthesis machine or PCR. These cells or automated procedures are then used to make many copies of the product.

genome of a food plant or of a sentient animal such as a dog or a chimpanzee.

Few people find fault with the use of GM bacteria to produce lifesaving pharmaceuticals such as insulin for diabetics, blood-clotting proteins for hemophiliacs, and clot-dissolving enzymes for stroke victims. GM plants, on the other hand, are bitterly opposed by some groups, especially in Europe. In some European countries, foods containing GM products must be labeled as such, kept separate from non-GM products, and monitored through the entire food production chain. Critics of GM foods worry that the presence of a GM protein in a common food might cause a severe allergic reaction in an unsuspecting consumer. Proponents counter that no adverse reactions have been authenticated in the United States, where millions of people have been eating GM foods for more than a decade.

Some environmentalists worry that engineering crops to be resistant to herbicides might promote increased use of herbicides, some of which could be harmful to the environment. Supporters of GMO technology say that herbicide-resistant plants would be good for soil health because farmers would not have to use soil-damaging tilling methods to uproot weeds if the weeds in a standing crop could be controlled by an herbicide spray instead. Many crops, including corn and cotton, have been engineered to produce Bt toxin, named after the bacterium (*Bacillus thuringiensis*) in which it was discovered. There were fears that the presence of this protein insecticide in corn pollen would be harmful to monarch butterflies, but recent studies have found no evidence of ill effects on this or other insect species.

Critics of genetic engineering have long argued that genes from genetically modified (GM) plants or animals could spread to wild species, potentially wreaking environmental havoc. Some have expressed concern that GMOs will escape from the bounds of farm fields, barnyards, and fish pens to contaminate natural ecosystems with genomes that have been altered by humans. Escaped GM salmon could threaten wild fish stocks not only by interbreeding with them and thereby reducing their natural diversity, but also by outcompeting them for resources and thereby driving them toward extinction.

Biology Matters

Have You Had Your GMO Today?

The use of genetically modified organisms (GMOs) in agriculture and food production has expanded dramatically since biotech crops were first commercially grown in 1996. So, too, has controversy over the risks and benefits of GMO foods. The United States grows more than 50 percent of the GMO crops that are raised worldwide; Canada grows about 6 percent.

Because the United States has no labeling requirements for GMO foods, many consumers are unaware that about 75 percent of all processed foods available in U.S. grocery stores may contain ingredients from genetically engineered plants. Breads, cereal, frozen pizzas, hot dogs, and soda are just a few of them. Corn syrup, derived from corn, is a common ingredient in many juices and sodas, and genetically modified corn syrup is used in most brand-name sodas. Soybean oil, cottonseed oil, and corn syrup are ingredients used extensively in processed foods. Soybeans, cotton, and corn dominate the 135 million acres of genetically engineered crops planted in the United

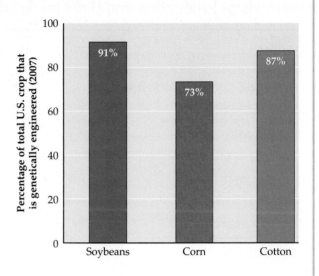

States in 2007, according to the U.S. Department of Agriculture (USDA). Through genetic engineering, these plants have been made to ward off pests and to tolerate herbicides used to kill weeds. Other crops, such as squash, potatoes, and papaya, have been engineered to resist plant diseases.

Of the world's 13 most important crop plants, 12 (all except corn) can mate and produce offspring with a wild plant species in some region where they are grown. If a crop plant is genetically engineered to be resistant to an herbicide, the potential exists for the resistance gene to be transferred (by mating) from the crop to the wild species. There is a risk that by engineering our crops to resist herbicides, we will unintentionally create "superweeds" resistant to the same herbicides.

In a recent study, scientists used a commercial-scale (400-acre) plot to examine the spread of genes for resistance to the herbicide Roundup. These resistance genes had been inserted into a variety of GM bent grass developed by the Monsanto and Scotts companies, which hope to use the engineered bent grass as turf in golf courses. Grasses are wind-pollinated, raising the concern that resistance genes could spread long distances if pollen blown by wind from GM bent grass were to fertilize any one of a dozen other grass species that can breed with bent grass. That fear appears to have been realized: DNA tests discovered the modified genes in non-GM bent grass located up to 13 miles away from a test plot; the genes were also found in other grass species growing wild up to 9 miles from the test plot. Representatives

from Monsanto and Scotts suggested that these results are not all that worrisome, since any plants that contained the Roundup resistance genes could be controlled by other herbicides.

The U.S. Department of Agriculture (USDA) has ordered that an environmental impact study be conducted on GM bent grass before the agency decides whether Scotts can sell it. This is the first time that the USDA has demanded such a review for a GM crop. Experts say a way around the problem is to create GM grasses with sterile pollen, making it much less likely that genes from such grasses could escape to wild species.

In some circles, the heated debate over GMOs tends to center on political and socioeconomic issues. The use of "terminator genes" in a GM plant can theoretically prevent that plant from making viable seed. Proponents say that "terminator technology" would be an effective barrier against the runaway spread of GMOs. Opponents of GMOs see this technology as a veiled attempt by seed companies to control the supply of seed, since farmers would have to buy seed from the company each year instead of saving their own.

Critics of genetic engineering point to the social costs of using such technology. They cite bovine growth hormone

(BGH), which is mass-produced by GM bacteria, as a case in point. Among its other effects, BGH increases milk production in cattle. Before the introduction of genetically engineered BGH in the 1980s, milk surpluses already were common. The use of BGH by large milk producers has created even larger milk surpluses, driving down the price of milk and forcing small producers of milk—the traditional family farms—out of business. As some see it, the lower milk prices for consumers are not worth the social cost of driving small dairy farms into bankruptcy. Others believe that all types of commercial enterprises, including family farms, should sink or swim on their own and unsuccessful businesses should not be rescued because of social sentiment. As these examples illustrate, the debate over the social dimensions of GMOs often gets caught up in a much larger discussion about the politics of food and the economics of modern agriculture.

With respect to altering human DNA, nearly everyone accepts that the use of genetic engineering in humans for the purpose of curing a horrible disease is ethical. What about the use of genetic engineering to make less critical changes, such as enhancing physical or mental traits? For example, if it were possible to do so, would it be ethical to alter the future intelligence, personality, looks, or sexual orientation of our children before birth? According to a 1990s March of Dimes survey, more than 40 percent of Americans would make such modifications if given the chance. Is it fair for parents to make such decisions on behalf of their children? As we shall see shortly, the genetic engineering of humans still faces substantial challenges, so there may be a few more years to resolve the dilemmas raised by such questions before the technology arrives at a clinic near you.

Concept Check

1. What are the roles of restriction enzymes and DNA ligases in DNA technology?
2. What is DNA fingerprinting?
3. Explain why the release of GMOs could pose potential risks to natural ecosystems.

Applying What We Learned

Human Gene Therapy

On September 14, 1990, 4-year-old Ashanthi DeSilva made medical history when she received intravenous fluid that contained genetically modified versions of her own white blood cells. She suffered from adenosine deaminase (ADA) deficiency, a genetic disorder that severely limits the ability of the body to fight disease and can make ordinary infections, such as colds or flu, lethal. The disorder is caused by a mutation in a single gene that affects the ability of white blood cells to fight off infections. Earlier, doctors had removed some of Ashanthi's white blood cells and added the normal ADA gene to them, in an attempt to fix the lethal genetic defect through genetic engineering. Ashanthi responded very well to the treatment and now leads an essentially normal life (Figure 15.13).

The treatment that Ashanthi received was the first gene therapy experiment on a person. Human gene therapy seeks to correct genetic disorders through genetic engineering and other methods that can alter gene function. The possibility of curing even the worst genetic disorders by reaching into our cells and restoring gene function, or turning off a troublesome gene, is a bold and captivating prospect.

As such, gene therapy has attracted much media attention—some of it, unfortunately, bordering on excess. Take Ashanthi's case: in addition to gene therapy, she received other treatments for ADA deficiency. Hence, contrary to some reports in the media, her remarkable good health cannot be attributed to gene therapy alone. Overall, more than 600 gene therapy experiments have been conducted worldwide, but there are relatively few success stories so far. Why? Has gene therapy been oversold?

Until recently, proponents of gene therapy could point to a pioneering French study and answer with a confident no. That study, published in 2000, described how researchers cured X-SCID—a disease similar to ADA deficiency that cripples the immune system—by inserting a healthy version of the gene into bone marrow cells. Of 11 children with X-SCID who were treated by gene therapy alone, 9 were cured. For the first time, scientists had achieved the holy grail of gene therapy: they had cured a human genetic disorder solely by fixing the gene that caused it.

Unfortunately, three of the children who were cured of X-SCID went on to develop leukemia, a form of cancer

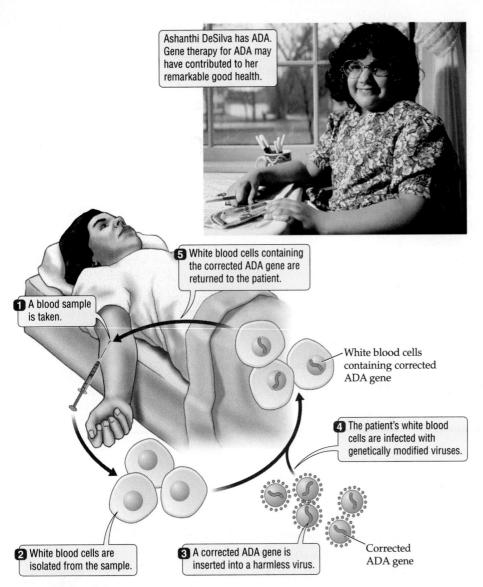

Ashanthi DeSilva has ADA. Gene therapy for ADA may have contributed to her remarkable good health.

5 White blood cells containing the corrected ADA gene are returned to the patient.

1 A blood sample is taken.

White blood cells containing corrected ADA gene

4 The patient's white blood cells are infected with genetically modified viruses.

2 White blood cells are isolated from the sample.

3 A corrected ADA gene is inserted into a harmless virus.

Corrected ADA gene

Figure 15.13 Gene Therapy for ADA Deficiency

ADA deficiency, a lethal disorder caused by a genetic defect in white blood cells, was the first genetic disorder to be treated by gene therapy.

Test your knowledge of how gene therapy can help ADA deficiency. 15.6

that strikes white blood cells. One of these children has died. What went wrong? By accident, the "good copy" of the gene was inserted near and promoted the expression of a gene that causes cells to divide rapidly, increasing the risk of cancers such as leukemia.

The three cases of leukemia followed close on the heels of another terrible incident: in 1999, a young man participating in a gene therapy experiment died from an allergic reaction to the virus used to deliver the engineered gene. The combined effect of these tragedies sent shock waves through the gene therapy field. Worldwide, many gene therapy trials were placed on hold, and there were calls to abandon gene therapy efforts altogether.

Scientists have pressed on, however, focusing their attention on the biggest hurdle: finding a way to deliver the engineered gene safely and effectively to just the right cells. Harmless viruses are often used for this purpose, but the use of viral vectors presents chal-

lenges. The human body defends itself so well against viruses that the recombinant viruses are often destroyed before they can deliver a "good copy" of the gene to enough of the target cells. Researchers at Cedars-Sinai Medical Center have developed a particularly stealthy version of a harmless virus that is almost invisible to the body's defense system. The new vector is being tested in clinical trials aimed at treating Parkinson disease, a devastating condition marked by progressive brain degeneration (see Chapter 9). Worrying about unforeseen consequences, the researchers applied the gene therapy to only one side of each patient's brain. One year later, brain scans revealed improved brain activity in the treated side of the brain while the untreated side showed a decline, compared to the pretrial status of brain function.

Scientists are also working on strategies to prevent the introduced gene from being inserted into the wrong tissues or in a wrong location within the DNA of the target cells (one that could increase the risk of cancer, for example). With respect to X-SCID, for instance, researchers are currently modifying the delivery virus so that it targets fewer types of cells and will therefore be less likely to cause cancer. Other scientists have focused on trying to understand how introduced genes are integrated into the DNA in human cells. This new understanding provides hope that the modified strategies now being proposed will make it possible to insert foreign genes into human cells in a safer, more controlled manner.

In just the past few years, RNA interference (RNAi) has emerged as a tremendously promising tool for conducting gene therapy. RNAi offers a potential cure for genetic disorders caused by the inappropriate activity, or overactivity, of a gene. In RNA interference (described in greater detail in the "Biology in the News" feature in Chapter 14, page 299), small chunks of RNA are able to silence genes that share nucleotide sequence similarity with them. Clinical trials are currently under way to test the effectiveness of RNAi for shutting off viral genes in people infected with hepatitis B or who have RSV pneumonia, a lethal viral infection. RNAi-based gene therapy is also being tested against macular degeneration, which is caused by overgrowth of blood vessels in the eye and is the leading cause of blindness in the elderly. Although viral vectors were employed in most of these trials, therapeutic RNA could also be delivered by direct injection into the cytoplasm or by packaging of the RNA into special lipids or polymers. These alternative methods for delivering RNAi are under intense investigation because if the delivery system works, there will be no need to insert any foreign DNA into the recipient's DNA.

Biology in the News

Mutants? Saviors? Modified Trees Eat Poisons; Rabbit Gene Helps Poplars Neutralize Chemicals Quickly

By Lisa Stiffler

Super trees that suck up and destroy toxic chemicals from the air and water faster than regular trees are the latest creation by scientists at the University of Washington. When the scientists stick a rabbit gene into poplar trees, the trees become dramatically better at eliminating a dozen kinds of pollutants commonly found on poisoned properties. The trees could prevent the need for digging up tons of soil or pumping out millions of gallons of water for treatment and disposal...

But while the poplars could benefit cleanup projects, they raise a multitude of ecological and ethical concerns. Many people are worried about transgenic organisms, in which a gene from one species is inserted into another, whether it's corn that produces a pig vaccine or a soybean that makes its own pesticide. There are concerns that mutant plants could spread, entering the food supply and threatening human health. Or they could interbreed with normal plants, transferring herbicide resistance to weeds, for example. No one can predict all of the potential side effects of a new gene on the host plant or other plants and animals.

When it comes to the pollution-consuming poplars, "it's really a question of trading some of the unknown risks of planting genetically modified trees with the positive environmental benefits," said Andrew Light, a UW professor of philosophy and public affairs. "This is a real dilemma for the environmental community." ... "It's commendable to be thinking about finding ways to reverse some of the pollution that has been caused in the past, but in doing so we have to make sure we don't cause new problems at the same time," said Doug Gurian-Sherman, a ... senior scientist with the Union of Concerned Scientists. "There are a lot of unknowns here," he said.

In the United States, more than 1,000 large areas—"Superfund sites"—are so contaminated with hazardous chemicals that they are off limits for commercial or public use until they can be cleaned up. Noxious waste has leached into the soil at these now-abandoned sites. The waste includes organic solvents, heavy metals such as mercury and lead, crude oil, and highly radioactive material. Since the 1980 law enacted to clean up these sites, only a small number have been decontaminated sufficiently, and the "Superfund" (created through federal taxes on the polluting industries) is widely believed to be completely inadequate for the cleanup.

Scientists have turned to biological systems to find effective, efficient, and cheaper ways of dealing with toxic waste. Bioremediation uses organisms to improve degraded environments. DNA technologists have also been exploring ways to enhance the natural ability of some organisms to eliminate hazardous chemicals.

The University of Washington scientists isolated a rabbit gene that codes for a liver enzyme that detoxifies hydrocarbons. The researchers cloned the gene into a plasmid vector and introduced it into *Agrobacterium tumefaciens*. This bacterium is a natural genetic engineer: it can insert parts of the plasmid DNA, and any gene embedded in it, into the nucleus of any host plant it infects. The bacterium was allowed to infect poplar tissue in a lab dish. The rabbit gene was transferred into the plant tissue, and 6-inch-tall poplar plants were raised from the GM tissues. The gene was strongly expressed in these poplars, and they proved extremely effective in converting various toxic pollutants into harmless by-products.

The GM poplars are currently being field tested, following strict USDA guidelines. Poplars grow quickly and produce relatively shallow but spreading root systems that could remove pollutants rapidly at soil depths up to 10 feet or so. They don't produce flowers until they are about 8 years old. By cutting down the GM trees before they start flowering, the UW team expects to keep them from interbreeding with wild trees.

Evaluating the News

1. What is bioremediation? Explain how the UW researchers developed poplars that could be used in bioremediation.

2. Do the benefits from the GM poplars outweigh potential risks? Would you object to the planting of a GM poplar in your neighborhood—say, to decontaminate a gasoline leak from an underground storage tank at the local gas station? Explain your reasoning.

3. Supporters of genetic engineering claim that GM plants are likely to be safer, in many ways, than crops bred through conventional methods. They claim that GM crops undergo extensive scrutiny and environmental impact studies, while non-GM crops do not. Are you persuaded by that argument? Should there be stricter oversight of new varieties of conventional crops, despite the economic cost?

SOURCE: Excerpted from the *Seattle Post-Intelligencer*, October 16, 2007.

Chapter Review

Summary

15.1 The Brave New World of DNA Technology

- Scientists can manipulate DNA using a variety of laboratory techniques. Because the structure of DNA is the same in all organisms, these techniques work in much the same way on the DNA of any species.
- A gene is said to be cloned if it has been isolated and many copies of it have been made in a suitable host cell, such as a bacterium.
- The polymerase chain reaction (PCR) produces millions of copies of DNA from a few molecules of template DNA.
- Genetic engineering introduces cloned DNA into cells, tissues, or whole organisms, to create genetically engineered (GM) organisms.

15.2 Working with DNA: Basic Techniques

- Restriction enzymes are used to break DNA into small pieces. Gel electrophoresis separates the resulting DNA fragments by size.
- Ligases are used to join pieces of DNA.
- Labeled DNA probes are used in DNA hybridization experiments to detect the presence of a particular gene in a sample of DNA.
- Automated machines greatly speed up the processes of sequencing and synthesizing DNA.
- To build a DNA library, a vector such as a plasmid is used to transfer DNA fragments from the donor organism to a host organism, such as a bacterium.
- To amplify a gene by PCR, primers (short segments of DNA that are complementary to the beginning and end of the target gene) are synthesized and used to produce billions of copies of the target gene in a few hours.

15.3 Applications of DNA Technology

- DNA fingerprinting, which creates a unique genetic profile for each individual, is widely used in criminal cases. PCR-based analysis is the most widely used method for DNA fingerprinting.
- In genetic engineering, a DNA sequence (often a gene) is isolated, modified, and inserted back into the same species or into a different species.
- Genetic engineering is used to alter the phenotype (especially the performance or productivity) of the genetically modified organism or to produce many copies of a DNA sequence, a gene, or a gene product.

15.4 Ethical and Social Dimensions of DNA Technology

- The use of DNA technology provides potential benefits, but also raises ethical questions and poses potential environmental risks.
- Genetically modified plants have been stiffly opposed by some people, especially in Europe. GM foods do not have to be labeled in the United States and are common in the American diet.

- Opponents of genetic engineering are concerned that genes from GM plants or animals could spread to wild species, potentially wreaking environmental havoc.
- Few people object to the use of gene therapy to cure severe genetic disorders. The technology, however, faces many challenges, especially in finding safe and effective means for introducing a cloned gene into a patient.

◉ Review and Application of Key Concepts

1. Discuss the extent to which our current ability to manipulate DNA differs from what people have done for thousands of years to produce a wide range of domesticated species, such as dogs, corn, and cows.

2. Judy and David are a couple considering whether to have children. Judy knows that her mother's sister died of sickle-cell anemia, which means Judy's grandparents passed the sickle-cell alleles on to her aunt, and Judy herself may have inherited the sickle-cell allele from them. The story on David's side is similar: two of his aunts have sickle-cell anemia. Describe in detail the DNA technology procedures that would enable Judy and David to know with certainty whether they carry the sickle-cell allele.

3. Define DNA cloning and describe how it is done.

4. Discuss some practical benefits of DNA cloning.

5. When a DNA library is made, tens to hundreds of thousands of fragments of an organism's DNA are stored in a large number of host organisms (one fragment per host cell). How is the library screened so that scientists can find a particular gene of interest?

6. What is genetic engineering? How is it accomplished? Select one example of genetic engineering and describe its potential advantages and disadvantages.

7. Is it ethical to modify the DNA of a bacterium? A single-celled yeast? A worm? A plant? A cat? A human? Give reasons for your answers.

8. Are some modifications to the DNA of humans not acceptable? Assuming you think so, what criteria would you use to draw the line between acceptable and unacceptable changes?

Key Terms

cloning (p. 311)
DNA cloning (p. 305)
DNA fingerprinting (p. 313)
DNA hybridization (p. 309)
DNA library (p. 311)
DNA primer (p. 312)
DNA probe (p. 309)
DNA sequencing (p. 310)
DNA technology (p. 305)
DNA vector (p. 311)
gel electrophoresis (p. 308)
gene therapy (p. 305)

genetic engineering (p. 305)
genetically modified organism
  (GMO) (p. 316)
genome (p. 305)
ligase (p. 308)
plasmid (p. 311)
polymerase chain reaction (PCR)
  (p. 312)
recombinant DNA (p. 305)
reproductive cloning (p. 314)
restriction enzyme (p. 307)
RFLP analysis (p. 315)

Self-Quiz

1. Which of the following cuts DNA at highly specific target sequences?
   a. DNA ligase
   b. DNA polymerase
   c. restriction enzymes
   d. RNA polymerase

2. A collection of an organism's DNA fragments stored in a host organism is called a
   a. DNA library.
   b. DNA restriction site.
   c. plasmid.
   d. DNA clone.

3. The pairing of complementary DNA strands from two different sources is called
   a. DNA replication.
   b. DNA hybridization.
   c. genetic engineering.
   d. DNA cloning.

4. Genetic engineering
   a. can be used to make many copies of recombinant DNA introduced into a host cell.
   b. can be used to alter the inherited characteristics of an organism.
   c. raises ethical questions in the minds of some people.
   d. all of the above

5. When DNA fragments are placed on an electrophoresis gel and subjected to an electrical current, the _____ fragments move the farthest in a given time.
   a. smallest
   b. largest
   c. PCR
   d. DNA library

6. A short, single-stranded sequence of DNA whose bases are complementary to a portion of the DNA on another DNA strand is called
   a. a DNA hybrid.
   b. a clone.
   c. a DNA probe.
   d. an mRNA.

7. Small loops of nonchromosomal DNA that are found naturally in bacteria are called
   a. plasmids.
   b. primers.
   c. vectors.
   d. clones.

8. If the DNA sequence at the beginning and end of a gene is known, which of the following methods could be used to produce billions of copies of the gene in a few hours?
   a. construction of a DNA library
   b. reproductive cloning
   c. therapeutic cloning
   d. PCR

Harnessing the Human Genome

The Genome Projects: A Crystal Ball for Life

In February 2001, the world witnessed a scientific milestone, the fruit of the combined efforts of thousands of researchers over a period of 15 years. For the first time in history, the DNA sequence of almost the entire human genome was available for perusal by anyone with a personal computer and access to the Internet. By 2003, much of the missing sequence information had been obtained, and the Human Genome Project, the quest to catalog the DNA sequence of humans, is now essentially complete. To many scientists, this represents one of the crowning achievements of modern biology, and many press conferences and articles touted it as such. But was it worth the approximately $3 billion price tag? What benefits can we expect as consumers and taxpayers? What type of ethical and social issues have been raised by this "crowning achievement"?

There is little doubt that the Human Genome Project has sped up our understanding of basic cell biology: we have a much clearer idea about how our DNA is organized and how many genes we have, and we can assign a specific function to more than 50 percent of our genes based on the sequence information. The accelerated pace of gene discovery has led to many new tests for alleles associated with genetic disorders and disease susceptibility. Genomewide analysis has shed light on our evolutionary history and also made it possible to identify regions of DNA that are essentially the same in all of us and other regions that are so different they can be used to create a DNA portrait of every person.

Recent improvements in DNA technologies, themselves spurred by the race to sequence the human genome, suggest that the greatest benefits of the Human Genome Project, and the many follow-up projects it has spawned, are just around the corner. Over the next ten years, we are likely

Figure C.1 The Bonds of DNA
The sequence of the human genome was deciphered by 2003. The canine genome sequence was first described in 2004. A pet boxer was chosen as the DNA donor, because this breed shows relatively little genetic variation among individuals compared to the 60 other dog breeds considered. Many purebreds are highly susceptible to specific genetic disorders, including certain cancers, deafness, blindness, and immune system problems. At least half of these disorders have counterparts in humans, so the analysis of the canine genome may uncover gene malfunction in humans as well, in addition to advancing veterinary medicine.

to witness big leaps in gene discovery, genome-based drug design, genetic screening for disease prevention as well as diagnosis, treatments customized to the patient's DNA, and new options for gene therapy—all based on the treasure trove of information about the human genome.

The field of **genomics** builds on genetics by seeking to understand the structure and expression of entire genomes and how they change during evolution. (As explained in Chapter 14, the genome of a eukaryote is all the genetic information contained in one set of its chromosomes.) Realizing the enormous potential of genomics in advancing knowledge, scientists have sequenced the genomes of a variety of nonhuman organisms, including many bacteria, yeasts and other fungi, algae and other protists, weedy experimental plants and vital crops such as wheat and rice, and animals ranging from nematode worms and fruit flies to the mouse. Many new genome projects have been launched in recent years (Figure C.1) because the cost of DNA sequencing, and the time it takes to sequence a genome, have both declined dramatically. The benefits of genome research are therefore likely to extend well beyond human health to affect areas such as wildlife conservation and agriculture.

Some persons have had their own specific DNA sequence determined. It cost about $1 million to sequence the genome of James Watson, one of the codiscoverers of the structure of DNA. According to plans announced by some companies at the forefront of genomics technology, in just a few years from now you could have your genome sequenced commercially for not much more than about $1,000.

What would you do with the complete sequence of the DNA in your nucleus? Knowing your complete DNA sequence amounts to having a blueprint for every biological process in your body. That would include all genetic characteristics and predispositions you might have—even your likely response to diet, exercise, alcohol, addictive substances, stress, various medications, and invasion by infectious organisms. The human genome contains about 3.3 billion base pairs of DNA, and although our genomes are 99.9 percent identical, the 0.1 percent that tends to be different has great bearing on our physical and behavioral attributes, including our susceptibility to certain genetic disorders and infectious diseases, certain aspects of personality, how we respond to the world around us—even our overall life span.

What would you want to know, if anything, about your genetic blueprint? Would you be comfortable having all or part of this information stored in a DNA database? Would you want to reveal any of it to a potential marriage partner, a physician, a team of research scientists, the government, an employer, an insurance company, or a court of law? The day is not far off when we, as a society, will have to make up our minds about these issues. In this essay we describe the genome projects, and their promise and potential in helping us understand the genetic basis of life and in improving human health and quality of life. We also consider some of the ethical and social dilemmas arising from this ability to gaze into the genetic crystal ball.

The Quest for the Human Genome

Efforts to understand how inheritance works have been under way for more than two millennia. There have been numerous intellectual and technical breakthroughs over the last 140 years, several of which we highlighted in earlier chapters of this unit. All of these discoveries paved the way for the focused drive to sequence the human genome (Figure C.2). By 1990, the National Institutes of Health (NIH) and the U.S. Department of Energy (DOE) had created an international consortium of sequencing laboratories or centers collectively known as the **Human Genome Project** (**HGP**). The HGP was originally scheduled to complete the "rough draft" of the human genome sequence by 2005. The DNA to be sequenced was obtained from several anonymous donors from diverse backgrounds. To ensure privacy, the identities of the donors were never associated with the DNA samples. DNA from several individuals was used in hopes of creating a reference genome that would represent humanity better than a single genome could.

In the early stages of the HGP, the decision was made to sequence the smaller genomes of several model organisms as a rehearsal for tackling the human genome. These smaller sequencing projects would enable scientists to improve on existing methods and develop the computer programs necessary for analyzing and knitting together huge quantities of sequence data. The selected genomes came from a bacterium (*Escherichia coli*), with 4.6 million base pairs of DNA; budding yeast (*Saccharomyces cerevisiae*), with 12 million base pairs; and a nematode worm (*Caenorhabditis elegans*), with 97 million base pairs.

Interestingly, however, the first complete genome sequence of a free-living organism was published in 1995 by a group not involved in the HGP. The genome of the *Haemophilus influenzae* bacterium, consisting of 1.8 million base pairs, was completely sequenced in a collaborative effort spearheaded by Craig Venter, a former NIH scientist who is now a billionaire entrepreneur. Venter's privately funded foray into genome sequencing planted the seeds of a controversy that continues to swirl around large-scale research efforts like the Human Genome Project: Can the private sector conduct such research more efficiently and cheaper than federally sponsored initiatives can? Is publicly funded genome research essential, no matter the cost, to ensure that this information remains in the public domain and is freely available to spur growth in basic knowledge and biomedical advances? Should companies be allowed to claim patents on genes discovered through genome analysis? So far, the answer to the first two questions appears to be yes, and judging by the agreements reached so far, the answer to the last question appears to be maybe.

A Preview of Our Blueprint

Greater than 99 percent of the human genome has been sequenced. The small portion that remains unsequenced has no protein-coding genes and is found mostly at the tips of chromosomes or at the centromeres. This type of DNA consists of distinctive stretches of nucleotides repeated over and over, sometimes thousands of times, and for technical reasons such repeated DNA is more difficult to sequence. However, we can browse the great majority of our genetic blueprint today, including the 1.5 percent of it that contains protein-coding genes. Analysis of the human genome sequence is still in the early stages, but the initial studies have led to some unexpected findings, especially in terms of the number of genes it takes to create the physical and behavioral complexity of humans.

Complexity in organisms is determined by more than just the number of genes

Prior to the initial publication of the human genome sequence, scientists had predicted that humans have at least 100,000 different genes. The first big surprise from reviewing the "rough draft" of the sequence was that the gene number needed revising: the new estimate was that the human genome contains between 30,000 and 40,000 genes. Today the estimate is lower still—20,000 to 30,000—with most genome experts betting on about 24,000 protein-coding genes.

Estimates of gene numbers are based largely on computer analyses of genome sequences, using programs that predict the beginnings and ends of previously unknown genes. Such analyses can be difficult, and they tend to miss genes whose final product is an RNA rather than a protein (refer to the refined definition of a gene in Chapter 13). Although the current estimate of protein-coding genes is close to definitive, the search for RNA-only genes will have to continue for several more years and will require experimental approaches instead of just computer-based analyses. Even so, it has become apparent that constructing and maintaining a human

Explore more about how the human genome was sequenced. ▶❚❚ C.1

Oswald Avery

Frederick Sanger

Figure C.2 Milestones in the Quest to Understand the Human Genome

Aristotle believes that parents pass biological information to their offspring.

Oswald Avery, Colin MacLeod, and Maclyn McCarty show that DNA is the transforming factor and the possible basis for heredity.

Frederick Sanger, and Allan Maxam and Walter Gilbert, develop two different methods for sequencing DNA. Sanger's approach eventually becomes the preferred technique.

Fourth century BC   **1865**   **1944**   **1953**   **1977**

Gregor Mendel performs his groundbreaking studies on inheritance in peas, defining the basic laws of heredity: equal segregation and independent assortment.

James Watson and Francis Crick discover the double helical structure of DNA.

Purple flowers

White flowers

Round, yellow peas

Round, green peas

Wrinkled, yellow peas

Wrinkled, green peas

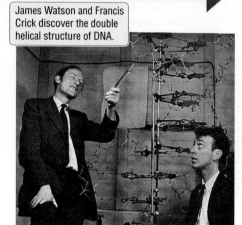

requires only 6 to 10 times the number of protein-coding genes needed by a bacterium such as *E. coli* (Table C.1).

Humans are clearly more complex than bacteria in that we are multicellular, have more than 200 different cell types organized into many organs and organ systems, which in turn enable us to display a wide range of complex behaviors. How can such a tremendous leap in structural and behavioral complexity be achieved with a mere tenfold increase in gene number? Insights from genome analysis, along with conventional genetic studies, have provided some explanations for this puzzle. First, we now understand that by mixing and matching the expression of a limited number of genes we can create a great variety of outputs, each of which could specify a unique cell type. It's as if we have three shirts and three pairs of pants, and by mixing and matching them in all combinations we create nine different wardrobe ensembles that will work for nine different occasions (in the biological context, nine different cell types would be specified).

All multicellular organisms express a unique combination of genes in each cell type. Less than a third of all the protein-coding genes in the human genome are actually expressed in most cell types in the human body. Most cells can also change their normal repertoire of gene expression in response to changes in the environment or important life events. Therefore, the composition of the different types of RNA found in one cell type can vary dramatically compared to another cell type or the same cell type faced with a different set of circumstances.

Let's assume that every gene can independently adopt two states, either on or off, so that it is either making the RNA it codes for or not making that RNA at all. (This supposition itself is an oversimplification, because in reality a gene can be expressed to various levels, producing large, medium, or small amounts of RNA.) A genome consisting of only three genes could generate eight different combinations in which these genes might be expressed, potentially creating

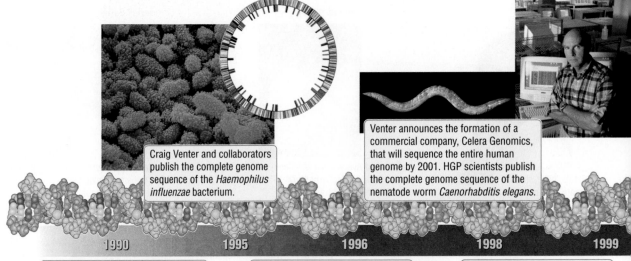

Craig Venter and collaborators publish the complete genome sequence of the *Haemophilus influenzae* bacterium.

Venter announces the formation of a commercial company, Celera Genomics, that will sequence the entire human genome by 2001. HGP scientists publish the complete genome sequence of the nematode worm *Caenorhabditis elegans*.

| 1990 | 1995 | 1996 | 1998 | 1999 |

NIH and the DOE establish the publicly funded Human Genome Project. An international consortium of sequencing centers is scheduled to completely sequence the human genome by 2005, but finishes in 2001.

The HGP publishes the complete genome sequence of budding yeast, *Saccharomyces cerevisiae*.

Both the public and commercial sequencing efforts pass the 1-billion-base-pair milestone. HGP scientists publish the first complete human chromosome sequence, that of chromosome 22.

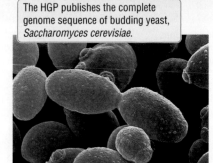

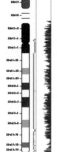

Table C.1

A Sample of the Genomes Sequenced to Date

Organism description	Scientific name	Date	Estimated genome size (millions of base pairs)	Predicted number of genes
Bacterium	*Haemophilus influenzae*	1995	1.8	1,740
Toxic shock bacterium	*Staphylococcus aureus*	2005	2.8	2,600
Bacterium	*Escherichia coli*	1997	4.6	3,240
Anthrax bacterium	*Bacillus anthracis* [… an-THRASS-iss]	2003	5.2	5,000
Budding yeast	*Saccharomyces cerevisiae*	1996	12	6,000
Fruit fly	*Drosophila melanogaster*	2000	180	13,600

Anthrax bacterium

Celera and collaborators publish the genome sequence of the fruit fly, *Drosophila melanogaster*. Venter and Francis Collins (leader of the HGP) agree to collaborate. Disagreement over the data release policy ends the short-lived collaboration between the HGP and Celera.

Since publication of the first-draft sequence, hundreds of laboratories around the world continue to fill in gaps in the sequence, proofread and correct the existing sequence, and confirm the identities of possible genes.

2000 **2001** **2005**

The Human Genome Project and Celera both publish complete draft sequences of the human genome in separate journals: HGP in *Nature*, Celera in *Science*.

The new field of comparative genomics depends on continuing efforts to sequence the genomes of other organisms such as the human, lab rat, *C. elegans* worm, and mustard plant. By June 2005, 266 complete genome sequences had been published, 80% from bacterial sources.

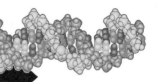

Table C.1

A Sample of the Genomes Sequenced to Date (*continued*)

Organism description	Scientific name	Date	Estimated genome size (millions of base pairs)	Predicted number of genes
Nematode worm	*Caenorhabditis elegans*	1998	97	19,100
Laboratory rat	*Rattus norvegicus* [… nore-*VAY*-juh-kuss]	2004	2,750	25,000
Flowering plant	*Arabidopsis thaliana* [uh-*RAB*-ih-*DOP*-siss *THAH*-lee-*AH*-nuh]	2000	125	25,500
Human	*Homo sapiens*	2001	3,200	20,000–30,000
Puffer fish	*Takifugu rubripes* [*TAHK*-ih-*FOO*-goo roo-*BRYPE*-eez]	2002	400	31,000

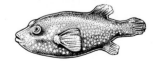

Puffer fish

eight unique cell types. The number of different possible combinations of gene products that can exist in an organism at any given time is determined by the formula 2^n, where n represents the number of genes. Because the formula is exponential, the number of different gene product combinations (that is, the number of different populations of RNAs or proteins) increases extremely rapidly as the number of genes increases. For example, imagine an organism that has only two genes. As Figure C.3 shows, this tiny genome could produce four ($2^2 = 2 \times 2$) different versions in which the two genes were either on or off. Similarly, a four-gene genome could have 16 ($2 \times 2 \times 2 \times 2 = 2^4$) different combinations of RNA types. And so on.

What about real-life organisms? The *E. coli* genome contains 3,240 genes (see Table C.1), which would yield $2^{3,240}$ different possible combinations of RNA types. This number is huge. But for contrast, consider the human genome: the number of ways in which RNA produced by 25,000 genes can be combined is $2^{25,000}$. This means that, even in such a simplified model of gene expression, the human genome can produce vastly larger combinations of gene products than can the genome of *E. coli*. A tenfold increase in gene number produces far more than a tenfold increase in the diversity of information that can be generated.

This calculation, however, understates the number of possibilities. A more realistic comparison of human and bacterial mRNA combinations would have to take into account a second factor—namely, the selective removal of introns from human mRNA products. As we saw in Chapter 13, introns must be removed from most eukaryotic mRNA transcripts before they can be used to make proteins. If different combinations of introns are removed, a single gene can produce many different mRNA transcripts. This variable removal of introns from eukaryotic mRNAs greatly increases the number of different mRNAs produced by eukaryotic genomes such as ours. Bacterial genes do not have introns, which means that each gene produces only one mRNA product.

A third factor to consider is the different mechanisms that humans and bacteria use for regulating gene expression (see Chapter 14). Genome analysis has revealed that a substantial amount of the noncoding DNA we previously dismissed as "junk DNA" actually has important roles in controlling gene expression. As we saw in Chapter 14, compared to prokaryotes, eukaryotes have much more noncoding DNA, some of which directly contributes to the greater complexity of gene expression seen in eukaryotes.

Finally, genomewide analyses in the past five years have shown us that small changes in the organization of the genome can produce dramatic shifts in the phenotype of the organism, most likely because such changes create novel patterns of gene expression. The chimpanzee and human genomes are about 98.8% similar, but these two primates obviously have large differences, especially in brain size and higher-order mental functions (cognition). Genome comparisons now show that the duplication of just a few genes, or a loss in the activity of a gene or two, or a few gene rearrangements (alteration in the position of the genes within the chromosome or even a shift to a different chromosome) can bring about large changes in phenotype. Some of the characteristics associated with the evolution of modern humans—a reorganization of vocal cords, an increased capacity for endurance running, decreased susceptibility to some infectious diseases, increased brain size—now appear to involve small changes in the organization of the genome, rather than the evolution of brand-new genes. Next we describe some of the other lessons learned from a comparison of genomes.

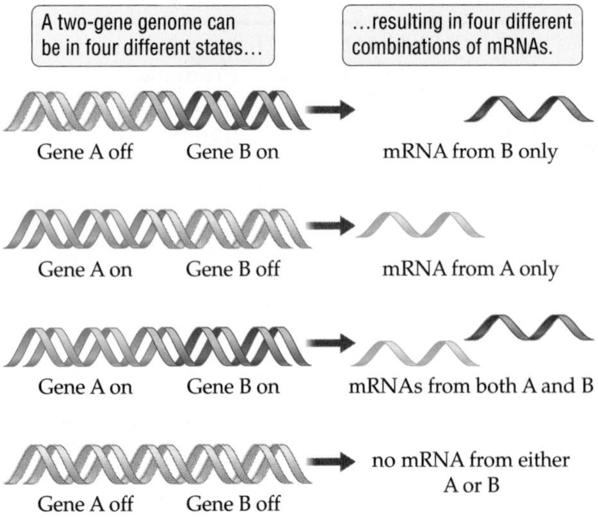

A two-gene genome can be in four different states…

…resulting in four different combinations of mRNAs.

Gene A off   Gene B on   mRNA from B only

Gene A on   Gene B off   mRNA from A only

Gene A on   Gene B on   mRNAs from both A and B

Gene A off   Gene B off   no mRNA from either A or B

Figure C.3 How Genes Can Produce Multiple RNAs

The number of combinations of mRNA transcripts an organism can produce rises exponentially as the number of genes increases. Two genes (shown here) could produce $2^2 = 2 \times 2 = 4$ different combinations of RNAs. Similarly, four genes would yield $2^4 = 16$ different mRNA combinations, eight genes would produce $2^8 = 264$ combinations, and so on.

Comparing the genomes of different organisms yields valuable insights

We can learn much about ourselves, our evolutionary past, and the organization of our genes and the DNA sequences that control their expression, by comparing the human genome with genome sequences from other organisms. These comparisons shed light on the common sets of genes that have been retained over evolutionary time. A gene is said to be conserved between any two groups of organisms if the nucleotide sequence of the gene is essentially the same in the genomes of these organisms. The homeotic genes described in Chapter 15 are examples of **conserved genes**, as is the *Noggin* gene illustrated in Figure C.4.

Roughly 60 percent of human genes are similar to genes found in fruit flies and worms. The proteins they encode participate in the core processes required for life, such as glycolysis and putting together the monomers that constitute the building blocks of life. Unsurprisingly, the genes that have been conserved over millions of years include those that code for DNA polymerases, proteins that regulate DNA transcription, enzymes involved in the processes of metabolism, and many of the receptors and kinases involved in cell signaling.

Evolutionary conservation across species also means that well-understood processes in other organisms may yield insight into similar but less understood processes in humans. Consider jet lag, the sleep disruption most of us experience when we fly across time zones in an airplane. This jumbling of the sleep–wake cycle is due to a disruption of the body's internal clock. The regulation of internal clocks is just one essential process that is likely to involve similar proteins in different species. In mammals, internal clocks dictate the daily rhythm of many physiological events, including sleep. Earlier studies in fruit flies had uncovered many so-called clock genes that control the timing of the fly's activities.

Comparisons between fruit fly clock gene sequences and the human genome have yielded several previously unknown candidate genes that seem to play a role in the daily rhythms of life. Furthermore, the human genome sequence has revealed the chromosomal locations of the known human clock genes, at least one of which is responsible for an inherited sleep disorder.

Studies like these that analyze and compare genomes from multiple organisms are part of a new field in biology called **comparative genomics**. The valuable insights to be gained from comparative genomics have motivated an explosion of efforts to sequence the genomes of a broad range of species. In March 2005, the NIH announced a new round of genome sequencing projects aimed at 12 more species, including a marmoset, a sea slug, a pea aphid, and three fungi (Figure C.5). Each selected organism represents a potential model for human disease or has significant economic importance. The sea slug *Aplysia californica* [uh-PLEEZ-yuh . . .], for example, has very large nerve cells and has been used as a model to study memory and its loss due to disease. Likewise, the marmoset—*Callithrix jacchus*, a Central and South American monkey—is an important model for human diseases such as multiple sclerosis. The pea aphid, *Acyrthosiphon pisum* [AY-ser-thoh-SYE-fun PEE-zum], is responsible for hundreds of millions of

Figure C.4 Gene Expression in a Mouse Embryo
This micrograph of a mouse embryo shows expression of a gene called *Noggin* (*Nog*). The greenish blue stain marks tissues in which *Noggin* mRNA is normally produced. These tissues include parts of the brain and the skeleton. The nucleotide sequence of *Nog* is strikingly similar in animals as diverse as sea anemones, frogs, mice, and humans. The gene controls various developmental pathways in these organisms, including brain and skeletal development in mammals. Mutations in the gene give rise to a variety of developmental effects in humans, including joint disorders such as symphalangism.

(a)
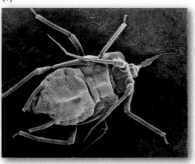

(b)

(c)

(d)
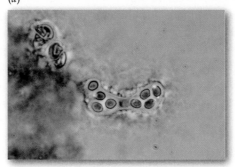

Figure C.5 More Organisms Will Be Sequenced
The NIH has scheduled projects to sequence the genomes of many more organisms, including the four shown here: (a) marmoset (*Callithrix jacchus*), (b) sea slug (*Aplysia californica*), (c) pea aphid (*Acyrthosiphon pisum*), (d) yeast (*Schizosaccharomyces octosporus*).

dollars of crop losses each year and is exceptionally resistant to pesticides. So far, more than 250 complete genome sequences have been published and are available for analysis, and the gold rush to sequence still more species shows no signs of abating.

Health Care for You Alone

Long before the attempt to sequence the human genome began, scores of genes were already known as culprits in human disease. The link between mutations in a given gene and the likelihood of coming down with a particular disease formed the basis for **genetic screening**. The early days of genetic screening involved looking for gene mutations in order to assess a person's future health risks. By the end of the twentieth century, people had the option of testing themselves, and often their unborn fetuses, for mutations linked to diseases such as breast cancer and genetic disorders such as cystic fibrosis and Huntington disease (see the "Science Toolkit" box in Chapter 11). With hundreds of such tests available, people were able to learn more about their children, their families, and themselves than they had ever thought possible.

As we emphasized in earlier chapters, however, many diseases, such as heart disease and diabetes (see the "Biology Matters" box in Chapter 11), are not caused by mutations in just one or two genes. Many malignant cancers, for example, require mutations in several different genes, resulting in a far more complex genetic profile than can be revealed by traditional tests of a few genes (see Interlude B). Knowledge of the human genome sequence and new genomic techniques have uncovered many more variations in an individual's genome, radically expanding the scope of genetic screening.

SNPs are a powerful means of characterizing individual genomes

Individual humans, no matter what their race or ethnicity, share 99.9 percent of their genome with all other humans. However, genome experts estimate that the human genome has about 30 million locations where different individuals do tend to vary, often in just a single nucleotide in a small stretch of DNA. For instance, a person might have a C–G base pair at one location on one chromosome, whereas the person sitting in the next seat in a classroom might have a T–A at exactly the same location. In that case, we would say that human DNA varies in terms of the nucleotide sequence found at that particular location, and the C–G version of the DNA and T–A version of the DNA are allelic versions of the DNA sequence at that location.

These differences in single base pairs, known as **single nucleotide polymorphisms (SNPs)** [SNIPS], account for the great majority of the 0.1 percent difference between one human and another. As such, SNPs are the main source of genomic variation (Figure C.6) among individuals, not only in human populations, but also in most other eukaryotes. Most of the estimated 30 million human SNPs are located in the noncoding regions between genes, rather than within the genes. Some SNPs *are* found within the protein-specifying region of a gene, or in regions that control the expression of a gene, and such SNPs are already familiar to us as allelic variations of a gene (see Chapter 10). Some of these SNPs may produce a distinct phenotype, while others may be phenotypically silent because they don't change the amino acid sequence of a protein enough to alter the phenotype.

What if we could associate the presence of a specific SNP, or a group of SNPs, with specific human traits, such as susceptibility to a disease like breast cancer, or a tendency toward drug addiction, or an allergic reaction to certain medications? If the relationship between having a specific set of SNPs and having a particular trait holds up in large numbers of individuals, we would have a means of testing individuals for that trait *before* they actually displayed the trait. For instance, if we would identify young women at risk for breast cancer on the basis of the presence of certain SNPs in their genome, these susceptible people could be forewarned. They could take measures to lower their risk, including frequent and more rigorous screening for the presence of precancerous cells or early-stage tumors, and they could consider even more drastic therapeutic interventions, depending on the degree of risk they faced.

Several large-scale projects are attempting to match specific SNPs with specific traits in humans, or to catalog SNPs in different human populations. All of these studies lean heavily on the knowledge of the human genome sequence, which serves as a basis for comparison. Even before the "rough draft" of the human genome sequence was released in 2001, scientists in Great Britain began laying the foundation for a massive database of human SNP profiles. Today that database relies on SNP profiles from hundreds of thousands of blood samples donated by adult volunteers. Physicians refer these volunteers to the project, whose staff record each volunteer's current health status and match it with that person's SNP profile. In addition, the volunteers are tracked over time so that changes in their health can also be matched with their SNP profiles. Many similar projects have been launched worldwide, including the International HapMap Project, which seeks to catalog similarities and differences in SNPs among people native to different parts of the world.

The completion of the Human Genome Project, followed by a concerted effort by laboratories across the world, has uncovered the identities and locations of more than 10 million human SNPs, and as we shall see in the next two sections, many of these SNPs have been linked to disease risk, or response to certain drugs, or vulnerability to adverse environmental conditions such as air pollution. Of course, for such breakthroughs to be useful in protecting us through genetic screening, we must have a rapid and affordable way to detect thousands of SNPs in an individual DNA sample. Recent refinements in the technology have made it possible to screen one patient's DNA for 100,000 SNPs in just a matter of hours.

SNP profiling can help identify the risk of getting certain diseases

Gene hunters looking for disease genes take the straightforward approach of comparing the DNA of those who have the genetic condition and those who don't, and asking what is different between the genomes of these two groups. The most comprehensive way to compare the DNA information in two sets of genomes is to look for those SNPs that are systematically and predictably different between the two groups. The task is easier if the two groups are relatively similar because that means the researcher has to sift through a smaller "haystack" of SNP differences, only a few of which are likely to be related to the genetic trait in question.

Groups that tend to marry among themselves, such as the orthodox communities of Ashkenazi Jews, have been a favorite among geneticists because such groups are likely to have less genetic variability and therefore present a smaller "haystack" of genetic differences. By looking for SNPs that are more common among orthodox Jews who live to be 100 years old, researchers have recently identified groups of genes that may play a part in longevity. Initial analysis shows that many of these genes are involved in fending off wear and tear caused by highly reactive by-products of metabolism, known as free radicals. Those who are less protected from such tissue damage appear to be more at risk for heart disease, cancer, and other conditions that grow more common with advancing years.

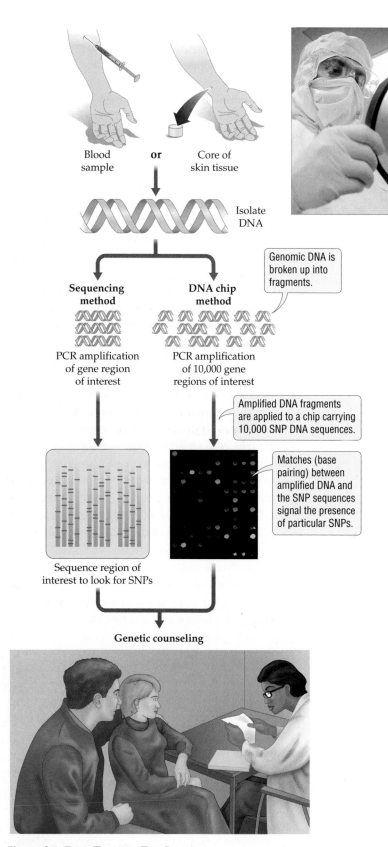

Figure C.6 From Tissue to Test Results

Using cells from tissues or body fluids, scientists can isolate a person's DNA and use a variety of methods to screen it for single nucleotide polymorphisms (SNPs). The inset photo shows a technician holding a DNA microarray plate containing ten thousand different SNPs.

In more than 50 instances, scientists have succeeded in linking distinctive SNPs to a specific genetic condition. These genetic conditions include complex polygenic diseases such as type 2 diabetes, schizophrenia, Crohn disease, multiple sclerosis, rheumatoid arthritis, and prostate cancer. In most instances, the SNPs are noncoding sequences that happen to be adjacent to genes associated with disease risk. Once researchers succeed in linking a SNP, or set of SNPs, to a disease, they have good odds of finding the gene variant (allele) that actually contributes to the disease. For example, researchers have linked the extremely high rates of obesity and diabetes among the Pima Indians of Arizona and New Mexico to specific SNPs that are unusually common in this group. Closer examination shows that these SNPs are close to genes involved in controlling metabolism. The investigators suggest that the gene variant that is unusually common among these desert tribes increases metabolic efficiency in those who possess it, and in bygone days it gave them a survival advantage in times of food scarcity. However, the ability to hoard a large proportion of the ingested calories as stored fat has a negative impact in modern times, when most people find themselves surrounded by an overabundance of calorie-dense food.

Similarly, researchers identified SNPs that were strongly associated with a high risk for Alzheimer's disease (AD), and later investigations showed that these SNPs were physically close to a gene known as *apoE*. This gene codes for a protein (apolipoprotein E) that transports various fatty substances, including cholesterol. There are three relatively common alleles of the gene: *apoE2*, *apoE3*, and *apoE4*. Those who are homozygous for the *apoE4* allele have eight fold higher odds of developing AD, and heterozygotes are about three times more likely to have AD, compared to individuals who lack the allele.

The leading cause of cognitive degeneration in the elderly, AD is characterized by the formation of plaques of protein in the brain, leading to extensive cell death (Figure C.7). Apolipoprotein E is thought to help in remodeling the plasma membranes of brain cells, but we don't yet understand why the *apoE4* version sometimes triggers damage leading to plaque formation. It is important to realize that having two copies of the *apoE4* does not necessarily condemn a person to AD. Many homozygotes do not go on to develop the disease, and nearly half of those who suffer from the disease lack the *apoE4* allele. It is clear that other genes, and environmental factors as well, have a role to play in the development of AD, and the *apoE4* allele is definitely not all there is to AD. A genetic test revealing that a person has two copies of the *apoE4* allele is no more than an indicator of increased risk. This is broadly true for many "tests" for genetic diseases, especially genetic conditions that are polygenic and are also influenced by environmental factors (see the "Biology Matters" box in Chapter 11 for some examples).

SNP profiling can lead to therapies customized for a particular patient

The SNPs you have represent your genetic makeup relative to that of the rest of humankind. A person's SNP profile could therefore indicate not only which diseases he or she might be prone to, but also how that person might respond to infectious agents such as viruses or bacteria, environmental toxins, foods that can cause allergies, and medications of all kinds. Scientists have found, for example, that about 10 percent of Europeans have extra copies of a SNP (named *CC3L1*) and those with this SNP profile appear to be resistant to HIV–AIDS. Children who are victims of abuse are more

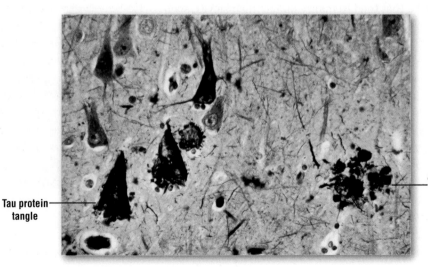

Figure C.7 Protein Plaques in Brain Tissue of Patient with Alzheimer's Disease

Alzheimer's disease is a progressive and fatal disorder that affects about 5 million Americans, most of them elderly. It is marked by the death of brain cells, as clumps of protein appear between and inside the cells. This micrograph shows protein deposits (tangles created by tau protein, and plaques formed by beta amyloid) that form in the brains of patients with Alzheimer's disease. The deposits are believed to disrupt communication between brain cells, resulting in loss of memory and other brain functions. Individuals who carry the *apoE4* allele are at increased risk for developing the disease.

Tau protein tangle

Amyloid plaque

vulnerable to lifelong harm if they carry certain SNPs than are children who have the more common version of those SNPs. In the same vein, some people are more vulnerable to posttraumatic stress disorder (PTSD) than others experiencing the same upsetting events, and that vulnerability has been linked to their SNP profile.

Physicians have noticed for a long time that people can respond to medicines in different ways: some people need higher doses of a drug than others, and some people are helped by a particular medication while others with the same disease show no benefit at all, for example. Beyond adjusting for size differences among men, women, and children, the pharmaceutical industry has traditionally designed drugs with the aim of treating the "average" patient. But the wide range of possible responses a patient can have to a particular drug means that it can be quite tricky to prescribe safe and appropriate therapies.

Why do people vary in their response to a certain drug therapy? There could be nongenetic causes, such as the presence of other drugs in the body or the patient's nutritional status. However, genetic variation is the most common root cause of variable drug response.

Two people can have the same condition, such as high blood pressure, but it could be caused by different alleles of the same gene or it could be due to mutations in different genes altogether. Medical researchers have found this to be true of drugs used to treat high blood pressure (hypertension). One class of antihypertensive drugs, known as beta blockers, are effective in lowering blood pressure in patients with a certain SNP profile. Other hypertensive patients, with a different SNP profile, do poorly on beta blockers but respond well to another family of drugs (called RAAS inhibitors).

Because of genetic variation, people can vary in how quickly chemicals, from caffeine to codeine, are cleared from the body. This diversity is illustrated, for example, by the many allelic variations that exist in genes coding for liver enzymes that modify ingested or injected chemicals. Several such genes have been identified, and SNP analysis shows that some of them are polymorphic in human populations. For example, an enzyme called CYP2D6 is required for the activation of painkillers such as codeine. Nearly 10 percent of the population is homozygous for an SNP that renders this enzyme inactive. These individuals fail to respond to codeine and do not get adequate pain relief from the drug.

One of the most important goals of SNP profiling is to uncover the genetic variety underlying human traits so that therapies can be tailored to the specific genetic problem a given patient has. This concept is often called **personalized medicine** or customized medicine. As the number of SNPs with known effects on gene function increases, SNP profiles are being used to optimize an individual's drug treatment plan. The DNA profile of a patient is often most readily generated by measurement of the activity of critical genes, initially identified by SNP analysis. For example, cancer specialists now use a DNA-based test (called Oncotype DX) to guide them in developing a treatment plan for a woman diagnosed with breast cancer. The test uses the DNA chip technology (see the "Science Toolkit" box in Chapter 14) to measure the activity of 21 different genes to help determine, for instance, whether a patient is likely to benefit from chemotherapy.

In the near future, it will become even more common for patients to sit down with their physicians, review their SNP profile, and find out what it reveals about their susceptibility to various diseases and how well a particular drug therapy might work for them. These methods will result in a personalized health program designed to head off future problems while simultaneously treating current illnesses as effectively as possible. In fact, submitting a blood sample for DNA chip analysis of hundreds of thousands of SNPs could become the first order of business during a baby's first visit to a pediatric clinic.

Ethical Issues Raised by Genetic Testing

Analysis of the human genome is already transforming the way we think about the many diseases caused by a combination of genetic and environmental factors. Understanding the genetic component of a disease enables physicians to predict more accurately the likelihood that a disease will occur in a particular individual, giving them the option either to prescribe preventive measures or to improve the course of treatment.

This is a brave new world of opportunity for improving human health and quality of life, and at some point it is likely to affect the lives of most young people living in the wealthier nations. At the same time, ethical and social issues stemming from these technologies have continued to grow. At the heart of the dilemma are the questions of *how* individuals should use the knowledge about their genomes, whether the rest of society should have any say in the matter, and what kinds of privacy rights individuals have with regard to their genomic information.

Explore how to get from tissue to test results.
C.2

New prenatal genetic screening methods provide added choices— and dilemmas—for parents

Learn more about prenatal screening. ►‖ C.3

Genetic screening for potential health problems does not have to wait until early childhood. While still pregnant, a woman can have numerous tests done on the genetic characteristics of her fetus. Depending on the outcome of a SNP profile (or a more traditional genetic test, as described in the "Science Toolkit" box in Chapter 11), she and her partner may or may not choose to terminate the pregnancy.

Such decisions are already altering the genotypes of babies being born today. In the United States, the number of Down syndrome babies is decreasing as a result of genetic screening (Figure C.8). According to some reports, the abortion rates for Down syndrome fetuses range from 50 percent up to 90 percent. And one study reported a 99 percent abortion rate of fetuses found to suffer from thalassemia [tha-luh-SEE-mee-uh], a genetic disorder of the blood. In some parts of Asia where boys are more highly valued than girls, large numbers of female fetuses are regularly aborted. Scientists who follow population trends are predicting a highly skewed sex ratio as fetuses allowed to survive become children and grow to adulthood in the next few decades.

What should prospective parents do if they discover that their fetus has a genetic disorder that will destine them and their child for a difficult life? A person with Down syndrome, for example, experiences mild to severe mental retardation along with problems, such as heart disease, that can require repeated and difficult surgeries. In addition, caring for a child with Down syndrome can be difficult for a family—financially, emotionally, and physically. For many parents, the answer is to terminate the pregnancy. Others argue that children with Down syndrome are among the most loving of people, with characteristically pleasant dispositions, able to enjoy life despite their impairments. Some individuals with Down syndrome hold part-time or full-time jobs and are able to live independently from their parents.

What about cystic fibrosis, another genetic disorder for which testing is available? Children with cystic fibrosis often have great difficulty breathing, requiring painful daily therapies to loosen the thick secretions in their lungs. In addition, these children often suffer from infections and pneumonias, experience digestive problems, and have trouble growing. Yet they have the potential to lead productive lives as adults. Babies with cystic fibrosis once routinely died as children; today many live into their forties or even their fifties.

When prenatal genetic screening reveals disorders such as these, the prospect of deciding whether to continue or terminate a pregnancy can be daunting. Many parents feel ill equipped to decide whether the quality of life that a child with a genetic disorder is likely to have will be worthwhile. Their choice is often complicated further by feelings of guilt and the question of how caring

(a)

(b)

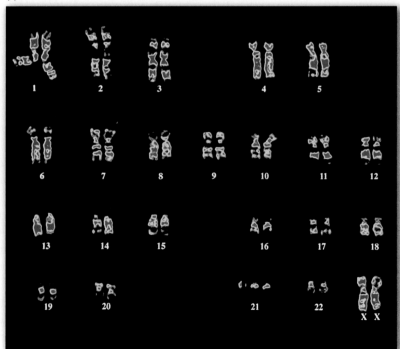

Figure C.8 A Disappearing Condition?

(a) This child shows the typical facial symptoms of Down syndrome, or trisomy 21. The major effect of the extra chromosome is mental retardation, but there are often other health problems as well. The number of Down syndrome babies being born is on the decline—a trend attributed to prenatal screening for the condition. (b) Individuals with Down syndrome carry two copies of all chromosomes except chromosome 21, for which they carry three copies, as shown.

for such a child might affect their lives and the lives of their other children.

Prenatal genetic screening, combined with other technologies, may encourage parents to choose embryos with the right genetics

In an even more extreme application of prenatal genetic screening, couples using assisted reproduction ("making test tube babies") can have their genetic choice of fetuses before a pregnancy has even begun, as we saw in Chapter 11 (see the discussion of preimplantation genetic diagnosis in the "Science Toolkit" box).

Let's return to the example of cystic fibrosis. Recall from Chapter 11 that cystic fibrosis is an autosomal recessive disorder: Individuals who have two copies of the defective allele develop cystic fibrosis. Those who have only a single defective allele are known as carriers; they are able to pass the allele for the disorder on to their children while remaining healthy themselves. When both parents are carriers, each of their children has a one in four chance of being homozygous for the defective allele and developing the disorder.

In 1994, one couple, both cystic fibrosis carriers, made reproductive history by undergoing in vitro fertilization ("in vitro" is Latin for "in glass," or "in a test tube"), having the resulting embryos tested for the cystic fibrosis allele, and then choosing which embryos to keep. Two homozygous embryos were rejected, and three embryos that were heterozygous were implanted in the mother-to-be. From these embryos, a single baby boy grew to term and was born. He is a carrier because he is heterozygous for the cystic fibrosis allele. He will pass the allele to his children, some of whom might get the disease if his future partner has cystic fibrosis or is a carrier—but he himself will not get the disease. Although these parents were interested only in the cystic fibrosis gene, there are no legal restrictions in place to stop parents from testing embryos before implantation for any number of genetic disorders or predispositions to disease using the growing diagnostic power of SNP profiling.

SNP profiling may affect individual privacy and freedom in new ways

Despite the exciting promise that detailed genetic profiling may hold, it raises a number of ethical problems. We have touched on some of these issues already, but there are others. For example, will such detailed biological profiles be used to discriminate against individuals? Health insurance companies might be tempted to raise premiums for people whose profiles show a high susceptibility to disease later in life. Individuals with a high risk of serious illness might even find themselves denied health and life insurance. Furthermore, the new field of behavioral genomics has revealed SNPs linked to a person's susceptibility to syndromes such as alcoholism, schizophrenia, and clinical depression. The revelation that certain people have a heightened susceptibility to drug addiction, for example, might cost them their jobs.

To prevent the potential misuse of SNP profiles, new guidelines for personal privacy must be established. To this end, about 5 percent of the budget for the Human Genome Project is devoted to studying the ethical, legal, and social issues surrounding the availability of genetic information. If we are to reap the benefits of this scientific achievement, addressing these issues will be at least as critical as the new quest of the genome researchers—discovering the function of every gene identified in the human genome and understanding how these genes interact.

◉ Review and Discussion

1. On the basis of what you have learned about cells, organelles, and metabolism, describe some of the genes you probably share with a tomato plant. What sets of genes do you think you do *not* share with plants?

2. What are SNPs, and what role do they play in genetic evaluations related to health issues?

3. Could knowing the entire genome sequence of the mosquito that carries malaria help us eradicate that deadly disease? Explain.

4. Do you think patients should be genetically screened before doctors issue them drug treatments? Why or why not? What ethical issues, if any, need to be considered?

5. As researchers discover more and more genes that predict certain tendencies—for example, toward obesity, high blood pressure, or depression—do you think doctors should screen their patients for some or all of these genes to help them adopt the healthiest lifestyles possible? Why or why not?

Key Terms

comparative genomics (p. C8)
conserved gene (p. C8)
genetic screening (p. C9)
genomics (p. C2)

Human Genome Project (HGP) (p. C3)
personalized medicine (p. C12)
single nucleotide polymorphism (SNP) (p. C9)

Key Concepts

- Evolution is change in the genetic characteristics of populations of organisms over time. For evolution to occur, there must be inherited differences among the individuals in a population. Populations evolve; individuals do not.

- Individuals with certain advantageous inherited characteristics may survive and reproduce at a higher rate than other individuals do—a process known as natural selection. The advantageous inherited characteristics of individuals that leave more offspring become more common in the following generation.

- Adaptations are features of an organism that improve its performance in its environment. Adaptations are products of natural selection.

- The great diversity of life on Earth has resulted from the repeated splitting of a single species into two or more species.

- When one species splits into two, the two species that result share many features because they evolved from a common ancestor.

- An enormous amount of evidence shows that evolution has occurred. One strong line of evidence comes from the fossil record, which allows biologists to reconstruct the history of life on Earth and shows how new species arose from previous species. Features of existing organisms, patterns of continental drift, and direct observations of genetic change also provide strong evidence for evolution.

Main Message: Strong evidence for evolution is provided by fossils, features of existing organisms, continental drift, and direct observations of genetic change.

A Journey Begins

The Galápagos [guh-*LAH*-puh-gohss] Islands, located off the west coast of South America, are isolated, encrusted with lava, and home to bizarre creatures found nowhere else on Earth: unthinkably huge tortoises, land-dwelling lizards that dive into the sea to scrape algae off of rocks for their dinner, and vampire finches that suck the blood of other birds.

No wonder Charles Darwin was so fascinated by the animals of the Galápagos. At the age of 22, he visited the islands as part of a remarkable round-the-world journey on the ship known as HMS *Beagle*, a trip that began in 1831 and ended 5 years later. During his travels as ship's naturalist, Darwin collected many animals and plants, but among the most interesting were those from the Galápagos.

Darwin came to find that these organism's were even more curious upon his return to England. For it was then that Darwin learned from taxonomists that many of his Galápagos specimens were new species and that many of these new species were confined to a single island. Darwin also realized that the islands as a whole were home to whole suites (related groups) of strange new species—for example, suites of tortoise species and finch species. The suites of finches, in particular, puzzled him. Some of the species were so strangely unfinchlike that at first, Darwin had had no idea they might be finches at all. Yet the entire suite of finches, once closely examined, appeared to consist of distinct yet similar forms. In fact, they were all notably similar to a finch species found on the South American mainland. **Why were there so many odd new finch species in the Galápagos?**

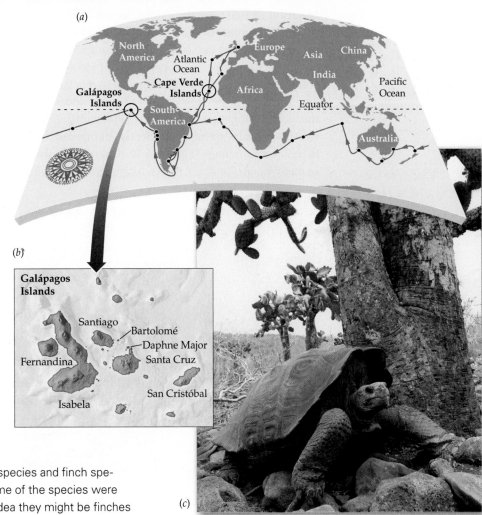

Charles Darwin's Voyage

(*a*) The course sailed by the *Beagle*. (*b*) The Galápagos Islands, located 1,000 kilometers to the west of Ecuador. (*c*) A giant Galapágos tortoise.

Charles Darwin

After pondering the oddities of the Galápagos, as well as considering many other biological observations, Darwin concluded that species were not, as was thought, the unchanging result of separate acts of creation by God. Instead, he would come to a bold new conclusion—one that would explain his odd little finches, as well as the diversity of all life on Earth. Species, he realized, had descended with modification from ancestor species; that is, they had changed over time. They had evolved. It was a revolutionary realization that he would unveil in 1859 by writing the book that shook the world: *The Origin of Species*.

It is hard to overstate the importance of Darwin's discovery. Evolution is biology's most powerful explanation for why living things are as they are, why they look, act, sound, breathe, grow—and do everything—as they do.

Earth teems with living things. One of the most striking aspects of the planet's many organisms is the beautiful fit that so many exhibit for life in their particular environments. A hawk with its broad wings and powerful muscles can soar easily through the sky, a flower's colorful petals and sweet scent quickly attract pollinators, and an insect's body looks so much like the leaves around it that it is almost perfectly hidden from the view of hungry predators. How did organisms come to be as they are, seemingly perfectly engineered to life in their surroundings (Figure 16.1)? And why are there so many kinds of animals, plants, fungi, and other organ-

isms? That is, why is there such a great diversity of life? And within that diversity, why do organisms share so many characteristics? The answer to all these questions is the same: evolution.

We open this chapter by defining the term "evolution" more fully. We then provide an overview of the mechanisms that cause evolution, the power of evolution to explain the characteristics of living things, the evidence that shows evolution is happening, and the impact of evolutionary thought on human society. We close the chapter by describing an example of evolution in action, explaining why there are so many odd species of finches in the Galápagos.

16.1 Biological Evolution: The Sum of Genetic Changes

Defined broadly, evolution is descent with modification, meaning changes in the form or behavior or some other characteristic of a lineage of living things over time. The term "evolution" can be used in nonbiological situations as well—for example, with cars, computers, or hats. In each of these cases, new items represent modified versions of previous items. But there is an important and fundamental difference between biological evolution and, say, the evolution of hats: hats change over time because of deliberate decisions made by their designers. As we will see throughout this chapter, biological evolution is not guided by a "designer" in nature (though humans can, and do, direct the course of evolution in some species—for example, when breeding dogs or crops).

Biological evolution can be defined in several ways. In this book, we define **evolution** as change in the genetic characteristics of a population of organisms over time. We take this approach so that we can use our knowledge of genetics to help us understand how evolution works. Note that evolution is defined at the level of the population, not the individual. Populations evolve, but individuals do not. That is because, overall, the genotypes of individuals do not change over the course of their lives. Mutations can and do occur within the cells of an individual, but most of the genes in any given cell do not mutate, and the mutations that do occur differ from cell to cell. In the next section we take a brief look at two mechanisms that cause evolution; we will examine these and other such mechanisms in greater detail in Chapter 17, where we discuss **microevolution**—small-scale evolutionary changes.

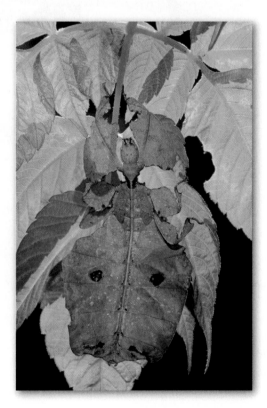

Figure 16.1 The Match between an Organism and Its Environment
These insects, known as leaf insects—a relative of stick insects—are a vivid example of the way in which many organisms seem so exquisitely designed to live in their particular environments. Not only do these *Phyllium giganteum* look like leaves, but they also act like them, remaining mostly still but beginning to move with a swaying motion if they perceive a gentle breeze.

Evolution can also be defined as the pattern of large-scale changes in life on Earth over long time periods. From this perspective, evolution is history; specifically, it is the history of the formation and extinction of species and higher taxonomic groups over time. Focusing on the rise and fall of different groups of organisms over time provides us with a grand view of the history of life on Earth, a view to which we will return in Chapter 19 when we discuss **macroevolution**—large-scale evolutionary changes.

16.2 Mechanisms of Evolution

Think of the students in a biology class. If they were lined up together, their many differences would make it easy to tell one person from the next. What is true for people is true for all organisms. Individuals of all species vary in many of their characteristics, including aspects of their **morphology** (form and structure), biochemistry, and behavior.

What causes organisms to be different from one another? As we learned in Unit 3, one thing that can cause individuals to be different from one another is the possession of different alleles for genes that influence their phenotypes. Such differences in the genetic makeup of individuals are essential for evolution. To see why, consider what would happen in a population in which (1) some individuals in the population were larger than others; (2) body size was an inherited characteristic; and (3) large individuals produced more offspring than small individuals. How might such a population change or evolve over time?

Here is one way that evolution could change this population: Body size is an inherited trait, so the offspring of large parents will themselves tend to be large. Furthermore, since large individuals produce more offspring than small individuals, from one generation to the next more and more of the offspring will be large. Over time, the average size of the population will increase and the alleles that cause large size will become more common. In other words, since we define evolution as change in the genetic characteristics of a population over time, the population will evolve.

As we have seen, evolution can occur if some individuals within a population survive and reproduce at a higher rate than other individuals. In nature, populations evolve—that is, the genetic characteristics of a population of organisms change over time—as a result of two processes: natural selection and genetic drift.

In many cases, certain individuals survive and produce more offspring because they have advantageous inherited traits—a process known as **natural selection**. How does natural selection work? Let's consider the preceding example, in which larger individuals were producing more offspring. Imagine that larger individuals were producing more offspring because being large was an advantage in survival or reproduction. In this case, the average size of individuals and the alleles for large size would increase in the population as a result of natural selection.

Biologists see such natural selection at work in nature all the time—for example, with garter snakes that eat highly poisonous rough-skinned newts (Figure 16.2). These newts produce one of the most potent neurotoxins known: tetrodotoxin [teh-*TROH*-doh-tox-in], or TTX. Newt toxicity varies from one geographic region to another, and the amount of TTX that a snake can tolerate is under genetic control. In regions where newts are highly toxic, snakes that tolerate more of the poison have an advantage over other snakes, and they have survived and reproduced more than snakes that cannot tolerate the toxin. As a result, snakes in such regions have evolved over time to become highly tolerant of the poison. Today, an individual snake from one of these regions can tolerate a quantity of TTX that would kill 25,000 mice.

Natural selection, a powerful force in natural populations, is the process also known as the survival of the fittest. How, then, does the process of natural selection or survival of the fittest work?

Organisms typically can produce more offspring than can survive to themselves reproduce. A butterfly can lay hundreds of eggs; a maple tree can produce

Figure 16.2 Natural Selection at Work Some garter snakes, like this one, have the genetic advantage of being able to survive eating toxic salamanders known as rough-skinned newts.

thousands of seeds. As a result, in natural populations, as organisms attempt to survive and reproduce, there can be fierce competition for food, shelter, and mates. Therefore, any individual with an advantageous inherited characteristic—for example, the ability to tolerate toxins—will be more likely to survive and reproduce and to pass those characteristics on. Meanwhile, any individual who lacks these advantageous characteristics or who has disadvantageous characteristics will be less likely to survive and reproduce and to pass those disadvantageous characteristics on. In this way, over time, natural selection can cause a population to evolve so that more and more individuals have the advantageous characteristics and fewer and fewer have the disadvantageous characteristics.

Note, however, that the process of natural selection does not create advantageous traits. That is, natural selection does not cause the mutations that give individuals advantageous traits—for example, the ability to tolerate more toxins. Mutations that cause advantageous traits happen randomly. If a mutation produces an advantageous trait, then natural selection can act upon it. Through the process of natural selection, individuals that have such advantageous traits will survive and reproduce more than others and so, over time, individuals with such advantageous traits will come to dominate a population.

Chance events can also cause some individuals to leave more offspring than others do. Consider again the example in which large individuals produce more offspring. This time, however, imagine that being large provides no advantage. Imagine instead that larger individuals just happen to produce more offspring than smaller individuals because the smaller individuals are killed by chance—say, as the result of a fire or flood. In this case, the average size of an individual and the alleles for large size will increase in the population; that is, the population will evolve, but this time the evolution will be a result of chance. This is **genetic drift**, a process in which the genetic makeup of a population fluctuates randomly over time, rather than being shaped in a nonrandom way by natural selection.

Biologists have observed genetic drift operating in many natural populations. Consider what happens when a windstorm causes trees to fall amid a population of forest wildflowers: some of the plants are crushed (and leave no offspring), while others survive (and leave offspring). In this case, whether a plant dies or survives is a matter of chance alone. As Figure 16.3 shows, such chance differences in survival can alter the genetic makeup of a population. Like natural selection, genetic drift can cause populations to evolve.

Although it is a simplification, we can summarize evolution as a two-step process: (1) individuals differ genetically for many characteristics, and (2) natural selection or genetic drift can cause individuals of one genotype to survive or reproduce better than individuals of other genotypes.

Explore the effects of genetic drift. ▶❙❙ 16.1

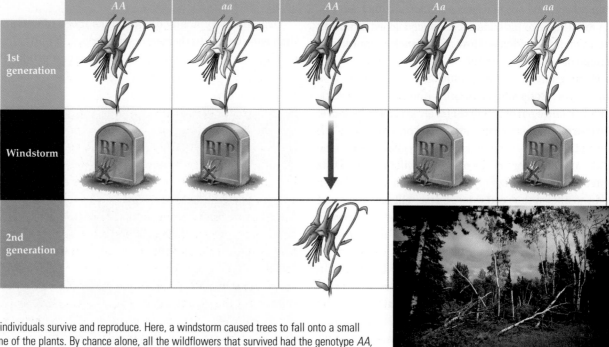

Figure 16.3 Genetic Drift
Chance events can determine which individuals survive and reproduce. Here, a windstorm caused trees to fall onto a small population of wildflowers, killing some of the plants. By chance alone, all the wildflowers that survived had the genotype *AA*, causing the proportion of *A* alleles in the population to change from 50 percent to 100 percent in a single generation.

Concept Check

1. What is the definition of evolution?
2. When populations evolve because some individuals with advantageous inherited characteristics survive and reproduce at a higher rate than other individuals, is that natural selection or genetic drift?

16.3 Evolution Can Explain Many Characteristics of Living Things

Life on Earth is distinguished by exquisite matches between organisms and their environments, by a great diversity of species, and by many puzzling examples of organisms that differ greatly in many respects yet share certain key characteristics. The fact that life evolves can explain all these features of life on Earth. In the following discussion we consider three of life's most striking features: (1) adaptations, or the match between organism and environment; (2) the diversity of life; and (3) the shared characteristics of life.

Adaptations result from natural selection

Some of the most striking features of the natural world are the complex characteristics of organisms and the often remarkable ways in which they are suited to their environments (see Figure 16.1). These characteristics of organisms that are advantageous features, increasing an organism's ability to survive or reproduce, are known as **adaptations**.

Adaptations are the products of natural selection. Here's the reasoning behind this statement: If their reproduction were not restricted in some way, all organisms would reproduce so much that their populations would outstrip the limited resources available to them. Because organisms produce more offspring than can survive, the individuals in a population must struggle for existence. In this struggle, the individuals whose inherited characteristics provide the best match to their environments leave more offspring than do other individuals. This is natural selection in action. Over time, natural selection leads to the accumulation of favorable features—adaptations—within the population.

There are many examples of adaptations. The snowshoe hare, a rabbitlike animal, grows a snow-white fur coat for winter, providing camouflage in the snow. In summer, it sports instead a dirt-brown coat that provides excellent camouflage on bare ground. Cacti are well known for their ability to survive on minute quantities of water—an excellent adaptation for life in the desert. Until Darwin came up with the idea of evolution by natural selection, it was very difficult to explain these sometimes astonishing matches between an organism and its environment—for example, the nearly perfect match of a leaf insect's body to the leaf-filled environment in which it lived. But natural selection explains exactly how such adaptations arise. In each generation, individuals with more advantageous features—for example, any leaf insect of the species *Phyllium giganteum* (see Figure 16.1) whose bodies are especially leaflike—are more likely to evade predators and thereby more likely to be able to continue devouring leaves to survive and reproduce. Over time, any given population slowly comes to be better and better adapted to its environment, resulting, after many generations, in organisms with features that are exquisite adaptations for surviving in their particular environments.

The diversity of life results from the splitting of one species into two or more species

Earth is home to millions of species. Why are there so many different kinds of living things? Here, too, evolution provides a simple, clear explanation: the diversity of life is a result of the repeated splitting of one species into two or more species—a process called **speciation**.

How do new species evolve? Speciation can happen in several different ways. One of the most important is the geographic separation, or geographic isolation, of populations. Consider two populations of a species that are isolated from each other, as by a mountain range or other barrier that prevents individuals from moving between the populations—or in the case of the Galápagos, for example, populations on separate islands, isolated from one another by the sea. The habitats in which the two populations of the original species reside will be different. Over time, natural selection may cause each population to become better adapted to its own particular environment on its own side of the barrier, leading to changes in the genetic makeup of both populations. Chance events may also cause genetic drift. In addition, such chance events may cause the genetic makeup of the populations to begin to diverge. Eventually, so many genetic changes may accumulate between the two populations that individuals from the two populations are no longer able to reproduce with each other. As we learned in Chapter 1, species are often defined in terms of reproduction: a species is a group of populations whose members can reproduce with each other but not with members of other such groups. In this way, geographic isolation can lead to the formation of new species.

Organisms share characteristics as a result of common descent

The natural world is filled with many seemingly puzzling examples of very different organisms that share certain characteristics that we would not imagine they should necessarily share. Consider the appendages we see in the wing of a bat, the arm of a human, and the flipper of a whale. They all have five digits and contain the same sets of bones (Figure 16.4a). But why should limbs that look so different and have such different functions share the same set of bones? Surely, if the best possible wing, arm, and flipper were designed from scratch, their bones would not be so similar. In addition, many living organisms also show the puzzling characteristic of having organs or other features that appear to be of no use to them. For example, many organisms have **vestigial organs** (reduced or degenerate parts whose function is hard to discern). Why do we humans have a reduced tailbone and the remnants of muscles for moving a tail? And why do some snakes have rudimentary leg bones but no legs (Figure 16.4b)? The answer to all these questions is the same: evolution.

Many similarities among organisms are due to the fact that the organisms evolved from a common ancestor. When one species splits into two, the two species that result share many features because they evolved from a common ancestor. Features of organisms related to one another through common descent are said to be **homologous** [hoh-MOLL-uh-guss]. For example, the wing of a bat, the arm of a human, and the flipper of a whale share the same set of homologous bones (see Figure 16.4a). Similarly, some snakes have rudimentary leg bones because they evolved from reptiles with legs (see Figure 16.4b), and humans have rudimentary bones and muscles for a tail because our (distant) ancestors had tails.

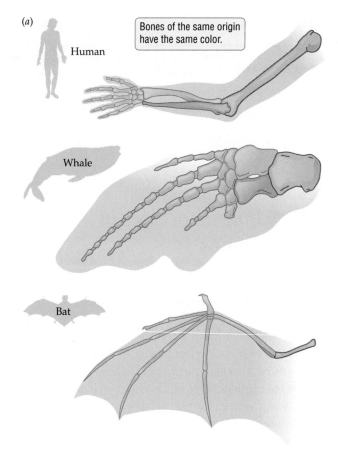

(a)

Bones of the same origin have the same color.

Human

Whale

Bat

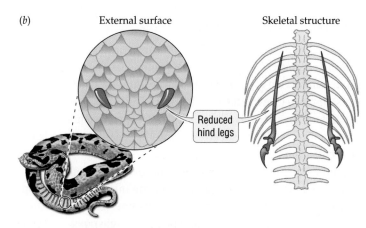

(b)

External surface

Skeletal structure

Reduced hind legs

Figure 16.4 Shared Characteristics

(a) The human arm, the whale flipper, and the bat wing are homologous structures, all of which have what are essentially a matching set of five digits and matching sets of arm bones that have been altered by evolution for different functions. (b) A python has rudimentary hind legs, as the external surface and the skeletal structure of the snake show.

Organisms can also share features as a result of **convergent evolution**, which occurs when natural selection causes distantly related organisms to evolve similar structures in response to similar environmental challenges. For example, the cacti found in North American

(a) (b) (c) (d) (e)

Figure 16.5 The Power of Natural Selection
Plants that grow in deserts often have fleshy stems (for water storage), protective spines, and reduced leaves. These three plants evolved from very different groups of leafy plants. They now resemble one another because of convergent evolution, driven by natural selection for life in a desert. Thus, their shared structures (fleshy stems, spines, reduced leaves) are analogous, not homologous. (a) *Euphorbia* [yoo-*FORE*-bee-uh] belongs to the spurge family and can be found in Africa. (b) *Echinocereus* [ee-*KYE*-noh-*SEER*-ee-uss] is a cactus found in North America. (c) *Hoodia,* a fleshy milkweed, can be found in Africa. Convergent evolution can be a powerful force shaping animals as well. Here we see how natural selection has caused two distantly related animals, sharks and dolphins, to look very similar. Sharks (d) are a kind of fish, related to other aquatic animals like skates and rays; dolphins (e) are mammals, as are cows, bears, and humans.

deserts share many convergent features with distantly related plants found in African and Asian deserts (Figure 16.5*a–c*). Similarly, although both sharks (a kind of cartilaginous fish) and dolphins (mammals) have bodies streamlined for aquatic life, these species are only very distantly related, and their overall similarities result from convergent evolution, not common descent (Figure 16.5*d–e*). When species share characteristics because of convergent evolution, not common descent, those characteristics are said to be **analogous** [uh-*NAL*-uh-guss].

Concept Check

1. Describe three aspects of the living world that evolution can explain.
2. What is another reason—besides descent from a common ancestor—that two species will share characteristics?

16.4 Strong Evidence Shows That Evolution Happens

Surveys taken during the past 10 years reveal that almost half of the adults in the United States do not believe that humans evolved from earlier species of animals. The results of these surveys are startling because evolution has been a settled issue in science for nearly 150 years. The vast majority of scientists of all nations, races, and creeds think that the evidence for evolution is very strong. In his landmark book, *The Origin of Species,* published in 1859, Charles Darwin argued convincingly that organisms are descended with modification from common ancestors. The scientific issues and questions being studied and debated by biologists today are not about whether evolution occurs, but concern the details of how it occurs.

Concept Check Answers

1. Adaptation, the great diversity of life on Earth, and the shared characteristics of life.
2. Convergent evolution. Such similarities are said to be analogous.

Biology Matters

Can't Live with 'Em, Can't Live without 'Em

Evolution is a natural process, but it also occurs in response to human interventions like the increasing development of antibacterial (and, to a lesser extent, antiviral) consumer products. Once largely limited to hospitals and other places at high risk for infections, antibacterial agents are now routinely added to soaps, lotions, dishwashing liquids, and other cleaning products. A recent study published in *Annals of Internal Medicine* estimated that 75 percent of liquid soaps and 29 percent of bar soaps available in the United States contain antibacterial ingredients. Furthermore, many additional products are now coated or impregnated with antibacterial agents, including tissues, cutting boards, toothbrushes, bedding, and children's toys.

So what's the problem? Wouldn't anyone prefer to avoid disease-causing bacteria? If it were that simple, the answer would be yes, but these antibacterial products also kill the beneficial bacteria that surround us in our environment and, according to the World Health Organization, contribute directly to the spread of antibiotic-resistant bacteria. Note that, as always, natural selection does not create advantageous traits. Natural selection does not cause antibiotic-resistant bacteria to arise out of nonresistant bacteria. Instead, once antibiotics are used, bacteria that are already resistant gain a huge advantage over nonresistant bacteria because only the resistant bacteria are not killed off by the antibiotics. They outsurvive and outreproduce the other bacteria and so, through the process of natural selection, these bacteria quickly come to dominate bacterial populations.

In 2000, the American Medical Association (AMA) advised consumers to avoid extensive use of "antibacterial soaps, lotions, and other household products" and also called for greater regulation of antibacterial products. In 2005, the Food and Drug Administration further announced that these soaps have no benefit over ordinary soap and water for ridding hands of bacteria. In fact, numerous studies have begun to show that the use of antibacterial consumer products doesn't decrease the frequency of illness in people that use them. The following guidelines (based on information from the American Society for Microbiology) will help you stay healthy and keep your home clean, while avoiding the risks associated with antibacterial products:

- Plain old soap and hot water remain the best for washing your hands, body, and dishes.
- Wash your hands thoroughly and often before you:
    Prepare or eat food
    Treat a cut or wound or tend to someone who is sick
    Insert or remove contact lenses

and after you:
    Use the bathroom
    Handle uncooked foods, particularly raw meat, poultry, or fish
    Change a diaper
    Blow your nose, cough, or sneeze
    Touch a pet, especially reptiles and exotic animals
    Handle garbage
    Tend to someone who is sick or injured

- Limit your use of antibacterial products to situations when you are most at risk, such as when you are unable to wash your hands.
- Use bleach to clean your bathroom.
- Use separate cutting boards for raw meat and foods that may not be cooked before eating (for example, fruits and vegetables).
- Wash all fruits and vegetables either in soapy water (rinse thoroughly, of course) or in one of the new fruit and vegetable washes.
- Wash all kitchen surfaces, dishes, and utensils in hot, soapy water. Make sure to rinse thoroughly. If possible, put everything (including cutting boards) in the dishwasher.
- Every time you run your dishwasher, throw in the kitchen sponge.
- Don't wipe counters with a sponge that's been sitting on your sink. This can deposit even more bacteria on countertops. Use paper towels or replace your dish rag every day with a clean one.

For example, biologists are still debating the relative importance of natural selection and other mechanisms of evolution (such as genetic drift), but they do not dispute whether evolution occurs. Today's scientific debate about the details of the workings of evolution can be compared to a dispute over what causes wars. Although we might argue about the causes, we all know that wars happen.

Why do scientists find the case for evolution so convincing? As we saw in Chapter 1, a scientific hypothesis must lead to predictions that can be tested, and hypotheses about evolution are no exception. Scientists have tested many predictions about evolution and have found them to be strongly supported by the evidence. Five lines of compelling evidence support evolution: (1) fossils, (2) traces of evolutionary history in existing organisms, (3) continental drift, (4) direct observations of genetic change in populations, and (5) the present-day formation of new species.

Evolution is strongly supported by the fossil record

Fossils are the preserved remains (or their impressions) of formerly living organisms. The fossil record allows biologists to reconstruct the history of life on Earth, and it provides some of the strongest evidence that species have evolved over time. For example, fossils document the fact of extinction and illustrate how the descendants of previously living species changed over time (Figure 16.6). The fossil record also reveals how environments have changed over time: fossils of whales have been found in the Sahara desert; and fossils of trees, dinosaurs, and tropical marine organisms have been found in Antarctica.

As we saw in Chapter 2, the evolutionary relationships among organisms—their pattern of descent from a common ancestor—can often be determined by comparison of their anatomical characteristics.

Fossils like this *Triceratops* skeleton show evidence of organisms now no longer alive

See how natural selection affects a population.

16.2

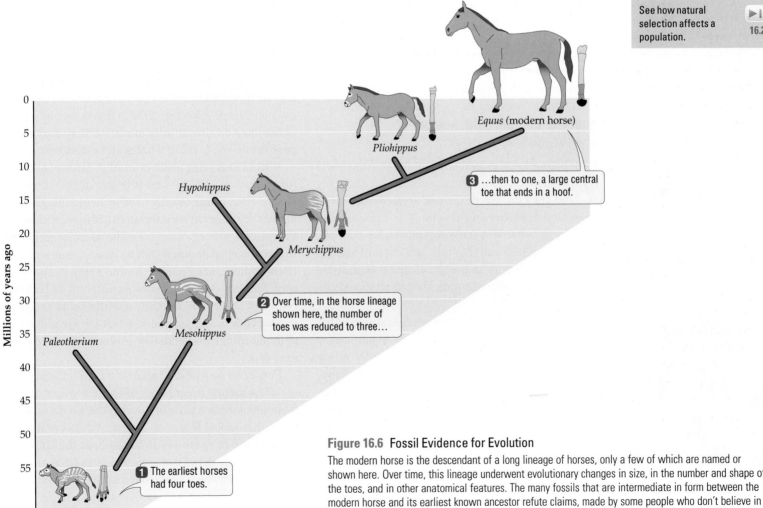

Equus (modern horse)

Pliohippus

3 ...then to one, a large central toe that ends in a hoof.

Hypohippus

Merychippus

2 Over time, in the horse lineage shown here, the number of toes was reduced to three...

Paleotherium

Mesohippus

1 The earliest horses had four toes.

Hyracotherium

Figure 16.6 Fossil Evidence for Evolution
The modern horse is the descendant of a long lineage of horses, only a few of which are named or shown here. Over time, this lineage underwent evolutionary changes in size, in the number and shape of the toes, and in other anatomical features. The many fossils that are intermediate in form between the modern horse and its earliest known ancestor refute claims, made by some people who don't believe in evolution, that the fossil record fails to document evolution.

Applying this technique to fossils, we find that the fossil record contains excellent examples of how major new groups of organisms arose from previously existing organisms. We will discuss one of these examples, the evolution of mammals from reptiles, in Chapter 19. Such fossils, which provide evidence for descent with modification by showing how new organisms evolved from ancestral organisms, exist for many other groups, including fishes, amphibians, reptiles, birds, and humans.

Finally, the times at which organisms appear in the fossil record match predictions based on evolutionary patterns of descent. For example, the evolutionary relationships among modern horses and their ancestors can be determined by comparison of the anatomical features of fossils. On the basis of those evolutionary relationships and the ages of fossils of horses thought to be ancestors to the modern horse, we can conclude that the modern horse (genus *Equus*) evolved relatively recently, about 5 million years ago (see Figure 16.6). From this estimate we can predict that *Equus* fossils should not be found in very old rocks—for example, in 30-million-year-old rocks—and so far they have not.

Organisms contain evidence of their evolutionary history

A major prediction of evolution is that organisms should carry within themselves evidence of their evolutionary past—and they do. We described some examples of such evidence earlier in this chapter—namely, the vestigial organs found in some organisms (for example, the "legs" of a snake and the "tail" of a human) and the remarkable similarity in design of limbs that differ greatly in function (for example, the bat wing and the human arm).

Patterns of growth in the very earliest stages of life can also provide evidence of organisms' evolutionary past. In animals, after a sperm and egg fuse, the newly conceived individual continues growing as an embryo. How these embryos grow and change in shape—that is, how they develop—can provide evidence of the species' evolutionary history. For example, anteaters and some whales do not have teeth as adults, but as embryos they do. Why should the embryos of these organisms develop teeth and then reabsorb them? Or consider the observation that the embryos of fishes, amphibians, reptiles, birds, and mammals (including humans), all develop gill pouches. In fishes, the gill pouches develop into gills that the adults use to "breathe" under water. But why should the embryos of organisms that breathe air develop gill pouches?

Our understanding of evolution provides an answer to these puzzles: similarities in patterns of development are caused by descent from a common ancestor. Fossil evidence suggests that anteater and whale embryos have teeth because anteaters and whales evolved from organisms with teeth. Similarly, fossil evidence indicates that the first mammals and the first birds each evolved from reptiles, although from different groups of reptiles. In addition, the first reptiles evolved from a group of amphibians. Even farther back in time, the first amphibians evolved from a group of fishes. So this line of evidence indicates that the embryos of air-breathing organisms such as humans, birds, lizards, and tree frogs all have gill pouches because all of these organisms share a common (fish) ancestor. In general, unless there is strong natural selection to remove anatomical features from the embryos of descendant groups (to remove teeth from whale embryos, or gill pouches from the embryos of air-breathing animals, for example), these features tend to remain by default. In the adults, however, they may be modified to serve other purposes (for example, gill pouches develop into gills in fishes, but into parts of the ear and throat in humans), or they may disappear (as do the teeth of whales and anteaters).

Within every organism is another piece of evidence for evolution—DNA—and it is one of the strongest pieces. DNA is universally used by living things as the hereditary or genetic material. In addition, with minor exceptions, all organisms use the same genetic code that we learned about in Unit 3. That is, they translate DNA codons into specific amino acids in the same way. The fact that organisms as different as bacteria, redwood trees, and humans use DNA and the same genetic code is further evidence that the great diversity of living things descended or evolved from a common ancestor.

Finally, as we saw in Chapter 2, the evolutionary relationships among organisms—their patterns of descent from a common ancestor—can often be determined from anatomical features. These patterns of descent can be used to make predictions about the similarity among organisms of molecules such as DNA and proteins. Biologists have predicted that the DNA sequences and proteins of organisms that share a more recent common ancestor should be more similar—and they

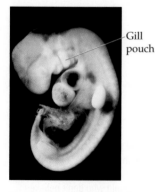

—Gill pouch

Human embryo with gill pouches

Figure 16.7 Independent Lines of Evidence Yield the Same Result

The pattern of evolutionary relationships among the animals shown here is based on the number of DNA sequence differences between the cytochrome *c* gene found in humans and that found in the other organisms. Cytochrome *c* is an enzyme that functions in aerobic respiration and is found in all eukaryotes. The pattern of evolutionary relationships shown here, in which humans are most closely related to rhesus monkeys and least closely related to moths, matches the pattern derived independently from anatomical features.

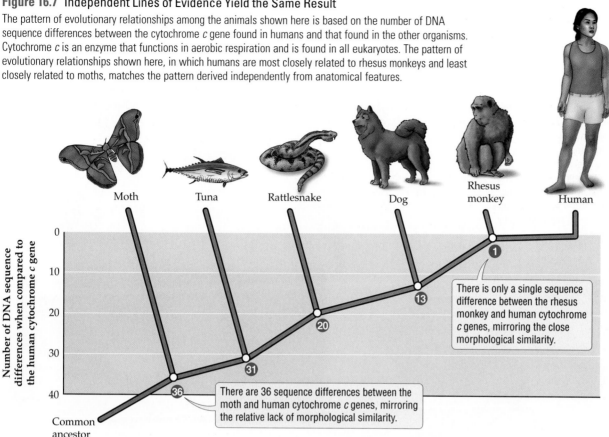

Moth   Tuna   Rattlesnake   Dog   Rhesus monkey   Human

Number of DNA sequence differences when compared to the human cytochrome *c* gene

0
10
20
30
40

Common ancestor

There is only a single sequence difference between the rhesus monkey and human cytochrome *c* genes, mirroring the close morphological similarity.

1

13

20

31

36

There are 36 sequence differences between the moth and human cytochrome *c* genes, mirroring the relative lack of morphological similarity.

---

are (Figure 16.7). If organisms were not related to one another by common descent, there would be no reason to expect that the degree of similarity found in DNA and proteins could be predicted from evolutionary relationships based on anatomical features. So we see these separate lines of evidence—anatomical features, and DNA and proteins—yielding the same result, providing strong evidence for evolution.

Continental drift and evolution explain the geographic locations of fossils

Earth's continents move over time, by a process called **continental drift**. Each year, for example, the distance between South America and Africa increases by about 3 centimeters, a little more than an inch. Although they are separating from one another now, about 250 million years ago South America, Africa, and all of the other landmasses of Earth had drifted together to form one giant continent, called **Pangaea** [pan-*JEE*-uh]. Beginning about 200 million years ago, Pangaea

slowly split up to form the continents we know today (see Figure 19.7).

We can use knowledge about evolution and continental drift to make predictions about the geographic locations where fossils will be found. For example, organisms that evolved when Pangaea was intact could have moved relatively easily between what later became widely separated regions, such as Antarctica and India. For that reason, we can predict that their fossils should be found on most or all continents. In contrast, the fossils of species that evolved after the breakup of Pangaea should be found on only one or a few continents, such as the continent on which they originated and any connected or nearby landmasses.

Predictions about the geographic distributions of fossils have proved correct, and they provide another important line of evidence for evolution. For example, today the lungfish *Neoceratodus fosteri* [nee-oh-sair-*AH*-toh-dus *FAW*-ster-ee] is found only in northeastern Australia, but its ancestors lived during the time of Pangaea, and fossils of those ancestors are found on

Figure 16.8 Once It Lived throughout the Earth

Ancestors of the freshwater lungfish *Neoceratodus fosteri* (inset) lived during the time of Pangaea (see map below). Fossils of these fish have been found on all continents except Antarctica. *N. fosteri,* which is currently found only in the orange-shaded region of northeastern Australia, is the only surviving member of its family. Places where fossils of *N. fosteri*'s ancestors have been found are indicated by the red dots.

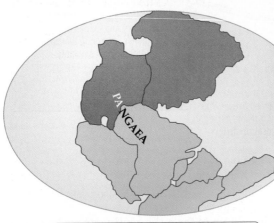

Portions of the supercontinent Pangaea began to drift apart about 200 million years ago.

all continents except Antarctica (Figure 16.8). At the other extreme, modern horses in the genus *Equus* first evolved in North America about 5 million years ago (see Figure 16.6), long after the breakup of Pangaea. The oldest *Equus* fossils are found only in North America, as we would predict. The land bridge that today connects North and South America was formed roughly 3 million years ago. As a result, we would predict that *Equus* fossils found in South America should be less than 3 million years old; and to date, all such discoveries have been less than 3 million years old.

Direct observation reveals genetic changes within species

In thousands of studies, researchers have observed populations in the wild, in agricultural settings, and in the laboratory changing genetically over time. Such observations provide direct, concrete evidence for evolution. Consider how farmers have altered the wild mustard, *Brassica oleracea* [BRASS-i-kuh oh-ler-ACE-ee-uh], to produce several distinct crops, all part of the same species (Figure 16.9). By allowing only individuals with certain characteristics to breed—a process called **artificial selection**—they have crafted enormous evolutionary changes within this species. The tremendous variation that we have produced within dogs, ornamental flowers, and many other species illustrates the power of artificial selection to produce evolutionary change. Natural selection can produce similar evolutionary changes, as shown by the often striking match between organisms and their environment (see Figure 16.1) and by case studies such as that of the medium ground finch, which we will examine shortly.

Formation of new species can be observed in nature and can be produced experimentally

Biologists have directly observed the formation of new species from previously existing species. The first experiment in which a new species was formed took

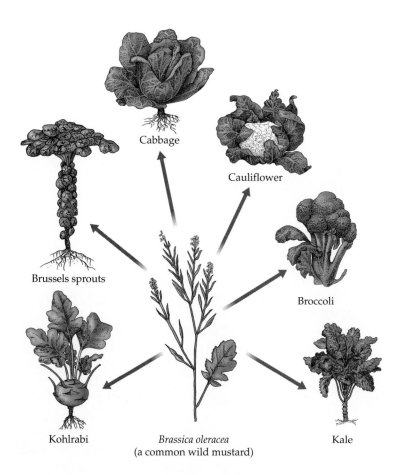

Brussels sprouts

Cabbage

Cauliflower

Broccoli

Kohlrabi

Brassica oleracea
(a common wild mustard)

Kale

Figure 16.9 Artificial Selection Produces Genetic Change
Humans have directed the evolution of the wild mustard *Brassica oleracea* (bottom center) to produce several different crop plants. Despite their obvious differences, all the plants shown here, both wild and domesticated, are members of the same species, *B. oleracea.*

place in the early 1900s, when the primrose *Primula kewensis* [PRIM-yoo-lah kee-WEN-siss] was produced. Scientists have also observed the formation of new species in nature. For example, two new species of salsify [SAL-suh-fee] plants were discovered in Idaho and eastern Washington in 1950. Neither of the new species had been found in those regions or anywhere else in the world in 1920. Genetic data reveal that both of the new species evolved from previously existing species, and field surveys indicate that this event occurred between 1920 and 1950. The two new species continue to thrive, and one of them has become common since its discovery in 1950.

16.5 The Impact of Evolutionary Thought

Before the theory of evolution was developed, adaptations often were taken as logical evidence for the existence of a creator. The existence of organisms that seemed well designed for their environments implied

to many people that there must be a supernatural designer—a God—much as the existence of a watch implies the existence of a watchmaker. In addition, from long tradition—dating from Plato, Aristotle, and other Greek philosophers more than 2,000 years ago—species were viewed as unchanging over time. Darwin's work shook these ideas to their very foundation. No longer could species be viewed as unchanging, nor could examples of apparent design in nature be offered as logical proof that God existed, since evolution by natural selection provided an alternative, scientific explanation for the morphology of organisms.

The evolution of species was a radical idea in the mid-nineteenth century, and the argument that the morphology of organisms could be explained by natural selection was even more radical. These ideas not only revolutionized biology but also had profound effects on other fields, ranging from literature to philosophy to economics.

The idea of Darwinian evolution had a profound effect on religion as well. Evolution was viewed initially as a direct attack on Judeo-Christian religion, and this presumed attack prompted a spirited counterattack by many prominent members of the clergy. Today, however, most religious leaders and most scientists view evolution and religion as compatible but distinct fields of inquiry (see the box on page 338). The Catholic Church, for example, accepts that evolution explains the physical characteristics of humans but maintains that religion is required to explain our spiritual characteristics. Similarly, although the vast majority of scientists accept the scientific evidence for evolution, many of those same scientists have religious beliefs. Overall, most scientists recognize that religious beliefs

Salsify

Biology on the Job

Evolution's Champion

Dr. Eugenie Scott is executive director of the National Center for Science Education (NCSE), a not-for-profit membership organization that provides information and resources for schools, parents, and concerned citizens working to keep evolution in public school science education. The NCSE educates the press and the public about the scientific, educational, and legal aspects of the evolution/creationism controversy. Dr. Scott has written a book about this controversy, entitled Evolution vs. Creationism.

Dr. Eugenie Scott

What do you do on a typical day at work? At work, I'm the "Dear Abby" of people who want to be sure that evolution continues to be taught in our public schools. People call me on the phone if they have questions about evolution, or when they feel the wrong people have been elected to their school board. I advise parents, teachers, legislators, and school boards about the difference between science and religion, and why religious views should not be taught as science. I also do a lot of public speaking at science teacher conferences, universities, and fund-raising events.

What are some of the major political and social issues at stake in the debate about evolution? There's a wide range of political and social issues at stake, including separation of church and state, public understanding of science, and the quality of science education. Evolution is central to the teaching of biology: without evolution as a central organizing principle, biology becomes a set of facts to memorize. Evolution is the periodic table of biology—it makes biology make sense, makes it hang together.

What do you like best about your job? I love my job because I think the issues are so important, and I really enjoy educating people about the evolution/creationism controversy. Another great thing is that I get to read a broad range of scientific material—I love science and I have to read broadly in order to counter claims that creationists make, not only about evolution, but also about physics, geology, and general aspects of biology.

The Job in Context

Creationism can take a variety of forms. One common version of creationism states that all species were created by God roughly 10,000 years ago and that they have not evolved since then. As scientific statements, these assertions are not correct. The scientific evidence indicates that (1) Earth is more than 4 billion years old; (2) life on Earth began about 3.5 billion years ago; and (3) evolution has occurred, continues to occur, and is responsible for the great diversity of life on Earth.

A surrogate for creationism, known as the intelligent design movement, doesn't insist that life began on Earth as recently as 10,000 years ago. Nevertheless, both intelligent design and creationism explain the diversity of life in terms of the actions of a supernatural creator. Such an explanation rings true to some people because their religious beliefs suggest that a supernatural being was directly or indirectly responsible for life on Earth. Science, however, is limited to natural, not supernatural, explanations. Furthermore, as we saw in Chapter 1, the scientific method is central to how science works: experiments are designed to test predictions made from a hypothesis, and if the results fail to support the hypothesis, it is modified or discarded. In contrast, creationism relies on supernatural, not natural, explanations, and does not use the scientific method to examine the validity of its ideas.

Like Dr. Scott, we think it is important for students—the future leaders of society—to understand evolution. If medical doctors, for example, had no understanding of evolution, they would not realize that overuse of antibiotics has the disastrous effect of causing bacteria to evolve resistance to those antibiotics (a topic we will explore further in Chapter 17). If creationists have their way and students are prevented from learning what science has to offer, we run the risk of having students unable to compete effectively in college classrooms or in today's global economy. When scientific understanding and nonscientific beliefs come into conflict, as illustrated by the conflict between evolutionary biology and creationism, the resulting debates can be enlightening. Such debates, however, should not be used as an excuse to keep students from being taught according to our best and most current scientific understanding of how the world works.

Keeping Evolution in the Classroom
This teacher is explaining how this skeleton evolved, and where it came from.

are up to the individual and that science cannot answer questions regarding the existence of God or other matters of religious import.

The emergence of evolutionary thought has also had an effect on human technology and industry. For example, an understanding of evolution has proved essential as farmers and researchers have sought to prevent or slow the evolution of resistance to pesticides by insects. Information about the evolutionary relationships among organisms can also be used to increase the efficiency of the search for new antibiotics and other pharmaceuticals, food additives, pigments, and many other valuable products.

Finally, in what may be a sign of changes that lie ahead, engineers have begun to use evolutionary principles to solve a variety of design problems in nonliving systems. In 2003, for example, a patent was granted for a new type of controller. A controller is a device like the cruise control on a car that can regulate processes such as car speed, heat output from a home furnace, or the access of information from a computer disk. Usually, such a discovery would be the direct brainchild of a human inventor. The controller patented in 2003, however, was "invented" by a computer program that mimicked the biological processes of mutation, mating, and natural selection. The program functioned much as natural selection would in the wild: from one generation to the next, the program selected the "digital individual" that functioned as the best controller, allowing that individual to produce the most "offspring." The end result of this process was a digital individual whose features outperformed the best controllers on the market at the time.

Concept Check

1. Name five pieces of evidence that life has evolved.
2. What is artificial selection?

Applying What We Learned

Evolution in Action

Why are there so many kinds of finches in the Galápagos Islands—species known today as Darwin's finches? Since Darwin's time, the Galápagos have provided a natural biological laboratory in which scientists have studied evolution, including the evolution of the islands' finches.

The climate of these islands is usually hot and relatively wet from January to May, and cooler and drier for the rest of the year. But in 1977, the wet season never arrived; very little rain fell during the entire year. On Daphne Major, a small island near the center of the Galápagos Islands, the lack of rain withered the plants that lived there (Figure 16.10). Soon the effects were also felt by a seed-eating bird, the medium ground finch. During the drought, the number of these birds on Daphne Major plummeted from 1,200 to 180.

The lack of rain not only caused many birds to die, but also induced evolutionary change in the population of medium ground finches. One effect of the drought, which prevailed until the wet season of 1978, was that the seeds available to the finches were larger than normal. The seed size increased because the finches had already eaten most of the smaller seeds by the time the drought began, and plants on the island produced few seeds during the drought. Large seeds can be difficult or impossible for birds with small beaks to crack open and eat. Suddenly, finches with larger beaks had an edge, and many more large-beaked than small-beaked finches survived the drought to contribute offspring to future generations. As a result, beak size in the population of medium ground finches evolved toward a larger size.

Galápagos Islands

Santiago   Bartolomé
          Daphne Major
Fernandina   Santa Cruz

Isabela   San Cristóbal

Before drought

After drought

Figure 16.10 A Drought Results in Rapid Evolutionary Change
The 1977 drought on Daphne Major in the Galápagos Islands had a dramatic effect on the plant life there, setting the stage for natural selection to cause rapid evolutionary change in birds that depended on the plants for food.

Medium ground finches on Daphne Major have been studied continually since 1972. As Figure 16.11 shows, the drought of 1977 had a strong but temporary effect on beak size: the average beak size of these birds shot up in the late 1970s but then declined over time. Beak size was not the only feature of the birds that was subject to bouts of natural selection. After

nearly 30 years of study, the body size of the birds had decreased, and their beak shape had changed from blunter to more pointed.

As the research on the medium ground finch shows, we live in an ever-changing world in which species are constantly being shaped by evolutionary forces. Here are natural selection and evolution in action. This research also provides a look at how quickly these finches and their beaks can evolve, a hint as to how the many odd finches of the Galápagos came to be.

The Galápagos are home to 13 species of finches (Figure 16.12). These finches are closely related to one another, they are found nowhere else on Earth, and they exhibit many behaviors that are unusual for finches. For example, whereas most finches around the world eat only seeds, different Galápagos finches are specialized to eat a variety of other foods. Some still eat seeds, but some use their unusually shaped beaks to eat insects, flowers, ticks, mites, leaves, and bird eggs—and some even to drink blood.

Why are there so many odd and unique finch species in the Galápagos? The answer seems to be that all the finches in the Galápagos Islands descended from a single species, most likely a finch that reached the islands (perhaps blown there by a storm) from the nearest mainland, South America, about 3 million years ago. Upon its arrival, this species found itself in a place where many kinds of birds, such as insect-feeding woodpeckers and warblers, were absent; it is likely that these birds were absent because the Galápagos Islands are geographically isolated and hence receive few immigrants. Over time, natural selection favored finches that developed new ways (for finches) to feed themselves. So although it is unusual for finches to feed on insects, the Galápagos finches that do so evolved to fill an ecological role that is usually taken by other birds, such as the absent woodpeckers and warblers. The end result of this process was that new species of finches evolved from the single species that originally colonized the islands. The odd behaviors of these newly evolved finch species make sense when viewed from the perspective of "evolution in action." In other words, these empty, isolated islands were a perfect place where whole suites of new species could and have evolved—tortoises, cacti, and Darwin's many curious finches.

(a)

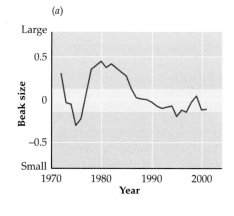

(b)

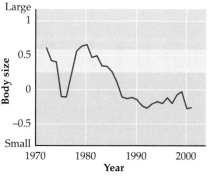

Figure 16.11 Evolutionary Change within a Species
Several morphological features of the medium ground finch population on Daphne Major fluctuated during a 30-year period. Although (a) beak size, (b) body size, and (c) beak shape all fluctuated over time, only body size and beak shape changed significantly from the beginning to the end of the study period. The light blue bands indicate the range of values that would be expected in the absence of evolutionary change.

(c)

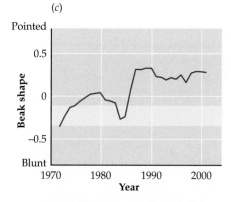

Medium ground finch

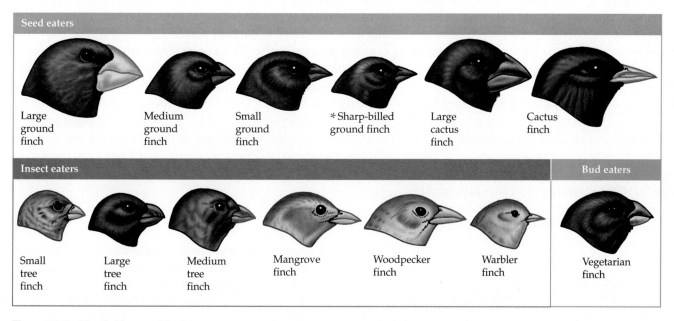

Figure 16.12 The Galápagos Finches
Thirteen unique species of finches evolved on the Galápagos Islands in isolation from other finches. Recent DNA evidence suggests that the warbler finch may actually comprise two different (but morphologically similar) species. The sharp-billed ground finch—the "vampire finch" mentioned at the beginning of this chapter—feeds on the blood of other birds in addition to eating seeds.

Empty-Stomach Intelligence

By Christopher Shea

Hunger makes the best sauce, goes the maxim. According to researchers at Yale Medical School, it may make quadratic equations and Kant's categorical imperative go down easier too. The stimulation of hunger, the researchers announced in the March issue of *Nature Neuroscience,* causes mice to take in information more quickly, and to retain it better—basically, it makes them smarter. And that's very likely to be true for humans as well.

A team led by Tamas Horvath ... had been analyzing the pathways followed in mouse brains by ghrelin, a hormone produced by the stomach lining, when the stomach is empty. To the scientists' surprise, they found that ghrelin was binding to cells not just in the primitive part of the brain that registers hunger (the hypothalamus) but also in the region that plays a role in learning, memory and spatial analysis (the hippocampus).

The researchers then put mice injected with ghrelin and control mice through a maze and other intelligence tests. In each case, the biochemically "hungry" mice—mice infused with ghrelin—performed notably better than those with normal levels of the hormone. The finding was startling, but "it makes sense," Horvath says. "When you are hungry, you need to focus your entire system on finding food in the environment." In fact, some biologists believe that human intelligence itself evolved because it made early hominids more effective hunters, gatherers and foragers.

Hormones are signaling molecules that are produced by cells in one part of an animal or plant's body and then travel to tissues in another part of the body where they trigger some kind of change or action. These Yale researchers are studying ghrelin—the "hunger hormone"—which is produced by the stomach and then travels to the brain, stimulating an organism to feel hungry and eat. Their studies focus on mice, but ghrelin is the hunger hormone in humans as well.

This new research suggests that we evolved to have our empty stomachs produce ghrelin not only to feel hungry but also to get smart. In other words, Horvath hypothesizes that natural selection has resulted in humans who not only feel hungry when ghrelin is released by their empty stomachs, but get smart as well. Though intriguing, is this idea even possible? Remember, we defined evolution as change in the genetic characteristics of a population of organisms over time. So, for humans to have evolved to react in a certain way to ghrelin, the human response to that hormone must have a genetic basis.

Ghrelin is being actively studied by researchers investigating the hormonal control of hunger and fullness. This work is particularly important now because more and more people are eating too much, leading to what some say is an epidemic of obesity. And these researchers have shown that the human ability to respond to ghrelin has a genetic basis. In fact, some mutations that have already been found appear to make people feel too hungry and overeat, causing a tendency to obesity.

Would there have been an advantage to being smarter when hungry? Throughout most of our history, humans (*Homo sapiens*) did not have access to malls, grocery stores, or Internet shopping. Instead, for many millennia we were hunter-gatherers who had to struggle to find food to survive. Probably being smarter when hungry would have been highly advantageous.

How, then, would natural selection have acted to create such an outcome? When their stomachs were empty, early humans whose brains were stimulated by ghrelin not only to feel hunger but to think smarter and more effectively would have survived and reproduced more successfully than those who did not receive such stimulation.

Ghrelin levels decrease in people after eating. So, if the researchers' hypothesis is correct, it may be useful not to eat too much before a biology test. But being too hungry, of course, can also make it hard to think. The Yale researchers suggest that students might not want to go into exams carbo-loaded (not enough ghrelin) or starving (too distracted), but mildly hungry, with the stomach just empty enough to produce the ghrelin required for maximum smartness.

Evaluating the News

1. What is another mechanism—besides natural selection—by which humans could have evolved to have ghrelin traveling to the hippocampus in the brain and making them smarter? Describe how the process would work. (*Hint:* Chance events can cause evolution via random genetic changes.)
2. Being smarter should always be an advantage in survival and reproduction. If ghrelin does make humans smarter, why wouldn't we have evolved to have ghrelin circulating in our brains constantly?
3. Neither students nor professional athletes are allowed to use performance-enhancing hormones like testosterone during competition. If ghrelin becomes commercially available, should students be allowed to use it when taking tests?

SOURCE: *New York Times,* December 10, 2006.

Chapter Review

Summary

16.1 Biological Evolution: The Sum of Genetic Changes

- Evolution can be broadly defined as change in the genetic characteristics of populations of organisms over time; such small-scale evolutionary changes are referred to as microevolution.
- Evolution can also be defined as the history of the formation and extinction of species over time; such large-scale evolutionary changes are referred to as macroevolution.

16.2 Mechanisms of Evolution

- Individuals in populations differ genetically in their morphological, biochemical, and behavioral characteristics.
- Natural selection or chance events can cause individuals with one set of inherited characteristics to survive and reproduce better than other individuals, producing evolutionary change.
- Natural selection is a mechanism by which some individuals, because they have particular inherited characteristics, consistently survive and reproduce better than other individuals. Thus, evolutionary changes wrought by natural selection are not random.
- Genetic drift is a process by which chance alone may cause individuals with one set of inherited characteristics to survive or reproduce better than other individuals. Thus, evolutionary changes caused by genetic drift are random.

16.3 Evolution Can Explain Many Characteristics of Living Things

- Evolution can explain three aspects of life on Earth: adaptations, the great diversity of species, and many seemingly puzzling examples in which otherwise dissimilar organisms share certain characteristics.
- Adaptations are characteristics of an organism that improve its performance in its environment. Adaptations result from natural selection.
- The diversity of life is a result of speciation, the repeated splitting of one species into two or more species.
- The shared characteristics of organisms can be due either to descent from a common ancestor or to convergent evolution. Shared characteristics that result from common descent are said to be homologous; those that result from convergent evolution are said to be analogous.

16.4 Strong Evidence Shows That Evolution Happens

- The fossil record provides clear evidence for the evolution of species over time. It also documents the evolution of major new groups of organisms from previously existing organisms.
- The extent to which organisms share characteristics is consistent with patterns of evolutionary relationships among them. Evidence for evolution includes remnant anatomical structures (vestigial organs), patterns of embryonic development, the universal use of DNA and the genetic code, and molecular (DNA and protein) similarities among organisms that are anatomically similar.
- As predicted by our understanding of evolution and continental drift, fossils of organisms that evolved when the present-day continents were all part of the supercontinent Pangaea have a wider geographic distribution than do fossils of more recently evolved organisms.
- In thousands of studies, researchers have observed genetic changes in populations over time, providing direct evidence of small evolutionary changes. Some of these studies involve artificial selection, in which people alter a population by choosing which individuals to breed; others involve observations of natural selection.
- Biologists have observed the evolution of new species from previously existing species.

16.5 The Impact of Evolutionary Thought

- Darwin's ideas on evolution and natural selection revolutionized biology, overturning the views that adaptations must be proof of God's existence and that species do not change over time. His ideas also had a profound effect on many other fields, including literature, economics, and religion.
- Evolutionary biology has many practical applications in agriculture, industry, and medicine.

⊚ Review and Application of Key Concepts

1. Explain what evolution is and why we state that populations evolve but individuals do not.

2. A population of lizards lives on an island and eats insects found in shrubs. Because the shrubs have narrow branches, the lizards tend to be small (so they can move effectively among the branches). A group of lizards from this population migrates to a nearby island where the vegetation consists mostly of trees; the branches of those trees are thicker than the shrubs on which the lizards used to feed. A few of the lizards that migrate to the new island are slightly larger than the others; large size is not a disadvantage for moving in the trees (because the branches are thicker) and is advantageous when males compete with other males to mate with females. Assuming that large size is an inherited characteristic, explain what is likely to happen to the average size of the lizards in their new home, and why.

3. How does evolution explain (a) adaptations, (b) the great diversity of species, and (c) the many examples in which otherwise dissimilar organisms share certain characteristics?

4. Why are scientists throughout the world convinced that evolution happens? Consider the five lines of evidence discussed in this chapter.

5. Although biologists agree that evolution occurs, they debate which mechanisms are most important in causing evolutionary change. Does this mean that the theory of evolution is wrong?

6. Genetic drift occurs when chance events cause some individuals in a population to contribute more offspring to the next generation than other individuals contribute. Are such chance events likely to have a greater effect in small or in large populations? (*Hint:* Examine Figure 16.3. Consider whether the proportion of the *A* allele in the population would be likely to change from 50 percent [5 of the 10 alleles originally present were *A* alleles] to 100 percent if there were 1,000 plants instead of 5 plants in the population.)

Key Terms

adaptation (p. 329)
analogous (p. 331)
artificial selection (p. 336)
continental drift (p. 335)
convergent evolution (p. 330)
evolution (p. 326)
fossil (p. 333)
genetic drift (p. 328)

homologous (p. 330)
macroevolution (p. 327)
microevolution (p. 326)
morphology (p. 327)
natural selection (p. 327)
Pangaea (p. 335)
speciation (p. 329)
vestigial organs (p. 330)

Self-Quiz

1. Which of the following provides evidence for evolution?
   a. direct observation of genetic changes in populations
   b. sharing of characteristics between organisms
   c. the fossil record
   d. all of the above

2. In natural selection,
   a. the genetic composition of the population changes at random over time.
   b. new mutations are generated over time.
   c. all individuals in a population are equally likely to contribute offspring to the next generation.
   d. individuals that possess particular inherited characteristics consistently survive and reproduce at a higher rate than other individuals.

3. Adaptations
   a. are features of an organism that hinder its performance in its environment.
   b. are not common.
   c. result from natural selection.
   d. result from genetic drift.

4. The fossil record shows that the first mammals evolved 220 million years ago. The supercontinent Pangaea began to break apart 200 million years ago. Therefore, fossils of the first mammals should be found
   a. on most or all of the current continents.
   b. only in Antarctica.
   c. on only one or a few continents.
   d. none of the above

5. The fact that the flipper of a whale and the arm of a human both have five digits and the same set of bones can be used to illustrate that
   a. genetic drift can cause the evolution of populations.
   b. organisms can share characteristics simply because they share a common ancestor.
   c. whales evolved from humans.
   d. humans evolved from whales.

6. The Galápagos Islands provide examples of
   a. microevolution only.
   b. macroevolution only.
   c. both micro- and macroevolutionary change.
   d. none of the above

7. Differences in survival and reproduction caused by chance events can cause the genetic makeup of a population to change at random over time. This process is called
   a. mutation.
   b. natural selection.
   c. macroevolution.
   d. genetic drift.

8. The splitting of one species into two or more species is called
   a. speciation.
   b. macroevolution.
   c. common descent.
   d. adaptation.

9. Features of organisms that are related to one another through common descent are
   a. convergent.
   b. homologous.
   c. divergent.
   d. analogous.

10. Artificial selection is the process by which
   a. natural selection fails to act in wild populations.
   b. humans prevent natural selection.
   c. humans allow only organisms with specific characteristics to breed.
   d. humans cause genetic drift in domesticated populations.

17 Evolution of Populations

Key Concepts

- Populations evolve when frequencies (proportions) of alleles change from generation to generation.

- Individuals within populations differ genetically as expressed in their behavior, morphology, and biochemistry. This genetic variation provides the raw material on which evolution can work.

- Mutations are the original source of genetic variation within populations. The genetic variation produced by mutations is increased by recombination.

- Four mechanisms can cause populations to evolve: mutation, gene flow, genetic drift, and natural selection.

- Some populations can evolve very rapidly (in months to a few years). This potential for rapid evolution has serious implications for the evolution of insect resistance to pesticides and bacterial resistance to antibiotics.

Evolution of Resistance

Our daily lives 150 years ago were considerably different from what they are today. There were not only no cell phones or laptops, but no phones or computers at all—not even electric lights, highways, or airplanes. In addition, life was often short and people frequently died of infection or disease. A healthy young person might fall, scrape a knee deeply, get an infection, and die shortly thereafter. Diseases like smallpox, malaria, cholera, tuberculosis, and the plague could strike with disastrous consequences, decimating towns or whole countries.

Today we are used to thinking that if we get an infection or disease, we can be cured. We expect this in part because of antibiotics, miracle drugs first introduced in the late 1930s. They worked like magic bullets, destroying bacteria and protecting people from all manner of once-deadly infection and disease. Around the same time, pesticides such as DDT began to be used to kill insects such as the mosquitoes that spread malaria and other fatal diseases. Insect pests are notoriously hard to kill, but the new poisons did an excellent job. In the late 1960s, the stunning success of antibiotics and pesticides led the U.S. surgeon general to testify confidently to Congress that it was time to "close the book on infectious diseases."

Unfortunately, the surgeon general was mistaken. Today, for every antibiotic now in use, there is one or more species of bacteria resistant to it. In addition, during the past 65 years the bacteria that cause rheumatic fever, staph infections, pneumonia, strep throat, tuberculosis, typhoid fever, dysentery, gonorrhea, and meningitis all have evolved resistance to many antibiotics. Some biologists worry that resistance may become such a widespread problem that bacterial diseases may again begin to ravage human populations as they did before the discovery of antibiotics.

Furthermore, bacteria are not the only disease agents affected by this trend. Most viruses, fungi, parasites, and insects that cause disease in humans or that attack crops or livestock have evolved resistance to our best efforts to destroy

Some Resistant Viruses and Organisms
HIV virus (top); Colorado potato beetle (middle); tuberculosis bacterium (bottom).

them. And the problem is accelerating. In insects, it used to take decades for pests to evolve resistance, but now resistance often appears in a few years or less.

Overall, the evolution of resistance sends a sobering message: the pests and killers we seek to destroy evolve resistance rapidly in response to our best efforts to defeat them. **How does this happen? What forces have led to the evolution of resistance in so many different organisms?**

Evolution was once thought to be a process too slow to observe, but over the last 80 years, in thousands of studies, biologists have observed and documented the evolution of populations. These studies show that populations evolve slowly in some cases and very rapidly in others (within a few generations, covering a time span of months to a few years). Similarly, new species form slowly in some cases but rapidly in others (in a year to a few thousand years).

In this chapter we describe evolutionary changes that take place in populations, sometimes over short time spans. In particular, we focus on how the frequencies (proportions) of alleles in populations can change from generation to generation. These changes in allele frequencies over time are referred to as **microevolution**, so called because they represent the smallest scale at which evolution occurs. An example of this type of evolutionary change would be a significant increase or decrease in the frequency of a particular allele in a particular population over several generations.

We begin our discussion of microevolution with definitions of two essential terms—"genotype frequency" and "allele frequency"—and with a brief look at what constitutes the raw material of evolution. With that information in hand, we then discuss four mechanisms that can cause allele frequencies to change over time: mutation, gene flow, genetic drift, and natural selection. We end the chapter by answering this question: how do organisms that cause disease evolve resistance to drugs, like antibiotics? We also consider some possible ways to combat infectious diseases more effectively by slowing the evolution of such resistance.

17.1 Alleles and Genotypes

If we want to know whether a population is evolving with regard to a particular trait, we need to determine the allele frequencies for the genes involved, since (as mentioned already) microevolution is defined as change in allele frequencies. **Allele frequency** refers to the proportion, or percentage in a population of a particular allele, such as an A or a allele. Similarly, **genotype frequency** refers to the proportion, or percentage, in a population of a genotype, such as the genotypes AA, Aa, or aa.

To see how we can calculate these proportions, imagine that flower color in a population of 1,000 plants is determined by a single gene that has two alleles: a dominant allele, R, for red flower color; and a recessive allele, r, for white flower color. If the population contains 160 RR individuals, 480 Rr individuals, and 360 rr individuals, then we can obtain the frequencies for the three genotypes (RR, Rr, and rr) by dividing their numbers by the total number of individuals in the population (1,000). The genotype frequencies for RR, Rr, and rr would be 0.16, 0.48, and 0.36, respectively. These genotype frequencies add up to 1.0—as they should, because RR, Rr, and rr are the only three genotypes possible and a frequency of 1.0 is equivalent to 100 percent. Hence, one way to check our genotype calculations is to make sure they add up to 1.0.

Allele frequencies can be computed by the following method, which we illustrate for the R allele (though we could have chosen the r allele instead). Our plant population has 1,000 individuals, each of which has two alleles of the flower color gene. Thus, the total number of alleles in the population equals 2,000. There are 160 RR individuals, each of which carries two R alleles, for a total of 320 R alleles. Rr individuals have one R allele each, for a total of 480 R alleles, and rr individuals have no R alleles. Thus there are 800 (320 + 480 + 0) R alleles in the population. Finally, we calculate the frequency of the R allele by dividing the number of R alleles by the total number of alleles in the population: (800 R alleles)/(2,000 total alleles) = 0.4. Since the gene for flower color has only two alleles, R and r, the sum of their frequencies must equal 1.0. Hence, the frequency of the r allele is

$1.0 - 0.4 = 0.6$. We can check this value by performing a calculation for the r allele similar to the one we did for the R allele.

17.2 Genetic Variation: The Raw Material of Evolution

Within a population, individuals differ in their behavior and morphology (Figure 17.1). As described in Unit 3, much of this variation is under genetic control. Organisms also vary greatly in biochemical traits that are under direct genetic control, such as the amino acid sequences of their proteins. The underlying cause of all inherited differences among individuals is variation in their DNA sequences. Such genetic differences among the individuals of a population are collectively referred to as **genetic variation**; this variation is important because it provides the raw material on which evolution can work.

Mutation is one source of genetic variation

Recall from Chapter 12 that mutations are changes in the sequence of an organism's DNA, and that mutations give rise to new alleles. Mutations that occur in an organism's gametes (as opposed to mutations in an animal's skin or a plant's leaf cells) can be passed on to the next generation. Such mutations are the original source of all genetic variation (though, as we shall see shortly, these DNA variants can then also be shuffled into new combinations through recombination). Gene

mutations are caused by various accidents, such as mistakes in DNA replication, collisions of the DNA molecule with other molecules, or damage from heat or chemical agents. As a result, mutations and the new alleles they produce do not appear because an organism needs them; instead, mutations occur at random and are not directed toward any goal.

Despite the efficiency with which repair proteins fix damage to DNA and correct errors in DNA replication, mutations occur regularly in all organisms. Humans, for example, have two copies each (one copy from each parent) of their approximately 25,000 genes. On average, two or three of these 50,000 gene copies have mutations that make them different from those of either parent.

Recombination provides another source of genetic variation

In sexually reproducing organisms, as we saw in Chapters 9 and 11, the different alleles produced by mutation are grouped in new arrangements by crossing-over, independent assortment of chromosomes, and fertilization; collectively, these three processes are known as **recombination**. Recombination causes offspring to have many new combinations of alleles that differ from those found in either parent. Thus, recombination greatly increases the genetic variation produced originally by mutation.

Genetic variation is random

All the genetic variation seen in natural populations—whether produced by individual mutations or through recombination—is random. That is, genetic variation does not arise because an organism needs a new or more advantageous trait or because an organism needs assistance in meeting the challenges of life in a new or changing environment. Instead, genetic variation simply happens. Only some of those random mutations and recombinations will, simply by chance, provide an individual with a new advantageous trait upon which natural selection can act.

Figure 17.1 Morphological Variation
The individuals within a population often vary greatly in many morphological traits, just as these monarch butterfly caterpillars do in color and banding pattern.

and fertilization.
assortment of chromosomes,
of crossing-over, independent
2. Recombination, in the form
tute microevolution.
quencies over time consti-
population. Changes in allele
allele (for example, A or a) in a
proportion or percentage of an
1. Allele frequencies are the

Concept Check

1. What are allele frequencies, and what roles do they play in micro-evolution?

2. In addition to mutation, what are other sources of genetic variation in populations?

17.3 Four Mechanisms That Cause Populations to Evolve

Evolution occurs when allele frequencies in a population change over time. Allele frequencies change primarily by four mechanisms: mutation, gene flow, genetic drift, and natural selection. Here we provide a brief overview of how these mechanisms work, saving more detailed discussion of each mechanism for the sections that follow.

Mutation and gene flow (the movement or exchange of alleles between populations) can introduce new alleles into a population and change allele frequencies, causing evolution. Although genetic drift is driven by chance events, it, too, can cause allele frequencies to change and hence evolution to occur, as we saw in Chapter 16. Finally, individuals that have certain alleles may have an advantage over individuals that do not have those alleles. In this case, natural selection causes the frequency of the favored alleles to increase, thereby contributing to evolutionary change.

If we know the genotype frequencies in a population, the box on the following page explains how we can use an equation called the Hardy–Weinberg equation to test whether these frequencies are changing, that is, whether evolution is occurring. We can use the Hardy–Weinberg equation to test whether or not gene frequencies are changing, regardless of the mechanisms that are causing allele frequencies to change.

Explore the Hardy-Weinberg equation.

▶❚❚

17.1

17.4 Mutation: The Source of Genetic Variation

As we have seen, mutation creates new alleles at random, thereby providing the raw material for evolution. In this sense, all evolutionary change depends ultimately on mutation.

We can also view mutation in another way: as a mechanism for changing allele frequencies in a population, thus causing evolution to occur. However, mutations occur so infrequently in any particular gene that they cause little direct change in the allele frequencies of populations. In contrast, allele frequencies in populations often change rapidly, indicating that mutation, acting alone, cannot be directly responsible for most evolutionary change.

Although mutations have little direct effect on allele frequencies, in some cases new mutations play a critical role in the evolution of populations. For example, HIV (the human immunodeficiency virus), which causes AIDS, has a high mutation rate and produces many new mutations, causing populations of the virus to evolve even within a single patient's body. Some of these mutations may allow the virus to resist new clinical treatments, making combating the virus difficult because it is a "moving target."

In another example, genetic evidence indicates that the resistance of the mosquito *Culex pipiens* [KYOO-leks PIP-yenz] to organophosphate pesticides was caused by a single mutation that occurred in the 1960s. Since that time, mosquitoes blown by storms or moved accidentally by people have carried the initially rare mutant allele from its place of origin in Africa or Asia to both North America and Europe. This mutant allele is highly advantageous to the mosquito: individuals that have the nonmutant allele die when exposed to organophosphate pesticides. When the mutant allele is introduced (by gene flow) into a population that is exposed to organophosphate pesticides, natural selection causes the mutant allele to increase rapidly in frequency, leading to the evolution of resistance within the new population.

Mutations like those that allow disease agents or pests to resist our best efforts to kill them are obviously beneficial to the organisms in which they occur. Most mutations, however, are either harmful to their bearers or have little effect. In general, the effect of a mutation often depends on the environment in which the organism lives. For example, certain mutations that provide houseflies with resistance to the pesticide DDT also reduce their rate of growth. Flies that grow more slowly take longer to mature and do not produce as many offspring in their lifetimes as do flies that grow at the normal rate. In the absence of DDT, such mutations are harmful. When DDT is sprayed, however, these mutations provide an advantage great enough to offset the disadvantage of slow growth. As a result, the mutant alleles have spread throughout housefly populations and can now be found globally.

Science Toolkit

Testing Whether Evolution Is Occurring in Natural Populations

Four mechanisms can cause evolutionary change in natural populations: mutation, gene flow, genetic drift, and natural selection. This sounds simple enough, but when biologists begin to study a population in nature, they may have little idea whether one or more of these four mechanisms are actually important in the study population. Fortunately, a quick genetic calculation can be performed, the results of which provide an initial indication of whether the population is evolving.

Let's assume we are dealing with a gene with just two alleles, A and a. And let's assign new letters to represent the allele frequencies: p for the frequency of A, and q for the frequency of a. (Adding two more letters—p and q—to the mix is not strictly necessary when calculating genotype frequencies, but it will prove very helpful shortly, when we calculate allele frequencies.) Because this gene has exactly two alleles, we know that the sum of the two allele frequencies is $p + q = 1$.

The quick genetic calculation to determine whether evolution is occurring in a population relies on the **Hardy–Weinberg equation**, which has the general form

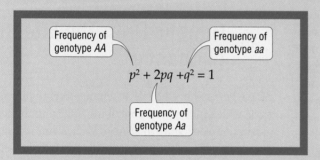

Frequency of genotype AA

Frequency of genotype aa

$$p^2 + 2pq + q^2 = 1$$

Frequency of genotype Aa

As described in the appendix, this equation is derived from the assumption that the population is *not* evolving—in other words, that mutation, gene flow, genetic drift, and natural selection do *not* cause allele frequencies to change. The Hardy–Weinberg equation gives us the genotype frequencies for a population that is not evolving. Because there are exactly three genotypes in this population, the three frequencies must sum to 1 (equivalent to 100 percent), as the Hardy–Weinberg equation shows.

The value of the Hardy–Weinberg approach is that we can use it to test whether the genotype frequencies in a real population match those predicted by the equation. If the actual frequencies differ considerably from the frequencies predicted by the Hardy–Weinberg equation, then the assumptions used to derive the equation must not be correct. Hence, we can conclude that one or more of the four evolutionary mechanisms (mutation, gene flow, genetic drift, or natural selection) are at work. As a result, the population is evolving and allele frequencies will change over time.

To find out whether genotype frequencies in a real population differ from those predicted by the Hardy–Weinberg equation, we must first determine the genotypes of individuals in the population; often this determination is made using the techniques described in Chapter 15. Let's assume that a population of 1,000 individuals has 460 individuals of genotype AA, 280 individuals of genotype Aa, and 260 individuals of genotype aa. Using this information, we calculate that the observed frequency of the A allele is $p = [(2 \times 460) + 280]/2,000 = 0.6$. And since $1.0 - 0.6 = 0.4$, we also know that q, the observed frequency of the a allele, is 0.4.

Now we can use the allele frequencies that we have calculated to determine whether the population is evolving. The observed frequency of the A allele is 0.6. If the population is not evolving, the Hardy–Weinberg equation will hold and the frequency of the AA genotype should be 0.36 (p^2), the frequency of Aa should be 0.48 ($2pq$), and the frequency of aa should be 0.16 (q^2). Therefore, since there are 1,000 individuals in the population, if the population is not evolving we expect to find 360 ($0.36 \times 1,000$) AA individuals, 480 ($0.48 \times 1,000$) Aa individuals, and 160 ($0.16 \times 1,000$) aa individuals. In fact, however, there are 460 AA individuals, 280 Aa individuals, and 260 aa individuals. That is, there are more AA and aa individuals, and fewer Aa individuals, than expected.

The differences between the actual and expected genotype frequencies just described are large. A biologist who obtained real data like those in this example would conclude that the actual genotype frequencies differed significantly from those in the Hardy–Weinberg equation, and that the population was evolving. Next, a researcher would begin to wonder what evolutionary mechanisms might be driving the population away from the predictions of the Hardy–Weinberg equation. Once you've finished reading this chapter, look again at the differences between the observed and the expected genotype frequencies here. Can you suggest one or more evolutionary mechanisms that might explain these differences?

Figure 17.2 Gene Flow

New alleles can be introduced into populations when individuals move from one population to another.

Population I is a large population containing birds of genotypes *AA*, *Aa*, and *aa*.

Population I

17.5 Gene Flow: Exchanging Alleles between Populations

Explore the process of gene flow.

▶II

17.2

When individuals move from one population to another, an exchange of alleles between populations, or **gene flow**, occurs (Figure 17.2). Gene flow can also occur when only gametes move from one population to another, as happens when wind or pollinators like insects transport pollen from one population of plants to another.

Gene flow can play a role similar to that of mutation by introducing new alleles into a population. Such introductions of new alleles can have dramatic effects. For example, in the case of the mosquito *Culex pipiens*, discussed in the previous section, a new allele that made the mosquito resistant to organophosphate pesticides spread by gene flow across three continents. This spread of a new mutant allele allowed billions of mosquitoes to survive the application of pesticides that otherwise would have killed them.

Because it consists of an exchange of alleles between one population and another, gene flow makes the genetic composition of different populations more similar. In this way, gene flow can counteract the effects of mutation, genetic drift, and natural selection, all of which can cause populations to become more different from one another. In some plant species, for example, neighboring populations live in very different environments, yet remain genetically similar. Natural selection, by favoring different alleles in the different environments, would

tend to make the populations differ genetically. The lack of genetic difference between such populations appears to be due to gene flow, which occurs at a rate high enough to cause the populations to remain genetically similar, despite the effects of natural selection.

17.6 Genetic Drift: The Effects of Chance

As we learned in Chapter 16, chance events may determine which individuals contribute offspring to the next generation (see Figure 16.3). As a result, chance events can cause alleles from the parent generation to be selected at random for inclusion in the next generation. The process by which alleles are selected at random over time is called **genetic drift**. This process leads to random changes in allele frequencies from generation to generation.

Genetic drift affects small populations

The chance events that cause genetic drift are much more important in small populations than in large populations. To understand why, consider what happens when a coin is tossed. It would not be all that unusual to get four heads in five tosses—that result has about a 15 percent chance of occurring. But it would be astonishing to get 4,000 heads in 5,000 tosses. Even though the frequency of heads would be the same in both cases (80 percent), the chance of getting many more, or many less, than the expected 50

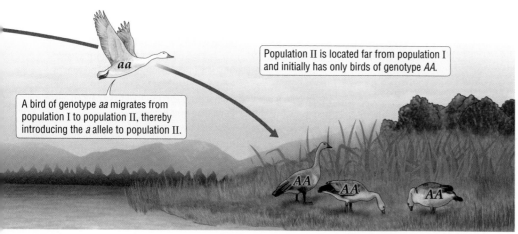

A bird of genotype *aa* migrates from population I to population II, thereby introducing the *a* allele to population II.

Population II is located far from population I and initially has only birds of genotype *AA*.

Population II

percent heads is much greater if the coin is tossed a few times than if it is tossed thousands of times.

In natural populations, the number of individuals in a population has an effect similar to the number of times a coin is tossed. Consider the small population of wildflowers shown in Figure 17.3. By chance alone, some individuals leave offspring and others do not. In this example, such chance events alter the allele frequencies of a gene with two alleles (*R* and *r*). The changes are rapid. One of the alleles (*r*) is lost from the population in just two generations. The other allele (*R*) reaches **fixation**, a frequency of 100 percent. When a population has many individuals, the likelihood that each allele will be passed on to the next generation greatly increases. If the population in Figure 17.3 had had many more individuals, it is unlikely that chance events could have caused such dramatic changes in so short a time.

Genetic drift also occurs in large populations, but in these cases its effects are more easily overcome by natural selection and other evolutionary mechanisms. In large populations, genetic drift causes little change in allele frequencies over time.

What types of chance events cause genetic drift? One important source of genetic drift is the random alignment of alleles during gamete formation, which causes (by chance alone) some alleles, but not others, to be passed on to offspring. Another source of genetic drift is chance events associated with the survival and reproduction of individuals. In this case, even though a particular genotype may increase in frequency from one generation to the next, it is important to remember that the increase is due to chance events (as in Figure 17.3), not to that genotype's having an advantage over other genotypes because of the alleles it carries (as would occur in natural selection).

Overall, genetic drift can affect the evolution of small populations in two ways:

1. It can reduce genetic variation within small populations because chance alone eventually causes one of the alleles to reach fixation. The fixation of alleles can happen rapidly in small populations, but in large populations it takes a long time.
2. It can lead to the fixation of alleles that are neutral, harmful, or beneficial. As emphasized in Chapter 16, only natural selection consistently leads to adaptive evolution.

Genetic bottlenecks can threaten the survival of populations

The importance of genetic drift in small populations has implications for the preservation of rare species. If the number of individuals in a population falls to very low levels, genetic drift may lead to a loss of genetic variation or to the fixation of harmful alleles, either of which can hasten the extinction of the species. When a drop in the size of a population leads to such a loss of genetic variation, the population is said to experience a **genetic bottleneck**. Genetic bottlenecks often occur in nature because of the **founder effect**, which results when a small group of individuals establishes a new population far from existing populations (for example, on an island).

Learn more about the three models of natural selection.
17.3

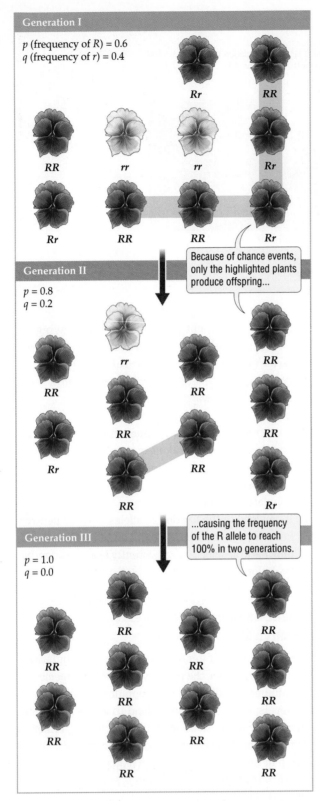

Generation I

p (frequency of R) = 0.6
q (frequency of r) = 0.4

Rr   RR

RR   rr   rr   Rr

Rr   RR   RR   Rr

> Because of chance events, only the highlighted plants produce offspring...

Generation II

p = 0.8
q = 0.2

rr   RR

RR   RR   RR

Rr   RR   RR

RR   Rr

> ...causing the frequency of the R allele to reach 100% in two generations.

Generation III

p = 1.0
q = 0.0

RR   RR

RR   RR

RR   RR   RR

RR   RR

RR   RR

Figure 17.3 Genetic Drift

In this small population of wildflowers, chance events determine which plants leave offspring without regard to which individuals are better equipped for survival or reproduction. Here, chance events cause the frequency of the *R* allele to increase from 60 percent to 100 percent in two generations. Note that this particular outcome is just one of many ways that genetic drift could have caused allele frequencies to change at random over time.

30 to 50 individuals in the 1980s. At that time, biologists discovered that male Florida panthers had low sperm counts and abnormally shaped sperm (Figure 17.4), probably as a result of the fixation of harmful alleles by genetic drift. The resulting low fertility in the males is thought to have contributed to the drop in population size. Panther numbers have increased to about 80 to 100 individuals in recent years, in part because of breeding programs designed to reduce the effects of genetic drift.

Although Florida panthers, northern elephant seals, and African cheetahs all show signs of having experienced genetic bottlenecks, each of these examples poses a problem: We don't know how much genetic variation was present before the population decreased in size. Hence, there is no way to be sure whether the observed low levels of genetic variation in these animals really were caused by a decrease in population size or were just a natural feature of the organism.

Recent studies on greater prairie chickens in Illinois avoided this problem by comparing the DNA of modern birds with the DNA of their pre-bottleneck ancestors, obtained from (nonliving) museum specimens. There were millions of greater prairie chickens in Illinois in the nineteenth century, but by 1993 the conversion of their prairie habitat to farmland had caused their numbers to drop to only 50 birds in two isolated populations (Figure 17.5). This drop in numbers caused a genetic bottleneck: the modern birds lacked 30 percent of the alleles found in the museum specimens, and they suffered poor reproductive success compared with other prairie chicken populations that had not experienced a genetic bottleneck (only 56 percent of their eggs hatched, versus 85 percent to 99 percent in other populations). From 1992 to 1996, 271 birds were introduced to Illinois from large populations in Minnesota, Kansas, and Nebraska, in order to increase both the size and the genetic variation of the Illinois populations. By 1997, the number of males in one of the two remaining Illinois populations had increased from a low of 7 to over 60 birds, and the hatching success of eggs had risen to 94 percent.

Elephant seal

Genetic bottlenecks are thought to have occurred in the Florida panther, the northern elephant seal, and the African cheetah. In the case of the endangered Florida panther, population sizes plummeted to about

(a)

Figure 17.4 Abnormally
Shaped Sperm in the Rare
Florida Panther
Florida panthers (*a*) have more
abnormal sperm (*b*) than do cats
from other cougar populations—
a possible effect of the fixation
of harmful alleles. Normal sperm
for comparison (*c*).

(b)   (c)

Illinois

| By 1993, only 50 greater prairie chickens remained in Illinois, causing both the number of alleles and the percentage of eggs that hatched to decrease. |

Pre-bottleneck
(1820)

Post-bottleneck
(1993)

	Illinois		Kansas	Minnesota	Nebraska
	Pre-bottleneck (1933)	Post-bottleneck (1993)	No bottleneck		
Population size	25,000	50	750,000	4,000	75,000 – 200,000
No. of alleles at 6 genetic loci	31	22	35	32	35
Percentage of eggs that hatch	93	56	99	85	96

In 1820, the grasslands
in which greater prairie
chickens live covered
most of Illinois.

In 1993, less than 1% of the
grassland remained, and the
birds could be found only in
these two locations.

Figure 17.5 A Genetic Bottleneck
The Illinois population of greater prairie chickens dropped
from 25,000 birds in 1933 to only 50 birds in 1993. This drop
in population size caused a loss of genetic variation and
a drop in the percentage of eggs that hatched. Here, the
modern, post-bottleneck Illinois population is compared with
the 1933 pre-bottleneck Illinois population, as well as with
populations in Kansas, Minnesota, and Nebraska that never
experienced a bottleneck.

17.7 Natural Selection: The Effects of Advantageous Alleles

Natural selection is a process by which individuals with particular inherited characteristics survive and reproduce at a higher rate than other individuals in a population. Natural selection acts by favoring some phenotypes over others, as when mosquitoes that are resistant to a pesticide survive at a higher rate and produce more offspring than do other mosquitoes. Although natural selection acts directly on the phenotype, not on the genotype, the alleles that code for forms of a trait favored by natural selection tend to become more common in the offspring generation than in the parent generation. For example, if large size is an inherited characteristic, and if natural selection consistently favors large individuals, then alleles that cause large size will tend to become more common in the population over time.

Even natural selection does not always lead to evolutionary change

Of the four mechanisms of evolution, natural selection is the only one that consistently improves the reproductive success of the organism in its environment. As shown by the study of the medium ground finch described in Chapter 16, natural selection can sometimes cause traits to evolve rapidly in response to changes in the environment.

However, even when natural selection favors one allele over another, it does not necessarily lead to evolutionary change. The other evolutionary mechanisms—genetic drift, gene flow, and mutation—may oppose its effects and prevent allele frequencies from changing. It also bears repeating that unless individuals within a population differ genetically, and unless some of them have beneficial mutations on which selection can act, natural selection is powerless. For example, if none of the mosquitoes in a population carry alleles for resistance to a pesticide, then natural selection cannot promote the evolution of resistance to that pesticide.

There are three types of natural selection

Natural selection can be divided into three types: directional selection, stabilizing selection, and disruptive selection. Despite these categories, it is important to remember that all types of natural selection operate by the same principle: individuals with certain forms of an

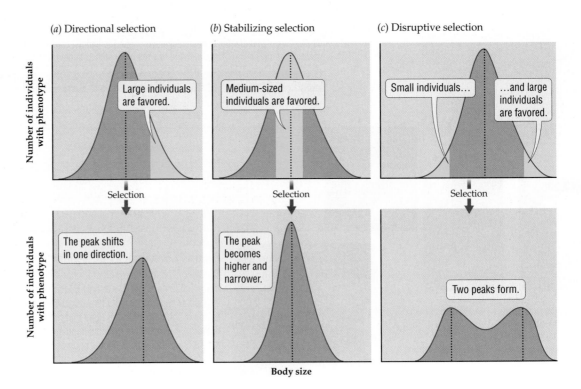

Figure 17.6 The Three Types of Natural Selection
Directional selection (a), stabilizing selection (b), and disruptive selection (c) affect phenotypic traits such as body size differently. The graphs in the top row show the relative numbers of individuals with different body sizes in a population before selection. The phenotypes favored by selection are shown in yellow. The graphs in the bottom row show how each type of natural selection affects the distribution of body size in the population.

inherited phenotypic trait tend to survive better and produce more offspring than do individuals with other forms of that trait.

In **directional selection**, individuals with one extreme of an inherited phenotypic trait have an advantage over other individuals in the population. For example, if large individuals produce more offspring than do small individuals, there will be directional selection for large body size (Figure 17.6a).

Directional selection can be illustrated by the rise and fall of dark-colored forms in various moth species. For example, dark-colored forms of the peppered moth were favored by natural selection when industrial pollution blackened the bark of trees in Europe and North America. The color of these moths is a genetically determined trait, and the allele for dark color is dominant to alleles for light color. Peppered moths are active at night and rest on trees during the day. The proportion of the dark-colored moths increased over time, apparently because they were harder for predators such as birds to find against the blackened bark of trees. The rise in the proportion

of dark-colored forms took less than 50 years. The first dark-colored moth was found near Manchester, England, in 1848. By 1895, about 98 percent of the moths near Manchester were dark-colored, and proportions of over 90 percent were common in other heavily industrialized areas of England—for example, nearby Liverpool (Figure 17.7).

Today, however, light-colored moths outnumber dark-colored moths. The reason, once again, is directional selection; but this time it was the light-colored moths that were favored, not the dark-colored moths. This turnaround was due to the Clean Air Act passed in England in 1956. As the air quality improved, the bark of trees became lighter, and light-colored moths became harder for predators to find than dark-colored moths. As a result, the proportion of dark-colored moths plummeted (see Figure 17.7).

In **stabilizing selection**, individuals with intermediate values of an inherited phenotypic trait have an advantage over other individuals in the population (Figure 17.6b). Birth weight in humans provides a classic example. Historically, light or heavy babies did not

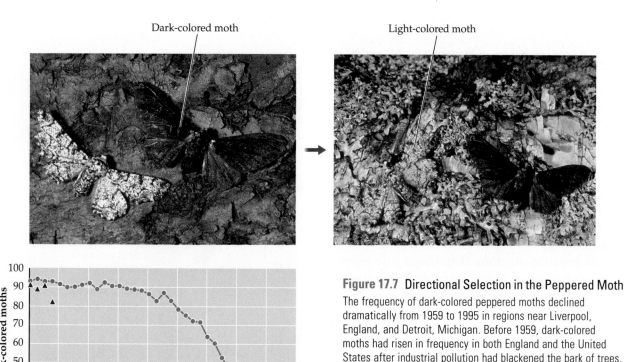

Figure 17.7 Directional Selection in the Peppered Moth
The frequency of dark-colored peppered moths declined dramatically from 1959 to 1995 in regions near Liverpool, England, and Detroit, Michigan. Before 1959, dark-colored moths had risen in frequency in both England and the United States after industrial pollution had blackened the bark of trees, causing dark-colored moths to be harder for bird predators to find than light-colored moths. A reduction in air pollution following clean-air legislation enacted in 1956 in England and in 1963 in the United States apparently removed this advantage, leading the dark-colored moths to decline in a similar way in the two regions. (No data were collected in Detroit for the 30-year period 1963–1993.)

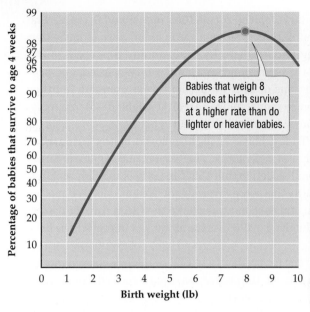

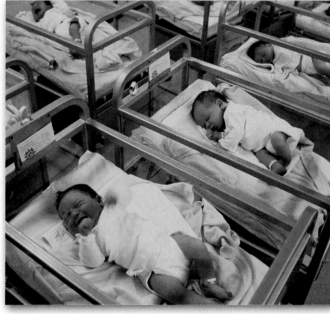

Figure 17.8 Stabilizing Selection for Human Birth Weight
This graph is based on data for 13,700 babies born from 1935 to 1946 in a hospital in London. In countries that can afford intensive medical care for newborns, the strength of stabilizing selection has been reduced in recent years; because of recent improvements in the care of premature babies and increases in the number of cesarean deliveries of large babies, a graph of such data collected today would be flatter (less rounded) at its peak than the graph shown here.

survive as well as babies of average weight, and as a result there was stabilizing selection for intermediate birth weights (Figure 17.8). By the late 1980s, however, selection against small and large babies had decreased considerably in some countries with advanced medical care, such as Italy, Japan, and the United States. This reduction in the strength of stabilizing selection was caused by advances in the care of very light premature babies and by increases in the use of cesarean deliveries for babies that are large relative to their mothers (and hence pose a risk of injury to mother and child at birth).

In **disruptive selection**, individuals with either extreme of an inherited phenotypic trait have an advantage over individuals with an intermediate phe-

Figure 17.9 Disruptive Selection for Beak Size
In African seed crackers, differences in feeding efficiencies may cause differences in survival. Among a group of young birds hatched in one year, only those with a small or large beak size survived the dry season, when seeds were scarce; all the birds with intermediate beak sizes died. Thus, natural selection favored both large-beaked and small-beaked birds over birds with intermediate beak sizes. Red bars indicate the beak sizes of young birds that survived the dry season; blue bars indicate the beak sizes of young birds that died.

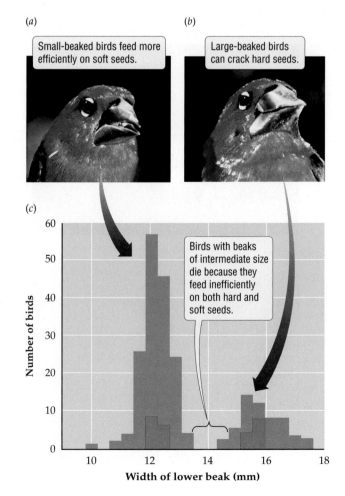

notype (Figure 17.6c). This type of selection is probably not common, but it appears to affect beak size within a population of African seed crackers, in which birds with large or small beaks survive better than birds with intermediate-sized beaks (Figure 17.9).

17.8 Sexual Selection: Where Sex and Natural Selection Meet

The males and females of many species differ greatly in size, appearance, or behavior. Many of these differences seem to be related to the ability of individuals to obtain mates. In lions, for example, males are considerably larger than females; and males fight, sometimes violently, for the privilege of mating with females. Since larger males are stronger and tend to be more successful in fights for females, natural selection may have favored large size in males, but not in females, thus causing the size difference between male and female lions.

When individuals differ in inherited characteristics that cause them to differ in their ability to get mates, a special form of natural selection, called **sexual selection**, is at work. Sexual selection favors individuals that are good at getting mates and, as with the lions we just described, it often helps explain differences between males and females in size, courtship behavior, and other features. However, characteristics that increase the chance of mating sometimes decrease the chance of survival. For example, male túngara frogs perform a complex mating call that may or may not end in one or more "chucks." Females prefer to mate with males that emit chucks, but frog-eating bats use that same sound to help them locate their prey (see Figure 18.5). As a result, an attempt to locate a mate can end in disaster.

In many species, the members of one sex—often females—are choosy about whether or not to mate. In such species, the choosy partner acts as the brake, while the suitor, who tries to convince its reluctant partner to mate, acts as the accelerator. In birds, for example, brightly colored males may perform elaborate displays in their attempts to woo a mate (Figure 17.10a). In other species, males may attract attention by other means, such as calling vigorously; females then select as their mates the males with the loudest calls.

For the process just described to make sense, we would expect that if the choosy partner bases her

(a)

(b)

Figure 17.10 Sexual Selection at Work
(a) A male peacock displays his magnificent tail for a peahen. (b) A male gray tree frog issues a call for mates.

(or, occasionally, his) choice of a mate on a trait such as color or calling vigor, then that trait should serve as a good indicator of the quality of the mate. A good-quality mate, for example, might be especially healthy and perhaps more likely to produce healthy offspring or to be better at guarding a nest of young or at gathering food. A series of recent studies has provided support for this hypothesis. In blackbirds, females choose males with orange beaks more often than males with yellow beaks. It turns out that orange-beaked males have had fewer infections than yellow-beaked males; by selecting males with orange beaks, the females are selecting males in good health. Similarly, in mice, females can tell from the odor of a male's urine how many parasites he is harboring, and they use this information to select their mates.

In the two examples just mentioned, females use a trait such as beak color or odor to tell how healthy the males currently are. In other cases, females seem to go beyond simply looking at a male's current bill of health to assess whether he has "good genes." Female gray tree frogs, for example, prefer to mate with males that give long mating calls (Figure 17.10b). In an elegant experiment, scientists collected unfertilized eggs from wild females, sperm from males with long calls, and sperm from males with short calls. The eggs of each female were then divided into two batches, and one was fertilized with sperm from long-calling males, the other with sperm from short-calling males. The offspring of long-calling male frogs grew faster, were bigger, and survived better than the offspring of short-calling male frogs. The long-calling males seem to have been genetically, and reproductively, superior to the short-calling males.

17.9 Putting It All Together: How Evolution in Populations Works

In this chapter we described how four mechanisms—mutation, gene flow, genetic drift, and natural selection—can cause allele frequencies to change in populations. With this information in hand, let's revisit our earlier description of how evolution works. In Section 16.2 we described how natural selection and genetic drift, oper-

ating on the genetic variation found in a population, can cause allele frequencies to change over time.

We can now more fully describe how evolution works, summarizing it as a three-step process. First, mutations and the genetic rearrangements caused by recombination occur at random. Second, these random events generate inherited differences in the characteristics of individuals in populations. And third, gene flow, genetic drift, and natural selection acting on that genetic variation cause allele frequencies in the population to change over time. By changing allele frequencies directly, mutation can also cause evolution to occur, although, as we have seen, this typically happens very slowly.

It is important to remember the role that chance can play in evolution, since three of the four mechanisms that cause allele frequencies to change in populations are influenced by chance events: (1) genes can mutate at random to produce new alleles; (2) alleles can be transferred by gene flow from one population to another (and the alleles that a migrating individual carries usually can be viewed as a random selection of the alleles in its population); and (3) alleles can increase in frequency in a given population as a result of genetic drift.

However, evolution in populations is by no means restricted to chance events. The fourth mechanism, natural selection, is not a random process. The directional selection exerted on peppered moths illustrates that point beautifully. Once pollution controls were imposed in the mid-twentieth century, light-colored moths were favored consistently—not by chance, but by the changing environmental conditions. This consistent advantage caused the proportion of dark-colored moths in the population to drop rapidly (see Figure 17.7). As this example shows, natural selection favors some individuals over others because they have particular characteristics that increase their ability to survive and reproduce in their environment. As a result, over time, natural selection improves the match between organisms and their environment—a topic we will discuss in more detail in the next chapter.

Concept Check

1. What are the four mechanisms by which populations can evolve?
2. Which of those mechanisms is the only one that can *consistently* lead a population toward adaptive evolution?

Learn about the crisis in antibiotic resistance.
17.4

Applying What We Learned

Halting the Spread of Antibiotic Resistance

The footprint of evolutionary change can be found throughout nature. We see it in battles between predator and prey, parasite and host, herbivore and plant. Natural selection favors individuals whose characteristics improve either their ability to consume others or their ability to avoid being consumed. Similar battles occur between people and organisms that we want to kill, such as fungi and insects that destroy crops and bacteria that cause disease. Unfortunately, our foes have proved hard to vanquish. In fact, our attempts to kill them are having the unintended effect of actually promoting the evolution of resistance.

Individual bacteria, fungi, and insects that are not resistant to our antibiotics or pesticides die, leaving behind ever larger proportions of individuals that are resistant—hence, our efforts to kill them impose selection for resistance to the very weapons we try to kill them with. This is how the magic-bullet antibiotics that once miraculously cured disease, as we described them at the beginning of the chapter, have grown to become less and less effective. The bacteria they were used to fight have undergone microevolution. Populations of bacteria all around the world, exposed again and again to antibiotics, have evolved resistance to those antibiotics via natural selection.

In 1941, for example, pneumonia could be cured in several days if patients took 40,000 units of penicillin per day. Today, however, even with excellent medical care and the administration of 24 million units of penicillin per day, a patient can still die from complications of this disease.

What is true of pneumonia is also true of other bacterial diseases. As a group, bacteria have many tactics for coping with once-lethal antibiotics. Some of these tactics provide resistance only to specific antibiotics, such as a bacterial cell's pump that removes a specific antibiotic, or a new enzyme that destroys a certain type of antibiotic, or a change in the bacterial cell wall that prevents an antibiotic from affecting its usual target (for example, a bacterium whose cell wall structure differs from that of other bacteria may be impervious to penicillin, which normally kills bacteria by attacking their cell walls). Other bacterial defense mechanisms provide resistance to many antibiotics. For example, some bac-

teria have developed pumps that remove many types of antibiotics, while other bacteria grow in layers that prevent the antibiotics from reaching cells that are not near the surface.

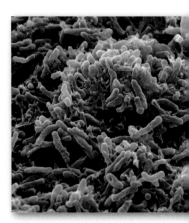

How do bacteria develop these and other resistance mechanisms so rapidly? In part, this happens because bacteria have such a short time between generations. As we saw in Chapter 3, bacteria reproduce by fission. Only two daughter cells are produced when each bacterium divides, but the time between divisions can be as little as 20 to 30 minutes—one bacterium can produce enormous numbers of offspring in just a few days. As a result of random mutation, some of these numerous offspring may end up carrying new mutations that, simply by chance, provide resistance to one or more antibiotics. Once such mutations occur, our widespread use of antibiotics causes alleles that provide resistance to increase rapidly in frequency through the process of natural selection.

This increase in resistant alleles is then made much worse by the fact that bacteria can transfer resistance genes within and among bacterial species with ease, often very quickly (Figure 17.11). Known as horizontal gene transfer, this kind of transfer of genes is much more common among bacteria than among other organisms, as we saw in Chapter 2. When genes that provide resistance are transferred or move from one bacterial species to another, it is especially troubling. That is because genes for resistance that evolve in a relatively harmless species of bacteria can be transferred to a highly dangerous species, creating the potential for a public health disaster.

Halting the spread of antibiotic resistance and preventing this sort of doomsday scenario will require concerted efforts on our part; this problem will not go away by itself. Consider the costs and benefits of the following actions:

- Devote greater resources to the study of the biology of bacteria and other disease agents, thereby improving our ability to design effective drugs or management strategies.
- Learn to live with bacteria, rather than seeking to annihilate them. For example, instead of using dis-

Biology Matters

You'll Just Have to Tough It Out

Being sick is never enjoyable, but the more often you take antibiotics, the more likely they won't work for you—or others—in the future. Every time we use antibiotics, we are applying selection pressure: killing off any nonresistant bacteria and giving an advantage to any resistant bacteria. That is, we are actually helping to speed the evolution of resistance to those very antibiotics. And remember, antibiotics don't even help with the most common ailments they are often used to treat, like a cold, flu, sore throat (except strep), bronchitis, most runny noses, and most earaches. Although doctors are prescribing antibiotics less than they did in the past, understanding when to use antibiotics, and when they *won't* help, will help you protect yourself and others. And don't forget—when you do need antibiotics, take them exactly as your health care provider prescribes, and continue to take them until they run out, even if you're feeling better.

The following information will help you care for yourself, and feel better, the next time you get sick:

Q: *If antibiotics will not help me, what will?*

A: Many over-the-counter products treat the symptoms of a viral infection. These include cough suppressants to help control coughing and decongestants to help relieve a stuffy nose. (Note, though, that such medications are thought to be of essentially no use and potentially harmful to infants.) Read the label and ask your pharmacist or doctor if you have questions about which will work best for you.

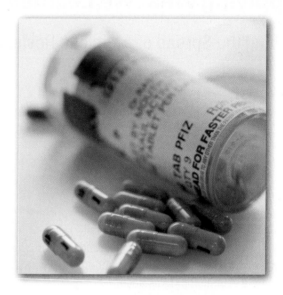

A cold usually lasts only a couple of days to a week. Tiredness from the flu may continue for several weeks. To feel better while you are sick,

- Drink plenty of fluids.
- Get plenty of rest.
- Use a humidifier—an electric device that puts water into the air.

Contact your doctor if

- Your symptoms get worse.
- Your symptoms last a long time.
- After feeling a little better, you develop signs of a more serious problem. Some of these signs are nausea, vomiting, high fever, shaking, chills, and chest pain.

SOURCE: U.S. Department of Health and Human Services.

infectants designed to kill bacteria, we could try to limit bacterial numbers by following safe cooking practices and by keeping our homes clean. Reducing our use of "antibacterial" disinfectants would be helpful because some genes that confer resistance to these products can also lead to antibiotic resistance (see Figure 17.11). Similarly, in combating bacterial diseases, we could search for drugs that make bacteria harmless but do not destroy them; because this approach is not lethal for the bacteria, resistance should evolve more slowly.

- Insist on prudent use of antibiotics in human, plant, and animal health care. Medical doctors and agriculturists frequently use antibiotics inappropriately. For example, the U.S. government estimates that half of the 100 million antibiotic prescriptions written by doctors each year are not necessary—often the conditions for which they are prescribed (such as colds and flu) are caused not by bacteria, but by other disease agents, such as viruses, that are not affected by antibiotics. Similarly, antibiotics are commonly used to increase the growth rates of farm animals—a practice that encourages the development of antibiotic-resistant strains of bacteria, including strains that attack people. Such inappropriate use of antibiotics encourages the evolution of antibiotic resistance in the many species of bacteria that are normally found in our bodies. It becomes possible, then, for these resistant, harmless bacteria to transfer genes for antibiotic resistance to other, harmful species of bacteria.

• Improve sanitation and decrease the spread of resistant bacteria from one person to another. This action is critically important in hospitals, where the abundant use of antibiotics has led to the emergence of highly resistant strains of bacteria capable of causing a variety of "hospital diseases," some of which can be lethal. It is also becoming increasingly important in other places where people come together—for example, schools—particularly among those who may have physical contact, like athletes.

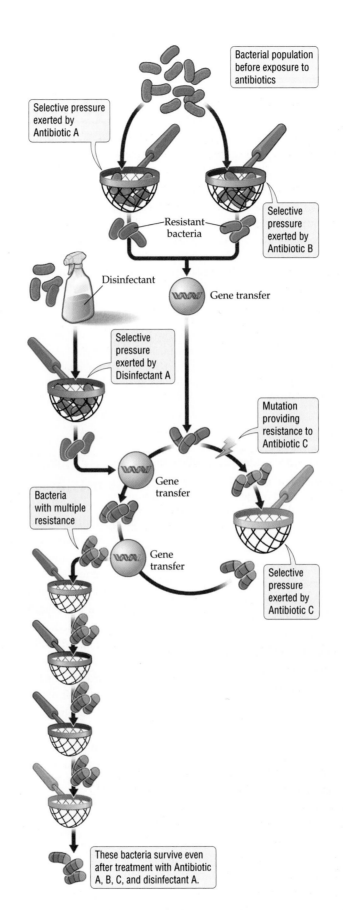

Figure 17.11 The Rise of Antibiotic Resistance
Before antibiotic use was common, most bacteria could be killed by antibiotics. Early use of antibiotics selected those few individuals that were resistant. Over time, new mutations and repeated episodes of selection—either by antibiotics (represented here by sieves) or by disinfectants—caused the frequency of resistant individuals to increase. In addition, transfer of genes within and among bacterial species allowed some bacteria to develop resistance to multiple antibiotics.

Antibiotic Runoff

One of the persistent problems of industrial agriculture is the inappropriate use of antibiotics. It's one thing to give antibiotics to individual animals, case by case, the way we treat humans. But it's a common practice in the confinement hog industry to give antibiotics to the whole herd, to enhance growth and to fight off the risk of disease, which is increased by keeping so many animals in such close quarters. This is an ideal way to create organisms resistant to the drugs. That poses a risk to us all.

A recent study by the University of Illinois makes the risk even more apparent. Studying the groundwater around two confinement hog farms, scientists had identified the presence of several transferable genes that confer antibiotic resistance, specifically to tetracycline. There is the very real chance that in such a rich bacterial soup these genes might move from organism to organism, carrying the ability to resist tetracycline with them. And because the resistant genes were found in groundwater, they are already at large in the environment.

There are two interdependent solutions to this problem, and hog producers should embrace them both. The first solution—the least likely to be acceptable in the hog industry—is to ban wholesale, herdwide use of antibiotics. The second solution is to continue to tighten the regulations and the monitoring of manure containment systems. The trouble is that there is no such thing as perfect containment.

The consumer, of course, has the choice to buy pork that doesn't come from factory farms. The justification for that kind of farming has always been efficiency, and yet, as so often happens in agriculture, the argument breaks down once you look at all the side effects. The trouble with factory farms is that they are raising more than pigs. They are raising drug-resistant bugs as well.

Although we tend to think of antibiotics as human medications, according to the Union of Concerned Scientists 70 percent of all antibiotics and related drugs used in the United States are given to livestock. An estimated 24 million pounds of antibiotics are used on livestock—pigs, cows, and chickens—each year.

In every exposure of bacteria to antibiotics—whether in humans or livestock—natural selection can weed out nonresistant bacteria and give an advantage to resistant bacteria. That is, every time antibiotics are used, there is the potential for antibiotic-resistant alleles to increase in frequency. Farmers use antibiotics to improve efficiency and cut costs because then they can raise more animals in closer quarters—conditions that, without widespread antibiotic use, would promote disease. In fact, however, this approach may be costing us all quite a lot in actual dollars. Treating antibiotic-resistant infections already costs billions of dollars each year in the United States alone.

One potentially huge cost described in this editorial could be incurred if resistance genes move from one kind of bacterium to another (in horizontal gene transfer). During such gene transfer, genes move not from one individual to its offspring, but from one individual to another individual of a different species. Such transfer is very common among bacteria. The real risk is movement of the resistance genes to a dangerous disease species—making the disease that much more difficult to treat. The groundwater around hog farms that contains such "transferable genes" could carry the organisms and their resistance genes anywhere.

Evaluating the News

1. How can raising healthier pigs be unhealthful for humans? Explain how herdwide applications of antibiotics could lead to increased antibiotic resistance in other kinds of bacteria, including bacteria that could cause disease in humans, via horizontal gene transfer.

2. Next time you're in a grocery store, check whether the packages of meat are from animals raised with antibiotics or without. (*Hint:* If they were raised without the use of antibiotics, the package will say so; otherwise, it will say nothing about antibiotics.) Compare the prices of meats raised with and without antibiotics; often those marked antibiotic-free are more expensive. Is it ethical for factory farms to use large-scale antibiotic treatment? On the other hand, would it be fair to farmers and consumers to outlaw such antibiotic use, if it meant that meats became more expensive?

3. A recent study found that a particular drug-resistant bacterium—methicillin-resistant *Staphylococcus aureus,* or MRSA—kills more Americans each year than even the AIDS virus. Common antibiotics, like methicillin and amoxicillin, are no longer effective against MRSA. Doctors are now using uncommon antibiotics to treat MRSA. How is natural selection likely to affect the effectiveness of those more uncommon antibiotics over time? If each use of an antibiotic helps promote resistance, by creating natural selection for resistant bacteria, what could doctors do to try to slow the evolution of additional resistance in MRSA?

Source: *New York Times* editorial, September 18, 2007.

Chapter Review

Summary

17.1 Alleles and Genotypes

- An allele's frequency is the proportion of that allele in a population. Calculating allele frequencies is an important part of determining whether a population is evolving.
- A genotype's frequency is the proportion of that genotype in a population.

17.2 Genetic Variation: The Raw Material of Evolution

- Individuals within populations differ in morphological, behavioral, and biochemical traits, many of which are under genetic control.
- Genetic variation provides the raw material on which evolution can work. The two sources of genetic variation are mutation and recombination.
- Mutations (changes in the sequence of an organism's DNA) give rise to new alleles.
- Recombination (crossing-over, independent assortment of chromosomes, and fertilization) causes the offspring of sexually reproducing organisms to have new combinations of alleles that differ from those found in either parent.
- Genetic variation is random and does not arise in response to an organism's needs for new or more advantageous traits.

17.3 Four Mechanisms That Cause Populations to Evolve

- Allele frequencies in populations can change over time as a result of mutation, gene flow, genetic drift, or natural selection.
- The Hardy–Weinberg equation can be used to test whether one or more of these four mechanisms is causing a population to evolve.

17.4 Mutation: The Source of Genetic Variation

- All evolutionary change depends ultimately on the production of new alleles by mutation.
- Mutations cause little direct change in allele frequencies over time.
- Mutations can, however, stimulate the rapid evolution of populations by providing new genetic variation on which natural selection, genetic drift, or gene flow can act.

17.5 Gene Flow: Exchanging Alleles between Populations

- Gene flow can introduce new alleles into a population, providing new genetic variation on which evolution can work.
- Gene flow makes the genetic composition of populations more similar.

17.6 Genetic Drift: The Effects of Chance

- Genetic drift causes random changes in allele frequencies over time.
- Genetic drift can cause small populations to lose genetic variation.

- Genetic drift can cause the fixation of harmful, neutral, or beneficial alleles.
- In a genetic bottleneck, a drop in population size causes genetic drift to have pronounced effects, including reduced genetic variation or the fixation of alleles. A genetic bottleneck can threaten the survival of a population.

17.7 Natural Selection: The Effects of Advantageous Alleles

- In natural selection, individuals that possess certain forms of an inherited phenotypic trait tend to survive better and produce more offspring than do individuals that possess other forms of that trait.
- Natural selection is the only evolutionary mechanism that consistently favors alleles that improve the reproductive success of the organism in its environment.
- There are three types of natural selection. In directional selection, individuals at one extreme of a trait have the advantage. In stabilizing selection, the advantage goes to individuals with intermediate values of a trait. In disruptive selection, individuals at both extremes have an advantage over individuals with an intermediate phenotype.

17.8 Sexual Selection: Where Sex and Natural Selection Meet

- Sexual selection occurs when individuals with different inherited characteristics differ in their ability to get mates.
- Sexual selection underlies many differences between males and females, such as differences in size and courtship behavior.
- Forms of a trait favored by sexual selection can lead to decreased survival.
- Sexual selection can occur when members of one sex—often females—are choosy about which individuals they will or will not mate with.

17.9 Putting It All Together: How Evolution in Populations Works

- Evolution can be summarized as a three-step process: (1) Mutations and the genetic rearrangements caused by recombination occur at random. (2) These random events then generate inherited differences in the characteristics of individuals in populations. (3) Finally, mutation, gene flow, genetic drift, and natural selection can cause allele frequencies to change over time.
- Of the four mechanisms of evolutionary change, three—mutation, gene flow, and genetic drift—are influenced by chance events. Natural selection, on the other hand, is a nonrandom process that favors individuals with particular characteristics that increase their ability to survive and reproduce in their environment.

Review and Application of Key Concepts

1. Select one of the four evolutionary mechanisms discussed in this chapter (mutation, gene flow, genetic drift, and natural selection), and describe how it can cause allele frequencies to change from generation to generation.

2. Summarize the role of genetic variation in evolution.

3. Describe how recombination increases the genetic variation originally produced by mutation.

4. Using your own words, define the following terms: gene flow, genetic drift, natural selection, sexual selection.

5. One way to prevent a small population of a plant or animal species from going extinct is to deliberately introduce some individuals from a large population of the same species into the smaller population. In terms of the evolutionary mechanisms discussed in this chapter, what are the potential benefits and drawbacks of transferring individuals from one population to another? Do you think biologists and concerned citizens should take such actions?

6. Consider the toads in question 2 of the Self-Quiz. How do the numbers of toads with genotypes *AA*, *Aa*, and *aa* compare with the numbers you would expect on the basis of the Hardy–Weinberg equation (see the box on page 351)? Discuss the factors that could cause any differences you find.

7. Explain the reasoning behind the following statement in the text (on page 361): "In fact, our attempts to kill them [bacteria] are having the unintended effect of actually promoting the evolution of resistance." If we shifted our efforts away from killing bacteria and toward reducing our exposure to them or slowing their growth, why might such a change in our approach slow the evolution of antibiotic resistance?

Key Terms

allele frequency (p. 348)
directional selection (p. 357)
disruptive selection (p. 358)
fixation (p. 353)
founder effect (p. 353)
gene flow (p. 352)
genetic bottleneck (p. 353)
genetic drift (p. 352)
genetic variation (p. 349)
genotype frequency (p. 348)
Hardy–Weinberg equation (p. 351)
microevolution (p. 348)
natural selection (p. 356)
recombination (p. 349)
sexual selection (p. 359)
stabilizing selection (p. 357)

Self-Quiz

1. A population of 1,500 individuals has 375 individuals of genotype *AA*, 750 individuals of genotype *Aa*, and 375 individuals of genotype *aa*. The genotype frequencies for genotypes *AA*, *Aa*, and *aa* are
   a. 0.33, 0.33, 0.33
   b. 0.25, 0.50, 0.25
   c. 0.375, 0.75, 0.375
   d. 0.125, 0.25, 0.125

2. A population of toads has 280 individuals of genotype *AA*, 80 individuals of genotype *Aa*, and 60 individuals of genotype *aa*. What is the frequency of the *a* allele?
   a. 0.24
   b. 0.33
   c. 0.14
   d. 0.07

3. A study of a population of the goldenrod *Solidago altissima* [soll-uh-DAY-goh al-TISS-ih-muh] finds that large individuals consistently survive at a higher rate than small individuals. Assuming that size is an inherited trait, the most likely evolutionary mechanism at work here is
   a. disruptive selection.
   b. directional selection.
   c. stabilizing selection.
   d. natural selection, but it is not possible to tell whether it is disruptive, directional, or stabilizing.

4. On the basis of the material in the box on page 351, if the frequency of the *A* allele is 0.7 and the frequency of the *a* allele is 0.3, what is the expected frequency of individuals of genotype *Aa* in a population that is not evolving?
   a. 0.21
   b. 0.09
   c. 0.49
   d. 0.42

5. Over time, a population of birds ranges in size from 10 to 20 individuals. If allele frequencies were observed to change in a random way from year to year, which of the following would be the most likely cause of the observed changes in gene frequency?
   a. stabilizing selection
   b. disruptive selection
   c. genetic drift
   d. mutation

6. Two large populations of a species found in neighboring locations with different environments are observed to become genetically more similar over time. Which evolutionary mechanism is the most likely cause of this trend?
   a. gene flow
   b. mutation
   c. natural selection
   d. genetic drift

7. Assume that individuals of genotype *Aa* are intermediate in size and that they leave more offspring than either *AA* or *aa* individuals do. This situation is an example of
   a. directional selection.
   b. disruptive selection.
   c. stabilizing selection.
   d. sexual selection.

8. The process by which differences in the inherited characteristics of individuals cause them to differ in their ability to get mates is most accurately called
   a. natural selection.
   b. reproductive success.
   c. mate choice.
   d. sexual selection.

CHAPTER 18 Adaptation and Speciation

Key Concepts

- An adaptation is a feature of an organism that improves the performance of the organism in its environment. Adaptations result from natural selection.

- The process by which natural selection improves the match between a population of organisms and its environment over time is called adaptive evolution. Through adaptive evolution, populations of organisms adjust to environmental change. This can take many thousands of years or occur in as short a time as a few years or even months.

- The biological species concept defines a species as a group of interbreeding natural populations that is reproductively isolated from other such groups.

- Speciation, the process by which one species splits to form two or more species, is usually a by-product of genetic differences between populations that are caused by factors such as natural selection or genetic drift.

- Speciation often occurs when populations of a species become geographically isolated. Such isolation limits gene flow between the populations, making the evolution of reproductive isolation more likely.

- New species can form in the absence of geographic isolation as well.

Cichlid Mysteries

The surface of Earth changes slowly but dramatically over time. Islands rise from the sea, new lakes form and old ones disappear, and mountains are thrust up to divide once-continuous landmasses. Such changes alter the environments in which species live, setting the stage for grand natural experiments in evolution.

No evolutionary experiment has been more wondrous than that of the cichlid fishes of Lake Victoria, the largest of the Great Lakes of East Africa. Lake Victoria first formed about 750,000 years ago, but geologic evidence indicates that the lake was nearly or completely dry as recently as 15,000 years ago. Whenever a lake forms or refills with water, it may be colonized by one or more fish species, some of which may then evolve to form new species unique to the new lake. Such a sequence of events has happened many times in lakes around the world, but nowhere more spectacularly than in Lake Victoria.

Until recently, Lake Victoria harbored over 500 species of cichlids, a greater number of fish species than is found in all the lakes and rivers of Europe combined. The cichlids are a diverse and colorful group of fishes, well known to the aquarium trade. Genetic evidence suggests that the cichlids of Lake Victoria originated about 100,000 years ago; thus they did not die out during the dry period 15,000 years ago (some may have survived in pools within the lake or in nearby rivers). Genetic data also indicate that the cichlids in Lake Victoria descended from two ancestor species from Lake Kivu, a smaller and older lake located about 275 kilometers from Lake Victoria. The finding that 500 species of fishes descended from two ancestor species in 100,000 years brings us to the first of our cichlid mysteries: **How did so many Lake Victoria cichlid species form in a relatively short period of time?**

The mystery deepens when we realize that many of the Lake Victoria cichlid species differ considerably from one another in color, jaw structure, and feeding specialization, yet genetically, all the species in the lake are extremely closely related to one another. **What evolutionary forces have caused some of these closely related species to differ so much?**

A Lake Victoria Cichlid

We discussed the great explanatory power of evolutionary biology in Chapter 16. Knowing that populations and species can evolve explains adaptation, the diversity of life, and the shared characteristics of life. In this chapter we return to two of these themes: adaptation and biodiversity. We examine the characteristics of adaptations and discuss how they are shaped by natural selection. Then we reconsider the concept of the species and how to define species effectively. We focus the remainder of the chapter on speciation, the process that generates the diversity of life.

18.1 Adaptation: Adjusting to Environmental Challenges

An **adaptation** is a feature of an organism that improves aspects of the performance of that organism in its environment—for example, the ability to grow quickly or avoid being eaten, or to otherwise survive or reproduce. Many adaptations result in what appears to be a remarkably well-designed match between the organism and its environment. As we saw in Chapter 16, adaptations do not result from any intentional "design," but rather from natural selection: over time, individuals with inherited characteristics that allow them to survive and reproduce better than other individuals replace those with less favorable characteristics. This process, which improves the match between organisms and their environment over time, is called **adaptive evolution**. A similar process can occur when people influence the way domesticated species such as dogs change over time (see the box on page 372).

There are many different kinds of adaptations

There are many striking examples of adaptations in the natural world. Consider the fascinating group known as the weaver ants. These ants actually construct nests out of living leaves. Building the nest requires the coordinated actions of many individual ants. Some draw the edges of leaves together while others weave them in place by moving silk-spinning larvae (immature ants) back and forth over the seam of the two leaves. These complex actions are not the result of conscious planning on the part of the ants; rather, they illustrate how a simple evolutionary mechanism—natural selection—can produce a complex behavioral adaptation (cooperative

Weaver ants at work

nest building, which benefits the ants by providing them with shelter).

In another example of an adaptation, the caterpillars (the larval or immature stage) of a certain moth species develop bodies with different shapes that depend on which part of their food plant (an oak tree) they feed on—the flowers or the leaves. Those that feed on flowers grow to resemble oak flowers; those that feed on leaves grow to resemble oak twigs (Figure 18.1a and b). In this way, the larvae develop so that they match the background they are feeding and living on, making them more difficult for predators to see.

Natural selection has also shaped some astonishing adaptations that facilitate reproduction. The flowers of some orchid species, for example, use chemical attractants and appearance to mimic female wasps, thereby attracting male wasps and fooling them into attempting to mate with the flowers (Figure 18.1c). In the course of these attempts, the insects become coated with pollen, which they then transfer from one plant to another.

All adaptations share certain key characteristics

Look carefully at the photograph and diagram of the eye of the four-eyed fish, *Anableps anableps,* in Figure 18.2. Although this fish really has only two eyes, they function as four, enabling the fish to see clearly through both air and water. The four-eyed fish is a surface feeder, so the ability to see above water helps it locate prey such as insects. Its unique eyes also allow it to scan simultaneously for predators attacking from above (such as birds) or below (such as other fish). The four-eyed fish can also walk on land, and it often escapes trouble by jumping out of the water. So, although it would be interesting to watch, this fish would make a poor choice for a home aquarium.

Though there are literally millions of examples of adaptations, the four-eyed fish and the other examples we've discussed so far illustrate their most important characteristics:

- Adaptations have the appearance of having been designed to match the organism to its environment (consider the caterpillars in Figure 18.1), but this close fit to the environment is the work of evolution and not a conscious designer.
- Adaptations are often complex (consider the eye of *Anableps anableps* and the nest-building behavior of weaver ants).

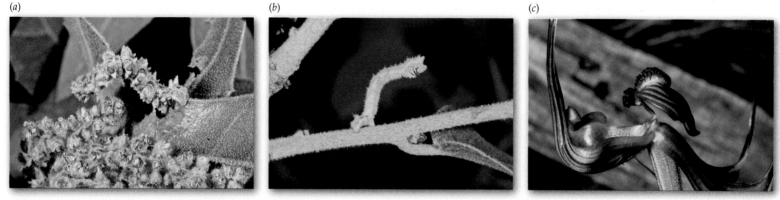

Figure 18.1 Evolution Can Lead to Astonishing Adaptations

Organisms can evolve to match their environments in striking ways. (*a, b*) Caterpillars of the moth *Nemoria arizonaria* [nee-*MORE*-ee-uh air-ih-zoh-*NAIR*-ee-uh] differ in shape depending on their diet: (*a*) Caterpillars that hatch in the spring resemble the oak flowers on which they feed. (*b*) Caterpillars that hatch in the summer eat leaves and resemble oak twigs. Experiments have demonstrated that chemicals in the leaves control the switch that determines whether the caterpillars will mimic flowers or twigs. (*c*) The flowers of this orchid have likewise evolved to become a striking match to one particular aspect of their environment: the females of a wasp species that can be found flying in the area. The match is so good that the orchids are able to achieve pollination by being "mated" by a fooled male wasp.

- Adaptations help the organism accomplish important functions, such as feeding, defense against predators, and reproduction. (This point is illustrated by all of the examples we have discussed.)

(*a*)

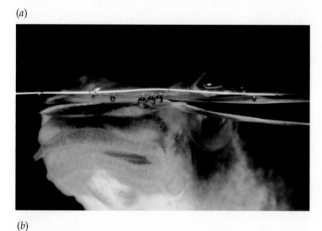

(*b*)

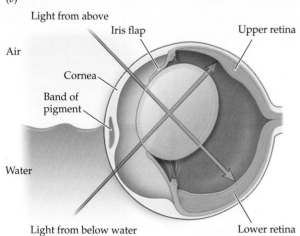

Populations can adjust rapidly to environmental change

Adaptive evolution can occur, and often does, over long periods of time. For example, as climates have slowly changed and ice ages have come and gone over the millennia, during those thousands of years populations of organisms have continued to evolve and adapt to their changing environments. But biologists have found that populations can undergo dramatic adaptive evolution surprisingly rapidly as well. For example, male guppies in the mountain streams of Trinidad and Venezuela have bright and variable colors that serve to attract females. But the bright colors that help the males succeed in attracting mates also make them easier for predators to find. How do guppy populations evolve in response to such conflicting pressures?

Field observations show that guppies from streams where few predators lurk are brightly colored, but guppies from streams with more predators are drab in

Figure 18.2 The Four-Eyed Fish, *Anableps anableps*

The four-eyed fish really has only two eyes, but each eye has a special design that lets it see clearly in both air and water. (*a*) The fish usually swims so that the band of pigment in the eye is level with the waterline, thus dividing the eye into above- and below-water halves. (*b*) The iris flap shields the upper pupil from glare off the surface of the water. When the fish looks above the water, light passes through a flattened lens and then strikes the lower retina (the retina is composed of photoreceptor cells that receive the image and nerve cells that send the image to the brain). Humans also have a flattened lens, which provides the best image when looking through air. When the fish looks below the water, light passes through a rounded lens before striking the upper retina. Ordinary fish also have a rounded lens, which provides the best image when looking through water.

Biology on the Job

Dog Show Judge Extraordinaire

Over the course of a single year in the United States, dogs of some 150 breeds, all created by artificial selection, participate in over 15,000 dog shows—some of them local and state shows, others national. Judging all these contestants requires the services of many dog judges, experts in the desired morphology and behavior of different dog breeds. Mr. Edd Bivin is so expert in his understanding of dog breed morphology that he was selected by a collection of dog judges, dog trainers, breeders, and kennel clubs as a "judging legend." The following excerpts are taken from an interview with Mr. Bivin by TheDogPlace, a Web site devoted to the interests of dog fanciers.

When, and why, did you decide to become a judge? I probably decided to become a judge when I was a kid. I showed good dogs [but I would just] get patted on the head. I believed in the sport and knew if I ever were to become a judge, I wouldn't pay any attention to where a dog comes from or the age of the individual handler. I would evaluate what I consider to be the best dog.

What do you most enjoy about judging? Obviously the dogs, and the people. I also consider judging to be a personal competition of Edd Bivin with himself. Every time I go in the ring I compete with myself, to do the best job I can do, on that day, within the circumstances with which I find myself.

Let's talk about the sport today. Are most breeds better than 10 years ago? No. Breeds are cyclical; they progress and they fall back. So I can't accept the term "most." I will tell you that there are many breeds that are better today and many breeds not as good as they were 10 years ago. A big part depends on who is directing breeding programs and pockets of interest around the country.

Judging a Dog Show

When you first look down the line [of dogs for show], what draws your eye? Balance and proportion. Carriage and outline. Outline and character.

Should showmanship and presentation be considered? Certainly. One should never miss a good animal with proper type and character. I become concerned about individuals applauding dogs or saying it's not a great such and such but it's a great "show dog." Dog shows are a format for the evaluation of breeding stock.

The Job in Context

When he judges a show, Mr. Bivin evaluates the degree to which a dog achieves the ideal standard for its breed. Judges such as Mr. Bivin are approved for selected breeds only; this specialization ensures that all judges thoroughly understand the physical form and character of the dog breeds that they evaluate.

When judges compare a dog with the ideal standard for that dog's breed, they become part of the process of artificial selection, which results in adaptations similar to those seen in adaptive evolution in nature. That is, prize-winning dogs are sought after as breeding stock, so they leave more offspring than do dogs that show poorly. So just as natural selection favors one thing in one population (say, brightly colored guppies in a predator-free river) and something else in another population (say, muted colors in a predator-filled river), dog show judges, by favoring one characteristic over another, are helping breeders artificially select for certain qualities in different breeds. For example, a huge Chihuahua is unlikely to win many prizes and be bred often, and an itsy-bitsy Great Dane will likely fair equally poorly in championship dog shows or the competition to become a highly desired breeder.

One might imagine that this process—coupled with the efforts of breeders, who also strive to improve the quality of the dogs over time—would ensure that dogs of a particular breed today were better in every way than dogs of 10 years ago. But instead, as Mr. Bivin notes, some are better, but many others are worse. Why? One possible reason is that, as we saw earlier in the chapter, there are always trade-offs in evolution. So while breeders may choose prize-winning dogs to breed because they show particularly good qualities in one area—coat color, leg length, and so on—they may be inadvertently promoting other, less desirable features that are part of the genetic makeup of those championship dogs. So just as natural selection does not result in perfect adaptation, continued artificial selection does not result in always better or perfect dogs.

comparison. Predators can influence guppy coloration because, with more predators present, a higher proportion of the most brightly colored guppies are likely to be eaten each generation, so they do not pass their genes on to the next generation. The color of populations of guppies can evolve very rapidly in response to predators. When guppies that live in areas with few predators, and so are brightly colored, are experimentally transferred to an area with many predators, or vice versa, the color patterns in each population can evolve to match the new conditions within 10 to 15 generations (14 to 23 months).

The ability to evolve rapidly in response to changing environmental conditions is not limited to guppies. For example, consider the soapberry bug, which feeds on the seeds contained in the fruit of the soapberry plant as illustrated in Figure 18.3a. In soapberry bug populations in Florida, beak length has evolved rapidly to match the size of the fruits that contain the seeds on which these insects feed (Figure 18.3b). Similarly, as we saw in Chapter 16, beak sizes in the medium ground finch evolved extremely rapidly in response to drought and the subsequent change in the size of available seeds. In addition, as we saw in Chapter 17, in only a few months or years viruses, bacteria, and insects can evolve resistance to our best efforts to kill them.

These examples illustrate an important point: while evolution by natural selection can improve the adaptations of organisms over very long periods of time, it can do so over surprisingly short periods of time as well.

18.2 Adaptation Does Not Craft Perfect Organisms

As impressive as the adaptations we see in nature may be, we do not mean to suggest that natural selection results in a perfect match between an organism and its environment. In many cases, genetic constraints, developmental constraints, or ecological trade-offs prevent further improvements in an organism's adaptation. In this section we look at these barriers to perfection.

Lack of genetic variation can limit adaptation

For an adaptation to evolve to be increasingly effective over time, there must be genetic variation for traits that can enhance the match between the organism and its environment. In some cases, the absence of such genetic variation places a direct limit on the ability of natural selection to cause adaptive evolution. For example, the mosquito *Culex pipiens* is now resistant to organophosphate pesticides, but this resistance is based on a single mutation that occurred in the 1960s, as we saw in Chapter 17. Before this mutation occurred, adaptation to these pesticides was not possible, and billions of mosquitoes died because their populations lacked genetic variation for resistance to the pesticides.

The multiple effects of developmental genes can limit adaptation

As we saw in Chapter 14, changes in genes that control development can have dramatic effects on the phenotype. Such a change in the developmental program of an organism often influences more than one part of its phenotype. Changes in genes that control development can have many effects, some of which may be advantageous while others may harm or kill the organism.

The multiple effects of developmental genes can limit the ability of the organism to evolve in certain directions, which in turn may limit what adaptive evolution can achieve. For example, the larval stages of some insects, such as beetles and moths, lack wings or well-developed eyes—two important adaptations that the adult forms

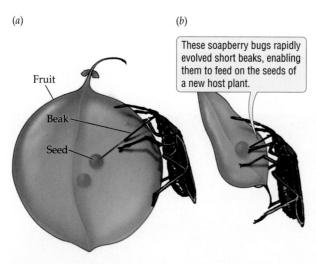

(a)

(b)

Fruit

Beak

Seed

These soapberry bugs rapidly evolved short beaks, enabling them to feed on the seeds of a new host plant.

Figure 18.3 Rapid Evolution in an Insect
(a) In Florida, soapberry bugs traditionally fed on seeds of a native plant species, the balloon vine. The bugs had to pierce to the center of the balloon vine's fruit to reach the seeds shown at the center.
(b) Over the past 30 to 50 years, some populations of soapberry bugs have evolved short beaks, enabling them to feed on seeds within the narrower fruit (shown as the narrower green silhouette with seeds at center) of an introduced species, the golden rain tree.

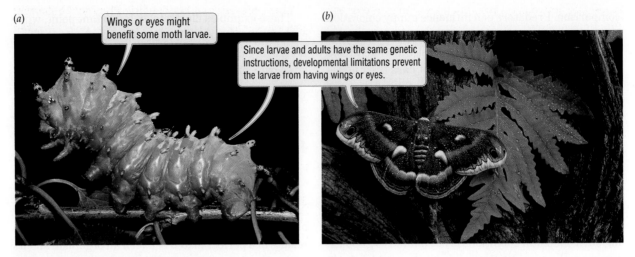

(a)

Wings or eyes might benefit some moth larvae.

Since larvae and adults have the same genetic instructions, developmental limitations prevent the larvae from having wings or eyes.

(b)

Figure 18.4 A Developmental Limitation
The larval form (*a*) of this moth does not have wings or eyes, two important adaptations that are found in the adult form (*b*).

of these insects have (Figure 18.4). Beetle and moth larvae have a wide range of lifestyles, so wings or well-developed eyes probably would benefit many of these larvae. Since the adult and larval forms of these insects carry exactly the same genetic instructions, it is likely that turning on or expressing the genes that control wing or eye production in adults would have other, extremely harmful effects in larvae, perhaps killing them. The lack of wings and eyes in beetle and moth larvae therefore is a result of developmental limitations.

Ecological trade-offs can limit adaptation

To survive and reproduce, organisms must perform many functions, such as finding food and mates, avoiding predators, and surviving the challenges posed by the physical environment. Within the realm of what is genetically and developmentally possible, natural selection increases the overall ability of the organism to survive and reproduce. However, the many and often conflicting demands that organisms face result in trade-offs

or compromises in their ability to perform important functions.

High levels of reproduction, for example, are often associated with decreased longevity. In some cases, such a trade-off is due to relatively subtle costs of reproduction: resources directed toward reproduction are not available for other uses, such as storing energy to help the organism survive a cold winter. In red deer, for example, females that reproduced the previous spring have a higher rate of death during winter than do females that did not reproduce. But costs associated with reproduction can sometimes be immediate and dramatic, as illustrated by the mating calls of the túngara frog described in Chapter 17 (Figure 18.5). In general, the widespread existence of

Figure 18.5 Does Love or Death Await?
Male túngara frogs face an ecological trade-off: the same type of call that is most successful at attracting females also makes it easier for predatory bats to locate calling males.

trade-offs between reproduction and other important functions ensures that organisms are not perfect, for the simple reason that it is not possible to be the best at all things at once.

Concept Check

1. Why do organisms show so many adaptations to life in their environment?
2. What is the range of time over which adaptive evolution has been shown to take place?

18.3 What Are Species?

Before we discuss the processes that have generated the great diversity of species on Earth, we must first define what a species is.

Species are often morphologically distinct

In a practical sense, species are usually defined in terms of morphology; that is, two organisms are classified as members of different species if they look sufficiently different. All of us use this type of definition, as when we distinguish bald eagles (Figure 18.6) from other birds by how they look.

Most species can be identified by morphological characteristics. That is, a group of organisms is often considered a separate and distinct species from all other organisms if its morphology is distinct enough. Indeed, morphology is the only way we have to identify and distinguish fossil species. For living species, however, a morphological definition does not always work well: Sometimes distinct and separate species have members that look exactly alike but cannot breed with each other. Conversely, different populations can vary phenotypically (sometimes dramatically) yet remain part of the same species because they can interbreed. What "holds a species together" and distinguishes it from other species?

Species are reproductively isolated from one another

In most cases, members of different species cannot reproduce with each other. When barriers to reproduction exist between species, the species are said to be **reproductively isolated** from one another. Barriers to reproduction are often divided into two types. Some barriers act before a male gamete (like a human sperm cell) and a female gamete (like a human egg cell) fuse to form a zygote; these are known as prezygotic barriers. Other barriers act after the formation of a zygote

(a)   *(b)*

Figure 18.6 In Most Species, Members Look Alike
Bald eagles that live in Alaska (*a*) look the same as bald eagles that live in Colorado (*b*). Although these birds live far apart, they remain phenotypically similar.

Table 18.1

Barriers That Can Reproductively Isolate Two Species in the Same Geographic Region

Type of barrier	Description	Effect
PREZYGOTIC BARRIERS		
Ecological isolation	The two species breed in different portions of their habitat, at different seasons, or at different times of the day.	Mating is prevented.
Behavioral isolation	The two species respond poorly to each other's courtship displays or other mating behaviors.	Mating is prevented.
Mechanical isolation	The two species are physically unable to mate.	Mating is prevented.
Gametic isolation	The gametes of the two species cannot fuse, or they survive poorly in the reproductive tract of the other species.	Fertilization is prevented.
POSTZYGOTIC BARRIERS		
Zygote death	Zygotes fail to develop properly and die before birth.	No offspring are produced.
Hybrid performance	Hybrids survive poorly or reproduce poorly.	Hybrids are not successful.

and are known as postzygotic barriers (Table 18.1). Although there are a wide range of barriers to reproduction, they all have the same overall effect: few or no alleles are exchanged between species. This restriction ensures that the members of a species share a common set of genes and alleles. Because members of a species exchange alleles with one another, but not with members of other species, they usually remain phenotypically similar to one another and different from members of other species.

This kind of reproductive isolation is the key to the most commonly used definition of a species, known as the **biological species concept**. The biological species concept defines a **species** as a group of populations that can interbreed but are reproductively isolated from other such groups. Note that reproductive isolation is not necessarily the same as geographic isolation; our definition of species includes populations that could interbreed if they were in contact with one another but do not interbreed because they have no opportunity to do so (for example, because they are too far away from one another, like our bald eagles from Alaska and Colorado).

The biological species concept has important limitations. For example, it is of no use when we are defining fossil species, since no information can be obtained about whether or not two fossil forms were reproductively isolated from each other. So, as mentioned earlier, fossil species are defined on the basis of morphology. Nor does our definition apply to organisms, such as bacteria and dandelions, that reproduce mainly by asexual means.

The biological species concept also fails to work well for the many plant and animal species that are obviously distinct and separate species—sometimes even looking quite different from one another—but that remain able to mate in nature to produce fertile offspring. Such distinct species that are able to interbreed in nature are said to **hybridize**, and their offspring are called **hybrids**. Although they can reproduce with each other, species that hybridize often look very different from one another (Figure 18.7) or are distinct ecologically. For example, they are usually found in different environments, or they differ in how they perform important biological functions, such as obtaining food.

Because of the limitations of the biological species concept, scientists have come up with alternative ways of defining species, each of which has its own limitations.

Gray oak tree

Gambel oak tree

Gray oak

Hybrids

Gambel oak

Figure 18.7 Some Species Interbreed Yet Remain Distinct
The gray oak and the gambel oak can mate to produce fertile hybrids. However, the two species remain phenotypically different, as is evident from their leaves.

The biological species concept, despite its limitations, remains the most useful definition for most biologists; and as a result, most biologists define species this way, so this is the definition we will use in this book. Thus, though it is useful to base species definitions on a single, simple characteristic (reproductive isolation), remember that the reality of a species in nature can be considerably more complicated.

18.4 Speciation: Generating Biodiversity

The tremendous diversity of life on Earth is caused by **speciation**, the process in which one species splits to form two or more species that are reproductively isolated from one another. The study of speciation is fundamental to understanding the diversity of life on Earth.

How do new species form? The crucial event in the formation of new species is the evolution of reproductive isolation, which requires that populations that once could interbreed diverge enough that they are no longer able to do so. As we saw in Chapter 17, however, populations within a species can be connected by gene flow, which tends to keep them genetically similar to one

another. How does reproductive isolation develop within a species, whose members interbreed and therefore share a common set of genes and alleles?

Speciation can be explained by the same mechanisms that cause the evolution of populations

Speciation is thought most often to be a secondary, incidental consequence of the evolution of populations. In essence, over time populations evolve genetic differences from one another—because of mutation, genetic drift, or natural selection—and these genetic differences sometimes result in reproductive isolation.

This idea can be illustrated by the results of an experiment with fruit flies. A population of flies was separated into smaller populations, all of which were placed in similar environments, except that some flies were fed maltose (a simple sugar) while others were fed starch (recall that starch is a polymer composed of many glucose molecules "stitched" together). Over time, the flies raised on these two different foods started to become reproductively isolated from each other (Figure 18.8). This reproductive isolation occurred simply because the flies adapted to living on two different kinds of food. That is, the fly populations changed genetically over time as the populations adapted to living on maltose or starch, respectively. The fly populations most likely

Figure 18.8 Selection Can Cause Reproductive Isolation

Fruit flies in an initial sample were separated into four populations and raised on two different kinds of food (starch or maltose) for several generations. In the experimental group, flies from the populations that had become adapted to feed on starch were then given the opportunity to mate with other flies that had adapted to feed on starch or with flies adapted to feed on maltose. As the mating frequencies show, scientists found that flies adapted to feed on starch preferred mating with other flies adapted to starch. Flies adapted to maltose likewise preferred other flies adapted to maltose. These preferences are the early stages of reproductive isolation that can eventually lead to speciation.

See speciation in action. ▶❚❚ 18.1

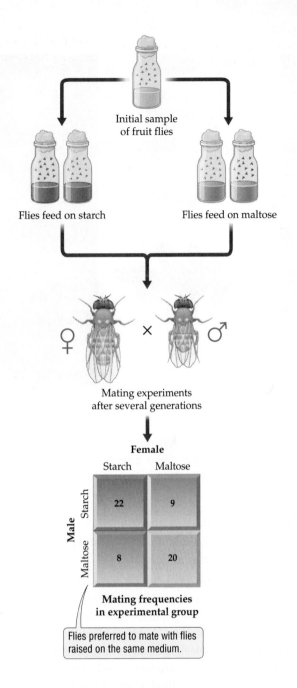

But the experiment illustrates how reproductive isolation can begin to evolve as a by-product of other evolutionary changes. In this case, the other change was adaptation to living on maltose or starch. In the wild, there is always the potential for this kind of evolution among populations of a single species, as populations adapt to living, for example, in different climates (bald eagles in Alaska versus Colorado) or to hunting different foods or avoiding different predators.

As suggested by Figure 18.8, natural selection can cause populations to diverge genetically when populations located in different environments face different selection pressures—in the case of our fruit flies, the selection to grow quickly on maltose versus starch. Populations can also diverge from one another as a result of mutation and genetic drift. In contrast, gene flow always operates to limit the genetic divergence of populations. For populations to accumulate enough genetic differences to cause speciation, the factors that promote divergence must have a greater effect than does the amount of ongoing gene flow.

Speciation can result from geographic isolation

A new species can form when populations of a single species become separated, or geographically isolated, from one another. This process can begin when a newly formed geographic barrier, such as a river or a mountain chain, isolates two populations of a single species. Such **geographic isolation** can also occur when a few members of a species colonize a region that is difficult to reach, such as an island located far outside the usual geographic range of the species. For example, as we saw in Chapter 16, Darwin's finches on the separate Galápagos islands were geographically isolated from one another and from finches on the South American mainland by ocean waters.

The distance required for geographic isolation to occur varies tremendously from species to species, depending on how easily the species can travel across any given barrier. Populations of squirrels and other rodents that live on opposite sides of the Grand Canyon—a formidably deep and large barrier for a rodent—have diverged considerably. Meanwhile, populations of birds—which can easily fly across the canyon—have not. In general, geographic isolation is said to occur whenever populations are separated by a distance that is great enough to limit gene flow.

However they arise, geographically isolated populations are essentially disconnected genetically. There is little or no gene flow between or among them. For this

changed genetically—that is, they evolved—in response to natural selection for the ability to grow and develop quickly on maltose versus starch. Those genetic changes had the side effect of causing reproductive isolation. That is, flies that were adapted to feed on maltose preferred to mate with flies that were also adapted to feed on maltose. The flies adapted to feed on starch likewise preferred to mate with flies adapted to feed on starch.

The experiment was not continued long enough for this budding reproductive isolation to become complete—that is, for speciation to occur. The flies from the maltose population and the starch population were still the same species and could still mate and reproduce with one another.

reason, mutation, genetic drift, and natural selection can more easily cause isolated populations to diverge genetically from one another. If the populations remain isolated long enough, they can evolve into new species. The formation of new species from populations that are geographically isolated from one another is called **allopatric speciation** (*allo*, "other"; *patric*, "country") (Figure 18.9).

Much evidence indicates that geographic isolation can lead to speciation. One line of evidence is the observation that in many groups of organisms, the number of species is greatest in regions where strong geographic barriers increase the potential for geographic isolation. Examples include species that live in mountainous regions (such as birds in New Guinea) or on island chains (such as plants in the Hawaiian Islands). One particularly interesting example of a possible case of speciation by geographic isolation is the recent discovery of the so-called hobbit people of the island of Flores in Indonesia. These very tiny but very humanlike creatures, also known as the "little people" of Indonesia, were discovered in 2004. Measuring 3 feet tall as adults, they lived as recently as 12,000 years ago on their isolated island along with giant komodo dragon lizards and pygmy elephants. The size of a typical human toddler, these people appeared to use stone tools and fire. Since their discovery, some scientists have continued to debate whether these little islanders are indeed a new and separate species of humans, *Homo floresiensis,* or whether they are simply small, misshapen individuals of our species, *Homo sapiens.* (For the latest on this controversy, see the "Biology in the News" feature on page 383.)

A second line of evidence for the importance of geographic isolation in speciation comes from cases in which individuals of a population found at one extreme end of a species' geographic range reproduce poorly with individuals of a population at the other end, even though individuals from both ends reproduce well with individuals from intermediate portions of the species' range. A special case of this phenomenon occurs when the populations loop around a geographic barrier; in this case, **ring species** form, in which populations at the two ends of the loop are in contact with one another, yet individuals from these populations cannot interbreed. Ring species have been found in salamander populations that loop around the mountains of the Sierra Nevada in California.

A third line of evidence comes from the results of laboratory experiments in which populations that are separated from one another develop reproductive isolation over time (see Figure 18.8).

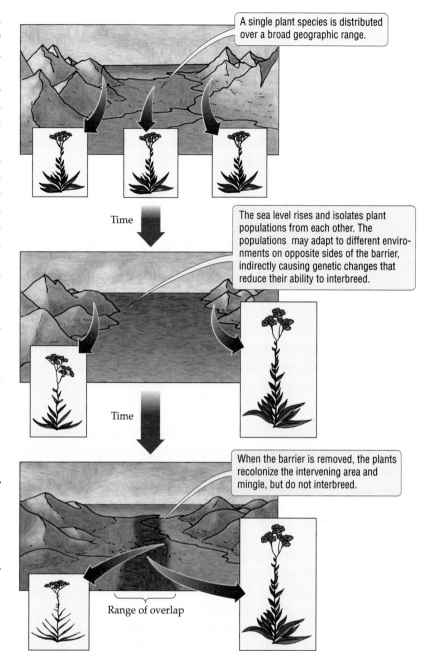

A single plant species is distributed over a broad geographic range.

Time

The sea level rises and isolates plant populations from each other. The populations may adapt to different environments on opposite sides of the barrier, indirectly causing genetic changes that reduce their ability to interbreed.

Time

When the barrier is removed, the plants recolonize the intervening area and mingle, but do not interbreed.

Range of overlap

Figure 18.9 Allopatric Speciation
New species can form when populations are separated by a geographic barrier, such as a rising sea.

Speciation can occur without geographic isolation

Most speciation events are thought to occur by allopatric speciation, because speciation requires the cessation of gene flow between populations and the potential for gene flow is much greater between populations whose geographic ranges overlap or are adjacent to one another than between populations that are geographically isolated from one another. Therefore,

Rhagoletis pomonella

at one time it was thought highly unlikely that new species could arise in any way other than by being geographically isolated—that is, via allopatric speciation. Today, however, it is well established that plants can form new species in the absence of geographic isolation, and recent work has provided convincing evidence that animals can as well. The formation of new species in the absence of geographic isolation is called **sympatric speciation**.

In plants, rapid changes in chromosome numbers can cause sympatric speciation. New plant species can form in a single generation as a result of **polyploidy**, a condition in which an individual has more than two sets of chromosomes, usually because of the failure of chromosomes to separate during meiosis (see Chapter 9). Polyploidy can also occur when a hybrid spontaneously doubles its chromosome number. Polyploidy is invariably fatal in people, but it is not lethal in many plant species. Although it does not kill them, however, a doubling of the chromosomes can lead to reproductive isolation because the chromosome number in the gametes of the new polyploid no longer matches the number in the gametes of either of its parents. Although relatively few plant species originate directly in this way, polyploidy has had a large effect on life on Earth: more than half of all plant species alive today are descended from species that originated by polyploidy. A few animal species also appear to have originated by polyploidy, including several species of lizards and fishes and one mammal (an Argentine rat).

Evidence is mounting that sympatric speciation can occur in animals by means other than polyploidy. For example, there is compelling evidence that new cichlid species have formed in the absence of geographic isolation in the lakes in Africa. Genetic data on some of these cichlids indicate that within the confines of Lake Bermin, 9 species have evolved; and within Lake Barombi Mbo, 11 species of cichlids have arisen. The formation of new species within these lakes provides strong evidence for sympatric speciation because there appears to be no way for fish to become geographically isolated in these small, environmentally uniform lakes. As a result, fish populations within these lakes cannot evolve into new species by living apart from one another in very different environments within the lake. That is, there appears to be no way that these fish species could have evolved by allopatric speciation (geographic isolation).

Strong evidence in support of sympatric speciation has also been found in other animals. Researchers believe that North American populations of the apple maggot fly, *Rhagoletis pomonella* [rag-oh-*LEE*-tiss poh-moh-*NELL*-uh], are in the process of diverging into new species, even though their geographic ranges overlap. Historically, *Rhagoletis* usually ate native hawthorn fruits, but in the mid-nineteenth century these flies were first recorded as pests on apples, an introduced nonnative species. *Rhagoletis* populations that feed on apples are now genetically distinct from populations that feed on hawthorns. Members of the apple and hawthorn populations mate at different times of the year and usually lay their eggs only on the fruit of their particular food plant. As a result, there is little gene flow between the apple and hawthorn populations. In addition, researchers have identified alleles that benefit flies that feed on one host plant but are detrimental to flies that feed on the other host plant. Natural selection operating on these alleles acts to limit whatever gene flow does occur. Over time, the ongoing research on *Rhagoletis* may well provide a dramatic case history of sympatric speciation.

18.5 Rates of Speciation

When speciation is caused by polyploidy or other types of rapid chromosomal change, new species form in a single generation. New species also appear to have formed relatively rapidly in the case of some African cichlid fishes. Scientists found that some 500 species of cichlids have existed in Lake Victoria, and genetic analyses indicate that they all descended from just two ancestor species. Genetic evidence also indicates that these 500 species evolved during the past 100,000 years. Thus, in roughly 100,000 years, 500 fish species evolved from two species.

In most cases, speciation probably occurs more slowly. Among freshwater fishes the time required for speciation has been estimated to range from a minimum of 3,000 years (in pupfish) to over 9 million years (in characins [*KAIR*-uh-sinz], a group that includes carp, piranha, and many aquarium fish). In other groups of organisms, including fruit flies, snapping shrimp, and birds, the time required for speciation has been estimated to range from 600,000 to 3 million years. Furthermore, some populations can be geographically isolated for a long time without evolving reproductive isolation. North American and European sycamore trees, for example, have been separated for more than 20 million years, yet the two populations remain morphologically similar and can breed with each other.

18.6 Implications of Adaptation and Speciation

Adaptation is the means by which organisms adjust to the challenges posed by new or changing environments, and speciation is the means by which the diversity of life has come into being. Understanding both, therefore, is critical to understanding how evolution works.

Adaptation and speciation are also very important from an applied perspective. For example, to combat rapidly evolving disease agents—such as HIV, the virus that causes AIDS—we must have a detailed understanding of the new adaptations that enable them to overcome our best efforts to destroy them. Speciation has long been of practical importance to humans, as our development of domesticated crop and animal species—cows, corn, and dogs to name just a few—readily attests.

In addition, understanding the often slow pace of speciation gives us a strong incentive to stop the ongoing extinctions of species. Speciation can require hundreds of thousands to millions of years, yet humans are driving species extinct in just decades to hundreds of years. If we continue to drive species extinct at the present rate, it will take millions of years before new species can evolve to replace the large number of species currently being lost.

Concept Check

1. What key characteristic determines whether a group is a separate species or not, according to the biological species concept?
2. What role does geographic isolation play in speciation?
3. Why was sympatric speciation once thought to be unlikely?

Applying What We Learned

Rapid Speciation in Lake Victoria Cichlids

As we saw earlier in this chapter, the Lake Victoria cichlids formed new species at a rapid rate, resulting in an evolutionary expansion unmatched by any other group of vertebrates. The rapid speciation of these fishes brings us back to one of the cichlid mysteries introduced at the opening of this chapter: how did so many species form in just 100,000 years? The answer hinges on two key aspects of cichlid biology.

First, cichlids in Lake Victoria appear to use color as a basis for mate choice: females prefer to mate with males of a particular color (an example of sexual selection; see Section 17.8). Within Lake Victoria, many pairs of ecologically similar cichlid species differ from each other in color but little else. In these pairs of species, the males of one species tend to be blue, while the males of the other species are red or yellow. Females of a given species show a strong preference for mating with males of their own species. This preference appears to be based on the color of the males: when

researchers experimentally changed the quality of the light so that the females could not see the difference between blue and red/yellow males, the female preference for mating with males of their own species broke down. This research suggests that if males in two populations of a cichlid species happened to have different colors, female preferences for mating with males of a certain color could cause these populations to become reproductively isolated, thus setting the stage for the formation of new species.

In addition, cichlids have unusual jaws that can be modified relatively easily over the course of evolution to specialize on new food items. This feature of their biology causes the cichlids in Lake Victoria to vary greatly in form and feeding behavior (Figure 18.10). How do their easily modified jaws relate to speciation? If female mate choice caused two populations to begin to be reproductively isolated from each other, the resulting lack of gene flow could allow the populations to specialize on

Figure 18.10 Variation among the Lake Victoria Cichlids
The four species shown here illustrate some of the differences in morphology and feeding behavior found among Lake Victoria cichlids.

Haplochromis chilotes
(feeds on insects)

Haplochromis macrognathus
(feeds on other fishes)

Astatotilapia elegans
(generalized bottom feeder)

Macropleurodus bicolor
(feeds on snails and other mollusks)

different sources of food, which would make it increasingly likely that they would continue to diverge and form new species.

However, although cichlids have evolved new species extremely rapidly, they cannot keep up with the even faster rate of extinction they are undergoing. In the past 30 years, roughly 200 Lake Victoria cichlid species have disappeared. Some, we know, were driven to extinction by an introduced predatory fish species, the Nile perch. But many species that are rarely eaten by Nile perch have also vanished. What might have driven these species to extinction?

Here, too, the answer may be related to female mate choice. As the experiments just described showed, female cichlids need to be able to see the colors of males to distinguish males of their own species from males of closely related species. However, pollution due to human activities has caused the water of Lake Victoria to become cloudy. If females were unable to distinguish the colors of potential mates in the murky water, repro-

ductive barriers between species might break down. As reproductive barriers lost their effectiveness, species that once were distinct might interbreed freely and become more similar to each other. If continued long enough, such interbreeding could "reverse" the speciation process and cause species to go extinct. Pollution would not be likely to have this effect if the cichlid species were separated by barriers to reproduction that occurred *after* mating took place, such as gametic isolation or zygote death (see Table 18.1).

Finally, remember that speciation in cichlids depends on the ability of females to recognize differences in the colors of males; when cloudy water impairs that ability, new species cannot form. Thus, pollution from human activities appears to have two profound effects: it halts the formation of new cichlid species while simultaneously causing existing species to go extinct. We must reduce that pollution if we are not to destroy one of nature's most amazing evolutionary experiments, the cichlids of Lake Victoria.

Study Says Bones Found in Far East Are of a Distinct Species

BY JOHN NOBLE WILFORD

[This article describes new findings on the recently discovered fossil remains of the extinct so-called hobbit people from the island of Flores in Indonesia. Some scientists have argued that they are a newly discovered species of human.]

In the continuing debate over the origin of the extinct "little people" of Indonesia, a team of scientists says it has found evidence in three wrist bones that these people were members of a distinct [new human] species rather than humans with a physical disorder.

Critics disputed the research, saying it did not present clear evidence for the existence of a separate species, known as *Homo floresiensis*.

The discovery, in a cave on the island of Flores, of skeletal remains of the diminutive people with unusually small heads was a sensation when it was announced three years ago. Some scientists contended that these were more likely to be modern humans who suffered a developmental disorder that causes the head and brain to be much smaller than average.

In the new study, scientists led by Matthew W. Tocheri, an anthropologist of human origins at the Smithsonian Institution, examined wrist bones from the skeletons and found them to be primitive and shaped differently from the wrist bones of modern humans.

This evidence, the scientists wrote, indicated that the individuals were not modern humans "with an undiagnosed pathology or growth defect." Rather they represent a species that descended from an ancestor that branched off from the human lineage at least 800,000 years ago, the scientists concluded.

Dr. Tocheri and his colleagues said the distinct species emerged from ancestors "that migrated out of Africa before the evolution of the shared, derived wrist morphology that is characteristic of modern humans, Neanderthals and their last common ancestor."

Dr. Robert B. Eckhardt, a professor of developmental genetics at Pennsylvania State University and one of several critics of the new-species designation, took issue with the new research. Dr. Eckhardt noted, in particular, that there was "a lot of variation in the form of wrist bones." Some variations, he said, are normal and others occur "as the result of various pathologies, such as from injuries or from anomalies of development."

As is typical with any kind of declaration made by scientists about the evolution of humans or their close relatives, the discovery of a possible new species of humans, *Homo floresiensis,* has been met with excitement, skepticism, and controversy.

Today our species is the only species of the genus *Homo* still alive. Earlier in history, however, there were other species, including *Homo habilis* and *Homo erectus* and possibly *Homo floresiensis.* The possibility that another species of humans, *Homo floresiensis,* was alive at the same time as modern humans a mere 12,000 years ago raises many questions.

Did *Homo sapiens* and the "little people" of Flores interact? Besides size and the wrist bone shape, did the two have other differences? If a *Homo sapiens* had mated with a little person of Flores, could they have produced healthy, fertile offspring? That is, were they separate species by our definition?

People have always been curious about the life of humans in our ancient, evolutionary past. Now the little people of Flores—by suggesting that there may have been more kinds of humans than had been guessed—have offered yet another possible window onto the nature of human life so long ago.

Evaluating the News

1. According to the biological species concept, could differences in wrist bones enable us to say that the "little people" of Flores Island are a separate species from modern-day humans? What information would we need about these people and about modern-day humans to figure that out?

2. If Dr. Tocheri and colleagues are right and these *Homo floresiensis* specimens represent a new and distinct species, and if the species were restricted to Flores Island, what would be the likely way in which they speciated: allopatric speciation or sympatric speciation? What other island species do you know of that speciated in the same way?

3. The "little people" of Indonesia died out, possibly as a result of a volcanic explosion on their island, thousands of years ago. But they raise the question of what it would mean to learn that there was another species of human alive today. If we were to discover another species of human, how would or should we behave toward them? Should they have all the rights and privileges of *Homo sapiens*? Should a new species of human be treated more like we treat other species—as pets, zoo animals, food, or work animals? If another human species viciously attacked or mauled a *Homo sapiens,* would it, like a dog, be killed, or would it simply be punished like a *Homo sapiens*?

SOURCE: *New York Times,* September 21, 2007.

Chapter Review

Summary

18.1 Adaptation: Adjusting to Environmental Challenges

- Adaptations result in an apparent match between organisms and their environment but are not caused by intentional design.
- Adaptive evolution is the process by which the fit between organisms and their environment is improved over time.
- Adaptations help organisms accomplish important functions, such as mate attraction and predator avoidance.
- Adaptations can evolve over very long and very short periods of time.

18.2 Adaptation Does Not Craft Perfect Organisms

- Adaptive evolution can be limited by genetic constraints: lack of genetic variation gives natural selection little or nothing on which to act.
- Adaptive evolution can be limited by developmental constraints: the multiple effects of developmental genes can prevent the organism from evolving in certain directions.
- Adaptive evolution can be limited by ecological trade-offs: conflicting demands faced by organisms can compromise their ability to perform important functions.

18.3 What Are Species?

- Species are often morphologically distinct from one another, but morphology is not always a reliable way to distinguish some species.
- The biological species concept defines species as a group of interbreeding natural populations that is reproductively isolated from other such groups.
- The biological species concept has important limitations. It does not apply to fossil species (which must be identified by morphology), to organisms that reproduce mainly by asexual means, or to organisms that hybridize extensively in nature.

18.4 Speciation: Generating Biodiversity

- The crucial event in the formation of a new species is the evolution of reproductive isolation.
- Speciation usually occurs as a by-product of the genetic divergence of populations from one another caused by natural selection, genetic drift, or mutation.
- Speciation usually occurs when populations are geographically isolated from one another long enough for reproductive isolation to evolve. Most new species are thought to arise by this process, which is called allopatric speciation.
- Speciation can also occur without geographic isolation. This process, called sympatric speciation, acts when part of a population diverges genetically from the rest of the population.
- Polyploidy is one way that many plants evolve new species during a single generation.

18.5 Rates of Speciation

- Speciation occurs rapidly in some cases, but it requires hundreds of thousands to millions of years in other cases.

18.6 Implications of Adaptation and Speciation

- Adaptations are the means by which organisms adjust to challenges posed by new or changing environments.
- Speciation is the means by which the diversity of life has come into being.
- Adaptation and speciation influence such practical matters as how we fight diseases and develop domesticated species.

◉ Review and Application of Key Concepts

1. Select an organism (other than humans) that is familiar to you. List two adaptations of that organism. Explain carefully why each of these features is an adaptation.

2. What is adaptive evolution? Apply your understanding of adaptive evolution to organisms that cause infectious human diseases, such as bacterial species that cause plague or tuberculosis. How do our efforts to kill such organisms affect their evolution? Are the evolutionary changes we promote usually beneficial or harmful for us? Explain your answer.

3. Imagine that a species legally classified as rare and endangered is discovered to hybridize with a more common species. Since the two species interbreed in nature, should they be considered a single species? Since one of the two species is common, should the rare species no longer be legally classified as rare and endangered?

4. Should species that look different and are ecologically distinct, such as the oaks in Figure 18.7, be classified as one species or two? These oak species hybridize in nature. Should species that hybridize in nature be considered one species or two?

5. High winds during a tropical storm blow a small group of birds to an island previously uninhabited by that species. Assume that the island is located far from other populations of this species, and that environmental conditions on the island differ from those experienced by the birds' parent population. Is natural selection or genetic drift (or both) likely to influence whether the birds on the island form a new species? Explain your answer.

6. Hundreds of new species of cichlids evolved within the confines of a large lake in Africa, known as Lake Victoria. Some of these species live in different habitats within the lake and rarely encounter one another. Would you consider such species to have evolved with or without geographic isolation?

7. How can new species form by sympatric speciation? Why is it harder for speciation to occur in sympatry than in allopatry?

Key Terms

adaptation (p. 370)
adaptive evolution (p. 370)
allopatric speciation (p. 379)
biological species concept (p. 376)
geographic isolation (p. 378)
hybrid (p. 376)
hybridize (p. 376)

polyploidy (p. 380)
reproductive isolation (p. 375)
ring species (p. 379)
speciation (p. 377)
species (p. 376)
sympatric speciation (p. 380)

Self-Quiz

1. Species that have overlapping geographic ranges but do not interbreed in nature are said to be
   a. geographically isolated.
   b. reproductively isolated.
   c. influenced by genetic drift.
   d. hybrids.

2. Which of the following evolutionary mechanisms acts to slow down or prevent the evolution of reproductive isolation?
   a. natural selection
   b. gene flow
   c. mutation
   d. genetic drift

3. The splitting of one species to form two or more species most commonly occurs
   a. by sympatric speciation.
   b. by genetic drift.
   c. by allopatric speciation.
   d. suddenly.

4. The time required for populations to diverge to form new species
   a. varies from a single generation to millions of years.
   b. is always greater in plants than in animals.
   c. is never less than 100,000 years.
   d. is rarely more than 1,000 years.

5. Adaptations
   a. match organisms closely to their environment.
   b. are often complex.
   c. help the organism accomplish important functions.
   d. all of the above

6. Prezygotic and postzygotic barriers to reproduction have the effect of
   a. reducing genetic differences between populations.
   b. increasing the chance of hybridization.
   c. preventing speciation.
   d. reducing or preventing gene flow between species.

7. Evidence suggests that sympatric speciation may have occurred or may be in progress in three of the following four cases. Select the exception.
   a. apple maggot fly
   b. squirrels on opposite sides of the Grand Canyon
   c. cichlid fishes
   d. polyploid plants (or their ancestors)

8. The diploid number of chromosomes in plant species A is 8; the diploid number in plant species B is 16. If plant species C originated when a hybrid between A and B spontaneously doubled its chromosome number, what is the most likely number of diploid chromosomes in C?
   a. 8
   b. 12
   c. 24
   d. 48

CHAPTER 19

The Evolutionary History of Life

Key Concepts

- Macroevolution is large-scale evolutionary change, including evolution above the species level.

- The fossil record documents the history of life on Earth and provides clear evidence of evolution.

- Early photosynthetic organisms released oxygen to the atmosphere as a waste product, thereby setting the stage for the evolution of the first eukaryotes, and later the first multicellular organisms.

- The Cambrian explosion was an astonishing increase in animal diversity that occurred 530 million years ago, when large forms of most of the major living animal phyla appeared suddenly in the fossil record.

- The colonization of land by the first plants, fungi, and animals marked the beginning of another major increase in the diversity of life.

- The history of life can be summarized by the rise and fall of major groups of protists, plants, and animals. This history has been greatly influenced by continental drift, mass extinctions, and adaptive radiations.

Puzzling Fossils in a Frozen Wasteland

Antarctica is a crystal desert, an ice-covered land in which heat and liquid water are very scarce. Few organisms can survive the extreme cold and lack of available water, and most of those that can are small and live near the sea. The entire continent has only two species of flowering plants (Antarctic hair grass and Antarctic pearlwort), and its largest terrestrial animal is a fly 5 milli-meters long.

In the interior of the continent the organisms are even smaller: in most places the only living things are microscopic bacteria and protists, including some that survive in a state of suspended animation (frozen but alive in the ice). Some of the interior valleys have little ice and hence seem a little less forbidding. But those valleys are so dry and cold that they support no visible life. There, the only organisms found on land are photosynthetic bacteria and lichens that spend their entire lives in a narrow zone just under the translucent surface of certain types of rocks.

Despite the nearly lifeless appearance of this continent, fossils reveal that Antarctica used to be very different from today's frozen landscape. Where life now maintains an uncertain foothold, the land was once bordered by tropical reefs and later covered with forests. At different times, ferns, freshwater fishes, large amphibians, aquatic beetles, and trees as tall as 22 meters thrived in what is now among the harshest of environments. Dinosaurs once roamed these lands, and millions of years later mammals and reptiles were pursued by the terrorbird, a fast-running flightless bird that stood 3.5 meters tall.

Early explorers and scientists were amazed when they discovered these fossils of ancient life forms in Antarctica. The fossils showed that life used to be rich and abundant where now it barely exists. These scientists and explorers were left with a simple question: **What happened?**

The Antarctic Landscape Then
An artist's reconstruction of Antarctica over 190 million years ago, full of lush, abundant life including the massive dinosaur *Glacialisaurus hammeri*.

Earth abounds with life. About 1.5 million species have been described, and millions more await discovery. (Most estimates for the total number of species on Earth range from 3 million to 30 million, as we saw in Interlude A). Though these numbers are large, the species alive today are thought to represent far less than 1 percent of all the species that have ever lived.

In previous chapters of this unit we discussed how the diversity of life arose, focusing on the mechanisms that drive the small-scale evolution of populations (microevolution) and lead to the formation of new species (speciation). In this chapter we broaden our scope to discuss large-scale evolutionary changes across the vast history of life on Earth. This we call macroevolution.

Macroevolution is large-scale evolution, including evolution above the species level. Examples of macroevolutionary change above the species level include the origin of new genera or higher taxonomic groups, like phyla. Macroevolution also includes increases or decreases in the number of members in such groups (for example, the number of species in a genus). In addition, macroevolutionary changes include evolutionary radiations that bring new groups to prominence and mass extinctions that greatly alter the diversity of life on Earth. Macroevolution emphasizes how life on Earth has changed over time. Such emphasis on the large scale leads us to think of evolution as a historical process—specifically, as the history of the formation and extinction of species and higher taxonomic groups over time.

We begin this chapter with a look at how the history of life on Earth is documented in the fossil record, followed by a summary of major events in the history of life. We then consider some of the forces that increase and decrease biodiversity over the long term—continental drift, mass extinctions, and adaptive radiations—and examine their effects with regard to the rise of our own group, the mammals.

19.1 The Fossil Record: A Guide to the Past

Fossils are the preserved remains or impressions of individual organisms that lived in the past (Figure 19.1). In many fossils, portions of the bodies of dead organisms are replaced with minerals. Petrified wood is a good example. In such fossils, the original body form is preserved but the fossil contains material not found in the living organism. Fossils are often found in sedimentary rock (rock that consists of layers of

sediments that have hardened), but they can also form in a few other situations. Insects, for example, have been found in amber, the fossilized resin or pitch (like pine tree pitch) of a tree (see Figure 19.1e), and many mammals, including mammoths and a 5,000-year-old man (see the photograph on page 25), have been found in melting glaciers.

As we saw in Chapter 16, the fossil record documents the history of life and is central to the study of evolution. Fossils provided the first compelling evidence that past organisms were unlike organisms alive today, that many forms have disappeared from Earth completely, and that life has evolved through time.

The relative depth or distance from the surface of Earth at which a fossil is found is referred to as its order in the fossil record. The ages of fossils correspond to their order: usually, older fossils are found in deeper, older rock layers. The order in which organisms appear in the fossil record agrees with our understanding of evolution based on other evidence, providing strong support for evolution. For example, analyses of the morphology, DNA sequences, and other characteristics of living organisms indicate that bony fishes gave rise to amphibians, which later gave rise to reptiles, which still later gave rise to mammals. That is exactly the order in which fossils from these groups appear in the fossil record. The fossil record also provides excellent examples of the evolution of major new groups of organisms, such as the evolution of mammals from reptiles (look ahead to Figure 19.13).

Although knowing the order of various organisms and groups of organisms in the fossil record is very helpful, it can provide only *relative* ages of fossils. That is, it can reveal which fossils are older than others. In some cases, we can approximate a fossil's age better by using **radioisotopes**, which are unstable, radioactive forms of elements that decay to more stable forms at a constant rate over time. For example, for a given amount of the radioisotope carbon-14 (^{14}C), half of the total decays to the stable element carbon-12 every 5,730 years. By measuring the amount of ^{14}C that remains in a fossil, scientists can estimate the age of the fossil. Carbon-14 can be used to date only relatively recent fossils (too little ^{14}C remains to date fossils formed more than 70,000 years ago), but elements such as uranium-235, which has a half-life of 700 million years, can be used to date much older materials. If, as commonly occurs, a fossil does not contain any radioisotopes, methods like carbon or uranium dating allow us to determine an approximate date for the fossil by dating rocks found above and below the fossil.

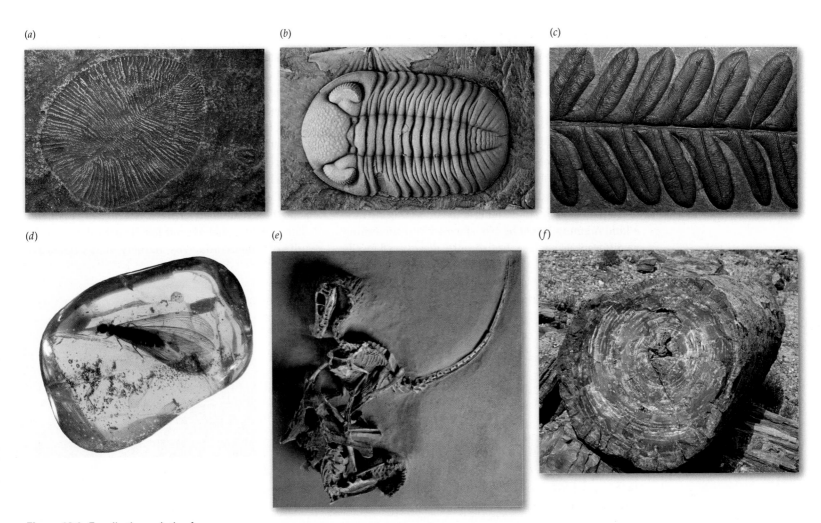

(a) (b) (c)

(d) (e) (f)

Figure 19.1 Fossils through the Ages

(a) Soft-bodied animals, such as the ones that left these fossils, dominated life on Earth 600 million years ago (mya). (b) A fossil of a trilobite [TRYE-loh-byte] that lived in the Devonian period (410–355 mya). Note the rows of lenses on each eye. (c) This leaf of a 300-million-year-old seed fern was found near Washington, DC. The fossil formed during the Carboniferous [kar-buh-NIFF-er-uss] period (355–290 mya). The great forests of this period led to the formation of the fossil fuels (oil, coal, and natural gas) that we use today as sources of energy. (d) A fossil of a *Velociraptor* entangled with a *Protoceratops*, which bit down on the predator's claw, locking both in a death grip. (e) This 20-million-year-old termite is preserved in amber, the fossilized resin of a tree. (f) Here we see how what was once solid wood has fossilized into solid rock.

The fossil record is not complete

The fossil record shows clearly that there have been great changes in the groups of organisms that have dominated life on Earth over time. As we will see throughout this chapter, these changes have been caused by the extinction of some groups and the expansion of other groups. Although many fossils have been found, however, the fossil record still has many gaps. Most organisms decompose rapidly after death; hence, very few form fossils. Even if an organism is preserved initially as a fossil, a variety of common geologic processes (such as erosion and extreme heat or pressure)

can destroy the rock in which it is embedded. And fossils can be difficult to find. Given the unusual circumstances that must occur for a fossil to form, remain intact, and be discovered by scientists, a species could evolve, thrive for millions of years, and become extinct without our ever finding evidence of its existence in the fossil record.

Still, although gaps in the fossil record remain, each year new discoveries fill in some of those gaps. One evolutionary event that has long been of interest is the evolution of seagoing whales from land-dwelling mammals. This event is one gap in the fossil record that has recently begun to be filled in by newly discovered fossils.

Fossils reveal that whales are closely related to a group of hoofed mammals

Let's take a closer look at what recent fossil discoveries revealed about one group of mammals, the whales. The origin of whales has long puzzled biologists. Most mammals live on land, and it is hard to imagine how a land mammal could be transformed into something as different as a whale. But recently discovered fossils provide a glimpse of how that transformation occurred (Figure 19.2a).

The bone structure of an early whale ancestor, *Pakicetus* [PACK-uh-SEET-uss], suggests that it probably spent most of its time on land. However, *Pakicetus* shared features (such as unusual bones in its inner ear) with modern whales and with the more whalelike creatures shown in Figure 19.2a. Over many generations, the ancestors of whales became increasingly similar to modern whales: their legs became smaller, and their overall shape took on the streamlined form of a fully aquatic mammal.

The recently discovered fossils also confirm the results of genetic analyses, namely, that whales are

(a)

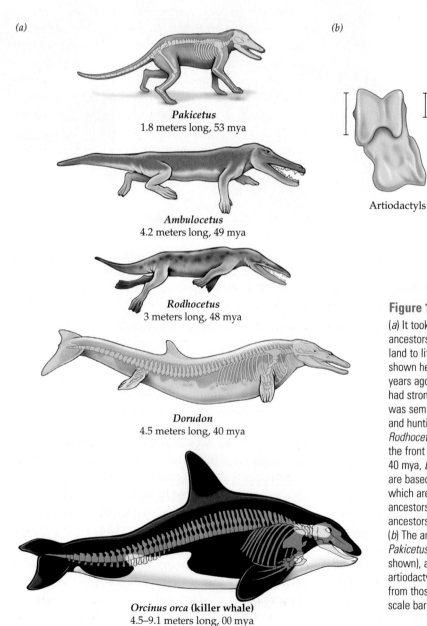

Pakicetus
1.8 meters long, 53 mya

Ambulocetus
4.2 meters long, 49 mya

Rodhocetus
3 meters long, 48 mya

Dorudon
4.5 meters long, 40 mya

Orcinus orca (killer whale)
4.5–9.1 meters long, 00 mya

(b)

This anklebone, from an extinct, hyena-like animal, is typical of non-artiodactyls.

Artiodactyls   Primitive whales

Figure 19.2 Shape-Shifters
(*a*) It took roughly 15 million years for whale ancestors to make the transition from life on land to life in water. The oldest whale ancestor shown here, *Pakicetus*, lived on land 53 million years ago. *Ambulocetus* [am-byoo-lo-SEET-uss] had strong, well-developed legs and probably was semiaquatic, living at the water's edge and hunting much as a crocodile does today. In *Rodhocetus*, the body is more streamlined, and the front legs are shaped more like flippers. By 40 mya, *Dorudon* was fully aquatic. The drawings are based on reconstructed fossil skeletons, which are superimposed on two of the whale ancestors. Compare this sequence of whale ancestors with *Orcinus*, a modern toothed whale. (*b*) The anklebones of two whale ancestors, *Pakicetus* (shown here) and *Rodhocetus* (not shown), are similar in shape to those of artiodactyls (hoofed mammals), but very different from those found in most other mammals. Each scale bar represents 1 cm.

most closely related to the artiodactyls [are-tee-oh-DACK-tulz]—even-toed, hoofed mammals such as camels, cows, pigs, deer, and hippopotamuses. In all artiodactyls, the anklebone has an unusual shape, in which both the top and bottom surfaces of the bone resemble those of a pulley (Figure 19.2*b*). In 2001, the anklebones of several whale ancestors, including *Pakicetus* and *Rodhocetus* [ROH-doh-SEET-uss], were discovered. These early whales had anklebones with the same unusual shape that the artiodactyl anklebones had. Since this shape is an adaptation for running on land, it is highly unlikely that whale ancestors developed such bones as a result of convergent evolution. Instead, these new fossils strongly suggest that whale ancestors had such bones because they shared a (recent) common ancestor with the artiodactyls.

19.2 The History of Life on Earth

In this section we focus on three of the main events in the history of life on Earth: the origin of cellular organisms, the beginning of multicellular life, and the colonization of the land. Figure 19.3 provides a sweeping overview of this history.

The first single-celled organisms arose at least 3.5 billion years ago

Our solar system and Earth formed 4.6 billion years ago. The oldest known rocks on Earth (3.8 billion years old) contain carbon deposits that hint at life. The first solid evidence for life, however, comes from 3.5-billion-year-old fossilized mats called stromatolites [stro-MATT-uh-lytes], which resemble similar mats formed by present-day bacteria. Although the earliest forms of life may have arisen between 4 billion and 3.5 billion years ago, fossils of that age have yet to be found. (See Chapter 3 for more discussion of the origin of life.)

Eukaryotes first appear in the fossil record at about 2.1 billion years ago. After the origin of prokaryotes 3.5 billion years ago, it took well over a billion years for the first eukaryotes to evolve. During this long period, the evolution of eukaryotes may have been limited in part by low levels of oxygen in the atmosphere. Chemical analyses of very old rocks indicate that Earth's atmosphere initially contained almost no oxygen. Roughly 2.75 billion years ago, however, some groups of bacteria evolved the ability to conduct photosynthesis, which releases oxygen

as a waste product. As a result, the oxygen concentration in the atmosphere increased over time (Figure 19.4).

Eukaryotic cells are larger than most prokaryotic cells. Oxygen and other materials spread more slowly through a large cell than through a small cell. Overall, because of their relatively large size, eukaryotic cells would not have been able to get enough oxygen to meet their needs until the atmospheric concentration of oxygen reached at least 2 percent to 3 percent of present-day levels. Once those levels were reached, about 2.1 billion years ago, the first single-celled eukaryotes—organisms that resembled some modern algae—evolved (see Figure 19.4). When oxygen levels reached their current levels, the evolution of larger and more complex multicellular organisms also became possible, as we shall see.

Oxygen was toxic to many early forms of life, and as the oxygen concentration in the atmosphere increased, many early prokaryotes went extinct or became restricted to environments that lack oxygen. Because the biologically driven increase in the oxygen concentration of the atmosphere drove many early organisms extinct while simultaneously setting the stage for the origin of multicellular eukaryotes, this increase in oxygen was one of the most important events in the history of life on Earth.

Multicellular life evolved about 650 million years ago

All early forms of life evolved in water. During the Precambrian period, about 650 million years ago (mya), the number of organisms appearing in the fossil record increased. At that time, much of Earth was covered by shallow seas, which were filled with plankton (protists, small multicellular animals, and single-celled and multicellular algae that float freely in the water). Later in the Precambrian, by about 600 mya, larger, soft-bodied multicellular animals had evolved (see Figure 19.1*a*). These animals were flat and appear to have crawled or stood upright on the seafloor, probably feeding on living plankton or their remains. No evidence indicates that any of these animals preyed on the others. Many of these early multicellular animals may have belonged to groups of organisms that are no longer found on Earth.

Later, during the Cambrian period, there was an astonishing burst of evolutionary activity. During the early to middle Cambrian period (530 mya), there was a dramatic increase in the diversity of life, known as the **Cambrian explosion**, in which large forms of most of the major living animal phyla, as well as other phyla that have since become extinct, appear in the fossil record.

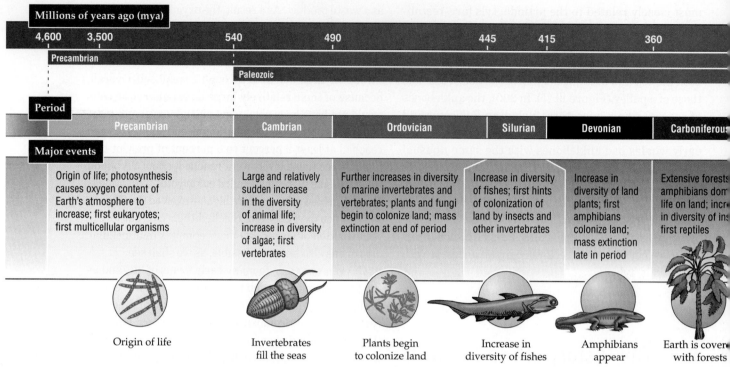

Millions of years ago (mya)							
4,600 3,500		540	490	445	415		360

Precambrian

Paleozoic

Period

Precambrian	Cambrian	Ordovician	Silurian	Devonian	Carboniferous

Major events

Origin of life; photosynthesis causes oxygen content of Earth's atmosphere to increase; first eukaryotes; first multicellular organisms	Large and relatively sudden increase in the diversity of animal life; increase in diversity of algae; first vertebrates	Further increases in diversity of marine invertebrates and vertebrates; plants and fungi begin to colonize land; mass extinction at end of period	Increase in diversity of fishes; first hints of colonization of land by insects and other invertebrates	Increase in diversity of land plants; first amphibians colonize land; mass extinction late in period	Extensive forests amphibians dom life on land; incr in diversity of ins first reptiles

Origin of life	Invertebrates fill the seas	Plants begin to colonize land	Increase in diversity of fishes	Amphibians appear	Earth is cover with forests

Figure 19.3 The History of Life on Earth: The Geologic Timescale

The history of life can be divided into 12 major geologic time periods, beginning with the Precambrian (4,600–540 mya) and extending to the Quaternary (1.8 mya to the present). This timescale is not drawn to scale; to do so and to include the Precambrian would require extending the diagram off the book page to the left by more than 5 *feet*.

Test your knowledge of the major geologic time periods. ▶ǁ 19.1

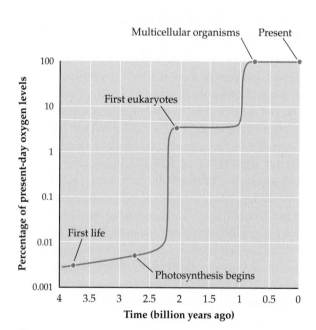

Figure 19.4 Oxygen on the Rise

The release of oxygen as a waste product by photosynthetic organisms caused its concentration in Earth's atmosphere to increase greatly over the last 3 billion to 4 billion years, facilitating the evolution of eukaryotes and multicellular organisms. Shown here are the hypothesized levels of oxygen at key points in the history of life, but still much remains unknown about the dynamics of the rise of oxygen over time.

The word "explosion" here refers to a rapid increase in the number and diversity of species, not to a physical explosion. And that is because, in geologic terms, the Cambrian explosion was extremely rapid, lasting only 5 million to 10 million years. It was a blink of an eye in geologic terms, when compared, for example, with the 1.4 billion (1,400 million) years it took for eukaryotes to evolve from prokaryotes.

The Cambrian explosion was one of the most spectacular events in the evolutionary history of life. It changed the face of life on Earth: from a world of relatively simple, slow-moving, soft-bodied scavengers and herbivores, suddenly there emerged a world filled with large, mobile predators in pursuit of herbivores with hard body coverings that defended them against those predators (Figure 19.5).

Colonization of land followed the Cambrian explosion

Because life first evolved in water, the colonization of land by living organisms posed enormous challenges. Indeed, many of the functions basic to life, including support, movement, reproduction, and the regulation of ions, water, and heat, must be handled very differently

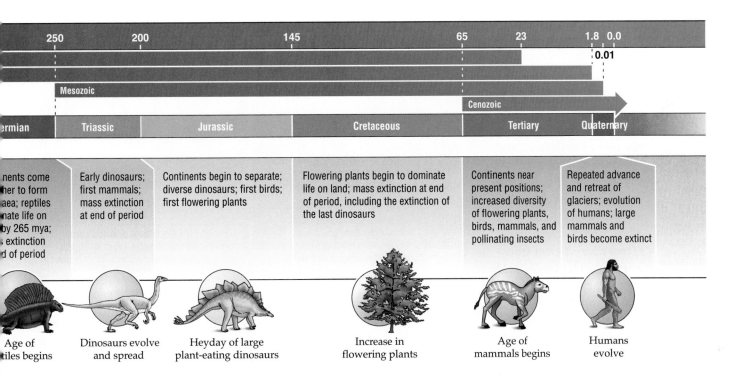

| | | 250 | | 200 | | 145 | | 65 | 23 | | 1.8 | 0.0 |

0.01

Mesozoic

Cenozoic

| ...rmian | Triassic | Jurassic | Cretaceous | Tertiary | Quaternary |

| ...nents come ...her to form ...aea; reptiles ...nate life on ...by 265 mya; ... extinction ...d of period | Early dinosaurs; first mammals; mass extinction at end of period | Continents begin to separate; diverse dinosaurs; first birds; first flowering plants | Flowering plants begin to dominate life on land; mass extinction at end of period, including the extinction of the last dinosaurs | Continents near present positions; increased diversity of flowering plants, birds, mammals, and pollinating insects | Repeated advance and retreat of glaciers; evolution of humans; large mammals and birds become extinct |

| Age of ...tiles begins | Dinosaurs evolve and spread | Heyday of large plant-eating dinosaurs | Increase in flowering plants | Age of mammals begins | Humans evolve |

on land than in water. Descendants of green algae were the first organisms to meet these challenges about 500 mya. These early terrestrial colonists had few cells and a simple body plan, but from them, land plants evolved and diversified greatly. By the end of the Devonian [deh-*voh*-nee-un] period (360 mya), Earth was covered with

plants. Like plants today, the plants of the Devonian included low-lying spreading species, short upright species, shrubs, and trees.

As new groups of land plants arose, they evolved a series of key innovations, including a waterproof cuticle, vascular systems, structural support tissues

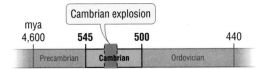

Cambrian explosion

mya
4,600   545   500   440

| Precambrian | Cambrian | Ordovician |

Figure 19.5 Before and After the Cambrian Explosion
The Cambrian explosion greatly altered the history of life on Earth.

Before the Cambrian explosion, Earth was dominated by soft-bodied scavengers and herbivores.

After the Cambrian explosion, life was dominated by more complex animals, including predators and well-defended herbivores.

Figure 19.6 The First Amphibians

(a) Amphibians probably evolved over long periods of evolutionary time from a lobe-finned fish ancestor like the one shown here. (b) This early amphibian was reconstructed from a 365-million-year-old (late Devonian) fossil.

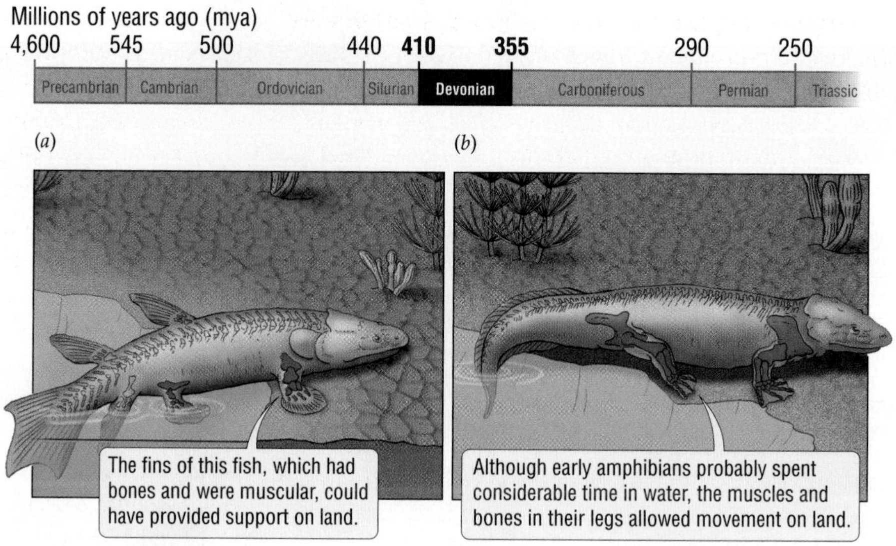

Learn more about the major events of each geologic time period. ►II 19.2, 19.3

Millions of years ago (mya)

| 4,600 | 545 | 500 | 440 | **410** | **355** | 290 | 250 |

| Precambrian | Cambrian | Ordovician | Silurian | Devonian | Carboniferous | Permian | Triassic |

(a)

(b)

The fins of this fish, which had bones and were muscular, could have provided support on land.

Although early amphibians probably spent considerable time in water, the muscles and bones in their legs allowed movement on land.

(wood), leaves and roots of various kinds, seeds, the tree growth form, and specialized reproductive structures. These and other important changes allowed plants to cope with life on land. Waterproofing, stems with efficient transport mechanisms, and roots, for example, were important features that helped plants acquire and conserve water while living on land.

Fungi are thought to have made their way onto land soon afterward, according to new studies. Scientists, for example, have found fossils of terrestrial fungi—both fossil hyphae and spores—that are 455 to 460 million years old.

The first definite fossils of terrestrial animals are of spiders and millipedes that date from about 410 mya, although there are hints of land animals as early as 490 mya. Many of the early animal colonists on land were predators; others, such as millipedes, fed on living plants or decaying plant material. Insects, which are currently the most diverse group of terrestrial animals, first appeared roughly 400 mya, and they played a major role on land by 350 mya.

The first vertebrates to colonize land were amphibians, the earliest fossils of which date to about 365 mya. Early amphibians resembled, and probably descended from, lobe-finned fishes (Figure 19.6). Amphibians were the most abundant large organisms on land for about 100 million years. In the late Permian period, the reptiles, which had evolved from a group of reptile-like amphibians, rose to become the most common verte-

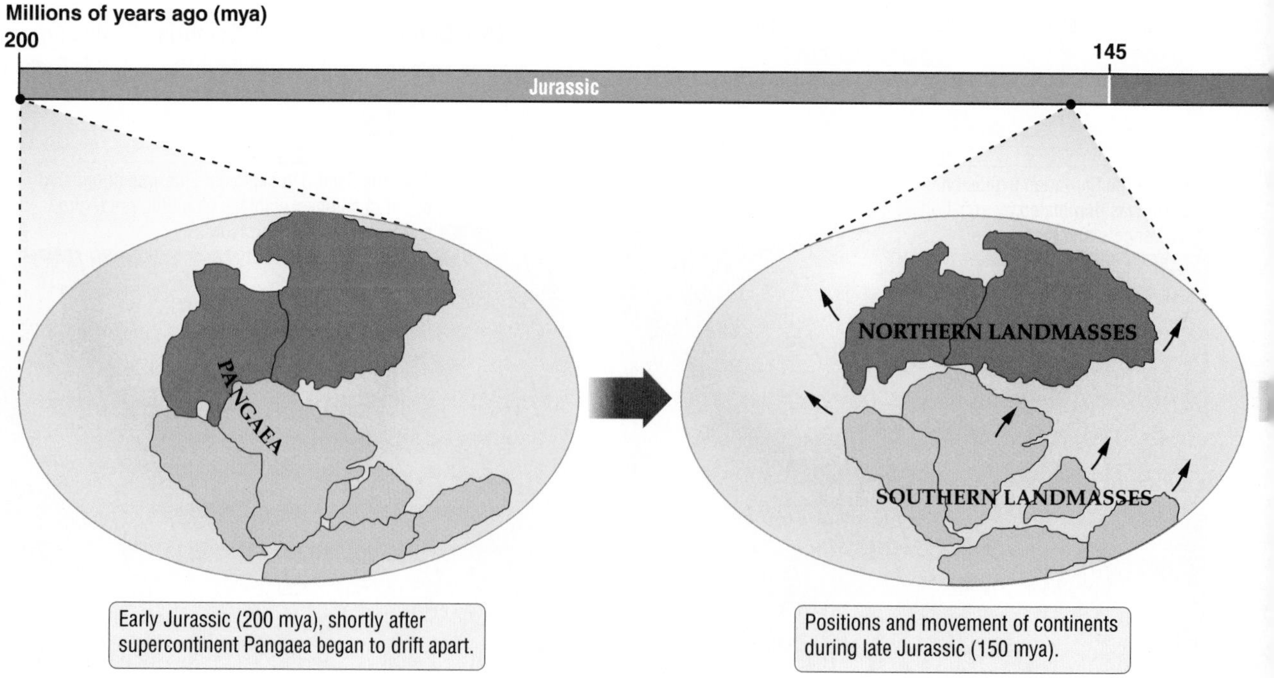

Millipede

Millions of years ago (mya)

| 200 | 145 |

Jurassic

Figure 19.7 Movement of the Continents over Time

The continents move over time, as shown by these "snapshots" illustrating the breakup of the supercontinent Pangaea. Earlier movements of the continents led to the gradual formation of Pangaea, a process that was complete by 250 mya.

Early Jurassic (200 mya), shortly after supercontinent Pangaea began to drift apart.

Positions and movement of continents during late Jurassic (150 mya).

brate group. Reptiles were the first group of vertebrates that could reproduce without returning to open water (for example, to lay eggs). As a result, reptiles were the first vertebrates that could fully exploit the available opportunities for terrestrial life. Reptiles, including the dinosaurs, dominated vertebrate life on land for 200 million years (265–65 mya), and they remain important today. Dinosaurs, in particular, arose about 230 mya and dominated the planet from about 200 mya to about 65 mya. Mammals, the vertebrate group that currently dominates life on land, evolved from reptiles roughly 220 mya (see Figure 19.3).

Concept Check

1. How does macroevolution differ from microevolution?
2. Why was it significant that early bacteria evolved the ability to carry out photosynthesis?
3. How did the Cambrian explosion change life on Earth?

19.3 The Effects of Continental Drift

The enormous size of the continents may cause us to think of them as immovable. But this notion is not correct. The continents move slowly relative to one another, and over hundreds of millions of years they travel considerable distances (Figure 19.7). This movement of the continents over time is called **continental drift**. The continents can be thought of as plates of solid matter that "float" on the surface of Earth's mantle, a hot layer of semisolid rock.

How can something as big as a continent move from place to place? Two forces cause the continental plates to move. First, hot plumes of liquid rock rise from Earth's mantle to the surface and push the continents away from one another. This process can cause the seafloor to spread, as it is doing between North America and Europe, which are separating at a rate of 2.5 centimeters per year. This process can also cause bodies of land to break apart, as is currently happening in Iceland and East Africa. Second, where two plates collide, one plate can sometimes slip underneath the other and begin to sink into the mantle below. As the now hidden end of the continental plate sinks down and slowly melts, it gradually pulls the rest of the plate down along with it, causing the rest of the plate to continue to move or drift.

Patterns of continental drift—most notably the breakup of the ancient supercontinent Pangaea—have had dramatic effects on the history of life. Pangaea began to break apart early in the Jurassic period (about 200 mya), ultimately separating into the continents we know today (see Figure 19.7). As the continents drifted apart, populations of organisms that

Concept Check Answers

1. Macroevolution is evolution above the species level, whereas microevolution happens within populations.
2. Because oxygen is a by-product of photosynthesis and increasing oxygen in the atmosphere set the stage for the evolution of eukaryotes, a group that includes all multicellular life.
3. Most of the major living animal phyla appeared, many of which were the world's first dangerous predators.

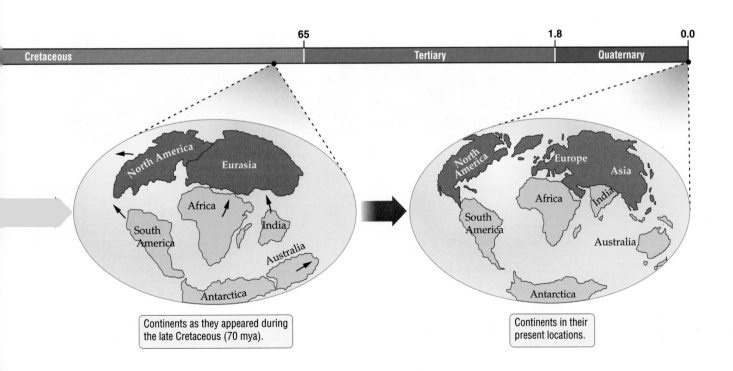

65   1.8   0.0

Cretaceous   Tertiary   Quaternary

Continents as they appeared during the late Cretaceous (70 mya).

Continents in their present locations.

once were connected by land became isolated from one another.

As we learned in Chapter 18, geographic isolation reduces or eliminates gene flow, thereby promoting speciation. The separation of the continents was geographic isolation on a grand scale, and it led to the formation of many new species. Among mammals, for example, kangaroos, koalas, and other marsupials unique to Australia evolved in geographic isolation on that continent, which broke apart from Antarctica and South America about 40 mya.

Continental drift also affects climate, which has a profound effect on the evolution of life by altering what kinds of organisms natural selection will favor. For example, consider animals adapted to life in a warm tropical climate. If continental drift moves the land those animals live on to a much colder climate, it will drastically alter which animals survive and thrive—that is, which are more likely to have an advantage in survival and reproduction. In addition, shifts in the positions of the continents alter ocean currents, and these currents have a major influence on the global climate. At various times, changes in the global climate caused by the movements of the continents have led to the extinctions of many species.

See continental drift in action. ▶❚❚ 19.4

Explore the causes of continental drift. ▶❚❚ 19.5

19.4 Mass Extinctions: Worldwide Losses of Species

As the fossil record shows, species have gone extinct throughout the long history of life. The rate at which this has happened—that is, the number of species that have gone extinct during a given period—has varied over time, from low to very high. At the upper end of this scale, the fossil record shows that there have been five **mass extinctions**, periods of time during which great numbers of species went extinct throughout most of Earth. Each of these upheavals left a permanent mark on the history of life, driving more than 50 percent of Earth's species to extinction. Figure 19.8 shows the effects of the five mass extinctions on animal life alone. Though difficult to determine, the causes of the five mass extinctions are thought to include such factors as climate change, massive volcanic eruptions, asteroid impacts, changes in the composition of marine and atmospheric gases, and changes in sea levels. In addition to the five mass extinctions revealed by the fossil record, we may be entering a sixth, human-caused mass extinction today (see the box on the next page and Interlude A).

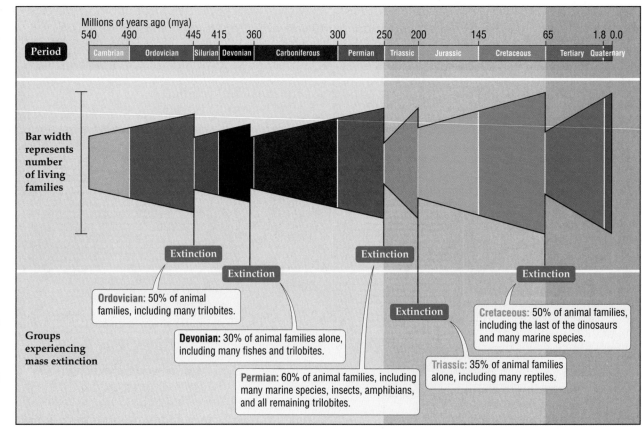

Figure 19.8 The Five Mass Extinctions

Each of these mass extinctions drastically reduced the diversity of marine and terrestrial animals, as shown here. Plant groups (not shown) were also severely affected. After each extinction, life again diversified.

Biology Matters

Is a Mass Extinction Under Way?

The International Union for the Conservation of Nature (IUCN), known also as the World Conservation Union, maintains what it calls its Red List, which identifies the world's threatened species. To be defined as such, a species must face a high to extremely high risk of extinction in the wild. The 2004 Red List contains 15,589 species threatened with extinction, of a total of approximately 57,000 species assessed. Because this assessment accounts for less than 3 percent of the world's 1.9 million described species, the total number of species threatened with extinction worldwide would actually be much larger.

Among the major species groups that were assessed for the 2004 Red List, the number of threatened species ranges between 12 percent and 52 percent, with, for example, 12 percent of birds, 23 percent of mammals, 32 percent of amphibians, and 52 percent of the plants known as cycads listed as threatened.

The Red List is based on an easy-to-understand system for categorizing extinction risk; it is also objective, yielding consistent results when used by different people. Because of these two attri-butes, the Red List is internationally recognized as an effective method to assess extinction risk. Table 1 shows some worldwide data for a few species groups that have been heavily assessed. In all four cases, the number of threatened species has risen since 1996–1998, and in the case of amphibians and gymnosperms the increase has been dramatic.

For types of organisms other than those listed in Table 1, extinction risk has been evaluated for only a few known species—for example, only 771 out of 950,000 described insect species. If the threatened species listed in the table do go extinct, and the percentage of species under threat in other taxonomic groups turns out to be similar to those listed, then the percentages of species that will go extinct will approach those in some of the previous mass extinctions.

Table 2 contains data on the numbers of species that the 2004 Red List identified as threatened in North America.

For more information about the 2004 Red List, visit www.iucnredlist.org.

Table 1

Group	Number of known species	Number of species evaluated for extinction risk	Number of threatened species in 1996–1998	Number of threatened species in 2006	Threatened species in 2006 (as a percentage of species evaluated)
VERTEBRATES					
Mammals	5,416	4,856	1,096	1,093	23%
Birds	9,934	9,934	1,107	1,206	12%
Amphibians	5,918	5,918	124	1,811	31%
PLANTS					
Gymnosperms	980	908	142	36	34%

Table 2

North America	Mammals	Birds	Reptiles	Amphibians	Fishes	Mollusks	Other	Plants	Total
Canada	16	19	2	1	24	1	10	1	**74**
United States	40	71	27	50	154	261	300	240	**1,143**

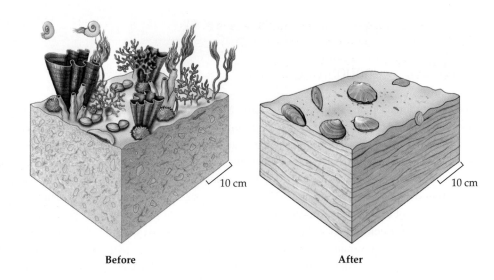

Figure 19.9 Before and After the Permian Mass Extinction

The Permian mass extinction wiped out nearly all of the rich variety of burrowing, bottom-dwelling, and swimming organisms that lived in the oceans before it occurred.

Before   After

10 cm   10 cm

Of the five previous mass extinctions, the largest occurred at the end of the Permian period, 250 mya. The **Permian extinction** radically altered life in the oceans (Figure 19.9). Among marine invertebrates (animals without backbones), an estimated 50 percent to 63 percent of existing families, 82 percent of genera, and 95 percent of species went extinct. The Permian mass extinction was also highly destructive on land. It removed 62 percent of the existing terrestrial families, brought the reign of the amphibians to a close, and caused the only major extinction of insects in their 400-million-year history (8 of 27 orders of insects went extinct).

What does it really mean for an entire family of animals to go extinct? As we learned in Chapter 2, all life is organized into kingdoms (Animalia, Plantae, and so on), within which are phyla, within which are classes, within which are orders, then families, genera, and finally species. So how significant a part of the living world is a single family? Today, extinction of an entire family such

as the Felidae would mean extinction of all the wild and domesticated cats, from lions to tigers to leopards to cougars to regular old kitty cats. Other familiar families include the Canidae (all wild and domesticated dogs) and the Ursidae (bears). And those are just single families. As described already, in some groups, such as the marine invertebrates, at least half the families disappeared altogether. Orders of insects were lost as well. Today, losing an entire order could mean losing the Lepidoptera (all butterflies and moths) or the Hymenoptera (including all bees, ants, and wasps).

Although not as severe as the Permian extinction, each of the other mass extinctions also had a profound effect on the diversity of life (see Figure 19.8). The best-studied mass extinction is the **Cretaceous extinction**, which occurred at the end of the Cretaceous period, 65 mya. At that time, half of the marine invertebrate species perished, as did many families of terrestrial plants and animals, including the dinosaurs (Figure 19.10). The Cretaceous mass extinction was probably caused at least in part by

(a)

(b)

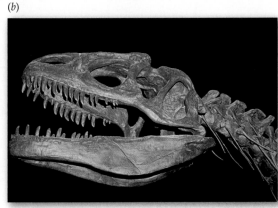

Figure 19.10 Gone for Good

(*a*) A dog sits next to a reconstruction of the head of a *Mapusaurus* dinosaur—what may have been the largest carnivore to have walked the Earth. (*b*) This *Allosaurus* skeleton reveals how sharp, pointed, large, and numerous a dinosaur's teeth could be.

the collision of an asteroid with Earth. A 65-million-year-old, 180-kilometer-wide crater lies buried in sediments off the Yucatán coast of Mexico; this crater is thought to have formed when an asteroid 10 kilometers wide struck Earth. An asteroid of this size would have caused great clouds of dust to hurtle into the atmosphere; this dust would have blocked sunlight around the globe for months to years, causing temperatures to drop drastically and driving many species extinct.

The effects of mass extinctions on the diversity of life are twofold. First, as noted earlier, entire groups of organisms perish, changing the history of life forever. Second, the extinction of one or more dominant groups of organisms can provide new ecological and evolutionary opportunities for groups of organisms that previously were of relatively minor importance, dramatically altering the course of evolution.

19.5 Adaptive Radiations: Increases in the Diversity of Life

After each of the five mass extinctions, some of the surviving groups of organisms diversified to replace those that had become extinct. These bursts of evolution were just as important to the future course of evolution as were the extinctions themselves. For example, when the dinosaurs went extinct 65 mya, the mammals diversified greatly in size and in ecological role (Figure 19.11). If mammals had not diversified to replace the dinosaurs, humans probably would not exist and the history of life over the past 65 million years would have been very different from what actually happened.

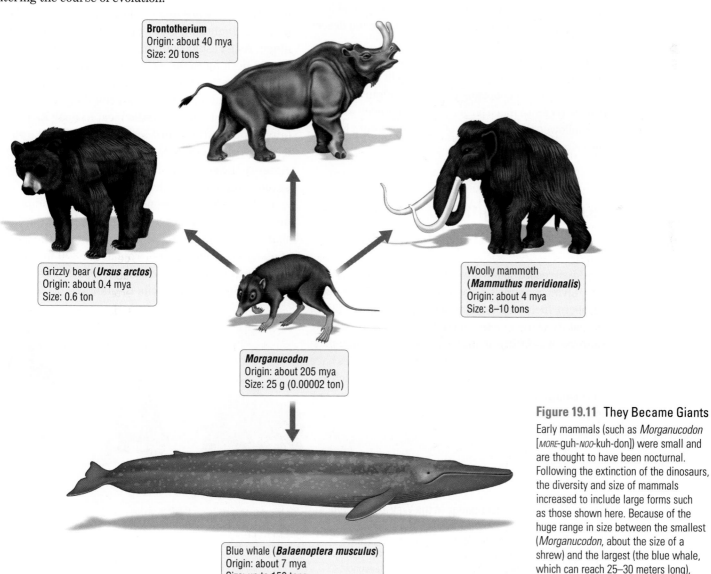

Brontotherium
Origin: about 40 mya
Size: 20 tons

Grizzly bear (**Ursus arctos**)
Origin: about 0.4 mya
Size: 0.6 ton

Woolly mammoth
(**Mammuthus meridionalis**)
Origin: about 4 mya
Size: 8–10 tons

Morganucodon
Origin: about 205 mya
Size: 25 g (0.00002 ton)

Blue whale (**Balaenoptera musculus**)
Origin: about 7 mya
Size: up to 150 tons

Figure 19.11 They Became Giants
Early mammals (such as *Morganucodon* [MORE-guh-NOO-kuh-don]) were small and are thought to have been nocturnal. Following the extinction of the dinosaurs, the diversity and size of mammals increased to include large forms such as those shown here. Because of the huge range in size between the smallest (*Morganucodon*, about the size of a shrew) and the largest (the blue whale, which can reach 25–30 meters long), none of these animals are drawn to scale.

When a group of organisms expands to take on new ecological roles and to form new species and higher taxonomic groups, that group is said to have undergone an **adaptive radiation**. Some of the great adaptive radiations in the history of life occurred after mass extinctions, as when the mammals diversified to replace the dinosaurs. In such cases, the adaptive radiations may have been caused by the release from competition that occurs after a dominant group of organisms (such as the dinosaurs) goes extinct.

In other cases, adaptive radiations have occurred after a group of organisms has acquired a new adaptation that allows it to use its environment in new ways. The first terrestrial plants, for example, possessed adaptations that allowed them to thrive on land, a new and highly challenging environment. The descendants of those early colonists radiated greatly, forming many new species and higher taxonomic groups that were able to live in a broad range of new environments (from desert to arctic to tropical regions).

The term "adaptive radiation" can refer to relatively small evolutionary expansions, as seen in the radiation of finches in the Galápagos Islands (see Figure 16.12). It can also refer to much larger expansions, such as those that followed the movement of vertebrates onto land or the origin of flowering plants.

19.6 The Origin and Adaptive Radiation of Mammals

The fossil record shows that mammals evolved from reptiles. Living mammals differ from living reptiles in many respects, including the way they move (Figure 19.12), the nature of their teeth, and the structure of their jaws. However, it is difficult to draw the line between mammals and reptiles in the fossil record. Some fossil species have features intermediate between the two groups; such fossils beautifully illustrate an evolutionary shift from one major group of organisms to another.

The mammalian jaw and teeth evolved from reptilian forms in three stages

Let's examine the evolution of mammals from reptiles from the perspective of two traits that are easily observed in fossils: the nature of their teeth and the structure of their jaws.

Compared with reptiles, mammals have complex teeth and jaws. For example, mammalian teeth differ considerably from one portion of the jaw to another: some teeth are specialized for tearing (incisors), others for hunting or defense (canines), still others for grinding (molars). In contrast, reptilian teeth change little in form or function from one position along the jaw to another (see Figure 19.13*a*).

The reptilian jaw has a hinge at the back for the attachment of muscles that simply snap the top and bottom of the jaw together. In mammals, the hinge has moved forward and is controlled by strong cheek muscles, including some positioned in front of the hinge. As a result, mammals actually have jaws that can be both more powerful and more precisely controlled than the simple snap-shut jaws of reptiles. To see why, think of how well we could close a door by pulling it shut with a handle located near its hinges (analogous to the reptilian method) versus pulling it shut with a handle located away from the hinges, closer to the actual position of a doorknob (analogous to the mammalian method).

How did these differences in the teeth and jaws of mammals and reptiles arise? The fossil record shows that these changes arose gradually, over the course of about 80 million years (from about 300 mya to 220 mya). During this time, there were three key steps in the transition from reptile to mammal, as shown in

The limbs of reptiles extend from the sides of their bodies.

Reptiles on the evolutionary line leading to mammals, such as this cynodont, had limbs positioned partially underneath their bodies.

The limbs of mammals are even more vertically oriented.

Figure 19.12 A Gradual Change in the Way They Move
The legs of most living reptiles stick out to the sides of their bodies, giving them a sprawling gait. Over time, the legs of mammal-like reptiles became positioned under the body, leading eventually to the vertical orientation of the legs and the upright gait of living mammals.

(a) Ancestral reptile (*Haptodus*)

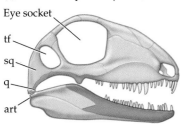

Eye socket
tf
sq
q
art

These reptiles had a temporal fenestra (tf), large jaw muscles, multiple bones in the lower jaw, and single-point teeth. They also had a single hinge (art/q) at the back of the jaw.

(b) Therapsid (*Biarmosuchus*)

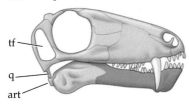

tf
q
art

Therapsids had a larger temporal fenestra (tf), large canine teeth, long faces, and a single hinge (art/q) at the back of the jaw.

(c) Early cynodont (*Procynosuchus*)

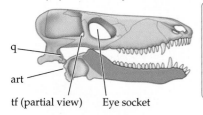

q
art
tf (partial view)   Eye socket

In early cynodonts, the temporal fenestra (here only partial view is shown) was further enlarged, allowing for very powerful jaw muscles.

(d) Cynodont (*Thrinaxodon*)

The dentary (the major jaw bone, colored in red) became enlarged, and the back teeth had multiple cusps.

(e) Advanced cynodont (*Probainognathus*)

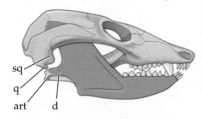

sq
q
art   d

In advanced cynodonts, complex teeth with multiple cusps enhanced chewing. The jaw had two hinges, art/q and d/sq.

(f) *Morganucodon* (early mammal)

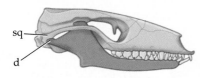

sq
d

Morganucodon had typical mammalian teeth. The jaw had two hinges, but the reptilian art/q hinge was reduced (not visible in this diagram).

(g) Tree shrew (*Tupaia*)

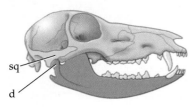

sq
d

In tree shrews and other mammals, the teeth are highly specialized. The lower jaw is composed of a single bone, and the jaw has one hinge (d/sq).

Figure 19.13 From Reptile to Mammal

Over an 80-million-year period, the jaws and teeth of the reptilian ancestors of mammals gradually changed to resemble those of living mammals. In addition to those shown here, there are dozens of other fossil species of mammal-like reptiles—species with features that are intermediate to those of reptiles and mammals. When fossils from all of these species are lined up next to one another, the transition from reptile to mammal appears very smooth. Here, red shows the size and position of the dentary bone, which ultimately formed the entire lower jaw in mammals. The muscles that close the jaw pass through an opening called the temporal fenestra (tf); a larger temporal fenestra allows for larger and more powerful jaw muscles. In reptiles (*a*), the hinge of the jaw is formed by the articular/quadrate (art/q) bones; in mammals (*g*), the hinge is formed by two entirely different bones, the dentary/squamosal (d/sq). Advanced cynodonts and early mammals had two hinges, the reptilian (art/q) hinge and the mammalian (d/sq) hinge.

Figure 19.13. First, a group of reptiles evolved to have an opening in the bones behind the eye called the temporal fenestra (Figure 19.13*a*). In the living animal, a muscle passed through this opening and increased the power with which the jaw could be closed. Second, a group of reptiles known as the therapsids [thuh-*RAP*-sidz] evolved a larger temporal fenestra (Figure 19.13*b*) (and hence more powerful jaw muscles), and their teeth showed the first signs of specialization.

The third step occurred when jaws very similar to mammalian jaws arose in one subgroup of the therapsids, the early **cynodonts** [*SYE*-noh-donts] (Figure 19.13*c*). In these animals—the last in a long line of mammal-like reptiles—the teeth became still more specialized, and the hinge of the jaws moved forward, using a different set of bones than in other reptiles (Figure 19.13*c–e*). Changes in the jaw hinge are particularly clear in some cynodont species (see Figure 19.13*e*) that had jaws with both a reptilian hinge (in reduced form) and a mammal-like hinge. Over time, mammals lost the reptilian hinge (Figure 19.13*g*), the bones of which evolved to become bones in the inner ear.

Mammals increased in size after the extinction of the dinosaurs

There were many species of mammal-like reptiles in the early Triassic period, 245 mya. By 200 mya, however, the mammal-like reptiles had declined as other reptiles, most notably the dinosaurs, came to dominate Earth. Although the mammal-like reptiles became extinct, they left behind the first mammals as their descendants. The earliest mammals, which were small,

Helpful to know

"Cynodont" comes from Greek roots meaning "dog-toothed." In cynodonts, the canine teeth (from *canis*, Latin for "dog") were prominent, though not as large as in some of their predecessors.

Biology on the Job

Geerat Vermeij: Sight Unseen

In addition to being a world-class biologist, Geerat J. Vermeij [GHEER-aht ver-MAY] is blind. Currently, he is Distinguished Professor of Geology at the University of California, Davis, where he studies the history of life on Earth. He has written over a hundred scientific papers and five books, including Evolution and Escalation: An Ecological History of Life *and an autobiography entitled* Privileged Hands: A Scientific Life. *Dr. Vermeij has been blind since he was 3 years old, when his eyes were removed to prevent a rare disease from causing neural damage.*

You have a legendary ability to learn about the ecology of a place and the history of life by touching, smelling, and listening to nature. Most scientists rely primarily on visual observations. To what extent does your use of senses other than sight allow you to discover new things about the natural world? While in the field, I use every sense available to me to observe and to keep track of my surroundings. My hands are essential in quickly identifying desirable specimens and potentially dangerous animals; I carefully listen to patterns of water movement for oncoming waves or other sudden changes. My hands, fingers, fingernails, and tactile extensions such as pins are essential for making detailed observations on shells or other natural objects. For the kinds of comparative research I do, it is essential to have a good working memory, so that I can compare one place with another I may have visited long before. Observation and comparison are at the heart of my science, done while I think about what I am observing in relation to previous experiences.

How did you become interested in studying the history of life? My interests in science and natural history date back to my earliest childhood. I began collecting shells seriously at age 10, at which point I knew in general what I wanted to do with my life.

Your research often takes you to remote locations in the field. What sorts of places have you visited? I have been fortunate to have visited every continent except Antarctica, and numerous islands ranging from Iceland and the Aleutians to many Caribbean islands, Hawaii, the Society Islands, Guam, and Palau. In all these places I have carried out extensive fieldwork. I have collected fossils in places ranging from Panama and Florida to the Aleutians.

What are some of the more unusual or hair-raising situations that you have experienced in the field? While aboard the R.V. *Alpha Helix* in Indonesia and the Philippines, we were stopped three times by military patrols, one of which came close to arresting my wife and me in the belief we were Taiwanese poachers. In the northern Mariana Islands, I jumped from a small boat onto a slippery lava shore in substantial surf and ended up scarring various parts of my body. I have also been stung by stingrays in Panama and bitten by moray eels at Mooréa in the Society Islands.

Is the history of life important for the general public? Why do you feel this way? The history of life should interest the public at large for several reasons. First, it provides the most direct evidence of evolution and reveals how organisms over the past 3.5 billion years have reacted to and created major environmental changes. This history provides the only long-term record of the consequences of extinction and diversification and enables us to study species invasion over the long run. Moreover, it places human actions and capacities in the larger perspective of the earlier history of life. For the imagination, the history of life transports us back in time to a very unfamiliar world, yet one in which all the principles and processes that govern life today were also in force.

The Job in Context

In over 30 years of studying fossil shells, Dr. Vermeij has used his hands to examine how the history of life in the oceans was influenced by ongoing battles between predator and prey. What he has found is that life has become ever tougher, more vicious, and more competitive over time. Dr. Vermeij's results show that over long periods of time, predators such as crabs and shell-drilling snails evolved increasingly sophisticated ways to crush, pry open, or drill through the shells of their prey. And the prey responded in kind, developing tougher, better-defended shells. Dr. Vermeij realized that this evolutionary "arms race" had proceeded in a single direction for hundreds of millions of years, shaping the history of life in the oceans. His hands revealed a fundamental story about life in the oceans, a story that no sighted biologist had discovered before him.

Shell Collection of Geerat Vermeij

rodent-sized organisms, evolved about 220 mya—in geologic terms, not long after the first dinosaurs, which arose 230 mya.

Throughout the long reign of the dinosaurs, most mammals remained small. Many appear to have been nocturnal (active at night) because they had large eye sockets, as do many living nocturnal organisms. By being nocturnal and small, early mammals may have been to dinosaurs what a mouse is to a lion: hard to notice and too small to eat.

Fossil and genetic evidence suggests that several of the orders of living mammals diverged from one another between 100 and 85 mya, well before the extinction of the dinosaurs. But most of the major radiations within these and other groups of mammals did not occur until after the dinosaurs went extinct (65 mya). After the dinosaurs were gone, the mammals radiated greatly to include many new forms that were large and active by day (see Figure 19.11). Some land mammals reached enormous sizes. An example is the extinct Beast of Baluchistan, which was over three times as large as an elephant. Other mammals (such as whales; see Figure 19.2) became specialized for life in water, while still others (the bats) became specialized for flight and hunting at night. One group of mammals, the primates, became specialized for life in trees and evolved especially large brains. We will examine the evolution of this group—the group to which we humans belong—in more detail in Interlude D.

19.7 An Overview of the Evolutionary History of Life

As we have seen, the history of life on Earth can be summarized as the rise and fall of major groups of protists, plants, and animals (see Figure 19.3). These broad patterns in the history of life are caused by the extinction or decline of some groups and the origin or expansion of other groups. Taken together, mass extinctions and adaptive radiations have been largely responsible for shaping macroevolutionary change. Continental drift also plays an important role, in that the movement of the continents can promote both mass extinctions and adaptive radiations.

In addition to offering this broad view of the evolutionary history of life, we want to emphasize two important and related concepts: how major new

groups of organisms evolve, and the difference between microevolution (small-scale evolutionary change within populations) and macroevolution (large-scale evolution—for example, the origin of higher taxonomic groups like genera or families or broad patterns of evolutionary change, like mass extinctions, over the history of life).

Major new groups of organisms evolve from existing groups

The Nobel Prize–winning geneticist François Jacob [YAH-kohb] described evolution as similar to "tinkering"—that is, as a process in which things are not built entirely from scratch, but by tinkering, as a tinkerer in a workshop does. Such a tinkerer uses existing forms—that is, forms of life that are already available—which are then modified and adjusted slightly. This view of evolution is especially relevant when we consider the origins of major new groups of organisms. New groups do not arise from scratch, but rather as modifications of organisms that already exist.

The origin of mammals from reptiles provides an excellent example of this process. Mammals have novel features, such as specialized teeth and powerful jaws, but the fossil record reveals that these features arose as gradual modifications of the teeth and jaws found in the reptiles from which they evolved (see Figure 19.13). These modifications occurred in the absence of a plan or design; instead, over time, bones and teeth that served one purpose in reptiles gradually changed and found new uses in mammals.

Macroevolution differs from microevolution

Microevolution is evolution on a small scale. One of the most powerful forces causing microevolutionary change is natural selection. As we saw in Chapter 18, natural selection can result in the often exquisite adaptations of organisms to their current environment. Can such natural selection provide a complete explanation of how macroevolution occurs; for example, can it explain the rise and fall of higher taxonomic groups? The answer is no.

In macroevolution, other larger forces can come into play. For example, asteroids can cause macroevolutionary change in the form of mass extinctions that remove entire groups of organisms. And such mass extinctions can remove groups without regard to the force of natural selection. In fact, these extinctions can wipe out

Oyster drill

groups seemingly at random—even those that have evolved unique and highly advantageous adaptations. A group of predatory gastropods (snails and their relatives), for example, went extinct in the Triassic mass extinction, shortly after they had evolved the ability to drill through the shells of other gastropods. The ability to drill through shells had opened up a major new way of life for these organisms (see the box on page 402). If these predatory gastropods had not gone extinct, they probably would have thrived and formed many new species that had the ability to drill through shells (as did another group of shell-drilling gastropods that evolved 120 million years later). As this example shows, even species that have highly beneficial adaptations, adaptations that evolved via natural selection, don't always survive over the long term.

Overall, broad patterns in the history of life cannot be predicted solely from an understanding of small-scale evolution, or how populations evolve—that is, microevolution. For full comprehension of the history of life on Earth, we must consider large-scale factors such as mass extinctions, adaptive radiations, and continental drift, all of which can have a tremendous effect on evolution above the species level—that is, macroevolution.

Concept Check

1. Name two ways in which mass extinctions have affected the history of life on Earth.
2. What did François Jacob mean when describing evolution as a tinkerer?

Applying What We Learned

When Antarctica Was Green

What had happened in Antarctica? Why were there fossils of lush tropical life in such an icy, barren place?

Antarctic fossils of dinosaurs, forest trees, and tropical marine organisms are vivid testimony to the fact that we live in a dynamic world. These fossils reveal great changes over time, ranging as they do from Cambrian marine organisms to early land plants to birds and mammals. The very different organisms that have lived in Antarctica at different times illustrate the broad changes in the history of life described in this chapter, such as the Cambrian explosion, the colonization of land, and

The Antarctic Landscape Now
Only two species of flowering plants are found on the continent, one of which is Antarctic hair grass.

the periods of domination by amphibians, reptiles, and mammals.

The Antarctic fossils also show the striking contrast between the diverse life forms that once lived in Antarctica and the few that live there today. The small number of present-day organisms in Antarctica is due in part to continental drift. As Pangaea broke apart, ocean currents were rerouted, causing Earth's climate to grow colder and ice caps to form at the poles. This process was especially pronounced in Antarctica, which experienced an ever-colder climate as it moved toward its present position over the South Pole. Once Antarctica had separated from Australia and South America, about 40 mya, the organisms on Antarctica were trapped there. As the climate of Antarctica became increasingly cold, most Antarctic species perished.

However, the same continental movements that brought destruction to Antarctica sowed the seeds of evolutionary diversity elsewhere. The rerouting of ocean currents that contributed to the formation of the Antarctic ice cap also produced the largest differences in temperature between the poles and the tropics that Earth has ever known. The wide range of new habitats that resulted from these temperature differences helped set the stage for adaptive radiations in many organisms, including humans.

Fossil Called Missing Link from Sea to Land Animals

BY JOHN NOBLE WILFORD

Scientists have discovered fossils of a 375-million-year-old fish, a large scale creature not seen before, that they say is a long-sought missing link in the evolution of some fishes from water to a life walking on four limbs on land.

The skeletons have the fins, scales and other attributes of a giant fish, four to nine feet long. But on closer examination, the scientists found telling anatomical traits of a transitional creature, a fish that is still a fish but has changes that anticipate the emergence of land animals—and is thus a predecessor of amphibians, reptiles and dinosaurs, mammals and eventually humans.

In the fishes' forward fins, the scientists found evidence of limbs in the making. There are the beginnings of digits, proto-wrists, elbows and shoulders. The fish also had a flat skull resembling a crocodile's, a neck, ribs and other parts that were similar to four-legged animals known as tetrapods.

Other scientists said that in addition to confirming elements of a major transition in evolution, the fossils were a powerful rebuttal to religious creationists, who have long argued that the absence of such transitional creatures are a serious weakness in Darwin's theory.

In an interview, Dr. Shubin, an evolutionary biologist, let himself go. "It's a really amazing, remarkable intermediate fossil," he said. "It's like, holy cow."

Discovered in Canada just 600 miles from the North Pole, this spectacular fossil fish find was first described by a team of scientists led by Neil Shubin, a biologist at the University of Chicago, in a paper in the journal *Science,* the most prestigious scientific journal in the United States. Researchers had uncovered several skeletons of the missing-link fish in far northern Canada and named the creature *Tiktaalik* at the suggestion of the elders of the First Nation people (Canada's Native American peoples) of Canada's Nunavut Territory. The name means "large, shallow-water fish" and it describes the fossil exactly. The fossil illustrates just the kind of transitional form expected to be found between the forms we see in Figure 19.6.

This fossil is particularly exciting because, while very obviously a fish, it shows some features that are simply not seen in fish—for example, a neck and proto-wrists. These features simply are not found in animals that are completely adapted for life in water. It is particularly interesting because transitional forms are so rarely found. First, organisms are rarely preserved as fossils. When they die, nearly all organisms, if not eaten, otherwise decay or are destroyed. The conditions have to be just right for a living creature to be preserved, undisturbed as an intact fossil. In addition, for a transitional fossil to be found, it must be preserved. Heat, the movement of the continents, weathering of exposed rocks, and many other forces conspire to destroy any fossil that is formed.

On top of all that, most of the world's rocks remain hidden from view, either buried at some level beneath the Earth's surface where we live or covered over with dirt and greenery. Finding new fossils at all is no easy matter, but finding particular transitional forms is quite a special feat indeed. It is expensive and often impractical to uncover rocks (for example, if there are buildings or farms or forests on top of them), and so much remains unknown. The fossil fungi mentioned earlier in the chapter were found, as so many are, incidentally. The important fossil spores and hyphae were discovered along a roadside, where crews had cut away at the rock to build the road.

Tiktaalik may have been found in part because Dr. Shubin and colleagues had searched the world for exactly the right conditions: the right age of rocks (about 375 million years old, dating to just about the same time that vertebrates appeared on land) and a likely area for such a transition to have taken place, where there were long ago, gentle, shallow waters.

Evaluating the News

1. Look at the timeline of the history of life (Figure 19.3) and find where *Tiktaalik* belongs. Identify another major transition in the history of life, for which biologists could find transitional forms.
2. *Tiktaalik* is a transitional form that could be described as a fish with some characteristics of a land-dwelling vertebrate. Given the major transition you chose in question 1, what would you expect your transitional form to be like? To look like? What particular characteristics might it have?
3. Some antievolutionists argue that an inability to find some "missing links" in the fossil record suggests that evolution may not occur. What is wrong with such an argument? (*Hint:* Why might we expect that most living things would not be seen in the known fossil record?)

SOURCE: *New York Times,* April 6, 2006.

Chapter Review

Summary

19.1 The Fossil Record: A Guide to the Past

- The fossil record documents the history of life on Earth.
- Fossils reveal that past organisms were unlike living organisms, that many species have gone extinct, and that the dominant groups of organisms have changed significantly over time.
- The order in which organisms appear in the fossil record is consistent with our understanding of evolution gained from other kinds of evidence, including morphology and DNA sequences. Sometimes the approximate age of a fossil can be determined through analysis of radioisotopes.
- Although the fossil record is not complete, it provides excellent examples of the evolution of major new groups of organisms.
- Recently discovered fossils show that whales are most closely related to artiodactyls (a group of hoofed mammals), supporting the results of genetic analyses.

19.2 The History of Life on Earth

- The first single-celled organisms resembled bacteria and evolved at least 3.5 billion years ago.
- About 2.75 billion years ago, some groups of bacteria evolved the ability to conduct photosynthesis, which releases oxygen as a waste product.
- The release of oxygen by photosynthetic bacteria caused oxygen concentrations in the atmosphere to increase. Rising oxygen concentrations made possible the evolution of single-celled eukaryotes about 2.1 billion years ago. Multicellular eukaryotes followed about 650 mya.
- Life on Earth changed dramatically during the Cambrian explosion (530 mya), when large predators and well-defended herbivores suddenly appear in the fossil record.
- The land was first colonized by plants (about 500 mya), fungi (about 460 mya), and invertebrates (about 410 mya), which were followed later by vertebrates (about 365 mya).

19.3 The Effects of Continental Drift

- Continental drift has had profound effects on the history of life on Earth.
- The separation of the continents over the past 200 million years has led to geographic isolation on a grand scale, promoting the evolution of many new species.
- At different times, climate changes caused by the movements of the continents have led to the extinctions of many species.

19.4 Mass Extinctions: Worldwide Losses of Species

- There have been five mass extinctions during the history of life on Earth.
- The extinction of some groups and the survival of others greatly alter the subsequent course of evolution.
- The extinction of a dominant group of organisms can provide new opportunities for other groups.

19.5 Adaptive Radiations: Increases in the Diversity of Life

- The history of life has been heavily influenced by adaptive radiations, in which a group of organisms diversifies greatly and takes on new ecological roles.
- Adaptive radiations can be caused by the release from competition that follows a mass extinction.
- Adaptive radiations can also occur when a group of organisms evolves a new adaptation that allows them to fill new ecological roles.

19.6 The Origin and Adaptive Radiation of Mammals

- Mammals evolved from reptiles over the course of about 80 million years (300 to 220 mya). During this time, a group of reptiles gradually evolved mammalian features such as vertically oriented legs, specialized teeth, and powerful jaws.
- The first mammals evolved from cynodonts, the last of a line of mammal-like reptiles. Mammals and dinosaurs originated at roughly the same time (230 mya for dinosaurs, 220 for mammals). Not long after, about 200 mya dinosaurs became the dominant land vertebrates.
- Throughout the long reign of the dinosaurs (200–65 mya), most mammals remained small and nocturnal. Following the extinction of the dinosaurs, the mammals radiated to include many species that were large and active by day.

19.7 An Overview of the Evolutionary History of Life

- Major new groups of organisms do not evolve from scratch, but rather arise through a series of modifications of existing organisms.
- Macroevolution cannot be predicted solely from an understanding of the mechanisms that cause microevolution.

⦿ Review and Application of Key Concepts

1. The fossil record provides clear examples of the evolution of new groups of organisms from previously existing organisms. Describe the major steps of one such example.

2. How did the evolution of photosynthesis affect the history of life on Earth?

3. What is the Cambrian explosion, and why was it important?

4. Life arose in water. Explain why the colonization of land represented a major evolutionary step in the history of life. What challenges—and opportunities—awaited early colonists of land?

5. Mass extinctions can remove entire groups of organisms, seemingly at random—even groups that possess highly advantageous adaptations. How can this be?

6. Evidence from the fossil record indicates that it usually takes 10 million years for adaptive radiation to replace the species lost during a mass extinction. Discuss this observation in light of your understanding of the speciation process (see Chapter 18). What does it suggest about the consequences of the losses of species that are occurring today?

7. Is macroevolution fundamentally different from microevolution? Can macroevolutionary patterns be explained solely in terms of microevolutionary processes? Does any evolutionary mechanism that we have studied link macroevolution and microevolution?

Key Terms

adaptive radiation (p. 400)
Cambrian explosion (p. 391)
continental drift (p. 395)
Cretaceous extinction (p. 398)
cynodont (p. 401)

macroevolution (p. 388)
mass extinction (p. 396)
Permian extinction (p. 398)
radioisotope (p. 388)

Self-Quiz

1. Continental drift
   a. can occur when liquid rock rises to the surface and pushes the continents away from one another.
   b. no longer occurs today.
   c. has led to the geographic isolation of many populations, thus promoting speciation.
   d. both a and c

2. The fossil record
   a. documents the history of life.
   b. provides examples of the evolution of major new groups of organisms.
   c. is not complete.
   d. all of the above

3. Mass extinctions
   a. are always caused by asteroid impacts.
   b. are periods of time in which many species go extinct worldwide.
   c. have little lasting effect on the history of life.
   d. affect only terrestrial organisms.

4. The Cambrian explosion
   a. caused a spectacular increase in the size and complexity of animal life.
   b. caused a mass extinction.
   c. was the time during which all living animal phyla suddenly appeared.
   d. had few consequences for the later evolution of life.

5. The history of life shows that
   a. biodiversity has remained constant for about 400 million years.
   b. extinctions have little effect on biodiversity.
   c. macroevolution is greatly influenced by mass extinctions and adaptive radiations.
   d. macroevolution can be understand solely in terms of the evolution of populations.

6. _____ are radioactive forms of elements that decay to more stable forms over time.
   a. X-rays
   b. Carbon-12 and carbon-14
   c. Radioisotopes
   d. Adaptive radiations

7. Large-scale evolution characterized by the rise and fall of major groups of organisms is called
   a. macroevolution.
   b. microevolution.
   c. mass extinction.
   d. adaptive radiation.

8. Which of the following terms most specifically describes what occurs when a group of organisms expands to take on new ecological roles, forming new species and higher taxonomic groups in the process?
   a. speciation
   b. evolution
   c. mass extinction
   d. adaptive radiation

9. Early bacteria evolved the ability to carry out photosynthesis, changing levels of atmospheric oxygen. This was significant for the evolutionary history of life because
   a. the decreased atmospheric oxygen set the stage for the evolution of the eukaryotes.
   b. the increased atmospheric oxygen set the stage for the evolution of eukaryotes.
   c. the increased atmospheric oxygen led to life's first mass extinction.
   d. the decreased atmospheric oxygen led to life's first mass extinction.

10. Which of the following are thought to have appeared first appeared on land?
    a. plants
    b. animals
    c. mammals
    d. fungi

Humans and Evolution

Who Are We? Where Do We Come From?

Imagine the commotion that would result if a group of Neandertals were to stroll down the street today. Neandertals had large, arching ridges above their eyes; a low, sloping forehead; almost no chin; and a face that, compared with ours, looked as if it had been pulled forward. Without shirts, they would be even more striking. Slightly shorter than most humans of today, Neandertals had thick necks, were heavily boned, and, as indicated by markings on the skeletons where their muscles were attached, were very strong (Figure D.1).

Neandertals take their name from the Neander Valley in eastern Germany (*Tal* is German for "valley"). A fossilized skeleton discovered there in 1856 came to be known as the Neandertal Man. At first, these bones were the subject of great debate: some people thought them to be those of a bear, while others argued that they were the remains of an ancient human. Finally, after similar fossils were found in other places, scientists became convinced that the bones were indeed those of ancient human relatives.

The Neandertal fossils shook our understanding of ourselves, for they provided dramatic proof that different forms of humans once existed. The fossils of our human and humanlike ancestors provide an eerie sense of recognition, for they reveal creatures that were like us, yet not like us. What do such fossils tell us about who we are and where we come from? Do findings from the fossil record match results from analyses of the human genome? In this essay we examine questions like these as we explore our evolutionary past and speculate about our evolutionary future. We also consider how human actions affect evolution, both in ourselves and in other species.

Figure D.1 An Artist's Interpretation of a Neandertal's Appearance

We Are Apes

Within the Linnaean hierarchy of all living things, we human beings—*Homo sapiens*—are members of the animal kingdom. Within the many phyla of the animal kingdom, we are members of the phylum known as the chordates, which includes all animals with a backbone. Within the chordates, we are members of the class known as the mammals. We share with all other mammals certain unique features, including body hair (which provides insulation in many mammals) and milk produced by mammary glands. Within the mammals, we are part of the order known as the primates. Like all other **primates**, we have flexible shoulder and elbow joints, five functional fingers and toes, thumbs that are **opposable** (that is, they can be placed opposite other fingers), flat nails (instead of claws), and brains that are large in relation to our body size. Within the primates we are members of the family known as the hominids, and within that family we belong to the genus *Homo*. The genus *Homo* comprises a number of human species, including our own species, *Homo sapiens*, as well as other extinct humans, which we'll learn more about later in the chapter.

As we think about the type of animal we are, we begin to address the questions posed at the beginning of this essay: Who are we? Where do we come from? For several hundred years we viewed ourselves as very distinct from other animals. Although we classified ourselves as primates, we placed ourselves in one family, the hominid family, and the species most similar to us—chimpanzees, gorillas, and orangutans—in a separate family, the pongid family. The decision not to place any other living species in our family both reflected and reinforced our view of ourselves as different from other animals.

After the publication of Darwin's book *The Origin of Species,* people realized that as animals, we descended from earlier animals. Even so, we continued to view ourselves as distinct from other animals. For example, before the early 1960s, when genetic analyses were first used to determine evolutionary relationships among primates, it was widely believed that the evolutionary lineage leading to humans split from the lineage leading to chimpanzees about 30 million years ago. Although scientists recognized that we share a common ancestor with chimpanzees and other apes, they believed that humans had evolved separately from the apes for tens of millions of years.

As we shall see, in the past 45 years or so, genetic analyses and a series of spectacular fossil discoveries have changed that view. Scientists now believe that the human–chimpanzee divergence occurred much more recently, about 5 to 7 million years ago (Figure D.2). Similarly, a combination of genetic analyses and fossil discoveries suggest that the evolutionary lineage leading to humans diverged from the lineage leading to gorillas about 7–8 mya, and from the lineage leading to orangutans about 12–16 mya.

In fact, genetic analyses published in the past few years suggest that our relationship to the apes is so close

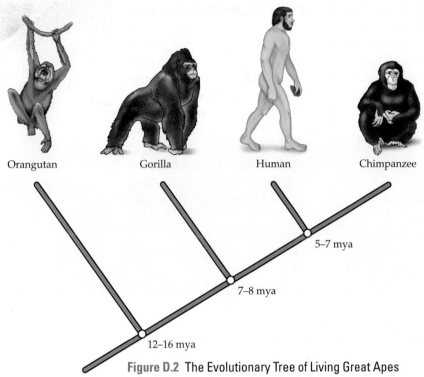

Orangutan   Gorilla   Human   Chimpanzee

5–7 mya

7–8 mya

12–16 mya

Figure D.2 The Evolutionary Tree of Living Great Apes
This tree, based on recent genetic analyses and fossil discoveries, shows that humans are most closely related to chimpanzees.

as to be entangled. Our DNA is very similar to the DNA of apes, especially chimpanzees. On average, the DNA sequence of a human differs from the DNA sequence of a chimpanzee at about 1 percent of the nucleotide bases. In addition, our chromosomes contain a complex mix

of DNA. Most of our DNA is closer in sequence to that of chimpanzees than to that of other apes, but in some regions our DNA is more like that of gorillas. In still other chromosomal regions, the chimpanzees and gorillas are more closely related to each other than either is to us.

The broad conclusion that emerges from studies published over the last 40 years is that we are not just closely related to apes, we *are* apes. Some commentators have suggested that if space aliens existed and observed life on Earth, they would classify humans as the "fourth chimpanzee" (there are three living species of chimpanzees). Be that as it may, we share many characteristics with apes, especially chimpanzees, including the use of tools, a capacity for symbolic language, and the performance of deliberate acts of deception (Figure D.3). These similarities—as well as our many differences—make the story of how humans evolved all the more interesting. We begin that story by describing the origin of the first **hominids**, the family that contains humans and our now extinct, humanlike ancestors.

Hominid Evolution: From Climbing Trees to Walking Upright

Primates are thought to have originated 65 to 80 million years ago from small nocturnal mammals, similar

Figure D.3 Our Closest Living Relative
We share many things, including rather specific behaviors, with our closest living relatives, chimpanzees.

to tree shrews, that ate insects and lived in trees. The fossil evidence of primate origins is sketchy, however, and the first definite primate fossils are 56 million years old. These early primates resembled modern lemurs (see Figure 2.11). Over time, the primates diversified greatly, eventually giving rise to the first hominids, roughly 5 to 7 million years ago.

As the hominids evolved, later species came to have large brains compared to their body size, an upright walking posture, and complex toolmaking behaviors. Of these traits, our intelligence and toolmaking abilities, and the cultures associated with them, are central to what it means to be human. In an evolutionary sense, however, the increases in our intelligence and toolmaking abilities were secondary changes: they occurred relatively late in our evolutionary history and were part of a general trend toward large brain size in primates.

The first big step in human evolution was the shift from being quadrupedal (moving on four legs) to being **bipedal** (walking upright on two legs)—a change that occurred long before hominids evolved large brains. Many skeletal changes accompanied the switch to walking upright, including the loss of opposable toes (the big toe is opposable in all primates except

humans). Try touching your little toe with the big toe on the same foot.

It is not necessary—or even a particularly good idea—for an organism living in the trees to try to walk upright. Moreover, for organisms that live primarily in trees, the loss of opposable toes that accompanied bipedalism would be a handicap, since such toes can helpfully grasp branches during climbing. On the ground, however, walking upright provides several advantages, including freeing the hands for carrying objects or making and using tools. In addition, walking upright elevates the head instantly, allowing the walker to see farther and over more things. It is likely that the evolution of an upright posture was linked to a switch from life in the trees to life on the ground—a change that probably occurred 5 to 8 million years ago.

The shift to life on the ground was probably not sudden or complete. The skeletal structure of some of the oldest fossil hominids (dating from 4.4 million years ago) indicates that they walked upright. However, foot bones and fossilized footprints that are 3 to 3.5 million years old show that the hominids living at that time still had partially opposable big toes (Figure D.4), perhaps because they continued to use trees some of the time.

Tree shrew

(a) Foot bones of early hominid

Partially opposable big toe

(b)

(c)

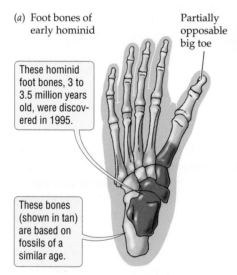

These hominid foot bones, 3 to 3.5 million years old, were discovered in 1995.

These bones (shown in tan) are based on fossils of a similar age.

Figure D.4 Early Hominids Had Partially Opposable Big Toes
(a) Fossilized hominid foot bones show that some hominids living 3 to 3.5 million years ago walked upright but had partially opposable big toes. That is, the big toe could be placed opposite (as is done when we use our thumb and pointer finger to pick up a pencil) some but not all of the toes. (b) These footprints of two early hominids walking upright, side by side, were found in Africa. (c) Chimpanzees, the closest living relatives of humans, have fully opposable big toes.

Figure D.5 The Million-Dollar Skull

(*a*) Paleontologist Michel Brunet is holding what has been called the find of the century: a nearly complete skull of the oldest known hominid, *Sahelanthropus tchadensis.* For this discovery, Dr. Brunet received a $1 million Dan David Prize; three of these prizes are awarded each year for achievements that have a major scientific, technological, cultural, or social effect on our world. (*b*) An artist's reconstruction of this earliest hominid species shows its distinctly apelike characteristics. Like *Homo sapiens*, however, this species was bipedal; that is, it walked upright on two legs. The finding of the fossil remains of other species nearby—including crocodiles and amphibious mammals, as well as rodents usually found in forests and grasslands—suggest that *S. tchadensis* lived in a forest near a lake.

The earliest known hominid is *Sahelanthropus tchadensis* [sah-heel-*AN*-throh-puss chuh-*DEN*-siss]. A 6- to 7-million-year-old skull of this species was discovered by Michel Brunet in 2002 (Figure D.5*a*). Found in the Sahel region of the African nation of Chad (hence its scientific name), this species gives scientists insights into the early evolution of the hominid family. *S. tchadensis* was apelike in most respects. It had a massive browridge (Figure D.5*b*)—larger even than a modern-day gorilla's—and the smallest brain of any known hominid, at a volume of 350 cubic centimeters, smaller even than an average modern gorilla's brain (about 500 cubic centimeters). The fossil pictured in Figure D.5a shows just how radically brain size evolved in the hominids, since a typical

modern human has a brain volume of about 1,400 cubic centimeters. Yet at the same time, this earliest hominid is thought to have been bipedal, walking on two legs.

Other early hominids include *Ardipithecus ramidus* [ahr-dih-*PITH*-uh-kuss *RAM*-uh-duss] (4.4 million years old) and several *Australopithecus* [*AW*-strah-loh-*PITH*-uh-kuss] species that are 3.0 to 4.2 million years old. *Ardipithecus* and *Australopithecus* are also thought to have walked upright. Their brains were still relatively small, and their skulls and teeth were more similar to those of other apes than to those of humans. Compare the skull of *Australopithecus afarensis*—one of the early *Australopithecus* species—with the skull of *Homo sapiens sapiens* in Figure D.6.

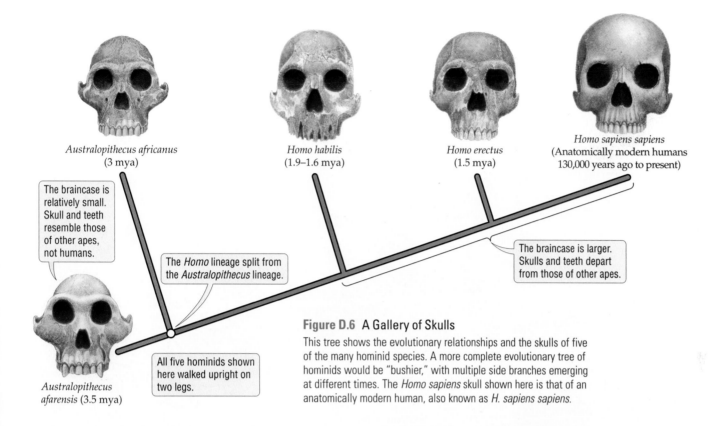

Australopithecus africanus (3 mya)

Homo habilis (1.9–1.6 mya)

Homo erectus (1.5 mya)

Homo sapiens sapiens (Anatomically modern humans 130,000 years ago to present)

The braincase is relatively small. Skull and teeth resemble those of other apes, not humans.

The *Homo* lineage split from the *Australopithecus* lineage.

The braincase is larger. Skulls and teeth depart from those of other apes.

All five hominids shown here walked upright on two legs.

Australopithecus afarensis (3.5 mya)

Figure D.6 A Gallery of Skulls

This tree shows the evolutionary relationships and the skulls of five of the many hominid species. A more complete evolutionary tree of hominids would be "bushier," with multiple side branches emerging at different times. The *Homo sapiens* skull shown here is that of an anatomically modern human, also known as *H. sapiens sapiens*.

Evolution in the Genus *Homo*

The oldest *Homo* fossil fragments were found in Africa and date from 2.4 million years ago (mya), suggesting that the earliest members of the genus *Homo* originated in Africa 2 to 3 mya. More complete early *Homo* fossils exist from the period from 1.9 to 1.6 mya; these fossils have been given the species name *Homo habilis* [HAB-uh-liss]. The oldest *H. habilis* fossils resemble those of *Australopithecus africanus,* a slightly more recent hominid than *A. afarensis* (see Figure D.6). In more recent *H. habilis* fossils, the face is not pulled forward as much, and the skull is more rounded. In these and other ways, more recent *H. habilis* specimens have features that are intermediate between those of *A. africanus* and *Homo erectus,* a species that evolved after *H. habilis.* Thus, *H. habilis* fossils provide an excellent record of the evolutionary shift from ancestral hominid characteristics (in *Australopithecus*) to more recent ones (in *H. erectus*).

Taller and more robust than *H. habilis, H. erectus* also had a larger brain and a skull more like that of modern humans (Figure D.7). It is likely that by 500,000 years ago *H. erectus* could use, but not necessarily make, fire. In addition, *H. erectus* probably

See the hominid species timeline. ▶‖ D.1

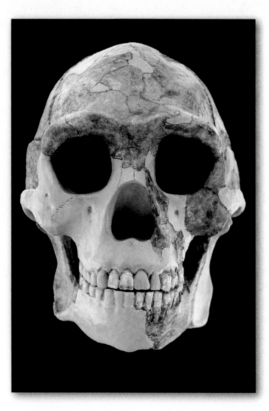

Figure D.7 Peking Man

The close resemblance of *Homo erectus* to a modern human can be seen in this photograph of a cast of an *H. erectus* fossil known as Peking Man. The fossil was found in China in a limestone cave. During World War II the fossil was being moved to the United States but it disappeared en route and has never been found.

Figure D.8 The Power of Teamwork
Both *Homo habilis* and *Homo erectus* likely worked in groups using stone tools when dealing with large kills.
This artist's reconstruction shows a family group harvesting the different parts of their prey.

hunted large species of game animals (Figure D.8). The evidence to support the latter conclusion includes the remarkable discovery in Germany of three 400,000-year-old spears, each about 2 meters long and designed for throwing with a forward center of gravity (like a modern javelin).

It was long thought that *H. habilis* gave rise to *H. erectus,* which then spread from Africa about 1 mya and later evolved into *Homo sapiens*. This simple picture has become more complicated with subsequent fossil discoveries. Some evidence now suggests that the fossils labeled *H. habilis* represent two different species, and there is debate over which of these species gave rise to *H. erectus*. In addition, it now appears that *H. erectus* or an earlier form of *Homo* migrated from Africa much

longer ago than was previously thought. *Homo* fossils dating from 1.9 to 1.7 mya have been found in Java, the Central Asian republic of Georgia, and China. Finally, in 2004, what may be a new miniature species of *Homo* was discovered on an Indonesian island (see the "Biology in the News" feature in Chapter 18). Described by its discoverers as *H. floresiensis,* these "little people" appear to have lived on that island 95,000 to 12,000 years ago. Other fossil evidence indicates that *H. erectus* lived on nearby islands from 1 million to 25,000 years ago, while *H. sapiens* lived in the same general region from 60,000 years ago to the present.

Overall, current research on *H. habilis, H. erectus,* and other early *Homo* species indicates that there were more species of *Homo* than was once thought, and that several of these species existed in the same places and times. A complete hominid evolutionary tree would therefore be much "bushier" than the version shown in Figure D.6. More research and evidence will be necessary before general agreement is reached regarding the number of early *Homo* species and their evolutionary relationships.

The Origin and Spread of Modern Humans

All humans alive today are *Homo sapiens*. More specifically, we are all what are known as anatomically modern humans, also known as *Homo sapiens sapiens,* a group that arose some 130,000 years ago (see skull, Figure D.6). But before anatomically modern humans arose, there existed earlier humans known as "archaic" *Homo sapiens*. Who were these early humans?

The fossil record indicates that these archaic *H. sapiens* bore features intermediate between those of *Homo erectus* and those of the anatomically modern *H. sapiens sapiens.* The first archaic *H. sapiens* originated between 400,000 and 300,000 years ago; some of these species are thought to have persisted until as recently as 30,000 years ago. Their fossils have been found in Africa, China, Java, and Europe. These ancestors of anatomically modern humans developed new tools and new ways of making tools, used new foods, built complex shelters, and controlled the use of fire. The long-distance exchange of goods between populations of humans, as well as toolmaking and other technologies, continued to improve and develop with the origin of anatomically modern humans about 130,000 years ago (Figure D.9).

What became of the archaic *H. sapiens*? Early populations of archaic *H. sapiens* eventually gave rise to both

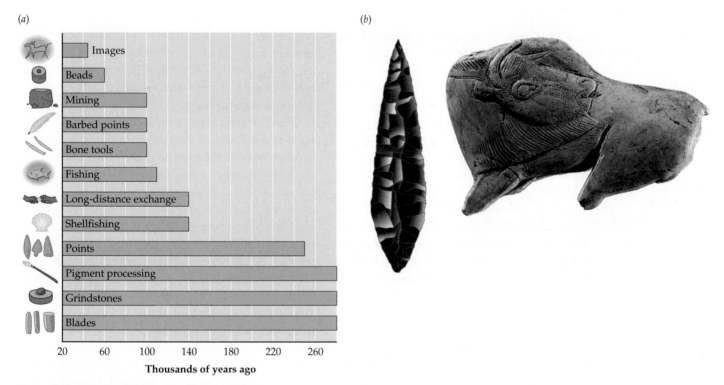

Figure D.9 Advanced Stone Age Tools and Art

(*a*) Over the past 300,000 years, first the archaic *Homo sapiens* and later the anatomically modern humans (*H. sapiens sapiens*) developed a rich set of new tools and technologies. The archaeological record shows that these tools and technologies did not appear suddenly together, but rather developed over a long period of time and a broad geographic area, first in Africa and then in other parts of the world. (*b*) This spearhead and the figurine of a bison date from 40,000 to 10,000 years ago.

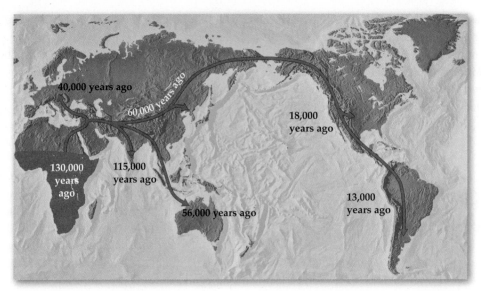

Figure D.10 Migration from Africa

The earliest known fossil and archaeological specimens of *Homo sapiens sapiens*, the first anatomically modern humans, come from Africa. The dates provided give the age of the earliest evidence that anatomically modern humans lived in different regions of the world. These dates are continually challenged by new evidence that scientists must then work to confirm. For example, a recent discovery suggests that modern humans may actually have lived in North America as early as 50,000 years ago.

Neandertals (an advanced type of archaic *H. sapiens* that lived from 230,000 to 30,000 years ago) and us—that is, anatomically modern humans (*Homo sapiens sapiens*). Though the oldest fossils of anatomically modern humans date from 130,000 years ago and have been found in Africa (Figure D.10), more recent fossils of anatomically modern humans have been found in such places as Israel (115,000 years ago), China (60,000 years ago), Australia (56,000 years ago), and the Americas (18,000 to 13,000 years ago).

There has been considerable controversy over exactly how anatomically modern humans arose from archaic *H. sapiens* and how we came to be found all across the globe. Two conflicting hypotheses have been proposed. According to the **out-of-Africa hypothesis** (Figure D.11*a*), anatomically modern humans first evolved in Africa from unknown populations of archaic *H. sapiens,* which evolved, also in Africa, from *H. erectus.* They then spread from Africa to the rest of the world, completely replacing (and not breeding with) all other *Homo* populations, including *H. erectus,* the Neandertals, and *H. floresiensis.* In contrast, the **multiregional hypothesis** (Figure D.11*b*) proposes that modern humans evolved over time from *H. erectus* populations located throughout the world.

According to this hypothesis, regional differences among human populations developed early, but worldwide gene flow caused these different populations to evolve modern characteristics simultaneously and to remain a single species.

Which of these hypotheses is correct? Let's consider some of the evidence. According to the multiregional hypothesis, when different populations of early humans came into contact, extensive gene flow should have caused them to become more similar to one another. If this were so, we would not expect different types of early humans to coexist in the same geographic region yet remain distinct for long periods of time. In fact, however, Neandertals and more modern humans coexisted in western Asia for about 80,000 years. Even as recently as 25,000 to 12,000 years ago, *H. sapiens* may have shared some parts of its range with *H. floresiensis* and *H. erectus.* This evidence from the fossil record indicates that modern humans overlapped in time yet remained distinct from *H. erectus, H. floresiensis,* and Neandertal populations. This finding calls into question the extensive gene flow assumed by the multiregional hypothesis, because if there were indeed such gene flow, it should have been impossible to maintain such distinct lineages side-by-side.

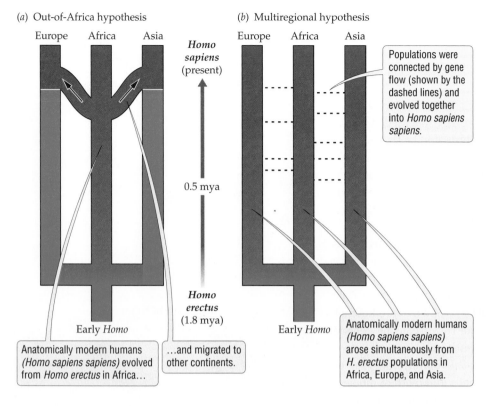

(a) Out-of-Africa hypothesis

Europe   Africa   Asia

Homo sapiens (present)

0.5 mya

Homo erectus (1.8 mya)

Early *Homo*

Anatomically modern humans (*Homo sapiens sapiens*) evolved from *Homo erectus* in Africa...

...and migrated to other continents.

(b) Multiregional hypothesis

Europe   Africa   Asia

Populations were connected by gene flow (shown by the dashed lines) and evolved together into *Homo sapiens sapiens*.

Early *Homo*

Anatomically modern humans (*Homo sapiens sapiens*) arose simultaneously from *H. erectus* populations in Africa, Europe, and Asia.

Figure D.11 The Origin of Anatomically Modern Humans (*a*) According to the out-of-Africa hypothesis, *Homo sapiens sapiens*, or anatomically modern humans (red), originated in Africa within the past 200,000 years, then migrated to Europe and Asia, replacing *Homo erectus* (blue) and archaic *H. sapiens* populations. (*b*) In contrast, the multiregional hypothesis proposes that populations of *H. erectus* in Africa, Asia, and Europe evolved simultaneously into anatomically modern humans (*H. sapiens sapiens*).

The best fossil evidence for the shift from archaic to modern *H. sapiens* comes from Africa, providing some support for the out-of-Africa hypothesis. Recent analyses of human DNA sequences are also consistent with the out-of-Africa hypothesis. However, the "complete replacement" part of the hypothesis may not be correct. Fossils have been found that some scientists interpret as showing a mix of Neandertal (a kind of archaic *H. sapiens*) and modern human characteristics. Similarly, some genetic studies (in which evolutionary trees were constructed on the basis of DNA sequences from living humans) indicate that genes from archaic *H. sapiens* populations outside Africa may have contributed to the genetic makeup of modern humans, suggesting that anatomically modern humans may instead have coexisted, even mated, with archaic *H. sapiens* like Neandertals.

In summary, many scientists think that anatomically modern humans arose in Africa and spread from there to other parts of the world. However, the details of the origin of modern humans continue to be a controversial issue; this debate is especially active concerning the extent to which early *Homo sapiens* interbred with, and hence did not completely replace, more ancient *Homo* populations.

The Evolutionary Future of Humans

What can our evolutionary past, the focus of this unit, tell us about our evolutionary future? Will we go the way of science fiction stories and evolve into beings with huge brains? Such a direction appears unlikely because for the last 75,000 years human brain size appears to have decreased slightly, not increased. The average brain size of Neandertal fossils (75,000 to 35,000 years ago) and anatomically modern human fossils (35,000 to 10,000 years ago) was about 1,500 cubic centimeters, and the brain size of an average human today is about 1,350 cubic centimeters. Most of this difference is due to body size: both Neandertals and early modern humans were larger than living humans, and large organisms have bigger brains than small organisms. Although our brains may not be shrinking much (relative to body size), there is no evidence that they are getting larger. But what changes can we expect as a result of ongoing genetic drift, gene flow, and natural selection in humans? (To review these concepts, see Chapter 17.)

To answer this question, let's consider several recent and current aspects of human societies. Before the development of agriculture about 10,000 years ago, human populations were small and widely scattered. On a large geographic scale, these populations were isolated from one another by geographic barriers. For example, 30,000 years ago the chance that an African would have met an Australian was virtually nil. Early human populations were probably isolated on a much smaller geographic scale as well, as illustrated by the fact that before the twentieth century, people from nearby valleys separated by the rugged mountains of New Guinea had little contact with one another and spoke different languages. Even today, 823 different indigenous languages are spoken in New Guinea.

Translated into evolutionary terms, the conditions of early human populations were exactly those under which genetic drift should be important: population sizes were small, and there was probably little gene flow among populations. We would thus be led to predict that genetic drift has played a major role in causing genetic differences among human populations.

Some evidence supports this claim: analyses of the rate of evolution of the skulls and teeth of modern humans indicate that genetic drift has been a more important factor in our evolution than has natural selection. Because the human population is now large and mobile, however, genetic drift is less likely to play a major role in future human evolution. Instead, high rates of gene flow among human populations could substantially reduce existing differences over time (Figure D.12), even for traits such as skin color that appear to have resulted from natural selection rather than genetic drift.

What about the role of natural selection? Because of our reliance on tools and technology, there is now relatively little selection pressure on many characteristics (such as poor vision) that might have been greatly disadvantageous at previous times in our evolutionary history. This does not mean that natural selection will have no effect on us in the future. Infectious diseases take a terrible toll each year in human death and suffering, so there remains strong selection pressure for the evolution of increased disease resistance in human populations. In a recent genetic analysis, for example, scientists found that alleles providing resistance to malaria increased in frequency so rapidly in populations exposed to malaria that their rise is best explained by natural selection—not by genetic drift, gene flow, or mutation.

Finally, at various times and places in recent history, humans have been tempted to direct our own evolution—a topic we shall return to later in this essay.

Figure D.12 Gene Flow in Our Future
Gene flow among human populations could reduce some of the features that distinguish different groups of people. Future humans might look something like this computer composite image, which was formed from photographs of eight Afro-Caribbean models, eight Caucasian models, and eight Japanese models.

The Effects of Humans on Evolution

Reports in the news media make it clear that humans profoundly affect their environment, but rarely do we hear mention of how we affect evolution. In fact, it is precisely because we have such a large effect on the environment that we also have a large effect on evolution.

People have profound effects on microevolution in other species

Humans greatly affect how allele frequencies in populations of other species change over time. In many cases, we exert strong natural selection on other species, and they change genetically in response to our actions. For example, when we apply pesticides to kill insects that spread disease or eat our crops, those insects rapidly

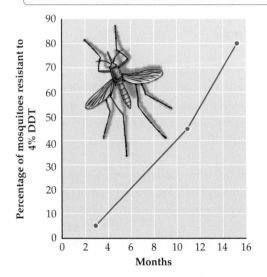

In one year, the percentage of mosquitoes that were resistant to a standard dose of DDT rose from 5% to 80%.

Figure D.13 Directional Selection for Pesticide Resistance

Over a 2-year period, mosquitoes were captured from a population at different times and sprayed with the pesticide DDT. Use of DDT in the area where the mosquitoes were captured caused the population to evolve resistance rapidly, thereby limiting the effectiveness of the pesticide. Mosquitoes were considered resistant if they were not killed by a standard dose of DDT (4 percent DDT) in 1 hour.

evolve resistance to our best efforts to exterminate them (Figure D.13). Similarly, disease-causing bacteria and viruses rapidly become resistant to new drug therapies. The evolution of resistance has large social costs in terms of human suffering and lost food production. It also costs us a great deal of money: at least $33 billion to $50 billion per year in the United States alone (see the box on page D13).

Rapid evolution caused by human actions is not limited to the evolution of resistance in disease or pest organisms. Intense fishing efforts, for example, have caused fish to evolve slower growth rates and thinner bodies, allowing them to slip through nets more easily. Similarly, in hatchery populations of salmon, natural selection is producing males that are smaller and return from the sea earlier than in the wild (since such males survive better at the hatcheries).

In addition to exerting strong selection on other species, our actions can change patterns of genetic drift and gene flow in natural populations, again affecting microevolution. For example, when we modify or destroy the habitat in which a species lives, the number of individuals in a population is often reduced, sometimes dramatically. With such population reductions, genetic drift can reduce genetic variation and cause the fixation of harmful alleles, as in the greater prairie chicken (see Figure 17.5). Similarly, when we destroy portions of what once were large regions of continuous habitat, as when we convert native grasslands to farmlands (Figure D.14), distances between remnants of the original habitat are increased, thereby reducing gene flow between natural populations.

In general, human actions alter large regions of Earth: we cut down forests; drain wetlands; plow grasslands;

(a)

(b)

Figure D.14 Human Effect on the Landscape

When habitats are changed from native grasslands and wetlands to farmlands, the impact on natural populations can be severe. For example, the number of species that can be found in natural habitats such as (*a*) the protected wetlands in this Michigan state park is considerably greater than the number to be found in (*b*) Michigan farmland.

Biology Matters

Humans: The World's Dominant Evolutionary Force?

Human impact on world ecology has "increased to the point where humans may be the world's dominant evolutionary force," according to a study reported in *Science* magazine in 2001. A wide range of organisms have evolved resistance to human efforts to kill or control them, and this evolution is costing vast sums of money. For example, as HIV (the virus that causes AIDS) has evolved, AZT, a relatively inexpensive drug used to treat it, has become less and less effective, leading to higher costs both for treating patients with newer and more powerful drugs, and for developing new drugs to keep up with the evolving virus. It is estimated that a pharmaceutical company must spend roughly $150 million to develop one new drug.

The following table estimates the annual costs of human-induced evolution in the United States, in several areas. These areas include resistance in insects and in two disease agents: HIV and *Staphylococcus aureus* (a bacterium that can cause food poisoning, toxic shock syndrome, and infections of the skin and other soft tissues). And although HIV is much better known as a deadly infection, scientists announced in 2007 that the spread of antibiotic-resistant

Staphylococcus aureus, also known as MRSA (short for methicillin-resistant *S. aureus*), kills more people each year even than AIDS. This point was brought home to many people after a number of widely publicized deaths, including that of an otherwise healthy high school student in Virginia in 2007, triggering a wave of nationwide concern over increasing antibiotic resistance. The total annual cost of these newly evolved problems, including items not cited in the table presented here, may exceed $100 billion.

Scientists are devising ways to slow human-induced evolution. For instance, one method used to thwart microbes' resistance to medicinal drugs is combination therapy, in which a patient takes several drugs in combination so that almost all of the microbes are killed, leaving few to reproduce and evolve. Another is direct observation therapy, in which health care providers administer all drug doses to patients to ensure that they take the drugs as frequently and as long as necessary to succeed in controlling the disease. Although some of these methods are labor-intensive, they are likely to cost society far less in the long run.

Some areas of increased spending due to resistance	Annual cost (in billions of dollars)
Additional pesticide use (to combat resistant insects)	1.2
Loss of crops	2–7
Treatment of patients infected with resistant *S. aureus*:	
Patients infected outside hospitals	14–21
Hospital infections: penicillin-resistant *S. aureus*	2–7
Hospital infections: methicillin-resistant *S. aureus*	8
HIV drug resistance	6.3
Total cost	**33–50**

SOURCE: S. R. Palumbi, Humans As the World's Greatest Evolutionary Force, *Science 293* (2001): 1786–1790.

add chemicals to the air, water, and soil; transport species to new environments; kill large predators; and harvest species for food, pets, and clothing. These activities cause profound changes in the environments where species live. Because species evolve in response to their environments, the changes we make to those environments alter the course of evolution in many species.

Although our effects on evolution are pervasive and have large costs for human societies, all is not hopeless. By understanding how we affect evolution, we can take steps to limit the negative effects we have on ourselves and on other species. We can use our knowledge of resistance, for example, to implement strategies designed to slow the rate at which it evolves (see Chapter 17). Similarly, we can design ways of moderating the effects of genetic drift by taking actions to increase population sizes in rare or endangered species. As we shall see in the next section, it is also within our power to halt what may become a sixth mass extinction, in which large numbers of species go extinct worldwide.

People also alter macroevolutionary patterns

In addition to affecting microevolution, humans have profound effects on the rise and fall of major taxonomic groups of organisms. Our effects on macroevolution are most easily seen in human-caused extinctions of species. Large numbers of species are in danger of extinction as a result of habitat destruction, introductions of invasive species, and overharvesting. For many different types of organisms, threats such as these are causing a dramatic increase in extinction rates. For example, the extinctions of birds and mammals known to have occurred in the last 400 years suggest that human actions have increased extinction rates in these groups by 100 to 1,000 times the usual (or "background") rates found in the fossil record.

Are we currently in the midst of another mass extinction? If not, are we at risk of experiencing one in the near future? A comparison of current extinction rates with those of prior mass extinctions indicates that we probably have not entered a sixth mass extinction yet (since we have currently driven far fewer than 50 percent of the known species on Earth extinct, whereas more than 50 percent of species perished in each previous mass extinction). Scientists estimate, however, that if the current trends in our behavior continue, extinction rates over the next 100 years will be at least 10,000 times the background rate for mammals, birds, plants, and many other organisms. From current rates of deforestation, for example, scientists estimate that virtually all of the world's tropical forests will be cut down in the next 50 years. If this happens, many of the species that live in tropical forests will go extinct. Since over 50 percent of the world's species live in tropical forests, the single action of removing tropical forests may cause extinctions of species on a scale that does rival the "Big Five" mass extinctions of the past.

When we drive species extinct, we alter the course of evolution and change the history of life on Earth forever. We don't know yet whether humans will cause a sixth mass extinction. If we do, however, it will be unique: It will be caused by one species, not many, and not by an environmental event. And the species causing it—*Homo sapiens*—will be one that could understand the consequences of its actions and could have prevented the mass extinction from occurring. It takes millions of years to recover from mass extinctions, as we saw in Chapter 19. Therefore, if we cause a mass extinction, our descendants—if we survive the mass extinction ourselves—will live in a biologically impoverished world for a long time to come.

People sometimes attempt to control human evolution

In addition to being able to affect other species, humans have sought to control our own evolution. In the United States, for example, the early part of the twentieth century saw the creation of a **eugenics** [yoo-*JENN*-icks] **movement**, an effort to breed better humans by encouraging the reproduction of people with "good" genes and discouraging the reproduction of people with "bad" genes (Figure D.15). The eugenics movement tried to achieve its goals by passing sterilization laws, marriage laws, and immigration laws; we shall focus here only on the sterilization laws.

In 1907, the state of Indiana passed the first sterilization law in the United States, allowing state officials to sterilize people deemed unfit to breed. Two years later, California passed an even stricter law, making it legal for officials there to castrate males or to remove the ovaries of women who were "feeble-minded." This vague term, popular at the time, was sometimes used to refer to people with mental handicaps, but also people with social behaviors held to be deviant, such as drunkenness or promiscuity. Similar fates could await prisoners considered to have sexual or moral perversions or

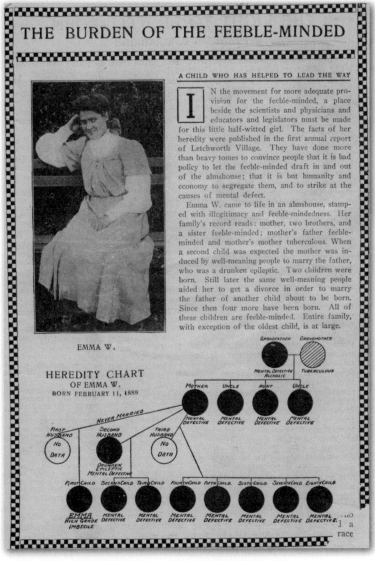

Figure D.15 A 1913 News Article Promoting Eugenics

The scientists and government officials involved with the eugenics movement believed that they were benefiting humankind by halting the spread of "bad" genes and lifting "the burden of the feeble-minded," as described in this pamphlet, which reads, "It is but humanity and economy to segregate them." The heredity chart at the bottom illustrates how the "mental defects" are passed from generation to generation.

anyone who had more than three criminal convictions. By 1940, 33 states had passed forced sterilization laws (few as severe as the one in California), and by 1960, over 60,000 people in the United States had been sterilized without their consent.

Sterilization laws gradually passed out of favor, in part because of public revulsion against similar (and more extreme) practices conducted by the Nazis in Germany. Members of the public also realized that it was difficult to establish who really was feebleminded. For example, government officials argued that Carrie Buck, the first woman forcibly sterilized in Virginia, should be prevented from having more children because she and her daughter were feebleminded—but later examination of school records showed that her daughter had been on the honor roll in first grade.

Even if society had found eugenics morally acceptable, is it likely that the movement could have been effective in its goal of breeding better humans? The answer is probably no. Eugenics laws aimed to reduce the occurrence of conditions such as so-called feeblemindedness. It was assumed that these conditions were inherited, and hence that if people with a condition held to be undesirable were prevented from having children, the frequency of the condition would decrease over time. However, it can take many years to remove an allele from a population, which could make it difficult for sterilization laws to have their intended effect. So-called feeblemindedness, for example, was assumed to be a recessive condition, and calculations made in the 1920s by opponents of the eugenics movement showed that it would take about 250 years of forced sterilization to reduce the frequency of feebleminded individuals in the population from 1 percent to ¼ percent. For sterilization laws to work as intended, society would have had to tolerate forced sterilizations for hundreds of years.

In addition, we now know that many of the traits of interest to the eugenics movement are not under simple genetic control; instead, they are strongly influenced by the environment or controlled by many genes. It takes even longer to reduce the frequency of traits under complex control than to reduce single-gene traits. Therefore, to make rapid changes in the human population, we would have to take drastic actions, such as preventing all carriers of any "bad" allele from reproducing, even if the carrier did not have any "undesirable" characteristics. Furthermore, because mutation would continually generate new copies of "bad" alleles, we would have to continually test the entire population for undesirable alleles and prevent anyone who carried even one of these alleles from reproducing. What would it be like to live under such a system?

◉ Review and Discussion

1. As discussed in this essay, most biologists classify humans as apes. Does the declaration that you are an ape make you view yourself or your life any differently?

2. Fossil evidence indicates that in the relatively recent past (about 30,000 years ago), anatomically modern humans, or *Homo sapiens sapiens,* may have shared the planet with at least three other distinct hominids: *H. erectus,* archaic *H. sapiens* like the Neandertals, and *H. floresiensis.* These species differed from us considerably, yet they also shared many similarities with us, and some may even have had the capacity for language. If one or more of these species were alive today, how would that affect the world as we know it?

3. Imagine what the world would be like if half of the species familiar to you were no longer alive. Given that our actions may be what cause a sixth mass extinction, resulting in the loss of 50 percent of the species on Earth, do we have an ethical responsibility to prevent that extinction from occurring?

4. In some instances, to protect the greater social good, human society limits the rights of an individual (for example, by placing a person who has committed a crime in prison). Do you think an ethical case can be made for limiting the right of people who carry genes with harmful effects to reproduce? Ethics aside, what do you think it would be like to live in a society in which everyone was tested for the presence of harmful alleles and anyone found to carry one was not allowed to have children?

Key Terms

bipedal (p. D4)
eugenics movement (p. D14)
hominid (p. D3)
multiregional hypothesis (p. D9)

opposable (p. D2)
out-of-Africa hypothesis (p. D9)
primate (p. D2)

CHAPTER 20 The Biosphere

Key Concepts

◉ Ecology is the study of interactions between organisms and their environment. All ecological interactions occur in the biosphere, which consists of all living organisms on Earth together with the environments in which they live.

◉ Climate has a major effect on the biosphere. Climate is determined by incoming solar radiation, global movements of air and water, and major features of Earth's surface.

◉ The biosphere can be divided into large terrestrial and aquatic areas, called biomes, based on their climatic and ecological characteristics.

◉ Terrestrial biomes are usually named for the dominant plants that live there. The locations of terrestrial biomes are determined by climate and by the actions of humans.

◉ Aquatic biomes are usually characterized by the environmental conditions that prevail there, especially salt content. They are heavily influenced by the surrounding terrestrial biomes, by climate, and by human activity.

◉ Because components of the biosphere interact with one another in complex ways, human actions that affect the biosphere can have unexpected side effects.

The Tapestry of Life

Take a walk outside, find a place where plants grow, and look around. If it is winter and cold where you live, at first glance you may not see much. But if you look carefully, you will find hidden signs of life everywhere. These signs lurk behind the bark of trees, in footprints left in snow, and under rocks, locked within the seeds and buds of plants waiting to burst forth come spring. If you walk outside in spring, summer, or fall, the wild riot of life is plain to see: it floats or buzzes by you in the air, crawls around your feet, spreads its leaves over you, and scampers nearby.

The rich abundance of life that surrounds you is but a small fraction of the biosphere: all organisms on Earth plus their environments (see Chapter 1). The biosphere includes grasslands, deserts, tropical forests, streams, and lakes, to name just a few of its many parts. It also includes the bottoms of our feet and the bacteria that grow there, as well as hydrothermal vents on the bottom of the ocean and the bacteria that grow there.

The biosphere is very complex. To put this statement in perspective, consider the challenge researchers faced in trying to understand how cells store and copy the billions of bits of genetic information contained in our DNA (see Chapter 12). Although challenging, problems such as understanding how DNA is copied pale before the difficulty of understanding the biosphere. An organism may have billions of bases of DNA, but there are many trillions of individual organisms on Earth, and they interact with one another and their environment in a vast number of ways.

The biosphere is not only complex, but also extremely important to us—our very lives depend on it. Interactions between organisms and their environment affect the quality of the air we breathe, the water we drink, the food we eat. Human actions have an enormous impact on the biosphere, sometimes producing surprising effects. Consider the seemingly harmless practice in which ships take up water in one port—to provide ballast (weight) to stabilize the ship's motion—and discharge it in another. Each time this is done, hundreds of aquatic species are moved from one part of the world to another. This is probably how the zebra mussel got from Eurasia

A Troublesome Invader
The introduction of the zebra mussel from Eurasia to North America, probably in ship ballast water, has caused a number of problems. In this picture, a worker is using a high-pressure hose to remove mussels from the inside surface of a giant pipe.

to North America, where it now clogs waterways, fouls pipes and boat hulls, outcompetes some native species, and causes billions of dollars of economic damage.

When people began the practice of taking up ballast water in one port and discharging it in another, no one suspected that doing so would affect native species and cause economic harm. Zebra mussels, for example, were not a major problem in Eurasia. **Why does the zebra mussel cause more trouble in North America than in its original home? And how can we prevent this and other similar problems in the future?** To begin to answer questions like these, we must understand how organisms interact with one another and with their environment.

A view of Earth from space highlights the beauty and fragility of the **biosphere**, which consists of all organisms on Earth together with the physical environments in which they live. Because the biosphere supplies us with food and raw materials, we depend on it for all aspects of human society. In this unit we discuss ecology, the branch of science devoted to understanding how the biosphere works.

Ecology can be defined as the scientific study of interactions between organisms and their environment, where the environment of an organism includes both biotic (other organisms) and abiotic (nonliving) factors. Ecologists are interested in how the two parts of the biosphere—organisms and the environments in which they live—interact with and affect each other. In the chapters of this unit, our study of ecology covers several levels of the biological hierarchy: individual organisms, populations, communities, ecosystems, and the biosphere (see Figure 1.11).

All ecological interactions, at whatever level they occur, take place in the biosphere, the focus of this chapter. We begin by discussing why ecology is important and what types of information ecologists must have to understand how the biosphere works. We go on to discuss climate and other factors that shape the biosphere. Then we take a brief look at the variety of terrestrial and aquatic biomes that those factors give rise to.

20.1 Why Is Ecology Important?

The science of ecology helps us understand the natural world in which we live. Beyond enriching our lives intellectually, such an understanding is becoming increasingly important because we are changing the biosphere in ways that can be expensive—and in some cases, difficult or even impossible—to fix. Consider species such as zebra mussels that are accidentally or deliberately brought by people to new geographic regions. In the United States, people have introduced thousands of non-native species, some of which have become pests (**invasive species**) that collectively cause an estimated $120 billion in economic losses each year—a huge cost, similar to the costs of smoking ($150 billion per year) and obesity ($90 billion per year). By studying the ecology of invasive species, we can understand how people help them spread to new regions, why they increase dramatically in abundance, how they affect natural communities, and how they cause economic disruption—all of which can be helpful in limiting the damage that these species cause.

As another example of human impact on the biosphere, consider chlorofluorocarbons (CFCs), which are synthetic chemicals used as refrigerants, in aerosol sprays, and in foam manufacture. CFCs rise to the upper reaches of Earth's atmosphere, where their chlorine atoms react with and destroy a gas called ozone (O_3). Our atmosphere possesses an ozone-containing layer that absorbs more than 90 percent of the most powerful ultraviolet radiation (UV-B), so relatively few of these DNA-damaging rays from the sun normally make it through. However, chlorine atoms released from CFCs destroy the ozone and therefore the UV-screening properties of this atmospheric layer.

When ultraviolet rays strike a molecule of CFC, they cause the chlorine atoms to break away. Free chlorine atoms interact with ozone molecules, triggering a chain reaction that results in the destruction of thousands of additional ozone molecules, as one of the by-products (chlorine monoxide, ClO) reacts with yet more ozone molecules.

$$Cl \quad + \quad O_3 \quad \longrightarrow \quad ClO \quad + \quad O_2$$
(Chlorine atom)   (Ozone)   (Chlorine monoxide)   (Oxygen gas)

Chlorine monoxide reacts with free oxygen atoms, which are routinely produced in small amounts as the high energy of ultraviolet rays breaks up some of the oxygen gas (O_2) in the upper atmosphere. The free

(a) September 1979

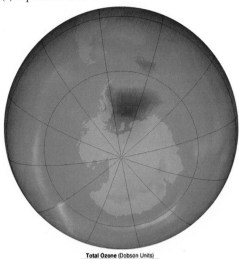

(b) September 2006

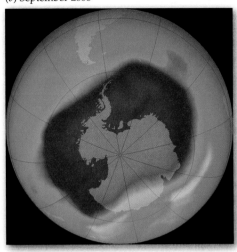

Total Ozone (Dobson Units)
110 220 330 440 550

Figure 20.1 The Antarctic Ozone Hole

These computer-generated maps of the southern end of Earth depict the total concentration of ozone in the air column above Antarctica for the months of (a) September 1979 and (b) September 2006. Ozone concentrations were measured (in Dobson units) by instruments aboard NASA's *Aura* satellite. The ozone "hole" refers to the extreme thinning of the ozone layer to values below 150 Dobson units (deep purple color). At 29 million square miles, the 2006 ozone hole above Antarctica was a record setter. The September 1979 image represents near-normal conditions, prior to significant loss of ozone in the upper atmosphere. The ozone hole waxes and wanes from year to year because the concentration of CFCs over the polar regions is influenced in complex ways by weather conditions.

oxygen atoms (O) interact with chlorine monoxide (ClO) to regenerate the destructive chlorine atom:

$$\text{ClO} + \text{O} \longrightarrow \text{Cl} + \text{O}_2$$

(Chlorine monoxide) (Oxygen atom) (Chlorine atom) (Oxygen gas)

The free chlorine atom released in this second reaction begins another round of reactions with ozone, as summarized by the first of the two equations here. A single CFC molecule can lead to the destruction of about 100,000 molecules of ozone. As ozone is depleted through this reaction, the thickness of the protective layer decreases. The complex airflow patterns of our upper atmosphere tend to funnel CFCs (and also other atmospheric pollutants) over the polar regions. The resulting ozone depletion creates a "hole" in the protective ozone layer above the poles (Figure 20.1). Low temperatures in the upper atmosphere promote the chemical processes that generate the highly reactive chlorine atoms. CFC accumulation and ozone destruction are more severe over Antarctica than over the Arctic Ocean because of the wind patterns that prevail in the southern hemisphere.

Decreases in the ozone layer may have many consequences, including more skin cancers in people, reduced yields for some crops, and reduced populations of phytoplankton, the small photosynthetic aquatic organisms on which all other aquatic organisms depend for food. Decreases in phytoplankton can result in decreased fish populations, which in turn lead to decreased catches by people and hence economic losses.

In the "Science Toolkit" box in Chapter 25, we describe how the discovery of the relationship between CFC pollution and ozone depletion led to an international treaty (the Montreal Protocol) to phase out the production and use of CFCs. Since the signing of the treaty in 1987, the rate of ozone depletion has declined, but the ozone layer is far from restored. At the twentieth anniversary of the treaty, in 2007, the member nations of the United Nations made a firmer commitment to the goals of the treaty, with developed nations pledging to phase out CFCs completely by 2020 and developing nations promising to do so by 2030.

As the examples of invasive species and the ozone hole illustrate, the actions people take can affect natural systems, which in turn affect us. Ecologists study how natural systems work, and they also analyze how we are affecting life on Earth and what the consequences of those actions are likely to be. As such, ecology is an important area of biology both because it gives us a deeper understanding of the world around us and because it can help us fix current environmental problems and prevent future ones.

20.2 Interactions with the Environment

All organisms interact with their environment. These interactions go both ways: organisms affect their environment (as when a beaver builds a dam that blocks the flow of a stream and creates a pond or lake), and the environment affects organisms (as when an extended drought limits the growth of plant species that the beaver depends on for food). Because the interactions go both ways, the organisms and physical environments of the biosphere can be thought of as forming a web of interconnected relationships.

Thinking of the biosphere as an interconnected web can also help us understand the consequences of human

Figure 20.2 An Explosion in Numbers

Red kangaroo numbers increased 166 times when their dingo predators were removed from the rangelands located south of the world's longest fence (red line).

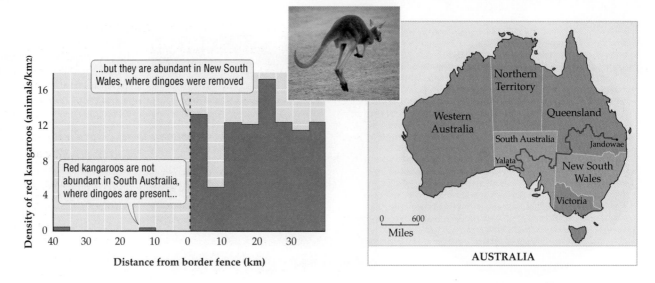

...but they are abundant in New South Wales, where dingoes were removed

Red kangaroos are not abundant in South Austrailia, where dingoes are present...

Distance from border fence (km)

Density of red kangaroos (animals/km2)

AUSTRALIA

Dingo

actions. Consider what happened when people used fencing, poison, and hunting to remove dingoes from a large region of rangeland in Australia. Dingoes, a type of wild dog, were removed to prevent them from eating sheep. As Figure 20.2 shows, in areas where dingoes were removed, the population of their preferred prey, red kangaroos, increased dramatically (166 times). Kangaroos decrease the food available to sheep because kangaroos and sheep like to eat the same plants. In addition, in times of drought, kangaroos resort to a behavior not found in sheep: they dig up and eat belowground plant parts. This behavior has the potential to change the numbers and types of plant species found in rangeland, thereby further increasing the effect that kangaroos have on sheep.

When people removed dingoes, the subsequent effects were not what they expected or desired, because red kangaroos outcompeted their sheep for food. With the advantage of hindsight, the negative side effects of removing dingoes seem predictable, because changes that affect one part of the biosphere (such as removal of dingoes) can have a ripple effect to produce changes elsewhere (such as increases in red kangaroos and decreases in the food available to sheep).

The examples given in this section illustrate how natural systems can be viewed as an interconnected web. To understand such connections further, we need to learn more about the physical factors that affect the biosphere.

Concept Check

1. Can ecological studies have economic value? Explain, using an example.

2. How have human activities contributed to the hole in the ozone layer? What are the likely consequences for humans and the natural world if the ozone hole grows larger?

20.3 Climate Has a Large Effect on the Biosphere

Weather refers to the temperature, precipitation (rainfall and snowfall), wind speed and direction, humidity, cloud cover, and other physical conditions of Earth's lower atmosphere at a specific place over a short period of time. **Climate** is more inclusive in area and time frame: it refers to the prevailing weather conditions experienced in a region over relatively long periods of time (30 years or more). Weather, as we all know, changes quickly and is hard to predict. Climate is more predictable.

Climate has major effects on ecological interactions because organisms are more strongly influenced by climate than by any other feature of their environment. On land, for example, whether a particular region is desert, grassland, or tropical forest depends primarily on such features of climate as temperature and precipitation. In the next section we look at the factors that determine the climate of a given geographic region.

Incoming solar radiation shapes climate

Sunlight strikes Earth directly at the equator, but at a slanted angle near the North and South poles. This difference causes more solar energy to reach the equator, and the regions on either side of it (the tropics), than the poles. Tropical regions receive 2.5 times the solar radiation that reaches polar regions, making them much warmer than the poles. Tropical regions show small seasonal fluctuations in temperature, so organisms that live there experience a relatively warm, stable climate throughout the year. Generally speaking, sunlight and warmth promote photosynthesis and thereby increase the productivity of plants and other light-dependent producers. Productivity is a measure of the energy that producers are able to store in the form of biological material, or biomass. Consumers, including all animals and all types of decomposers, are dependent on the productivity of producers.

Wind and water currents affect climate

Near the equator, intense sunlight heats moist air, causing the air to rise from the surface of Earth. Warm air rises because heat causes it to expand and therefore to be less dense, or lighter, than air that has not been heated. The warm, moist air cools as it rises. Because cool air cannot hold as much water as warm air can, much of the moisture is "wrung out" from a cooling air mass and falls as rain (Figure 20.3).

Usually, cool air sinks. The cool air above the equator, however, cannot sink immediately, because of the warm air rising beneath it. Instead, the cool air is drawn to the north and south, tending to sink back to Earth at about 30° latitude. Part of the air mass flows back toward the equator, and as it does so, it absorbs moisture from Earth's surface. By the time it reaches the equator, the air is once more warm and moist, so it rises, repeating the cycle.

Earth has four giant **convection cells** in which warm, moist air rises and cool, dry air sinks (see Figure 20.3). Two of the four convection cells are located in tropical regions and two in polar regions, where they generate relatively consistent wind patterns. In temperate regions (roughly 30° to 60° latitude), winds are more variable, and there are no stable convection cells. Precipitation occurs when cool, dry air from polar regions collides with warm, moist air moving north, so that most temperate regions receive ample moisture.

The winds produced by the four giant convection cells do not move straight north or straight south, relative to the land spinning below. Instead, because of Earth's rotation, these winds appear to curve as they travel

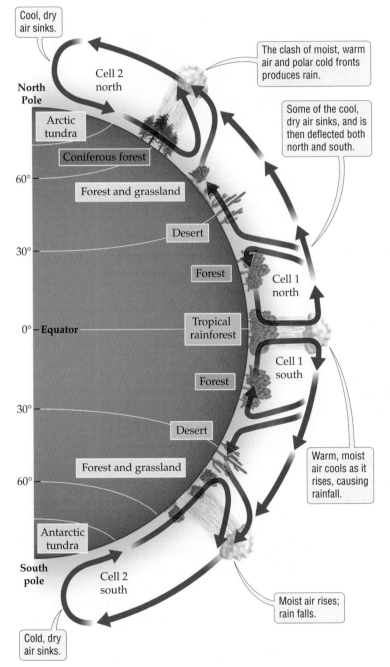

Cool, dry air sinks.

Cell 2 north

North Pole

Arctic tundra

Coniferous forest

60°

Forest and grassland

Desert

30°

Forest

Cell 1 north

0° Equator

Tropical rainforest

Cell 1 south

Forest

30°

Desert

Forest and grassland

60°

Antarctic tundra

South pole

Cell 2 south

Cold, dry air sinks.

The clash of moist, warm air and polar cold fronts produces rain.

Some of the cool, dry air sinks, and is then deflected both north and south.

Warm, moist air cools as it rises, causing rainfall.

Moist air rises; rain falls.

Figure 20.3 Earth Has Four Giant Convection Cells

Two giant convection cells are located in the Northern Hemisphere and two in the Southern Hemisphere. In each convection cell, relatively warm, moist air rises, cools, and then releases moisture as rain or snow. The cool, dry air then sinks to Earth and flows toward regions where warm air is rising. Some of the descending air heads poleward at about 30°, and because these wind currents are dry, most of the world's deserts are found at these latitudes.

Learn more about these convection cells. ▶❚❚ 20.1

Figure 20.4 Prevailing Winds Are Determined by Global Patterns of Air Circulation

The four giant convection cells determine the basic pattern of air circulation on Earth. Earth's rotation causes winds to curve to the east or west. The direction in which they curve depends on their latitude, but for any given geographic region on Earth, the winds usually blow from a consistent direction. The photo shows westerly wind flow across a wheat field in North Dakota.

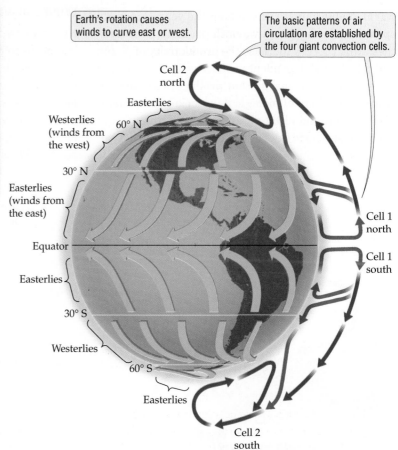

Earth's rotation causes winds to curve east or west.

The basic patterns of air circulation are established by the four giant convection cells.

Cell 2 north

Easterlies

Westerlies (winds from the west)

60° N

30° N

Easterlies (winds from the east)

Equator

Cell 1 north

Cell 1 south

Easterlies

30° S

Westerlies

60° S

Easterlies

Cell 2 south

near Earth's surface (Figure 20.4). Winds traveling toward the equator appear to curve to the right. To an Earth-bound observer, these winds seem to blow from the east; hence, such winds are called easterlies ("from the east"). Similarly, winds that travel toward the poles curve to the left; and because they seem to blow from the west, these winds are called westerlies. At any given location the winds usually blow from a consistent direction; these patterns of air movement are known as prevailing winds. In southern Canada and much of the United States, for example, winds blow mostly from the west, so storms in these regions usually move from west to east.

Ocean currents also have major effects on climate. The rotation of Earth, differences in water temperatures between the poles and the tropics, and the directions of prevailing winds all contribute to the formation of ocean currents. In the Northern Hemisphere, ocean currents tend to run clockwise between the continents; in the Southern Hemisphere, they tend to run counterclockwise (Figure 20.5).

Ocean currents carry a huge amount of water and can have a great influence on regional climates. The Gulf Stream, for example, moves 25 times as much water as is carried by all the world's rivers combined. Without the warming effect of the water carried by this current, the climate in countries such as Great Britain and Norway would be subarctic instead of temperate. Overall, the Gulf Stream causes cities in western Europe to be warmer than cities at similar latitudes in North America, as illustrated by a comparison of Rome with Boston, Paris with Montréal, and Stockholm with Kuujjuaq in Québec, Canada (Figure 20.5).

The major features of Earth's surface also shape climate

The climate of a place may also be affected by the presence of large lakes, the ocean, and mountain ranges. Heat is absorbed and released more slowly by water than by land. Because they retain heat comparatively well, large lakes and the ocean moderate the climate of the surrounding lands. Mountains can also have a large effect on a region's climate. For example, mountains often produce a **rain shadow** effect, in which little precipitation falls on the side of the mountain that faces away from the prevailing

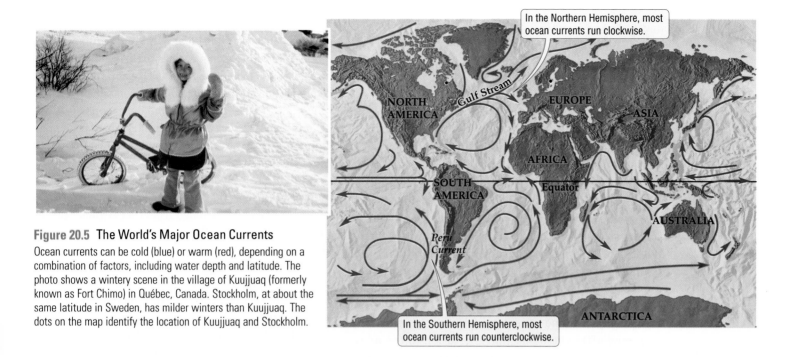

Figure 20.5 The World's Major Ocean Currents
Ocean currents can be cold (blue) or warm (red), depending on a combination of factors, including water depth and latitude. The photo shows a wintery scene in the village of Kuujjuaq (formerly known as Fort Chimo) in Québec, Canada. Stockholm, at about the same latitude in Sweden, has milder winters than Kuujjuaq. The dots on the map identify the location of Kuujjuaq and Stockholm.

In the Northern Hemisphere, most ocean currents run clockwise.

In the Southern Hemisphere, most ocean currents run counterclockwise.

winds (Figure 20.6). In the Sierra Nevada of North America, five times as much precipitation falls on the west side of the mountains (which faces toward winds that blow in from the ocean) as on the east side, where the lack of precipitation contributes to the formation of deserts. Mountain ranges in northern Mexico, South America, Asia, and Europe also create rain shadows.

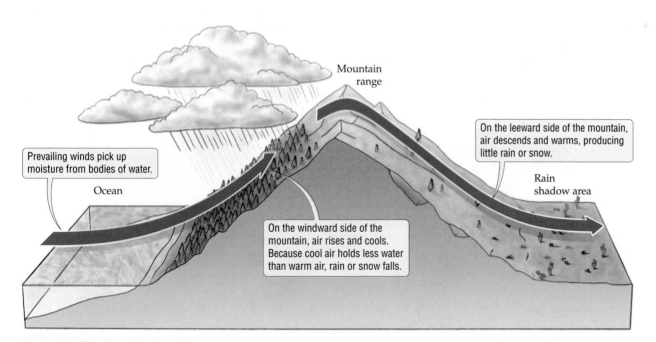

Prevailing winds pick up moisture from bodies of water.

On the windward side of the mountain, air rises and cools. Because cool air holds less water than warm air, rain or snow falls.

On the leeward side of the mountain, air descends and warms, producing little rain or snow.

Figure 20.6 The Rain Shadow Effect
The side of a high mountain that faces the prevailing winds (the windward side) receives more precipitation than the side of the mountain that faces away from the prevailing winds (the leeward side). The leeward side is therefore said to be in a rain shadow.

Learn more about a rain shadow.
20.2

20.4 Life on Land

Now that we've discussed factors that influence climate, let's explore the effects of climate on the biosphere in more detail.

The biosphere can be divided into biomes

Ecologists categorize large areas of the biosphere into distinct regions, called **biomes**, based on the unique climatic and ecological features of each such region. Because climate strongly influences the life-forms that can live in a particular place, each biome is associated with a characteristic set of plant and animal species. A biome may encompass more than one ecosystem and typically stretches across large swaths of the globe.

Biomes on land—terrestrial biomes—are usually named after the dominant vegetation in the area. Aquatic biomes are classified on the basis of physical and chemical features, such as salt content. Earth's major terrestrial biomes include **tundra**, **boreal forest**, **temperate deciduous forest**, **grassland**, **chaparral**, **desert**, and **tropical forest** (Figure 20.7). In this section we'll delve into each of these biomes in some detail. In Section 20.5 we'll take a closer look at aquatic biomes.

The location of terrestrial biomes is determined by climate and human actions

Climate is the single most important factor controlling the potential (natural) location of terrestrial biomes. The climate of the area—most important, the temperature and the amount and timing of precipitation—allows some species to thrive and prevents other species from living there. Overall, the effects of temperature and moisture on different species cause particular biomes to be found under a consistent set of conditions (Figure 20.8). Terrestrial biomes change from the equator to the poles, and from the bottom to the top of mountains.

Climate can exclude species from a region directly or indirectly. Species that cannot tolerate the climate of a region are directly excluded from that region. Species that can tolerate the climate but are outperformed by other organisms that are better adapted to that climate are excluded indirectly.

Although climate places limits on where biomes can be found, the actual extent and distribution of biomes in the world today are very strongly influenced by human activities. We will return to the effects of humans on natural biomes later in this chapter and also when we discuss global change in Chapter 25. First, let's tour seven great terrestrial biomes of our planet to examine the distinctive features of each.

The tundra is marked by cold winters and a short growing season

The word "tundra" comes from *tunturi*, which means "treeless plain" in the language of the Sami people of Finland. The arctic tundra covers nearly one-fourth of Earth's land surface, encircling the North Pole in a vast sweep that includes half of Canada and Alaska and sizable portions of northern Europe and Russia. A similar

Identify the major terrestrial biomes.

20.3

Figure 20.7 Major Terrestrial Biomes

This map shows the potential distribution of some of the major terrestrial biomes on Earth. The actual distribution of biomes is heavily influenced by human activities.

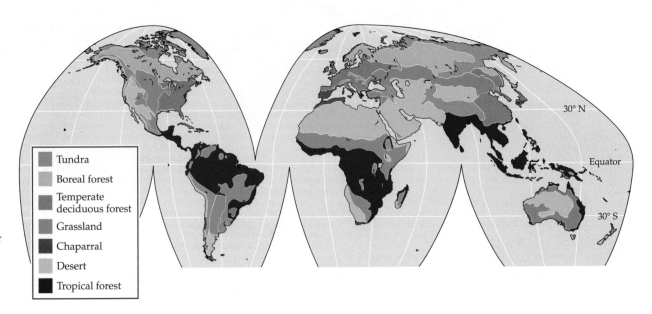

- Tundra
- Boreal forest
- Temperate deciduous forest
- Grassland
- Chaparral
- Desert
- Tropical forest

30° N

Equator

30° S

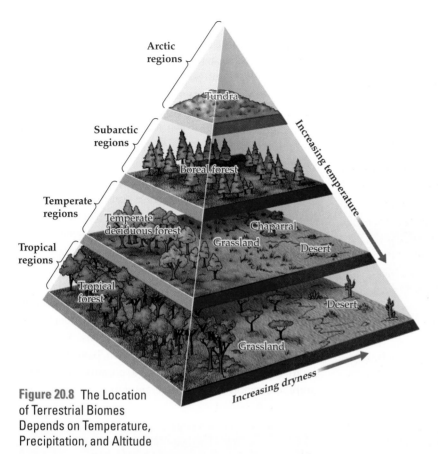

Figure 20.8 The Location of Terrestrial Biomes Depends on Temperature, Precipitation, and Altitude

(Labels on pyramid: Arctic regions; Subarctic regions; Temperate regions; Tropical regions; Increasing temperature; Tundra; Boreal forest; Temperate deciduous forest; Chaparral; Grassland; Desert; Tropical forest; Grassland; Desert; Increasing dryness)

Test your knowledge of biomes and climate zones.
20.4

food source for herbivores such as reindeer (known as caribou in North America). Rodents such as lemmings, voles, and arctic hare provide food for carnivores like arctic foxes and wolves. Bear and musk ox are among the few large mammals. Insects abound in the summer, along with the migratory birds that feed on them. There are few amphibians and even fewer reptiles.

A few coniferous species dominate in the boreal forest

The boreal forest (from *borealis*, "northern") is the largest terrestrial biome. It is also known as the taiga [*TYE*-guh], based on the Mongolian word for coniferous forests. It includes the subarctic landmass immediately south of the tundra, covering a broad belt of Alaska, Canada, northern Europe, and Russia, approximately between 60° and 50° north latitude. Winters in the boreal forest are nearly as cold as in the tundra and last about half the year. Summers are longer and warmer than in the tundra, with temperatures reaching as high as 30°C (86°F) in some boreal forests of

habitat, known as the alpine tundra, is encountered above the tree line in high mountains.

Winter temperatures in the arctic tundra can dip to –50°C (–58°F), although they average about –34°C (–29°F). Even in summer, temperatures don't climb much above 12°C (54°F) and freezing weather is not uncommon. The land is frozen for 10 months of the year, thawing to a depth of no more than a meter (about 3 feet) during the short summer. Below these surface layers of soil is **permafrost**, permanently frozen soil that may be quarter of a mile deep. Precipitation in the arctic tundra ranges from 15 to 25 centimeters (up to 10 inches) per year—lower than in many of the world's deserts. However, evaporation is low because of the cold. In the summer, ice melt and the thawing of the upper layers of soil create an abundance of bogs, ponds, and streams, which are prevented from draining because of the underlying permafrost.

Trees are absent or scarce in the tundra (Figure 20.9) because of the short growing season and because the permafrost is a barrier to the deep taproots of woody plants. The vegetation is dominated by low-growing flowering plants, such as grasses, sedges, and members of the heath family. The rocky landscape is covered in mosses and lichens, which are an important

Figure 20.9 Tundra in Denali National Park, Alaska

Tundra is found at high latitudes and high elevations, and it is dominated by low-growing shrubs and nonwoody plants that can cope with a short growing season.

Figure 20.10 Boreal Forest, Banff National Park, Alberta
Boreal forests are dominated by coniferous (cone-bearing) trees that grow in northern or high-altitude regions with cold, dry winters and mild summers.

the world. Typically, the soil is thin and nutrient-poor. Rainfall is low in most boreal forests, but it may be high in some areas, most notably the Pacific Coast rainforests of North America. Because evaporation rates are low in these cold, northern latitudes, plants in the boreal forest generally receive adequate moisture during the growing season.

Conifers, which are cone-bearing trees with needle-like leaves, dominate the boreal forest vegetation (Figure 20.10). Spruce and fir are the most common species in the North American boreal forest; pine and larch are abundant in Scandinavia and Russia. Broad-leaved species such as birch, alder, willow, and aspen are also found, especially in the southern ranges of this biome. Plant diversity is relatively low, except in the rainforests of the Pacific Coast, where the seaside climate is milder and soils are richer. The large herbivores of the boreal forest include elk and moose. Small carnivores, such as weasel, wolverine, and marten, are common. Larger carnivores include lynx and wolves. Bears, which are omnivores, are found throughout the world's boreal forests.

Temperate deciduous forests have fertile soils and relatively mild winters

Temperate deciduous forests are a familiar forest type for most people who live in North America. They also constitute the dominant biome in large parts of Europe and Russia and parts of China and Japan. Temperate deciduous forests occur in regions with a distinct winter, with subfreezing weather that may last 4 or 5 months. However, winter temperatures are not as harsh as in the arctic and subarctic biomes, and summers are typically much warmer. Temperatures commonly range from lows of –30°C (–22°F) in winter to summertime highs above 30°C (86°F). Annual rainfall varies from 60 centimeters (nearly 24 inches) to more than 150 centimeters (59 inches), and precipitation is distributed evenly through much of the year.

Deciduous trees—that is, trees that drop their leaves in the cold season—are the dominant vegetation in this biome (Figure 20.11). These forests display greater species diversity than the tundra and boreal biomes. Oak, maple, hickory, beech, and elm are common in these woodlands, which also harbor an understory of shade-tolerant shrubs and herbaceous plants forming a ground cover. Coniferous species, such as pine and hemlock, also occur, but they are not the dominant trees. The fauna includes squirrel, rabbit, deer, raccoon, beaver, bobcat, mountain lion and bear. There are many fishes, and amphibians and reptiles are common.

Figure 20.11 Temperate Deciduous Forest, Poconos Mountains, Pennsylvania
Temperate deciduous forests are dominated by trees and shrubs adapted to relatively rich soil, snowy winters, and moist, warm summers.

Grasslands appear in regions with good soil but relatively little moisture

The grassland biome is characteristic of regions that receive about 25 to 100 centimeters (about 10–40 inches) of precipitation annually. The moisture levels are insufficient for vigorous tree growth but they are not as low as those in a desert. Grasslands are found in both temperate and tropical latitudes. The prairies of North America, the steppe in Russia and central Asia, and the pampas of South America are examples of temperate grasslands. Tropical and subtropical grasslands are known as savanna. Grasses dominate in this biome, although the landscape may be dotted with shrubs and small trees, as in the African savanna (Figure 20.12a). Soils in some grasslands, especially the prairies of North America and the pampas of Argentina, are exceptionally deep and fertile. As a result, most of these areas have been converted to agriculture, and today they are some of the most productive grain-growing regions of the world.

Before they succumbed to the plow, the grasslands of central North America formed the largest stretch of grassland biome in the world. The northern and central parts of the Great Plains receive moderately good rainfall—about 100 centimeters annually—enough to support the growth of "hat-high" grasses (like big bluestem, averaging about 6 feet in height). The tallgrass prairies, as these grasslands are called, once stretched from Manitoba in Canada, through the Dakotas and Nebraska, south to Kansas and Oklahoma. They also extended into neighboring states to the east, as far as the western edge of Indiana.

Only about 1 percent of the original prairie remains, most of it in protected areas. Grasses, and herbaceous plants such as coneflowers and shooting stars, predominate in these grasslands. That predominance is maintained

(a) A savanna in Kenya

(b) Northern mixed grassland

(c) Western short grassland

Figure 20.12 Grassland

Grasslands are common throughout the world and are dominated by grasses, although scattered trees are found in some such as the tropical grasslands known as the Savanna. (a) A lone Acacia tree in the Lewa Conservancy Savanna in Kenya. (b) Purple cone flowers (*Echinacea purpurea*) in a mixed grassland, Theodore Roosevelt National Park, North Dakota. (c) Buffalo grass (*Buchloe dactyloides*) in a short grassland.

by prairie fires, which destroy shrubs and trees but not the underground roots and stems of prairie plants. Burrowing rodents like voles and prairie dogs (a type of ground squirrel) aerate the soil, thereby improving growing conditions for the root systems of prairie plants. There are many butterflies and other insects, and the greatest diversity of mammals in North America is found here.

Rainfall declines farther west, and the prairie is replaced by mixed grassland (Figure 20.12b), and then short grassland along the eastern foothills of the Rocky Mountains from Montana, Wyoming, and Colorado, south to New Mexico. Drought-resistant grasses, such as buffalo grass, are the dominant vegetation (Figure 20.12c), and bison and pronghorn are the larger herbivores of the short grasslands.

Chaparral is characteristic of regions with wet winters and hot, dry summers

The chaparral [SHAP-uh-RAL] is a shrubland biome dominated by dense growths of scrub oak and other drought-resistant plants (Figure 20.13). It is found in regions with a "Mediterranean climate," characterized by cool, rainy winters and hot, dry summers. Annual rainfall ranges from 25 to 100 centimeters (about 10–40 inches). Chaparral occurs in regions of southern Europe and North Africa bordering the Mediterranean Sea, and also in coastal California, southwestern Australia, and along the west coast of Chile and South Africa.

The soil is relatively poor in these habitats, and most species are adapted to hot, dry conditions. Many chaparral plants have thick, leathery leaves that reduce water loss. Low moisture and high temperatures in summer make the chaparral exceptionally susceptible to wildfires. Common vegetation in the California chaparral includes scrub oak, pines, mountain mahogany, manzanita, and the chemise bush. The California quail is an iconic bird of the chaparral. Rodents such as jackrabbits and gophers are common, and there are many species of lizards and snakes. Mammals include deer, peccary (a piglike animal), lynx, and mountain lion.

The scarcity of moisture shapes life in the desert

The desert biome makes up one-third of Earth's land surface. It is characterized by less than 25 centimeters (10 inches) of precipitation annually. The defining feature of a desert is the scarcity of moisture, not high temperatures. Antarctica, which receives less than 2 centimeters of precipitation per year, is the largest cold desert in the world. The Sahara desert in northern Africa is the largest hot desert. Because desert air lacks moisture, it cannot retain heat and therefore it cannot moderate daily temperature fluctuations. As a result, temperatures can be above 45°C (113°F) in the daytime, then plunge to near freezing at night.

Figure 20.13 Chaparral, North of San Francisco, California
Chaparral is characterized by shrubs and small, nonwoody plants that grow in regions with winter rain and hot, dry summers.

Figure 20.14 Desert, near Phoenix, Arizona
Deserts form in regions with low precipitation, usually 25 centimeters (10 inches) per year or less. The photo shows the saguaro cactus (tall green columns) and other plants in the Sonoran Desert.

Desert plants have small leaves; the reduced surface area minimizes water loss. Succulents, such as cacti, store water in their fleshy stems or leaves (Figure 20.14). Some desert plants produce enormously long taproots that are able to reach subsurface water. Most animals in the desert are nocturnal, hiding in burrows during the heat of the day and emerging at night to feed. Jackrabbits have large ears that act like a radiator to help dissipate heat. Desert mammals, such as the desert fox, have light-colored fur, which deflects some of the radiant energy from the sun. The kangaroo rat loses very little water in urine because its kidneys enable it to recover water with exceptional efficiency. The rodent's respiratory passages wring moisture from exhaled air by cooling it. A similar mechanism enables the camel's nose to recover moisture from the air it breathes out.

Tropical forests have high species diversity

The tropical forest biome is characterized by warm temperatures and about 12 hours of daylight all through the year. Rainfall may be heavy throughout the year or occur only during a pronounced wet season. Tropical rainforests, which are tropical forests that remain wet all year round, may receive in excess of 200 centimeters (80 inches) of rain annually. Because organic matter decomposes rapidly at warm temperatures, there is little leaf litter in tropical forests. Soils in this biome tend to be nutrient-poor for two main reasons: First, a large percentage of nutrients are locked up in the living tissues (biomass) of organisms, especially large trees. Second, heavy rains tend to leach out soil nutrients, depleting certain minerals in particular.

The abundance of sunshine and moisture makes tropical rainforests the most productive terrestrial habitats. They are also hotspots for diversity of life-forms (biodiversity). Tropical rainforests currently occupy about 6 percent of Earth's land area but harbor nearly 50 percent of its plant and animal species (Figure 20.15). South America has the largest tracts of tropical rainforest, especially in Brazil, Peru, and Bolivia. Large areas of Southeast Asia and equatorial Africa are also covered in tropical rainforests. However, these forests are under severe threat from logging and clearing for livestock grazing and other types of agricultural activity. More than half of the original tropical rainforests has been lost. About 2 acres of tropical rainforest are lost every second. These forests represent an enormous sink for CO_2 and their loss is likely to worsen global warming (see Chapter 25).

Figure 20.15 Tropical Forest, El Yunque National Forest, Puerto Rico

Tropical forests form in warm regions with seasonal, or year-round, rain. Tropical rainforests, which receive abundant moisture throughout the year, are some of the most productive ecosystems on Earth. They have a rich diversity of trees, vines, and shrubs.

20.5 Life in Water

Life evolved in water billions of years ago, and aquatic ecosystems cover about 75 percent of Earth's surface. Aquatic biomes are shaped primarily by the physical characteristics of their environment, such as salt content, water temperature, water depth, and the speed of water flow. Within an aquatic biome, we can recognize various habitats, or ecological zones—defined by their nearness to the shore, water depth, and the depth to which light penetrates. The bottom surface of any body of water, for example, is classified as the **benthic zone**.

Two main types of aquatic biomes can be distinguished on the basis of salt content: freshwater and marine. Lakes, rivers, and wetlands are examples of ecosystems within the freshwater biome. Estuaries, coral reefs, the coastal region, and the open ocean are examples of ecosystems within the marine biome.

Aquatic biomes are influenced by terrestrial biomes and climate

Aquatic biomes, especially lakes, rivers, wetlands, and the coastal portions of marine biomes, are heavily influenced by the terrestrial biomes that they border or through which their water flows. High and low points of the land, for example, determine the locations of lakes and the speed and direction of water flow. In addition, when water drains from a terrestrial biome into an aquatic biome, it brings with it dissolved nutrients (such as nitrogen, phosphorus, and salts) that were part of the terrestrial biome. Rivers and streams carry nutrients from terrestrial environments to the ocean, where they may stimulate large increases in the abundance of phytoplankton.

Aquatic biomes also are strongly influenced by climate. Climate helps determine the temperature, depth, and salt content of the world's oceans, for example. Such physical conditions of the ocean have dramatic effects on the organisms that live there, and hence climate has a powerful effect on marine life. Consider the El Niño [ell-*NEEN*-yoh] events that are often reported in the news. These events begin when warm waters from the west deflect the cold Peru Current along the Pacific coast of South America (Figure 20.16). The results of this change are spectacular, including dramatic decreases in numbers of fish, die-offs of seabirds, storms along the Pacific coast of North America that destroy underwater "forests" of a brown alga called kelp, crop failures in Africa and Australia, and drops in sea level in the western Pacific that kill huge numbers of coral reef animals.

Aquatic biomes are also influenced by human activity

Like terrestrial biomes, aquatic biomes are strongly influenced by the actions of humans. Portions of some aquatic biomes, such as wetlands and estuaries, are often destroyed to allow for development projects. Rivers, wetlands, lakes, and coastal marine biomes are negatively affected by pollution in most parts of the world. Aquatic biomes also suffer when humans destroy or modify the terrestrial biomes they occupy. For example, when forests are cleared for timber or to make room for agriculture, the rate of soil erosion increases dramatically because trees are no longer there to hold the soil in place. Increased erosion can cause streams

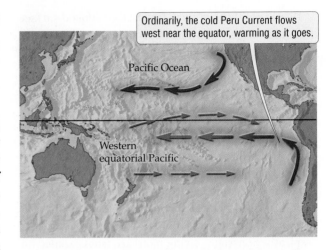

Ordinarily, the cold Peru Current flows west near the equator, warming as it goes.

Pacific Ocean

Western equatorial Pacific

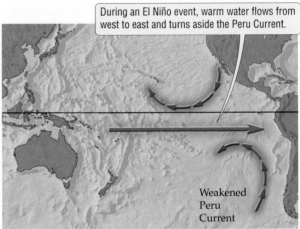

During an El Niño event, warm water flows from west to east and turns aside the Peru Current.

Weakened Peru Current

Figure 20.16 El Niño Events
During an El Niño event, winds from the west push warm surface water from the western Pacific to the eastern Pacific. The resulting changes in sea surface temperatures cause changes in ocean currents (shown here in blue for cold, red for warm). El Niño events cause many additional changes (not shown here), altering wind patterns, sea levels, and patterns of precipitation throughout the world.

and rivers to become clogged with silt, which harms or kills invertebrates, fishes, and many other species.

Next, we take a closer look at six ecosystems that are commonly found within the freshwater or marine biome: **lake**, **river**, **wetland**, **estuary**, **coastal region**, and **oceanic region**.

Lakes, rivers, and wetlands are part of the freshwater biome

Lakes are standing bodies of fresh water of variable size, ranging from thousands of square meters to thousands of square kilometers (Figure 20.17*a*). The productivity of a lake, and the abundance and distribution of its life-forms, is strongly influenced by nutrient concentrations, water depth, and the extent to which the lake water is

(a) Lake

(b) River

(c) Wetland

Figure 20.17 Lakes, Rivers, and Wetlands

(a) Lakes are standing bodies of fresh water of variable size, ranging from a few square meters to thousands of square kilometers. (b) Rivers are bodies of fresh water that move continuously in a single direction. (c) Wetlands are characterized by shallow waters that flow slowly over lands that border rivers, lakes, or ocean waters.

mixed. Northern lakes tend to be clear because they usually have low nutrient concentrations and therefore do not support vigorous growth of photosynthetic plankton (floating microscopic organisms). Lakes with higher nutrient concentrations appear more turbid because plankton thrive in such lakes. In temperate regions, seasonal changes in temperature cause the oxygen-rich water near the top of a lake to sink in the fall and the spring, bringing oxygen to the bottom of the lake. In tropical regions, seasonal differences in temperature are not great enough to cause a similar mixing of water. This lack of mixing causes the deep waters of tropical lakes to have low oxygen levels and relatively few forms of life.

Rivers are bodies of fresh water that move continuously in a single direction (Figure 20.17b). The physical characteristics of a river tend to change along its length. At its source—whether glacier, lake, or underground spring—the current is stronger and the water colder. The waters in these upper reaches are highly oxygenated because O_2, like most other gases, dissolves more readily in cold water and the turbulence created by rapids and riffles causes more of the gas to dissolve in water. As they approach their emptying point, rivers become wider, slower, warmer, and less oxygenated.

Wetlands are characterized by standing water shallow enough that rooted plants emerge above the water surface. A bog is a freshwater wetland with stagnant, acidic, oxygen-poor water. Because bogs are nutrient-poor, their productivity and species diversity are low. In contrast, marshes and swamps are highly productive

Figure 20.18 Estuaries
Estuaries are tidal ecosystems where rivers flow into the ocean. They are usually classified as part of the marine biome. This photo shows a salt marsh at sunrise in Rachel Carson Wildlife Refuge, Maine.

wetlands. A marsh is a grassy wetland (Figure 20.18), but swamps are dominated by trees and shrubs.

Estuaries and coastal regions are highly productive parts of the marine biome

The marine biome is the largest biome on our planet, since it includes the open oceans that cover about three-fourths of Earth's surface. This biome is critically important for us, for all other biomes, and for our planet

in general. Simply because of the sheer size of this biome, photosynthetic plankton in the oceans release more oxygen (and absorb more CO_2) through photosynthesis than do all the terrestrial producers put together. Much of the rain that falls on terrestrial biomes comes from water evaporated off the surface of the world's oceans.

Estuaries are the shallowest of the marine ecosystems. An estuary is a region where a river empties into the sea. It is marked by the constant ebb and flow of fresh water and salt water, and all organisms that thrive on its bounty have to be able to tolerate daily and seasonal fluctuations in salinity (salt levels). The water depth fluctuates with ocean tides and river floods, but most of the time it is shallow enough that light can reach the bottom. The plentiful light, the abundant supply of nutrients delivered by the river system, and the regular stirring of nutrient-rich sediments by water flow create a rich and diverse community of photosynthesizers. Grasses and sedges are the dominant vegetation in most estuaries (Figure 20.18b). The producers provide food and shelter for a varied and prolific community of invertebrates and vertebrates, including crustaceans, shellfish, and fishes. The abundance and diversity of life make estuaries one of the most productive ecosystems on our planet.

The coastal region stretches from the shoreline to the edge of the continental shelf, which is the undersea extension of a continent (Figure 20.19). The coastal region is among the most productive marine ecosystems because of the ready availability of nutrients and

Figure 20.19 Ecological Zones in the Marine Biome
This diagram depicts a cross-sectional view of the land and water, progressing from the shoreline toward the open ocean (oceanic region). The coastal region stretches out to sea as far as the continental shelf, which is the underwater extension of a continent's rim. Productivity, and the abundance of life forms, declines with increasing water depth because sunlight, which producers need for photosynthesis, diminishes with depth. The sunlit zone, or photic zone, extends to a depth of 200 meters (656 feet). The waters are dimly lit up to 1,000 meters, but lie in complete darkness at greater depths. Productivity also decreases with distance away from the shore because nutrient levels typically decline farther out to sea. The well-lit waters of the open ocean are less productive than the well-lit regions of coastal areas, and the deepest layers of the ocean (the abyssal zone) are typically the least productive of all aquatic habitats.

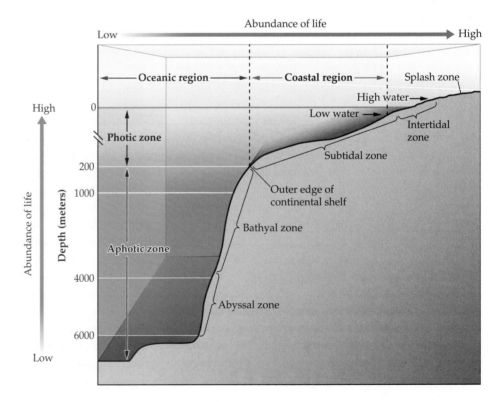

oxygen. Nutrients delivered by rivers and washed off the surrounding land accumulate in the coastal region. Nutrients that settle to the bottom are stirred up by wave action, tidal movement, and the turbulence produced by storms. The nutrients support the growth of photosynthetic producers, which inhabit the well-lit upper layers of coastal waters to a depth of about 80 meters (260 feet). The vigorous mixing by wind and waves also adds atmospheric oxygen to the water. A majority of Earth's marine species live in the coastal region. Not surprisingly, most of the world's highly productive fisheries are also located along coasts.

The **intertidal zone** is the part of the coast that is closest to the shore, where the ocean meets the land. This ecological zone spans the uppermost reaches of ocean waves during the highest tide and the shoreline that remains submerged at the lowest of low tides. In other words, the intertidal zone extends from the highest tide mark to the lowest tide mark.

The intertidal zone is a challenging environment for plants and animals because they must cope with being submerged and exposed to dry air on a twice-daily basis, in addition to being pounded by surf and sand (Figure 20.20a). The organisms that inhabit the upper regions of the intertidal zone are subject to predation by shorebirds and other animals when the tide is out. Despite these challenges, a diverse community of seaweeds, worms, crabs, sea stars, sea anemones, mussels, and other species is adapted to living in this zone.

The benthic zone of coastal regions may lie as deep as 200 meters (656 feet) below the water surface. The coastal benthic zone is a relatively stable habitat, rich in sediments containing the dead and decaying remains of organisms (detritus). The detritus forms the basis of a food web that supports a wide variety of consumers, including sponges, worms, sea stars, sea fans, sea cucumbers, and many fishes (Figure 20.20b).

The productivity of the oceanic region is limited by nutrient availability

The open ocean, or the **oceanic region**, begins about 40 miles offshore, where the continental shelf, and therefore the coastal region, ends. The open oceans form a vast, complex, interconnected ecosystem that we know relatively little about. Although they are well-lit and have sufficient oxygen, the surface layers of the open ocean are much less productive than estuarine and coastal waters because they are relatively nutrient-poor. That's because detritus tends to settle on the seafloor and the nutrients locked in it are not readily stirred up from the great depths of the open ocean.

(a)

(b)

Giant tube worms (*Riftia pachyptila*) on a hydrothermal vent in the abyssal zone of the deep oceans.

Figure 20.20 The Marine Biome

(a) Intertidal zones are found in coastal regions where the tides rise and fall on a daily basis, periodically submerging a portion of the shore. (b) Benthic zones, located on the bottom surfaces of lakes, rivers, wetlands, estuaries, coastal regions, and oceanic regions, are home to a variety of consumers, many of which feed on dead organisms drifting down from the upper, better-lit zones.

Where the continental shelf ends, the seafloor plunges steeply to a depth of approximately 6,000 meters (almost 20,000 feet). The cold, dark waters at these great depths constitute the **abyssal zone**. Few organisms can survive at the great pressures and low temperatures (about 3°C, or 37°F) of the abyssal zone. However, complex communities of archaeans and invertebrates are known to be associated with hydrothermal vents in geologically active regions of the deep oceans. A submarine hydrothermal vent is a crack in the seafloor that releases hot water containing dissolved minerals. Certain archaeans can extract energy from these minerals (see Figure 3.5), and these single-celled prokaryotes form the basis of a food web that supports invertebrates such as giant tube worms (Figure 20.20b), clams, and shrimps.

Concept Check

1. Which aspects of the environment help shape the climate in a region? How does climate affect life-forms?

2. What are some of the physical conditions that influence life in aquatic biomes?

Applying What We Learned

The Human Impact on the Biosphere

The discharge of ships' ballast water opened the way for zebra mussels to colonize the waterways of North America. These mussels have harmed native species and caused serious economic problems, including the shutdown of a Ford Motor Company plant and the closure of a Michigan town's power plant when the mussels choked off pipes supplying cooling water to the furnaces.

Species like the zebra mussel that become major pests in a new environment are referred to as invasive species. Such species often cause much less trouble in their native environment. Zebra mussels, for example, come from Eurasia, where they are held in check by a wide range of predators and parasites. In North America, fewer predators and parasites attack zebra mussels, so the numbers of these mussels were able to increase explosively after their arrival.

What is true of zebra mussels is true of other invasive species: when species are introduced by people to novel environments, the population size of a non-native species may increase greatly because its usual predators, parasites, and competitors are not present. When this occurs, the introduced species may harm the native species of its new home and use large amounts of resources or space. In Interlude A we saw how native species are harmed by invasive species such as the Nile perch in Lake Victoria and the brown tree snake on Guam. When an invasive species has a major and disruptive effect on its new environment, people can suffer great economic losses, as seen with the zebra mussel and dozens of other troublesome invaders.

The effects of humans on the biosphere are not limited to introductions of invasive species. When we control the flow of rivers or eliminate (or even reduce the numbers of) a predator such as the dingo (see page 412), we may also cause outcomes that we did not expect or want. Thinking of the biosphere as an interconnected web suggests that when we change one component of the biosphere, we recognize that our actions may have unintended and undesirable effects on other parts of the biosphere. What's more, global patterns of air and water movement can cause actions in one area to have consequences that reach distant locations. Such long-distance effects are especially evident with pollution: though emitted in one country or continent, pollution can cause problems in other countries or continents (for example, sulfur dioxide emitted in Great Britain can cause acid rain in Scandinavia, as we shall see in Chapter 24).

People can—indeed, to feed and house ourselves we must—take actions that affect the natural world. Given the view of the biosphere as an interconnected web, what do you think people should do when we consider taking actions whose possible outcomes we either don't understand well or suspect may have harmful side effects? Should we let fear or lack of knowledge paralyze us into inaction? How can we use what we know about how the biosphere works to solve current environmental problems and prevent future ones? Although there are no easy answers to such questions, we'll examine how people have begun to address such issues at the close of this unit, in Interlude E.

Invasion of the Cuban Treefrogs

BY RALPH MITCHELL

As the rainy season approaches, all types of amphibians that thrive in the wet conditions and breed in our ponds and puddles will make an appearance.

Amphibians such as frogs and toads are part of our great Florida ecosystem and offer an interesting chorus at night as well as ... some free insect control as they feed.

There is, however, one treefrog that is a problem, not only as a nuisance, but also as an aggressive invader that actually eats native frogs, lizards and other small native creatures—the Cuban treefrog! ...

Like all of our invasive species, Cuban treefrogs came from somewhere else, namely Cuba, the Cayman Islands and the Bahamas.

Sometime in the 1920s, the Cuban treefrog made its appearance in the Florida Keys. Known as great stowaways and able to hitchhike on plants, cars and boats, the Cuban treefrog has expanded its range all the way to Cedar Key, Gainesville and Jacksonville.

... Cuban tree frogs are pushing out and eating native treefrog species such as squirrel treefrogs and green treefrogs. Even the Cuban treefrog tadpoles outcompete native frog tadpoles. These frogs sleep in trees, nooks and crannies during the day and have even been found buried in soil.

What can you do to reduce Cuban treefrog impacts in your own yard?

Eliminate eggs and breeding sites where possible. Also make sure to maintain your swimming pool so it is not an attractive breeding site. Check for and eliminate stagnant sources of water in such things as buckets and old coolers. Encourage the presence of native wildlife that will actually eat Cuban treefrogs such as rat snakes, black racers, garter snakes, owls, crows and some wading birds.

Cuban tree frogs were first noticed in the United States about 80 years ago, when they were brought accidentally to the Florida Keys from Cuba, probably in vegetable crates. Since then, they have migrated north, decimating native species as they go. The invader attempts to eat anything that moves and fits into its mouth. It eats so many native frogs that it can take over their range. It also secretes a toxic chemical that covers its skin and causes painful stings to anyone who picks it up.

The females lay up to 15,000 eggs, and their tadpoles are such voracious eaters that tadpoles of many native species are losing ground to them. Cuban tree frogs are common in fields, ditches, swamps, and ponds, and they don't shy away from suburbia. They sleep through the day and emerge at night to chomp on landscaping plants and lie in wait for insects near porch lights and window wells. They can enter homes through open doors, pipes, and vents. Toilets appear to be their favorite indoor habitat. They manage to enter transformers and other electrical equipment, leading to short-circuits and blackouts that exasperate home owners and businesses and cost utility companies hundreds of thousands of dollars each year.

Before 2004, the Cuban tree frog had never been seen north of Jacksonville, Florida. Since then, it has been spotted in Georgia and South Carolina, sparking alarm throughout the southeast. Because little is known about how cold-tolerant the frogs are, scientists worry that the frog may be able to survive north of Florida.

Evaluating the News

1. Why do you suppose Cuban tree frogs are such a nuisance in Florida, but not on their home islands, such as Cuba? Do all, or most, non-native species become pests? Why or why not?
2. Consider the following hypothetical situation: An introduced ornamental plant species escapes from cultivation and can grow in the wild. It harms native plant species—outcompeting and eventually replacing them—but causes little economic damage. Should such a species be considered a problem species, one that we should spend money to control? Or should we spend money only on controlling introduced species that harm human economic interests?
3. Many introduced species cause problems, some costing society huge sums of money. Should governments spend money to reduce the number of introduced species (for example, by inspecting vegetable crates more thoroughly) and to control those that are already here? If so, should states handle this problem, or should it be a federal responsibility? At either the state or federal level, would you be willing to pay for government efforts to control invasive species by having your taxes increase?

SOURCE: *Charlotte Sun-Herald*, Port Charlotte, FL, May 27, 2007.

Chapter Review

Summary

20.1 Why Is Ecology Important?

- Ecology is the scientific study of interactions between organisms and their environment. It helps us understand the natural world around us and our relationship with it.
- Ecology is an important area of biology because an understanding of how natural systems work can help us address current environmental problems and predict the effects of future human actions on the biosphere.

20.2 Interactions with the Environment

- Organisms affect—and are affected by—their environment, which includes not only their physical surroundings but also other organisms. As a result, a change that affects one organism can affect other organisms and the physical environment.
- It is helpful to think of organisms and their environment as forming a web of interconnected relationships.
- Because components of the biosphere depend on one another in complex ways, human actions that affect the biosphere can have surprising side effects.

20.3 Climate Has a Large Effect on the Biosphere

- Weather describes the physical conditions of Earth's lower atmosphere in a specific place over a short period of time. Climate describes a region's long-term weather conditions.
- Climate depends on incoming solar radiation. Tropical regions are much warmer than polar regions because sunlight strikes Earth directly at the equator, but at a slanted angle near the poles.
- Climate is strongly influenced by four giant convection cells that generate relatively consistent wind patterns over much of the Earth.
- Ocean currents carry an enormous amount of water and can have a large effect on regional climates.
- Regional climates are also affected by major features of Earth's surface, as when mountains create rain shadows.

20.4 Life on Land

- Climate is the most important factor controlling the potential (natural) location of terrestrial biomes. Climate can exclude a species from a region directly (if it finds the climate intolerable) or indirectly (if by other species in the region outcompete it for resources).
- Human activities heavily influence the actual location and extent of terrestrial biomes.

- Some of the major terrestrial biomes are tundra, boreal forest, temperate deciduous forest, grassland, chaparral, desert, and tropical forest.

20.5 Life in Water

- Aquatic biomes are usually characterized by physical conditions of the environment, such as temperature, salt content, and water movement.
- Aquatic biomes are strongly influenced by the surrounding terrestrial biomes, by climate, and by human actions.
- Two major aquatic biomes are recognized on the basis of salt content: the freshwater and marine biomes. The freshwater biome includes lakes, rivers, and wetlands; estuaries, the coastal region, and the open ocean are ecosystems that lie within the marine biome.

⊙ Review and Application of Key Concepts

1. This chapter suggests that the organisms and environments of the biosphere can be thought of as forming an "interconnected web." Why do ecologists think of the biosphere this way? Give an example illustrating the interconnections between organisms and their environment.

2. Explain in your own words how global patterns of air and water movement can cause local events to have far-reaching ecological consequences. Give an example that shows how local ecological interactions can be altered by distant events.

3. Name 7 of the major terrestrial biomes. How many of those biomes are located within 100 kilometers (about 60 miles) of your home? Describe the chief climatic and ecological characteristics of one terrestrial biome and explain how those characteristics affect which life-forms are found in that biome.

4. Using examples, explain the following statement: The defining characteristic of a desert is low moisture, not high temperature.

5. What accounts for the high productivity of estuaries and coastal regions? Explain why the open ocean has lower productivity than these two regions of the marine biome.

Key Terms

Self-Quiz

1. Which of the following is not a level of the biological hierarchy commonly studied by ecologists?
   a. ecosystem
   b. individual
   c. organelle
   d. population

2. Earth has four stable regions ("cells") in which warm, moist air rises, and cool, dry air sinks back to the surface. Such cells are known as
   a. temperate cells.
   b. latitudinal cells.
   c. rain shadow cells.
   d. giant convection cells.

3. The biosphere consists of
   a. all organisms on Earth only.
   b. only the environments in which organisms live.
   c. all organisms on Earth and the environments in which they live.
   d. none of the above

4. What aspect(s) of climate most strongly influence the locations of terrestrial biomes?
   a. rain shadows
   b. temperature and precipitation
   c. only temperature
   d. only precipitation

5. Winds that blow from the west across warm waters in the Pacific Ocean become warm and moist. By analogy to what happens in a rain shadow, what do you think would happen if such warm, moist winds blew across the cold Peru Current (see Figure 20.5)?
   a. An El Niño event would occur.
   b. The winds would continue to pick up moisture from the ocean currents.
   c. The warm, moist winds would cool, causing rain to fall.
   d. The warm, moist winds would cool, but rain would not fall.

6. Which of the following represents a large area of the globe with unique climatic and ecological features?
   a. a population
   b. a community
   c. the biosphere
   d. a biome

7. The ozone layer in the upper atmosphere
   a. is depleted when it interacts with chlorine atoms.
   b. is created by the accumulation of human-made pollutants.
   c. has been increasing in thickness since the 1960s.
   d. is currently thickest over the Antarctic and thinnest over the equator.

8. Wetlands, ponds, and streams are common in summertime in the arctic tundra because
   a. the permafrost thaws completely, releasing large amounts of water.
   b. melted snow and ice is prevented from draining by the underlying permafrost.
   c. precipitation in the arctic is high, exceeding 80 inches per year.
   d. tundra trees grow best in wet, boggy ground.

9. Temperate deciduous forests
   a. experience cool, rainy winters and hot, dry summers.
   b. are maintained by frequent fires.
   c. are not dominated by coniferous trees.
   d. were the predominant biome in Great Plains states such as Nebraska before they were settled by Europeans.

10. Estuaries are highly productive because
    a. they are dominated by trees and shrubs, which have a large biomass.
    b. they receive nutrients delivered by rivers and stirred up by tide action.
    c. they lie between the high-tide mark and low-tide mark in the coastal region.
    d. archaeans that extract energy from minerals are especially abundant in this type of ecosystem.

CHAPTER **21** Growth of Populations

Key Concepts

◉ A population is a group of interacting individuals of a single species located within a particular area.

◉ Populations increase in size when birth and immigration rates exceed death and emigration rates, and they decrease when the reverse is true.

◉ A population that increases by a constant proportion from one generation to the next exhibits exponential growth.

◉ Eventually, the growth of all populations is limited by environmental factors such as lack of space, food shortages, predators, disease, and environmental deterioration.

◉ Different populations may exhibit different patterns of growth over time, including J-shaped curves (which result from exponential growth), S-shaped curves, population cycles, and irregular fluctuations.

◉ The world's human population is increasing exponentially. Rapid human population growth cannot continue indefinitely; either we will limit our own growth, or the environment will do it for us.

The Tragedy of Easter Island

Imagine standing at the edge of a cliff on Easter Island, in a remote corner of the South Pacific, looking into the long-abandoned quarry of Rano Raraku. Scattered about the grassy slopes of the quarry lie hundreds of huge, eerie statues that were carved from stone centuries ago. The scene is beautiful, yet also ghostly and disturbing. Some of the statues, known as Moai, stand upright but unfinished; they look almost as if the artists dropped their tools in midstroke. Others are complete, but lie fallen at odd angles. Hundreds more statues are scattered along the coast of Easter Island. Who carved these statues? Why were so many left unfinished? What happened to the people who made them?

The mystery deepens when we consider where Easter Island is and what it looks like today. Extremely isolated, the island is a small, barren grassland, just 166 square kilometers in area, with little water and little potential for agriculture. How could such a remote and forbidding place support a civilization capable of carving, moving, and maintaining these enormous stone statues?

The answers to these questions provide a sobering lesson for people today. Easter Island was not always a barren grassland; at one time most of the island was covered by forest. According to archaeological evidence, no humans lived on the island until about AD 400. At that time, about 50 Polynesians arrived in large canoes, bringing with them crops and animals to support themselves. These people developed a well-organized society capable of sophisticated technological feats, such as moving 15- to 20-ton stone statues long distances without the aid of wheels (they probably rolled the statues on logs).

Easter Island

Located 3,700 kilometers west of the coast of Chile and 4,200 kilometers east of the nearest major population center in Polynesia (Tahiti), Easter Island is small and extremely isolated.

By the year 1500, the population had grown to about 7,000 people. By this time, however, virtually all the trees on the island had been cut down to clear land for agriculture and (presumably) to provide the logs used to roll the statues from one place to another. The cutting of trees, along with other forms of environmental destruction, increased soil erosion and decreased crop production, leading to mass starvation.

With no large trees remaining on the island, the people could not build canoes to escape the ever-worsening conditions. The society collapsed, resorting to warfare, cannibalism, and living in caves for protection. The population also crashed; even 400 years later (in 1900) only 2,000 people remained on the island, less than one-third the number that had lived there in 1500.

What caused the events on Easter Island? Does the story of Easter Island provide a preview of what will happen to the whole human race? In this chapter we examine the characteristics of populations, including how and why populations increase or decrease in size—topics that are the main focus of population ecology. Toward the end of the chapter we return to the story of Easter Island and ask if we humans are using more resources than the planet can support.

R̲ecall from Chapter 20 that ecology is the study of interactions between organisms and their environment. **Population ecology** is concerned with questions that relate to how many organisms live in a particular environment, and why. The answers to such questions not only provide insight into the natural world, but also are essential for solving real-world problems, such as protecting rare species or controlling pest species. To set the stage for our study of the factors that influence how many individuals are in a population, we begin by defining what populations are. We then describe how populations grow over time, and we consider the limits to growth that all populations face, including the human population.

21.1 What Is a Population?

A **population** is a group of interacting individuals of a single species located within a particular area. The human population of Easter Island, for example, consists of all the people who live on the island.

Ecologists usually describe the number of individuals in a population by the **population size** (the total number of individuals in the population) or by the **population density** (the number of individuals per unit of area). To calculate population density, we divide the population size by the total area. To illustrate, let's return to the Easter Island example: in the year 1500, the population size was 7,000 people. If we divide that number by 166 square kilometers (the size of the island), we get a population density of 42 people per square kilometer (7,000/166 = 42). The density of people that lived on Easter Island in 1500 was therefore higher than the 31 people per square kilometer (80 per square mile) who lived in the United States in 2008.

Easter Island is an easy example for determining what constitutes a population: islands have well-defined boundaries, and human individuals are relatively easy to count. But often it is more difficult to determine the size or density of a population. Suppose a farmer wants to know whether the aphid population damaging a crop is increasing or decreasing (Figure 21.1). Aphids are

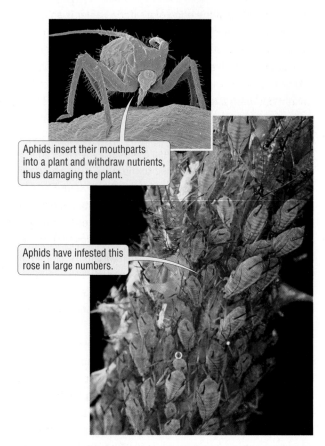

Aphids insert their mouthparts into a plant and withdraw nutrients, thus damaging the plant.

Aphids have infested this rose in large numbers.

Figure 21.1 Populations of Aphids Can Cause Extensive Crop Damage
Aphids are small insects with sucking mouthparts. They are pests on many plant species, which they infest in such large numbers that they can be difficult to count.

small and hard to count. More important, it is not obvious how the aphid population should be defined. What do we mean by "a particular area" in this case? Aphids can produce winged forms that can fly considerable distances, so how do we know which aphids to count? Should we count only the aphids in the farmer's field? What about the aphids in the next field over?

In general, what constitutes a population often is not as clear-cut as in the Easter Island example. Overall, the area appropriate for defining a particular population depends on the questions being asked and on aspects of the biology of the organism of interest, such as how far and how rapidly they move.

21.2 Changes in Population Size

All populations change in size over time—sometimes increasing, sometimes decreasing. In one year, abundant rainfall and plant growth may cause mouse populations to increase; in the next, drought and food shortages may cause mouse populations to decrease dramatically. Such changes in the population sizes of organisms can have important consequences for people. For example, an increase in the number of deer mice, carriers of hantavirus, is thought to have been responsible for the 1993 outbreak of this deadly disease in the southwestern United States.

Whether a population increases or decreases in size depends on the number of births and deaths in the population, as well as on the number of individuals that immigrate to (enter) or emigrate from (leave) the population. A population increases in size whenever the number of individuals entering (by birth and immigration) is greater than the number of individuals leaving (by death and emigration). We can express this relationship in equation form:

$$birth + immigration > death + emigration$$

The environment plays a key role in the increase or decrease of a population because birth, death, immigration, and emigration rates are affected by environmental factors. Consider how features of the environment caused monarch butterfly populations to fluctuate wildly in 2002. Each spring, monarchs make a spectacular migration that begins in the mountains west of Mexico City (where the butterflies overwinter) and ends in eastern North America (Figure 21.2). On January 13, 2002, the monarchs' overwintering sites were hit with an unusual storm that first drenched the butterflies with rain, then

Spring migration

Figure 21.2 Before the Crash
Huge numbers of monarch butterflies overwinter in mountains west of Mexico City and then migrate each spring to eastern North America. In 2002, 70 to 80 percent of the overwintering butterflies died in an unusual winter storm. (Not shown are smaller populations of monarchs found west of the Rockies that overwinter along the Pacific coast of California.)

Overwintering monarchs

subjected them to freezing cold. The combination of wet and cold proved lethal: an estimated 70 to 80 percent of the butterflies—roughly 500 million of them—died overnight, the worst die-off in 25 years.

Fortunately for the butterflies, this huge increase in winter death rates was followed by an equally spectacular rise in birth rates during the summer of 2002. Monarch birth rates shot up that year because it turned out to be a great summer for the monarch's primary food plant, milkweed. Because of this chance good fortune, butterflies that had survived the winter produced so many young in the summer of 2002 that monarch numbers quickly rebounded to almost the historic average.

Concept Check

1. What is population density? Explain why it can be difficult to measure.
2. What aspects of a population determine whether it increases or decreases in size?

Deer mouse

Concept Check Answers

1. Population density is the total number of interacting individuals of a single species, divided by the particular area they inhabit. It may be difficult to measure because individuals may be hard to detect, may move between populations, and may inhabit a complex, hard-to-define area.

2. Birth and immigration increase population size; death and emigration reduce it. Environmental factors influence these characteristics, so they have a strong impact on population size.

21.3 Exponential Growth

Like monarch butterflies, many organisms produce vast numbers of young. If even a small fraction of those young survive to reproduce, a population can grow extremely rapidly.

Exponential growth results in rapid population increases

An important type of rapid population growth is **exponential growth**, which occurs when a population increases by a constant proportion (λ) over a constant time interval, such as over the course of 1 year (Figure 21.3). We can represent exponential growth from one year to the next by the equation

$$N_{\text{next year}} = \lambda \times N_{\text{this year}}$$

where N is the number of individuals in the population and λ (lambda) is the proportional increase in population size, a constant multiplier that determines the population size from one year to the next. For example, if λ = 1.5 and the current population size is 40, then the population size in the following year will be 60 (40 × 1.5), and in successive years it will be 90 (60 × 1.5), and then 135 (90 × 1.5), and so on.

In exponential growth, the proportional increase (λ) is constant, but the numerical increase—the number of individuals added to the population—becomes larger with each generation. For example, the population in Figure 21.3 doubles every generation (that is, λ = 2). With respect to its *numerical* increase, however, the population increases vary: the population increases by only 1 individual between generations 1 and 2, but by 16 individuals between generations 5 and 6. When plotted on a graph, exponential growth forms a **J-shaped curve**, as seen in Figure 21.3.

The time it takes a population to double in size—the **doubling time**—can be used as a measure of how fast the population is growing. We like it when our bank accounts double rapidly, but when populations grow exponentially in nature, problems eventually result, as we shall see in the following sections.

Exponential growth often occurs when a species moves into a new area

Populations can increase exponentially, at least initially, when they migrate to, or are introduced into, a new area. Consider the following tale of woe: In 1839, a rancher in Australia imported from South America a species of

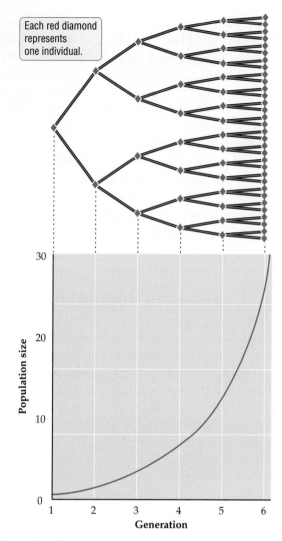

Each red diamond represents one individual.

Figure 21.3 Exponential Growth
In this hypothetical population, each individual produces two offspring, so the population increases by a constant proportion with each generation (in this case, where λ = 2, the population doubles). The number of individuals added to the population increases each generation, resulting in the J-shaped curve that is characteristic of exponential growth. Exponential growth curves are always J-shaped, regardless of the value of λ, though different λ values will change the curve's steepness.

Opuntia (prickly pear cactus) and used it as a "living fence" (a thick wall of this cactus is nearly impossible for human or beast to cross). Unlike a real fence, however, the *Opuntia* cactus did not stay in one place; it spread rapidly throughout the landscape. As the cactus spread, whole fields were turned into "fence," crowding out cattle and destroying good rangeland.

In about 90 years, *Opuntia* cacti spread across eastern Australia, covering more than 243,000 square kilometers (over 60 million acres) and causing great economic damage. All attempts at control failed until 1925, when scientists introduced a moth species, appropriately named *Cactoblastis cactorum*, whose caterpillars feed on

(a)

(b)

Figure 21.4 Blasting the Cactus

In Australia, the moth *Cactoblastis cactorum* was introduced in 1925 to halt the exponential growth of populations of an introduced cactus species. (*a*) Two months before release of the moth, *Opuntia* cacti were growing in a dense stand. (*b*) Three years after introduction of the moth, the same stand had been almost completely eliminated.

the growing tips of the cactus. This moth killed billions of cacti, successfully bringing the cactus population under control (Figure 21.4). Introducing non-native species to control a problematic invader (biological control) is fraught with risk because the control agent (the cactus moth, in this case) could multiply exponentially and become a problem itself. Fortunately, the cactus moth is a specialist feeder, preferring cacti, which are native to the Americas, over any other plant the Australian outback had to offer. The moth's success led to its own demise, as, after the prickly pear cactus population had been decimated, moth numbers plummeted because of lack of food. Today, both the cactus and the moth exist in eastern Australia in low numbers.

Overall, the *Opuntia* population in Australia increased exponentially at first; then it declined even more rapidly after introduction of the moth. Exponential growth has also been observed in other species introduced by people to new areas, as well as in species that have expanded naturally to new areas.

21.4 Limits to Population Growth

A giant puffball mushroom can produce up to 7 trillion offspring (Figure 21.5). If all of these offspring survived and reproduced at this same (maximal) rate, the descendants of a giant puffball would weigh more than Earth in just two generations. Humans and *Opuntia* cacti have much longer doubling times than giant puffballs, but given enough time, they, too, can produce an astonishing number of descendants. Obviously, however, Earth is not covered with giant puffballs, *Opuntia* cacti, or even humans. These examples illustrate an important general point: No population can increase in size indefinitely. Limits exist.

Growth is limited by essential resources and other environmental factors

The most obvious reason that populations cannot continue to increase indefinitely is simple: food and other resources become diminished. Imagine that a few bacteria are placed in a closed jar containing a source of food. The bacteria absorb the food and then divide, and their offspring do the same. The population of bacteria grows exponentially, and in short order the jar contains billions of bacteria. Eventually, however, the food runs out and metabolic wastes build up. All the bacteria die.

Figure 21.5 Will They Overrun Earth?
Given the number of spores it produces, a giant puffball mushroom has the potential to produce 7 trillion offspring in a single generation. However, relatively few of those spores land in a habitat suited for their growth. Large giant puffballs weigh 40 to 50 kilograms each; a medium-sized example is shown here.

See a simulation of unrestrained population growth. ▶II 21.1

See a simulation of restrained population growth.

▶❚❚

21.2

This example may seem extreme because it involves a closed system: no new food is added, and the bacteria and the metabolic wastes cannot go anywhere. In many respects, however, the real world is similar to a closed system. Space and nutrients, for example, exist in limited amounts. In the *Opuntia* example of the previous section, even if humans had not introduced the *Cactoblastis* moth, the cactus population could not have sustained exponential growth indefinitely. Eventually, the growth of the cactus population would have been limited by an environmental factor, such as a lack of suitable **habitat** (the type of environment in which an organism lives).

The growth pattern of some populations can be represented by an **S-shaped curve**. Such populations grow close to exponentially at first, but then stabilize at the maximum population size that can be supported indefinitely by their environment. This maximum population size is known as the **carrying capacity** (Figure 21.6). The growth rate of the population decreases as the population size nears the carrying capacity because resources such as food and water begin to be in short supply. At the carrying capacity, the population growth rate is zero.

In the 1930s, the Russian ecologist G. F. Gause carried out experiments on *Paramecium caudatum*, a common protist. He found that laboratory populations of paramecia increased to a certain size and then remained there (see Figure 21.6). In these experiments, Gause added

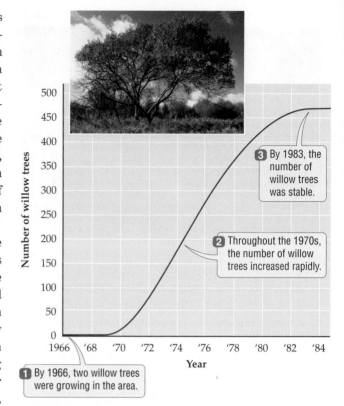

3 By 1983, the number of willow trees was stable.

2 Throughout the 1970s, the number of willow trees increased rapidly.

1 By 1966, two willow trees were growing in the area.

Figure 21.7 An S-Shaped Curve in a Natural Population
At a site in Australia, rabbits heavily grazed young willow trees, preventing willows from growing in the area. The rabbits were removed in 1954. By 1966, two willows had taken root in the area, presumably from seed blown in or carried in by animals. They increased rapidly in number, and the population then leveled off at about 475 trees.

new nutrients to the protists' liquid medium at a steady rate and removed the old solution at a steady rate. At first, the population increased rapidly in size. But as the population continued to increase, the paramecia used nutrients so rapidly that food began to be in short supply, slowing the growth of the population. Eventually the birth and death rates of the protists equaled each other and the population size stabilized.

In contrast to natural systems, there was no immigration or emigration in Gause's experiments. In natural systems, populations reach and remain at a constant population size when

$$\text{birth} + \text{immigration} = \text{death} + \text{emigration}$$

for extended periods of time.

Like laboratory populations of bacteria and paramecia, natural populations also experience limits (Figure 21.7). Their growth can be held in check by a number of environmental factors, including food

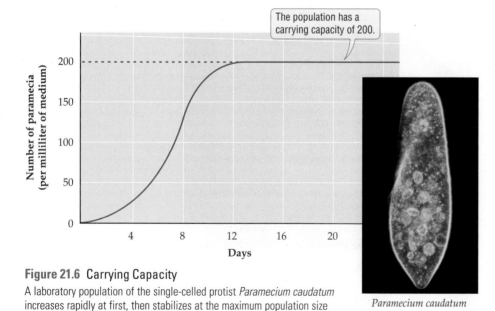

The population has a carrying capacity of 200.

Figure 21.6 Carrying Capacity
A laboratory population of the single-celled protist *Paramecium caudatum* increases rapidly at first, then stabilizes at the maximum population size that can be supported indefinitely by its environment—that is, at its carrying capacity. This growth pattern can be graphed as an S-shaped curve.

Paramecium caudatum

shortages, lack of space, disease, predators, habitat deterioration, weather, and natural disturbances. When a population has many individuals, birth rates may drop or death rates may increase; either effect may limit the growth of the population, and sometimes both effects occur. Let's take a brief look at how this works.

Any area contains a limited amount of food and other essential resources. Therefore, as the number of individuals in a population increases, fewer resources are available to each individual. As resources diminish, each individual, on average, produces fewer offspring than when resources are plentiful, causing the birth rate of the population to decrease.

In addition, when a population has many individuals, disease spreads more rapidly (because individuals tend to encounter one another more often), and predators may pose a greater risk (because many predators prefer to hunt abundant sources of food). Disease and predators obviously increase the death rate.

Large populations can also damage or deplete their resources. If a population exceeds the carrying capacity of its environment, it may damage that environment so badly that the carrying capacity is lowered for a long time. A drop in the carrying capacity means that the habitat cannot support as many individuals as it once could. Such habitat deterioration may cause the population to decrease rapidly (Figure 21.8).

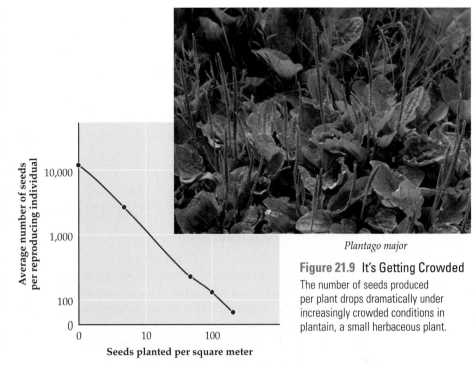

Plantago major

Figure 21.9 It's Getting Crowded
The number of seeds produced per plant drops dramatically under increasingly crowded conditions in plantain, a small herbaceous plant.

Some growth-limiting factors depend on population density; others do not

Food shortages, lack of space, disease, predators, and habitat deterioration—all these factors influence a population more strongly as it grows and therefore increases in density. The birth rate, for example, may decrease or the death rate may increase when the population has many individuals. When birth and death rates change as the density of the population changes, such rates are said to be **density-dependent**. In natural populations, the number of offspring produced (Figure 21.9) and the death rate are often density-dependent.

In other cases, populations are held in check by factors that are not related to the density of the population; such factors cause the population to change in a **density-independent** manner. Density-independent factors can prevent populations from reaching high densities in the first place. Year-to-year variation in weather, for example, may cause conditions to be suitable for rapid population growth only occasionally. Poor

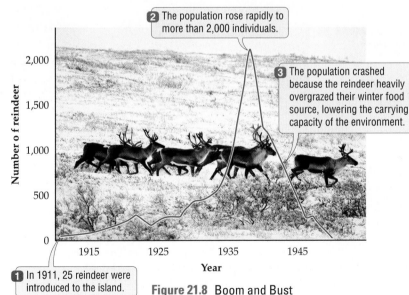

2 The population rose rapidly to more than 2,000 individuals.

3 The population crashed because the reindeer heavily overgrazed their winter food source, lowering the carrying capacity of the environment.

1 In 1911, 25 reindeer were introduced to the island.

Figure 21.8 Boom and Bust
When reindeer were introduced to Saint Paul Island, off the coast of Alaska, in 1911, their population increased rapidly at first and then crashed. By 1950, only eight reindeer remained.

weather conditions may reduce the growth of a population directly (by freezing the eggs of an insect, for example) or indirectly (by decreasing the number of plants available as food to that insect). Natural disturbances such as fires and floods also limit the growth of populations in a density-independent way. Finally, the effects of environmental pollutants such as DDT are density-independent; such pollutants can threaten natural populations with extinction.

21.5 Patterns of Population Growth

Different populations may exhibit a number of different growth patterns over time. We'll discuss four such patterns: J-shaped curves, S-shaped curves, population cycles, and irregular fluctuations.

Under favorable conditions, the population size of any species increases rapidly. An initial period of rapid population growth can be seen in growth patterns that are J-shaped (see Figure 21.3) or S-shaped (see Figure 21.7). With the J-shaped growth pattern, rapid population growth may continue until resources are depleted, causing the population size

to drop dramatically (see Figure 21.8). In contrast, with the S-shaped growth pattern, the rate of population growth slows as the population size nears the carrying capacity. Predators, disease, and other factors may then keep the population near the carrying capacity for a long time.

As we have seen, populations change in size over time, increasing at some times and decreasing at others. Even populations with an S-shaped growth pattern do not remain indefinitely at a single, stable population size; instead, they fluctuate slightly over time, yet remain close to the carrying capacity.

In some cases, the population sizes of two species change together in a tightly linked cycle. Such **population cycles** can occur when at least one of the two species involved is very strongly influenced by the other. The Canada lynx, for example, depends on the snowshoe hare for food, so lynx populations increase when hare populations increase and they decrease when hare populations drop (Figure 21.10).

In relatively few examples from nature do populations of two species show regular cycles like those of the hare and lynx. As illustrated dramatically by monarch butterflies, however, the populations of most species do rise and fall over time—just not as regularly as in Figure 21.10. **Irregular fluctuations** are far more common in

See predator-prey interactions. 21.3

See consumer-victim interactions. 21.4

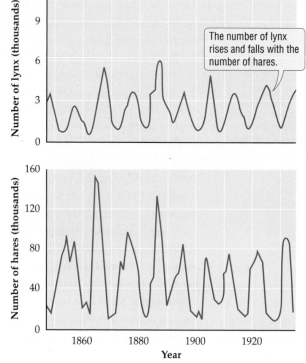

The number of lynx rises and falls with the number of hares.

Figure 21.10 Population Cycles

Populations of two species occasionally increase and decrease together. The Canada lynx depends on the snowshoe hare for food, so the number of lynx is strongly influenced by the number of hares. Experiments conducted in the early twentieth century indicate that hare populations are limited by their food supply and by their lynx predators. (The researchers supplemented their observations with population estimates based on the numbers of hare and lynx pelts sold by trappers to the Hudson's Bay Company, Canada.)

Science Toolkit

Population Counts Help Save Endangered Species

Populations of many species are threatened by human actions, some to the point of being in danger of extinction. One of the most important tools for people trying to save endangered species is the population count. Like the censuses used by demographers studying human populations, a population count reveals the number of individuals in a particular population. Results from such counts can prod organizations and legislatures into taking action, as happened in the case of the bald eagle.

Bald eagles are relatively easy to count: they have large, conspicuous nests to which they return year after year. The bald eagle was one of many bird species in the United States that were severely affected by DDT poisoning; birds with high levels of DDT in their bodies produced such fragile eggshells that they could not reproduce. By the early 1960s, population counts revealed that only 417 breeding pairs of bald eagles remained in the lower 48 states—a huge drop from the estimated 100,000 breeding pairs present in 1800.

The effect of DDT on these birds prompted a 1972 ban on its use within the United States. This ban gave populations of bald eagles the opportunity to bounce back from extinction. Today there are more than 6,400 breeding pairs in the lower 48 states. The encouraging results from ongoing population counts of bald eagles illustrate that when people recognize a problem and take decisive measures to fix it, populations sometimes can return from the brink of extinction. Furthermore, actions taken with one species in mind can give other species a chance to recover; this was the case for the DDT ban, which enabled other birds, such as the peregrine falcon, to recover from their perilously low numbers as well.

Bog alkaligrass

Not all species are as easy to count as bald eagles. Bog alkaligrass [*AL*-kuh-lye-*GRASS*] (*Puccinellia parishii* [puh-chuh-*NELL*-ee-uh pah-*REE*-shee]) is designated as an endangered plant species under state and tribal statutes in Arizona, New Mexico, and the Navajo Nation. This small, inconspicuous grass germinates in the winter and typically dies by late spring or early summer. Only 30 populations are known, which means that the species is vulnerable to extinction. Population counts are challenging because the number of seeds that germinate can fluctuate wildly: one year there may be millions of plants, and the next there may be few or none (the number depends on growing conditions). Because no plants are visible in some years, it is hard to know whether all existing populations have been located. Furthermore, depending on how many plants there are, different techniques must be used in different years to estimate the size of the 30 known populations.

Although more difficult to perform, population counts are as essential for bog alkaligrass as they are for bald eagles. They document which populations are increasing and which are decreasing—information that can help scientists and policy makers decide how best to protect the species from the threat of extinction.

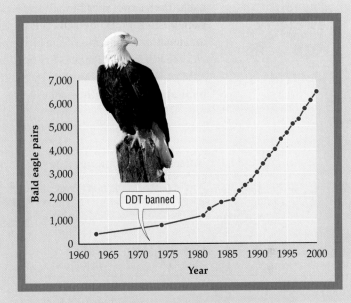

Recovery of the Bald Eagle after the Ban on Use of DDT

---

nature than is the smooth rise to a stable population size shown in Figure 21.7.

Finally, different populations of the same species may experience different patterns of growth. Understanding the reasons for these differences can provide critical information on how best to manage endangered or economically important species. The first step toward such an understanding is to perform population counts (see the "Science Toolkit" box) to determine whether different patterns of population growth are present. If there are different growth patterns, the next step is to figure out why.

Spotted owl

During the 1980s, forest managers needed to decide where and how much (if any) mature or old-growth forest could be cut without harming the rare spotted owl. For each owl population, researchers first gathered data on the birth rate and the amount of habitat used by each individual. The researchers then used these data to predict how the growth of spotted owl populations would be affected by the number, size, and location of patches of the bird's preferred habitat: old-growth forest. (Patches are portions of a particular habitat that are surrounded by a different habitat or habitats.) The amount—that is, the total area—and arrangement of old-growth forest patches was found to have a large effect on owl population growth rates (Figure 21.11).

21.6 Human Population Growth: Surpassing the Limits?

The human population is growing today at a spectacular rate (Figure 21.12). It took more than 100,000 years for our population to reach a billion people, but now it increases by a billion people every 13 years. Our use of resources and our overall impact on the planet have increased even faster than our population size. For example, from 1860 to 1991, the human population increased 4-fold, but our energy consumption increased 93-fold.

The global human population passed the 6.7 billion mark early in 2008. At present, the human population is growing exponentially, increasing by about 79 million people each year, or over 9,100 people per hour. These numbers are all the more sobering when we consider the following facts:

- More than 1.3 billion people live in absolute poverty.
- Two billion people lack basic health care or safe drinking water.
- More than 2 billion people have no sanitation services.
- Each year 14 million people, mostly children, die from hunger or hunger-related problems.

By the year 2025, the global human population is projected to increase to more than 8 billion people. Even if our birth rate dropped immediately from the current 2.8 children per female to a level that ultimately would allow the human population to simply replace itself, but not increase (about 2.1 children per female), the human population would continue to grow for at least another 60 years. The population will continue to increase long after birth rates drop, because a huge number of existing children have not yet had children of their own.

How did the human population increase so rapidly, apparently escaping the limits to population growth described in this chapter? There are several reasons. First, as our ancestors emigrated from Africa (see Figure D.10), they encountered and prospered in many kinds of new habitats. Few other species can thrive

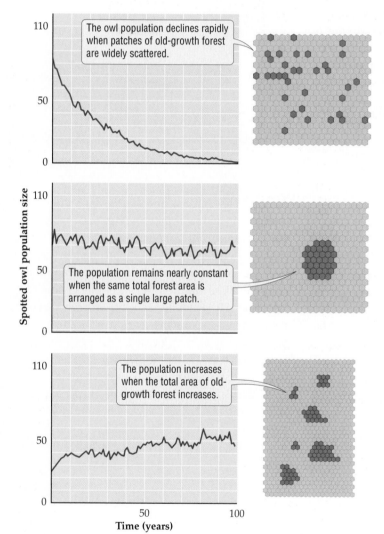

The owl population declines rapidly when patches of old-growth forest are widely scattered.

The population remains nearly constant when the same total forest area is arranged as a single large patch.

The population increases when the total area of old-growth forest increases.

Time (years)

Figure 21.11 Same Species, Different Outcomes
Different populations of the endangered spotted owl are predicted to show different patterns of growth over time, depending on the arrangement and area of their preferred habitat: old-growth forest. Patches of old-growth forest are shown in blue.

Learn more about human population growth.

▶ ‖
21.5

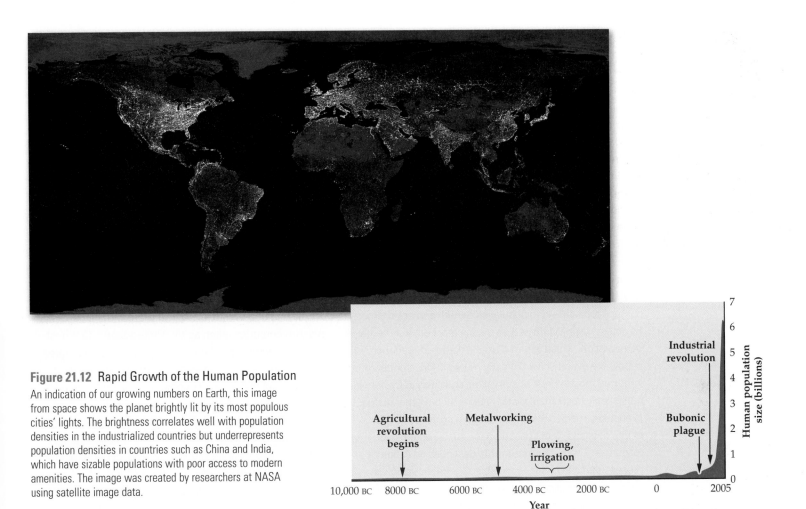

Figure 21.12 Rapid Growth of the Human Population
An indication of our growing numbers on Earth, this image from space shows the planet brightly lit by its most populous cities' lights. The brightness correlates well with population densities in the industrialized countries but underrepresents population densities in countries such as China and India, which have sizable populations with poor access to modern amenities. The image was created by researchers at NASA using satellite image data.

in places as different as grasslands, coastal environments, tropical forests, deserts, and arctic regions. Second, people increased the carrying capacity of the places where they lived. The development of agriculture, for example, enabled more people to be fed per unit of land area. More recently, our heavy use of fossil fuels and nitrogen fertilizers in the twentieth century led to great increases in crop yields. Finally, in the last 300 years death rates have dropped as a result of improvements in medicine, sanitation, and food storage and transportation. However, birth rates did not drop at the same time, so our population has continued to grow.

Viewed broadly, human inventiveness and technology have enabled us to sidestep limits to population growth for some time. Like all other populations,

however, our population cannot continue to increase without limit; ultimately we will be subject to the environmental factors that limit the growth of all species.

Concept Check

1. Can a population grow in an exponential manner indefinitely? Explain.
2. How is a population that shows an S-shaped growth curve different from one that exhibits a J-shaped growth curve?
3. A few seeds of plantain arrive in a vacant lot and start growing exponentially over the next three seasons. A late frost in the fourth season destroys most of the individuals in the population. Was the size of this population limited in a density-dependent manner or a density-independent manner? Explain.

Applying What We Learned

What Does the Future Hold?

As described at the opening of this chapter, the people who colonized Easter Island initially maintained a culturally rich and densely populated society. But their society did not persist. The people of Easter Island temporarily increased the carrying capacity of the island by cutting down the forest to create farm fields. Ultimately, however, cutting down the trees led to environmental deterioration, starvation, and the collapse of their civilization.

As on Easter Island, many of the problems facing humans today relate to population growth and environmental deterioration. More people means more environmental deterioration, which in turn makes it harder to feed the people we already have. Already much of Africa depends on imported food to prevent starvation, and cities in California persist only because of water imported from other states. In addition, many people think that our society, like that of Easter Island, is not based on the sustainable use of resources. The term **sustainable** describes an action or process that can continue indefinitely without using up resources or causing serious damage to the environment.

Many lines of evidence suggest that the current human impact on Earth is not sustainable. For example, water tables are dropping throughout the world, global fish populations have plummeted in response to over-harvesting, and if current rates of logging continue (14 million hectares, or 35 million acres, per year), scientists estimate that all tropical forests will be gone in 100 to 150 years.

One measure of sustainability is the **ecological footprint**, which is the area of productive ecosystems needed throughout the year to support a population and cope with its waste materials. Early results of calculating our ecological footprint are not encouraging. Scientists recently estimated that the average person's ecological footprint is 2.7 hectares (6.7 acres), which is almost 30 percent higher than the 2.1 hectares (5.2 acres) that could be sustained for each of the world's 6.7 billion people. The ecological footprint of individuals in some countries, such as the United States (9.4 hectares per person) and the United Kingdom (5.3 hectares per person), is three to five times what is sustainable. Overall, such estimates suggest that, since the late 1970s, people have been using resources faster than they can be replenished (Figure 21.13)—a pattern of resource use that, by definition, is not sustainable.

Will people limit the growth and impact of our global population, or will the environment do it for us? There are some hopeful signs: the growth rate of the human population has slowed in recent years, and people throughout the world are conscious of the risks of environmental degradation. But much remains to be done. To limit the growth and impact of the human population, we must address the interrelated issues of population growth rates, poverty, unequal use of resources, environmental deterioration, and sustainable development. It is especially important for people who live in North America, Japan, and Europe to address such issues, because people in these regions have such large ecological footprints.

Our hope for the future—for your future and the future of your children—lies in realistically assessing the problems we face, and then committing ourselves to take bold actions to address those problems (see Interlude E). In the end, it is up to all of us to help ensure that humankind does not repeat on a grand scale the tragic lessons of Easter Island.

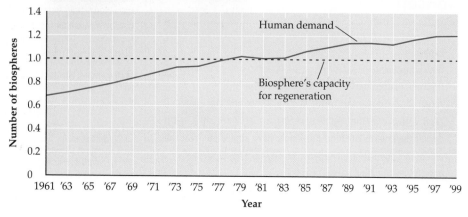

Figure 21.13 Our Rising Ecological Impact

The global ecological impact of people has increased steadily since 1961. This graph compares human demands on the biosphere in each year with the capacity of the biosphere to regenerate itself. One vertical unit on the graph represents the entire capacity of the biosphere to regenerate itself in a given year. Human demand has exceeded the biosphere's entire regenerative capacity since the late 1970s.

Biology in the News

Mosquitoes Bring West Nile Virus

BY MARY K. REINHART

Just as the flu season is winding down, mosquito season and the West Nile virus are ramping up. And in case evenings on your patio haven't convinced you, state and county health officials confirm that the pesky bloodsuckers are breeding faster than rabbits.

The state's first West Nile-positive mosquitoes were confirmed Friday in a batch from Clarkdale in Yavapai County, an early start for the cooler climes . . . "We've had some pretty heavy rains. That's creating backyard breeding," said Craig Levy, head of the Arizona Department of Health

Services' vector-borne and zoonotic diseases section . . . Levy said he was getting eaten alive in his backyard until he discovered the source: a neighbor's untended pool. Once the pool was cleaned up, Levy's mosquito problem was solved. "Being a little nosy may not be a bad thing," Levy said. "If you've got a problem, you need to find the source." The county has stepped up efforts to fine green-pool owners, and it issues dozens of citations each month.

Last year, it was late June before the first West Nile-packing mosquitoes were

confirmed, and they were in La Paz County. By October, the virus had infected 150 people and killed 11 of them, including six in Maricopa County. That put Arizona fifth in the country in the number of 2006 West Nile fatalities . . . Most people bitten by infected mosquitoes won't develop any symptoms, about 20 percent will develop flulike symptoms, and fewer than 1 percent develop encephalitis. Those most at risk are the elderly and people with compromised immune systems.

Deserts and mosquitoes usually do not go together, but Arizona has one species, *Culex tarsalis* [KYOO-leks TAR-suh-liss], that tolerates heat well and can increase rapidly in numbers if there is enough standing water for its young to develop into adults. (Mosquitoes lay their eggs in water, and their larvae are fully aquatic.) *Culex tarsalis*, like all other mosquito species, can breed in standing water found in wheelbarrows, old tires, or even bottles—the kinds of miscellaneous small sources of standing water found in towns and cities throughout the country. But in some parts of Arizona, *C. tarsalis* mosquitoes recently hit the "double jackpot" with two additional sources of standing water.

A view from the air shows that the Phoenix area is covered with swimming pools. A well-maintained pool has too many chemicals for mosquitoes to breed in it. But of 600,000 residential pools in the state, health department officials estimate that 10,000 can support mosquitoes. These abandoned pools are like miniature backyard swamps—perfect homes for mosquito larvae.

In addition, many homeowners have created lush landscapes in their yards, which in Phoenix's dry climate require large amounts of water to maintain. With irrigation canals, sprinklers, and lawn misters, there seems to be no shortage of human-related water sources for mosquitoes to breed in. Altogether, there is so much standing water in the Phoenix area that mosquitoes have become a problem. Their populations have increased greatly in size, which in turn has contributed to the sharp increase in the number of West Nile cases reported in Arizona.

Evaluating the News

1. *Culex tarsalis* was not common in Arizona at the turn of the twentieth century. Explain why populations of this mosquito have increased in recent decades. What steps can residents take to reduce the carrying capacity for this species in the Arizona desert?

2. In some desert regions, homeowners and farmers apply so much water to their yards and fields that the humidity has steadily increased over the past 20 years. One consequence is that evaporative cooling systems (called "swamp coolers") don't work very well any more, causing people to switch to more expensive cooling systems (such as standard air conditioners). Do you think this trend of modifying local climates to such an extent is good, bad, or not significant? Why?

3. People who live in Phoenix or other cities located in deserts—such as Tucson, Arizona; El Paso, Texas; Las Cruces, New Mexico; and Las Vegas, Nevada—must water their lawns and gardens heavily if they want to create lush landscapes. Far too little rain falls in these areas to contribute significantly to the maintenance of these landscapes, so water must be taken from rivers, pumped out of underground reservoirs, or imported from other states. Such high levels of water use in a desert are not sustainable, because water is used more rapidly than it is replenished.
   a. Do you think it is appropriate to use large amounts of water on plants in lawns and gardens that otherwise could not live in a desert?
   b. Should cities encourage people to conserve water, and if so, how should they do that?

SOURCE: *East Valley Tribune*, Phoenix, AZ, May 5, 2007.

Chapter Review

Summary

21.1 What Is a Population?

- A population is a group of interacting individuals of a single species located within a particular area.
- Two basic concepts used in studying populations are population size (the total number of individuals in the population) and population density (the number of individuals per unit of area).
- What constitutes an appropriate area for determining a population depends on the questions of interest and the biology of the organism under study.

21.2 Changes in Population Size

- All populations change in size over time.
- Populations increase when birth and immigration rates are greater than death and emigration rates, and they decrease when the reverse is true.
- Because birth, death, immigration, and emigration rates are all affected by environmental factors, the environment plays a key role in changing the size of populations.

21.3 Exponential Growth

- A population grows exponentially when it increases by a constant proportion from one generation to the next. Exponential growth produces a J-shaped curve.
- The doubling time is one measure of how fast a population is growing.
- Populations may grow at an exponential rate when organisms are introduced into or migrate to a new area.

21.4 Limits to Population Growth

- Because the environment contains a limited amount of space and resources, no population can continue to increase in size indefinitely.
- Some populations increase rapidly at first and then level off and stabilize at the carrying capacity, the maximum population size that their environment can support. This growth pattern is represented by an S-shaped curve.
- Density-dependent environmental factors limit the growth of a population more strongly when the density of the population is high. Such factors include food shortages, diminishing space, disease, predators, and habitat deterioration.
- Density-independent factors, such as weather and natural disturbances, limit the growth of populations without regard to their density.

21.5 Patterns of Population Growth

- Different populations (including those of the same species) can exhibit different patterns of growth over time, including J-shaped curves, S-shaped curves, population cycles, and irregular fluctuations.
- Two populations of different species can change together in tightly linked cycles when one or both species are strongly influenced by the other.
- In natural systems, a growth pattern of irregular fluctuations is much more common than an S-shaped growth pattern or tightly linked cycles.

- Understanding why different populations have different patterns of growth can provide critical information on how best to manage endangered species.

21.6 Human Population Growth: Surpassing the Limits?

- The global human population is growing exponentially, increasing by 1 billion people every 13 years.
- So far, we have been able to postpone dealing with limits to our population growth by increasing the carrying capacity of our environment. We have accomplished this through inventiveness and technology, in particular through agriculture and the use of fossil fuels.
- Our use of resources is now increasing even faster than our rate of population growth. The same environmental factors that limit other species' growth will eventually limit ours, unless we take steps to limit it ourselves before that happens.

◉ Review and Application of Key Concepts

1. Explain why it can be difficult to determine what constitutes a population.

2. Populations increase in size when birth and immigration rates are greater than death and emigration rates. Keeping this basic principle in mind, what actions do you think a scientist or policy maker might take to protect a population threatened by extinction?

3. Assume that a population grows exponentially, increasing by a constant proportion of 1.5 per year. If the population initially contains 100 individuals, it will contain 150 individuals in the next year. Graph the number of individuals in the population versus time for the next 5 years, starting with 150 individuals in the population.

4. Population growth cannot increase indefinitely.

   a. What environmental factors prevent unlimited growth?

   b. Why is it common for populations of species that enter a new region to grow exponentially for a period of time?

5. Describe the difference between density-dependent and density-independent factors that limit population growth. Give two examples of each.

6. Different populations of a species can have different patterns of population growth. Explain how an understanding of the causes of these different patterns can help managers protect rare species or control pest species.

7. List five specific actions that you can take to limit the growth or impact of the human population.

Key Terms

carrying capacity (p. 436)
density-dependent (p. 437)
density-independent (p. 437)
doubling time (p. 434)
ecological footprint (p. 442)
exponential growth (p. 434)
habitat (p. 436)
irregular fluctuations (p. 438)

J-shaped curve (p. 434)
population (p. 432)
population cycle (p. 438)
population density (p. 432)
population ecology (p. 432)
population size (p. 432)
S-shaped curve (p. 436)
sustainable (p. 442)

Self-Quiz

1. A group of interacting individuals of a single species located within a particular area is
   a. a biosphere.
   b. an ecosystem.
   c. a community.
   d. a population.

2. A population of plants has a density of 12 plants per square meter and covers an area of 100 square meters. What is the population size?
   a. 120
   b. 1,200
   c. 12
   d. 0.12

3. A population that is growing exponentially increases
   a. by the same number of individuals each generation.
   b. by a constant proportion each generation.
   c. in some years and decreases in other years.
   d. none of the above

4. In a population with an S-shaped growth curve, after an initial period of rapid increase, the number of individuals
   a. continues to increase exponentially.
   b. drops rapidly.
   c. remains near the carrying capacity.
   d. cycles regularly.

5. The growth of populations can be limited by
   a. natural disturbances.
   b. weather.
   c. food shortages.
   d. all of the above

6. Factors that limit the growth of populations more strongly at high densities are said to be
   a. density-dependent.
   b. density-independent.
   c. exponential factors.
   d. sustainable.

7. The maximum number of individuals in a population that can be supported indefinitely by the population's environment is called the
   a. exponential size.
   b. J-shaped curve.
   c. sustainable size.
   d. carrying capacity.

8. A population that initially has 40 individuals grows exponentially with an (annual) proportional increase (λ) of 1.6. What is the size of the population after 3 years? (*Note*: Round down to the nearest individual.)
   a. 16
   b. 163
   c. 192
   d. 102,400

CHAPTER 22 Interactions among Organisms

Key Concepts

- Organisms interact with one another in many different ways, the three most important of which are mutualism, exploitation, and competition.

- In mutualism, two species interact for the benefit of both species. Mutualism evolves when the benefits of the interaction outweigh the costs for both species.

- In exploitation, one species in the interaction benefits (the consumer) while the other is harmed (the species that is eaten). Species that are eaten by other organisms have evolved elaborate ways of defending themselves against their consumers.

- In competition, two species that share resources have a negative effect on each other. Competition can cause greater differences to evolve between species.

- Mutualism, exploitation, and competition help determine where organisms live and how abundant they are.

- Changes in interactions among organisms can change the community of species that live in an area and the underlying nature of the ecosystem.

Suicidal Mantises and Gruesome Parasites

Praying mantises have been seen to walk to the edge of a river, throw themselves in, and drown shortly thereafter. If they are rescued from the water, they will immediately throw themselves back in. What causes them to do this?

This bizarre behavior appears to be driven not by the mantises themselves, but by a parasitic worm (*Gordius*). Less than a minute after the mantis lands in the water, a worm emerges from its anus. This worm attacks and infects terrestrial insects, such as praying mantises, but it also depends on an aquatic host for part of its life. The worm has performed a neat trick: it has evolved the ability to cause its insect host to jump into the river, an act that kills the insect but increases the chance that the worm will eventually reach its aquatic host.

Moving from the bizarre to the gruesome, examine the photographs here. The fungus that killed the ant first grew throughout the ant's entire body, dissolving portions of its body and using them for food. Eventually, the fungus sprouted reproductive structures (indicated by arrows), which allowed it to spread and attack other ants. Fungi attack many other species, including crops such as the corn plant shown. If you have trouble empathizing with ants or corn, the third photograph shows the effects of one of the many parasites that attack people. Hundreds of millions of people are disabled every year by more than a thousand different types of parasitic organisms that afflict humans.

Parasites, such as the worms that plague praying mantises and the fungi that riddle the bodies of ants, are organisms that live in or on other organisms (known as their hosts). They obtain nutrients from their hosts, often causing them harm but not immediate death. **But can a parasite alter the behavior of its host organism? How would a parasite benefit from behavioral changes in its host?** In this chapter, we examine interactions among species that share a habitat and see how changes in such interactions can change communities and ecosystems. We will see that the malevolent actions of parasites are just one of the many, and often complex, interspecies relationships that exist in biological communities.

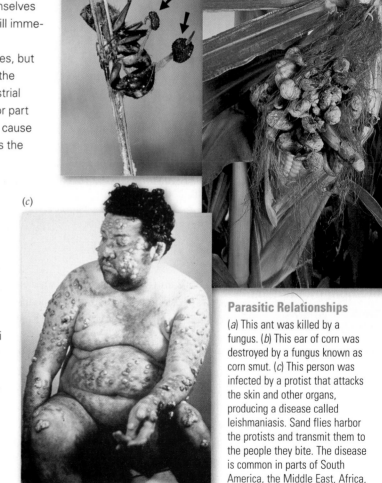

(a)

(b)

(c)

Parasitic Relationships
(a) This ant was killed by a fungus. (b) This ear of corn was destroyed by a fungus known as corn smut. (c) This person was infected by a protist that attacks the skin and other organs, producing a disease called leishmaniasis. Sand flies harbor the protists and transmit them to the people they bite. The disease is common in parts of South America, the Middle East, Africa, and Central and South Asia.

447

Citrus mealybug

As we saw in the previous two chapters, ecology focuses on interactions between organisms and their environment. An organism's environment includes the other organisms that live there. All the interacting species in a defined area together make up a biological community. The subject of this chapter—interactions among organisms—is a central theme in community ecology.

Interactions among organisms have huge effects on natural communities. For example, as we saw in Chapter 21, the moth *Cactoblastis cactorum*, by feeding on the cactus *Opuntia*, caused *Opuntia* populations to crash throughout a large region of Australia. Overall, interactions among organisms have an influence at every level of the biological hierarchy at which ecology is studied.

The millions of species on Earth can interact in many different ways. In this chapter we classify interactions among organisms into three broad categories, based on whether the interaction is beneficial (+) or harmful (–) to each of the interacting species. These three categories represent some of most common and most important kinds of ecological interactions:

1. *+/+ interactions*, in which both species benefit (mutualism)
2. *+/– interactions*, in which one species benefits and the other is harmed (exploitation)
3. *–/– interactions*, in which both species may be harmed (competition)

Each type of interaction plays a key role in determining where organisms live and how abundant they are. We also discuss how changes in interactions among organisms can alter ecological communities.

See mutualism in action. 22.1

Helpful to know

"Symbiosis" (from *sym*, "with"; *biosis*, "life") describes some (but not all) mutualistic and parasitic interactions. Individuals of one species, for at least part of their life cycle, must live close to, in, or on individuals of another species for the interaction to be considered symbiotic.

22.1 Mutualism

Mutualism (+/+ interaction) is an association between two species in which both species benefit. Mutualism is common and important to life on Earth: many species receive benefits from, and provide benefits to, other species. These benefits increase the survival and reproduction of both of the interacting species.

Mutualism can occur when two or more organisms of different species live together—an association known as **symbiosis**. Insects such as aphids and mealybugs, both of which feed on the nutrient-poor sap of plants, often have a mutualistic, symbiotic association with bacteria that live within their cells. The bacteria receive food and a home from the insects, and the insects receive nutrients that the bacteria (but not the insects) can synthesize from sugars in the plant sap. Such symbiotic associations can be amazingly complex. Scientists have recently discovered that a second bacterial species lives within the bacteria that live inside of citrus mealybug cells; it is not yet clear whether this second species benefits or harms the bacteria in which they live.

This open question illustrates an important point: although some symbiotic associations are clearly mutualistic, benefiting both organisms (a +/+ interaction), in many cases of symbiosis one species harms rather than helping the other species in the association (a +/– interaction). This sort of interaction is true for many parasites, which spend all or most of their lives within their hosts, deriving benefits from their hosts yet harming rather than benefiting their hosts. By some estimates, nearly half of all species on Earth are parasites.

There are many types of mutualism

Nature abounds with varieties of mutualism; here we describe only some of the most common types. In gut inhabitant mutualism, organisms that live in an animal's digestive tract receive food from their host and benefit the host by digesting foods, such as wood or cellulose, that the host otherwise could not use. The interaction between a mealybug and the first bacterial species living in its gut is an example of this type of mutualism. So, too, is the interaction between termites and the bacteria that inhabit their guts, which enable the termites to digest wood. The interaction between humans and some of the bacteria that live in our intestines, which help us digest and absorb certain nutrients, is yet another example.

In seed dispersal mutualism, an animal, such as a bird or mammal, eats a fruit that contains plant seeds, and then later defecates the seeds far from the parent plant. Such dispersal by animals is the primary way that many plant species reach new areas of favorable habitat. For example, most of the plant species that live on isolated oceanic islands (those that are farther than 1,000 kilometers from land) are thought to have arrived there by bird dispersal of their seeds.

Mutualism in which each partner has evolved to alter its behavior to benefit the other species is called behavioral mutualism. The relationship between certain shrimps and fishes is a good example of behavioral mutualism (Figure 22.1). Shrimps of the genus *Alpheus* live in an environment with plenty of food but little shelter. They dig burrows to hide in, but they see poorly, so they are vulnerable to predators when they leave their burrows to feed. These shrimps have formed a fascinating relationship with some goby fishes in the genera *Cryptocentrus* and *Vanderhorstia*. When a shrimp ventures out of its burrow to eat, it keeps an antenna in contact with an individual goby with which it has formed a special relationship. If a predator or other disturbance causes the fish to make a sudden movement, the shrimp darts back into the burrow. The goby acts as a "Seeing Eye" fish for the shrimp, warning it of danger. In return, the shrimp shares its burrow with the goby, thereby providing the fish with a safe haven.

In pollinator mutualism, an animal, such as a honeybee, transfers pollen (which contains sperm) from one flower to the female reproductive organs (carpels) of another flower of the same species. These animals are known as **pollinators**, and without them many plants could not reproduce. To ensure that pollinators come to their flowers, plants offer a food reward, such as pollen or nectar. Both species benefit from the interaction. Pollinator mutualism is important in both natural and agricultural ecosystems. For example, the apples we buy at the supermarket are available only because honeybees pollinate the flowers of apple trees, enabling the trees to produce their fruit.

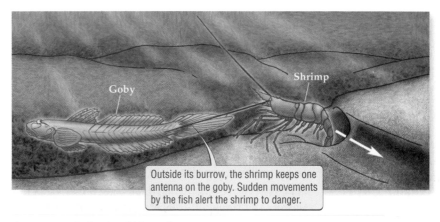

Outside its burrow, the shrimp keeps one antenna on the goby. Sudden movements by the fish alert the shrimp to danger.

Figure 22.1 Behavioral Mutualism
Each *Alpheus* shrimp builds a burrow for shelter, which it shares with a goby fish. The fish provides an early-warning system to the nearly blind shrimp when the shrimp leaves the burrow to feed.

of the reproductive parts of a flower, see Figure 22.6.) When the moth larvae hatch, they feed on the seeds of the yucca plant.

In this mutualism, the plant gets pollinated (a reproductive benefit provided to the plant by the moth) and the moth eats some of its seeds (a food benefit provided to the moth by the plant). In fact, plant and pollinator each depend absolutely on the other—the yucca is the moth's only source of food, and this moth

Honeybee

Mutualists are in it for themselves

Although both species in a mutualism benefit from the relationship, what is good for one species may come at a cost to the other. For example, a species may use energy or increase its exposure to predators when it acts to benefit its mutualistic partner. From an evolutionary perspective, mutualism evolves when the benefits of the interaction outweigh the costs for both species. Even in mutualism, however, the interests of the two species may be in conflict.

Consider the pollinator mutualism between the yucca plant and the yucca moth. A female yucca moth collects pollen from yucca flowers, flies to another group of flowers, and lays her eggs at the base of the carpel in a newly opened flower. After she has laid her eggs, the female moth climbs up the carpel and deliberately places the pollen she collected earlier onto the stigma of the flower. By this act she fertilizes the eggs of that second yucca plant (Figure 22.2). (For a review

Figure 22.2 Pollinator Mutualism
The yucca and the yucca moth depend on each other for survival.

is the only species that pollinates the yucca—so this association is mutualistic, not parasitic. But there are costs for both species. Let's examine these costs more closely.

In a cost-free situation for the plant, the moth would transport pollen but would not destroy any of the plant's seeds. In a cost-free situation for the moth, the moth would produce as many larvae as possible, and they would consume many of the plant's seeds. In actuality, an evolutionary compromise has been reached: the moth usually lays only a few eggs per flower, and the plant tolerates the loss of a few of its seeds. Yucca plants have a defense mechanism that helps keep this compromise working: if a moth lays too many eggs in one of the plant's flowers, the plant can selectively abort that flower, thereby killing the moth's eggs or larvae.

Mutualism is everywhere

Mutualism is very common. Most of the plant species that dominate forests, deserts, grasslands, and other biomes are mutualists. For example, most plant species have mutualistic associations with fungi, called **mycorrhizae**. The fungi help the plant roots absorb nutrients and water from the soil, and the plant provides the

Figure 22.3 Mycorrhizae
Fungal hyphae surround a plant root and penetrate some of its cells, helping the plant roots absorb mineral nutrients and water from the soil and allowing carbohydrates to be transported from the plant to the fungus.

fungi with carbohydrates produced by photosynthesis (Figure 22.3).

As mentioned earlier, many animal species are pollinators involved with plants in pollinator mutualisms. Other examples of mutualisms involving animals include the spectacular reefs found in tropical oceans (Figure 22.4). These reefs are built by corals (soft-bodied animals), most of which house photosynthetic algae—their mutualistic partners—inside their bodies. The corals provide

Figure 22.4 The Home a Mutualism Built
The great diversity of life in tropical reefs depends on corals, many of which benefit from mutualistic associations with algae.

the algae with a home and several essential nutrients, such as phosphorus, and the algae provide the corals with carbohydrates produced by photosynthesis.

Mutualism can determine the distribution and abundance of species

Mutualism can influence the **distribution** (the geographic area over which a species is found) and **abundance** (number of individuals of a species in a defined habitat) of organisms in two ways. First, because each species in a mutualism survives and reproduces better where its partner is found, the two species strongly influence each other's distribution and abundance. For example, because the yuccas and yucca moths described earlier depend absolutely on each other, each species is found only where the other is present.

Second, a mutualism can have indirect effects on the distribution and abundance of species that are not part of the mutualism. It can affect how a community is organized and what types of species are found in it, where, and in what numbers. Coral reefs, for example, are unique habitats that are home to many different organisms, including certain species of fishes, mollusks, crustaceans, and echinoderms (such as sea stars). Because the corals that build the reefs depend on their mutualisms with algae, the many other species that live in coral reefs depend on those mutualisms indirectly.

Concept Check

1. The yucca moth pollinates the yucca and depends on the plant for food. Is the relationship between the moth and the yucca a mutualism or an exploitation? Is this a cost-free interaction for both moth and yucca? Explain.
2. Using an example, describe how a mutualism between two species can affect species not directly involved in the relationship.

22.2 Exploitation

Exploitation (+/− interaction) includes a variety of interactions in which one species (the consumer) benefits and the other (the species that is eaten, or "food organism") is harmed. The consumers in such interactions can be classified into three main groups:

1. **Herbivores** are consumers that eat plants or plant parts.

2. **Predators** are animals (or, in rare cases, plants) that kill other animals for food; the animals that are eaten are called **prey**.
3. **Parasites** are consumers that live in or on the organisms they eat (which are called **hosts**). An important group of parasites are **pathogens**, which cause disease in their hosts.

These three major types of +/− interactions are very different from one another. For example, whereas predators (such as wolves) kill their food organisms immediately, herbivores (such as cows) and parasites (such as fleas) usually do not. Although the three types of exploitation have obvious and important differences, in this section we look at some general principles that apply to all three.

Consumers and their food organisms can exert strong selection pressure on each other

The presence of consumers in the environment has caused many species to evolve elaborate strategies to avoid being consumed. Many plants, for example, produce spines and toxic chemicals as defenses against herbivores. Some plants rely on **induced defenses**, responses that are directly stimulated by an attack from herbivores. Spine production is an induced defense in some cactus species: an individual cactus that has been partially eaten, or grazed, is much more likely to produce spines than is an individual that has not been grazed (Figure 22.5).

Many prey organisms have evolved bright colors or striking patterns that warn potential predators that they are heavily defended, usually by chemical means

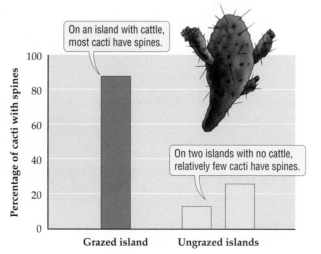

On an island with cattle, most cacti have spines.

On two islands with no cattle, relatively few cacti have spines.

Percentage of cacti with spines

Grazed island   Ungrazed islands

Figure 22.5 Spines on Some Cacti Are an Induced Defense

On three islands off the coast of Australia, the percentage of cacti with spines is higher on the island that has cattle than on the two islands that do not. Field and laboratory experiments show that grazing by cattle directly stimulates the production of spines in this species of cactus.

Figure 22.6 Warning Coloration Can Be Highly Effective

(*a*) The bright colors of this poison dart frog warn potential predators of the deadly chemicals contained in its tissues. (*b*) An inexperienced blue jay vomits after eating a brightly colored monarch butterfly.

(*a*)

(*b*)

defenses (immune systems) to help them fight off the effects of microbial diseases and parasitic infections.

Interactions among species can drive evolutionary change in the interacting species—a concept known as **coevolution**. Put another way, two species that interact may trigger evolutionary change in each other as a consequence of their interactions. For example, the many ways in which producers have evolved to protect themselves against consumers indicate that consumers often apply strong selection pressure on their food organisms. Selection occurs in the other direction as well. If a plant or prey species evolves a particularly powerful defense against attack, its consumers, in turn, experience strong selection pressure to overcome that defense. Many defenses work against all consumers except a few species that have evolved the ability to overcome them. Consider the rough-skinned newt, whose skin contains unusually large amounts of the potent neurotoxin TTX (tetrodotoxin)—enough to kill 25,000 mice. The newt is so toxic that only one predator, the garter snake, can tolerate its poison well enough to eat the newt and survive (see Figure 16.2).

Consumers can alter the behavior of the organisms they eat

The bizarre story of the praying mantises that jump to their deaths in rivers (at the opening of this chapter) provides a dramatic example of how consumers can alter the behavior of their food organisms. But exploitation can alter the behavior of food organisms in more subtle ways as well.

Predators can be a driving force that provides advantages to animals living or feeding in groups. In some cases, several prey individuals acting together may be able to thwart attacks from predators (Figure 22.7).

(Figure 22.6*a*). Such warning coloration can be highly effective. Blue jays, for example, quickly learn not to eat monarch butterflies, which are brightly colored and have tissues that contain cardiac glycosides; these chemicals cause nausea in birds (and people) (Figure 22.6*b*) and, at high doses, sudden death from heart failure. Other prey have evolved to avoid predators by being hard to find or hard to catch. In addition, the potential hosts for parasites have evolved molecular

Figure 22.7 Come and Get Us

Although a single musk ox may be vulnerable to predators such as wolves, a group that forms a circle makes a difficult target.

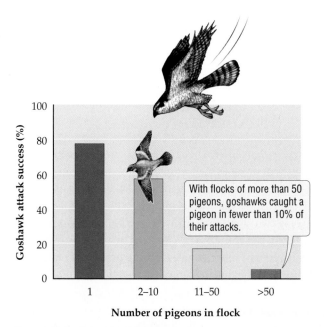

Figure 22.8 Safety in Numbers
The success of goshawk attacks on wood pigeons decreases greatly when there are many pigeons in a flock.

[Graph: Goshawk attack success (%) vs. Number of pigeons in flock. Callout box reads: "With flocks of more than 50 pigeons, goshawks caught a pigeon in fewer than 10% of their attacks."]

Large groups of prey may also be able to provide better warning of a predator's attack. Because more individuals can watch for predators, a large flock of wood pigeons detects the approach of a goshawk [GOSS-hawk] (a predatory bird) much sooner than a single pigeon does. The success rate of goshawk attacks drops from nearly 80 percent when attacking single pigeons to less than 10 percent when attacking flocks of more than 50 birds (Figure 22.8).

Consumers can restrict the distribution and abundance of their food organisms

The American chestnut used to be a dominant tree species across much of eastern North America. Within its range, anywhere from one-quarter to one-half of all trees were chestnuts. They were capable of growing to large size: trunks up to 10 feet in diameter were noted by settlers in colonial times. In 1900, however, a fungus that causes a disease called chestnut blight was introduced into the New York City area. This fungus spread rapidly, killing most of the chestnut trees in eastern North America. Today the American chestnut survives throughout its former range only in isolated patches, primarily as sprouts that arise from the base of otherwise dead trunks. With few exceptions, the new sprouts die back from reinfection with the fungus before they can grow large enough to generate new seeds.

The effect of chestnut blight on the American chestnut shows how a consumer (the fungus) can limit the distribution and abundance of its food organism (the chestnut): in this case, a formerly dominant tree species was virtually eliminated from its entire range. The effects of consumers on the distribution and abundance of the species they eat are also shown by what can happen when a food organism is separated from its consumers. As we saw in Chapter 20, nonnative species introduced by people to new regions sometimes disrupt the ecological communities there. A series of recent studies suggests that some introduced species are able to increase rapidly in number in their new areas in part because they have many fewer parasites there than in their original homes (Figure 22.9).

Consumers can drive their food organisms to extinction

Laboratory experiments with protists and with mites have shown that predators can drive their prey extinct (see the box on the next page for a description of how experiments are used in ecology). Exploitation can

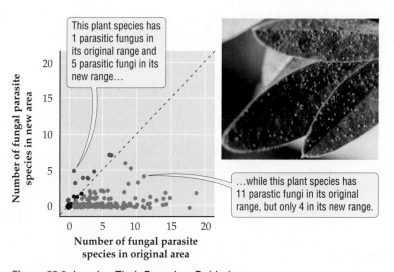

[Graph: Number of fungal parasite species in new area vs. Number of fungal parasite species in original area. Callout box reads: "This plant species has 1 parasitic fungus in its original range and 5 parasitic fungi in its new range…" and "…while this plant species has 11 parasitic fungi in its original range, but only 4 in its new range."]

Figure 22.9 Leaving Their Parasites Behind
Most introduced plant species have fewer fungal parasites in their new homes than in their original homes. Each point on the graph represents a different plant species that has been introduced to a new area. Points below the diagonal line represent plants with fewer fungal parasites in their new home than in their old home; points above the diagonal line represent the opposite. Points falling on the diagonal line indicate plants showing no difference in the number of fungal parasite species between their new and old ranges. The photograph shows the perennial bush clover (*Lespedeza capitata*) infected with the pathogenic rust fungus (*Uromyces lespedezae-procumbentis*); each separate "bump" is an individual fungus that infected the leaf from a separate microscopic spore and is now producing its own spores.

Science Toolkit

Answering Ecological Questions with Experiments

Like all other scientists, ecologists observe nature and ask questions about their observations. They then rely on three approaches to answer these questions: experiments, additional observations, and models. Observations are helpful if the questions concern events that cover large geographic regions or occur over long periods of time—experiments designed to answer such questions may not be feasible. Models, such as those that predict changes in the global climate, can also be used to answer questions that are hard to examine experimentally. Nevertheless, experiments are one of the most important parts of the modern ecologist's scientific toolkit.

In an **ecological experiment**, an investigator alters one or more features of the environment and observes the effect of that change. Experiments in ecology range from laboratory investigations to studies conducted outside in an artificial environment (such as an artificial pond) to field experiments conducted in a natural environment in which one or more factors are manipulated by the experimenter but all else is left as it is in the local environment. In any ecological experiment, the purpose is to examine the effects of different treatments—such as a high amount of pesticide, a low amount of pesticide, and no pesticide—on natural processes.

Field Experiment with Fishes

An ecologist studying lake fishes might perform experiments in laboratory aquariums, in natural lakes, or in artificial (human-constructed) ponds as seen here.

The "no pesticide" treatment is an example of a **control** (applied to a **control group**)—an important feature included in most ecological experiments (as well as most other kinds of experiments). The control group is subjected to the same environmental conditions as the experimental groups, except that the factor or factors being tested in the experiment are omitted. Including a control group in an experiment strengthens the argument that any effects seen in the experimental group (but not in the control group) result from the factor or factors under investigation.

When performing an experiment, an ecologist replicates each treatment (that is, performs it more than once), including the control. The advantage of replication is that as the number of **replicates** increases, it becomes less likely that the results are actually due to a lack of measurement or control of a variable in the study. Consider an experiment designed to test whether the presence of pesticides in a pond causes frogs to have more deformities. If the experiment were performed with only two ponds, one with detectable levels of pesticides and the other without, the results would be hard to interpret. Suppose that frog deformities were more common in the pond that contained pesticides. Although pesticides may have caused this result, the two ponds might have differed in many other ways, and one or more of these differences might have been the real cause of the result. On the other hand, if many ponds were used—that is, many replicates—it becomes much less likely that each pond with pesticides also contained something else that increased the chance of frog deformities.

By assigning treatments at random, ecologists also seek to limit the effects of unmeasured variables. Suppose an experiment was designed to test whether insects that eat plants decrease the number of seeds the plants produce. To test this idea, a natural area could be divided into a series of experimental plots. Each plot would receive one of two treatments: it would either be sprayed regularly with an insecticide, reducing the number of plant-eating insects, or be left alone. If plant-eating insects have much effect, seed production should be higher in plots that are sprayed. Usually, the decision of whether a particular plot will be sprayed is made at random at the start of the experiment, to make it less likely that the plots receiving a particular treatment will share other features that might influence seed production, such as high or low levels of soil nutrients.

Ecological experiments enable us to test whether we understand how nature works. The results may answer the question the experiment was designed to address, and they may also stimulate a whole new set of questions. As new questions lead to new discoveries, what we know about ecology constantly changes. Therefore, our understanding of ecology is, and always will be, a work in progress.

drive food organisms to extinction in natural systems as well. The effect of chestnut blight on the American chestnut provides one clear example: although the chestnut tree is not extinct through its entire range, many local populations have been driven to extinction. Similarly, *Cactoblastis* moths drove many populations of the *Opuntia* cactus in Australia extinct (see Figure 21.4). If a consumer eats only one species and then drives a population of the species it eats to extinction, the consumer must either locate a new population of food organisms or go extinct itself. This is exactly what happened to *Cactoblastis* in eastern Australia: the moth drove most populations of the cactus it eats extinct, and now both species are found in low numbers.

22.3 Competition

In **competition** (–/– interaction), each of two interacting species has a negative effect on the other. Competition is most likely when two species share an important resource, such as food or space, that is in short supply. When two species compete, each has a negative effect on the other because each uses resources (such as a source of food) that otherwise could have been used by its competitor. This is true even when one species is so superior as a competitor that it ultimately drives the other species extinct: until the inferior competitor actually becomes extinct, it continues to use some resources that could have been used by the superior competitor.

There are two main types of competition:

1. In **interference competition**, one organism directly excludes another from the use of a resource. For example, individuals from two species of birds may fight over the tree holes that they both use as nest sites.
2. In **exploitative competition**, species compete indirectly for a shared resource, each reducing the amount of the resource available to the other. For example, two plant species may compete for a resource that is in short supply, such as nitrogen in the soil.

Competition can limit the distribution and abundance of species

Competition between species often has important effects on natural populations. These effects, as shown by a great deal of field evidence, include limiting the distributions and abundances of species. Let's explore two examples.

Along the coast of Scotland, the larvae of two species of barnacles, *Semibalanus balanoides* [. . . buh-*LAY*-nus bah-luh-*NOY*-deez] and *Chthamalus stellatus* [thuh-*MAY*-lus stell-*AY*-tus], both settle on rocks on high and low portions of the shoreline. However, as adults, *Semibalanus* individuals appear only on the lower portion of the shoreline, which is more frequently covered by water, and *Chthamalus* individuals are found only on the higher portion of the shoreline, which is more frequently exposed to air (Figure 22.10).

In principle, the distributions of these two barnacles could have been caused either by competition or by environmental factors. In an experimental study, however, ecologists discovered that *Chthamalus* could thrive on low portions of the shoreline, but only when *Semibalanus* was removed. Hence, competition with *Semibalanus* ordinarily prevents *Chthamalus* from living low on the shoreline. This interaction is an example of interference competition because *Semibalanus* individuals often crush the smaller and more delicate *Chthamalus* individuals. The distribution of *Semibalanus*, on the other hand, depends mainly on

Learn more about competition.
▶❙❙
22.2

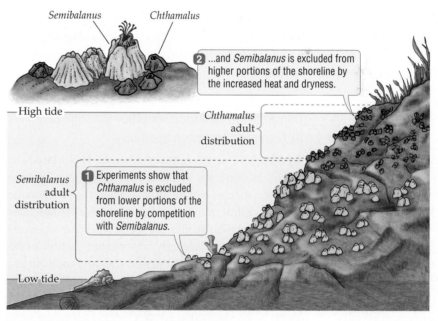

Figure 22.10 What Keeps Them Apart?
On the rocky coast of Scotland, the larvae of *Semibalanus* and *Chthamalus* barnacles settle on rocks on both high and low portions of the shoreline. However, adult *Semibalanus* barnacles are not found on high portions of the shoreline, and adult *Chthamalus* individuals are not found on low portions of the shoreline.

environmental factors: the increased heat and dryness found at higher levels of the shoreline prevent *Semibalanus* from surviving there.

A second case of competition affecting distribution and abundance concerns wasps of the genus *Aphytis* [ay-FYE-tus]. These wasps attack scale insects, which can cause serious damage to citrus trees. Female wasps lay eggs on a scale insect, and when the wasp larvae hatch, they pierce the scale insect's outer skeleton and then consume its body parts.

In 1948, the wasp *Aphytis lingnanensis* was released in southern California to curb the destruction of citrus trees caused by scale insects. A closely related wasp, *A. chrysomphali*, was already living in that region at the time. *A. lingnanensis* was released in the hope that it would provide better control of scale insects than *A. chrysomphali* did. *A. lingnanensis* proved to be a superior competitor (Figure 22.11), in most locations driving *A. chrysomphali* to extinction by exploitative competition. As hoped for, *A. lingnanensis* also provided better control of scale insects.

Although competition between species is very common, note that it does not always occur when two species share resources or space. This is especially true when the resources are abundant. Competition among leaf-feeding insects, for example, is relatively uncommon for this reason. A huge amount of leaf material is available for the insects to eat, and usually there are too few insects to cause their food to be in short supply. As long as their food remains abundant, little competition occurs.

Competition can increase the differences between species

As Charles Darwin realized when he formulated the theory of evolution by natural selection, competition between species can be intense when the two species are very similar in form. For example, birds whose beaks are similar in size eat seeds of similar sizes and therefore compete intensely, whereas birds whose beaks differ in size eat seeds of different sizes and compete less intensely. Intense competition between similar species may result in **character displacement**, in which the forms of the competing species evolve to become more different over time. By reducing the similarity in form between species, character displacement reduces the intensity of competition. As we saw in Chapter 17, however, species can evolve in this way only if their populations vary genetically for traits (in this case, beak size) on which natural selection can act.

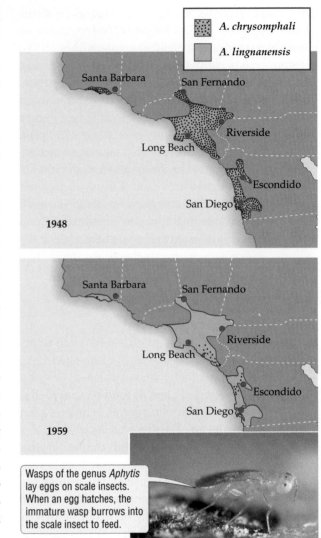

Figure 22.11 **A Superior Competitor Moves In**
After being introduced to southern California in 1948, the wasp *Aphytis lingnanensis* rapidly drove its competitor, *A. chrysomphali*, extinct in most locations. Both species of wasps prey on scale insects that damage citrus crops (such as lemons and oranges).

Wasps of the genus *Aphytis* lay eggs on scale insects. When an egg hatches, the immature wasp burrows into the scale insect to feed.

Some evidence for character displacement comes from observations that the forms of two species are more different when they live together than when they live in separate places. In the Galápagos Islands, for example, the beak sizes of two species of Galápagos finches, and hence the sizes of the seeds the birds eat, are more different on islands where both species live than on islands that have only one of the two species (Figure 22.12). Recent experiments with other groups, such as fishes and lizards, also suggest that character displacement is important in nature.

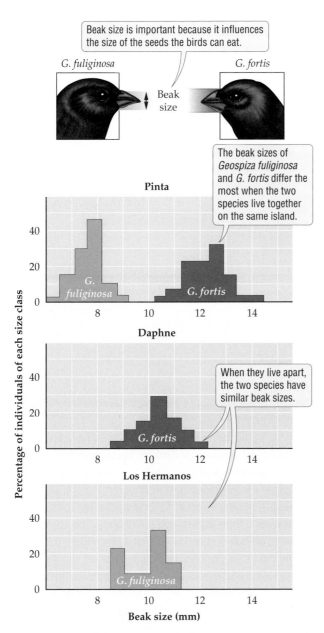

Beak size is important because it influences the size of the seeds the birds can eat.

G. fuliginosa   G. fortis

Beak size

Pinta

The beak sizes of *Geospiza fuliginosa* and *G. fortis* differ the most when the two species live together on the same island.

G. fuliginosa

G. fortis

Daphne

When they live apart, the two species have similar beak sizes.

G. fortis

Los Hermanos

G. fuliginosa

Beak size (mm)

Percentage of individuals of each size class

Figure 22.12 Character Displacement
Competition for resources may cause the competing species to become more different over time. Competition between two species of Galápagos finches, the small ground finch (*Geospiza fuliginosa*) and the medium ground finch (*Geospiza fortis*), may be the driving force that causes the beak sizes of these birds to be more different when they live on the same island (Pinta) than when they live apart (on Daphne, or Los Hermanos, island).

desert shrubs may become more abundant. These changes in the abundances of grasses and shrubs can change the physical environment. The rate of soil erosion may increase because shrubs do not stabilize soil as well as grasses do. Ultimately, if overgrazing is severe, the ecosystem can change from a dry grassland to a desert.

Changes in interactions among organisms can have complex effects on natural communities. Recall what happened when people removed dingoes from rangelands in Australia to prevent them from eating sheep (see Figure 20.2). The removal of this predator caused other, unintended changes, including increases in the number of red kangaroos and decreases in the availability of plants that sheep like to eat.

Mutualism can have similar large effects. For example, the fire tree (Figure 22.13) was brought to Hawaii by Portuguese immigrants for use as an ornamental plant and as firewood. The fire tree forms a mutualistic association with bacteria that can convert nitrogen gas (N_2) from the air into ammonium (NH_4^+) in the soil, a form of nitrogen that plants can use. The fire tree has escaped cultivation and has invaded ecological communities such as those found on recent volcanic deposits. Because of its mutualism, the fire tree causes four times the usual amount of nitrogen to enter volcanic site ecosystems. In essence, the mutualism enables the fire tree to fertilize itself, which helps it

Figure 22.13 A Tree That Doesn't Need Fertilizer
The fire tree (*Myrica faya*) was brought to Hawaii by people, but it has since escaped from cultivation. The fire tree forms a mutualism with bacteria that provides the tree with extra nitrogen. The mutualism gives this tree an advantage over other species, enabling it to invade ecological communities such as recent volcanic deposits and exclude other species from those sites. An invading fire tree, with bright green foliage, is seen in the foreground in this photograph. The larger tree with gray-green foliage is the native 'ohi'a lehua (*Metrosideros polymorpha*).

22.4 Interactions among Organisms Shape Communities and Ecosystems

Throughout this chapter we have seen how interactions among organisms help determine their distribution and abundance. Interactions among organisms also have large effects on the communities and ecosystems in which those organisms live.

When dry grasslands are overgrazed by cattle, for example, grasses may become less abundant and

grow rapidly and exclude other species from colonizing volcanic sites.

In the examples discussed in this section, a change in an interaction between organisms had a ripple effect, altering the abundance of individuals, the community of species living in an area, and even, in the case of the dry grasslands, converting one ecosystem (dry grassland) into another (desert shrubland). In general, interactions among organisms can affect all levels of the biological hierarchy at which ecology can be studied: the individual organisms involved in an interaction, populations of those organisms, the communities in which those organisms live, and whole ecosystems.

Concept Check

1. Explain why the plants that herbivores graze are not driven to extinction in natural communities.
2. Why do prey species often live in herds or flocks, but predators rarely do so?
3. *Chthamalus* (a barnacle that lives high on the shoreline) and *Semibalanus* (a larger, less delicate barnacle that lives in the low intertidal zone) exhibit interference competition. How would the survival and reproduction of a colony of *Chthamalus* be affected if it were relocated to the low intertidal zone (a) in the absence of *Semibalanus*, and (b) intermixed with *Semibalanus*.

Applying What We Learned

Parasites Often Alter Host Behavior

Parasites affect their hosts in ways that range from merely annoying (for example, fleas) to downright deadly (such as fungal parasites of ants). In addition, many parasites cause their hosts to perform unusual or even bizarre behaviors that harm the host but benefit the parasite. Recall the parasitic worms described at the opening of this chapter that cause praying mantises to throw themselves into rivers and drown. Similarly, the protist *Toxoplasma gondii* causes its rat host to become more curious and less fearful. Such changes make infected rats easier prey for cats, the other host of the protist. As these examples show, some parasites cause broad changes in their hosts' behavior, such as making the host less cautious or causing it to move from one habitat to another.

Other parasites cause much more specific changes in their hosts' behavior. A parasitic wasp called *Hymenoepimecis* [HYE-men-oh-EPP-ee-MEE-sis] attacks the spider *Plesiometa argyra* [PLEZ-ee-oh-MEE-tuh ahr-JEER-uh]. A female wasp stings the spider into temporary paralysis and then lays an egg on its body. The spider recovers quickly and builds normal webs (Figure 22.14a) for the next week or two. During this period, the wasp egg hatches, and the wasp larva feeds by sucking body fluids from the spider. Then, one evening, the larva injects a chemical into the spider, causing the spider to spin a unique "cocoon web" (Figure 22.14b). In response to the injected chemical, the spider performs many repetitions of one part of its normal web-building process, suppressing the other parts. The wasp has evolved the ability to cause a very particular change in how the spider builds its web.

What adaptive benefit does the wasp gain in altering the behavior of the spider in this way? As soon as the spider finishes the web, the larva kills and consumes the spider. The larva then spins a cocoon, in which it will complete its development. The larva uses the spider's altered web as a strong support from which to hang its cocoon, protecting itself from being swept away by heavy rains. In effect, the wasp not only consumes the spider, but also forces the spider to build it a safe haven.

(a)

(b)

Figure 22.14 My Parasite Made Me Do It

(a) A typical web of the spider *Plesiometa argyra*. (b) A "cocoon web," produced by a *P. argyra* spider infected by a parasitic wasp that alters the spider's web-spinning behavior. The wasp's cocoon can be seen hanging down from the center of the cocoon web.

River Parasite Eats at Children: Neglected Scourge in Africa Is Cheap and Easy to Treat, but the "Pennies Cannot Be Found"

By Colleen Mastony

Flowing through the shantytowns and yam fields of this dust-choked region, the River Uke glimmers like a mirage, tiny white diamonds of sunlight dancing on its surface. As the temperature rises to 100 degrees, wiry boys run to the river and leap into its waters.

Ask the people of Nasarawa, and they say the river is the center of their lives. But the water hides a debilitating scourge: schistosomiasis, a disease spread by microscopic parasites that live in the river, burrow through skin and slowly infect organs, stunting children's growth and sometimes causing death.

The solution, experts say, lies with just one dose, once a year, of about three white pills called praziquantel. Studies show that a single dose—at a cost of 20 cents—can reverse up to 90 percent of the damaging health effects of schistosomiasis within six months of treatment.

But while Nigeria profits handsomely from its oil industry and giant pharmaceutical corporations donate millions every year to treat more prominent diseases in developing countries, no one has stepped forward to help mass-produce and distribute praziquantel, which costs 7 cents per pill to manufacture.

"The pennies cannot be found," said Frank Richards, a doctor who heads a program to study the disease at the Atlanta-based Carter Center.

More than 1,000 different parasites can attack people. The boys in this article were at risk from a parasitic flatworm (*Schistosoma*) that can cause seizures or paralysis, and typically damages organs such as the liver, intestines, and lungs. Over 200 million people are infected by this worm, which causes the disease schistosomiasis.

The infection cycle begins when an infected person urinates or defecates into water, contaminating the water with the worm's eggs. The eggs hatch, and the parasites grow and develop inside snails. Next, the parasite leaves the snail and enters the water, where it can survive for about 48 hours. If the parasite comes into contact with the skin of a person in the water, the worm burrows through and begins to grow inside the blood vessels. Within weeks, the worms produce eggs, and the vicious cycle is ready to begin again.

In most villages in Nigeria, a third of the children have heavy infestations of the worm. The warning sign of infestation is clear: the children have blood in their urine. Children with lots of worms are stunted in their growth, do poorly in school, and suffer from ongoing medical problems.

Schistosomiasis can be treated with a drug that kills virtually all of the worms in the body. Not-for-profit organizations purchase the drug and distribute it for free to infected children. The pills are relatively inexpensive (only 7 cents apiece), but they are made by only three companies. These companies make 89 million of the pills each year, but unfortunately, nearly five times that number would be needed each year to treat the 200 million people infected with schistosomiasis.

Evaluating the News

1. In recent years, schistosomiasis has declined sharply in China, mostly because of the successful control of snails through the use of molluscicides (antisnail pesticides) and biological controls (snail predators). Explain why reducing snail populations would be expected to reduce the incidence of schistosomiasis in humans.

2. Do you think governments should provide subsidies to companies making relatively inexpensive drugs that can combat human parasites in developing nations? Or should market forces alone be used to address these diseases?

3. Former Costa Rican president and winner of the 1987 Nobel Peace Prize Oscar Arias Sánchez estimated recently that universal health care could be provided to everyone in the world's developing nations for an annual cost equal to 12 percent of the combined annual military budgets of those nations. He also noted that in developing nations, there are about 20 soldiers for each physician, although the chances of dying from malnutrition and preventable diseases are 33 times greater than the chances of dying from a war. Developing nations are not unique in spending much more money on military expenditures than on solutions to global health and environmental problems. Why do human societies consistently behave in this way? Should we—and could we—do things differently?

Source: *New York Times,* November 2, 2004.

Chapter Review

Summary

22.1 Mutualism

- Mutualism is an association between two species in which both species benefit (+/+ interaction).
- Symbiosis is an association of two species that live together. Symbiosis may or may not be mutualistic.
- There are many types of mutualism, including gut inhabitant, seed dispersal, behavioral, and pollinator mutualisms.
- Mutualism evolves when the benefits of the interaction to both partners are greater than its costs for both partners.
- Mutualism is very common in nature. Most plant species form mycorrhizae, a type of mutualism in which plant roots associate with a fungus and both organisms benefit.
- Mutualism helps determine the distribution and abundance of the mutualist species, as well as other species that depend directly or indirectly on the mutualist species.

22.2 Exploitation

- In exploitation (+/− interaction), one species (the consumer) benefits and the other (the species that is eaten) is harmed.
- Consumers include herbivores, predators, and parasites.
- Consumers can be a strong selective force, leading their food organisms to evolve various ways to avoid being eaten. Many plants have evolved induced defenses, such as the growth of spines, that are directly stimulated by attacking herbivores.
- Food organisms, in turn, exert selection pressure on their consumers, which evolve ways to overcome the defenses of the species they eat.
- Consumers can restrict the distribution and abundance of the species they eat, in some cases driving their food organisms to extinction.

22.3 Competition

- In competition (−/− interaction), each of two interacting species has a negative effect on the other.
- In interference competition, one species directly excludes another species from the use of a resource.
- In exploitative competition, species compete indirectly, each reducing the amount of a resource available to the other species.
- Competition can have a strong effect on the distribution and abundance of species.
- Competition between similar species may result in character displacement, in which the forms of the competing species evolve to become more different over time.

22.4 Interactions among Organisms Shape Communities and Ecosystems

- Interactions among organisms affect individuals, populations, communities, and ecosystems.

◉ Review and Application of Key Concepts

1. Mutualism typically has costs for both of the species involved. Why, then, is mutualism so common?

2. Consumers affect the evolution of the organisms they eat, and vice versa. Explain how this interaction occurs, and illustrate your reasoning with an example described in the text.

3. How can a species that is an inferior competitor have a negative effect on a superior competitor?

4. How do ecological interactions affect the distribution and abundance of organisms?

5. Rabbits can eat many plants, but they prefer some plants over others. Assume that the rabbits in a grassland containing many plant species prefer to eat a species of grass that happens to be a superior competitor. If the rabbits were removed from the region, which of the following do you think would be most likely to happen?
   a. The plant community would have fewer species.
   b. The plant community would have more species.
   c. The plant community would remain largely unchanged.

   Explain and justify your answer.

Key Terms

abundance (p. 451)	induced defense (p. 451)
character displacement (p. 456)	interference competition (p. 455)
coevolution (p. 452)	mutualism (p. 448)
competition (p. 455)	mycorrhizae (p. 450)
control (p. 454)	parasite (p. 451)
control group (p. 454)	pathogen (p. 451)
distribution (p. 451)	pollinator (p. 449)
ecological experiment (p. 454)	predator (p. 451)
exploitation (p. 451)	prey (p. 451)
exploitative competition (p. 455)	replicate (p. 454)
herbivore (p. 451)	symbiosis (p. 448)
host (p. 451)	

Self-Quiz

1. Which of the following statements about consumers is true?
   a. They cannot drive the species they eat to extinction.
   b. They are not important in natural communities.
   c. They can apply strong selection pressure on their food organisms.
   d. They cannot alter the behavior of their food organisms.

2. In what type of interaction do species directly confront each other over the use of a shared resource?
   a. interference competition
   b. exploitative competition
   c. exploitation
   d. distribution competition

3. Interactions among species
   a. do not influence the distribution or abundance of organisms.
   b. are rarely beneficial to both species.
   c. have a strong influence on communities and ecosystems.
   d. cannot drive species to extinction.

4. The advantages received by a partner in a mutualism can include
   a. food.
   b. protection.
   c. increased reproduction.
   d. all of the above

5. The shape of a fish's jaw influences what the fish can eat. Researchers found that the jaws of two fish species were more similar when they lived in separate lakes than when they lived together in the same lake. The increased difference in jaw structure when the fishes live in the same lake may be an example of
   a. warning coloration.
   b. character displacement.
   c. mutualism.
   d. exploitation.

6. Which of the following statements about symbiosis is *not* correct?
   a. Symbiosis is an association in which two or more organisms of different species live together.
   b. Symbiosis almost always involves species that benefit each other.
   c. Mutualism can occur between the species in a symbiotic association.
   d. One species in a symbiotic association may harm the other species.

7. Experiments with the barnacle *Semibalanus balanoides* showed that
   a. where this species was found on the shoreline was not influenced by physical factors.
   b. competition with *Chthamalus* restricted *Semibalanus* to high portions of the shoreline.
   c. competition with *Chthamalus* restricted *Semibalanus* to low portions of the shoreline.
   d. this species restricted *Chthamalus* to high portions of the shoreline.

8. The American chestnut used to be a dominant tree species in eastern North America, but it is now virtually gone from its entire range. This species is much less common than it once was because
   a. a consumer (an introduced fungus) nearly drove it extinct.
   b. other tree species outcompeted it.
   c. insect herbivores evolved the ability to overcome the tree's defenses against herbivore attack.
   d. it could not form mycorrhizae, because acid rain killed its fungal partner.

CHAPTER 23 Communities of Organisms

Key Concepts

- A community is an association of populations of different species that live in the same area.

- Food webs describe the feeding relationships within a community.

- Keystone species play a critical role in determining the types and abundances of species in a community.

- All communities change over time. As species colonize new or disturbed habitat, they tend to replace one another in a directional and fairly predictable process called succession. Communities also change over time in response to changes in climate.

- Communities can recover rapidly from some forms of natural and human-caused disturbance, but recovery from other forms of human-caused disturbance may take thousands of years.

The Formation of a New Community

The origin of new habitat—as when an island rises out of the sea or a new lake is formed—marks the beginning of a huge and exciting natural experiment. What organisms will colonize the new habitat first? How will those organisms interact and evolve over time? Will the new habitat come to have unusual communities of organisms? Or will the communities of the new habitat be similar to other, nearby ecological communities?

The outcome of such grand experiments can be spectacular, especially when the new communities are located far from existing ones. Consider what has happened on the Hawaiian Islands, a remote chain of volcanic islands, the most recent of which (Hawaii) was formed about 600,000 years ago.

The Hawaiian Islands are so isolated (they lie 4,000 kilometers from the nearest continent) that they have been colonized by relatively few species. Over time, however, the few species that have colonized the islands have evolved to form many new species. As a result, the communities of organisms on Hawaii are very different from those anywhere else on Earth. Such a series of events is not restricted to Hawaii; unusual communities often form when a new habitat is located far from existing communities.

Today, many of the unique communities on Hawaii are threatened by various human activities—including habitat destruction and the introduction of nonnative species. People brought beard grass, for example, to Hawaii as forage for cattle. By the late 1960s, beard grass had invaded the seasonally dry woodlands of Hawaii Volcanoes National Park. Before that time, fires had occurred there every 5.3 years on average, and each fire had burned an average of 0.25 hectare (about five-eighths of an acre). Since the introduction of beard grass, fires have occurred at a rate of more than one per year, and the average burn area of each fire has increased to more than 240 hectares (about 600 acres). The fires are now so frequent and intense that the seasonally dry woodlands that once thrived in the park have disappeared.

A Natural Experiment
The Hawaiian Islands are part of a chain of volcanic islands that rose from the sea over the past 70 million years. As newly formed islands were colonized by species from the mainland and as new species evolved on the islands, unique new ecological communities like this one formed.

463

Why did the introduction of beard grass increase the frequency and size of fires in Hawaii Volcanoes National Park? Is there something about Hawaii and other island communities that makes them particularly vulnerable to human disturbance? These questions are relevant to communities of all types, not just those on islands. In this chapter we consider how and why ecological communities change over time; and how, why, and whether they can bounce back from disturbances, including those created by people.

An ecological **community** is an association of populations of different species that live in the same area. There are many different types of communities, ranging from those found in grasslands and forests to those found in the digestive tract of a cow or deer (Figure 23.1). Most communities contain many species and, as we learned in Chapter 22, the interactions among those species can be complex. Ecologists seek to understand how interactions among organisms influence natural communities.

Ecologists also study how human actions affect communities. People are having a profound effect on many different kinds of ecological communities. When we cut down tropical forests, we destroy entire communities of organisms, and when we give antibiotics to a cow, we alter the community of microorganisms that live in its digestive tract. To prevent such actions from having effects that we do not anticipate or want, we must understand how ecological communities work. In this chapter we describe the factors that influence what species are found in a community. We pay particular attention to how communities change over time and how they respond to disturbance, including disturbance caused by people.

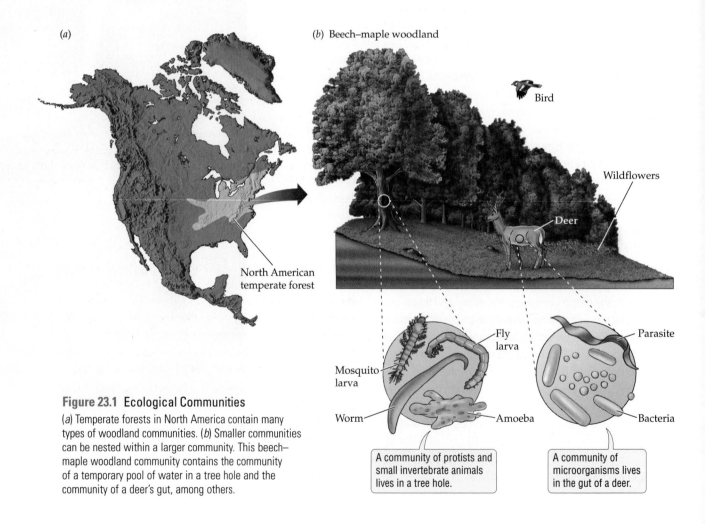

(a)

(b) Beech–maple woodland

Bird

Wildflowers

Deer

North American temperate forest

Mosquito larva

Fly larva

Worm

Amoeba

Parasite

Bacteria

Figure 23.1 Ecological Communities
(a) Temperate forests in North America contain many types of woodland communities. (b) Smaller communities can be nested within a larger community. This beech–maple woodland community contains the community of a temporary pool of water in a tree hole and the community of a deer's gut, among others.

A community of protists and small invertebrate animals lives in a tree hole.

A community of microorganisms lives in the gut of a deer.

23.1 The Effects of Species Interactions on Communities

Communities vary greatly in size and complexity, from the community of microorganisms that inhabits a small temporary pool of water, to the community of plants that lives on the floor of a forest, to a forest community that stretches for hundreds of kilometers. Communities can also be nested within one another, as Figure 23.1 shows. Whatever its size or type, an ecological community can be characterized by its composition, or diversity. The **diversity** of a community has two components: the number of different species that live in the community (also known as species richness), and the relative abundances of those species (how common a species is compared to others in the community). Figure 23.2 compares the diversity of two communities that have the same number of species; because community A is dominated by a single, highly abundant species, it is considered less diverse than community B, in which all species are equally common.

Overall, ecological communities are influenced by the individual species that live in them, by interactions among those species, and by interactions between those species and the physical environment.

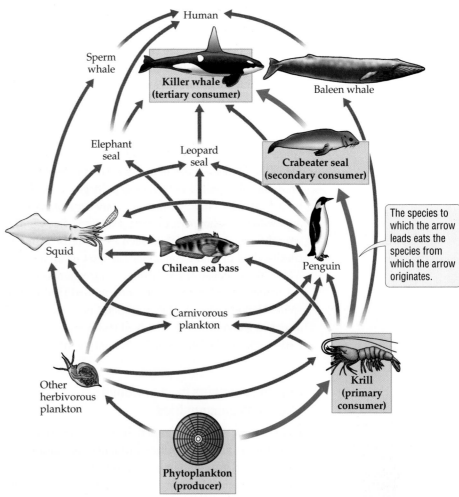

The species to which the arrow leads eats the species from which the arrow originates.

Figure 23.3 A Food Web
Food webs summarize the movement of food through a community. This illustration is a simplified version of the food web in the Antarctic Ocean. Food webs are composed of many specific sequences, known as food chains, that show one species eating another. To make it easier to follow a single sequence of feeding relationships, one of the food chains in this food web is highlighted with red arrows and orange boxes.

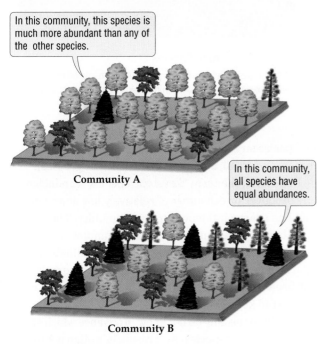

In this community, this species is much more abundant than any of the other species.

Community A

In this community, all species have equal abundances.

Community B

Figure 23.2 Which Community Has Greater Diversity?
Communities A and B have the same four species of trees. However, community A is dominated by a single species, whereas all four species are equally represented in community B. Therefore, ecologists consider community A less diverse than community B.

Food webs consist of multiple food chains

One important aspect of a community is which species eat which other species. These feeding relationships can be illustrated by **food chains**, each of which describes a single sequence of who eats whom in a community. We can summarize the movement of energy and nutrients through a community by connecting the different food chains to form a **food web**, which describes the interconnected and overlapping food chains of a community (Figure 23.3).

As we saw in Chapter 1, food webs and the ecological communities they describe are based on a foundation of producers. **Producers** are organisms that use energy from an external source, such as the sun, to produce their own food without having to eat other organisms

Learn more about this tangled food web.
23.1

or their remains. On land, photosynthetic plants, which harvest energy from the sun, are the major producers. In aquatic biomes, a wide range of organisms serve as producers, including phytoplankton in the oceans, algae in intertidal zones and lakes, and bacteria in deep-sea hydrothermal vents.

Consumers are organisms that obtain energy by eating all or parts of other organisms or their remains. Important groups of consumers include decomposers (which we'll discuss in Chapter 24) and the herbivores, predators, and parasites (including pathogens) described in Chapter 22. **Primary consumers** are organisms, such as cows or grasshoppers, that eat producers. **Secondary consumers** are organisms, such as humans or birds, that feed on primary consumers as part or all of their diet. This sequence of organisms eating organisms that eat other organisms can continue: a bird that eats a spider that ate a beetle that ate a plant is an example of a **tertiary consumer**. In the food chain highlighted in Figure 23.3, the killer whale is a tertiary consumer.

Keystone species have profound effects on communities

Interactions among organisms such as mutualism, exploitation, and competition influence the number of species found in a community, as shown by the coral reef, dingo, and barnacle examples discussed in Chapters 20 and 22. In addition, certain species have a disproportionately large effect, relative to their own abundance, on the types and abundances of the other species in a community; these influential species are called **keystone species**.

In an experiment conducted along the rocky Pacific coast of Washington State, ecologist Robert Paine demonstrated that the sea star *Pisaster ochraceus* [pih-ZASS-ter oh-KRAY-see-us] is a keystone species in its intertidal-zone community. He removed sea stars from one site and left an adjacent, undisturbed site as a control. In the absence of sea stars, all of the original 18 species in the community except mussels disappeared (Figure 23.4). When the sea stars were present, they ate the mussels, thereby keeping the number of mussels low enough that the mussels did not crowd out the other species.

Pisaster is a predator, but organisms other than predators can also be keystone species. Plants such as fig trees, herbivores such as snow geese and elephants, and pathogens such as the distemper virus that kills lions have been found to be keystone species. In addition, humans often function as a keystone species: we have a large effect on interactions among other species, includ-

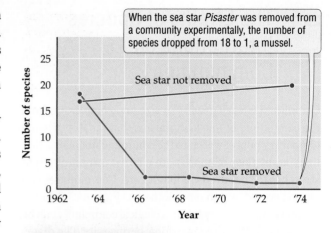

When the sea star *Pisaster* was removed from a community experimentally, the number of species dropped from 18 to 1, a mussel.

A *Pisaster* sea star feeding on a mussel.

Figure 23.4 A Keystone Species

The sea star *Pisaster ochraceus* is a predator that feeds on mussels and thereby prevents the mussels from crowding out other species in their community.

ing species far more abundant than we are (for example, insects and small, abundant plants such as grasses).

In general, the term "keystone species" can include any producer or consumer of relatively low abundance that has a large influence on its community. The most abundant or dominant species in a community (such as the corals in a coral reef or the mussels in Paine's intertidal zone) also have large effects on their communities, but because their abundance is not low, they are not considered keystone species.

Finally, we usually do not know in advance which species are keystone species. As a result, it is often not until after people remove a species from a community, and then observe large changes in that community, that a species is discovered to have been a keystone species. When people removed rabbits from a region in England, for example, grasslands with a variety of plant species, including many nongrasses, were converted (unintentionally) into grasslands consisting primarily of just a few species of grasses. This change occurred because the rabbits were no longer present to hold the grasses in check. In the absence of rabbits, the grasses crowded out other plant species.

23.2 Communities Change over Time

All communities change over time. The number of individuals of different species in a community often changes as the seasons change. For example, although butterflies might be abundant in summer, we would not find any flying in a North Dakota field in the middle of winter. Similarly, every community shows year-to-year changes in the abundances of organisms, as we saw in Chapter 21. In addition to such seasonal and yearly changes, communities show broad, directional changes in species composition over longer periods of time.

Succession establishes new communities and replaces disturbed communities

A community may begin when new habitat is created, as when a volcanic island like Hawaii rises out of the sea. New communities may also form in regions that have been disturbed, as by a fire or hurricane. Some species arrive early in such new or disturbed habitat. These early colonists tend to be replaced later by other species, which in turn may be replaced by still other species. These later arrivals replace other species because they are better able to grow and reproduce under the changing conditions of the habitat.

The process by which species in a community are replaced over time is called **succession**. In a given location, the order in which species will replace one another is fairly predictable (Figure 23.5). Such a sequence of species replacements sometimes ends in a **climax community**, which, for a particular climate and soil type, is a community whose species are not replaced by other species. But in many—perhaps most—ecological communities, **disturbances** such as fires or windstorms occur so frequently that the community is constantly changing in response to a disturbance event, and a climax community never forms.

Primary succession occurs in newly created habitat, as when an island rises from the sea or when rock and soil are deposited by a retreating glacier. In such a situation, the process begins with a habitat that contains no species. The first species to colonize the new habitat usually have one of two advantages over other species: either they can disperse more rapidly (and hence reach the new habitat first), or they are better able to grow and reproduce under the challenging conditions of the newly formed habitat.

In some cases of primary succession, the first species to colonize the area alter the habitat in ways that enable later-arriving species to thrive. In other cases, the early colonists hinder the establishment of other species. An experimental study of primary succession was conducted on marine intertidal communities on

See this transition from shrubs to trees. ▶II 23.2

Figure 23.5 An Example of Succession
When strong winds cause sand dunes to form at the southern end of Lake Michigan, succession often leads to a community dominated by black oak. Succession on such dunes occurs in three stages and forms communities of black oak that have lasted up to 12,000 years. Under different local environmental conditions, succession on Michigan sand dunes can lead to the establishment of stable communities as different as grasslands, swamps, and sugar maple forests (inset).

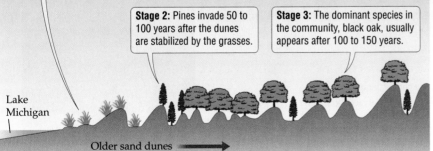

Stage 1: Bare sand is first colonized by dune-building grasses, such as marram grass, which spread rapidly and stabilize the moving sand of the dunes.

Stage 2: Pines invade 50 to 100 years after the dunes are stabilized by the grasses.

Stage 3: The dominant species in the community, black oak, usually appears after 100 to 150 years.

Lake Michigan

Older sand dunes ➡

the rocky coast near Santa Barbara, California, where researchers created new habitats by placing concrete blocks along the shoreline. The first species of algae to colonize the concrete blocks initially inhibited other species—the ones that ultimately replaced them—from establishing themselves. In cases like this one, the early colonists eventually lose their hold on a habitat by being more susceptible than later species to a particular feature of the environment, such as disturbance, grazing by herbivores, or extremes of heat or cold.

Secondary succession is the process by which communities recover from disturbance, as when natural vegetation recolonizes a field that has been taken out of agriculture, or when a forest grows back after a fire (Figure 23.6). In contrast to primary succession, habitats undergoing secondary succession often have well-developed soil containing seeds from species that usually predominate late in the successional process. The presence of such seeds in the soil can considerably shorten the time required for the later stages of succession to be reached.

Communities change as climate changes

Some groups of species stay together for long periods of time. For example, an extensive plant community once stretched across the northern parts of Asia, Europe, and North America. As the climate grew colder during the past 60 million years, plants in these communities migrated south, forming communities in Southeast Asia and southeastern North America that are similar to one another—and similar in composition to the community from which they originated.

Although groups of plants can remain together for millions of years, the community located in a particular place changes as the climate of that place changes. The climate at a given location can change over time for two reasons: global climate change and continental drift.

Consider first the climate of Earth as a whole, which changes over time. What we experience today as a "normal" climate is warmer than what was typical during the previous 400,000 years. Over even longer periods of time, the climate of North America has changed greatly (Figure 23.7), causing dramatic changes in the plant and animal species that live there. For example, fossil evidence indicates that 35 million years ago, the areas of southwestern North America that are now deserts were covered with tropical forests. Historically, changes in the global climate have been due to relatively slow natural processes, such as the advance

Learn more about climate changes over time. ►❚❚ 23.3

(a)

(b)

Figure 23.6 Secondary Succession

Forests in the United States grow back after being cut by people, blown down by windstorms, or burned by fire. These photographs, taken in different locations, show the stages of regrowth following the large fire that struck Yellowstone National Park in 1988. (*a*) Shortly after the fire. (*b*) Four years later, lodgepole pine saplings are growing in a stand of trees killed by the fire. (*c*) A mature lodgepole pine forest (not burned in the fire).

(c)

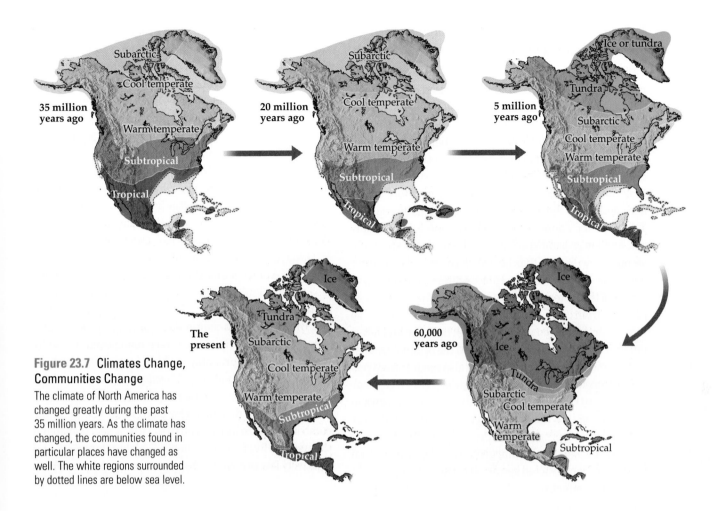

Figure 23.7 Climates Change, Communities Change
The climate of North America has changed greatly during the past 35 million years. As the climate has changed, the communities found in particular places have changed as well. The white regions surrounded by dotted lines are below sea level.

and retreat of glaciers. However, evidence is mounting that human activities are now causing rapid changes in the global climate (see Chapter 25).

Second, as the continents move slowly over time (see Figure 19.7), their climates change. To give a dramatic example of continental drift, 1 billion years ago Queensland, Australia, which is now located at 12° south latitude, was located near the North Pole. Roughly 400 million years ago, Queensland was at the equator. Since the species that thrive at the equator and in the Arctic are very different, continental drift has resulted in large changes in the communities of Queensland over time.

Concept Check

1. What is a keystone species?
2. Describe the distinctive characteristics of species that tend to be the first to colonize a new habitat.
3. Compare primary succession with secondary succession.

23.3 Recovery from Disturbance

Ecological communities are subject to many natural forms of disturbance, such as fires, floods, and windstorms. Following a disturbance, secondary succession can reestablish the previously existing community. In this way communities can and do recover from some forms of disturbance. Depending on the community, the time required for recovery varies from years to decades or centuries.

Communities have been exposed over long periods of time to natural forms of disturbance, such as windstorms. In contrast, people may introduce entirely new forms of disturbance, such as the dumping of hot wastewater into a river. Human actions may also alter the frequency of an otherwise natural form of disturbance—for example, causing a dramatic increase or decrease in the frequency of fires or floods.

Communities can recover from some human-caused disturbances

Can communities recover from disturbances caused by people? For some forms of human-caused disturbance, the answer is yes. Throughout the eastern United States, for example, there are many places where forests were cut down and the land used for farmland; years later, the farmland was abandoned. Second-growth forests have grown on these abandoned farms, often within 40 to 60 years after farming stopped. Second-growth forests are not identical to the forests that were originally present. The sizes and abundances of the tree species are different, and fewer plant species grow beneath the trees of a second-growth forest than beneath a virgin forest (one that has never been cut down). However, the second-growth forests of the eastern United States already have recovered partially from cutting. If current trends continue, over the next several hundred years there will be fewer and fewer differences between such forests and the original forests.

In some cases, communities can also recover from pollution. Consider Lake Washington, which is a large, clear lake in Seattle. As the city of Seattle grew, raw sewage was dumped into the lake. This practice declined after 1926 and was stopped by 1936. Beginning in 1941, however, treated sewage was discharged into Lake Washington from newly constructed sewage treatment plants.

A major effect of discharging sewage—treated or not—into Lake Washington was the addition of extra nutrients to the lake, in the form of nitrogen and phosphorus in the sewage. As a result, the numbers of algae soared, decreasing the clarity of the water. As the algae reproduced and then died, their bodies provided an abundant food source for bacteria, whose populations also increased. Bacteria use oxygen when they consume dead algae, so concentrations of dissolved oxygen in the water decreased. The lowered oxygen concentrations killed invertebrates and fishes. These events illustrate how a body of water can be degraded by **eutrophication** [YOO-truh-fih-KAY-shun] (from *eu*, "well"; *troph*, "nourished"), a process in which enrichment of water by nutrients (often from sewage or runoff from fertilized agricultural fields) causes bacterial populations to increase and oxygen concentrations to decrease.

By the early 1960s, Lake Washington was so degraded that it was referred to in the local press as "Lake Stinko" (Figure 23.8). From 1963 to 1968, the dumping of sewage into the lake was reduced, and virtually no sewage was dumped after 1968. Once inputs of sewage had stopped, algae populations declined, oxygen concentrations increased, and Lake Washington returned to its former, clear state.

Other, even larger bodies of water, including Lake Erie in North America and the Black Sea in Eurasia, have also shown signs of recovery from disturbances caused by people. In the 1980s, high nutrient inputs caused the Black Sea to become eutrophic [yoo-TROH-fick]. Conditions in the Black Sea were made even worse by the introduction (probably in ships' ballast water) of the North American comb jelly, *Mnemiopsis* [NEM-ee-OP-sis], a voracious predator that ate the plankton that anchovies and other fishes depend on for food (Figure 23.9). When comb jellies were introduced to the Black Sea, they found themselves in an environment with abundant food and relatively few predators. As a result, comb jelly numbers

(a)

(b)

Figure 23.8 Lake Stinko No More

In the early 1960s, Lake Washington (in Seattle, Washington) was so polluted that it was dubbed "Lake Stinko" by the local press (a). Nutrient flow from raw or treated sewage ceased in 1968, which stopped eutrophication and restored the clarity of the water, as seen in a 2004 photo (b).

Figure 23.9 Changes in the Black Sea

In the late 1980s, the accidental introduction of the comb jelly *Mnemiopsis* wreaked havoc on the food webs of the Black Sea, causing a decline in the abundance of many species and the collapse of the anchovy fishing industry. Conditions began to improve after the accidental introduction of *Beroe*, a predator capable of controlling *Mnemiopsis* populations.

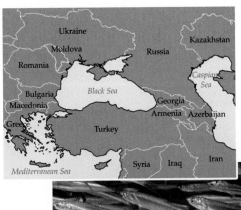

Anchovies swimming | A *Beroe* jelly eating another comb jelly

exploded, causing anchovy populations to crash and the Turkish anchovy fishing industry to collapse. A mere 10 years later, however, the situation had improved: through reductions of nutrient inputs into the Black Sea and the accidental introduction (again, probably in ballast water) of a predator that eats *Mnemiopsis*, the sea had become less eutrophic, *Mnemiopsis* numbers had plummeted, and anchovy populations had recovered.

People can cause long-term damage to communities

It is encouraging that complex ecological communities such as those in Lake Washington and the Black Sea can recover rapidly from disturbances caused by people. However, communities do not always recover so quickly from human-caused disturbances, as a few examples will show.

Northern Michigan once was covered with a vast stretch of white pine and red pine forest. Between 1875 and 1900, nearly all of these trees were cut down, leaving only a few scattered patches of virgin forest. The loggers left behind large quantities of branches and sticks, which provided fuel for fires of great intensity. In some locations, the pine forests of northern Michigan have never recovered from the combination of logging and fire.

Throughout South America and Southeast Asia, large tracts of tropical forest have been converted into grasslands by a combination of logging and fire. Scientists estimate that it will take tropical forest communities hundreds to thousands of years to recover from such changes.

Finally, in some areas of the American Southwest, overgrazing by cattle has transformed dry grasslands into desert shrublands (Figure 23.10). How do cattle cause such large changes? Grazing and trampling by cattle decrease

(a)

(b)

This dry grassland is in southern New Mexico.

This nearby, former dry grassland is now a desert.

Figure 23.10 Overgrazing Can Convert Grasslands into Deserts

(*a*) More than 200 years ago, large regions of the American Southwest were covered with dry grasslands. (*b*) Most of these grasslands have been converted into desert shrublands, in large part because of overgrazing by cattle.

Biology on the Job

Designing with Nature

When planting a garden, constructing a pond, building a road, or planning for the long-term growth of a city, we think about how to alter outdoor space for our use. All such activities involve changing existing ecological communities and establishing new ones—a subject we discuss with Carlyn Worstell, a landscape architectural designer who works at the Bronx Zoo.

What is landscape architecture? Landscape architecture is the design of outdoor spaces for use by people. I was drawn to it because of my interest in nature and the environment.

What are the foremost concerns in designing a landscape for a zoo? Well-being and safety of the animals, and creating an environment that will elicit the most natural behaviors from the animals. We also have to consider the aesthetic quality, and such urban design concerns as accessibility and the flow of people through a space. Also, an exhibit must appeal to children and other visitors to provide the best opportunities for education.

Are some ecological communities more difficult to re-create than others? Why or why not? Desert or tropical communities are difficult in New York. To create a more realistic-looking environment, we sometimes use plants that might be native to New York but look like or are pruned to resemble plants native to the habitat we're trying to re-create. Indoor enclosures can be challenging because they must be climate-controlled yet energy-efficient. That way we can both protect wildlife and conserve Earth's natural resources.

What has been your most challenging or memorable assignment? I am currently working on an indoor LEED (leadership and energy efficient) design in a historic building from 1911. It's a landmark building so the outside façade cannot be changed, but we are trying to build an extensive Madagascar exhibit inside. To conserve energy we need to use low lighting levels, but the plants need high light to live. This is a tough one!

The Job in Context

To do her job well, Carlyn Worstell must understand the animals she works with and the communities they live in. When she helps build an exhibit, she mimics a natural community, building it from scratch. This process leaps over many of the typical steps in ecological succession—machines are used to move earth, and animals and plants are brought

A Pair of Snow Leopards at the Bronx Zoo
These endangered animals are native to the high mountains of Central Asia. Only about 5,000 of them survive from Mongolia to Nepal. About 600 snow leopards live in zoos around the world. A number of zoos have succeeded in breeding these beautiful cats in captivity.

from other locations to form a human-made ecological community. But succession and other aspects of ecological change cannot be completely circumvented. The plants in the exhibit must be given time to establish themselves before the animals are introduced—otherwise, the plants will not be able to tolerate the tendency of animals to dig them up, scratch against them, and sometimes eat them. Zoo staff must also prevent colonization by unwanted species (such as a plant species that might outcompete those in the exhibit), they must monitor the quality of the water, and they must treat or replace animals or plants that are not healthy. Like any ecological community, a zoo exhibit is a dynamic place—it is a challenging and rewarding job to help ensure that it continues to serve the needs of the animals that live there.

the abundance of grasses in the community. With less grass to cover the soil and hold it in place, the soil dries out and erodes more rapidly. Desert shrubs thrive under these new soil conditions, but grasses do not. These changes in soil characteristics can make it very difficult to reestablish grasses, even when the cattle are removed.

Grazing can have a dramatic effect on dry grasslands, but how do you think the effects of grazing would compare to the effects of an atomic bomb? The first aboveground explosion of an atomic bomb occurred on July 16, 1945, at the Trinity site in New Mexico (Figure 23.11). Fifty years later, dry grasslands destroyed by the bomb blast (but never grazed) had recovered. In contrast, nearby dry grasslands that had been heavily grazed (but not destroyed by the bomb blast) had not recovered, even though they had not been grazed since the time of the blast. The plant community recovered more rapidly from the effects of a nuclear explosion than from the effects of grazing—a dramatic example of how strongly ecological interactions can affect natural communities.

Figure 23.11 The Trinity Site
The world's first aboveground atomic explosion occurred in July 1945 at the Trinity site in New Mexico. Fifty years later, researchers learned that grasslands destroyed by the explosion had recovered more rapidly than grasslands that had been overgrazed by cattle.

Changes in communities affect their value to humans

Human actions have the power to change communities rapidly across large geographic regions. Disturbances not caused by people can also occur rapidly and affect large geographic regions. For example, most scientists think that the impact of a large asteroid contributed to the sudden extinction of many species, including the last of the dinosaurs, 65 million years ago (see Chapter 19). However, changes in communities caused by people are different in one important sense: we can consider the consequences of our actions and use that information to decide what to do. Many people think our unique ability to control our actions brings with it the responsibility to use our power to change communities wisely.

One argument for taking such responsibility is that human actions that degrade or destroy ecological communities have ethical implications. When people disrupt communities, our actions kill individual organisms, alter communities that may have persisted for thousands of years, and threaten species with extinction. According to results from surveys in North America and Europe, many people think that such effects of human actions on individuals, communities, and species are ethically unacceptable.

Another argument is that human-caused changes can reduce the aesthetic value of a community. Tropical forests, for example, have unique aesthetic value to many people. When our actions cause tropical forests to be destroyed and replaced by introduced grasses, we deprive current and future generations of experiencing the beauty of those forests.

Finally, often when people change a community, its economic value is reduced. Economic value is lost, for example, when human actions convert grasslands into deserts, in part because there is more plant material to support grazing in grasslands than in deserts. In general, when our actions cause long-term damage to ecological communities, we run the risk of harming our own long-term economic interests.

Overall, communities change constantly in the face of both natural and human-caused disturbances. Can human societies manage Earth's changing communities while maintaining the aesthetic and economic value that those communities provide for us? One way to achieve such an objective is to design the indoor and outdoor spaces that we use in ways that reduce their environmental impact (see the box on page 472). We'll return to efforts to reduce the human impact on the environment in Interlude E, which focuses on how to build a sustainable society.

Applying What We Learned

Introduced Species: Taking Island Communities by Stealth

The Hawaiian Islands are the most isolated chain of islands on Earth. Because the islands are so remote, entire groups of organisms that live in most communities never reached them. For example, there are no native ants or snakes in Hawaii, and there is only one native mammal (a bat, which was able to fly to the islands).

The few species that did reach the Hawaiian Islands found themselves in an environment that lacked most of the species from their previous communities. The sparsely occupied habitat and the lack of competitor species resulted in the evolution of many new species. For example, genetic evidence indicates that the many different species of Hawaiian silversword [SIL-ver-SORD] plants found on the islands today (Figure 23.12) all arose from a single ancestor. The new silversword species evolved very different forms, enabling them to live in a number of different habitats. Because other groups of plants and animals on the islands also evolved many new species, the Hawaiian Islands have many unique natural communities.

Since the arrival of people on the islands, about 1,500 to 2,000 years ago, Hawaii's unique communities have been threatened by habitat destruction, overhunting, and introduced species. Of these threats, the effects of introduced species can be the easiest to overlook because such species often wreak their havoc quietly, behind the scenes. Introduced Argentine ants, for example, may drive native insects to extinction, but it can take years before even a trained biologist realizes what the introduced ants have done.

Island communities are particularly vulnerable to the effects of introduced species. Relatively few species colonize newly formed islands, and those species then evolve in isolation. For this reason, species on islands may be ill equipped to cope with new predators or competitors that are brought by people from the mainland. In addition, introduced species often arrive without the predator and competitor species that held their populations in check on the mainland. On islands the potential exists for populations of introduced species to increase dramatically and become invasive.

In some cases, introduced species can destroy entire communities. For example, if most of the native plants are not adapted to fire, an introduced species that alters the frequency or intensity of fire can have devastating effects. Recall the description at the opening of this chapter of what happened after beard grass was introduced to Hawaii Volcanoes National Park: the frequency of fires in the park increased by more than 5 times, and the burn area of an average fire increased by more than 960 times.

Why did the introduction of beard grass have these effects? As beard grass grows, it deposits a large amount of dry plant matter on the ground. This material catches fire easily, and the fires burn much hotter than they would in the absence of beard grass. Beard grass recovers well from large, hot fires, but the native trees and shrubs of the seasonally dry woodland do not. As a result, former woodlands have now been converted into open meadows filled with beard grass and other, even more fire-prone, introduced grasses.

The Hawaiian dry woodland community has been destroyed, probably forever. Because there is no hope of restoring the native community, ecologists are now trying to construct a new community that is tolerant of fire yet contains native trees and shrubs. This is a difficult challenge, and it is uncertain whether the effort will succeed. If not, what was once woodland is likely to remain indefinitely as open meadows filled with introduced grasses.

Figure 23.12 Great Diversity from a Single Ancestor
Hawaiian silverswords are found only in the Hawaiian Islands. Genetic evidence indicates that this diverse group of plant species evolved from a single ancestor. Although the three silversword species shown here are closely related, they live in very different habitats and differ greatly in form.

Biology in the News

Shrinking Number of Fish Is Bad Sign

By Ray Grass

It's not easy trying to put things back together once they've been tampered with. Take fish, for example. In some areas, non-native fish have pushed out the natives. The trick now is to get the native fish back on a competitive level, which isn't easy . . .

It seems like each fish ever brought into Utah back in the late 1800s and early 1900s was dropped in Utah Lake, everything from bass to carp to catfish. Eventually, the June sucker all but vanished. Now, the Division of Wildlife Resources' Logan Fisheries Experiment Station is working to return the fish to the Utah Lake. Once found in large numbers, the odd-looking fish is on the verge of extinction.

. . . Why all the trouble for fish that few will ever see or catch? Native fish that are unique to Utah are often referred to as the "canary in the mine." If they are having trouble, after surviving for centuries, then it's a good indication that there's something wrong with the water source, and water is critical to humans. They are telling us something is wrong with this habitat, and we should be sitting up and paying attention.

A study conducted by the U.S. Geological Survey (USGS) revealed that about one-fourth of all fishes in 12 western states are non-native, and invaders were identified in more than half of the streams studied. Non-native species had completely replaced native species in about 11 percent of the 400,000 miles of streams surveyed. According to Scott Bonar, a USGS scientist, non-native fishes are found "across the landscape in all habitat types." These non-native species are not necessarily exotic fishes from distant lands. Most are North American species. A majority were intentionally introduced for recreational fishing, as bait for larger game fish, or for mosquito control in the battle against West Nile disease (see the "Biology in the News" feature in Chapter 21). What are the consequences when humans alter the species composition of an ecological community? In many cases, the community remains similar to what it once was, but some species are driven to extinction while others are catapulted to new dominance. Even in the less extreme cases, often more species are lost than are gained, so the number of species in the community drops.

Sometimes, however, human actions appear to have led to an increase in the number of species. Working with several other scientists, Scott A. Smith made use of a detailed biological survey that had been conducted just before the Panama Canal was built. "We knew which species were where," Smith said. "Then we were able to sample [the same two] rivers and examine how community composition had changed over time." They found that, at least for the fishes in the Rio Chagres and the Rio Grande, the effect of building the canal has been positive: no species in either river went extinct, and some species from the Rio Chagres successfully colonized the Rio Grande, and vice versa.

These results surprised some ecologists, especially those who think of communities as being full, or "saturated." If a community is saturated, all possible food sources and microenvironments are being used by species in the community. According to this view, colonization events should occur only if newcomers replace one or more of the existing species, thereby driving local species extinct. Smith's results showed, however, that the number of species in the two rivers in Panama actually rose rather than fell. Smith concluded that at least some communities consist of "neutral" groups of species, in which species are not so precisely adapted and the community is not so tightly knit that new species will necessarily drive some of the existing species extinct.

Evaluating the News

1. Given Smith's results, should the citizens of Utah be concerned about the decline of the June sucker and other native species? Explain your view.
2. Imagine you're a policy maker who must decide what to do about a complex global environmental problem, such as global warming or declining fish populations in the ocean. Would it be best to take decisive action now to prevent possible future harmful effects of human actions? Bear in mind that taking decisive action now would risk spending money needlessly if the problem should turn out to be less serious than is feared. Or would it be better to proceed cautiously, spending limited sums to gather more information, but waiting until that information becomes available before taking decisive action? The risk with this second course of action is that the problem will get worse in the interim and may cost much more to fix than it would have cost if decisive action had been taken earlier.

SOURCE: *Deseret Morning News*, Salt Lake City, UT, October 4, 2007.

Chapter Review

Summary

23.1 The Effects of Species Interactions on Communities

- Communities can be described by food webs, which summarize the interconnected food chains describing who eats whom in a community.
- Producers obtain their energy from an external source, such as the sun. Consumers get their energy by eating all or parts of other organisms. Primary consumers eat producers, and secondary consumers feed on primary consumers.
- A keystone species has a large effect on the composition of a community relative to its abundance.
- Keystone species alter the interactions among organisms in a community, and they change the types or abundances of species in the community.

23.2 Communities Change over Time

- All communities change over time.
- Directional changes that occur over relatively long periods of time have two main causes: succession and climate change.
- Primary succession occurs in newly created habitat. Secondary succession occurs in communities recovering from disturbance.
- The climate at a given location can change because of global climate change or continental drift, leading to changes in ecological communities.

23.3 Recovery from Disturbance

- Communities can recover from some forms of natural and human-caused disturbance. The time required for recovery varies from years to decades or centuries.
- Degradation of water bodies resulting from eutrophication can be reversed if the sources of nutrient enrichment are removed.
- It can take hundreds to thousands of years for communities to recover from some forms of human-caused disturbance.
- Community change caused by people is unique in that we can consider the consequences of our actions and use that information to decide what to do. We can choose to act in ways that do not reduce the aesthetic or economic value of ecological communities.

⊙ Review and Application of Key Concepts

1. Describe how each of the following factors influences ecological communities: (a) species interactions, (b) disturbance, (c) climate change, (d) continental drift.

2. What is the difference between primary and secondary succession?

3. Provide an example of how the presence or absence of a species in a community can alter a feature of the environment, such as the frequency of fire.

4. Consider two forms of human disturbance to a forest:
   a. All trees are removed, but the soils and low-lying vegetation are left intact.
   b. The trees are not removed, but a pollutant in rainfall alters the soil chemistry to such an extent that the existing trees can no longer thrive.

   Which form of disturbance do you think would require the longest recovery time before a healthy forest community could once again be found at the site? Explain your assumptions in answering this question, and justify the conclusion you reach.

5. Do you think it is ethically acceptable for people to change natural communities so greatly that it takes thousands of years for the communities to recover? Why or why not?

6. Explain the difference between a keystone species and one of the most abundant or dominant species of a community.

Key Terms

climax community (p. 467)	keystone species (p. 466)
community (p. 464)	primary consumer (p. 466)
consumer (p. 466)	primary succession (p. 467)
disturbance (p. 467)	producer (p. 465)
diversity (p. 465)	secondary consumer (p. 466)
eutrophication (p. 470)	secondary succession (p. 468)
food chain (p. 465)	succession (p. 467)
food web (p. 465)	tertiary consumer (p. 466)

Self-Quiz

1. A low abundance species that has a large effect on the composition of an ecological community is called a
   a. predator.
   b. herbivore.
   c. keystone species.
   d. dominant species.

2. Organisms that can produce their own food from an external source of energy without having to eat other organisms are called
   a. suppliers.
   b. consumers.
   c. producers.
   d. keystone species.

3. Ecological communities
   a. cannot recover from disturbance.
   b. can recover from natural but not human-caused disturbance.
   c. can recover from all forms of disturbance.
   d. can recover from some, but not all, forms of natural and human-caused disturbance.

4. Which of the following was *not* caused by the introduction of beard grass to Hawaii?
   a. an increase in the growth of native trees and shrubs
   b. an increase in the frequency and intensity of fire
   c. the decline of native trees and shrubs
   d. the conversion of dry woodlands into grasslands

5. A directional process of species replacement over time in a community is called
   a. global climate change.
   b. succession.
   c. competition.
   d. community change.

6. A community whose species are not replaced by other species is known as a _____ community.
   a. primary succession
   b. climax
   c. competitive
   d. disturbance-based

7. A single sequence of feeding relationships describing who eats whom in a community is a
   a. life history.
   b. keystone relationship.
   c. food web.
   d. food chain.

8. The process in which the enrichment of water by nutrients causes bacterial populations to increase and oxygen concentrations to decrease is called
   a. eutrophication.
   b. disturbance.
   c. fertilization.
   d. nutrient loading.

Key Concepts

⊙ An ecosystem consists of a community of organisms together with the physical environment in which those organisms live. Energy, materials, and organisms can move from one ecosystem to another.

⊙ Energy enters an ecosystem when producers capture it from an external source, such as the sun. A portion of the energy captured by producers is lost as metabolic heat at each step in a food chain. As a result, energy cannot be recycled within an ecosystem.

⊙ Earth has a fixed amount of nutrients. If nutrients were not recycled between organisms and the physical environment, life on Earth would cease. Human activities affect the cycling of some nutrients.

⊙ Ecosystems provide humans with essential services, such as nutrient recycling, at no cost. Our civilization depends on these and many other ecosystem services.

Main Message: Materials are recycled in ecosystems, but energy flows through them in one direction. Ecosystems provide humans and other organisms with essential services.

Is There a Free Lunch?

Next time you drink a glass of water, think for a moment about where your water comes from. If you're like many of us, you may not know. Does it come from surface waters, such as rivers, lakes, or reservoirs? Or does it come from deep underground? Whatever its source, the delivery of safe drinking water can make for an interesting story.

Consider New York City. The 8 million people who live there get about 90 percent of their water from the Catskill Watershed, with the remainder coming from the Croton Watershed. Together, these watersheds store 580 billion gallons of water in 19 reservoirs and 3 controlled lakes. Over 1.3 billion gallons of this water is delivered to New York each day. The water flows by gravity to the city in a vast set of pipes, some of them large enough to drive a bus through.

For years, New Yorkers drank high-quality water, essentially for free: their water was kept pure by the root systems, soil microorganisms, and natural filtration processes of forests in the Catskill and Croton watersheds. By the late 1980s, however, pollutants such as sewage, fertilizers, pesticides, and oil had begun to overwhelm these purification processes, causing the quality of the water to decline. New York had a problem, one that could be very expensive to fix.

The standard way of dealing with surface water contamination is to build a water treatment plant. Faced with violations of Environmental Protection Agency (EPA) water quality regulations, New York readily agreed to build such a plant for the Croton Watershed, for an estimated cost of $300 million. But the city balked at the price tag for treating the much larger supply from the Catskill Watershed. Early estimates put the cost of that treatment plant at $6 billion to $8 billion, plus another $300 million per year for its operation. Could a less expensive solution be found?

The answer turned out to be yes. For a projected cost of $1.5 billion, in the early 1990s the city embarked on an ambitious but simple plan: protect the environment of the watershed so that natural systems could resume supplying the city with clean water. The city is buying land that borders rivers and streams in the Catskills, and protecting the land from development to minimize the flow of fertilizers, pesticides, and other pollutants into the water. The city is also building new storm sewers

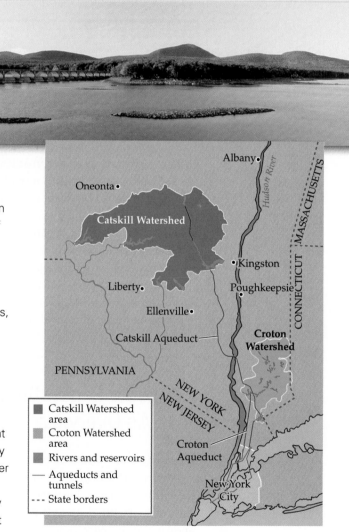

New York City's Water Supply System
The Ashokan Reservoir, located in the Catskill Forest Preserve near Woodstock, New York, is one of several reservoirs that supply water to New York City.

and septic systems, upgrading existing sewage treatment plants, and providing funds to encourage environmentally friendly forms of development.

New York City is not alone. Cities throughout the world are struggling to cope as increasing development causes natural water purification systems to lose their effectiveness. And in at least one case, a private company has taken actions similar to those taken in New York. The bottled-water company Perrier Vittel became concerned in the late 1980s that pollutants would threaten the quality of its water. These concerns were driven home a few years later when the company had to pull its water from the shelves because of contamination with benzene, a carcinogen found in gasoline. Rather than getting its water from a new source—as some other companies had done when faced with similar issues— Perrier Vittel decided to invest in the environment. To protect its water, the company spent $9 million to acquire land and to reach agreements with farmers to reduce fertilizer and pesticide use.

As the New York City and Perrier Vittel examples suggest, there really can be a free lunch, at least as far as the availability of clean, fresh water is concerned. **How does an ecosystem deliver clean, fresh water? What other essential services do ecosystems provide—not just for us, but for other life-forms as well? In what ways can human actions jeopardize these "free services"?** Before we can tackle these questions, we must understand how a healthy ecosystem works. In particular, we must examine how energy and materials are exchanged between the nonliving and living components of an ecosystem, and how energy and materials move through the chain of life-forms that makes up the living part of an ecosystem. In this chapter we will see how our individual lives are bound to the living and nonliving world around us, and how our everyday actions can affect that intricate network of relationships.

To survive, all organisms need energy to fuel their metabolism and materials to construct and maintain their bodies. For their energy needs, most organisms depend directly or indirectly on solar energy, an abundant supply of which reaches Earth each day. By contrast, materials such as the carbon, hydrogen, oxygen, and other elements of which we are made are added to our planet in relatively small amounts (in the form of meteoric matter from outer space). Earth, therefore, has an essentially fixed amount of materials for organisms to use. This simple fact means that for life to persist, natural systems must recycle materials. In this chapter we focus on these two essential aspects of life—energy and materials—as we discuss ecosystem ecology, the study of how energy and materials are used in natural systems.

24.1 How Ecosystems Function: An Overview

An **ecosystem** consists of communities of organisms together with the physical environment they share. The assemblage of interacting organisms—prokaryotes, protists, animals, fungi, and plants—constitute the **biotic** world within an ecosystem ("biotic" means "having to do with life"). The physical environment that surrounds the biotic community—the atmosphere, water, and Earth's crust—is the **abiotic** part of an ecosystem. An ecosystem is therefore the sum of its biotic and abiotic components. Ecosystems don't always have sharply defined physical boundaries. Instead, ecologists recognize an ecosystem by the distinctive ways in which it functions, especially the means by which energy is acquired by the biotic community.

Ecosystems may be small or very large: a puddle teeming with protists is an ecosystem, as is the Atlantic Ocean. Smaller ecosystems can be nested inside larger, more complex, ecosystems. In fact, global patterns of air and water circulation (see Chapter 20) may be viewed as linking all the world's organisms into one giant ecosystem, the biosphere. Whether they are large or small, ecosystems can be challenging to study because organisms, energy, and materials often move from one ecosystem to another.

Figure 24.1 gives a broad overview of how ecosystems work. First, examine the orange and red arrows in the figure, which show the movement of energy through the ecosystem. At each step in a food chain (see Figure 23.3 for an example of a food chain), a portion of the energy captured by producers is lost as metabolic heat (red arrows). Metabolic heat is a by-product of chemical reactions within a cell. Cellular respiration, the chemical process that most cells use to extract energy from food molecules (see Chapter 7), is responsible for much of the metabolic heat released by any oxygen-utilizing organism. Organisms lose a lot of energy as metabolic heat, as revealed by the fact that a

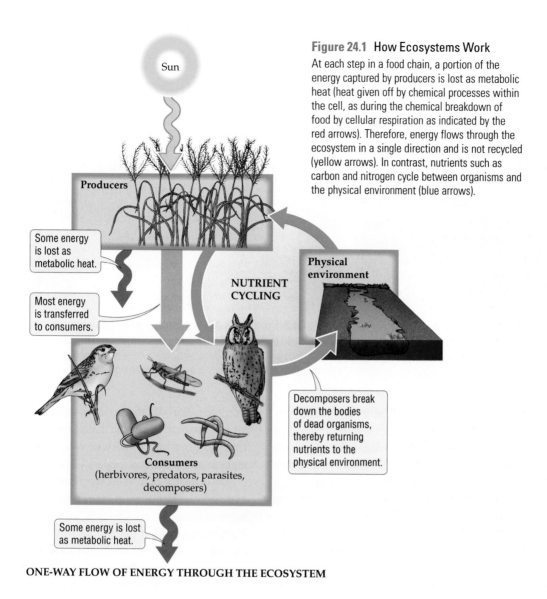

Figure 24.1 How Ecosystems Work
At each step in a food chain, a portion of the energy captured by producers is lost as metabolic heat (heat given off by chemical processes within the cell, as during the chemical breakdown of food by cellular respiration as indicated by the red arrows). Therefore, energy flows through the ecosystem in a single direction and is not recycled (yellow arrows). In contrast, nutrients such as carbon and nitrogen cycle between organisms and the physical environment (blue arrows).

Sun

Producers

Some energy is lost as metabolic heat.

Most energy is transferred to consumers.

NUTRIENT CYCLING

Physical environment

Decomposers break down the bodies of dead organisms, thereby returning nutrients to the physical environment.

Consumers
(herbivores, predators, parasites, decomposers)

Some energy is lost as metabolic heat.

ONE-WAY FLOW OF ENERGY THROUGH THE ECOSYSTEM

small room crowded with people rapidly becomes hot. Because of this steady loss of heat, energy flows in only one direction through ecosystems: it enters Earth's ecosystems from the sun (in most cases) and leaves them as metabolic heat.

In contrast to energy, **nutrients**—the chemical elements required by living organisms—are largely recycled between living organisms and the physical environment. While Earth receives a constant stream of light energy from the sun, it has a constant and finite pool of chemical elements distributed through the land, water, and air. These elements may pass from rocks and mineral deposits to the soil and water and then to producers, consumers, and decomposers—and back again to the abiotic world. However, Earth does not acquire more of these substances on a daily basis, the way it receives light energy, and at most, the elements are moved within and between ecosystems.

As Figure 24.1 shows (blue arrows), nutrients are absorbed from the environment by producers, cycled among consumers for varying lengths of time, and eventually returned to the environment when the ultimate consumers—decomposers—break down the dead bodies of organisms. Ecologists and earth scientists use the term "nutrient cycle" or "biogeochemical cycle" to describe the passage of a chemical element through the abiotic and biotic worlds.

The physical, chemical, and biological processes that link the biotic and abiotic worlds in an ecosystem are known as ecosystem processes. Energy capture through photosynthesis, release of metabolic heat, decay of biomass through the action of decomposers, and movement of nutrients from living organisms to the abiotic world around them are all examples of ecosystem processes. The activity of producers, in particular, profoundly influences ecosystem processes and therefore the

Helpful to know

In discussing the biology of humans and other animals, "nutrients" refers to vitamins, minerals, essential amino acids, and essential fatty acids. In the ecosystem context of this chapter, however, "nutrients" has a more restricted meaning, referring only to the essential elements required by producers.

(a)

(b)

Figure 24.2 Producers Are the Energy Base in an Ecosystem
(a) In tropical rainforests, the abundant producers (plants) store a lot of chemical energy, which in turn is available to consumers. (b) In deserts, because of the sparse plant life, relatively little chemical energy is available for consumers.

characteristics of an ecosystem (Figure 24.2). Ecologists often demarcate an ecosystem according to the types of producers it contains and the community of consumers that these producers support. A duckweed-covered pond, a saltwater marsh, a tallgrass prairie, and an oak–maple woodland are all examples of ecosystems that can be defined on the basis of the specific types of producers that capture energy and supply it to a characteristic assemblage of consumers.

To understand how ecosystems work, ecologists focus on two ecosystem processes in particular: the one-way flow of energy and the cycling of nutrients. As we'll describe, they study how organisms capture energy and nutrients from the environment, transfer energy and nutrients to one another, and, ultimately, return nutrients to the environment. We turn now to a more detailed look at the first of these steps, energy capture.

24.2 Energy Capture in Ecosystems

Life cannot exist without a source of energy to support it. Most life on Earth depends directly or indirectly on the capture of solar energy by producers. In some unusual habitats, such as deep-sea vents and hot springs, the main producers are prokaryotes that can extract chemical energy from inorganic molecules such as compounds of

iron and sulfur. Our discussion of energy capture will focus on ecosystems that depend mainly on solar energy because these are common in both terrestrial (land-based) and aquatic (saltwater and freshwater) regions of the world.

The energy captured by plants and other photosynthetic organisms is stored in their bodies in the form of chemical compounds, such as carbohydrates. Herbivores (which eat plants and other producers), predators (which eat herbivores and other predators), and decomposers (which consume the remains of dead organisms) all depend indirectly on the solar energy originally captured by plants and other producers (such as photosynthetic bacteria and algae).

To better understand these points, imagine that all the plants in a terrestrial ecosystem suddenly vanished. Although bathed in sunlight, herbivores and predators alike would starve because they could not use that energy to produce food. This line of thinking also helps us understand why some environments can support more animals than other environments can. In a tropical forest, for example, many plants can capture energy from the sun (see Figure 24.2a). As a result, a large amount of energy from the sun is stored in chemical forms that can be used as food by animals. In contrast, in environments with few plants (such as arctic or desert regions, as shown in Figure 24.2b), relatively little energy is captured from the sun. Hence, less food is available in such environments, and fewer animals can live there.

Assessing the overall amount of energy captured by plants is an important first step in determining how a terrestrial ecosystem works: it influences the amount of plant growth and hence the amount of food available to other organisms. Each of these factors, in turn, influences the type of terrestrial ecosystem found in a region and how that ecosystem functions.

The rate of energy capture varies across the globe

The amount of energy captured by photosynthetic organisms, minus the amount they lose as metabolic heat, is called **net primary productivity (NPP)**. Although NPP is defined in terms of energy, it is usually easier to estimate it as the amount of new **biomass** (the mass of organisms) produced by the photosynthetic organisms in a given area during a specified period of time. In a grassland ecosystem, for example, ecologists would estimate NPP by measuring the average amount of new grass and other plant matter produced in a square meter each year. Such NPP estimates based on biomass can be converted to units based on energy.

NPP is not distributed evenly across the globe. On land, NPP tends to decrease from the equator toward the poles (Figure 24.3a). This decrease occurs because the amount of solar radiation available to plants also decreases from the equator toward the poles (as we saw in Chapter 20). But there are many exceptions to this general pattern. For example, there are large regions of very low NPP in northern Africa, central Asia, central Australia, and the southwestern portion of North America. Each of these regions is a site of one of the world's major deserts.

The low NPP in deserts emphasizes the fact that sunlight alone is not sufficient to produce high NPP; water is also required. In addition to water and sunlight, productivity on land can be limited by temperature and the availability of nutrients in the soil. The most productive terrestrial ecosystems are tropical rainforests and farmland; the least productive are deserts and tundra (including some mountaintop communities).

The global pattern of NPP in marine ecosystems (Figure 24.3b) is very different from that on land. There is little tendency for NPP to decrease from the equator toward the poles. Instead, the general pattern relates to distance from shore: the productivity of marine ecosystems is often high in ocean regions close to land, but relatively low in the open ocean, which is, in essence, a marine "desert." Calling the open ocean a desert makes sense largely because nutrients needed by aquatic photosynthetic organisms are in short

supply there. Nutrients released by the death and decay of organisms tend to settle on the deep-ocean floor, in which case they are not immediately available to photosynthetic producers, which live near the surface. Aquatic ecosystems that experience **upwelling**,

(a)

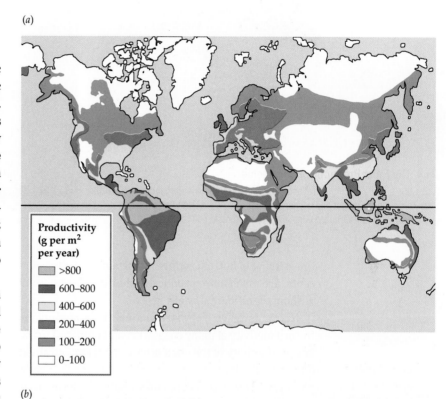

Productivity
(g per m^2
per year)
- >800
- 600–800
- 400–600
- 200–400
- 100–200
- 0–100

(b)

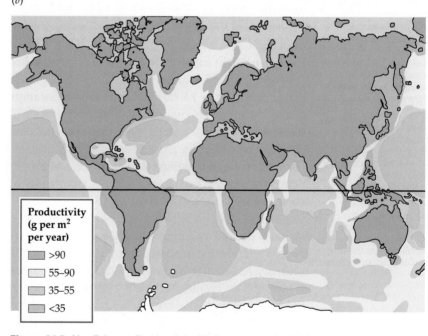

Productivity
(g per m^2
per year)
- >90
- 55–90
- 35–55
- <35

Figure 24.3 Net Primary Productivity Varies across the Globe

Net primary productivity varies greatly among both (a) terrestrial ecosystems and (b) marine ecosystems. Net primary productivity is measured as grams of new biomass made by plants or other producers each year in a square meter of area.

(a)

(b)

Figure 24.4 The Most Productive Ecosystems
Wetlands like this estuary in Virginia (*a*) and coral reefs like this one in Indonesia (*b*) can harness energy from sunlight and oxygen and translate it into high net primary productivity on par with terrestrial systems like tropical forests and farmlands.

the stirring of bottom sediments by air and water currents, are much more productive than similar habitats without significant upwelling.

Nutrients delivered by streams and rivers account for the productivity of many coastal areas, compared to the low productivity of the open oceans. Nutrients drained off the land stimulate the growth and reproduction of phytoplankton, the small photosynthetic producers that form the foundation of aquatic food webs. Estuaries—regions where rivers empty into the sea—are some of the most productive habitats on the planet precisely because the rich nutrient supply supports large populations of producers, which in turn nourish large populations of consumers (Figure 24.4*a*). Wetlands such as swamps and marshes can also match the productivity levels of tropical forests and farmland. They trap soil sediments rich in nutrients and organic matter, thereby promoting the growth of flooding-tolerant plants and phytoplankton, which in turn feed a complex community of consumers.

As on land, the NPP in aquatic ecosystems can be strongly limited by sunlight and temperature. Coral reefs, which abound in warm, sunny, relatively shallow seas, are among Earth's most productive ecosystems (Figure 24.4*b*), rivaling tropical forests and farmland. Even though nutrients are relatively scarce in these waters, this scarcity is offset by a mutually beneficial relationship (mutualistic symbiosis) between photosynthetic protists (algae) and their animal hosts (coral polyps). The algal and animal partners share the bounty

they secure: the algae capture carbon through photosynthesis, and the polyps supply nitrogen from the microscopic consumers (zooplankton) and other small marine life they trap. This relationship thrives best in warm, sunny, and clear seas.

Concept Check

1. Compare the transfer of energy with the movement of nutrients in an ecosystem.
2. What is NPP? What factors limit NPP in terrestrial ecosystems?
3. Explain why the open ocean has lower NPP than coastal areas.

24.3 Energy Flow through Ecosystems

As described in Chapter 23, organisms capture energy in two major ways, depending on whether they are producers or consumers. Producers get their energy from abiotic sources, such as the sun. The organisms at the bottom of a food web, such as plants, algae, and photosynthetic bacteria, are producers. Consumers get their energy by eating all or parts of other organisms or their remains. Consumers include decomposers, such as bacteria and fungi, that break down the dead bodies of organisms, as well as the herbivores, predators, and parasites described in Chapter 22.

An energy pyramid shows the amount of energy transferred up a food chain

Energy from the sun is stored by plants and other producers in the form of chemical compounds. This stored energy can be transferred from one organism to another up a food chain, but eventually every unit of energy captured by producers is lost as heat. Therefore, energy cannot be recycled.

To illustrate, let's follow the fate of energy from the sun after it strikes the surface of a grassland. A portion of the energy captured by grasses is transferred to the herbivores that eat the grasses, and then to the predators that eat the herbivores. The transfer of energy from grasses to herbivores to predators is not perfect, however. When a unit of energy is used by an organism to fuel its metabolism, some of that energy is lost from the ecosystem as unrecoverable heat, as we have seen. Therefore, energy moves through ecosystems in a single direction: proceeding up a food chain (for example, from grass to grasshopper to bird, as shown in Figure 24.5), portions of the energy originally captured by photosynthesis are steadily lost. Because of this steady loss of energy, more energy is available at a lower level, than a higher level, in a food chain.

The amounts of energy available to organisms in an ecosystem can be represented by a pyramid. Each level of an energy pyramid corresponds to a step in a food chain and is called a **trophic level** (see Figure 24.5). The grass–grasshopper–bird example shown in the figure has four trophic levels: grass is on the first trophic level, the grasshopper is on the second, the insect-eating bird is on the third, and the bird-eating bird is on the fourth. On average, roughly 10 percent of the energy at one trophic level is transferred to the next trophic level. The energy that is not transferred to the next trophic level is the energy that is not consumed (for example, when we eat an apple, we eat only a small part of the apple tree), is not taken up by the body (for example, we cannot digest the cellulose that is contained in the apple), or is simply lost as metabolic heat.

Secondary productivity is highest in areas of high NPP

The rate of new biomass production by *consumers* is called **secondary productivity**. As we've seen, because consumers depend on producers for both energy and materials, secondary productivity is highest in ecosystems with high net primary productivity. A tropical forest, for example, has a much higher NPP than tundra has. For this reason, per unit of area there are many more herbivores and other consumers in tropical forests than in tundra, and hence secondary productivity is much higher in tropical forests than in tundra.

In natural ecosystems, new biomass made by plants and other producers is consumed either by herbivores or by decomposers. In some ecosystems, 80 percent of the biomass produced by plants is used directly by decomposers. Eventually, since all organisms die, all biomass made by producers, herbivores, predators, and

Learn more about the idealized energy pyramid.

24.1

Figure 24.5 An Idealized Energy Pyramid

Of each 10,000 kilocalories (kcal; 1 kilocalorie = 1,000 calories) of energy from the sun captured by producers, primary consumers harvest only about 10 percent. Roughly 10 percent of the energy at each trophic level is then transferred to the next trophic level.

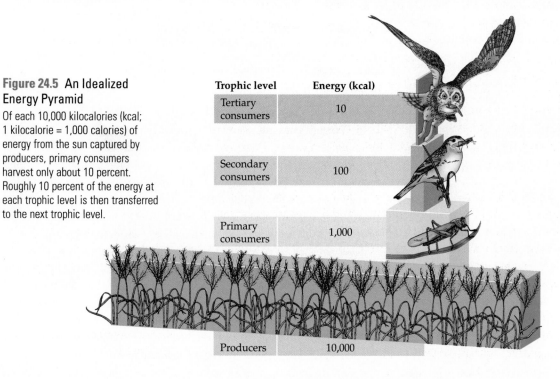

Trophic level	Energy (kcal)
Tertiary consumers	10
Secondary consumers	100
Primary consumers	1,000
Producers	10,000

Biology on the Job

From Corn to Cars

With gasoline prices soaring and dependence on oil produced in other countries a concern, many nations are looking for alternative sources of automotive fuel. Some countries have turned to ethanol, a form of alcohol that can be made from organic materials. Brazil, for example, invested heavily in ethanol fuels beginning in the 1970s; now, 40 percent of the fuel used to drive cars in that country is ethanol, compared with 3 percent in the United States. Ben P. Sever is a lawyer at a company in the United States that makes ethanol for use as automotive fuel.

What are your day-to-day tasks? What does your job entail? I draft and revise contracts and advise the company on contract negotiations. Sometimes I perform negotiations that lead to contracts with shippers, corn suppliers, enzyme suppliers, and people buying the ethanol. The ethanol we sell is blended with gasoline for motor vehicle fuel.

Do you have a scientific background? Not at all. I have a BS in political science, and I began my career in securities and corporate law. It would be handy to have a scientific background, since otherwise you face a pretty steep learning curve. But I learn faster on the job than in the classroom anyway. If you are good student—not necessarily of academic things but of life—you'll do well no matter what subject your degree is in, or what you go into.

What do you like most about your job? I believe in the product we produce, and I think it is environmentally sound. Ethanol is good for farmers and it decreases the nation's dependence on foreign oil. Also, I get to look at a variety of legal issues, so every day brings something new.

What future do you see for ethanol-based alternative fuels? Traditionally, ethanol fuels are 10 to 15 percent ethanol blended with gasoline. Emerging E-85 fuel is 85 percent ethanol. A standard automobile can be modified to run on that, and although there aren't many gas stations that currently sell it, there will be. In terms of the environmental implications of operating an ethanol production plant, you don't have to be very concerned about wastes or air emissions. This is true because part of the corn or biomass that you are using is turned into ethanol, and leftover wastes are sold to feed companies. The carbon dioxide that is produced when you make ethanol can be sold to companies that collect it.

The Job in Context

When fully combusted, ethanol is a clean-burning fuel—only carbon dioxide and water are produced. Although carbon dioxide contributes to global warming (see Section 25.4), use of ethanol fuels can help reduce emissions of other pollutants into the atmosphere and could therefore reduce smog in many large cities.

Despite these potential benefits, efforts to expand the use of ethanol as an automotive fuel have been criticized. Some experts say diversion

Switchgrass, an Up-and-Coming Alternative Source of Automotive Fuel
New ethanol fuels may prove to be an important alternative source of energy. Switchgrass (*Panicum virgatum*), a drought-tolerant native of the Great Plains of North America, is being promoted as one of the most promising sources of cellulosic ethanol. The perennial grass can grow on marginal lands with little or no need for fertilizer. Its root system is almost as deep as the aboveground parts are tall, and it is exceptionally good at binding soil and preventing soil erosion.

of the corn crop for ethanol production contributed to the recent spike in food prices that put many of the world's poor at risk for malnutrition and starvation. Further, studies show that current agricultural practices require more energy to produce ethanol fuels than is gained from those fuels. Energy must be expended to grow the corn (including energy used to run farm machinery and to manufacture fertilizers and pesticides), to transport the corn, to culture the microorganisms that produce the ethanol, and to transport the ethanol to the pump.

Over the last 20 years, steady progress has been made in reducing the units of energy expended to produce corn ethanol. Rapid progress is also being made in finding nonfood sources of commercial ethanol, and cellulosic ethanol appears especially promising. Decomposers such as bacteria and fungi are used to degrade cellulose (the major component of plant cell walls) to glucose, which is then fermented to ethanol under low-oxygen conditions (see Chapter 8). Cornstalks, wood waste, rice hulls, and a native grass known as switchgrass (*Panicum virgatum*) are being investigated as more efficient sources of ethanol than corn kernels.

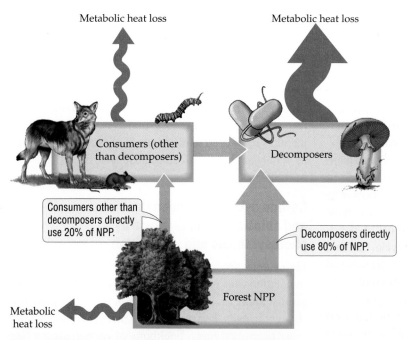

Figure 24.6 Decomposers Consume Most of the NPP
Decomposers such as bacteria and fungi use more than 50 percent of net primary productivity in ecosystems of all types. In this forest, 80 percent of NPP is used directly by decomposers, and the remaining 20 percent is used by other consumers (such as herbivores and predators).

parasites is consumed by decomposers (Figure 24.6). In some instances, people bypass the decomposers, as when we use crops or agricultural refuse to produce fuels (see "Biology on the Job").

24.4 Nutrient Cycles

Nutrients—chemical elements such as carbon, hydrogen, oxygen, and nitrogen—are used by producers and other organisms to construct their bodies. Producers obtain these and other essential nutrients from the soil, water, or air in the form of ions such as nitrate (NO_3^-) or inor-

ganic molecules such as carbon dioxide (CO_2). Consumers obtain them by eating producers or other consumers. Because these nutrients are essential for life, their availability and movement through ecosystems influence many aspects of ecosystem function.

Nutrients are transferred between organisms (the biotic community) and the physical environment (the abiotic world) in cyclical patterns called **nutrient cycles** or biogeochemical cycles. Figure 24.7 provides a general description of how nutrient cycles work. Nutrients can be stored for long periods of time in abiotic reservoirs such as rocks, ocean sediments, or fossilized remains of organisms. The nutrients stored in such abiotic reservoirs are not readily accessible to producers. Weathering of rocks, geologic uplift, human actions, and other forces can move nutrients back and forth between such reservoirs and **exchange pools**—abiotic sources such as soil, water, and air where nutrients *are* available to producers.

Once captured by a producer, a nutrient can be passed from the producer to an herbivore, then to one or more predators or parasites, and eventually to a decomposer. Decomposers break down once-living tissues into simple chemical components, thereby returning nutrients to the physical environment. Without decomposers, nutrients could not be repeatedly reused, and life would cease because all essential nutrients would remain locked up in the bodies of dead organisms.

Abiotic conditions, especially temperature and moisture, influence the length of time it takes for a nutrient to cycle from the biotic community to the exchange pool

Helpful to know

"Reservoir" in this chapter has two meanings: It can refer to a body of water held aside for use by people, as in the New York City water supply. More generally, a reservoir is any place or substance that holds something (not necessarily water) in storage, as is meant with regard to nutrient cycling.

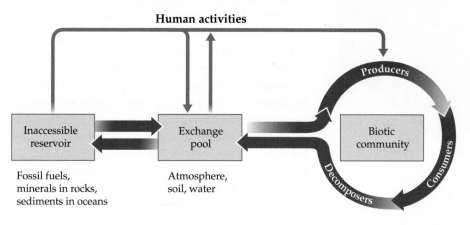

Human activities

Figure 24.7 Nutrient Cycling
Nutrients cycle among reservoirs inaccessible to producers, exchange pools available to producers, and living organisms. Human activities, such as fossil fuel use and fertilizer synthesis, can move nutrients between inaccessible reservoirs and exchange pools, altering their availability to producers and changing nutrient cycles.

and back again to a producer. Warmer temperatures, for instance, increase the activity of decomposers, which in turn speeds up nutrient release from biomass. By influencing such processes as weathering and runoff, rainfall may control nutrient release from inaccessible reservoirs and also determine if significant amounts of nutrients are lost from one ecosystem (a forest community, for example) to another (such as a stream).

The length of time it takes for a nutrient to cycle from a producer to the physical environment and back to another producer also varies from one type of nutrient to another. Some chemical elements are transported more quickly and over longer distances than others, depending on whether they have an atmospheric cycle. Carbon, hydrogen, oxygen, nitrogen, and sulfur are all essential nutrients that have atmospheric cycles. Nutrients that have an **atmospheric cycle** can exist as a gas under natural conditions, which means they can be released into the atmosphere, or be absorbed from it, by the biotic and abiotic components of an ecosystem. Once in the atmosphere, these nutrients can be transported by wind from one region of Earth to another. When nutrients are transported long distances in this way, they can affect nutrient cycles in distant ecosystems.

Phosphorus is one of the few major nutrients that does not have an atmospheric cycle and displays an exclusively sedimentary cycle instead. A nutrient is said to have a **sedimentary cycle** if it is not commonly found as a gas under natural conditions and moves mostly through land and water, rather than the atmosphere, in its passage through an ecosystem. Nutrients with a sedimentary cycle tend to move slowly and are not dispersed as widely as those with an atmospheric cycle.

Carbon cycling between the biotic and abiotic worlds is driven by photosynthesis and respiration

Living cells are built mostly from organic molecules—molecules that contain carbons bonded to hydrogen atoms. Next to oxygen, carbon is the most plentiful element in cells. Every one of the main macromolecules in an organism has a backbone of carbon atoms (see Chapter 4). The transfer of carbon within biotic communities, between living organisms and their physical surroundings, and within the abiotic world is known as the global **carbon cycle** (Figure 24.8).

Although carbon makes up a large part of biomass, the element is not abundant in the atmosphere. Carbon, in the form of CO_2 gas, makes up only about 0.04 percent of Earth's atmosphere, although that percentage has been creeping upward every year for the last 100 years or so. The oceans represent the largest store of carbon on our planet. Most of this carbon is dissolved inorganic carbon (such as bicarbonate ions, HCO_3^-), and a minor portion is represented by the carbon held in the biomass of marine organisms. Earth's crust generally contains only about 0.038 percent carbon by weight, but it also has regions with carbon-rich sediments and rocks formed from the remains of ancient marine and terrestrial organisms (see Figure 24.8). The deep layers of carbon-containing rocks and sediments constitute an inaccessible reservoir of carbon in natural ecosystems.

Some of the organic matter from ancient organisms has been transformed by geologic processes into deposits of fossil fuel, such as petroleum, coal, and natural gas. When we extract these fossil fuels and burn them to meet our energy needs, the carbon locked in these deposits for several hundred million years is released into the atmosphere as carbon dioxide (shown by a red arrow in Figure 24.8). Humans today have the power to change the global carbon cycle but, as we discuss in Chapter 25, altering an ecosystem process that is so vital, large-scale, and still poorly understood might prove dangerous for natural ecosystems and for human welfare.

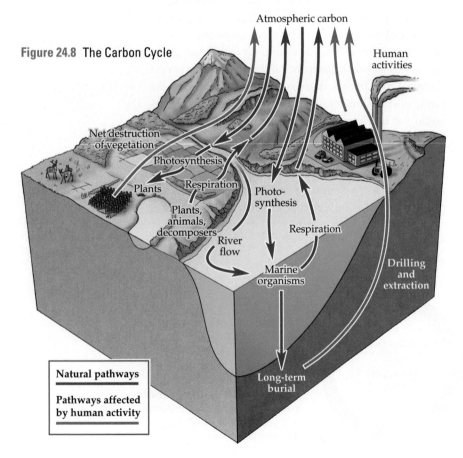

Figure 24.8 The Carbon Cycle

Atmospheric carbon

Human activities

Net destruction of vegetation

Photosynthesis

Plants

Respiration

Plants, animals, decomposers

Photo-synthesis

Respiration

River flow

Marine organisms

Drilling and extraction

Long-term burial

Natural pathways

Pathways affected by human activity

Living organisms, in both aquatic and terrestrial ecosystems, acquire carbon mostly through photosynthesis. Aquatic producers, such as photosynthetic bacteria and algae, can absorb dissolved carbon dioxide (in the form of bicarbonate or carbonate ions) and convert it into organic molecules using sunlight as a source of energy. Plants, the most important producers in terrestrial ecosystems, absorb CO_2 from the atmosphere and transform it into food with the help of sunlight and water. The extraction of energy from food molecules, by producers and consumers alike, releases CO_2 and returns it to the abiotic world. Cellular respiration is the main pathway by which most organisms extract energy from organic molecules, although some organisms, including many decomposers, use other pathways that do not depend on molecular oxygen.

Although decomposers release a great deal of the carbon contained in the dead organisms they live on, some of it typically remains in the ecosystem as partially decayed organic matter. Leaf litter, humus, and peat are examples of partially decayed organic matter that form an important store of soil carbon. The top layers of soil in northern coniferous forests have large amounts of organic matter because decomposition is slow in these cold, wet regions. The vast peat deposits found in the arctic tundra and northern boreal forests represent a large carbon store in the biosphere. Because decomposition is faster at warmer temperatures, in terrestrial ecosystems typical of the tropics the soil carbon levels are lower and carbon cycles more rapidly.

Biological nitrogen fixation is the most important source of nitrogen for biotic communities

Nitrogen is a key component of amino acids, proteins, and nucleic acids such as DNA. It is therefore an essential element for all living organisms. The fact that this nutrient is not naturally abundant in soil and water limits the growth of producers in most ecosystems that are unaltered by human actions.

Earth's atmosphere is the largest store of nitrogen in the biosphere: N_2 gas (molecular nitrogen) makes up 78 percent of the air we breathe. Energy from lightning converts some of the atmospheric N_2 into nitrogen compounds that mix with rainwater to form nitrate ions (NO_3^-), as shown in Figure 24.9. Nitrate ions are the most common form of nitrogen in soil and water, although smaller amounts of ammonium ions (NH_4^+) also occur naturally. Nitrate is highly soluble and can therefore be easily lost from an ecosystem through runoff.

Lightning contributes only a small amount of the nitrogen found in soil and water, with much of it originating in a remarkable metabolic process, known as biological nitrogen fixation, that can be carried out only by some prokaryotes. Waste produced by animals, and the death and decay of organisms, adds ammonium ions to soil and water (see Figure 24.9). Bacteria common in most ecosystems rapidly convert ammonium ions to nitrate, which is why ammonium ions do not accumulate at high levels in most habitats. Denitrifying bacteria, common in oxygen-poor environments, convert nitrate into molecular nitrogen (N_2), or into N_2O (nitrous oxide), returning nitrogen to the vast atmospheric pool.

The conversion of molecular nitrogen (N_2) into ammonium ions (NH_4^+) by bacteria is known as biological **nitrogen fixation.** Some nitrogen-fixing bacteria live free in the soil, and some fix nitrogen in a mutualistic symbiosis with certain plants, including legumes (members of the bean family), alder trees, and an aquatic fern called *Azolla*. The plant host obtains ammonium from its nitrogen-fixing partner and supplies to the prokaryote food energy derived from photosynthesis. Nitrogen fixed by bacteria is added to soil or water when these organisms, and any host species they

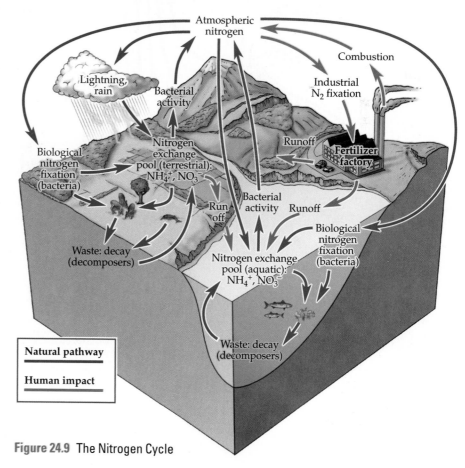

Figure 24.9 The Nitrogen Cycle

might associate with, die and decompose. Consumers acquire nitrogen when they eat plants that have acquired nitrogen from the soil or through their association with nitrogen-fixing bacteria. The action of decomposers transfers nitrogen from the biotic community to the abiotic component of an ecosystem.

Like carbon, nitrogen is said to have an atmospheric cycle because the element passes through a gaseous phase during its cycling through an ecosystem. Like all other nutrients with an atmospheric cycle, nitrogen can be transported long distances in the biosphere. Humans manufacture fertilizer through industrial nitrogen fixation, a process that combines nitrogen gas (N_2) with hydrogen gas (H_2) to make ammonium compounds. The manufacturing process relies on high temperature and pressure generated by the burning of fossil fuels. Applying nitrogen fertilizer to crop plants boosts their productivity, but the runoff can stimulate NPP in aquatic ecosystems in ways that are harmful to the biotic community. Excess nitrogen can disturb terrestrial communities as well, sometimes decreasing diversity (species number and abundance). This shift in

diversity has been noted in some grassland communities, in which a few species that responded exceptionally vigorously to supplemental nitrogen were able to outcompete the other species.

Sulfur is one of several important nutrients with an atmospheric cycle

Sulfur is a component of certain amino acids, many proteins, and some polysaccharides and lipids. It is also found in other organic compounds that have important roles in metabolism. Sulfur displays an atmospheric cycle because it moves easily among terrestrial ecosystems, aquatic ecosystems, and the atmosphere. Sulfur enters the atmosphere from terrestrial and aquatic ecosystems in three natural ways (Figure 24.10): as sulfur-containing compounds in sea spray; as a metabolic by-product (the gas hydrogen sulfide, H_2S) released by some types of bacteria; and, least important in terms of overall amount, as a result of volcanic activity.

About 95 percent of the sulfur that enters the atmosphere from the world's oceans does so in the form of strong-smelling sulfur compounds (such as dimethyl sulfide) that are breakdown products of organic molecules made by phytoplankton. The odorous compounds are lofted into the air by wave action and contribute to the smell we associate with seaside air. Hydrogen sulfide is another smelly gas. It is generated by metabolic reactions in bacteria that inhabit oxygen-poor environments such as swamps and sewage.

Sulfur enters terrestrial ecosystems through the weathering of rocks and as atmospheric sulfate (SO_4^{2-}) that mixes with water and is deposited on land as rain. Sulfur enters the ocean in stream runoff from land and, again, as sulfate lost from the atmosphere and falling as rain. Once in the ocean, sulfur cycles within marine ecosystems before being lost in sea spray or deposited in sediments on the ocean bottom. Like most other nutrients with atmospheric cycles, sulfur cycles through terrestrial and aquatic ecosystems relatively quickly. As we'll see shortly, human activities can also add sulfur to the atmosphere, often with adverse consequences for biotic communities and human economic interests.

Learn more about the sulfur cycle.
▶❚❚
24.2

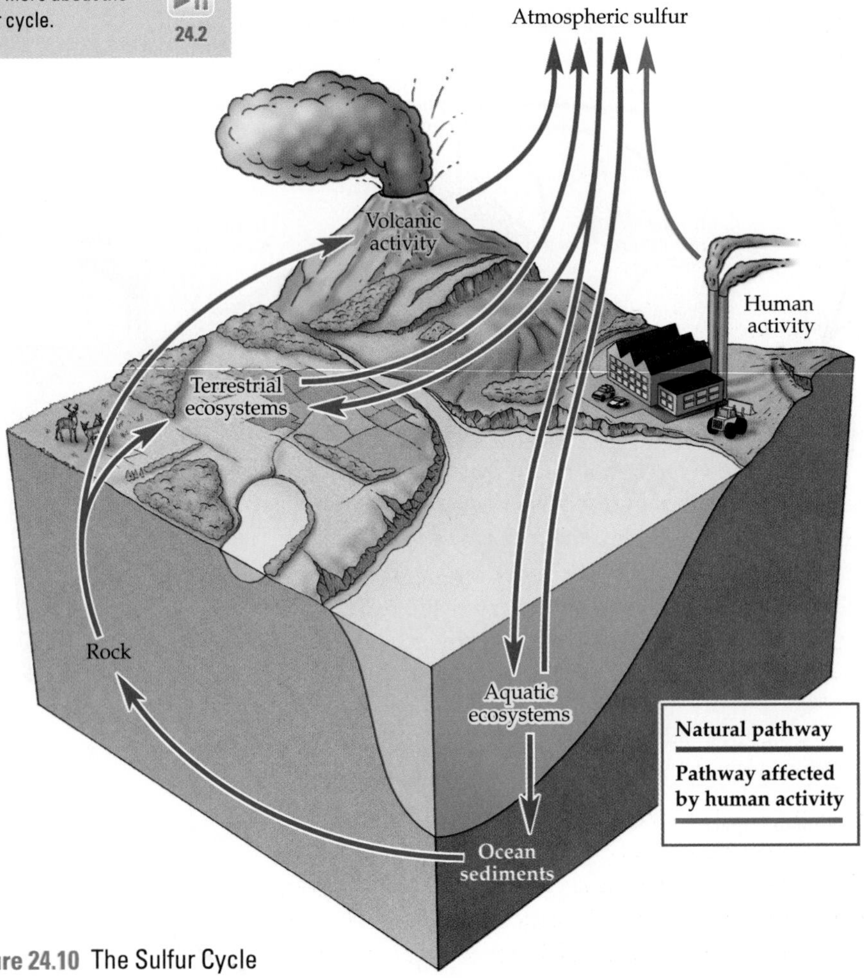

Figure 24.10 The Sulfur Cycle

Phosphorus is the only major nutrient with a sedimentary cycle

Phosphorus is a vital nutrient because it is a component of crucial macromolecules, such as DNA. Phosphorus is important to ecosystems because it strongly affects net primary productivity, especially in aquatic ecosystems.

Figure 24.11 The Phosphorus Cycle

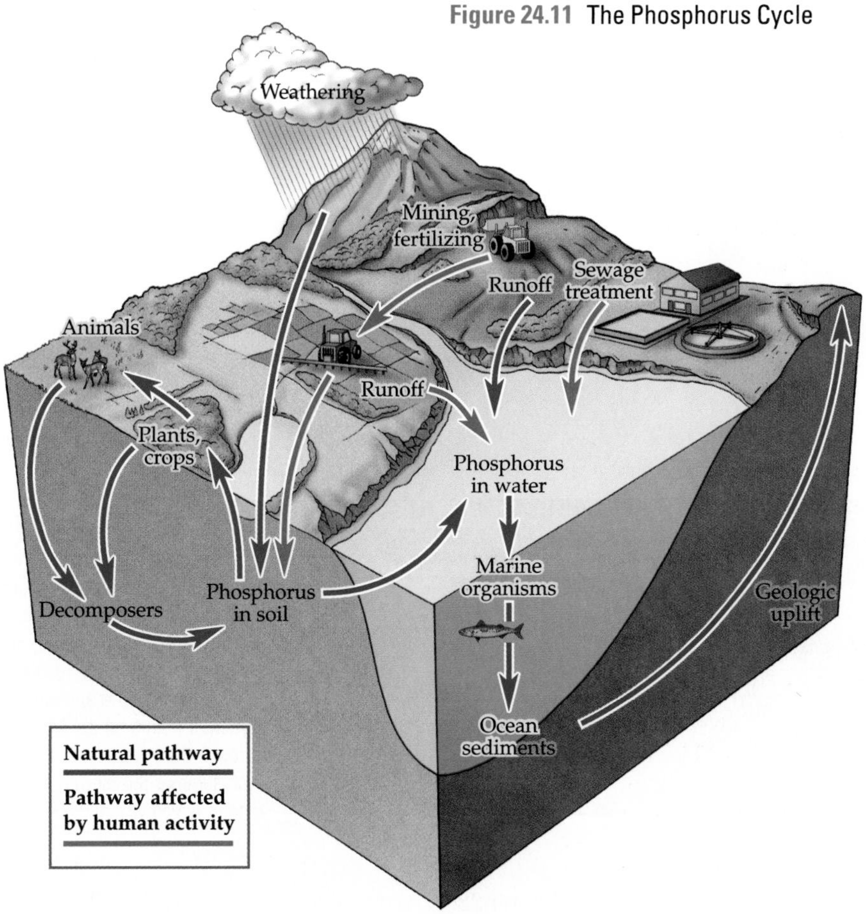

Explore the phosphorous cycle. ▶ II 24.3

Figure 24.11 The Phosphorus Cycle

NPP usually increases, for example, when phosphorus is added to lakes. As we will see in the next section, such an increase in productivity can have undesirable effects, including eutrophication, which leads to the death of aquatic plants, fishes, and invertebrates (see Figure 23.8 and look ahead to Figure 24.12).

Among the major nutrients that cycle within ecosystems, phosphorus is the only one with a sedimentary cycle (Figure 24.11). The reason is that soil conditions usually do not allow bacteria to carry out the chemical reactions required for the production of a gaseous form of phosphorus (phosphine, PH_3). Nutrients like phosphorus first cycle within terrestrial and aquatic ecosystems for variable periods of time (from a few years to many thousands of years); then they are deposited on the ocean bottom as sediments. Nutrients may remain in sediments, unavailable to most organisms, for hundreds of millions of years. Eventually, however, the bottom of the ocean is thrust up by geologic forces to become dry land, and once again the nutrients in the sediments may be available to organisms. Sedimentary nutrients usually cycle very slowly, so they are not replaced easily once they are lost from an ecosystem.

24.5 Human Actions Can Alter Ecosystem Processes

Humans have been disrupting ecological communities for many hundreds, perhaps thousands, of years. The history of the moai-building Polynesian settlers of Easter Island (see Chapter 21) demonstrates that such disruptions can produce tragic consequences for humans. The disruptions of preindustrial times are dwarfed, however, by the scale of the ecological changes that humans have brought about in the past 200 years. In this section we examine some of the ecosystem processes that are readily altered by human activities, reserving a broader discussion of the human impact on the biosphere for the next chapter.

Human activities can increase or decrease NPP

Human activities can change the amount of energy captured by ecosystems on local, regional, and global scales. For example, rain can cause fertilizers to wash from a farm field into a stream, which then flows into a lake. The addition of extra nutrients to lakes or streams or offshore

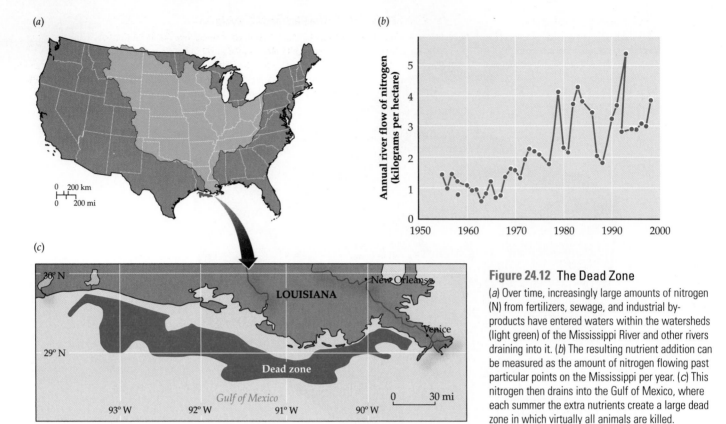

Figure 24.12 The Dead Zone
(*a*) Over time, increasingly large amounts of nitrogen (N) from fertilizers, sewage, and industrial by-products have entered waters within the watersheds (light green) of the Mississippi River and other rivers draining into it. (*b*) The resulting nutrient addition can be measured as the amount of nitrogen flowing past particular points on the Mississippi per year. (*c*) This nitrogen then drains into the Gulf of Mexico, where each summer the extra nutrients create a large dead zone in which virtually all animals are killed.

waters leads to **eutrophication**, as we saw in Chapter 23. In a eutrophic lake, NPP increases because the added nutrients cause photosynthetic algae to become more abundant, so they absorb more energy from the sun.

It is not necessarily a good thing when human actions increase NPP. In some instances, nutrient enrichment has increased NPP to such an extent that large bodies of water have become nearly devoid of animal life. For example, each summer, large amounts of nitrogen and phosphorus are brought by the Mississippi and other rivers to the Gulf of Mexico (Figure 24.12). Algae populations (and hence NPP) increase in these nutrient-rich waters, which float on top of the colder, saltier waters of the gulf. As the algae die, they drift into deeper waters, where bacteria decompose their bodies, using oxygen in the process. As a result, oxygen levels in these deeper waters drop so low that virtually all animals die, creating a large "dead zone" (see Figure 24.12*c*). This dead zone threatens to diminish the fish and shellfish industry in the gulf, which produces an annual catch worth about $500 million. In the summer of 2002, the dead zone reached its largest size ever—about 22,000 square kilometers (8,500 square miles)—covering an area greater than the state of Massachusetts.

Human activities can also change NPP on land. For example, NPP decreases when logging and fire convert tropical forest to grassland. Globally, scientists estimate that human activities leading to such land conversions have decreased NPP in some regions while increasing it in other regions, but the net effect is a 5 percent decrease in NPP worldwide.

Human activities can alter nutrient cycles

Human activities can have major effects on nutrient cycles. Ecologists have shown, for example, that the clear-cutting of a forest, followed by spraying with herbicides to prevent regrowth, causes the forest to lose large amounts of nitrate, an important source of nitrogen for plants (Figure 24.13).

On a larger geographic scale, nutrients such as nitrogen and phosphorus used to fertilize crops can be carried by streams to a lake or an ocean hundreds of kilometers away, where they can increase NPP and cause the body of water to become eutrophic. Finally, on a still larger geographic scale, many human activities, such as shipping crops and wood to distant locations, transport nutrients around the globe. Many human activities also release chemicals into the air, where global wind patterns move them over long distances.

When people alter atmospheric nutrient cycles, the effects are often felt across international borders.

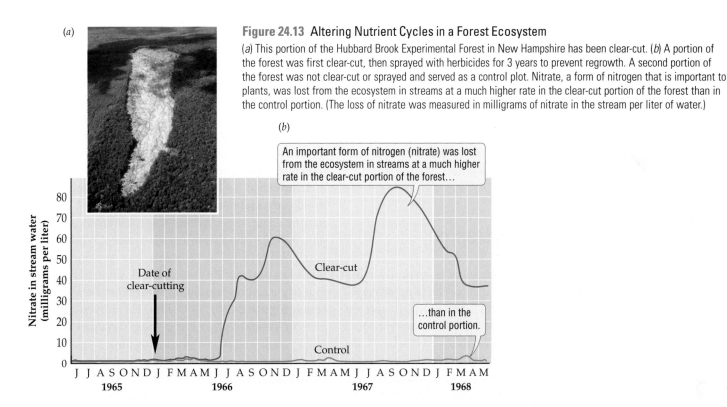

(a)

Figure 24.13 Altering Nutrient Cycles in a Forest Ecosystem
(a) This portion of the Hubbard Brook Experimental Forest in New Hampshire has been clear-cut. (b) A portion of the forest was first clear-cut, then sprayed with herbicides for 3 years to prevent regrowth. A second portion of the forest was not clear-cut or sprayed and served as a control plot. Nitrate, a form of nitrogen that is important to plants, was lost from the ecosystem in streams at a much higher rate in the clear-cut portion of the forest than in the control portion. (The loss of nitrate was measured in milligrams of nitrate in the stream per liter of water.)

(b)

> An important form of nitrogen (nitrate) was lost from the ecosystem in streams at a much higher rate in the clear-cut portion of the forest…

> …than in the control portion.

Consider sulfur dioxide (SO_2), which is released into the atmosphere when we burn fossil fuels such as oil and coal. The burning of fossil fuels has altered the sulfur cycle greatly: annual human inputs of sulfur into the atmosphere are more than one and a half times the inputs from all natural sources combined.

Most human inputs of sulfur into the atmosphere come from heavily industrialized areas such as northern Europe and eastern North America. Once in the atmosphere, SO_2 combines with oxygen and water to become sulfuric acid (H_2SO_4), which then returns to the land in rainfall. Rainfall normally has a pH of 5.6, but sulfuric acid (as well as nitric acid, HNO_3, caused by nitrogen-containing pollutants) has caused the pH of rain to drop to values as low as 2 or 3 in the United States, Canada, Great Britain, and Scandinavia (see Chapter 4 for a review of pH). Rainfall with a low pH is called **acid rain**.

Acid rain can have devastating effects on human-made structures (such as statues) and on natural ecosystems. Acid rain has drastically reduced fish populations in thousands of Scandinavian and Canadian lakes. Much of the acid rain that falls in these lakes is caused by sulfur dioxide pollution that originates in other countries (such as Great Britain, Germany, and the United States). Acid rain has also caused extensive damage to forests in North America and Europe (Figure 24.14).

The international nature of the acid rain problem has led nations to agree to reduce sulfur emissions. In the United States, annual sulfur emissions were cut by nearly 40 percent between 1980 and 2001 (see Figure 24.14). Such reductions are a very positive first step, but the problems resulting from acid rain will be with us for a long time: acid rain alters soil chemistry and therefore has

Figure 24.14 Acid Rain
Acid rain can damage many ecosystems, such as this spruce forest killed by acid rain in the Jizerské Mountains of the Czech Republic. As the graph shows, however, the amount of sulfur dioxide, a major contributor to acid rain, emitted into the atmosphere each year in the United States has fallen by more than 6.5 million tons since 1980.

effects on ecosystems that will last for many decades after the pH of rainfall returns to normal levels.

Concept Check Answers

2. Nutrients in fertilizer and sewage are delivered by the Mississippi to the Gulf of Mexico, where they stimulate NPP and an algal bloom. Oxygen-consuming decomposers feed on algae that die and settle to deeper waters; dissolved oxygen is depleted as the decomposer population soars, depriving invertebrates and fishes, which die in large numbers.

1. Energy transfer decreases from one trophic level to the next-higher level in an energy pyramid because the biomass at each trophic level loses a substantial amount of energy as heat, so less energy is available to the organisms at the next-higher trophic level.

Concept Check

1. Does the amount of energy transferred up a food chain increase, decrease, or stay the same? Explain.

2. How does runoff from farm fields and outflow from sewage treatment facilities contribute to the "dead zone" in the Gulf of Mexico?

24.6 Ecosystem Design

Throughout the world, much effort is currently being devoted to the design and construction of ecosystems. One reason for these efforts is the desire to restore or replace ecosystems that have been degraded or destroyed by human activities. In the Netherlands, for example, where the land has been so heavily modified by people for so long that few natural ecosystems remain, intensive work is under way to rebuild some of the original ecosystems. In many other countries, similar efforts are in progress to restore heavily damaged ecosystems, such as native prairie ecosystems in the United States.

People also seek to design and build ecosystems for economic reasons. There is often a potential for conflict between developers who want to build homes or industrial parks and environmental activists who seek to prevent such development in order to protect natural ecosystems. Some argue that such conflict between developers and environmentalists could be avoided if we played a type of "zero sum" game: if an ecosystem such as a wetland is destroyed or degraded by development, a comparable ecosystem should be built to replace the damaged natural ecosystem.

In principle, such a policy could work if people could build ecosystems with properties similar to those of natural ecosystems. Are we able to do this? Results to date indicate that we have much to learn. With respect to our ability to build ecosystems from scratch, consider what happened with Biosphere II, an experimental 3-acre enclosure containing several artificial ecosystems (Figure 24.15). Biosphere II—named in reference to Biosphere I, or Earth—was designed to house eight people and be self-sustaining for a 2-year period. Despite the expenditure of more than $200 million on this project, it had to be shut down early because of a slew of problems, including pest population outbreaks, the extinction of all pollinators, the extinction of 19 of 25 vertebrate species, a drop in oxygen levels (from 21 percent to 14 percent, the level usually found at elevations of 5,300 meters, or 17,500 feet, at which most people develop altitude sickness), and spikes in the concentration of nitrous oxide high enough to impair human brain function.

With respect to efforts to build specific ecosystems, a National Academy of Sciences report on wetland restoration concluded that we can build ponds and cattail marshes, but we cannot replace fens, bogs, and many other complex, species-rich wetland ecosystems. Similar challenges confront efforts to restore terrestrial ecosystems. Restored prairies, for example, can resemble native prairies in some respects after as few as 5 to 10 years. However, even 30 to 50 years after restoration work has begun, critical differences may remain in the way nutrients are cycled in restored versus native prairie ecosystems. Most ecosystems are complex, and we often do not know enough about fundamental ecosystem processes to replicate those processes in a human-built ecosystem.

Figure 24.15 Biosphere II: A Lesson in Humility
Despite an expenditure of over $200 million, the Biosphere II experiment, which began in 1991 and was intended to run for 2 years, had to be shut down early. The artificial ecosystems in Biosphere II, such as the ocean zone shown in the inset underwater view, could not duplicate the services found in natural communities, and hence were not self-sustaining.

Applying What We Learned

The Economic Value of Ecosystem Services

New York City had a compelling economic reason to restore the ability of forest ecosystems to purify the city's water supply: faced with spending an estimated $6 to $8 billion on a water treatment plant, planners chose instead to spend roughly $1.5 billion to purchase land, update sewer and septic systems, and promote environmentally friendly forms of development.

More recent estimates put the price tag for the treatment plant at closer to $4 billion, while costs for restoring the ecosystem's water purification services are now expected to be over $2 billion. Still, the New York City example shows that what is good for the environment can also be sound economic policy. Is this an isolated example? Or do ecosystems provide people with services that have economic value in other cases as well?

Consider the floods that struck the western United States in 1996 and 1997. Damage amounted to billions of dollars in the states of Nevada, California, Oregon, and Washington (Figure 24.16). Most news reports about the floods and associated mud slides said they were caused by unusually large amounts of rainfall and snowfall. But a huge flood is not always something that just happens, beyond our control. Some human actions prevent ecosystems from responding as they normally would to heavy rainfall, thereby helping to set the stage for a flood.

How can human actions increase the chance of flooding? People often build dikes or divert the flow of rivers to protect homes or industrial areas located in what were once floodplains. By preventing rivers from overflowing into floodplains, we reduce the ability of the ecosystem to handle periods of heavy rainfall. Floodplains normally function as huge sponges: when streams and rivers overflow, floodplains absorb the excess water, thereby preventing even more severe floods from occurring farther downstream. By building on floodplains and attempting to control floods, we unintentionally make it more likely that when a flood does occur, it will be a big one.

(a)

(b)

Figure 24.16 Flood Devastation in the Pacific Northwest
(*a*) Following heavy flooding in Oregon City during the 1990s, people used boats and float tubes to get around. (*b*) Mud slides can kill people, destroy homes, contaminate stream ecosystems, and have undesirable aesthetic effects.

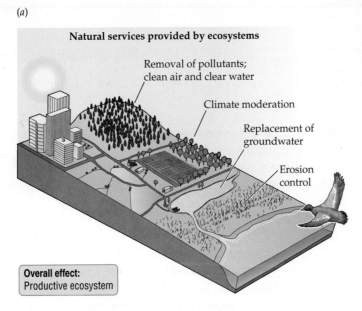

(a)

Natural services provided by ecosystems

Removal of pollutants;
clean air and clear water

Climate moderation

Replacement of
groundwater

Erosion
control

Overall effect:
Productive ecosystem

(b)

Damage to ecosystems resulting in loss of natural services

Pollution

More severe climate

Depletion of
groundwater

Severe runoff
and erosion

Overall effect: Loss of
ecosystem productivity

Figure 24.17 Services Provided by Ecosystems

(*a*) Ecosystems provide humans with many essential services for free. These same services are also essential to maintaining the productivity of ecological communities. Only a few of the many ecological services provided by ecosystems are diagrammed here. (*b*) These services may be lost when ecosystems are damaged by human actions, with potentially devastating consequences for natural communities as well as human well-being.

Like the forests that filter pollutants from New York City's water, floodplains provide us with a free service: they act as safety valves for major floods, preventing even larger floods. Ecological communities provide us with many such **ecosystem services** (Figure 24.17*a*), including removal of pollutants from the air by plants, pollination of plants by insects (essential for many crops), maintenance of breeding grounds for shellfish and fish in marine ecosystems, prevention of soil erosion by plants, screening of dangerous ultraviolet light by the atmospheric ozone layer, moderation of the climate by the ocean, and, as we have seen throughout this chapter, nutrient cycling. Such ecosystem services are essential for maintaining healthy ecological communities, and they also provide people with enormous economic benefits.

What is the total economic value of ecosystem services? In one sense, their value is infinite: we cannot live without the clean air and water they provide. But even in a more restricted sense, the monetary value of ecosystem services is enormous, as illustrated by the billions of dollars that New York City would need to spend to replace just one of those services: water purification. Globally, the value of ecosystem services provided by lakes and rivers alone has been estimated by some researchers to be a staggering $1.7 trillion each year. Other examples of the value of ecosystem services include the world's fish catch (valued at between $50 billion and $100 billion per year) and the billions of dollars' worth of crops that could not be produced if plants were not pollinated by insects.

Human civilization depends on many essential services provided by ecosystems. At present, though we can design landscapes that incorporate aspects of natural ecosystems, we cannot duplicate with technology what ecosystems provide us for free. Although we usually have only a rough idea of the economic value of particular ecosystem services, we do know that when we destroy or degrade ecosystems (see Figure 24.17*b*), we place ourselves—and our wallets—at peril.

Hitting the Squids: Deep-Sea Squid and Octopi Full of Human-Made Chemicals

Human-made chemicals have snuck on down into the ocean depths, showing up in the tissues of deep-sea cephalopods, says new research. In a study to be published in the journal *Marine Pollution Bulletin*, researchers found various persistent organic pollutants—including PCBs and DDT—in nine species of octopi, squid, cuttlefish, and nautiluses. "The fact that we detected a variety of pollutants in specimens collected from more than 3,000 feet deep is evidence that human-produced chemicals are reaching remote areas of the open ocean, accumulating in prey species, and therefore available to higher levels of marine life," says study coauthor Michael Vecchione. "Contamination of the deep-sea food web is happening, and it is a real concern."

More than 80 percent of the streams and lakes in urban and agricultural areas of the United States have detectable levels of human-made chemicals. Contaminants include over-the-counter painkillers such as ibuprofen, hormones from birth control pills, triclosan (the active ingredient in many antibacterial soaps) and DEET (the active ingredient in the most widely used insect repellents), and a variety of carbon-based chemicals known as persistent organic pollutants (POPs). In most lakes and rivers, these chemicals occur at relatively low concentrations, less than 1 part per billion in most cases. Some of these pollutants, however, have been shown to harm aquatic life even at low concentrations. Researchers are also investigating the possibility that certain combinations of pollutants may be more toxic to organisms than is any single pollutant on its own.

Of the approximately 30,000 chemicals in commercial use in the United States, about 400 are classified as POPs. Because they are exceptionally stable, POPs stay for a long time once they enter the environment. A breakdown product of the pesticide DDT can be detected in the body tissues of most Americans, even though its use was prohibited in the United States in 1972. Polychlorinated biphenyls (PCBs), found in sealants, paints, and other industrial products, are another class of POPs that are widespread in our environment, despite being banned in the 1970s.

As the news article indicates, many pollutants are transported widely throughout the globe—and can be detected in the tissues of animals inhabiting ecosystems we tend to think of as pristine. The researchers studied nine species of cephalopods, an invertebrate group that includes octopus, squid, cuttlefish, and nautilus. They found a wide variety of POPs, some at relatively high levels, in each species. Cephalopods live on a wide variety of primary and secondary consumers, including crustaceans (such as the tiny, shrimplike krill), mollusks (such as clams), and fishes (such as anchovies). POPs have been detected in cephalopod prey such as krill, which are filter feeders that ingest particles of organic debris, as well as phytoplankton.

As secondary consumers, cephalopods consume a large biomass of krill and other herbivorous species, so contaminants present in the prey build up in the tissues of these predators. Cephalopods themselves are food for many marine mammals, including dolphins, killer whales, and sperm whales. Very high concentrations of POPs (about 80 parts per million) have been found in the blubber of whales and seals. The pollutants have also been reported in arctic peoples who hunt whales and seals, thereby acting as quaternary (fourth-level) consumers. According to a Canadian study, Inuit have five times the concentration of "legacy contaminants," such as DDT and PCBs, than other Canadians have.

Evaluating the News

1. The detection of POPs in humans and other organisms suggests that we cannot release chemicals into the environment and expect them to go away. From what you've learned about the transfer of nutrients in ecosystems, explain why there is no "away." Describe how the chemicals that wash off our pesticide-sprayed lawns can wind up on our dinner plates.

2. The hundreds of POPs in wide use are the basis of a multibillion-dollar industry. The health and environmental risks associated with most of these POPs are currently unknown. Should these chemicals be banned until they are proven harmless? Or should we decide what to do only after more information is available about whether these chemicals really are a threat? Defend your answer.

3. Hormones from oral contraceptives have been shown to disrupt reproduction in some fish species, but it is unclear if the "hormone pollution" harms people. Sewage facilities could filter such hormones from wastewater, but the cost of such technology is high and, in the view of many, not worth it. What is your view? Who should bear the cost of the technology, if it is implemented? Would a tax on oral contraceptives be justified? What about other chemicals, such as the POP-laced products that nearly all of us have used at some point in our lives?

SOURCE: *Grist Magazine,* Seattle, WA, June 13, 2008

Chapter Review

Summary

24.1 How Ecosystems Function: An Overview

- Energy and materials can move from one ecosystem to another.
- A portion of the energy captured by producers is lost as metabolic heat at each step in a food chain. Therefore, energy moves through ecosystems in just one direction.
- Nutrients are recycled in ecosystems. They pass from the environment to producers to various consumers, then back to the environment when the ultimate consumers—decomposers—break down the bodies of dead organisms.

24.2 Energy Capture in Ecosystems

- Most life on Earth depends on energy that is captured from sunlight by producers and stored in their bodies as chemical compounds.
- Energy capture in an ecosystem is measured as net primary productivity (NPP); assessing the amount of NPP in an area is an important first step in determining the type of ecosystem found there and how it functions.
- On land, NPP tends to decrease from the equator toward the poles. In marine ecosystems, NPP tends to decrease from relatively high values where the ocean borders land to low values in the open ocean (except where upwelling provides scarce nutrients to marine organisms). Aquatic ecosystems on land (such as wetlands) can also show high NPP.
- Human activities can increase or decrease NPP on local, regional, and global scales.

24.3 Energy Flow through Ecosystems

- A portion of the energy captured by producers can be transferred from organism to organism up a food chain.
- Once an organism uses energy to fuel its metabolism, that energy is lost from the ecosystem as heat.
- The amounts of energy available to organisms at different trophic levels in an ecosystem can be represented in the form of a pyramid, with each successive level harvesting only 10 percent of the energy from the level below.
- Secondary productivity is highest in areas of high net primary productivity.

24.4 Nutrient Cycles

- Nutrients are cycled between organisms and the physical environment.
- Decomposers return nutrients from the bodies of dead organisms to the physical environment.
- Nutrients that enter the atmosphere easily (in gaseous form) have atmospheric cycles, which occur relatively rapidly and can transfer nutrients between distant parts of the world.
- Nutrients that do not enter the atmosphere easily (such as phosphorus) have sedimentary cycles, which usually take a long time to complete.
- Carbon is the most abundant nutrient in living organisms, if we disregard water. In most ecosystems, the biotic community gains carbon from the abiotic environment through photosynthesis carried out by producers such as bacteria, plants, and algae.

- Photosynthetic producers convert atmospheric CO_2 to carbohydrate. Consumers gain carbon when they feed on producers or other consumers. Producers and consumers release CO_2 into the air when they extract energy from carbohydrates through cellular respiration.
- Decomposers unleash much of the carbon locked inside the remains of dead organisms and release it as CO_2. Partially decayed organic matter, such as leaf litter and peat, is an important reservoir of soil carbon, especially in the colder regions of the world.
- Nitrogen is a key component of vital macromolecules such as proteins and DNA. The atmosphere, which is 78 percent N_2, is the largest store of nitrogen in the biosphere although N_2 is inaccessible to producers.
- In biological nitrogen fixation, N_2 is converted to NH_4^+ by bacteria. Some bacteria fix nitrogen in a mutualistic relationship with plants such as legumes.
- Producers such as plants can absorb NH_4^+ or NO_3^- from the nitrogen exchange pool. Consumers acquire nitrogen when they feed on producers or other consumers.
- In industrial nitrogen fixation, N_2 is converted to ammonium compounds in the manufacture of fertilizer.
- Sulfur displays an atmospheric cycle. It enters land and water through weathering of rocks. Sulfur-containing compounds produced by phytoplankton are lofted into the atmosphere in sea spray. Atmospheric SO_4^- mixes with water and rains to the ground as sulphuric acid (H_2SO_4).
- Phosphorus is a component of vital macromolecules such as DNA. Phosphorus inputs usually boost NPP in an ecosystem. Phosphorus has a sedimentary cycle because it does not generally enter the atmosphere.

24.5 Human Actions Can Alter Ecosystem Processes

- Human activities can alter nutrient cycles on local, regional, and global scales.
- The addition of extra nutrients (especially nitrogen and phosphorus) to lakes, streams, and coastal waters can lead to eutrophication. Nutrient enrichment triggers a population explosion among algae; when dead algae settle into deeper waters, they stimulate a population explosion among oxygen-using decomposers, which can kill shellfish and fish by depriving them of oxygen.
- Human inputs to the sulfur cycle exceed those from all natural sources combined, creating problems of international scope, such as acid rain.

24.6 Ecosystem Design

- People seek to design and restore ecosystems for aesthetic and economic reasons.
- Attempts to create self-sustaining ecosystems from scratch or restore damaged ecosystems have not fully succeeded. Humans do not yet know enough about the fundamental, complex processes of ecosystems to build ecosystems that function like those in the natural world.

Review and Application of Key Concepts

1. Some people think the current U.S. Endangered Species Act should be replaced with a law designed to protect ecosystems, not species. The intent of such a law would be to focus conservation efforts on what its advocates think really matters in nature: whole ecosystems. Given how ecosystems are defined, do you think it would be easy or hard to determine the boundaries of what should and should not be protected if such a law were enacted? Give reasons for your answer.

2. What prevents energy from being recycled in ecosystems?

3. What essential role do decomposers play in ecosystems?

4. Explain why human alteration of nutrient cycles can have international effects.

5. Describe some key ecosystem services and discuss the extent to which human economic activity depends on such services.

Key Terms

abiotic (p. 480)
acid rain (p. 491)
atmospheric cycle (p. 486)
biomass (p. 483)
biotic (p. 480)
carbon cycle (p. 486)
ecosystem (p. 480)
ecosystem services (p. 494)
eutrophication (p. 490)
exchange pool (p. 485)

net primary productivity (NPP) (p. 483)
nitrogen fixation (p. 487)
nutrient (p. 481)
nutrient cycle (p. 485)
secondary productivity (p. 485)
sedimentary cycle (p. 486)
trophic level (p. 485)
upwelling (p. 483)

Self-Quiz

1. The amount of energy captured by photosynthesis, minus the amount lost as metabolic heat, is
   a. secondary productivity.
   b. consumer efficiency.
   c. net primary productivity.
   d. photosynthetic efficiency.

2. The movement of nutrients between organisms and the physical environment is called
   a. nutrient cycling.
   b. ecosystem services.
   c. net primary productivity.
   d. a nutrient pyramid.

3. Free services provided to humans by ecosystems include
   a. prevention of severe floods.
   b. prevention of soil erosion.
   c. filtering of pollutants from water and air.
   d. all of the above

4. Each step in a food chain is called
   a. a trophic level.
   b. an exchange pool.
   c. a food web.
   d. a producer.

5. What type of organisms consume 50 percent or more of the NPP in all ecosystems?
   a. herbivores
   b. decomposers
   c. producers
   d. predators

6. Sources of nutrients that are available to producers, such as soil, water, or air, are
   a. called essential nutrients.
   b. called exchange pools.
   c. considered eutrophic.
   d. called limiting nutrients.

7. Which of the following is the most representative term for an organism that gets its energy by eating all or parts of other organisms or their remains?
   a. fungus
   b. predator
   c. consumer
   d. producer

8. Nutrients that cycle between terrestrial and aquatic ecosystems and are then deposited on the ocean bottom
   a. have a short cycling time.
   b. have an atmospheric cycle.
   c. are more common than nutrients with a gaseous phase.
   d. have a sedimentary cycle.

CHAPTER 25 Global Change

Key Concepts

◉ The effects of human actions on the world's lands and waters are thought to be the main causes of the current high rate of species extinctions.

◉ Human activities have added natural and synthetic chemicals to the environment, and these additions in turn have altered how natural chemicals cycle through ecosystems.

◉ Human inputs to the global nitrogen cycle now exceed those of all natural sources combined. If unchecked, these changes to the nitrogen cycle are expected to have negative effects on many ecosystems.

◉ The concentration of carbon dioxide (CO_2) gas in the atmosphere is increasing at a dramatic rate, largely because of the burning of fossil fuels. Increased CO_2 levels are expected to have large but unpredictable effects on ecosystems.

◉ Increased concentrations of CO_2 and certain other gases in the atmosphere are predicted to cause a rise in temperatures on Earth. Most scientists think that such global warming is occurring, but its extent and consequences remain uncertain.

◉ Because global change caused by humans is expected to have large, negative consequences for many species, including our own, most ecologists think we must learn to use Earth's ecosystems in a sustainable fashion.

Devastation on the High Seas

Nearly 75 percent of Earth's surface is covered with oceans. The oceans are so deep and so vast that many scientists once thought people could never drive marine species extinct. No matter how much we overhunted a species or polluted local portions of its habitat, it was thought there would always be places where the species could thrive. Now it seems this assumption was wrong.

Consider the white abalone's tale of woe. This large marine shellfish once was common along 1,200 miles of the California coast. It lives on rocky reefs in relatively deep water (25 to 65 meters or deeper). The fact that it lives in deep water protected it for a while: white abalone is delicious to eat, but people first hunted other species of abalone that live in shallower waters and hence are easier to find. When the shallow-water species became rare, fishermen turned to the white abalone. After only 9 years of commercial fishing, the fishery collapsed. This species, which once covered the seafloor with up to 4,000 individuals per acre, is now on the verge of extinction.

In general, when people begin fishing in a new region, the number of fish they catch drops sharply—typically by 80 percent—within the first 15 years. In 1958, for example, having depleted fishing grounds in the western Pacific, the Japanese fleet still was able to catch many large predatory fishes in portions of the Indian Ocean, southern Pacific, and Atlantic. Six years later, the regions that yielded many fish in 1958 gave much lower returns, and some 20 years later many of them were no longer viable as fishing grounds.

Overall, people have had a large negative effect on fish populations worldwide. Recent studies indicate that 66 percent of the world's marine fisheries are in trouble because of overfishing. Furthermore, over the past 45 years the catch has included fewer large predatory fishes and more invertebrates and small fishes that feed on plankton. The reason is that large predatory fishes are preferred by fishermen worldwide. As a result, populations of these fishes plummeted. In addition to reducing populations of individual species, people have altered the food webs of ocean communities (by removing top predators).

The White Abalone and Its California Coastal Habitat

The near-extinctions of species and the other effects of human actions on marine ecosystems are just a few of the changes we are making to the biosphere. Other examples include changes in the global sulfur cycle (see Chapter 24), effects on the locations of biomes (see Chapter 20), and the extinctions or declines of many species worldwide (see Interlude A). **What are the consequences of such changes for life on Earth, including us? The sea and sky and land have changed in uncountable ways for billions of years, so why should we care that humans are now changing the biosphere?** We consider such large and crucial issues in this chapter. We will see that the message of ecology is that good stewardship of our planet is in our own self-interest.

Statements by politicians, talk show hosts, and others can give the impression that worldwide change in the environment, or **global change**, is a controversial topic. Such statements cause many in the general public to think that global change may not really be occurring, or cause them to wonder whether anything really needs to be done about it.

This impression of controversy is unfortunate because we know with certainty that global change is occurring. Invasions of non-native species have increased worldwide (see the examples in Chapters 21 and 23), large losses of biodiversity have occurred (see Interlude A), and pollution has altered ecosystems throughout the world (see Chapter 24). These are important types of global change that we know with certainty are happening today.

Although these three types of global change are caused by people, the biosphere has always changed over time. As we saw in Chapter 23, the continents move, the climate changes, and succession and natural forms of disturbance change the composition of communities. Even in the absence of human actions, we know that ecological communities face—and always have faced—global change.

In this chapter we describe how people have influenced global change. We first discuss two types of global change that we know have occurred and that we know are caused by people: changes in land and water use, and changes in the cycling of nutrients through ecosystems. We finish with a look at climate change caused by global warming and its possible consequences for our future.

25.1 Land and Water Transformation

People make many physical and biotic changes to the land surface of Earth, which collectively are referred to as **land transformation**. Such changes include the destruction of natural habitat to allow for resource use (as when a forest is clear-cut for lumber), agriculture, or urban growth. Land transformation also includes many human activities that alter natural habitat to a lesser degree, as when we graze cattle on grasslands.

Similarly, **water transformation** refers to physical and biotic changes that people make to the waters of our planet. For example, we have drastically altered the way water cycles through ecosystems. People now use more than half of the world's accessible fresh water, and we have altered the flow of nearly 70 percent of the world's rivers. Since water is essential to all life, our heavy use of the world's waters has many and far-reaching effects, including changing where water is found and altering which species can survive at a given location.

Many of our effects on the lands and waters of Earth are local in scale, as when we cut down a single forest or pollute the waters of one river. However, such local effects can add up to have a global impact.

There is ample evidence of land and water transformation

Aerial photos, satellite data, changing urban boundaries, and local instances of the destruction of natural habitats show how humanity is changing the face of Earth (Figure 25.1). Together, many such lines of evidence show that land and water transformation is occurring, is caused by human actions, and is global in scope.

To estimate the total amount of land that has been transformed by people, the effects of many different human activities must be added together for every acre of the world. This task may seem nearly impossible, but with the aid of satellites and other new technologies, we can now measure our total impact on Earth for the first time in history. Although researchers are just beginning this task, one reasonable estimate is that people have substantially altered one-third to one-half of Earth's land surface. The exact amount of land transformed by humans is not yet known, but we know that we have altered a large percentage of it. Though we cannot yet measure our total impact

(a)

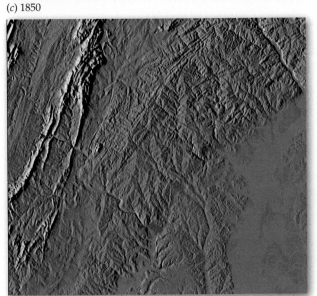

(b)

Figure 25.1 Examples of Land Transformation

(*a*) A clear-cut forest in Washington State. (*b*) An open-pit copper mine in Arizona. (*c,d*) Change in the boundaries (in red) of urban regions near Baltimore, Maryland, and Washington, DC, between 1850 (*c*) and 1992 (*d*).

(*c*) 1850

(*d*) 1992

See a city growing over time. ▶❚❚ 25.1

on the waters of Earth (in part because we know so little about the deep ocean), global problems with water pollution and observed declines in many aquatic populations make it clear that we have transformed Earth's waters.

In modifying land and water for our own use, we have had dramatic effects on many ecosystems. Examples of human effects on ecosystems include the ongoing destruction of tropical rainforests (Figure 25.2) and the conversion of once vast grasslands in the American Midwest to cropland. Half of the world's wetlands, from

Figure 25.2 Disappearing Forests

The graph, based on data from the United Nation's Food and Agriculture Organization (FAO), shows that forest cover has shrunk in most regions of the world except Europe. Asia's relatively good standing is due largely to extensive reforestation efforts in China over the past few years. The inset shows cattle grazing on land previously covered by Amazonian rainforest, in Para, Brazil.

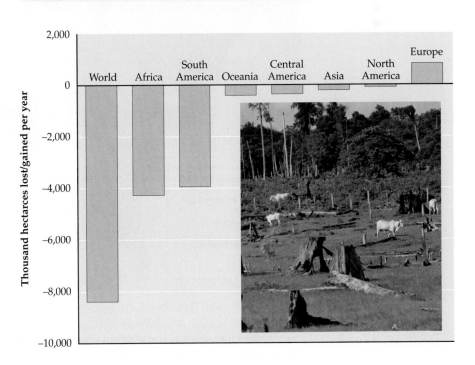

mangrove swamps to northern peat bogs, have been lost in the last 100 years. During the 200-year period beginning in the 1780s, wetlands declined in every state in the United States. Thanks to protective legislation, widespread public outreach, and initiatives that encourage landowners and public groups to conserve and restore wetlands, the loss of wetland was sharply curtailed in the last decade of the twentieth century.

About 50 percent of the world's population lives within 3 miles of the coast, making coastal ecosystems extremely vulnerable to human activity (Figure 25.3). Estuaries, saltwater marshes, mangrove swamps, and coastal shelf waters are among the most productive ecosystems on Earth. Most of the world's coastline is under siege today from urban development, sewage, nutrient runoff from farm fields, chemical pollution, and unsustainable harvesting of shellfish and fish.

Land and water transformation has important consequences

Many ecologists think that land transformation and water transformation are—and will remain for the immediate future—the two most important components of global change. There are several reasons for this assessment.

First, as we alter land and water to produce goods and services for an increasing number of people, we use a very large share of the world's resources. Estimates suggest that humans now control (directly and indirectly) roughly 30 to 35 percent of the world's total net primary productivity (NPP) on land (see Chapter 24 for a definition of NPP). By controlling such a large portion of the world's land area and resources, we have reduced the amount of land and resources available to other species, causing some to go extinct. Water transformation has similar effects. As we saw at the opening of this chapter, when people overfish or pollute Earth's waters, we may cause dramatic changes in the abundances and types of species found in the world's aquatic ecosystems (see also Chapters 23 and 24).

The transformation of land and water has other effects as well. One of these effects is change in local climate. For example, when a forest is cut down, the local temperature may increase and the humidity may decrease. Such climatic changes can make it less likely that the forest will regrow if the logging stops. In addition, as we'll see shortly, the cutting and burning of forests increases the amount of carbon dioxide in the atmosphere—an aspect of global change that may alter the climate worldwide.

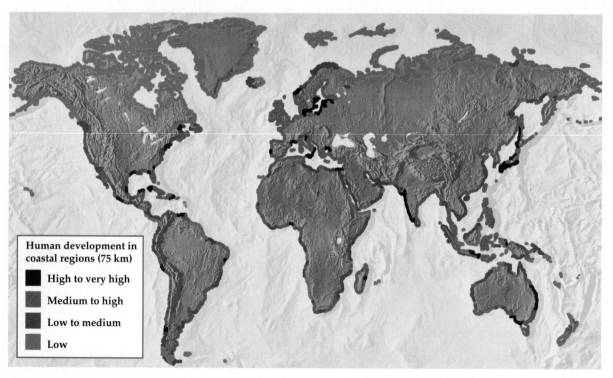

Human development in coastal regions (75 km)

■ High to very high
■ Medium to high
■ Low to medium
■ Low

Figure 25.3 Coastal Ecosystems Are Heavily Affected by Human Populations
This map shows the impact of humans on Earth's coastal shelf, which harbors some of the most productive ecosystems on the planet. The most extensively developed coastal regions show the largest impact, including the eastern coasts of the United States and China, much of Southeast Asia, and parts of Europe.

25.2 Changes in the Chemistry of Earth

Life on Earth depends on, and is heavily influenced by, the cycling of nutrients in ecosystems. Net primary productivity often depends on the amount of nitrogen and phosphorus available to producers, and the amount of sulfuric acid in rainfall has many effects on ecological communities, as we saw in Chapter 24. The nitrogen and phosphorus that stimulate NPP and the sulfur in acid rain are just two of many examples of naturally occurring chemicals that cycle through ecosystems.

Synthetic chemicals (artificial chemicals made by humans) and other substances that humans release into the air, water, and soil, may also cycle through ecosystems. Like nutrients that organisms gain from food or by direct absorption from their environment, chemicals released by humans can bioaccumulate in an individual. **Bioaccumulation** of a substance refers to the deposition of the substance within an organism at concentrations higher than in the surrounding abiotic environment. Substances that bioaccumulate tend to be stable chemicals that are deposited within cells and tissues faster than they can be eliminated (in urine, for instance).

Many synthetic chemicals—especially organic molecules found in pesticides, plastics, paints, solvents, and other industrial products—tend to bioaccumulate in cells and tissues. Long-lived organic molecules of synthetic origin that bioaccumulate in organisms, and that can have harmful effects, are broadly classified as **persistent organic pollutants (POPs)**. Some of the most damaging POPs that are widespread in our biosphere include different types of PCBs (polychlorinated biphenyls) and dioxins. Because many of these pollutants have an atmospheric cycle (see Chapter 24), they can be transported over vast distances across the globe to contaminate food chains in remote places where the chemicals have never been used (see the "Biology in the News" feature in Chapter 24).

Heavy metals such as mercury, cadmium, and lead can also bioaccumulate in a wide variety of organisms, both producers and consumers. Mercury is a toxic chemical that occurs naturally in small amounts. However, the concentration of this heavy metal in air, water, and soil has increased threefold since the industrial revolution, mainly from the burning of coal and the incineration of mercury-containing waste. Mercury enters the food chain when bacteria absorb it from soil or water and convert it to an organic form known as methylmercury. Methylmercury is much more toxic than inorganic forms of mercury, in part because the organic form bioaccumulates more readily, being stored in muscle tissues of shellfish, fish, and humans. Methylmercury bioaccumulated by bacteria is passed on to consumers, such as zooplankton (microscopic aquatic animals), that feed on mercury-accumulating bacteria. In this way, the methylmercury is progressively transferred to other consumers throughout the food web (see Chapter 24).

The increase in the tissue concentrations of a bioaccumulated chemical at successively higher trophic levels in a food chain is known as **biomagnification**. Bioaccumulation and biomagnification might seem similar at first glance, but whereas bioaccumulation is the accumulation of a substance in individual organisms within a trophic level, biomagnification is the increase in tissue concentrations of a chemical as organic matter is passed from one trophic level to the next in a food chain.

Chemicals that show biomagnification are resistant to degradation in the body or in the environment and are not easily excreted by animals, usually because they bind to macromolecules such as proteins or fats. PCBs, for example, are hydrophobic molecules that combine with fat and become locked within fatty tissues. Predators in the next trophic level acquire the chemical when they eat the fatty tissues of the prey species. Because predators consume large quantities of prey, and lose little if any of the chemical, its concentration builds up in their tissues over time. That is why top predators—those that feed at the end of a food chain—usually have the highest tissue concentration of biomagnified chemicals. Figure 25.4 illustrates the 25 million–fold biomagnification of PCBs that has been recorded in some northern lakes. An important aspect of biomagnification is that pollutants that may be present in

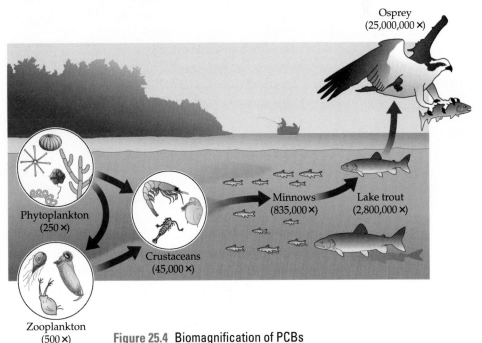

Figure 25.4 Biomagnification of PCBs

Science Toolkit

From Science to Public Policy and Back: Repairing the Ozone Layer

Human societies often have to make tough choices regarding scientific issues such as genetic engineering, gene therapy, cloning, causes of cancer, effects of pollutants on public health, and the environmental impacts of human actions. The background science can be complex, so scientists must provide input to the policy makers who evaluate possible courses of action and their consequences. In any particular case, policy makers must consider questions such as, Should science be allowed to progress unhindered, regardless of the consequences? How can we prevent a new technology from having unintended and undesirable consequences? Who should pay to fix problems caused by existing technologies or policies?

We can illustrate how science influences public policy (and vice versa) with what is shaping up to be a major environmental success story: efforts to repair the damage that humans caused to the ozone layer. In 1974, Mario Molina and F. Sherwood Rowland wrote an influential scientific article suggesting that chlorofluorocarbons in aerosol spray cans damaged the ozone layer; for this work, Molina and Rowland shared the 1995 Nobel Prize in Chemistry with Paul Crutzen, who performed similar research. Various scientists described the threats posed by CFCs to members of the public, and almost immediately, sales of aerosol products began to drop as people worried that ozone loss would cause skin cancer and other health problems. Many policy makers and members of the business community balked at taking action, citing high projected costs and arguing that the science was too weak to support restrictions on ozone production. By 1978, however, CFC-containing aerosols were banned in the United States, and other countries soon followed suit.

After 2 years of tough negotiations, the Montreal Protocol, an international treaty designed to halt the production and use of CFCs and other chemicals that harm the ozone layer, was signed in 1987. New scientific evidence indicates that the treaty is working. CFC emissions measured at Earth's surface have dropped, and satellite data show that the atmosphere is now losing ozone less rapidly than between 1979 and 1996. Although we are not gaining ozone yet, results like these provide an early sign that the ozone layer has begun to recover. If current trends continue, scientists expect the ozone layer to recover completely in about 50 years.

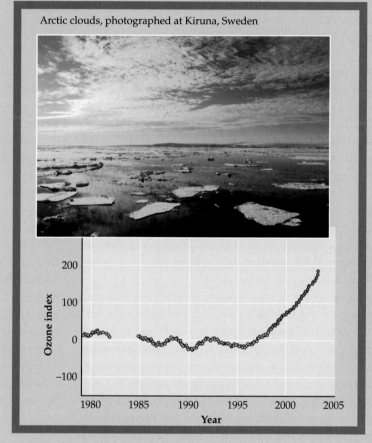

Arctic clouds, photographed at Kiruna, Sweden

The Ozone Layer Begins to Recover

In this arctic sky and elsewhere in the world, the destruction of the ozone layer has begun to slow down. The "ozone index" in the graph would have remained close to zero if the rate at which ozone was destroyed had not changed from the rate observed between 1979 and 1996. Instead, by 1999, as shown by the points drawn in red, the index was significantly greater than zero, indicating that the rate of ozone destruction was decreasing. The air samples used to calculate the index were taken at latitudes 30° to 50° north, at an altitude of 35 to 45 kilometers.

minuscule amounts in the abiotic environment, such as the water in a lake, can build up to damaging, even lethal, concentrations in the top predators in the food chain.

The pesticide DDT, whose effects on bald eagles were described in the "Science Toolkit" box in Chapter 21, is an example of a POP that is bioaccumulated and bio- magnified along a food chain. Until its use was banned in 1972, DDT was extensively sprayed in the United States to control mosquitoes and protect crops from insect pests. The pesticide ended up in lakes and streams, where it was taken up by plankton, such as algae, which were in turn ingested by zooplankton. As the pesticide moved up the food chain, from zooplankton to shellfish to birds of prey such as ospreys and bald eagles, its tissue concentrations increased by hundreds of thousands of times. DDT disrupts reproduction in a variety of animals, but predatory birds were especially hard-hit. The chemical interferes with calcium deposition in the developing egg, resulting in thin, fragile eggshells that break easily.

DDT is an example of an **endocrine disrupter**, a chemical that interferes with hormone function to produce negative effects, such as reduced fertility, developmental abnormalities, immune system dysfunction, and increased risk of cancer. Bisphenol A (found in plastics that many water bottles are made of) and phthalates (found in everything from soft toys to cosmetics) are examples of endocrine disrupters that can be readily detected in the tissues of most Americans. In laboratory animals, bisphenol increases the risk of diabetes, obesity, reproductive problems, and various cancers; phthalate exposure is associated with lowered sperm counts and defects in the development of the male reproductive system. There are many reports of endocrine disrupters reducing fertility and causing developmental abnormalities in wild populations, especially among amphibians and reptiles. What is less clear at present is whether endocrine disrupters are harming human health to a significant degree. On the other hand, there is no assurance that long-term exposure to multiple endocrine disrupters, even at low doses of each, is necessarily safe for us.

Some of the POPs that we have poured into our environment have toxic effects on organisms, possibly including humans. Other POPs damage ecosystem processes on such a broad scale that there is little doubt about their negative impact on humans, and on a whole host of organisms from producers to top predators. The addition of **chlorofluorocarbons** [*KLOR*-oh-*FLOR*-oh . . .] **(CFCs)** to the atmosphere is one of the most wide-ranging changes that humans have made to the chemistry of Earth. CFCs have caused a decrease in the thickness of the atmospheric ozone layer across the globe, and contributed to the ozone hole above Antarctica (see Chapter 20). Because the ozone layer shields the planet from harmful ultraviolet light (which can cause mutations in DNA), damage to the ozone layer poses a serious threat to all life. Fortunately, the international community responded quickly to this threat, and the ozone layer has recently begun to show signs of a recovery (see the "Science Toolkit" box). Clearly, in some cases we have succeeded in slowing down or undoing the harm caused by chemical pollution or the alteration of nutrient cycles (the mitigation of acid rain, which we discussed in Chapter 21, is another

example). However, in other cases, such as those of the global nitrogen cycle and the global carbon cycle, great challenges lie ahead.

Concept Check

1. Describe some of the causes and consequences of the degradation of coastal ecosystems.

2. Compare bioaccumulation and biomagnification. What are some distinctive characteristics of chemicals that tend to bioaccumulate?

25.3 Changes in Global Nutrient Cycles

There is a large amount of nitrogen in Earth's atmosphere, where N_2 gas makes up 78 percent of the air we breathe. However, organisms cannot use N_2 directly to build biologically important molecules such as proteins and nucleic acids. Instead, the nitrogen in N_2 gas must be converted to other forms, such as ammonium (NH_4^+) or nitrate (NO_3^-), before it can be used by any eukaryote (plant, animal, protist, or fungus). The conversion of N_2 to NH_4^+, called **nitrogen fixation**, is accomplished by several species of bacteria and, to a much lesser degree, by lightning (see Chapter 24). The amount of nitrogen that cycles among organisms is much smaller than the amount found in gaseous form in the atmosphere.

Human technology is also capable of fixing nitrogen. In recent years, the amount of nitrogen fixed by human activities has exceeded the amount fixed by all natural processes combined (Figure 25.5). Much of this nitrogen fixation by humans is the result of the industrial

Learn more about the nitrogen cycle. ▶❚❚ 25.2

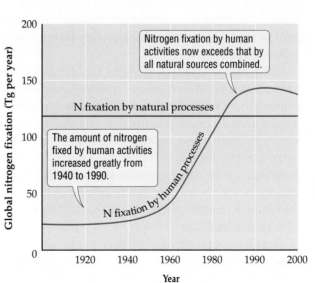

Figure 25.5 Human Impact on the Global Nitrogen Cycle Nitrogen is fixed naturally by bacteria and by lightning at a rate of about 130 teragrams (Tg) per year (1 Tg = 1,012 grams, or 1.1 million tons). Human activities such as the production of fertilizers now fix more nitrogen than do all natural sources combined.

(a)

(b)

Figure 25.6 A Nitrogen Addition Experiment

Native grasslands in Minnesota often have 20 to 30 plant species per square meter. (*a*) No nitrogen was added to this control plot, and it lost no species between 1984 and 1994. (*b*) Researchers added nitrogen to this nearby experimental plot during the same time period. Most of the native species disappeared from this plot, and an introduced species, European quack grass, became dominant.

production of fertilizers. A significant portion of the fertilizer that is applied on farm fields is broken down by bacterial action and released. Other major sources of nitrogen fixation include car engines, in which heat from combustion converts some of the N_2 found in the air to nitrogen monoxide (NO) and nitrogen dioxide (NO_2). These gases enter the atmosphere as engine exhaust, combine with oxygen and water in the air, and then fall to the ground as nitrate (NO_3^-) dissolved in rain. Biological nitrogen fixation by bacteria, including those that have a mutualistic relationship with certain plants, accounts for most of the nitrogen available to producers in natural ecosystems. The fact that human inputs of nitrogen to ecosystems exceed natural inputs tells us that our activities have greatly changed the global nitrogen cycle.

The potential effects of changing the nitrogen cycle are far-reaching. When nitrogen is added to terrestrial communities, NPP usually increases, but the number of species often decreases (Figure 25.6). The reason for the loss in diversity is that the species best able to use the extra nitrogen outcompete other species. For example, the addition of nitrogen to grasslands in the Netherlands that historically were poor in nitrogen resulted in the loss of more than 50 percent of the species from some of those communities.

Similarly, when nitrogen is added to nitrogen-poor aquatic ecosystems, such as many ocean communi-

ties, productivity increases but species are lost (see Chapter 23). In general, an increase in productivity caused by the addition of nitrogen is not necessarily a good thing for the ecosystem.

Vast amounts of carbon are found throughout the biosphere, and this carbon cycles readily among organisms, soils, the atmosphere, and the ocean (see Figure 24.8). Next we focus our discussion on one portion of the global carbon cycle that has been altered by human activities: the concentration of carbon dioxide gas (CO_2) in the atmosphere.

Atmospheric carbon dioxide levels have risen dramatically

Although CO_2 makes up less than 0.04 percent of Earth's atmosphere, it is far more important than its low concentration might suggest. As we saw in earlier chapters, CO_2 is an essential raw material for photosynthesis, on which most life depends. CO_2 is also the most important of the atmospheric gases that contribute to global warm-

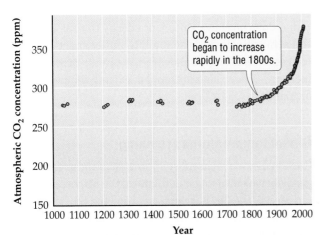

Figure 25.7 Atmospheric CO₂ Levels Are Rising Rapidly
Atmospheric CO_2 levels (measured in parts per million, or ppm) have increased greatly in the past 200 years. The red circles show results from direct measurements of the concentration of CO_2 in the atmosphere. The green circles indicate CO_2 levels measured from bubbles of air trapped in ice that formed many hundreds of years ago.

ing. Therefore, scientists took notice in the early 1960s when new measurements showed that the concentration of CO_2 in the atmosphere was rising rapidly.

Scientists have directly measured the concentration of CO_2 in the atmosphere since 1958. By measuring CO_2 concentrations in air bubbles trapped in ice for hundreds to hundreds of thousands of years, scientists have also estimated the concentration of CO_2 in both the recent and relatively distant past (Figure 25.7). For ice formed recently, direct measurements of CO_2 in the air match estimates from ice bubbles, giving us confidence that the ice bubble measurements for the past are accurate. Both types of measurements show that CO_2 levels have risen greatly during the past 200 years. Overall, of the current yearly increase in atmospheric CO_2 levels, about 75 percent is due to the burning of fossil fuels. Logging and burning of forests are responsible for most of the remaining 25 percent.

The recent increase in CO_2 levels is striking for two reasons. First, the increase happened quickly: the concentration of CO_2 increased from 280 parts per million (ppm) to 380 ppm in roughly 200 years. Measurements from ice bubbles show that this rate of increase is greater than even the most sudden increase that occurred naturally during the past 420,000 years. Second, although the concentration of CO_2 in the atmosphere has ranged from about 200 to 300 ppm during the past 420,000 years, CO_2 levels are now higher than those estimated for any time during this period. Global CO_2 levels have changed very rapidly in recent years and have reached concentrations unmatched in the last 420,000 years.

Increased carbon dioxide concentrations have many biological effects

Learn more about the carbon cycle. ▶ll 25.3

An increase in the concentration of CO_2 in the air can have large effects on plants (Figure 25.8). Many plants increase their rate of photosynthesis and use water more efficiently, and therefore grow more rapidly, when more CO_2 is available. When CO_2 levels remain high, some plant species keep growing at higher rates, but others drop their growth rates over time. As CO_2 concentrations in the atmosphere rise, species that maintain rapid growth at high CO_2 levels might outcompete other species in their current ecological communities or invade new communities.

Differences in how individual species respond to higher CO_2 levels may cause changes to entire communities. However, it is difficult (at best) to predict exactly how communities will change under higher CO_2 levels. Increased CO_2 levels in the atmosphere have contributed to warming Earth's climate, as we shall see in the following section. As both temperatures and CO_2 levels change, many different competitive and exploitative interactions may also change, but usually in ways that will not be known in advance. As we learned in Chapter 22, when interactions among species change, entire communities can change dramatically.

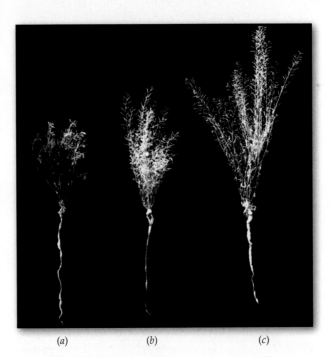

(a)   (b)   (c)

Figure 25.8 High CO₂ Levels Can Increase Plant Size
These three *Arabidopsis thaliana* plants all have the same genotype but were grown under different CO_2 concentrations: (a) 200 ppm, a level similar to that found roughly 20,000 years ago; (b) 350 ppm, the level found in 1988; and (c) 700 ppm, a predicted future level. Notice that plants grew larger at higher CO_2 concentrations.

25.4 Climate Change

Some gases in Earth's atmosphere, such as carbon dioxide (CO_2), water vapor (H_2O), methane (CH_4), and nitrous oxide (N_2O), absorb heat that radiates from Earth's surface to space. These gases are called **greenhouse gases** because they function much as the walls of a greenhouse or the windows of a car do: they let in sunlight but trap heat. Figure 25.9 illustrates how these gases contribute to the **greenhouse effect** that warms the surface of the Earth. About one-third of the solar radiation received by Earth is bounced back by the upper layers of the atmosphere. The rest is absorbed by the land and oceans, and to a lesser degree by the air. The warmed Earth emits some of its heat as long wavelengths of energy, known as infrared radiation. Some infrared radiation escapes Earth's atmosphere, but a good deal is absorbed by the greenhouse gases. Much of the heat absorbed by greenhouse gases is effectively trapped on Earth because when it is reemit-

ted, it does not have sufficient energy to pass through the atmosphere and escape into outer space. As the concentration of greenhouse gases in the atmosphere increases, more heat is trapped, raising temperatures on Earth.

Global temperatures are rising

Carbon dioxide is the most important of the greenhouse gases because so much of it enters the atmosphere. As far back as the 1960s, scientists predicted that the ongoing increases in atmospheric CO_2 concentrations would cause temperatures on Earth to rise. This aspect of global change, known as **global warming**, has provoked controversy in both the media and the political arena. Although year-to-year variation in the weather can make it hard to persuade everyone that the climate really is getting warmer, the overall trend in the data has convinced the great majority of the world's climatologists and other Earth scientists (Figure 25.10). A 2007 report

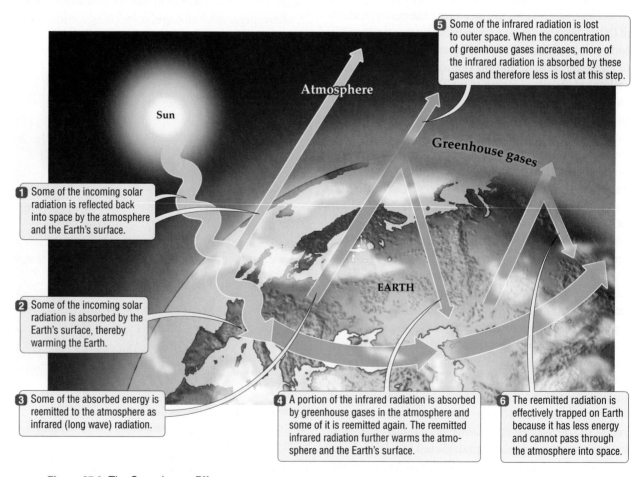

Figure 25.9 The Greenhouse Effect

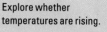

Explore whether temperatures are rising.

25.4

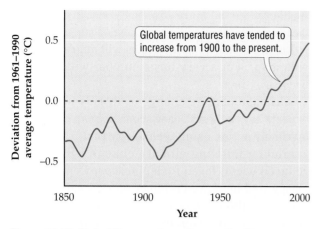

Figure 25.10 Global Temperatures Are on the Rise
Global air temperatures are plotted relative to the average temperature between 1961 and 1990 (dashed line). Portions of the curve below and above the dashed line represent lower-than-average and higher-than-average temperatures, respectively.

from the United Nations–sponsored Intergovernmental Panel on Climate Change (IPCC) concluded that global surface temperatures rose by an average of 0.74°C between 1906 and 2005, with land warming more than the oceans, and higher rates of warming in the northern latitudes compared to the more tropical and equatorial regions of the planet. The IPCC also concluded that the increase in global temperatures since the mid-twentieth century is very likely a result of human-caused increases in the concentration of CO_2 and other greenhouse gases in the atmosphere.

The IPCC's conclusion is based on statistical analyses showing that the recent rise in global temperatures represents a significant trend, not just ordinary variation in the weather. Global warming and changes in our biosphere that are caused by global warming are broadly known as **climate change**. Consistent with the warming trend, satellite images show that arctic sea ice has been declining by 2.7 percent per decade since 1978. Sea levels rose by an average of 1.8 millimeters per year between 1961 and 1993, and they have been rising by an average of 3.1 millimeters per year since then. Thermal expansion—the increase in volume as water warms up—has contributed to sea level rise, and the melting of glaciers and polar ice has added to it. Rainfall patterns have changed in some regions of the world—there is more rain in the eastern United States and northern Europe, less in parts of the Mediterranean, northeastern and southern Africa, and parts of South Asia. Studies show that recent temperature increases have also changed the biotic (living) component of ecosystems. For example, as temperatures increased in Europe during the twentieth century, dozens of bird and butterfly

species shifted their geographic ranges to the north. Similarly, plants in northern latitudes have increased the length of their growing season as temperatures have warmed since 1980.

Studies published since 1995 indicate that the warming since 1950 has been caused largely by human activities. For example, computer simulations performed on data for the second half of the twentieth century were able to predict the observed 0.1°C rise in temperature of the top 2,000-meter layer of the world's oceans only when human activities (such as greenhouse gas emissions) were included in the computer's calculations. Overall, an increasing amount of scientific evidence suggests that global warming is already happening, that it is affecting ecological communities, and that it is caused in large part by human activities.

What will the future bring?

Because there is no end in sight to the rise in CO_2 levels, the current trend of increasing global temperatures seems likely to continue. How will increased temperatures affect life on Earth? Not surprisingly, the effects will depend on how much, and how fast, global warming occurs.

Computer models predict that by the end of the twenty-first century, average temperatures on Earth will have risen anywhere from 1.1°C to 6.4°C (2°F–11.5°F) above the average global temperatures that prevailed between 1980 and 1990. The projections are based on a "business-as-usual" scenario, with no checks on the current trends in greenhouse gas emissions. The broad range reflects best estimates based on differing assumptions about some aspects of climate change that scientists are still uncertain about. The most optimistic climate models project a *minimum* increase of 1.1°C by the end of the century. According to the most pessimistic models, the increase *could* be as high as 6.4°C, although a 4°C (7.2°F) increase is more probable.

What are the implications of such temperature increases for ecosystems and for human well-being? Even an optimistic 1.8°C (3.2°F) increase in surface temperatures is likely to raise sea levels by as much as 0.38 meter (about 1.2 feet) and reduce ocean pH by at least 0.14 pH unit. Summer sea ice is likely to disappear entirely by the end of the century. Extreme weather, including hurricanes, floods, and severe drought, is expected to become more common. Many species are expected to become extinct. Agricultural productivity is expected to increase in the northern latitudes but decrease in most other parts of the world.

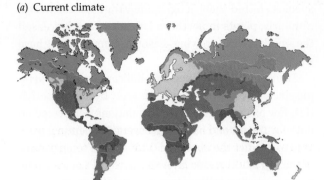

(a) Current climate

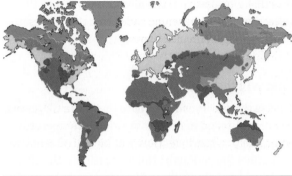

(b) Future climate (4°C increase in temperature)

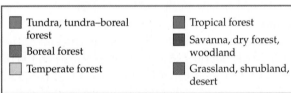

- Tundra, tundra–boreal forest
- Boreal forest
- Temperate forest
- Tropical forest
- Savanna, dry forest, woodland
- Grassland, shrubland, desert

Figure 25.11 Biomes on the Move
If Earth's temperatures warm by an average of 4°C, the distribution of forests, grasslands, deserts, and other biomes could be altered by global climate change.

A global temperature increase of 4°C or so will intensify the severity of the effects described above. Sea levels, for instance, are likely to rise as much as 0.59 meter (almost 2 feet). That may not sound like much, but combined with storm surges, it is enough to wipe out some island nations and destroy many coastal communities across the globe. The world's agricultural systems will be severely strained, and it is unlikely that there will be enough food to nourish the extra 4 billion to 5 billion who are expected to join the human population between now and the end of this century.

As illustrated in Figure 25.11, a 4°C rise in global surface temperatures will wreak large-scale alterations in Earth's biomes. Some species will migrate, others will adapt, but a very large number will probably become extinct.

Although climate change is already under way, experts say the worst-case scenarios can be averted by timely action using technology that is currently available. Battling climate change will require reduced use of fossil fuels, increased energy efficiency, and increased reliance on renewable energy such as cellulose-based ethanol and solar power. Innovative carbon capture methods have been developed, and more are under development, to reduce atmospheric CO_2 levels. In one such strategy, carbon dioxide from factories and power plants is turned into oil by algae, and the oil is converted to biodiesel. Improvements in waste management—reducing the release of greenhouse gases by landfills, for example—will be necessary. Agricultural practices will have to change, placing greater emphasis on sustainability, improved manure management to minimize the release of methane, and fertilizer application techniques that reduce the emission of N_2O (nitrous oxide, a greenhouse gas that is released in the breakdown of synthetic fertilizers). Halting deforestation in tropical countries and increasing forest cover worldwide will be crucial.

Some of the renewable-energy technologies may need government support (such as tax credits) to enable them to compete with conventional energy sources. New regulations, such as higher fuel economy standards for vehicles and stringent energy efficiency codes for appliances, are needed; and these will require political will, resting on public support. Efforts to curb global warming will have social and economic costs, but any delay will most likely lead to even greater costs in the future.

Concept Check

1. How do human activities contribute nitrogen to the environment? How do human inputs of nitrogen compare with natural inputs?
2. Explain how increased CO_2 levels contribute to global warming.
3. Which predicted effects of global warming are already apparent?

Concept Check Answers

1. Humans contribute through fertilizer, which is manufactured by conversion of N_2 into NH_4^+ and NO_3^-; and through combustion, which converts N_2 into gases such as NO and NO_2. Human inputs substantially exceed natural inputs.
2. CO_2 is a greenhouse gas. Increase in atmospheric CO_2 levels increases global temperatures by intensifying the greenhouse effect, which is caused by the trapping of infrared radiation by greenhouse gases in Earth's atmosphere.
3. Shrinking of polar ice sheets and glaciers; sea level increases; change in rainfall patterns predicted by climate models; poleward migration of many warmweather species; earlier blooming of many northern plants.

Applying What We Learned

A Message of Ecology

The science of ecology has important and timely messages for humanity, such as the one we learned in Chapter 21: No population can continue to increase without limit. Although related, the message of this chapter is more complex. As one ecologist wrote, "We are changing the world more rapidly than we are understanding it." In a very real sense, the world is in our hands. What we do to change it will determine our future and the future of all other species on Earth.

As we saw in this chapter, human activities have had profound effects on life on Earth. Most scientists are convinced that people are causing global change at a rate and intensity unmatched by natural patterns of change. Depending on the actions we take, global change has the potential to have even greater effects in the future.

As scientists, we believe that the main message provided by our knowledge of how people have changed the planet is that we must reduce the rate at which humans alter Earth's ecosystems. Such a change in our behavior not only will be good for other species but is in our own self-interest. Our entire civilization depends on the many services that ecosystems provide to us at no cost. If we continue to ignore the effects of our actions on these natural systems, ultimately we will harm ourselves.

To reduce our effects on natural systems, we must limit the growth of the human population and, equally important, we must use Earth's resources more efficiently. Simply put, we must strive to have a sustainable impact on Earth—that is, our impact should be one that can continue indefinitely without causing serious environmental damage and without using up resources faster than they are replenished. For example, if we want the world's oceans to continue to provide their bounty, we must stop harvesting fish populations more rapidly than they can regenerate. Otherwise, their numbers will crash (Figure 25.12). In general, to have a sustainable impact, we must stop altering natural systems in ways and at speeds that lead to short-term gain but result in long-term damage.

To achieve the goal of having a sustainable impact on the planet, we must anticipate the effects of our actions before they have disastrous consequences. No other species is capable of such forethought. Will we use that capability? Will we be bold enough, creative enough, and intelligent enough to take responsibility for our impact on Earth? As the cases described throughout Unit 5 suggest, there is hope that the answers to these questions will be yes. For example, our response to the threat that CFCs posed to the ozone layer shows that people can face reality and solve challenging environmental problems. The conversion to a sustainable society is an even bigger challenge, but the first steps to meet that challenge have already been taken, as we shall see in Interlude E.

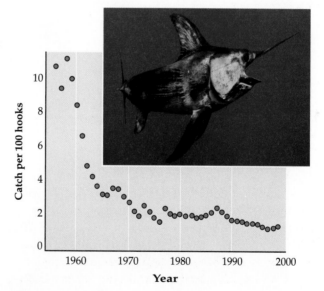

Figure 25.12 **Declining Fish Populations**
The catch of large predatory fishes (such as swordfish and tuna) declines rapidly once people begin to harvest an area intensively. This graph shows catches of these fishes in the tropical Atlantic; similar patterns were found in 14 other ocean regions.

Biology Matters

What's the Size of Your Footprint?

Each person on Earth has an ecological footprint—the acreage of productive land and water required to supply the resources we use and to dispose of the wastes we generate. At present, the average footprint of a person on Earth is 2.3 hectares (ha; 5.7 acres), which is greater than the amount of biologically productive land available per person on a sustainable basis (1.9 ha, or 4.7 acres). The figure of 2.3 ha per person is for an "average" person—the ecological footprint is much higher in some countries, such as the United States (9.7 ha, or 24 acres) and Canada (8.8 ha, or 21.7 acres), and much lower in others (for example, 0.5 ha, or 1.2 acres, in Bangladesh). As the world population grows, the amount of biologically productive land available per person will decline, increasing the speed at which the Earth's resources are consumed.

Although wealthier countries do tend to have higher average footprint sizes, there is considerable variation; for example, average European footprints are substantially lower than American and Canadian footprints. As countries such as China and India develop, their average footprints will probably grow. Currently, the size of a person's footprint is most closely correlated with the size of that person's residence and the amount of traveling one does, especially by car or airplane.

What is your ecological footprint? If you are a typical college student, your footprint is probably close to the U.S. average of 9.7 ha (24 acres). In other words, it would take about five planet Earths to support the human population if everyone on Earth enjoyed the same standard of living that you do.

Water Managers Told: Plan Now for Crisis

BY DAVID PERLMAN

California and Bay Area cities must start planning now for new and costly systems to control increasing runoff from urban storms, springtime floods from swollen rivers and rising sea levels as they invade lowlands, all as a result of global warming, climate scientists and water experts warn. Climate change, they say, will result in thinner winter snowpacks in the Sierra and other Western mountains. As snowpacks melt earlier each spring, the meltwater will increase river flows and raise new threats of floods. Even a small rise in sea levels could threaten cities and farmland in low-lying areas...

... "The challenge is daunting," said Paul C. Milly of the U.S. Geological Survey ... "Patterns of change are complex, uncertainties are large; and the knowledge base changes rapidly." Another report ... [from] researchers headed by Tim P. Barnett ... warns of "a coming crisis in water supply for the western U.S.," largely due to what the scientists call human-caused climate changes ... Those changes, according to Barnett's team, have been altering the West's river flows, water temperatures and snowpacks for 50 years, according to records the team has analyzed. In California, runoff from the Sierra is starting earlier, while droughts already threaten Arizona, the scientists say ... Barnett [maintains] that models of warming in the future mean "a coming crisis in water supply for the Western United States."

Burning fossil fuels adds CO_2 to the atmosphere, which increases global temperatures by trapping more of the sun's heat. If we take a "business as usual" approach to global warming (that is, if we postpone taking serious action to mitigate climate change and fail to invest heavily in alternative fuels and energy conservation), temperatures in California are likely to increase by 7°C to 10°C by the end of the century, and the state's economy will take a big hit. If fossil fuel use does not decrease, heat waves could occur six to eight times more often in Los Angeles, and heat-related deaths could increase by 500 to 700 percent. Air pollution worsens at warmer temperatures, and a 2008 Stanford University study estimates that each 1°C rise in temperature will lead to 1,000 additional deaths, mostly from respiratory illnesses such as asthma.

The predicted effects of a 7°C to 10°C increase include an 89 percent reduction in snow in the Sierra Nevada by 2050, with almost all the precipitation falling as rain instead of snow. Less snowfall in this mountain range means less snowmelt in the spring, which means less water for agriculture, hydroelectricity, and urban use. Climate models predict the annual runoff from snowmelt will take place earlier, perhaps peaking as early as February instead of May. Rapid release of snowmelt, combined with increased severity of winter storms, is likely to produce intense flooding in early spring or late winter. With the snowpacks diminishing from year to year, the runoff from snowmelt will decline sharply by mid-century, plunging California into an acute water crisis. By 2030, sea level is projected to rise by 6.5 inches in San Francisco Bay. Coastal flooding and storm surges will bring destruction to some famed seaside communities, such as Malibu, and to California's productive estuaries.

California has become a world leader in the battle against global warming, enacting measures such as the Global Warming Solutions Act, which aims to reduce the emissions of greenhouse gases to 1990 levels by the year 2020. But even if aggressive actions are taken now to reduce global warming, temperatures in California will still rise by 4°C to 6°C. California's $32 billion agriculture industry, which employs one out of every 10 Californians, is especially vulnerable. A big worry is the Napa and Sonoma wine harvest, which experts say could be hit hard by even a slight temperature rise. Other crops sensitive to such temperatures include tomatoes, onions, and most fruit. Because most crops in the state depend on irrigation, California agriculture would be hit hard by water scarcity, especially because the state is home to nearly 37 million people, whose demands for water compete with those of thirsty crops. State agencies and private groups are now developing plans to cope with the floods, increased fire risk, water shortage, reduced crop yields, and harm to human health that are the expected outcomes of global warming in even the best-case scenario.

Evaluating the News

1. Explain the relationship between a 7°C to 10°C rise in annual temperatures and water scarcity in a place such as California.
2. Would you be willing to pay a gasoline tax to help fund aggressive actions to reduce global warming? If so, how much tax per gallon would you be willing to pay—50 cents, one dollar, two dollars? If not, why not?

SOURCE: *San Francisco Chronicle*, February 1, 2008.

Chapter Review

Summary

25.1 Land and Water Transformation

- Many lines of evidence show that human activities are changing land and water worldwide.
- Land and water transformation has caused extinctions of species and has the potential to alter local and global climate.

25.2 Changes in the Chemistry of Earth

- Human activities are changing the way many chemicals, both natural and synthetic, are cycled through ecosystems.
- Bioaccumulation is the tendency of some chemicals to be deposited in the tissues of an organism at concentrations higher than those found in the surrounding environment. Methylmercury and PCBs are among the many persistent organic pollutants (POPs) that are bioaccumulated.
- Biomagnification is the increase in tissue concentrations of pollutants as biomass is transferred from one trophic level to another. Fishes, birds, and mammals that feed at the highest trophic levels in a food chain tend to accumulate the highest tissue concentrations of biomagnified chemicals.
- The release of chlorofluorocarbons (CFCs) into the atmosphere thinned the ozone layer over Earth, posing a serious threat to all life.

25.3 Changes in the Global Nutrient Cycles

- Nitrogen requires fixation—conversion from N_2 gas to ammonium (NH_4^+)— before it can be used by producers. In nature, most nitrogen fixation is performed by N_2-fixing bacteria.
- Human activities (fertilizer production, release of car exhaust) fix more nitrogen than all natural sources combined.
- The extra nitrogen fixed by human activities has altered the global nitrogen cycle, leading to increases in productivity that can cause losses of species from ecosystems.
- Concentrations of atmospheric CO_2 have increased greatly in the past 200 years and are higher now than in the past 420,000 years. These CO_2 increases are caused by the burning of fossil fuels and the destruction of forests.
- Increased CO_2 concentrations can alter the growth of plants in ways that will probably cause changes in many ecological communities.

25.4 Climate Change

- Carbon dioxide and other greenhouse gases in the atmosphere trap heat that radiates from Earth's surface. As the concentration of greenhouse gases increases, average temperatures on Earth are expected to rise.
- Human activities have contributed to the global warming that has occurred in the past 100 years.
- Some predicted effects of global warming are already being witnessed; these include the melting of polar ice sheets, sea level rise, changed rainfall patterns, and northward migration of some species.
- Although the amount of global warming that will occur in the twenty-first century is uncertain, if high-end predictions are correct, the social, economic, and environmental costs will be very large.

◉ Review and Application of Key Concepts

1. Summarize the major types of global change caused by human activities. What consequences do such types of global change have for species other than humans?

2. Compare examples of human-caused global change with examples of global change not caused by people. What is different or unusual about human-caused global change?

3. Producers such as algae and plants require nitrogen to grow, and nitrogen is often in limited supply in both aquatic and terrestrial ecosystems. People are adding considerable amounts of nitrogen to ecosystems. Is that a good thing? Explain why or why not.

4. How does the current atmospheric CO_2 concentration compare with concentrations over the past 420,000 years? How do scientists know what Earth's CO_2 concentrations were hundreds of thousands of years ago?

5. The future magnitude and effects of global warming remain uncertain. Do you think we should take action now to address global warming, despite those uncertainties? Or do you think we should wait until we are more certain what the ultimate effects of global warming will be? Support your answer with facts already known about global warming.

6. What changes to human societies would have to be made for people to have a sustainable impact on Earth?

7. How does your ecological footprint compare with that of the average American or Canadian? You may want to use one of the many ecological footprint "calculators" that are now available on the Internet to arrive at an estimate. On the basis of your self-evaluation, do you think your impact on Earth is sustainable? If not, what changes could you make so that your impact would be sustainable?

Key Terms

bioaccumulation (p. 505)
biomagnification (p. 505)
chlorofluorocarbons (CFCs) (p. 507)
climate change (p. 511)
endocrine disrupter (p. 507)
global change (p. 502)
global warming (p. 510)

greenhouse effect (p. 510)
greenhouse gas (p. 510)
land transformation (p. 502)
nitrogen fixation (p. 507)
persistent organic pollutants (POPs) (p. 505)
water transformation (p. 502)

Self-Quiz

1. Which of the following is most directly responsible for global warming?
   a. increased CO_2 concentration in the atmosphere
   b. melting sea ice
   c. persistent organic pollutants
   d. loss of species diversity

2. CO_2 absorbs some of the _____ that radiates from the surface of Earth to space.
   a. ozone
   b. infrared energy
   c. ultraviolet light
   d. smog

3. The conversion of N_2 gas to ammonia by bacteria is known as
   a. biological nitrogen fixation.
   b. fertilizer production.
   c. nitrogen cycling.
   d. denitrification.

4. The concentration of CO_2 in the atmosphere is now about 380 ppm, a level that is roughly _____ the levels of 200 years ago.
   a. the same as
   b. 300 percent higher than
   c. 30 percent higher than
   d. 30 percent lower than

5. Human activities can alter the natural cycling of nitrogen
   a. through biomagnifications of methylmercury.
   b. through pollution controls that reduce acid rain.
   c. by preventing fertilizer runoff from farm fields.
   d. by burning fossil fuels in automobile engines.

6. The release of chlorofluorocarbons into the atmosphere caused what aspect of the global environment to change?
   a. the carbon cycle
   b. the ozone layer of the atmosphere
   c. acid rain
   d. the sulfur cycle

7. Most scientists think that three of the following four statements related to global warming are correct. Select the exception.
   a. The concentration of greenhouse gases in the atmosphere is not increasing.
   b. Dozens of species have shifted their geographic ranges to the north.
   c. Plant growing seasons are longer now than they were before 1980.
   d. Human actions, such as the burning of fossil fuels, contribute to global warming.

8. Compared with the middle of the twentieth century, fish catches worldwide now include more small fishes and invertebrates and fewer large predators. What has caused this trend?
   a. People no longer want to eat large predators such as tuna and swordfish.
   b. Pollution has reduced the abundance of large predators (but not the abundance of small fishes and invertebrates).
   c. Net primary productivity in the oceans has declined, so most marine ecosystems cannot support large predators.
   d. Global populations of large predators have been reduced by overfishing.

Building a Sustainable Society

The State of the World

Each year, representatives of nations and corporations give speeches and produce reports that summarize what they've done in the past year and where they are headed for the upcoming year. If such an update could be provided for Earth, it would tell us how the planet's air, water, soil, and living organisms changed in the previous year. No one makes anything close to a complete version of such a report, and indeed, no one could; we do not even know how many species there are on Earth, let alone the current status of each of those species and the environments in which they live.

Although we cannot give a complete "State of the World" report, we do know how some pieces of the planetary puzzle are changing over time. Some of that news is good. As described in Unit 5, populations of some endangered species (such as the bald eagle) are increasing in size, sulfur emissions that cause acid rain have decreased by about 40 percent in the United States after peaking to their highest levels in 1974, and the ozone layer is showing early signs of recovery. Other news is bad. Nitrogen pollution continues to have negative effects on ecosystems worldwide, populations of many species are in serious decline, and global CO_2 levels continue to rise rapidly.

In addition to being able to provide particular bits of good and bad news, we know enough about Earth to make an overall assessment of our effect on the planet (Figure E.1). Unfortunately, that assessment indicates that the current human impact on the biosphere is not sustainable. As we'll see, people are using and damaging many of Earth's resources more rapidly than they can be renewed.

Although scientific evidence indicates that our current impact is not sustainable, there are many hopeful signs for the future. Five aspects of human society—education, individual action, research, government, and business—have already begun to contribute to the formation of a sustainable society. In this essay we first describe some of the evidence suggesting that the current human impact on the biosphere is not sustainable. With that material as background, we turn to our main focus: sources of hope for the future, and case studies that provide clues as to how to build a sustainable society.

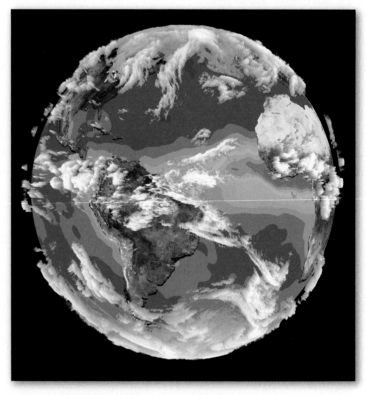

Figure E.1 Assessing the State of the World
New tools enable us to monitor Earth's vital signs in unprecedented detail, as in this computer image made using four different types of satellite data. Fires over land are shown in red. The large plume that extends from Africa over the Atlantic (and ranges in color from red to orange to yellow to green) was caused by the burning of vegetation and by windblown dust.

Main Message: Aware that the current human impact on the global environment cannot be sustained, many individuals, corporations, and governments are taking innovative actions to help build a sustainable society.

The Current Human Impact Is Not Sustainable

Many different lines of evidence suggest that the current human impact on the biosphere is not **sustainable**. What does this mean? An action or process is sustainable if it can be continued indefinitely without resources being used up or serious damage being caused to the environment. To begin with a simple example, modern societies depend on fossil fuels such as oil and natural gas to power our vehicles, heat our homes, and generate electricity. Although fossil fuels provide abundant energy now, our use of these fuels is not sustainable: they are not renewable, and hence supplies will run out, perhaps sooner rather than later (Figure E.2). Already, the volume of new sources of oil

discovered worldwide has dropped steadily from over 200 billion barrels during the period from 1960 to 1965, to less than 30 billion barrels during 1995 to 2000. In 2007, the world used about 31 billion barrels of oil, but only 5 billion barrels of new oil were discovered in that year.

Actions that cause serious damage to the environment are also considered unsustainable, in part because our economies depend on clean air, clean water, and healthy soils. But what constitutes "serious" environmental damage? One way to tell if an action causes serious damage is to see whether it disrupts important features of an ecosystem. As we have seen, many human actions have such effects. Human inputs to the global nitrogen and sulfur cycles, for example, now exceed all natural inputs combined (see Chapters 24 and 25). Such changes to the world's nutrient cycles

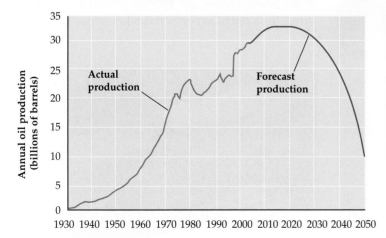

Figure E.2 Running Out of Oil
Many experts predict that the annual global production of oil will peak, then decline, sometime before 2020.

E2

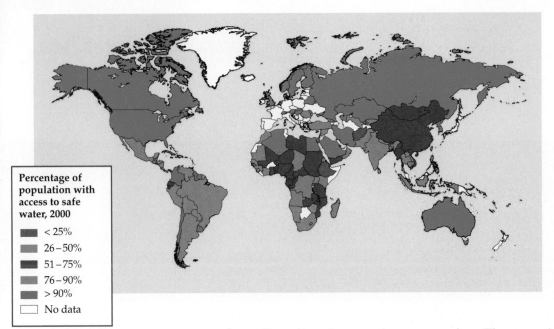

Percentage of
population with
access to safe
water, 2000

■ < 25%
■ 26–50%
■ 51–75%
■ 76–90%
■ > 90%
☐ No data

Figure E.3 Water Quality
Varies across the Globe

Overall, countries in the
Northern Hemisphere have much
greater access to safe water
than do those in the Southern
Hemisphere, especially Africa.

Declining water resources are a serious problem

People currently use over 50 percent of the world's annual supply of available fresh water, and demand is expected to rise over time. Many regions of the world already experience problems with either the amount of water available or its quality and safety (Figure E.3). Declining water resources are a serious issue today, and experts are worried that matters may get much worse.

To illustrate the problem, let's look at water pumped from underground sources, or **groundwater**. We use groundwater to drink, to irrigate our crops, and to run our industries. How does the rate at which people use groundwater compare with the rate at which it is replenished by rainfall?

can cause large disruptions to ecosystems, as manifested by such global problems as acid rain (see Section 24.5), the Antarctic ozone hole (see Figure 20.1), rising CO_2 concentrations (see Section 25.3), and lethal drops in oxygen levels in lakes and oceans (see Figures 23.8 and 24.12).

Let's consider the human use of two important resources: water and forests.

The answer is that we often use water in an unsustainable way: we pump it from **aquifers** (underground bodies of water, sometimes bounded by impermeable layers of rock) much more rapidly than it is renewed.

In Texas, for example, more than 6 million acre-feet of water were removed in 1995 from the vast Ogallala aquifer (an acre-foot is enough water to cover an acre of land with a foot of water). That amount—which would cover a football field with a wall of water more than 1,000 miles high—is more than 20 times the 0.3 million acre-feet of new water each year. Water has been pumped from the Ogallala aquifer faster than it is replenished for 100 years, causing the Texan portion of the aquifer to lose half its original volume. If that rate of use were to continue, in another 100 years the water would be gone, and many of the farms and industries that depend on it would collapse.

Texas is not alone. Pumping has caused groundwater levels to drop at rates of 1 to 3 feet per year in many regions of the world (Figure E.4a). Rapid drops in groundwater levels (about 1 meter per year) in China pose a severe threat to its recent agricultural and economic gains; and

(a)

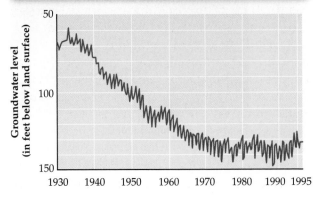

Figure E.4 Declining Groundwater Levels

(a) Groundwater levels steadily declined in Memphis, Tennessee, during the twentieth century because groundwater was pumped out more rapidly than it could be renewed. (b) The effects of subsidence can be quite visible, as in the sinkhole that developed in the yard of this home in Nixa, Missouri.

(b)

A stand of old-growth forest in the Pacific Northwest.

Old-growth forest is shown in red.

1620

Each dot represents 25,000 acres of old-growth forest.

1920

Figure E.5 The Destruction of Old-Growth Forests
From 1620 to 1920, vast regions of old-growth forests in the United States were cut down to produce lumber and to make room for agriculture, housing, and industry.

at current rates of use, large agricultural regions in India will completely run out of water in 5 to 10 years. In addition, because there is less water underground to provide support, the land surface may sink when groundwater levels drop. Such land **subsidence** [sub-*SYE*-denss] (Figure E.4*b*) can force people to stop pumping long before the water runs out. In Mexico City, for example, pumping has caused land within the city to sink by an average of 7.5 meters (more than 24 feet) since 1900, damaging buildings, destroying sewers, and causing floods.

Global deforestation continues at an alarming rate

For centuries, people have cut down trees at rapid rates. Much of Europe, for example, was deforested from AD 900 to 1500. Forest losses have approached 100 percent in some cases, as in Kenya and in São Paulo State, Brazil. Today we continue to cut some forests at unsustainable rates. Forested area in developing regions of the world declined by 9 percent from 1980 to 1995. These regions include most of the world's tropical forests, which are currently being cut at the rate of 14 million hectares (35 million acres) per year. If this rate of loss were to continue, the world's tropical forests would be gone in 100 to 150 years.

The news is not all bad for global forests, however. In industrial regions of the world, including Europe, the United States, Canada, and Japan, forested area has been increasing in the past three decades. Even so, this good news must be qualified: these increases were mainly due to growth in young forests, not to an increase in the area covered by older, more mature forests (which often harbor unique species). Few old-growth forests remain in industrial regions of the world (Figure E.5), and their regeneration takes hundreds of years. As a result, even if all harvesting of old-growth forests were to stop immediately, it would take several hundred years before the area covered by such forests began to increase.

Sources of Hope for the Future

As we all know, if we start with a fixed amount of money in the bank and consistently spend more than we earn, the funds in the bank account will shrink. Similarly, if people consistently use resources more rapidly than they are renewed, nature's "bank accounts" will shrink. For resources such as forests, we can think of nature's bank account as consisting of the total amount of forest in the world. Let's consider how the amount of forest changes over time.

Each year, total forest area can decrease as a result of natural disturbances (such as fire or windstorms) and human actions (including logging) that remove forests, and it can increase by natural growth and by human actions (such as planting trees) that regenerate forests. Currently, increases and decreases due to natural factors roughly balance one another, but logging is causing the global area of tropical forests to decrease each year.

For forests, or for any other resource, sustainable use requires that we not use resources faster than they are replenished. Unfortunately, data collected during the past few decades suggest that we are using many resources, including forests, more rapidly than they are being replenished. As a result, nature's "bank accounts" are shrinking. These findings are discouraging, for they suggest that future generations will have fewer resources than we have today.

All is not bad news, however. Thinking in terms of nature's bank accounts also provides a source of hope: we can alter our actions to avoid using resources faster than they are replenished. This attitude—along with the realization of the dire consequences if we continue to use resources unsustainably—has prompted many to embrace the monumental challenge of helping to change our societies and economies to make our collective impact on Earth sustainable. Hope is being provided in many arenas of human society: education, individual action, research, government, and business.

Figure E.6 "Worthless" Desert?
Desert regions, such as the Sonoran Desert of Arizona pictured here, are a vital ecosystem that is home to the likes of jackrabbits, cacti, wrens, hawks, prairie dogs, Gila monsters, rattlesnakes, and coyotes—all of which can be seen as valuable in their own right.

Awareness and understanding are the first steps to solving environmental problems

Efforts to build a sustainable society depend on education in part because people cannot solve problems they don't know about. Many people are aware, in general terms, that environmental problems exist, but often they are not aware of the extent of those problems or the amount of scientific evidence demonstrating how serious those problems are. For example, after listening to a presentation about the status of the world's oceans, one judge for the U.S. Ninth Circuit Court said, "I thought I knew a lot about the environment. But I was staggered by what I didn't know." In addition to playing a critical role in informing people about the scope of environmental problems, education can help people explore connections between the economy and the environment—showing, for example, that a "jobs versus the environment" trade-off is not always necessary (as we shall see later in this essay).

Education also plays a key role in shaping attitudes about nature. As various commentators have written, we will save only what we love, and we love only what we know. In many human societies, the dominant view has been that a species or community is worth saving only to the extent that it provides direct benefits to people. The appreciation for natural communities that can develop from education provides a powerful alternative view: communities can be worth saving for their own sake. For example, some people who live in desert regions refer to the land that surrounds their city or town as "worthless." Such views can change when a person spends time in the desert, learns more about it, and comes to appreciate its beauty and unique value (Figure E.6).

Because it is fundamental to how we view the world and to our knowledge of particular problems, education is central to all efforts to build a sustainable society. Education has already produced great changes. People are far more aware of environmental issues today than they were 30 or 40 years ago, and they place more value on solving environmental problems than they once did. One national poll in the United States asked people whether they agreed with the statement, "Protecting the environment is so important that requirements and standards cannot be too high, and continuing environmental improvements must be made regardless of cost." The percentage of people who agreed rose from 45 percent in 1981 to 74 percent in 1990. Polls conducted in other nations have obtained similar results, indicating that environmental issues are of worldwide concern.

Individual actions can have a ripple effect

Each day we make many choices that affect the environment. For example, we can choose to purchase energy-efficient cars and appliances—or not. Similarly, we can refuse to buy throwaway products, such as disposable cameras—or not. Many people do make choices with the environment in mind, and as a result, "green" products that minimize impacts to the environment are becoming increasingly common. As one example of this trend, sales of organic food in the United States skyrocketed from $180 million in 1986 to about $17 billion in 2007. Similarly, consumers in many parts of the industrial world can now opt to receive electricity generated from renewable sources, such as wind or solar power (although consumers still must pay extra for renewable energy).

In addition to using their wallets as leverage, some individuals take other actions to help support the conversion to a sustainable society. For example, individuals, corporations, and local governments have established green roof projects in Europe, North America, and Japan. A **green roof** is a 2- to 4-inch-thick "living"

rooftop; it has a layer of soil or other material in which plants grow, under which there are one or more layers that absorb water and prevent roots and water from damaging the underlying roof structure (Figure E.7). Green roofs are well established in countries such as Germany, where more than 13 million square meters of rooftops had green roof systems by 2002.

Individuals who build green roofs enjoy a number of benefits, including reduced stormwater flow, decreased heating and cooling bills (because the roofs insulate the building), and reduced levels of dust and pollutants (because they are absorbed by the plants). As part of a series of steps designed to convert a 600-acre, $2 billion assembly plant to sustainable forms of manufacturing, managers at the Ford Motor Company decided to install green roofs on 454,000 square feet of roofing (see Figure E.7). These roofs, which are covered with the ground-cover sedum, can absorb an inch of rainfall with no runoff. Other benefits include reduced energy costs and increased absorption (by the plants) of CO_2, the most important greenhouse gas. Some green roof

Chicago City Hall

A residential rooftop garden in Tokyo, Japan

A school in Unterensingen, Germany, combining solar panels with a green roof

Figure E.7 Green Roofs around the World

Green roofs in Germany, the United States, Japan, and other countries differ in appearance. They tend to be similar, however, in their basic design, which includes a plant layer, a soil or soil-like matrix, an absorbent layer, a drainage layer, and a protective membrane, as illustrated here for the green roof design used at the Ford Rouge Center, located near Dearborn, Michigan.

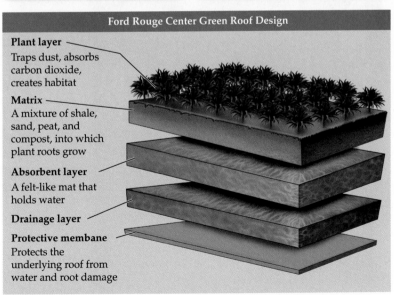

Ford Rouge Center Green Roof Design

Plant layer
Traps dust, absorbs carbon dioxide, creates habitat

Matrix
A mixture of shale, sand, peat, and compost, into which plant roots grow

Absorbent layer
A felt-like mat that holds water

Drainage layer

Protective membane
Protects the underlying roof from water and root damage

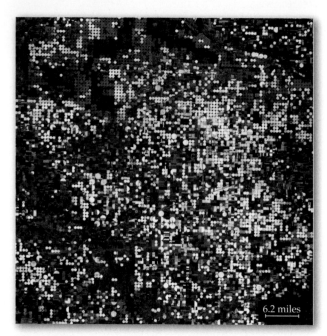

Figure E.8 Putting Endangered Species on Top
As part of their plan to redevelop an abandoned meatpacking plant in Nashville, Tennessee, a husband-and-wife team installed a green roof on one of the plant's buildings. Their green roof uses species found in Tennessee's endangered Cedar Glade community, shown here. Cedar Glade is home to many rare species, including the Tennessee coneflower, the first plant species listed under the Endangered Species Act.

projects, such as one initiated by a husband-and-wife team in Nashville, Tennessee, offer additional benefits by providing a place to grow rare or endangered species (Figure E.8).

Efforts by individuals to build a sustainable society can start small— sometimes in a person's backyard— and then grow into something larger. Such was the case with green roofs in the city of Portland, Oregon: a green roof was started in 1997 on a garage by one person, and that effort has now snowballed into several large green roof projects sponsored by the city. Efforts by one person

Figure E.10 A Polka-Dotted Landscape
Satellite images provide a unique perspective on our world, as seen in this image of irrigated farmland near Garden City, Kansas. The red and white circles are farmland watered by circular or "center pivot" irrigation systems; red indicates healthy vegetation.

can even blossom into national movements. In 1977, for example, Wangari Maathai (winner of the 2004 Nobel Peace Prize) founded the Green Belt Movement (GBM) in Kenya (Figure E.9). The GBM, initially a small organization with no staff or funds, began by planting seven trees in a small park in Nairobi. By 2003, the GBM had gone on to plant more than 20 million trees in Kenya. At present, the movement has more than 3,000 tree nurseries and has provided jobs for thousands of people.

New research tools are being used to measure human impacts

Documenting the human impact on the planet is a monumental task: we must understand our impact, acre by acre, across the entire globe. Now, for the first time in history, we have the research tools needed to accomplish this seemingly impossible task. Satellite images, for example, enable us to monitor Earth in unprecedented detail (Figure E.10). The preliminary results from such monitoring efforts show that we are causing serious damage to the world's ecosystems. But our recently acquired ability to recognize how much we are changing the planet provides a powerful source of hope: we can use that information to motivate change and to guide our efforts to build a sustainable society.

Figure E.9 One Tree at a Time
In 1977, Wangari Maathai founded the Green Belt Movement, initially a small organization with no staff or funds. The GBM has grown and has gone on to plant millions of trees in Kenya. Ms. Maathai is pictured here after she received the 2004 Nobel Peace Prize for her efforts toward sustainable development, democracy, and peace.

Consider the new research capabilities unveiled by NASA in April 2003: a satellite system able to generate the world's first consistent and continuous measurements of net primary productivity (NPP; see Figure 24.3) on a global scale (Figure E.11). Such measurements show how much new plant growth is occurring throughout the world, thereby providing a snapshot of Earth's "metabolism." Early results from this satellite system are fascinating—showing, for example, the speed with which land plants respond

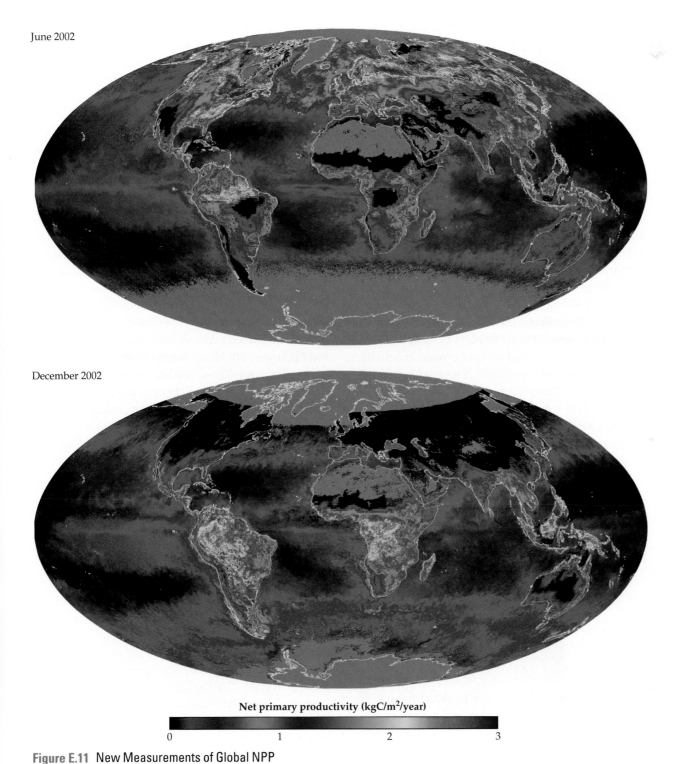

June 2002

December 2002

Net primary productivity (kgC/m²/year)

0   1   2   3

Figure E.11 New Measurements of Global NPP

Data to construct these images were collected by NASA's new MODIS satellite system. NPP is expressed as kilograms of new growth produced per square meter per year.

to changing weather conditions. (NPP rises and falls considerably in a matter of days.) Researchers can now use such data to monitor how changes in climate affect plant growth, to measure the rate at which deserts are expanding globally, and to analyze the effects of droughts on ecosystems. By monitoring plant growth in real time, such data could also be used to improve crop production forecasts and to help ranchers decide when to move cattle from one pasture to another.

Finally, research in fields other than the sciences can help build a sustainable society. For example, traditional measures of a nation's economic output, such as **gross domestic product (GDP)**, keep track of goods produced but do not consider the social and environmental costs that result from producing those goods. Such costs include billions of tax dollars spent to clean up polluted waters and soils, billions of dollars in medical bills for lung conditions caused by air pollution, and long-term problems caused by unsustainable practices that increase GDP in the short run but harm our environment and economy in the long run. To account for such costs, economists are developing measures of economic output, such as the **index of sustainable economic welfare**, that include a wider range of the benefits and costs of economic activities than does the traditional GDP. The goal is to create a new system of accounting that provides a realistic measure of the true long-term health of our economy, thereby enabling us to recognize and correct problems before a crisis occurs.

Government has an important role to play

Efforts to build sustainable societies require government action. Treaties among governments have reduced emissions of pollutants that cross national boundaries, as illustrated by international agreements to curb the production of CFCs that damage the ozone layer (see the box in Chapter 25, page 506). Considerable potential remains for governments to take other actions to promote a sustainable society, such as initiating major efforts to develop new sources of energy or enacting legal reform so that tax laws would penalize polluters and provide incentives for environmentally friendly practices. Some governments provide such incentives; in Germany, for example, tax breaks are given to families who use thatch roofs, which are made of renewable plant materials rather than asphalt-derived materials.

Governments can also play an essential role by placing limits on resource use. Although many fisheries are in trouble, as we saw in Chapter 25, that is not the case for the lobster fishery in Maine, where the catch of lobsters was stable for many years and then increased in recent years (Figure E.12). The lobster fishery in Maine is regulated by strict laws that forbid the catch of several categories: (1) females bearing eggs, (2) small individuals (which probably have not yet reproduced), and (3) large individuals (each of which can produce many offspring). These and other regulations, which help prevent lobsters from being caught more rapidly than they can reproduce, allowed the catch of lobsters to remain roughly stable from 1950 to 1990. From 1990 to 2000, the lobster catch increased dramatically, sug-

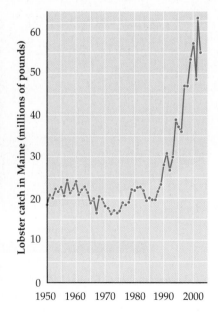

Figure E.12 A Sustainable Fishery

Statewide, the catch of lobsters in Maine was fairly stable from 1950 to 1990, indicating that lobster harvests there were sustainable during that time period. In the early 1990s, the catch increased dramatically, in part because of declining populations of fishes that eat lobster eggs and young lobsters.

gesting an increase in the lobster population. Declining populations of fish species, such as flounder and cod, that feed on lobster eggs and immature lobsters may have contributed to this dramatic increase.

As a final example, consider how government regulations reduced sulfur dioxide (SO_2) emissions from power plants by nearly 40 percent in the United States (see Figure 24.14). Reductions in SO_2 emissions are required by the Acid Rain Program, established by Congress in 1990 as an amendment to the Clean Air Act. Reduced SO_2 emissions have already decreased acid rain, and additional reductions mandated by the law are expected to save more than $50 billion per year in medical costs by 2010. These impressive results were achieved by an innovative approach to pollution control: the cap-and-trade system.

Here's how a **cap-and-trade system** works (Figure E.13). First, the government sets a nationwide limit, or cap, on the amount of a pollutant that can be added to the atmosphere each year. In the case of SO_2, the cap for 2010 is 8.95 million tons, a decrease of nearly 50 percent from 1980 levels. Second, each factory is given a certain number of "emission allowances" in each year, which allow it to emit a specified number of tons of SO_2. At the end of the year, the factory must provide government officials with enough allowances to cover its emissions for the year. Unused allowances can be sold, traded, or saved (banked) for future use. Factories that cannot declare enough in allowances to cover their yearly emissions are fined and must use future allowances to cover the shortfall, just as we would use future earnings to repay present debts.

The cap-and-trade system reduced SO_2 emissions at a fraction of the estimated costs. When the system was proposed in the 1980s, it was estimated that reaching the 2010 caps would cost the industry over $12 billion. By 1998, however, updated estimates placed the total cost of meeting the 2010 caps at $0.87 billion, significantly less than was originally feared.

Why has the cap-and-trade system worked so well? First, the cap restricts total emissions. So even if the industry grows, total emissions must go down, resulting in both health and environmental benefits. Second, although the government limits total emissions, it does not specify how the reductions are to be achieved. This lack of direction frees companies to seek the most cost-effective way to meet the caps. In addition, because the total number of allowances is limited, emission allowances are scarce, and hence valuable. Simple economics then takes over: because allowances are worth money and can be bought and sold, the profit motive stimu-

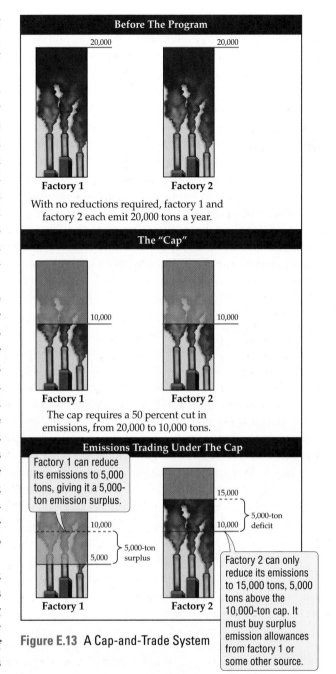

Figure E.13 A Cap-and-Trade System

lates the development of new ways to reduce emissions. Such market forces may soon provide a new source of hope for reducing CO_2 emissions: the European Union recently established the world's first international trading scheme for carbon dioxide emissions, which began as scheduled in 2005.

Business participation is crucial

Successful conversion to a sustainable economy will be an ongoing and complex process. For this process to work, business must play a central role—for two reasons. First, corporations use large amounts of resources and emit large quantities of pollutants. Hence, there is enormous potential for corporations to reduce the environmental impact of their activities. Second, as we've seen with the cap-and-trade system for SO_2, the profit motive can help drive the rapid development of innovative new technologies that reduce the human impact on the environment.

Given current patterns of investment, a growing number of corporations anticipate huge profits if they are well positioned to help societies convert to sustainable forms of development. For example, the energy giant British Petroleum (BP) is investing $1 billion in alternative sources of energy, such as fuel cells, solar power, and wind power. Similarly, ABB, a Swiss company with revenues of $24 billion per year, sold its large electric power plants in 1999 as part of a move to become an industry leader in small-scale and renewable sources of energy. ("Small-scale" here refers to power sources that serve a large building or a cluster of homes.)

What other steps are businesses taking to help build a sustainable society? Consider global warming. Most scientists think global warming is real and is caused at least in part by human actions, such as fossil fuel use, that cause CO_2 levels to rise. Although some politicians and members of the public argue against the scientific consensus, many corporations view climate change as a reality that threatens both society and their business interests. Such companies include economic powerhouses such as the energy firms Shell and BP, the chemical company DuPont, and the aluminum manufacturer Alcoa.

To lessen the impact of projected global warming, these and other large corporations have set targets that are more ambitious than actions required by the **Kyoto Protocol**, an international agreement designed to reduce global warming by reducing CO_2 emissions. Signed in 1997 and ratified by 141 countries (but not by the United States, the world's largest producer of CO_2 emissions), the Kyoto Protocol calls for industrial nations to reduce CO_2 emissions by an average of 7 percent from 1990 levels. BP has already beat that goal: by 2001, after just 4 years of effort, it had reduced CO_2 emissions by 10 percent from 1990 levels at no net cost to the company. Other companies have reduced emissions of CO_2 and other greenhouse gases by even greater levels, as shown in Figure E.14.

Finally, firms such as the Home Depot, McDonald's, and Alcoa seek to document the total environmental impact of their products, from manufacture to disposal. This approach, called **life cycle engineering**, gives corporations the information they need to develop new business practices that cause less environmental damage, often saving money at the same time.

Life cycle engineering, the cost-free reductions in CO_2 emissions achieved by BP, the success of cap-and-trade systems, and savings from green roofs all suggest that what is good for the environment can also be good for the bottom line. As a result, some economists think we can protect the environment without causing economic harm or large job losses. Even if these economists are only partly right, in some cases we will be able to avoid the tough choices required by a "jobs versus the environment" trade-off. Even when such trade-offs are necessary, it is clear that people can draw on a range of motivating factors—including the profit motive, concern for the environment, and concern for human welfare—in efforts to build a sustainable society.

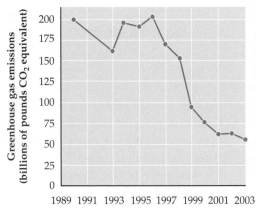

Figure E.14 Declining Greenhouse Gas Emissions

Greenhouse gas emissions at the chemical company DuPont declined by 72 percent from 1990 to 2003. In the graph, DuPont's total greenhouse gas emissions—including CO_2, nitrous oxide, and several types of CFCs—are expressed as the number of pounds of CO_2 emissions that would have the same effect on global warming if they had been released.

◉ Review and Discussion

1. Some critics object to cap-and-trade systems because they don't think it's right for a company that does not meet the cap to be able to purchase emission allowances and thereby continue to pollute at a high level. Other people counter that because the cap-and-trade system reduces *total* emissions, it doesn't matter if some companies continue to pollute as usual. What do you think?

2. Describe in detail two lines of evidence from this essay (or elsewhere in Unit 5) indicating that the impact of people on the environment is not sustainable.

3. Many new technologies and products that reduce human impact on the environment (such as renewable energy sources, lumber certified not to be from old-growth forests, or fish raised using low-impact aquaculture) cost more than standard technologies, at least initially. Would you be willing to pay more for use of such technologies or products? Why or why not?

4. Do you think governments should switch from traditional measures of economic productivity, such as the gross domestic product (GDP), to new measures that incorporate the environmental and social costs of producing goods (such as the index of sustainable economic welfare)?

5. Many corporations are investing large sums of money in products and technologies that reduce the human impact on the environment. Why? What unique features of corporations enable them to rapidly develop innovative new products?

Key Terms

aquifer (p. E3)
cap-and-trade system (p. E10)
green roof (p. E6)
gross domestic product (GDP) (p. E9)
groundwater (p. E3)
index of sustainable economic
   welfare (p. E9)

Kyoto Protocol (p. E11)
life cycle engineering (p. E11)
subsidence (p. E4)
sustainable (p. E2)

Appendix: The Hardy–Weinberg Equilibrium

In this appendix we describe the conditions under which populations do not evolve. Specifically, we discuss the conditions for the Hardy–Weinberg equation, a formula that allows us to predict genotype frequencies in a hypothetical nonevolving population. As described in Chapter 17 (see the box on page 351), this equation provides a baseline with which real populations can be compared in order to figure out whether evolution is occurring.

A population can evolve as a result of mutation, gene flow, genetic drift, or natural selection. Put another way, a population does *not* evolve when the following four conditions are met:

1. There is no net change in allele frequencies due to mutation.
2. There is no gene flow. This condition is met when new alleles do not enter the population via immigrating individuals, seeds, or gametes.
3. Genetic drift does not change allele frequencies. This condition is met when the population is very large.
4. Natural selection does not occur.

The Hardy–Weinberg equation is derived from the assumption that all four of these conditions are met. In reality, these four conditions are rarely met completely in natural populations. However, many populations meet these conditions well enough that the Hardy–Weinberg equation is approximately correct, at least for some of the genes within the population.

To derive the Hardy–Weinberg equation, consider a hypothetical population of 1,000 moths. The dominant allele for orange wing color (W) has a frequency of 0.4, and the recessive allele for white wing color (w) has a frequency of 0.6. What we seek to do now is predict the frequencies of the WW, Ww, and ww genotypes in the next generation for a population that is not evolving.

If mating among the individuals in the population is random (that is, if all individuals have an equal chance of mating with any member of the opposite sex), and if the four conditions just described are also met, we can use the approach described in the accompanying figure to predict the genetic makeup of the next generation. This approach is similar to mixing all the possible gametes in a bag and then randomly drawing one egg and one sperm to determine the genotype of each

The Hardy–Weinberg Equation

When mating is random and certain other conditions are met, allele and genotype frequencies in a population do not change. p = frequency of the W allele; q = frequency of the w allele.

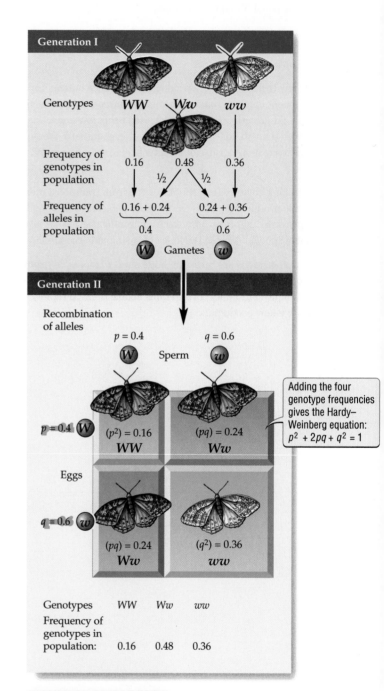

Adding the four genotype frequencies gives the Hardy–Weinberg equation: $p^2 + 2pq + q^2 = 1$

Conclusion:
Genotype and allele frequencies have not changed.

offspring. With such random drawing, the allele and genotype frequencies in our moth population do not change from one generation to the next, as the figure shows.

Because the *WW*, *Ww*, and *ww* genotypes are the only three types of zygotes that can be formed, the sum of their frequencies must equal. As the figure shows, when we sum the frequencies of the three genotypes, we get the Hardy–Weinberg equation:

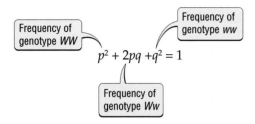

Frequency of genotype *WW*

Frequency of genotype *ww*

$$p^2 + 2pq + q^2 = 1$$

Frequency of genotype *Ww*

In this equation, the frequency of the *W* allele is labeled p and the frequency of the *w* allele is labeled q.

In general, once the genotype frequencies of a population equal the Hardy–Weinberg frequencies of p^2, $2pq$, and q^2, they remain constant over time if the four specified conditions continue to be met. A population in which the observed genotype frequencies match the Hardy–Weinberg predicted frequencies is said to be in Hardy–Weinberg equilibrium.

Table of Metric–English Conversion

Common conversions

Length		To convert	Multiply by	To yield
nanometer (nm)	0.000000001 (10^{-9}) m	inches	2.54	centimeters
micrometer (µm)	0.000001 (10^{-6}) m	yards	0.91	meters
millimeter (mm)	0.001 (10^{-3}) m	miles	1.61	kilometers
centimeter (cm)	0.01 (10^{-2}) m			
meter (m)	—	centimeters	0.39	inches
kilometer (km)	1,000 (10^{3}) m	meters	1.09	yards
		kilometers	0.62	miles
Weight (mass)				
nanogram (ng)	0.000000001 (10^{-9}) g	ounces	28.35	grams
microgram (µg)	0.000001 (10^{-6}) g	pounds	0.45	kilograms
milligram (mg)	0.001 (10^{-3}) g			
gram (g)	—	grams	0.035	ounces
kilogram (kg)	1,000 (10^{3}) g	kilograms	2.20	pounds
metric ton (t)	1,000,000 (10^{6}) g (=10^{3} kg)			
Volume				
microliter (µl)	0.000001 (10^{-6}) l	fluid ounces	29.57	milliliters
milliliter (ml)	0.001 (10^{-3}) l	quarts	0.95	liters
liter (l)	—			
kiloliter (kl)	1,000 (10^{3}) l	milliliters	0.034	fluid ounces
		liters	1.06	quarts
Temperature				
degree Celsius (°C)	—	To convert Fahrenheit (°F) to Celsius (°C): $°C = \frac{5}{9}°F - 32°$ To convert Celsius (°C) to Fahrenheit (°F): $°F = \frac{9}{5}°C + 32°$		

PERIODIC TABLE OF THE ELEMENTS

Legend:

1	— Atomic number
H	— Symbol
Hydrogen	— Name
1.00794	— Average atomic mass

- ☐ Metals
- ☐ Metalloids
- ☐ Nonmetals

1 1A	2 2A	3 3B	4 4B	5 5B	6 6B	7 7B	8	9 8B	10	11 1B	12 2B	13 3A	14 4A	15 5A	16 6A	17 7A	18 8A
1 **H** Hydrogen 1.00794																	2 **He** Helium 4.002602
3 **Li** Lithium 6.941	4 **Be** Beryllium 9.012182											5 **B** Boron 10.811	6 **C** Carbon 12.0107	7 **N** Nitrogen 14.0067	8 **O** Oxygen 15.9994	9 **F** Fluorine 18.9984032	10 **Ne** Neon 20.1797
11 **Na** Sodium 22.98976928	12 **Mg** Magnesium 24.3050											13 **Al** Aluminum 26.9815386	14 **Si** Silicon 28.0855	15 **P** Phosphorus 30.973762	16 **S** Sulfur 32.065	17 **Cl** Chlorine 35.453	18 **Ar** Argon 39.948
19 **K** Potassium 39.0983	20 **Ca** Calcium 40.078	21 **Sc** Scandium 44.955912	22 **Ti** Titanium 47.867	23 **V** Vanadium 50.9415	24 **Cr** Chromium 51.9961	25 **Mn** Manganese 54.938045	26 **Fe** Iron 55.845	27 **Co** Cobalt 58.933195	28 **Ni** Nickel 58.6934	29 **Cu** Copper 63.546	30 **Zn** Zinc 65.409	31 **Ga** Gallium 69.723	32 **Ge** Germanium 72.64	33 **As** Arsenic 74.92160	34 **Se** Selenium 78.96	35 **Br** Bromine 79.904	36 **Kr** Krypton 83.798
37 **Rb** Rubidium 85.4678	38 **Sr** Strontium 87.62	39 **Y** Yttrium 88.90585	40 **Zr** Zirconium 91.224	41 **Nb** Niobium 92.90638	42 **Mo** Molybdenum 95.94	43 **Tc** Technetium [98]	44 **Ru** Ruthenium 101.07	45 **Rh** Rhodium 102.90550	46 **Pd** Palladium 106.42	47 **Ag** Silver 107.8682	48 **Cd** Cadmium 112.411	49 **In** Indium 114.818	50 **Sn** Tin 118.710	51 **Sb** Antimony 121.760	52 **Te** Tellurium 127.60	53 **I** Iodine 126.90447	54 **Xe** Xenon 131.293
55 **Cs** Cesium 132.9054519	56 **Ba** Barium 137.327	57 **La** Lanthanum 138.90547	72 **Hf** Hafnium 178.49	73 **Ta** Tantalum 180.94788	74 **W** Tungsten 183.84	75 **Re** Rhenium 186.207	76 **Os** Osmium 190.23	77 **Ir** Iridium 192.217	78 **Pt** Platinum 195.084	79 **Au** Gold 196.966569	80 **Hg** Mercury 200.59	81 **Tl** Thallium 204.3833	82 **Pb** Lead 207.2	83 **Bi** Bismuth 208.98040	84 **Po** Polonium [209]	85 **At** Astatine [210]	86 **Rn** Radon [222]
87 **Fr** Francium [223]	88 **Ra** Radium [226]	89 **Ac** Actinium [227]	104 **Rf** Rutherfordium [261]	105 **Db** Dubnium [262]	106 **Sg** Seaborgium [266]	107 **Bh** Bohrium [264]	108 **Hs** Hassium [277]	109 **Mt** Meitnerium [268]	110 **Ds** Darmstadtium [271]	111 **Rg** Roentgenium [272]							

6 Lanthanides

58 **Ce** Cerium 140.116	59 **Pr** Praseodymium 140.90765	60 **Nd** Neodymium 144.242	61 **Pm** Promethium [145]	62 **Sm** Samarium 150.36	63 **Eu** Europium 151.964	64 **Gd** Gadolinium 157.25	65 **Tb** Terbium 158.92535	66 **Dy** Dysprosium 162.500	67 **Ho** Holmium 164.93032	68 **Er** Erbium 167.259	69 **Tm** Thulium 168.93421	70 **Yb** Ytterbium 173.04	71 **Lu** Lutetium 174.967

7 Actinides

90 **Th** Thorium 232.03806	91 **Pa** Protactinium 231.03588	92 **U** Uranium 238.02891	93 **Np** Neptunium [237]	94 **Pu** Plutonium [244]	95 **Am** Americium [243]	96 **Cm** Curium [247]	97 **Bk** Berkelium [247]	98 **Cf** Californium [251]	99 **Es** Einsteinium [252]	100 **Fm** Fermium [257]	101 **Md** Mendelevium [258]	102 **No** Nobelium [259]	103 **Lr** Lawrencium [262]

We have used the United States system as well as the system recommended by the International Union of Pure and Applied Chemistry (IUPAC) to label the groups in this periodic table. The system used in the United States includes a letter and a number (1A, 2A, 3B, 4B, etc.), which is close to the system developed by Mendeleev. The IUPAC system uses numbers 1–18 and has been recommended by the American Chemical Society (ACS). Elements with atomic numbers 112 and higher have been reported, but not yet fully authenticated.

Answers to Self-Quiz Questions

Chapter 1
1. *b*
2. *a*
3. *c*
4. *d*
5. *c*
6. *b*
7. *a*
8. *d*

Chapter 2
1. *c*
2. *c*
3. *a*
4. *d*
5. *c*
6. *c*
7. *d*
8. *d*
9. *c*
10. *c*

Chapter 3
1. *d*
2. *c*
3. *b*
4. *a*
5. *b*
6. *a*
7. *a*
8. *c*
9. *b*
10. *a*

Chapter 4
1. *a*
2. *c*
3. *d*
4. *a*
5. *c*
6. *b*
7. *c*
8. *b*
9. *d*
10. *c*

Chapter 5
1. *b*
2. *a*
3. *c*
4. *d*
5. *b*
6. *d*
7. *b*
8. *a*
9. *b*
10. *c*

Chapter 6
1. *d*
2. *b*
3. *c*
4. *d*
5. *b*
6. *a*
7. *d*
8. *d*
9. *b*
10. *c*

Chapter 7
1. *c*
2. *a*
3. *b*
4. *c*
5. *d*
6. *a*
7. *c*
8. *d*
9. *d*
10. *c*

Chapter 8
1. *d*
2. *d*
3. *d*
4. *a*
5. *b*
6. *b*
7. *c*
8. *a*
9. *b*
10. *c*

Chapter 9
1. *b*
2. *a*
3. *a*
4. *b*
5. *d*
6. *c*
7. *c*
8. *c*
9. *d*
10. *a*

Chapter 10
1. *a*
2. *c*
3. *b*
4. *d*
5. *d*
6. *a*
7. *d*
8. *d*

Chapter 11
1. *c*
2. *b*
3. *c*
4. *c*
5. *a*
6. *c*
7. *a*
8. *d*

Chapter 12
1. *c*
2. *c*
3. *b*
4. *d*
5. *a*
6. *c*
7. *d*
8. *d*

Chapter 13
1. *b*
2. *c*
3. *b*
4. *d*
5. *a*
6. *d*
7. *b*
8. *c*

Chapter 14
1. *b*
2. *c*
3. *b*
4. *a*
5. *c*
6. *d*
7. *b*
8. *b*
9. *d*

Chapter 15
1. *c*
2. *a*
3. *b*
4. *d*
5. *a*
6. *c*
7. *a*
8. *d*

Chapter 16

1. d
2. d
3. c
4. a
5. b
6. c
7. d
8. a
9. b
10. c

Chapter 17

1. b
2. a
3. b
4. d
5. c
6. a
7. c
8. d

Chapter 18

1. b
2. b
3. c
4. a
5. d
6. d
7. b
8. c

Chapter 19

1. d
2. d
3. b
4. a
5. c
6. c
7. a
8. d
9. b
10. a

Chapter 20

1. c
2. d
3. c
4. b
5. c
6. d
7. a
8. b
9. c
10. b

Chapter 21

1. d
2. b
3. b
4. c
5. d
6. a
7. d
8. b

Chapter 22

1. c
2. a
3. c
4. d
5. b
6. b
7. d
8. a

Chapter 23

1. c
2. c
3. d
4. a
5. b
6. b
7. d
8. a

Chapter 24

1. c
2. a
3. d
4. a
5. b
6. b
7. c
8. d

Chapter 25

1. a
2. b
3. a
4. c
5. d
6. b
7. a
8. d

Answers to Review Questions

Chapter 1

1. **Observation:** Huge numbers of fish were being found dead; their bodies, covered with bleeding sores, were found floating by the millions in the estuaries of North Carolina.

   Hypothesis: On the basis of a previous experience in which laboratory fish had died suddenly after exposure to local river water, Dr. Burkholder hypothesized that *Pfiesteria*, a protist found in high numbers in the tanks containing the dead lab fish, was also responsible for the fish die-offs in local estuaries.

   Experiment: Dr. Burkholder isolated samples of *Pfiesteria* and exposed healthy fish to the samples. The fish were quickly killed, upholding the prediction that Dr. Burkholder had formulated on the basis of her hypothesis.

2. The characteristics of living things can be used to make a prediction. For example: "If nanobes are living things, it should also be true that they are made of cells."

3. All living organisms share all of the following characteristics: they are built of cells; they reproduce themselves using DNA; they develop; they capture energy from their environment; they sense their environment and respond to it; they show a high level of organization; they evolve. Prions do not share many of these characteristics and therefore cannot be considered living organisms. Prions are not built of cells; they are protein molecules. Prions do not reproduce themselves using DNA; rather, they reproduce by causing other molecules to mimic their shape. Because prions lack these essential characteristics of life, they are not alive.

4. Levels of the biological hierarchy from smallest to largest (with examples): molecule (DNA); cell (bacteria); tissue (muscle tissue); organ (heart); organ system (stomach, liver, intestines in digestive system); individual organism (human); population (field mice in one field); community (different species of insects living in a forest); ecosystem (river ecosystem); biome (the arctic tundra, coral reefs); biosphere (Earth).

5. Energy flows from the sun to photosynthesizing organisms such as grasses. The grasses, which are producers, use the sun's energy to produce chemical energy in the form of sugars and starches. Antelope, which are consumers, feed on the grasses to produce energy for their own use. Lions, also consumers, then eat the antelope. Ticks, consumers as well, feed on both the antelope and the lions.

6. a. The grasses are producers because they capture sunlight and convert it to energy. The antelope, lions, and ticks are consumers because they eat either plants or other organisms that derive energy from plants.

   b.

Chapter 2

1. By analyzing the similarities and differences in both structural and behavioral features between an unknown organism and known organisms, biologists can determine how closely or distantly related the unknown organism is to an organism already on the tree of life and can therefore place the new organism on the tree. The most useful features for this purpose are those known as shared derived features.

2. Both family trees and evolutionary trees demonstrate how different members of a group are biologically related. Common ancestors shared by individuals can be traced using both family and evolutionary trees. An evolutionary tree is a hypothesis—a best educated guess—but typically a family tree is built from known facts.

3. Unique structural or behavioral features that have evolved in a group's most recent common ancestor and are then shared by the descendant species of that ancestor are known as shared derived features. The panda's thumb and the human thumb are not shared derived features, because this trait (an opposable thumb) did not evolve in these different species from their most recent common ancestor. Rather, pandas and humans each evolved opposable thumbs independently; such features are referred to as convergent features.

4. Biologists view an evolutionary tree as a hypothesis because it is the best approximation of the relationships of organisms to one another, given what we know today. They continue to study and reevaluate these relationships as new information becomes available.

5. Using an evolutionary tree like the one in Figure 2.4, which shows the relationship of crocodilians to birds, biologists are able to speculate about the behavior of dinosaurs. Both crocodilians and birds are dutiful parents: they build nests and defend their young. Parental behavior is therefore presumed to be a shared derived feature of crocodilians and birds inherited from, and exhibited in, their most recent common ancestor. Crocodilians and birds share their most recent common ancestor with dinosaurs, so we can hypothesize that dinosaurs, too, exhibited parental behavior.

6. Figure 2.4 shows that birds and crocodilians share a common ancestor with dinosaurs. If both birds and crocodilians are known to either sing or make chirping vocalizations, it can be assumed—given the relationship between these groups—that this behavior is a shared derived feature inherited from their common ancestor. Because these two groups share their most recent common ancestor with dinosaurs, one could conclude that singing (or chirping) dinosaurs existed.

7. From the smallest to the largest, the groupings of the Linnaean hierarchy are species, genus, family, order, class, phylum, kingdom.

8. DNA enables us to see connections or separations between species that are not expressed in structural or behavioral features, thereby changing our understanding of certain organisms' placement on the tree of life and their relationships to one another. For example, organisms that exhibit similar traits may be genetically very distantly related, while organisms that appear very different on the outside may have very similar DNA. Through the study of DNA, biologists have begun to hypothesize that, at its base, the tree of life is structured more like an interconnected web than like a tree.

   DNA studies of different organisms have resulted in the discovery of bacterial DNA in archaeans and in eukaryotes. This finding raises the question of how these three distinct lineages—Bacteria, Archaea, and Eukarya—came to share similar DNA. Dr. W. Ford Doolittle suggests that throughout the early history of life, organisms within the three lineages were freely exchanging genes, thereby passing genetic information horizontally—in a *horizontal gene transfer*—as well as vertically to their descendants. Therefore, we can attribute this new understanding of the shape of the tree of life to the study of DNA.

Chapter 3

1. The three major systems used to categorize living organisms are the evolutionary tree of life, the Linnaean hierarchy, and the system of three domains.

2. The two kingdoms that make up the prokaryotes are Bacteria and Archaea. Some factors that contributed to the success of prokaryotes are their simple and efficient structure, their uncomplicated reproduction, their ability to reproduce rapidly and prolifically, the diverse ways in which they obtain nutrition, and their ability to persist in extreme environments.

3. Unlike many other eukaryotes, *Giardia lamblia* has two nuclei, no chloroplasts, and no mitochondria. Biologists theorize that this unique cell structure is a result of one experiment in engulfment that has gone on over time—a conclusion supporting the idea that eukaryotic cells first formed from prokaryotic cells engulfing other prokaryotic cells.

   Slime molds provide biologists with insight into the early evolution of multicellularity. Because these protists live their lives in two phases—as independent, single-celled organisms, and then as individual members of a multicellular society—they are studied by biologists who hope to understand the evolutionary transition from single-celled organisms to multicellular organisms.

4. When cells began to work together in a coordinated fashion, some became specialized in order to aid the new multicellular organisms in survival. Sponges, which are among the simplest of animals, are loose collections of specialized cells. One of the earliest animal groups to develop true tissues—specialized, coordinated collections of cells—was the Cnidaria. Cnidarians evolved stinging cells used to stun their prey, as well as nervous tissues, musclelike tissues, and digestive tissues. With cells working together as tissues, animals were able to evolve organs: body parts composed of different tissues organized to carry out specialized functions. Flatworms were one of the earliest groups to evolve true organs.

5. Viruses are not classified in any kingdom or domain, because they occupy a gray area between living organisms and nonliving matter. Viruses lack many of the qualities present in living organisms. They cannot reproduce outside of a host organism, and they lack a clear evolutionary relationship to any single group.

6. When plants colonized land, they were forced to evolve in ways that enabled them to persist in an environment where they were no longer surrounded and supported by water. To deal with the problem of obtaining and retaining water, plants evolved root systems (which enable them to absorb water and nutrients from soil) and the waxy covering over their stems and leaves known as the cuticle (which prevents their tissues from drying out when exposed to sun and air).

   No longer floating in water, plants had to meet the challenge of gravity. They evolved rigid cell walls composed of cellulose, which gave them enough rigidity to grow up and into the air. In conjunction with strong cell walls, plants evolved vascular systems—networks of specialized tissues that can transport fluids—extending from their roots throughout their bodies. When a plant is properly hydrated, these systems act like a water-filled balloon and enable the plant to remain and grow upright.

7. Members of the kingdom Plantae contain chloroplasts. Plant cells contain an organelle that is lacking in animal cells: the chloroplast. Chloroplasts enable plants to produce their own food from carbon dioxide and sunlight. As photosynthesizing organisms and producers, plants provide an important source of food for nearly all organisms on land and are the basis of terrestrial food webs.

8. Plants have developed many characteristics that utilize the mobility of animals. Animals are attracted to various features that plants have evolved in order to reproduce or spread their seeds. Some plants—specifically, angiosperms—evolved flowers, specialized reproductive structures that produce nectar, a sugary liquid used as food by some animals. Animals attracted by nectar may transport pollen between very distant plants, or from flower to flower.

Angiosperms have evolved another food that attracts animals for seed distribution: fruits. Fruits develop from the ovary surrounding an angiosperm embryo. Animals eat fruits and later excrete the seeds, usually in a location far from the parent plant where the embryo will not be in competition with its parent for food, water, or sunlight. The animal's feces may also provide a nutrient-rich environment for germination of the new plant.

Interlude A

1. Terry Erwin fogged a single rainforest tree with insecticide and found more than 1,100 species of beetles living in its canopy. He estimated that 160 of these species were likely to be specialists on a single tree species. There are an estimated 50,000 or so tropical tree species. Therefore, if the tree species Erwin was studying were typical, beetle species in tropical trees would number 8,000,000 ($50,000 \times 160$). Beetle species are thought to make up about 40 percent of all arthropod species. If that is the case, then the total number of arthropod species in tropical tree canopies should be 20 million. Many scientists believe that the total number of arthropod species in the canopy is double the number found in other parts of tropical forests, suggesting that there are another 10 million arthropod species in other parts of the tropical forest environment. So Erwin estimated the total number of arthropod species in the tropics at 30 million. Such estimates are difficult because they require many assumptions, any of which could be wrong.

   There are many possible ways to try to estimate the number of species remaining on Earth. Erwin began with fogging and counting the beetles in a single tree. Another possible method would be for a group of scientists to try to find and count all the species—plants, animals, fungi, prokaryotes, protests—of all kinds in one very small area. For example, they could scour a 100 foot by 100 foot patch of forest. As Erwin did, scientists could try to extrapolate from there to the species count for the entire world.

2. We should take seriously the claims about the extinctions of species around the world because, even though we do not know the exact number of species or their rate of extinction, reasonable estimates of these numbers are invariably quite high. The lack of precise data does not lessen the actual problem of species extinction.

3. Biodiversity has fluctuated ever since life began, declining rapidly during mass extinctions. Five previous mass extinctions (about 440, 350, 250, 206, and 65 million years ago) are thought to have been caused by natural forces, including climate changes, volcanic eruptions, changes in sea level, and atmospheric dust from the collision of Earth with an asteroid. After each of these events, biodiversity recovered slowly, over millions of years. Some scientists believe that the current rapid expansion of human populations could lead to another mass extinction.

4. As human populations grow, cities, suburbs, and commercial areas expand and natural habitats are destroyed and degraded, driving out other species. Human population growth has increased pollution, further threatening biodiversity. The expansion of human populations has also increased the introduction of non-native species, which constitute another threat to biodiversity.

5. Human beings require places to live, just as other species do. If we have a right to exist, then other species must as well. Where human rights end and those of other species begin will depend on how we, as a society, decide to live and to what degree we agree to either protect or encroach on the habitats and lives of other species.

6. Compromise between development and habitat protection for endangered species can be a better outcome than prolonged conflict between staunch environmentalists and determined developers that results in one party winning and the other losing. In such compromises, parties can often come to a speedier, less costly, and more practical solution that provides benefits for humans and other species as well.

Chapter 4

1. Monomers are small molecules that serve as repeating units in a larger molecule (macromolecule). Macromolecules that contain monomers as building blocks are known as polymers. Lipids, such as triglycerides and sterols, are macromolecules, but since they are not built from discrete units that are repeated multiple times, they are not usually regarded as polymers.

2. The pH of pure water should be 7. Units on the pH scale represent the concentration of free hydrogen ions in water. In the presence of a base, the pH of a solution will be above 7, indicating that there are more hydroxide ions than hydrogen ions, so the solution is basic. In the presence of an acid, the pH will be below 7, indicating more free hydrogen ions than hydroxide ions, so the solution is acidic. Pure water has equal amounts of hydrogen and hydroxide ions and is therefore neutral.

3. A hydrogen bond is a noncovalent bond created by the electrical attraction between a hydrogen atom with a partial positive charge and any other atom that has a partial negative charge. Hydrogen bonds are weaker than ionic bonds, which are weaker than covalent bonds. Water molecules are polar: the region around the oxygen atom is slightly negative, and the regions around the two hydrogen atoms are slightly positive. This property provides for the formation of hydrogen bonds between water molecules, since each partially charged hydrogen atom in a water molecule is attracted to any atom with a partial negative charge, including the oxygen atom of a nearby water molecule.

4. Each carbon atom can form strong covalent bonds with up to four other atoms, including other carbons, creating large molecules containing hundreds, even thousands, of atoms. These molecules play many different roles critical to life.

5. Cells use carbohydrates as a readily available energy source. Some carbohydrates, such as the cellulose found in plant cell walls, have

structural functions. Nucleic acids such as DNA and RNA, which carry genetic information, are polymers of nucleotides. Some nucleotides act as energy carriers. Proteins make up the physical structures of organisms, as well as the enzymes that catalyze biochemical reactions. Lipids such as triglycerides are common means of long-term energy storage, and lipids such as phospholipids are important components of cell membranes.

Chapter 5

1. The plasma membrane, which is a feature of all cells, provides a necessary boundary between a cell and its surrounding environment. The plasma membrane is selectively permeable, controlling what gets in and what flows out. Both prokaryotic and eukaryotic cells also contain DNA, cytosol, and ribosomes. DNA contains the information for producing the proteins needed by each cell. Cytosol is the watery medium in which biochemical reactions take place. Ribosomes are the workbenches for producing proteins.

2. The major components of a plasma membrane are a phospholipid bilayer and an assortment of proteins. The phospholipid molecules are oriented so that their hydrophilic heads are exposed to the watery environments both inside and outside the cell. Their hydrophobic fatty acid tails are grouped together inside the membrane away from the watery surroundings. Some membrane proteins extend all the way through the phospholipid bilayer and act as gateways for the passage of selected ions and molecules into and out of the cell. Other membrane proteins are used by the cell to detect changes in and signals from the environment outside the cell. Proteins that are not anchored to structures within the cell are free to move sideways within the phospholipid bilayer. This freedom of movement supports what is known as the fluid mosaic model, which describes the plasma membrane as a highly mobile mixture of phospholipids and proteins. This mobility is essential for many cellular functions, including movement of the cell as a whole and the ability to detect external signals.

3. Chloroplasts are found in photosynthetic cells of algae and plants. Each chloroplast has two concentric membranes and an internal network of membranes (thylakoids) that contain the light-absorbing green pigment, chlorophyll. Chloroplasts harness light energy to convert carbon dioxide (CO_2) to sugars, splitting water molecules (H_2O) and releasing oxygen gas (O_2) in the process of photosynthesis. Mitochondria are found in nearly all cell types in all eukaryotes, both producers and consumers. Each mitochondrion has two membranes, the inner of which is thrown into many folds (cristae) and contains proteins and other components that enable the organelle to generate the energy carrier, ATP, through the oxygen-dependent process of cellular respiration. In this process, the chemical energy of organic molecules (such as sugars) is converted into energy stored within ATP molecules, and CO_2 and H_2O are released, as oxygen (O_2) is consumed.

4. The nucleus houses the cell's DNA; it directs the activities of the cell and determines which proteins will be produced by the cell in response to messages received from both inside and outside the cell. The endoplasmic reticulum is the site where proteins and lipids are manufactured. The Golgi apparatus directs proteins and lipids produced in the ER to their final destinations. Lysosomes, which are found in animal cells, contain enzymes that assist in breaking down macromolecules for cell use. Vacuoles, found in plant and fungal cells, contain enzymes that break down substances and can store nutrients for later use by the cell. Mitochondria use chemical reactions to transform the energy from many different molecules into ATP, the universal cellular fuel. Chloroplasts, found in plant cells, capture energy from sunlight and convert it into chemical energy to be used by the cell.

5. A fibroblast cell uses microfilaments—twisted polymers of the protein actin—to move. Microfilaments can change rapidly in length, and a fibroblast cell is able to "crawl" as different arrangements of microfilaments alter the structure of the leading and trailing parts of the cell. In the leading parts of the fibroblast cell—the pseudopodia—microfilaments are aligned pointing outward in a forward direction. As the filaments lengthen, they push against the plasma membrane, thereby extending the pseudopodia in the direction the cell is moving. Microfilaments at the trailing end of the cell tend not to be well organized. While the well-aligned filaments in the pseudopodia expand, the filaments in the trailing end of the fibroblast shrink, so the cell appears to be pulling up its rear end as it moves forward. A bacterium with a flagellum moves quite differently from a fibroblast. Each bacterial flagellum is composed of twisted cables of proteins that spin like a boat's propellers. The rotary action of the flagella pushes the cell forward in a fluid medium.

6. Both mitochondria and chloroplasts exhibit characteristics that have led scientists to hypothesize that they are descendants of primitive prokaryotic cells. The two organelles have their own DNA (which resembles prokaryotic DNA), are able to manufacture some of their own proteins, have ribosomes that resemble prokaryotic ribosomes, and reproduce independently of the cell by dividing in two in a manner similar to the way prokaryotic cells divide.

Chapter 6

1. The phospholipid bilayer that constitutes the plasma membrane blocks large and hydrophilic molecules from entering or leaving the cell. Channel proteins and passive carrier proteins in the plasma membrane allow certain molecules or ions to cross the membrane by moving down their concentration gradients. Active carrier proteins can transport molecules or ions against their concentration gradients.

2. Osmosis is the passive diffusion of water molecules across a selectively permeable membrane. A cell must maintain an optimum

balance of water and solutes in the cytosol because most metabolic processes that are crucial for life can only proceed under these conditions. An animal cell, such as a red blood cell, will shrink in a hypertonic environment because it will lose water through osmosis; there will be a net flow of water molecules out of the cell toward the region of higher solute concentration.

3. Exocytosis is the release of substances from a cell when transport vesicles fuse with the plasma membrane from the cytosolic side and release their contents into the space outside the cell. Endocytosis is the opposite process, in which substances are brought into a cell after becoming enclosed in vesicles that bud inward from the plasma membrane. The cell can import specific substances through receptor-mediated endocytosis, in which specialized receptor proteins on the outer surface of the plasma membrane recognize surface characteristics of the material to be brought into the cell.

4. In animals, leakproof barriers are provided by tight junctions, which bind cells together with protein strands. Anchoring junctions consist of protein hooks that hold neighboring cells together. Gap junctions are cytoplasmic tunnels, created by protein complexes, that directly connect adjacent animal cells. In plants, plasmodesmata provide cytoplasmic connections through cell walls.

5. Slow-acting cell signals affect target cells far away from their source, and they usually travel in the bloodstream (in animals) or in the sap (in plants). Fast-acting signals are short-lived and travel only between neighboring cells.

Chapter 7

1. The second law of thermodynamics holds that systems tend toward disorder. In a living system such as a cell, the order maintained by chemical reactions is counterbalanced by the release of heat energy (disorder) into the surroundings.

2. The reactions of photosynthesis obtain carbon from the air as CO_2, using light energy to combine that CO_2 with water to synthesize sugars. Organisms ultimately break down these sugars for energy and release the stored carbon back into the atmosphere as CO_2. The burning of fossil fuels by humans adds extra carbon dioxide to the atmosphere and in recent decades has contributed to an increase in the average temperature at Earth's surface (global warming).

3. Anabolism refers to those metabolic processes that manufacture larger molecules from smaller units. Anabolic pathways need an input of energy. Catabolic processes break down macromolecules, releasing energy and small organic molecules. Photosynthesis is an anabolic process because it creates large molecules, such as sugars, from smaller units such as carbon dioxide and water. It is driven by an input of light energy.

4. Concentrating enzymes and their substrates in a compartment, as with components of the citric acid cycle in the mitochondrial matrix, can increase catalytic efficiency by making collisions between enzymes and their substrates more likely. Another factor that increases efficiency is the sequential arrangement of enzymes for a series of reactions, as found in the arrangement of enzymes for ATP synthesis on the inner membrane of the mitochondrion.

5. According to the induced fit model of enzyme action, the binding of the substrate to the active site of an enzyme further molds the active site to create a more stable interaction between an enzyme and its substrate. The bound substrate reshapes the binding site of the enzyme slightly, the way your hand shapes a properly-sized glove when you put it on.

Chapter 8

1. Photosynthesis occurs in chloroplasts, uses light energy, synthesizes sugars from carbon dioxide and water, and releases oxygen as a byproduct. Cellular respiration occurs in mitochondria, releases energy from organic molecules such as sugars in an oxygen-dependent process, generating carbon dioxide and water as byproducts. Photosynthesis is an anabolic process, while cellular respiration is a catabolic process. Photosynthesis occurs only in producers (algae and plants), while cellular respiration takes place in both producers and consumers.

2. The transfer of electrons down an electron transport chain (ETC) produces a proton gradient in both chloroplasts and mitochondria. The protons move down that gradient through a membrane channel protein known as ATP synthase. The movement of the protons releases energy, which is used by ATP synthase to phosphorylate ADP to form ATP.

3. In the light reactions of photosynthesis, light energy is first captured by chlorophyll. The energy excites electrons, which travel down the electron transport chain (ETC) in the thylakoid membrane of the chloroplast. This process produces a proton gradient that drives ATP synthesis. The electrons removed from chlorophyll are handed over to $NADP^+$, which combines with a proton to form NADPH. Electrons released by the splitting of water molecules (photolysis) replace the electrons stripped from chlorophyll, generating oxygen gas as a byproduct. In the Calvin cycle, the ATP and NADPH made by the light reactions help power the incorporation of CO_2 into sugars (carbon fixation).

4. When protons pass through ATP synthase, their energy is used to convert ADP into ATP. Drugs that allow protons to bypass ATP synthase will inhibit the production of ATP and such drugs are therefore deadly poisons.

5. Electrons are removed from water in photosystem II, and then transferred to electron-accepting pigments and proteins in the thylakoid membrane. The electrons pass from the electron transport chain (ETC) to photosystem I. From there, the electrons pass to another ETC and finally to $NADP^+$, which is reduced to NADPH.

Chapter 9

1. Interphase consists of the G_1, S, and G_2 phases. During the gap phases (G_1 and G_2), the cell has time to grow and make the proteins it will need for the next phase. Between the two gap phases is the S phase, during which the DNA of the cell is replicated. During mitosis, the duplicated DNA molecules (sister chromatids) separate from each other and are delivered to the opposite ends of the parent cell; during cytokinesis, the cytoplasm of the parent cell is split in two, creating two daughter cells. Cells in the G_0 stage are highly differentiated and do not normally divide.

2. A horse cell would have a total of 128 DNA molecules in the G_2 phase prior to mitosis. At the end of meiosis I, each daughter cell would contain 64 DNA molecules in the form of 32 pairs of linked sister chromatids.

3.

	Mitotic cell division	Meiosis
1. In humans, the cell undergoing this type of division is diploid.	True	True
2. The daughter cells have half as many chromosomes as the parent cell.	False	True
3. A total of four daughter cells are produced when one parent cell undergoes this type of division.	False	True
4. In male animals, this type of division directly gives rise to sperm.	False	True
5. Stem cell self-renewal involves this type of division.	True	False
6. All daughter cells are genetically identical to the parent cell.	True	False
7. This type of division involves two nuclear divisions.	False	True
8. This type of division involves two rounds of cytokinesis.	False	True
9. Maternal and paternal homologues pair up to form bivalents at some point during this type of division.	False	True
10. Sister chromatids separate from each other at some point during this type of cell division.	True	True

4. A human male has a karyotype of 44 non–sex chromosomes (called autosomes, as noted in Chapter 10) in homologous pairs, plus one X and one Y sex chromosome. A female has 44 autosomes plus two X chromosomes. Therefore, the haploid gamete of the female would contain 22 autosomes plus an X chromosome, and a male haploid gamete would contain 22 autosomes plus either an X or a Y sex chromosome.

5. If gametes of sexually reproducing organisms were produced by mitosis, the offspring of each succeeding generation would have double the number of chromosomes of the parent generation.

Interlude B

1. When a proto-oncogene mutates to form an oncogene, some cellular processes may become uncontrolled and thereby increase the risk of cancer.

2. Cell signals must promote the activity of proto-oncogenes in order to promote cell division, and/or they must inactivate or reduce the activity of tumor suppressor genes in order for cell division to occur.

3. Colon cancer begins with a benign growth of cells called a polyp, which is commonly caused by mutations that inactivate both copies of a tumor suppressor gene and/or transform a proto-oncogene into an oncogene. In many patients, loss of a part of chromosome 18 deletes two other tumor suppressor genes, allowing for more aggressive and rapid cell division in the tumor. In a majority of cases, the protective activity of the tumor suppressor gene *p53* is lost, which removes all remaining cell division controls and enables the cancer cells of the now malignant tumor to metastasize (spread to other parts of the body).

4. The possibility of carcinogens in many commonly consumed products should persuade people to lessen their risk of cancer by taking more care in choosing what products to use and what lifestyle choices to make. Studies should be done to identify unknown carcinogens and determine what levels constitute a health risk to consumers.

5. You will have your own opinion on this issue, but all 50 states in the United States place some restrictions on the sale and advertising of tobacco products (for example, a ban on the sale of cigarettes to minors, and a ban on the use of cartoon characters in tobacco advertising). Most states restrict smoking in public places and require warning labels about the harmful effects of tobacco. Public health experts have proposed additional regulations, which include a ban on candy-flavored tobacco products that might be attractive to children; a ban on the labeling of cigarettes as "light" and "mild" because of concerns that such products might be incorrectly perceived as safer; limits on the tar and nicotine content in cigarettes; and, larger and more informative warning labels.

6. It is important to provide sufficient information about cancer risks so that consumers can make informed decisions about their product purchases. Food labels could be simplified to provide more concise information about possible cancer risks, and public health programs could raise public awareness of the cancer risks present in certain foods and other products.

Chapter 10

1. Genes are the basic units of inheritance; as such, they carry genetic information for specific traits. Genes are composed of DNA and are located on chromosomes. Most genes govern a genetic trait by influencing the production of specific proteins.

Mendel's theory of inheritance can be summarized as follows: (1) Alternative versions of genes, called alleles, cause variation in inherited traits. (2) Offspring inherit one copy of a gene from each parent. (3) An allele is dominant if it determines the phenotype of an organism even when paired with a different allele. (4) The two copies of a gene separate during meiosis and end up in different gametes. (5) Gametes fuse without regard to which alleles they carry. (6) The alleles of one genetic trait are assorted independently from the alleles of another genetic trait during gamete formation (modern genetics recognizes that this is uniformly true only of genes that lie on different chromosomes).

2. A sexually reproducing organism contains two copies of each gene because it gets one copy from each parent. If an individual is homozygous for a particular gene—say, it has genotype *gg*—each parent of that individual must have had at least one copy of the *g* allele. Therefore, the mother could have had genotype *Gg* or genotype *gg*; the same is true of the father.

3. New alleles arise when genes mutate. A mutation is any change in the DNA that makes up a gene. When a mutation occurs, the new allele that results may contain instructions for a protein with a form different from that of the protein specified by the original allele. By specifying different versions of proteins, the different alleles of a gene cause hereditary differences among organisms.

4.

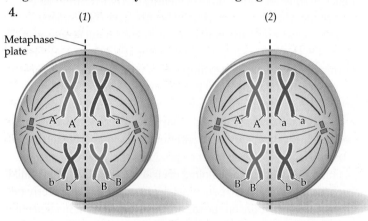

(1)   (2)

Metaphase plate

The diagrams shown here illustrate metaphase I of meiosis for an individual of genotype *AaBb*; the DNA has already replicated, so each of the four chromosomes consists of two identical chromatids (see Chapter 9). In these diagrams, the maternal chromosomes are shown in red and the paternal chromosomes are in blue. In diagram 1, by chance one maternal chromosome lined up to the right of the metaphase plate, while the other maternal chromosome lined up to the left of the plate. When the chromosomes line up in this way, the gametes have genotype *Ab* or *aB*; if you are not clear why this is so, see Figure 9.10 for more information. Alternatively, and also by chance, the chromosomes could have lined up as shown in diagram 2, causing the gametes to have genotype *AB* or *ab*. Overall, genes on different chromosomes are inherited independently of each other (in this case, producing gametes of genotype *AB*, *Ab*, *aB*, or *ab*) because the chromosomes on which they are found line up at random on the spindle microtubules during meiosis.

5. To determine the genotype of the purple-flowered plant (*PP* or *Pp*), you could cross it with a white-flowered plant (*pp*). When either genotype is crossed with the homozygous recessive (the white-flowered plant), the resulting phenotypes of the offspring will indicate whether the purple-flowered parent plant is heterozygous or homozygous. A plant with the genotype *Pp*, when crossed with a plant possessing the *pp* phenotype, will produce white-flowered offspring 50 percent of the time and purple-flowered offspring 50 percent of the time (diagram 1). A dominant homozygous plant (*PP*), when crossed with a recessive homozygous plant (*pp*), will always produce purple-flowered offspring (diagram 2).

(1)

	P	p
p	Pp	pp
p	Pp	pp

(2)

	P	P
p	Pp	PP
p	Pp	PP

6. Although identical twins are genetically identical, their phenotypes can differ because environmental factors can alter the effects of genes. A twin who is well nourished in childhood, for example, may grow up to be tall, while one who is malnourished in childhood may grow up to be short. Similarly, exposure to different amounts of sunlight will cause their skin color to differ. The phenotypes of twins can differ radically if one of them is exposed to environmental factors that trigger the onset of a genetic disorder to which both are predisposed, but the other is not.

7. Dominant alleles for lethal human diseases may be uncommon because a person carrying such an allele would be likely to develop the disease and perish before producing any offspring, so the frequency of such an allele in the human population would be low. Conversely, individuals possessing a recessive allele for a lethal genetic disorder can live as carriers, unaffected by the gene, and pass the allele on to their children, who, depending on the genetic makeup of their other parent, might develop the disease, or might become carriers in turn, remaining unharmed by the allele and passing it on to the next generation.

Sample Genetics Problems

1. a. *A* and *a*
   b. *BC*, *Bc*, *bC*, and *bc*
   c. *Ac*
   d. *ABC*, *ABc*, *Abc*, *AbC*, *aBC*, *aBc*, *abC*, and *abc*
   e. *aBC* and *aBc*

2. a. Genotype ratio: 1:1. Phenotype ratio: 1:1.

	A	a
a	Aa	aa

   b. Genotype ratio: 1:0. Phenotype ratio: 1:0.

	B
b	Bb

c. Genotype ratio: 1:1. Phenotype ratio: 1:1.

	AB	Ab
ab	AaBb	Aabb

d. Genotype ratio: 1BBCC:1BBCc:2BbCC:2BbCc:1bbCC:1bbCc. Phenotype ratio: 6:2, reduced to 3:1.

	BC	Bc	bC	bc
BC	BBCC	BBCc	BbCC	BbCc
bC	BbCC	BbCc	bbCC	bbCc

e. Genotype ratio: 1AABbCC:2AABbCc:1AABbcc:1AAbbCC:2AAbbCc: 1AAbbcc:1AaBbCC:2AaBbCc:1AaBbcc:1AabbCC:2AabbCc:1Aabbcc. Phenotype ratio: 6:2:6:2, reduced to 3:1:3:1.

	ABC	ABc	AbC	Abc	aBC	aBc	abC	abc
AbC	AABbCC	AABbCc	AAbbCC	AAbbCc	AaBbCC	AaBbCc	AabbCC	AabbCc
Abc	AABbCc	AABbcc	AAbbCc	AAbbcc	AaBbCc	AaBbcc	AabbCc	Aabbcc

3.

	S	s
S	SS	Ss
s	Ss	ss

Genotype ratio: 1SS:2Ss:1ss.
Phenotype ratio: 3 healthy:1 sickle-cell anemia.
Each time two Ss individuals have a child there is a 25% chance that the child will have sickle-cell anemia.

4. 100% of the offspring should be chocolate Labs.

5. a. NN and Nn individuals are normal; nn individuals are diseased.
   b.

	N	n
N	NN	Nn
n	Nn	nn

Genotype ratio: 1NN:2Nn:1nn.
Phenotype ratio: 3 normal:1 diseased.

   c.

	N	n
N	NN	Nn

Genotype ratio: 1:1. Phenotype ratio: 2 healthy:0 diseased.

6. a. DD and Dd individuals are diseased; dd individuals are normal.
   b.

	D	d
D	DD	Dd
d	Dd	dd

Genotype ratio: 1DD:2Dd:1dd.
Phenotype ratio: 3 diseased:1 normal.

   c.

	D	d
D	DD	Dd

Genotype ratio: 1:1. Phenotype ratio: 2 diseased:0 healthy.

7. The parents are most likely BB and bb. The white parent must be bb. The blue parent could potentially be BB or Bb, but if it were Bb, we would expect about half of the offspring to be white. Therefore, if the cross yields many offspring and all are blue, it is extremely likely that the blue parent's genotype is BB.

8. The allele for green fruit pods is dominant. Since each parent breeds true, that means each parent is homozygous. When a homozygous recessive parent is bred to a homozygous dominant parent, the F_1 generation will exhibit only the dominant phenotype. Therefore, the phenotype of the F_1 generation—green fruit pods—is produced by the dominant allele.

Chapter 11

1. A gene is a small region of the DNA molecule in a chromosome. Genes are located on chromosomes.

2. Human females have two X chromosomes, while human males have one X and one Y chromosome; therefore, human males have only one copy of each gene that is unique to either the X or the Y chromosome. As a result, patterns of inheritance for genes located on the X chromosome may differ between males and females. A mother can pass an X-linked allele, such as one for a genetic disorder, to her male or female offspring. A male can pass an X-linked allele only to his female offspring (because his male offspring receive his Y chromosome, not his X chromosome).

3. Genes located on different chromosomes separate into gametes independently of one another during meiosis; hence, such genes are not linked. If the genes for the traits shown in Figure 11.4 were inherited independently of each other, Morgan would have obtained approximately equal numbers of flies for each of the four genotypes shown in the figure. Since the numbers of the two parental genotypes outnumbered the other two genotypes by a wide margin, Morgan concluded that the genes must be located on the same chromosome. Because they are physically connected to each other, they are inherited together, or linked.

4. Crossing-over occurs when genes are physically exchanged between homologous chromosomes during meiosis. Part of the chromosome inherited from one parent is exchanged with the corresponding region from the other parent. Two genes that are far apart on a chromosome are more likely to be recombined by crossing-over than are two genes that are close to each other. Understanding this, one can assume that genes A and C are more likely to be separated into different gametes because of crossing-over than are genes A and B.

5. Nonparental genotypes may arise because of crossing-over. The exchange of genes that takes place during crossing-over makes possible the formation of gametes with combinations of alleles that differ from those found in either parent (see Figure 9.10).

6. Relatively few human genetic disorders are caused by inherited chromosomal abnormalities, probably because most large changes in the chromosomes kill the developing embryo. Genetic disorders caused by single-gene mutations appear to be more common, because the survival rate of embryos with single-gene mutations is higher.

Sample Genetics Problems

1. a. A male inherits his X chromosome from his mother, since his Y chromosome must come from his father. His mother does not have a Y chromosome to give him, and he must have one in order to be male.

   b. No, she does not have the disorder. If she has only one copy of the recessive allele, her other X chromosome must then have a copy of the dominant allele. She is a carrier, but she does not have the disorder herself.

   c. Yes, he does have the disorder. The trait is X-linked, he has only one X chromosome, and that X chromosome carries the recessive, disorder-causing allele. His Y chromosome does not carry an allele for this gene, so it cannot contribute to the male's phenotype relative to this trait.

   d. If the female is a carrier of an X-linked recessive disorder, her genotype is $X^D X^d$, where D is the dominant allele and d is the recessive, disorder-causing allele. This means she can produce two types of gametes relative to this trait: X^D and X^d. Only the X^d gamete carries the disease-causing allele.

   e. None of their children will have the disorder, because the mother will always contribute a dominant, non-disorder-causing allele to each child. However, all of the female children will be carriers, because their second X chromosome comes from their father, who has only one X chromosome to contribute, and it carries the disorder-causing allele.

2. a. 50% chance of aa cystic fibrosis genotype

	A	a
a	Aa	aa
a	Aa	aa

   b. 0% chance of aa cystic fibrosis genotype

	A	A
A	AA	AA
a	Aa	Aa

   c. 25% chance of aa cystic fibrosis genotype

	A	a
A	AA	Aa
a	Aa	aa

   d. 0% chance of aa cystic fibrosis genotype

	A	A
a	Aa	Aa
a	Aa	Aa

3. a. 50% chance of Huntington's disease genotype, Aa

	A	a
a	Aa	aa
a	Aa	aa

   b. 100% chance of Huntington's disease genotype, AA or Aa

	A	A
A	AA	AA
a	Aa	Aa

   c. 75% chance of Huntington's disease genotype, AA or Aa

	A	a
A	AA	Aa
a	Aa	aa

   d. 100% chance of Huntington's disease genotype, Aa

	A	A
a	Aa	Aa
a	Aa	Aa

4. a. 0% chance of a child with hemophilia

	X^a	Y
X^A	$X^A X^a$	$X^A Y$
X^A	$X^A X^a$	$X^A Y$

   b. 50% chance of a child with hemophilia, $X^a X^a$ or $X^a Y$

	X^a	Y
X^A	$X^A X^a$	$X^A Y$
X^a	$X^a X^a$	$X^a Y$

   c. 25% chance of a child with hemophilia, $X^a Y$

	X^A	Y
X^A	$X^A X^A$	$X^A Y$
X^a	$X^A X^a$	$X^a Y$

   d. 50% chance of a child with hemophilia, $X^a Y$

	X^A	Y
X^a	$X^A X^a$	$X^a Y$
X^a	$X^A X^a$	$X^a Y$

   e. No, male and female children do not have the same chance of getting the disease. Male children are more likely to have hemophilia, because they do not possess a second allele for this trait to mask a recessive allele that they may inherit.

5. The terms "homozygous" and "heterozygous" refer to pairs of alleles for a given gene. Since a male has only one copy of any X-linked gene, it does not make sense to use these pair-related terms.

6. Although neither the mother nor the father expresses the trait in question, some of their children do, so the disease-causing allele (d) is recessive and is carried by both parents. The disease-causing allele is located on an autosome. If it were on the X chromosome, the father would express the gene, since we have already

determined that he must carry one recessive copy of the gene and he would not have another copy of the gene to mask this recessive allele. Both individuals 1 and 2 of generation I have the genotype Dd.

7. Designate the dominant, X-linked allele D and the recessive normal allele d. According to the Punnett squares shown in (a) and (b) (below), males are not more likely than females to inherit a dominant, X-linked genetic disorder.

    a. There are two possible Punnett squares, depending on whether the affected female has genotype $X^D X^d$ or genotype $X^D X^D$:

	X^d	Y
X^D	$X^D X^d$	$X^D Y$
X^d	$X^d X^d$	$X^d Y$

or

	X^d	Y
X^D	$X^D X^d$	$X^D Y$

    b. This cross is $X^d X^d \times X^D Y$, which gives the following Punnett square:

	X^D	Y
X^d	$X^D X^d$	$X^d Y$
X^d	$X^D X^d$	$X^d Y$

8. The disorder allele is a recessive allele, located on the X chromosome. We know the allele is recessive because individual 2 in generation II carries the allele but does not have the condition. If the allele were located on an autosome, the parents in generation I would be of genotype AA (the male) and aa (the female). In this case, none of the individuals in generation II could have the condition—yet two of them do have the condition implying that the allele is on a sex chromosome. Finally, we know that the allele is on the X chromosome, because otherwise only males could get the condition.

9. a. If the two genes are completely linked:

	AB	ab
aB	$AaBB$	$aaBb$

   b. If the two genes are on different chromosomes:

	AB	Ab	aB	ab
aB	$AaBB$	$AaBb$	$aaBB$	$aaBb$

Chapter 12

1. Frederick Griffith's experiment with two strains of bacteria and mice showed that harmless strain R bacteria could be transformed into deadly strain S bacteria when exposed to heat-killed strain S bacteria. This finding suggested that genetic material from the heat-killed strain S bacteria had somehow changed living strain R bacteria into strain S bacteria.

    Oswald Avery, Colin MacLeod, and Maclyn McCarty isolated and tested different compounds from the bacteria in Griffith's experiments and found that only DNA from heat-killed strain S bacteria was able to transform harmless strain R bacteria into deadly strain S bacteria. This finding led to the conclusion that DNA, not protein, is the genetic material.

    Alfred Hershey and Martha Chase studied a virus that consists only of a DNA molecule surrounded by a coat of proteins. By using radioisotopes to selectively label either the DNA or the protein portion of the virus, Hershey and Chase showed that the DNA, but not the proteins, entered a bacterium to take over the bacterial cell and produce the next generation of viruses.

2. The three main components of a nucleotide from a DNA molecule are the sugar deoxyribose, a phosphate group, and a nitrogenous base. The base is what makes one type of nucleotide different from others. The four bases are adenine (A), cytosine (C), guanine (G), and thymine (T). When covalent bonds link nucleotides, one strand of the double helix that is a DNA molecule is formed.

3. The two strands that make up the double helix of the DNA molecule are held together by hydrogen bonds between the nucleotides' nitrogenous bases.

4. The genetic information of the alleles is contained in the sequence of the nitrogenous bases—adenine (A), cytosine (C), guanine (G), and thymine (T)—found within the segment of DNA that constitutes each allele. At any genetic locus, different alleles differ in the sequence of bases they contain. Therefore, the DNA segments of the two codominant alleles A^1 and A^2 differ in the sequence of bases found in each allele.

5. The double helical structure of DNA and the base-pairing rules theorized by Watson and Crick suggested a simple way that genetic material could be copied. Because A pairs only with T and C pairs only with G, each strand of DNA contains the information needed to produce the complementary strand. DNA replication involves separation of the two strands of the helix, each of which then serves as a template for the construction of its complementary strand, resulting in two identical copies of the original DNA molecule.

6. The sequence of bases in DNA is the basis of inherited variation. A change in the sequence of bases in DNA, whether because of an error during replication or because of exposure to a mutagen, is a mutation. Such a change could result in a new allele that would encode a new version of the protein encoded by the gene in which the mutation occurred. If the new allele produced a protein that did not function properly (or at all), serious damage could be done to the cell, and consequently to the organism; such an allele could cause a genetic disorder.

7. DNA is repaired by protein complexes that include enzymes. When DNA is being replicated, enzymes check for and immediately correct mistakes in pair bond formation. Mistakes that escape this process, called mismatch errors, are caught and corrected by repair proteins.

    DNA repair is essential for cells to function normally because DNA is constantly being damaged by chemical, physical, and biological agents. If none of this damage were repaired, genes that encoded proteins critical to life would eventually cease to function, thereby disabling the production of those proteins and killing the cell, and ultimately the organism.

Chapter 13

1. A gene is a DNA sequence that contains information for the synthesis of one of several types of RNA molecules used to make proteins. A gene stores information in its sequence of nitrogenous bases.

2. Genes control the production of a variety of RNA products including mRNA, rRNA, tRNA. Messenger RNA (mRNA) encodes the amino acid sequence of proteins, ribsomal RNA (rRNA) is an essential component of ribosomes (the site of protein synthesis), and transfer RNA (tRNA) carries amino acids to the ribosomes during protein construction. Therefore, each of the RNA products specified by genes functions in the synthesis of proteins. Proteins are essential for many functions that support life. In cells and organisms, proteins provide structural support, transport materials through the body, and defend against disease-causing organisms. Enzymes are a class of proteins that speed up chemical reactions.

3. Genes commonly contain instructions for the synthesis of proteins. Each gene is composed of a segment of DNA on a chromosome and consists of a sequence of the four bases: adenine (A), cytosine (C), guanine (G), and thymine (T). The sequence of bases specifies the amino acid sequence of the gene's protein product. Through transcription and translation, proteins are produced from the information stored in genes. In transcription, mRNA is synthesized directly from the sequence of bases in one DNA strand inside the nucleus of a cell. Translation occurs in the cytoplasm and converts the sequence of bases in an mRNA molecule into the sequence of amino acids in a protein. Proteins, by their many and various functions, influence the phenotype of an individual.

4. For a protein to be made in eukaryotes, the information in a gene must be sent from the gene, which is located in the nucleus, to the site of protein synthesis, on a ribosome. This transfer of information requires an intermediary molecule because DNA does not leave the nucleus but ribosomes are located in the cytoplasm. In eukaryotes, a newly formed mRNA molecule usually must be modified before it can be used to make a protein. The reason is that most eukaryotic genes contain internal sequences of bases (introns) that do not specify part of the protein encoded by the gene. DNA sequences copied from introns must be removed from the initial mRNA product if the protein encoded by the gene is to function properly.

5. RNA splicing is a step in RNA processing that involves the removal of introns from a newly transcribed mRNA and the joining together of the remaining exons to create the mature, export-ready form of the mRNA. RNA splicing is not known to occur in prokaryotes, but is common among eukaryotes. The great majority of our protein-coding genes produce mRNA that must undergo RNA splicing before they are allowed to exit the nucleus.

6. Messenger RNA is the product of transcription, and is a version of the genetic information stored in a gene. The mRNA moves from the nucleus to the cytoplasm, where it binds with a ribosome to guide the construction of a protein.

   Ribosomal RNA is a major component of ribosomes. Translation occurs at ribosomes, which are molecular machines that make the covalent bonds linking amino acids to form a particular protein.

   Transfer RNA molecules carry the amino acids specified by the mRNA to the ribosome. At the ribosome, a three-base sequence (anticodon) on the tRNA binds by means of complementary base pairing with the appropriate codon on the mRNA. Each tRNA molecule carries the amino acid specified by the mRNA codon to which its tRNA anticodon can bind.

7. If a tRNA molecule does not function properly because of a mutation, each protein that it helps to build will be altered in some way. By failing to bind properly with the mRNA codons of many different genes, a mutant tRNA may significantly affect the structure of many different protein products. Because their structure is altered, the function of these protein products may be impaired. Since proteins are key components of many metabolic reactions, changing the function of many different proteins can result in a series of metabolic disorders.

8. A mutation is any alteration in the information coded within an individual's DNA. Sometimes the effects of that change can be detected as a change in the inherited characteristics of that individual (a phenotypic change). But in other cases, the mutation may be "silent," with no outward sign that a mutation has occurred. Most mutations are neutral in their impact on the individual, neither benefiting it nor harming it. Some mutations have harmful effects, and rarely a mutation might produce a change that enhances the individual's ability to survive and reproduce in a particular environment.

   A very large number of our genes carry information for the construction of specific proteins. That information is carried in the form of a sequence of chemical units, called nitrogenous bases, that in turn specify the sequence in which the amino acids in a protein are strung together (proteins are built from amino acids, and each unique protein has a unique sequence of amino acids). When the base sequence of a protein-coding gene changes, the amino acid sequence of its protein is altered as well. Every protein has special chemical and biological properties that are critical to its function, and most of those properties stem from the precise sequence of amino acids in it. If the amino acid sequence of a protein changes because of a mutation in the gene that codes for it, the biological function of that protein may change as well.

Chapter 14

1. Two features enable cells to pack an enormous amount of DNA into a very small space: the thinness of the DNA molecule and a highly organized, complex packing system. Each portion of a chromosome

which contains one DNA molecule, consists of many tightly packed loops. Each loop is composed of a chromatin fiber consisting of many nucleosome spools, which are made of proteins called histones. A segment of DNA winds around each spool, and if that DNA were unwound it would reveal its double helical structure.

2. Prokaryotes have less DNA than eukaryotes have. All DNA in a prokaryote is located on one chromosome; in eukaryotes, by contrast, the DNA is distributed among several chromosomes. Eukaryotes have more genes than prokaryotes do, and genes constitute only a small portion of the eukaryotic genome. Most prokaryotic DNA encodes proteins, and very little of it is noncoding DNA and transposons. Functionally related genes in prokaryotes are grouped together on the chromosome; eukaryotic genes with related functions often are not located near one another.

3. The bacterium would begin expressing the gene responsible for encoding the enzyme that breaks down arabinose. Organisms can turn genes on and off in response to short-term changes in food availability or other features of the environment.

4. In multicellular organisms, different cell types express different genes by controlling transcription, along with other methods. By switching specific genes on or off, cells can vary their structure and perform specialized metabolic tasks, even though each has exactly the same genes (and alleles).

5. Homeotic genes are a class of master-switch genes that have a central role in the development of the body plan and in the differentiation of organs. The homeotic genes of multicellular animals first evolved hundreds of millions of years ago, and since then they have been used in similar ways to organize the body plans of animals as different as fruit flies and humans. Homeotic genes are an example of highly conserved genes, those that have a similar base sequence and general function among diverse groups of organisms. Many homeotic genes control early stages of development—events such as specification of the head and tail ends of an embryo or the demarcation of body segments—that tend not to change much over evolutionary time.

6. From gene to protein, the steps at which gene expression can be controlled are as follows: (1) The expression of tightly packed DNA can be prevented, in part because the proteins necessary for transcription cannot reach them. (2) Transcription can be regulated by regulatory proteins that bind to regulatory DNA, effectively switching a gene on or off. (3) The breakdown of mRNA molecules can be regulated such that mRNA is destroyed hours or weeks after it is made. (4) Translation can be inhibited when proteins bind to mRNA molecules to prevent their translation. (5) Proteins can be regulated after translation, either when the cell modifies or transports them or when they are rendered inactive by repressor molecules. (6) Synthesized proteins can be destroyed.

7. Regulatory DNA sequences such as the tryptophan operator switch genes on and off. The tryptophan operator controls whether or not the gene that encodes tryptophan is transcribed. If tryptophan is present, it binds to a repressor protein, which is then able to bind to the operator and prevent transcription because the presence of tryptophan indicates that the cell does not need to waste energy

by making more. If tryptophan is absent, the repressor is unable to bind to the operator, allowing the gene to be transcribed so that the tryptophan needed by the cell is produced.

8. A DNA chip consists of thousands of samples of DNA placed on a small glass surface. Often the DNA at each position on a chip corresponds to the DNA of one gene. When a gene is expressed, an mRNA copy of the information in that gene is produced. To study many genes at once, mRNA is isolated from the organism or cells being tested, labeled (as with a dye that glows red or green), and then washed over the DNA chip. The labeled mRNA can bind to the DNA representing the gene from which it was originally produced. Because the gene that corresponds to each location on the chip is known, results from this procedure can tell us which of the organism's genes produced mRNA—and hence, which of the organism's many genes were expressed (and which were not).

Chapter 15

1. To produce domesticated species, humans have manipulated the reproduction of other organisms, selecting for desirable qualities that, over time, have become standard in domesticated species. Although such selection practices do lead to changes in the DNA of organisms (that is, they lead to an increase in the frequencies of alleles that control the inheritance of the traits we select for), genetic engineering enables us to make much greater changes in a much shorter span of time. Using such methods, we can manipulate the DNA of organisms directly, and we can transfer genes from one species to another. Transfers of DNA from, say, a human to a bacterium (as is done in the production of human insulin) far exceed the scope of DNA transfers that occur in nature or are possible through conventional breeding of crops and farm animals. We can also selectively change specific DNA sequences—something we could never do before. Overall, we can now manipulate DNA with greater power and precision than we could when we domesticated species such as dogs, corn, and cows.

2. By using restriction enzymes and gel electrophoresis together, geneticists can examine differences in DNA sequences. Judy and David could be tested for the sickle-cell allele by use of the restriction enzyme *Dde*I, which cuts the normal hemoglobin allele into two pieces but cannot cut the sickle-cell allele. Their doctor might also want to use a DNA probe to test for the sickle-cell allele in these would-be parents. A DNA probe is a short, single-stranded segment of DNA with a known sequence, usually tens to hundreds of bases long. A probe can pair with another single-stranded segment of DNA if the sequence of bases in the probe is complementary to the sequence of bases in the other segment.

3. DNA cloning is the introduction of a DNA fragment into a host cell that can generate many copies of the introduced DNA. The purpose of DNA cloning is to multiply a particular DNA fragment, such as a specific gene, so that a large amount of this DNA is made available for

further analysis and manipulation. Bacteria are the most common host cells in DNA cloning. Two of the most common methods of cloning a gene are constructing a DNA library and using the polymerase chain reaction. To build a DNA library, a vector such as a plasmid is used to transfer DNA fragments from the organism whose gene is to be cloned to a host organism, such as a bacterium. To clone a gene by PCR, primers are synthesized, enabling DNA polymerase to produce billions of copies of the gene in a few hours.

4. The advantage of DNA cloning is that it is easier to study a gene and its function, and to manipulate that function for practical benefits, once you have many copies of it. Once a gene is cloned, it can be sequenced, transferred to other organisms, or used in various experiments. Today, many lifesaving pharmaceuticals, such as human insulin, human growth hormone, human blood-clotting proteins, and anticancer drugs, are manufactured by bacteria that have been genetically modified by inserting cloned human genes into them.

5. Bacterial colonies acting as hosts in the DNA library are screened by DNA hybridization to see if their DNA can pair with a probe for the gene of interest. Colonies whose DNA can pair with the probe contain all or part of the desired gene. It is often necessary to screen the colonies on many petri dishes before such a colony is found.

6. Genetic engineering is the permanent introduction of one or more genes into a cell, a certain tissue, or a whole organism, leading to a change in at least one genetic characteristic in the recipient. The organism receiving the DNA is said to be genetically modified (GM) or genetically engineered (GE). To create GM organisms, a DNA sequence (often a gene) is isolated, modified, and inserted back into the same species or into a different species.

7. and 8. These answers depend on your viewpoint, which we hope is also guided by scientific facts.

Interlude C

1. A human and a tomato plant would probably share sets of genes that control DNA replication, gene transcription and translation, cell division, glycolysis and cellular respiration, and some aspects of membrane structure and function (synthesis of phospholipids, for example). Humans would not share with plants sets of genes for proteins involved with vision, various sensory receptors, or nerve function.

2. Single nucleotide polymorphisms (SNPs) are differences in single base pairs that are found in the DNA of unrelated individuals. SNPs or groups of SNPs can be linked to the propensity to develop certain diseases, so SNP testing can reveal whether an individual is at risk for developing one of those diseases.

3. Knowing the genome of the mosquito that carries malaria may assist in the development of pesticides that could specifically disrupt certain biological processes in that species and be more effective at killing it.

4. Genetic screening of patients could help doctors diagnose diseases and assign proper treatments or preventive measures. On the other hand, if genetic screening of an individual showed the likelihood of a serious illness whose treatment would be costly, insurance companies might deny coverage. Genetic screening could raise concerns of privacy violation or increase the likelihood that unborn children inheriting "undesirable" genetic conditions would be aborted.

5. Some considerations: Because of privacy concerns, patients should have the option of whether or not to undergo genetic screening, but it would often be in the best interest of patients to know what preventive steps to take. Doctors should inform their patients that possible genetic dispositions toward certain medical conditions do not always lead to illness.

Chapter 16

1. Evolution is change in the genetic characteristics of a population over time, which can occur through mechanisms such as genetic mutation or natural selection. Since the genotypes of individuals do not change, a population can evolve but an individual cannot.

2. In the new habitat, larger lizards will have an advantage over smaller ones. The larger lizards of the species will therefore be more likely to pass the trait of large size on to their offspring, and the average size of lizards in the population will increase over time because of this selective advantage.

3. Each of these aspects of life on Earth can be explained by evolution. (a) Adaptations, which improve the performance of an organism in its environment, result from natural selection. (b) The diversity of life results from speciation, which occurs when one species splits to form two or more species. (c) Organisms can share puzzling characteristics because of common descent. Consider the wing of a bird, the flipper of a whale, and the arm of a human. Even though these appendages are used for very different purposes—and hence we would not expect them to be structurally similar—they are composed of the same set of bones. The reason is that birds, whales, and humans share a common ancestor that had these bones. Organisms can also share less puzzling characteristics because of convergent evolution resulting from similar selective pressures.

4. Overwhelming evidence indicates that evolution occurred and continues to occur. Support for evolution comes from five lines of evidence:

a. The fossil record provides clear evidence of the evolution of species over time and documents the evolution of major groups of organisms from previously existing organisms.

b. Organisms contain evidence of their evolutionary history. For example, scientists find that studies of proteins and DNA support the evolutionary relationships determined by anatomical data; that is, the proteins and DNA of closely related organisms are more similar than those of organisms that do not share a recent common ancestor. In this and many similar examples, the extent to which organisms share characteristics other than those used to determine evolutionary relationships is consistent with scientists' understanding of evolution.

c. Scientists' understanding of evolution and continental drift has enabled them to predict the geographic distributions of certain fossils, depending on whether the organisms evolved before or after the breakup of Pangaea.

d. Scientists have gathered direct evidence of small evolutionary changes in thousands of studies by documenting genetic changes in populations over time.

e. Scientists have observed the evolution of new species from previously existing species.

5. In any area of science, new pieces of information are continually being added to our knowledge. The debate among scientists as to which mechanisms of evolution are most important means only that evolution is not fully understood, not that it does not occur.

6. Genetic drift has a greater effect on smaller populations. If the plant population were larger, the likelihood that all plants of a certain genotype would die in a windstorm would be smaller, and the dramatic shift in the frequencies of the A and a alleles would therefore be less likely.

Chapter 17

1. **Mutation:** A nonlethal mutation of a particular allele can be inherited by offspring, thereby increasing the frequency of the mutant allele in a population over time.
Gene flow: The exchange of alleles between populations can change the frequencies at which alleles are found in the populations by introducing new alleles. Populations affected by gene flow tend to become more genetically similar to one another.
Genetic drift: Random events (such as chance events that influence the survival or reproduction of individuals) can cause one allele to become dominant in a small population. By chance alone, drift can lead to the fixation of alleles in small populations; if these alleles are harmful, the population may decrease in size, perhaps to the point of extinction.
Natural selection: If an inherited trait provides a selective advantage for individuals in a certain population, individuals with that trait will be more likely to reproduce, and the frequency of the allele for that trait will increase in succeeding generations. Likewise, an allele for a disadvantageous trait will be selected against, and will be found with decreasing frequency over time.

2. Genetic variation in a population provides the "raw material" on which evolution can work. In the absence of genetic variation, evolution cannot occur. Natural selection, for example, causes individuals of some genotypes to leave more offspring than individuals of other genotypes leave, but this sorting process cannot occur if all individuals have the same genotype.

3. Recombination (a term that refers collectively to fertilization, crossing-over, and the independent assortment of chromosomes) causes offspring to have combinations of alleles that differ from one another and from those found in their parents. Therefore, recom-

bination greatly increases the genetic variation on which evolution can act.

4. **Gene flow** is the exchange of alleles between populations. Gene flow makes populations more similar to one another in their genetic makeup. Genetic drift is a process by which alleles are sampled at random over time. **Genetic drift** can have a variety of causes, such as chance events that cause some individuals to reproduce and prevent others from reproducing. **Natural selection** is a process in which individuals with particular inherited characteristics survive and reproduce at a higher rate than other individuals. **Sexual selection** is a form of natural selection in which individuals with certain traits have an advantage in attracting mates, and consequently in passing those traits on to offspring.

5. The potential benefits include making the population larger, and therefore less susceptible to genetic drift, and providing an input of new alleles on which natural selection can operate. The potential drawbacks include the introduction of individuals with genotypes that are not well matched to the local environmental conditions of the smaller population. Throughout time, some species have gone extinct locally or worldwide. Extinction is a natural process. However, humans have greatly increased the rate at which populations and species become extinct. If numerous other populations of a species exist, it may not be worth introducing new members to the smaller population. If the smaller population is one of the few populations of that species left, however, it may be important to introduce new individuals in an attempt to enable the population to recover and survive.

6. Genotype frequencies for the original population:

$$AA: \frac{280}{280\ +\ 80\ +\ 60} = 0.67$$

$$Aa: \frac{80}{280\ +\ 80\ +\ 60} = 0.19$$

$$aa: \frac{60}{280\ +\ 80\ +\ 60} = 0.14$$

Allele frequencies for this population:

Frequency of A allele =

$$p\ =\ \frac{2(280)+80}{2(280+80+60)}\ =\ 0.76$$

Frequency of a allele =

$$q\ =\ \frac{2(60)+80}{2(280+80+60)}\ =\ 0.24$$

The Hardy–Weinberg equation predicts that the frequency of genotype AA should be
$$p^2 = (0.76)(0.76) = 0.58,$$
that the frequency of genotype Aa should be
$$2pq = (2)(0.76)(0.24) = 0.36,$$
and that the frequency of genotype aa should be
$$q^2 = (0.24)(0.24) = 0.06.$$

Note the sum of the genotype frequencies,

$$p^2 + 2pq + q^2 = 1.0.$$

These calculated genotype frequencies do not match those of the original population. This difference could be due to mutation, nonrandom mating, gene flow, a small population size, and/or natural selection.

7. Using an antibiotic drug to kill large numbers of bacteria tends to give a considerable reproductive advantage to bacteria that possess resistance to the drug. Since bacteria reproduce extremely rapidly, the entire population of bacteria will soon be resistant. Reducing human exposure to the bacteria would not enable resistant strains to have as great a degree of reproductive advantage over normal bacterial strains. Slowing the growth of bacteria would likewise help limit the reproductive advantage of resistant strains.

Chapter 18

2. Individuals with inherited traits that enable them to survive and reproduce better than other individuals replace those with less favorable traits. This process, by which natural selection improves the match between organisms and their environment over time, is called adaptive evolution. In our efforts to kill or control bacteria that cause infectious diseases, we are creating a new environment in which bacteria that cannot withstand antibiotics are eliminated. Often, some bacteria in the population are not killed by antibiotics; these bacteria reproduce, increasing the frequency of resistant bacteria in the population. These evolutionary changes in disease-causing bacteria are harmful to us because more and more of the diseases we encounter will be resistant to medical treatment.

3. Species that hybridize in nature may still be distinct species, because of a host of alleles that do not affect their ability to interbreed but may cause them to look different or to differ from each other ecologically. For this reason, many people would argue that the rare species is separate from the common species and should remain classified as rare and endangered.

4. Defining species by their inability to reproduce sexually with other species is convenient, but many alleles do not affect reproductive isolation yet could cause the two oaks to be different enough that they could be classified as separate species, even though they would be able produce hybrids.

5. Because this storm-blown population is now geographically isolated from other populations of its species, there will be little or no gene flow to such populations. As a result, genetic changes due to mutation, genetic drift, and natural selection will accumulate over time. Natural selection is likely to cause genetic change in the population because the new environment is different from the parent population's environment; genetic drift will probably also be important, because the island population is small (making drift more likely). As a by-product of genetic changes due to selection, drift, and mutation, the island population may become reproductively isolated from the parent population. If the island population remains isolated long enough, a sufficient number of genetic changes may accumulate for it to evolve into a new species.

6. Some of the cichlid populations of Lake Victoria may have had so little contact with one another that they can be said to have evolved into separate species in geographic isolation, even though they live in the same lake. Other populations may have evolved into new species in the absence of geographic isolation.

7. New plant species can form in the absence of geographic isolation as a result of polyploidy, a condition in which an individual has more than two sets of chromosomes. Strong evidence suggests that sympatric speciation occurred in the cichlids that live in Lake Bermin and Lake Barombi Mbo, and evidence is accumulating that apple and hawthorn populations of the apple maggot fly are diverging into two species, despite living in the same area. In apple maggot flies and cichlids, sympatric speciation is promoted by ecological factors (such as selection for specialization on different food items) and sexual selection.

There is a greater potential for gene flow between populations whose geographic ranges overlap than between populations that are geographically isolated from one another. Gene flow tends to cause populations to remain (or become) similar. Therefore, in the absence of geographic isolation, it can be difficult for genetic differences great enough to cause reproductive isolation to accumulate over time. As a result, sympatric speciation occurs less readily than allopatric speciation.

Chapter 19

1. One example of the evolution of one group of organisms from another is the emergence of mammals from reptiles. The emergence of the mammalian jaw and teeth illustrates the steps in this process. The first step was the development of an opening in the jaw behind the eye. Then, more powerful jaw muscles and specialized teeth appeared with the therapsids. Finally, a subgroup of these reptiles, the cynodonts, emerged with more specialized teeth and a more forward hinge of the jaws, completing this aspect of mammalian evolution.

2. The emergence of photosynthesis in ancient organisms gradually led to the buildup of O_2 in the atmosphere, which killed many organisms to which oxygen was toxic. However, the oxygen supplied by photosynthetic organisms made possible the evolution of eukaryotes and later multicellular life-forms.

3. The Cambrian explosion was a large increase in the diversity of life-forms over a relatively short time about 530 million years ago. Larger organisms of most phyla emerged during this period, setting the stage for the colonization of land.

4. The colonization of land led to another great increase in the diversity of life-forms. Life on land required different means of mobility and reproduction, adaptations to obtain and retain water, and ways to breathe in air rather than water. Early terrestrial organisms had the opportunity to expand into new types of largely unoccupied

habitat, which provided ample resources for those organisms able to survive the challenges of life on land.

5. A mass extinction event may be associated with rapid environmental changes that have no relation to the conditions that favored a particular adaptation. Organisms with wonderful adaptations can (and have) become extinct during mass extinction events.

6. Although speciation can happen within a single year, it often takes hundreds of thousands to millions of years to occur. It is not surprising, then, that it usually takes 10 million years for the number of species found in a region to rebound after a mass extinction event. The time required to recover from mass extinction events provides a powerful incentive for humans to halt the current, human-caused losses of species; otherwise, it will take millions of years for biodiversity to recover.

7. *Microevolution* refers to the changes in allele or genotype frequencies within a population of organisms. This concept is fundamentally different from *macroevolution*, which deals with the rise and fall of entire groups of organisms, the large-scale extinction events that cause some of these changes, and the evolutionary radiations that follow extinction events. Macroevolutionary changes cannot be predicted solely from an understanding of the evolution of populations. Of the evolutionary processes discussed in Unit 4, speciation occupies a "middle ground" between macroevolution and microevolution.

Interlude D

2. If other human species shared the world with us today, there would undoubtedly be instances of social or cultural friction between them and modern humans, just as there exist tensions between "different" ethnic groups today. In order for there to be peaceful cohabitation, societies would need to recognize the humanity of other human species and discuss ways of cooperation in sharing living space and resources.

3. It is generally recognized that people have an ethical responsibility to care for living things in the environment. The loss of so many species would have negative consequences not only for the environment, but for people as well. Human actions that could potentially lead to a mass extinction should therefore be seriously examined and measures taken to avert such an occurrence.

4. An ethical case for preventing people who carry harmful genes from reproducing would be very difficult to make. Deciding which harmful alleles might be detrimental to the societal good would be largely subjective, and mandatory testing would be a serious violation of human rights and privacy. Living under such a system would probably be very restrictive to personal freedom.

Chapter 20

1. The description of the biosphere as an "interconnected web" is apt because all organisms within it are connected by their interactions, as shown by examples ranging from various food chains to symbiotic relationships between species. The global spread of invasive species and their often detrimental effects on native ecosystems and the change in red kangaroo populations in response to the "dingo fence," are some of the case histories discussed in this chapter that underscore the interrelatedness of organisms throughout the biosphere.

2. Giant convection cells in the atmosphere and ocean currents carry the results of local events (such as a volcanic eruption or an oil spill) to distant areas around the globe. For example, oil spilled into an ocean current next to one continent's shore may be carried by that current and end up coating the shores of other continents. If shorebirds on those continents are killed when they become coated with oil from the spill, they may no longer keep populations of their food organisms under control, and they will no longer be available as a food source for their predators.

3. These are some of Earth's terrestrial biomes: tropical forest, temperate deciduous forest, grassland, chaparral, desert, boreal forest, and tundra. Identify the terrestrial biome in this list that is closest to where you live. Read section 20.5 for a description of the climatic and ecological characteristics of this biome.

4. The defining feature of a desert is the scarcity of moisture (less than 25 cm of precipitation), not high temperatures. Antarctica, which receives less than 2 centimeters of precipitation per year, is the largest cold desert in the world. The Sahara desert in northern Africa is the largest hot desert. Because desert air lacks moisture, it cannot retain heat and therefore it cannot moderate daily temperature fluctuations. As a result, temperatures may be above 45°C (113°F) in the daytime, plunging to near freezing at night.

5. An estuary, where a river empties into the sea, is a shallow marine ecosystem. The plentiful light, the abundant supply of nutrients delivered by the river system, and the regular stirring of nutrient-rich sediments by water flow creates a highly productive community of photosynthesizers. The coastal region, which stretches from the shoreline to the continental shelf, is among the most productive marine ecosystems because of the ready availability of nutrients and oxygen. Nutrients delivered by rivers and washed off the surrounding land accumulate in the coastal region. Nutrients that settle to the bottom are stirred up by wave action, tidal movement, and the turbulence produced by storms. The nutrients support the growth of photosynthetic producers, which inhabit the well-lit upper layers of coastal waters to a depth of about 80 meters (260 feet). The vigorous mixing by wind and waves also adds atmospheric oxygen to the water.

Chapter 21

1. A population may be difficult to define if the boundaries of its range are unclear, if its members move around frequently, or if its members are small and hard to count.

2. If a population is threatened with extinction, possible options for saving it might be to protect it from human disturbances, to treat diseases, to reduce the number of predators, to move the population to an area with greater food supply (limiting death or emigration), to introduce individuals from other populations of the species (increasing immigration), or to institute captive breeding programs (increasing the birth rate).

3. Coordinates for the graph, in the notation (x coordinate, y coordinate): (1, 150); (2, 225); (3, 337.5); (4, 506.3); (5, 759.4).

4. a. Some factors that limit population growth include available habitat, available food and water, disease, weather, natural disturbances, and predators.

   b. Species new to an area often do not have established predators. In addition, they have not yet reached the carrying capacity of their habitat.

5. A density-dependent factor is one whose intensity increases as the density of the population increases. An example of a density-dependent factor is an infectious disease that spreads more rapidly in densely populated areas. A density-independent factor is not affected by the density of the population. Temperature is a density-independent factor: if the temperature drops below what a certain plant species can tolerate, it does not matter how dense the plant population is; the plants will still die.

6. If the pattern of population growth is understood, managers may be able to manipulate the factors that most directly affect the population's growth rate. If a population of organisms were more successful because, for example, it had adequate access to water, a manager would be sure that nearby rivers were not drained off for agricultural needs.

7. Limit reproduction to no more than one child per parent; reduce the consumption of unnecessary goods; reuse and recycle items to promote the sustainable use of resources; work to develop and follow environmentally friendly policies and activities (for instance, using energy-efficient cars and lightbulbs); purchase goods that have a lower impact on the environment, including organically grown clothing and food items.

Chapter 22

1. Mutualism is common because its costs are outweighed by the benefits it provides. Yucca plants, for example, may lose a few seeds to the offspring of their moth pollinators, but they still end up with more seeds than if the moths had not pollinated them to begin with.

2. Organisms eaten by consumers are under selection pressure to develop defenses against those consumers. Likewise, consumers experience selection pressure to overcome the defenses of their food organisms. Adaptations that improve the survival of individuals in either group will therefore be likely to spread throughout the population. An example would be the poison of the rough-skinned newt, which can kill nearly all predators; garter snakes, however, have evolved the capacity to tolerate the toxin and eat the newt.

3. The inferior competitor is still using resources that the superior competitor needs, thereby possibly limiting the superior competitor's distribution or abundance.

4. A superior competitor can reduce the abundance of an inferior competitor in a particular habitat, and even restrict the other species' distribution, by competitively excluding it from a habitat. A consumer can have a similar effect on its food organisms. In contrast, each partner in a mutualism is likely to be found only where the other partner is present.

5. The plant community would have fewer species (a). When the rabbits were removed, the grass they prefer would no longer be eaten, so that grass would assert its dominance as the superior competitor and would probably drive some of the other grass species in the area to extinction.

Chapter 23

1. a. Food webs influence the movement of energy and nutrients through a community. Some species, called keystone species, have a disproportionately large effect, relative to their abundance, on the types and abundances of other species in the community.

   b. Disturbances such as fire occur so often in many ecological communities that the communities are constantly changing and hence may never establish climax communities (relatively stable end points of ecological succession). Depending on the type and severity of the disturbance, a given community may or may not be able to recover.

   c. Climate is a key factor in determining which organisms can live in a given area, so if the climate changes the community changes.

   d. As continents move to different latitudes, their climates—and therefore their communities—change.

2. Primary succession occurs in a newly created habitat that contains no species. Secondary succession is the process by which communities recover from disturbance.

3. The introduction of beard grass to Hawaii has increased the frequency and size of fires on the island. This change is due to the large amount of dry matter that the grass produces, which burns more easily and hotter than does the native vegetation. In this way, the presence of one species in the community has profoundly altered its disturbance pattern, leading to other large changes in the community.

4. The disturbance described in (b) would probably require a longer recovery time, assuming that no other disturbances, such as fire, were to occur. In disturbance (a), the soil and ground cover would be left intact, so new trees would be able to sprout according to natural succession, eventually growing to replace the trees that were removed. In (b), however, the pollutant would have damaged the soil, which would hinder the ability of the trees and ground vegetation to grow. The soil chemistry would need to return to normal before the forest vegetation would be able to grow back and thrive again.

5. Change is a part of all ecological communities. However, human-caused change is unique in that we can consider the impact of our actions, and we can decide whether or not to take actions that cause community change. Whether or not a particular change is viewed as ethically acceptable will depend on the type of change, the reason for it, and the perspective of the person evaluating the change. For example, a person might find it ethically acceptable to alter a region so as to produce a long-term source of food for the growing human population, yet not ethically acceptable to take actions that result in short-term economic benefit but cause long-term economic loss and ecological damage.

6. A keystone species is one that has a disproportionate effect, relative to its own abundance, on other species in a community. As such, keystone species often control the numbers of other species that otherwise would be dominant and far more abundant in a community.

Chapter 24

1. An ecosystem consists of a community of organisms together with the physical environment in which those organisms live. The organisms in an ecosystem interact with one another in various ways; organisms can also move from one ecosystem to another. For this reason, determining the boundaries of protection for a particular ecosystem would probably be difficult. Such a plan would require an understanding of the roles that certain organisms play in the overall function of an ecosystem.

2. Energy captured by producers from an external source, such as the sun, is stored in the bodies of producers in chemical forms, such as carbohydrates. At each step in a food chain, a portion of the energy captured by producers is lost from the ecosystem as metabolic heat. This steady loss prevents energy from being recycled.

3. Decomposers break down the tissues of dead organisms into simple chemical components, thereby returning nutrients to the physical environment so that they can be used again by other organisms.

4. Nutrients can cycle on a global level. When sulfur dioxide pollution, for example, enters the atmosphere in one area of the world, winds can move that pollution around the world, where it can affect other ecosystems.

5. Human economic activity is interwoven with several key ecosystem services. Pollination is essential for the productivity of both commer-

cial crops and backyard gardens. Floodplains act as safety valves for major floods, provided we do not build on them or separate them from the bodies of water they help control. Forests act as water filtration systems. We rely on nutrient cycling to keep us alive. When ecosystem services such as these are damaged, human economic interests are damaged as well.

Chapter 25

1. Major types of global change caused by humans include global warming, land and water transformation, and changes in the chemistry of Earth (for example, changes to nutrient cycles). By altering the conditions under which species live, all of these changes could result in increased dominance of certain types of species and the disappearance of others from various ecosystems.

2. Human-caused global changes often happen at a much more rapid rate than do changes due to natural causes. Continental drift and natural climate change happen much more slowly than do the measurable increases humans have caused in atmospheric carbon dioxide levels and nitrogen fixation. In addition, humans have a choice about the global changes we cause.

3. The addition of large amounts of nitrogen to the environment by human activities is not necessarily a good thing. When more nitrogen becomes available than would exist naturally, a few species may outcompete others for the extra nitrogen, causing the other species to disappear from their communities.

4. The present levels of atmospheric CO_2 are higher than any seen in the previous 420,000 years. Since the middle of the last century, atmospheric CO_2 levels have been rising at about 2 parts per million (ppm) per year, and stood at 385 ppm toward the end of 2008. Scientific instruments can directly measure the amount of carbon dioxide in the atmosphere. By measuring CO_2 levels in bubbles of air trapped in ancient ice, scientists can estimate the amount of CO_2 that was present in the atmosphere up to hundreds of thousands of years ago.

5. It would be prudent to take action on global warming sooner rather than later, despite present uncertainties as to its extent. There is already evidence of climate changes that are consistent with the predicted effects of global warming. In addition, the correlation of rising CO_2 levels and worldwide temperature increases suggests that these increases will continue in the future if carbon dioxide emissions are not reduced. If action is delayed too long, it may be too late to undo many of the effects of global warming.

6. For people to have a sustainable impact on Earth, we must reduce the rate of growth of the human population, and we must reduce the rate of resource use per person. To achieve these goals, many aspects of human society would have to change. For example, we would need to alter our view of nature from looking at it as a limitless source of goods and materials that can be exploited for short-

term economic gain to accepting nature's limits and seeking always to take only those actions that can be sustained for long periods of time. Many specific actions would follow from such a change in our view of the world, such as an increase in recycling; the development and use of renewable sources of energy; a decrease in urban sprawl; an increase in the use of technologies with low environmental impact (such as organic farming); and a concerted effort to halt the ongoing extinction of species.

7. Examples of actions you could take to make your impact on Earth more sustainable include the following: reducing the quantity of nonfood items purchased; reusing items until they are no longer usable; buying used items rather than getting everything new; recycling paper, plastic, glass, and metal; taking along reusable cloth bags when shopping; rarely using paper cups, plates, or towels; planting trees and other native plants, especially those that help feed native wildlife; reducing water use by not leaving water running when brushing teeth, by adjusting the water level of washing machines to match the size of the load, and by using water-saving fixtures; reducing fossil fuel use by choosing a fuel-efficient car and by using household heating and air-conditioning only as needed; using compact fluorescent lightbulbs and turning off lights that are not in use; supporting organic farmers by purchasing organically grown food.

Interlude E

1. Cap-and-trade systems remain controversial. To learn more about the issues involved, you might want to visit this webpage, maintained by the Environmental Protection Agency (EPA): www.epa.gov/airmarkt/cap-trade/index.html

2. **Deforestation:** The destruction of forests can have many negative consequences for both the environment and for humans. At the current rate of forest destruction, nearly all the world's tropical rainforests will be gone in 100 to 150 years. The rate of new forest growth is not keeping up with the rate of forest loss. In addition, old-growth forests take centuries to recover from logging, and they contain many unique species that are not found in second-growth forests. Some areas of the world have already lost virtually all of their original forested area. **Water use:** Humans use more than half of the world's fresh water, and population growth will add to the demand. In many parts of the world, more water is drawn out of underground aquifers than can be naturally replenished. The rate of water use in some agricultural areas of China and India is not sustainable and will result in water shortages and potential economic damage in only a few years.

3. Most surveys in North America and the UK show that a majority of consumers prefer "eco-friendly" products and services. However, only a minority are willing to pay more for "greener" goods and services. In one survey, focused on food and grocery items, only 10 percent said they would pay more for foods produced in a sustainable way, and that cost, quality, and health aspects carried more weight in their decision to pay more for groceries. The respondents who said they would be willing to pay more for sustainably produced groceries were disproportionately older, female, better educated, or better off economically. However, since 40 percent of the respondents were unable to explain what "sustainability" refers to, the researchers point to the need to communicate environmental concepts to the general public with greater clarity.

4. In contrast to the GDP, methods of measurement such as the index of sustainable economic welfare give a more realistic picture of the output and value of an economy, by taking into account the environmental costs of producing goods and services.

5. Corporations have the motive of making a profit, and if environmental restrictions are in place for certain products or activities, businesses will have an incentive to develop innovative and marketable solutions. In a market economy, competition and the availability of capital help corporations develop new products rapidly.

Glossary

abiotic Of or referring to the nonliving environment that surrounds the community of living organisms in an ecosystem; the atmosphere, water, and Earth's crust. Compare *biotic*.

abundance The number of individuals of a species in a defined habitat.

abyssal zone The deeper waters of the open ocean, beyond the continental shelf, at depths greater than 6,000 meters (nearly 20,000 feet).

acid A chemical compound that can give up a hydrogen ion. Compare *base* (definition 1), *buffer*.

acid rain Precipitation with an unusually low pH, compared to that of unpolluted rain (which has a pH of about 5.2). Acid rain is a consequence of the release of sulfur dioxide and other pollutants into the atmosphere, where they are converted to acids that then fall back to Earth in rain or snow.

actin The protein component of actin filaments, which are cytoskeletal elements that bring about movement in muscle tissue, and in bacterial flagella.

activation energy The small input of energy required for a chemical reaction to occur at a noticeable rate.

active carrier protein A cell membrane protein that, using energy from an energy-rich molecule such as ATP, changes its shape to transfer an ion or a molecule across the membrane. Compare *passive carrier protein*.

active site The specific region on the surface of an enzyme where substrate molecules bind.

active transport Movement of ions or molecules across a biological membrane that requires an input of energy. Compare *passive transport*.

adaptation An inherited characteristic that, because of natural selection, improves an organism's ability to survive and reproduce in its environment.

adaptive evolution The process by which natural selection improves the match between organisms and their environment over time.

adaptive radiation An evolutionary expansion in which a group of organisms takes on new ecological roles and forms new species and higher taxonomic groups.

adenosine triphosphate See *ATP*.

adult stem cell Also called *somatic stem cell*. A cell that retains the capacity for self-renewal and persists into adulthood. Compare *embryonic stem cell*.

aerobe An organism that requires oxygen to survive. Compare *anaerobe*.

aerobic Of or referring to a metabolic process or organism that requires oxygen gas. Compare *anaerobic*.

allele One of several alternative versions of a gene. Each allele has a DNA sequence that is somewhat different from that of all other alleles of the same gene.

allele frequency The proportion (percentage) of a particular allele in a population.

allopatric speciation The formation of new species from populations that are geographically isolated from one another. Compare *sympatric speciation*.

amino acid A nitrogen-containing, small organic molecule that has an amino group, a carboxyl group, and a variable R group, all covalently attached to a single carbon atom. Proteins are polymers of amino acids.

amniocentesis A procedure in which a needle is inserted through the abdomen into the uterus to extract a small amount of amniotic fluid from the pregnancy sac that surrounds the fetus; this fluid contains fetal cells that can be used to test for genetic disorders.

anabolic reaction See *biosynthetic reaction*.

anabolism Metabolic pathways that build macromolecules. Compare *catabolism*.

anaerobe An organism that can survive without oxygen. Compare *aerobe*.

anaerobic Of or referring to a metabolic process or organism that does not require oxygen gas. Compare *aerobic*.

analogous Of or referring to a characteristic shared by two groups of organisms because of convergent evolution, not common descent. Compare *homologous*.

anaphase The stage of mitosis during which sister chromatids separate and move to opposite poles of the cell.

anchoring junction A protein complex that acts as a "hook" between two animal cells or between a cell and the extracellular matrix.

angiosperms Also called *flowering plants*. One of four major groups of plants. Angiosperms have vascular tissues, seeds, flowers, and fruits. They include most plants on Earth today. Compare *gymnosperms*.

Animalia The kingdom made up of animals—multicellular, heterotrophic eukaryotes that have evolved specialized tissues, organs and organ systems, body plans, and behaviors.

antenna complex A disclike grouping of pigment molecules, including chlorophyll, that harvests energy from sunlight in the thylakoid membrane of a chloroplast.

anticodon A sequence of three nitrogenous bases on a transfer RNA molecule that base-pairs with a particular codon on an mRNA molecule. Compare *codon*.

apoptosis (pl. apoptoses) A form of programmed cell death that generally benefits the organism.

aquifer An underground body of water that is sometimes bounded by impermeable layers of rock.

Archaea One of the three domains of life, encompassing the microscopic, single-celled prokaryotes that arose after the Bacteria. The domain Archaea is equivalent to the kingdom Archaea. Compare *Bacteria, Eukarya*.

arthropods A group of animals characterized by a hard exoskeleton. Arthropods include millipedes, crustaceans, insects, and spiders.

artificial selection A process in which only individuals that possess certain characteristics are allowed to breed. Artificial selection is used to guide the evolution of crop plants and domestic animals in ways that are advantageous for people.

asexual reproduction The production of genetically identical offspring without the exchange of genetic material with another individual. Compare *sexual reproduction*.

atmospheric cycle A type of nutrient cycle in which the nutrient enters the atmosphere easily. Compare *sedimentary cycle*.

atom The smallest unit of a chemical element that still has the properties of that element.

atomic mass number The sum of the number of protons and neutrons found in the nucleus of an atom of a particular chemical element.

atomic number The number of protons found in the nucleus of an atom of a particular chemical element.

ATP Adenosine triphosphate, a molecule that is commonly used by cells to store energy and to transfer energy from one chemical reaction to another.

autosome Any chromosome that is not a *sex chromosome*.

Bacteria One of the three domains of life, encompassing the microscopic, single-celled prokaryotes that were the first organisms to arise. The domain Bacteria is equivalent to the kingdom Bacteria. Compare *Archaea, Eukarya*.

bacterial flagellum (pl. flagella) A ropelike structure built from actin proteins that, through rotary movements, propels a bacterium through fluid. Compare *eukaryotic flagellum*.

base 1. A chemical compound that can accept a hydrogen ion. Compare *acid, buffer*. 2. A nitrogen-containing molecule that is part of a nucleotide. See *nitrogenous base*.

base pair A pair of complementary nitrogenous bases that have formed hydrogen bonds with each other. Base pairs form the "rungs" of the DNA double helical "ladder." In DNA, adenine pairs with thymine (A–T) and cytosine pairs with guanine (C–G); in RNA, uracil replaces thymine.

benign tumor A relatively harmless cancerous growth that is confined to a single tumor and does not spread to other tissues in the body. Compare *malignant tumor*.

benthic zone The bottom surface of any body of water.

binary fission A form of asexual reproduction in which a cell divides to form two genetically identical daughter cells that replace the original parent cell.

bioaccumulation The deposition of a substance within an organism at concentrations higher than in the surrounding abiotic environment. Compare *biomagnification*.

biodiversity The variety of organisms on Earth or in a particular location, ranging from the genetic variation and behavioral diversity of individual organisms or species through the diversity of ecosystems.

biological hierarchy The nested series in which living things, their building blocks, and their living and nonliving surroundings can be arranged—from molecules at the lowest level to the entire biosphere at the highest level.

biological species concept The idea that a species is defined as a group of populations that can interbreed but are reproductively isolated from other such groups.

biology The study of life.

biomagnification An increase in the tissue concentrations of a bioaccumulated chemical at successively higher trophic levels in a food chain. Compare *bioaccumulation*.

biomass The mass of organisms per unit of area.

biome A large area of the biosphere that is characterized according to its unique climatic and ecological features. Terrestrial biomes are usually classified according to their dominant vegetation; aquatic biomes, on the basis of physical and chemical features.

biosphere All living organisms on Earth, together with the environments in which they live. Compare *ecosystem*.

biosynthetic reaction Also called *anabolic reaction*. A chemical reaction that manufactures complex molecules in living cells. Compare *catabolic reaction*.

biotic Of or referring to the assemblage of interacting organisms—prokaryotes, protists, animals, fungi, and plants—in an ecosystem. Compare *abiotic*.

bipedal Of or referring to an organism that walks upright on two legs.

bivalent The paternal and maternal members of a pair of homologous chromosomes aligned parallel to each other during meiosis I.

blastocyst A hollow ball of cells that forms the stage following a morula in mammalian development.

boreal forest Also known as *taiga*. A terrestrial biome dominated by coniferous trees that grow in northern or high-altitude regions with cold, dry winters and mild summers.

buffer A chemical compound that can both give up and accept hydrogen ions. Buffers can maintain the pH of water within specific limits. Compare *acid, base*.

Calvin cycle The sugar-manufacturing reactions in photosynthesis. This series of enzymatic reactions takes place in the stroma of the chloroplast and synthesizes sugars from carbon dioxide and water. Compare *light reactions*.

Cambrian explosion A major increase in the diversity of life on Earth that occurred about 530 million years ago, during the Cambrian period. The Cambrian explosion lasted 5 million to 10 million years, during which time large and complex forms of most living animal phyla appeared suddenly in the fossil record.

cancer A group of diseases caused by rapid and inappropriate cell division.

canopy The habitat in the branches of forest trees.

cap-and-trade system An approach to pollution control in which a government sets a nationwide limit, or cap, on the amount of a pollutant that can be added to the environment each year. Each factory that emits the pollutant is given a certain number of "emission allowances" in each year. Unused allowances can be sold, traded, or saved (banked) for future use.

carbohydrate Any of a class of organic compounds that includes sugars and their polymers, in which each carbon atom is usually linked to two hydrogen atoms and an oxygen atom. See also *sugar*.

carbon cycle The transfer of carbon within biotic communities, between living organisms and their physical surroundings, and within the abiotic world.

carbon fixation The process by which carbon dioxide is incorporated into organic molecules. Carbon fixation occurs in the chloroplasts of plants and results in the synthesis of sugars.

carcinogen A physical, chemical, or biological agent that causes cancer.

carrier In genetics, an individual that carries a disease-causing allele but does not get the disease.

carrying capacity The maximum population size that can be supported indefinitely by the environment in which the population is found.

catabolic reaction A chemical reaction that breaks down complex molecules to release energy for use by the cell. Compare *biosynthetic reaction*.

catabolism Metabolic pathways that take macromolecules apart. Compare *anabolism*.

catalyst A substance that speeds up a specific chemical reaction without being permanently altered in the process. Enzymes, which are usually proteins, are an example of biological catalysts.

cell The smallest self-contained unit of life, enclosed by a membrane.

cell communication Exchange of signals, which are usually diffusible molecules, between cells.

cell cycle A series of distinct stages in the life cycle of a cell that is capable of dividing. Cell division is the last stage in the cycle.

cell differentiation The process through which a daughter cell becomes different from its parent cell and acquires specialized functions.

cell division The final stage in the life of an individual cell, in which, a parent cell is split up to generate two (mitotic cell division) or four (meiotic cell division) daughter cells.

cell junction A structure that anchors a cell, or connects it with a neighbor, or creates communication passageways between two cells.

cell plate A partition, consisting of membrane and cell wall components, that appears during cytokinesis in dividing plant cells. The cell plate matures into a polysaccharide-based cell wall flanked on either side by the plasma membranes of the two daughter cells.

cell wall A polysaccharide support layer that lies outside the plasma membrane of the cells of many prokaryotes, fungi, and protists, and all plants.

cellular respiration A metabolic process that extracts chemical energy from organic molecules, such as sugars, to generate the universal energy carrier ATP, consuming oxygen and releasing carbon dioxide and water in the process. Cellular respiration has three phases: glycolysis, the Krebs cycle, and oxidative phosphorylation. ATP fuels a wide variety of cellular activities in every organism on Earth.

cellulose An extracellular polysaccharide (a type of carbohydrate) produced by plants and some other organisms that strengthens cell walls.

centromere A physical constriction that holds sister chromatids together.

centrosome A cytoskeletal structure in the cytosol that helps organize the mitotic spindle and defines the two poles of a dividing cell.

CFCs See *chlorofluorocarbons*.

channel protein A cell membrane protein that forms an opening through which specific ions or molecules can pass.

chaparral A terrestrial biome characterized by shrubs and small nonwoody plants that grow in regions with rainy winters and hot, dry summers.

character displacement A process by which intense competition between species causes the forms of the competing species to evolve to become more different over time.

chemical bond The attractive interaction that causes two atoms to associate with each other.

chemical compound An association of atoms of different chemical elements linked by ionic or covalent bonds.

chemical reaction A process that creates or rearranges chemical bonds between atoms.

chemoautotroph An organism that obtains its energy from chemicals and derives its carbon from carbon dioxide in the air. All chemoautotrophs are prokaryotes. Compare *chemoheterotroph*, *photoautotroph*.

chemoheterotroph An organism that obtains its energy from chemicals and derives its carbon from carbon-containing compounds found mainly in other organisms. All fungi and animals, as well as many protists and prokaryotes, are chemoheterotrophs. Compare *chemoautotroph*, *photoheterotroph*.

chlorofluorocarbons (CFCs) Synthetic chemical compounds whose release into the atmosphere can damage the ozone layer.

chlorophyll A green pigment that is used to capture light energy for photosynthesis.

chloroplast An organelle found in plants and algae that is the primary site of photosynthesis.

chorionic villus sampling (CVS) A procedure in which a flexible tube is inserted through a woman's vagina and into her uterus. The tip of this tube is placed next to the chorionic villi and a few cells are removed by gentle suction so that they can be tested for genetic disorders.

chromatid Either of two identical, side-by-side copies of a chromosome that are firmly linked at the centromere.

chromatin DNA complexed with packaging proteins. Chromosomes are made up of compacted chromatin.

chromosome Threadlike structures, each composed of a single DNA molecule packaged with proteins. Chromosomes achieve the highest level of compaction when prophase begins during mitosis or meiosis.

chromosome theory of inheritance A theory, supported by much experimental evidence, stating that genes are located on chromosomes.

cilium (pl. cilia) A hairlike structure found in some eukaryotes that uses a rowing motion to propel the organism or to move fluid over cells. Compare *eukaryotic flagellum*.

citric acid cycle See *Krebs cycle*.

class In reference to biological classification systems, the level in the Linnaean hierarchy above order and below phylum.

climate The prevailing weather conditions experienced in an area over relatively long periods of time (30 years or more). Compare *weather*.

climate change Significant and long-term changes in the average climatic conditions in our biosphere, such as global warming.

climax community A community, typical of a given climate and soil type, whose species are not replaced by other species. A climax community is the end point of succession for a particular location; in many cases, however, ongoing disturbances such as fire or windstorms prevent the formation of a stable climax community.

cloning Making a copy (of a gene, a cell, or a whole organism) that is genetically identical to the original.

coastal region An ecosystem of the marine biome defined as the area of the ocean stretching from the shoreline to the continental shelf, which is the undersea extension of a continent's edge.

codominance The situation in which the effects of both alleles in a pair are equally visible in the phenotype of a heterozygote. In other words, the influence of each codominant allele is fully displayed in the heterozygote, without being diminished or diluted by the presence of the other allele (as in *incomplete dominance*).

codon A sequence of three nitrogenous bases in an mRNA molecule. Each codon specifies either a particular amino acid or a signal to start or stop the translation of a protein. Compare *anticodon*.

coevolution The process by which interactions among species drive evolutionary change in those species.

community An association of populations of different species that live in the same area.

comparative genomics A field of scientific study that analyzes and compares the genomes of multiple species.

competition An interaction between two species in which each has a negative effect on the other.

complementary strand A strand of DNA whose sequence of bases can pair (according to base-pairing rules) with the sequence of bases found in a focal DNA strand.

concentration gradient A change in the concentration of molecules from one location to another.

conserved gene A gene whose genomic nucleotide sequence and general function are essentially the same in two or more diverse groups of organisms.

consumer An organism that obtains its energy by eating other organisms or their remains. Consumers include herbivores, carnivores, and decomposers. Compare *producer*.

continental drift The movement of Earth's continents over time.

control group A group of participants in an experiment that are subjected to the same environmental conditions as the experimental groups, except that the factor or factors being tested in the experiment are omitted.

convection cell A large and consistent atmospheric circulation pattern in which warm, moist air rises and cool, dry air sinks. Earth has four stable giant convection cells (two in tropical regions and two in polar regions) and two less stable cells (located in temperate regions).

convergent evolution Evolutionary change that occurs when natural selection causes distantly related organisms to evolve similar structures in response to similar environmental challenges.

convergent feature A feature shared by two groups of organisms not because it was inherited from a common ancestor, but because it arose independently in the two groups.

coupled reactions Paired reactions in which one provides the energy to make the other happen.

covalent bond A strong chemical linkage between two atoms based on the sharing of electrons. Compare *hydrogen bond, ionic bond.*

Cretaceous extinction A mass extinction that occurred 65 million years ago, wiping out many marine invertebrates and terrestrial plants and animals, including the last of the dinosaurs.

crista (pl. cristae) A fold in the inner mitochondrial membrane.

crossing-over Also called *recombination.* A physical exchange of chromosomal segments between paired paternal and maternal members of homologous chromosomes. Crossing-over takes place during prophase of meiosis I.

cuticle A waxy layer that covers aboveground plant parts, helping to prevent water loss and to keep enemies, such as fungi, from invading the plant.

CVS See *chorionic villus sampling.*

cynodont A member of a group of mammal-like reptiles from which the earliest mammals arose, roughly 220 million years ago.

cytokinesis The stage following mitosis, during which the cell physically divides into two daughter cells.

cytoplasm The contents of a cell enclosed by the plasma membrane but, in eukaryotes, excluding the nucleus. Compare *cytosol.*

cytoskeleton A complex network of protein filaments found in the cytosol of eukaryotic cells. The cytoskeleton maintains cell shape and is necessary for cell division and movement.

cytosol The water-based fluid component of the cytoplasm. In eukaryotes, the cytosol consists of all the contents of a cell enclosed by the plasma membrane, but excluding all organelles. Compare *cytoplasm.*

decomposer An organism that breaks down dead tissues into simple chemical components,

thereby returning nutrients to the physical environment.

deletion In genetics, a mutation in which one or more nucleotides are removed from the DNA sequence of a gene, or a piece breaks off from a chromosome and is lost. Compare *insertion, substitution.*

denaturation The destruction of a protein's three-dimensional structure, resulting in loss of protein activity.

density-dependent Of or referring to a factor, such as food shortage, that limits the growth of a population more strongly as the density of the population increases. Compare *density-independent.*

density-independent Of or referring to a factor, such as weather, that can limit the size of a population but does not act more strongly as the density of the population increases. Compare *density-dependent.*

deoxyribonucleic acid See *DNA.*

desert A terrestrial biome dominated by plants that grow in regions with low precipitation, usually 25 centimeters per year or less.

deuterostome Any of a group of animals, including sea stars and vertebrates, in which the second opening to develop in the early embryo becomes the mouth. Compare *protostome.*

development The process by which an organism grows from a single cell to its adult form.

diffusion The passive movement of a substance from areas of high concentration of that substance to areas of low concentration.

diploid Of or referring to a cell or organism that has two complete sets of homologous chromosomes (2*n*). Compare *haploid, polyploid.*

directional selection A type of natural selection in which individuals with one extreme of an inherited characteristic have an advantage over other individuals in the population, as when large individuals produce more offspring than do small and medium-sized individuals. Compare *disruptive selection, stabilizing selection.*

disaccharide A molecule made up of two *monosaccharides.* Some common examples are sucrose, lactose, and maltose. Compare *polysaccharide.*

disruptive selection A type of natural selection in which individuals with either extreme of an inherited characteristic have an advantage over individuals with an intermediate phenotype, as when both small and large individuals produce more offspring than do medium-sized individuals. Compare *directional selection, stabilizing selection.*

distribution The geographic area over which a species is found.

disturbance In an ecological community, an event, such as a fire or windstorm, that kills or damages some organisms in the community, thereby creating an opportunity for other organisms to become established.

diversity The composition of an ecological community, which has two components: the number of different species that live in the community and the relative abundances of those species.

DNA Deoxyribonucleic acid, a polymer of nucleotides that stores the information needed to synthesize proteins in living organisms.

DNA chip A small surface, or "chip," roughly the size of a dime on which thousands of samples of DNA are placed in a regimented order.

DNA cloning The introduction of recombinant DNA into a host cell (usually a bacterium) that can generate many copies of the introduced DNA.

DNA fingerprinting The use of DNA analysis to identify individuals and determine the relatedness of individuals.

DNA hybridization An experimental procedure in which DNA from two different sources bind to each other through complementary base pairing.

DNA library A collection of an organism's DNA fragments that is introduced into, and stored within, a host organism, such as a bacterium.

DNA polymerase The key enzyme that cells use to replicate their DNA. In DNA technology, DNA polymerase is used in the polymerase chain reaction to make many copies of a gene or other DNA sequence.

DNA primer A short segment of DNA used in PCR amplification that is designed to pair with one of the two ends of the gene being amplified by PCR.

DNA probe A short sequence of DNA (usually tens to hundreds of bases long) that is labeled with a radioactive or fluorescent tag and then allowed to hybridize with a target DNA sample, with the objective of determining if the two share sequence similarity.

DNA repair A three-step process in which damage to DNA is repaired. Damaged DNA is first recognized, then removed, and then replaced with newly synthesized DNA.

DNA replication The duplication, or copying, of a DNA molecule. DNA replication begins when the hydrogen bonds connecting the two strands of DNA are broken, causing the strands to unwind and separate. Each strand is then used as a template for the construction of a new strand of DNA.

DNA segregation The process by which the duplicated DNA of a dividing cell is sorted equally between two daughter cells.

DNA sequence The sequence, or order, in which the nitrogenous bases adenine (A), cytosine (C), guanine (G), and thymine (T) are arranged in a gene or a DNA fragment, or in an organism's genome.

DNA sequencing A procedure, usually automated, used to determine the sequence of nitrogenous bases in a DNA fragment.

DNA technology The set of techniques that scientists use to manipulate DNA.

DNA vector A loop or chain of DNA designed to serve as a "DNA vehicle." DNA vectors are useful in the cloning of DNA fragments and in the transfer of recombinant DNA from one cell to another.

domain A level of biological classification above the kingdom. The three domains are Bacteria, Archaea, and Eukarya.

dominant allele An allele that determines the phenotype (dominant phenotype) when it is paired with a *recessive allele* in a heterozygous individual.

double helix The structure of DNA, in which two long strands of covalently bonded nucleotides are held together by hydrogen bonds and twisted into a spiral coil.

doubling time The time it takes a population to double in size. Doubling time can be used as a measure of how fast a population is growing.

duplication In genetics, a type of mutation in which an extra copy of a gene or DNA fragment appears alongside the original, increasing the length of the chromosome.

ecological footprint The area of productive ecosystems needed throughout a year to support a population and cope with its waste materials. Ecological footprint is a measure of the demands that humans place on ecosystems relative to the capacity of these ecosystems to deliver resources and services on a sustainable basis.

ecology The scientific study of interactions between organisms and their environment.

ecosystem A community of organisms, together with the physical environment in which the organisms live. Global patterns of air and water circulation link all the world's organisms into one giant ecosystem, the *biosphere*.

ecosystem service An action or function of an ecosystem that provides a benefit to humans, such as pollination by insects or water filtration by wetlands.

electron A negatively charged particle found in atoms. The atoms of a particular element contain a characteristic number of electrons. Compare *proton*.

electron transport chain (ETC) A group of membrane-associated proteins and other molecules that can both accept and donate electrons. The transfer of electrons from one ETC component to another releases energy that can be used to drive protons across a membrane, and ultimately, in both chloroplasts and mitochondria, to manufacture ATP.

element In reference to chemicals, a pure substance made up of only one type of atom, with a characteristic number of protons. The physical world is made up of 92 naturally occurring elements.

embryonic stem cell A pluripotent stem cell in or derived from the inner cell mass, which exists only at the blastocyst stage in mammalian embryos. Compare *adult stem cell*.

endangered In reference to species in danger of extinction.

endocrine disrupter A chemical that interferes with hormone function to produce negative effects, such as reduced fertility, developmental abnormalities, immune system dysfunction, and increased risk of cancer.

endocytosis A process by which a section of a cell's plasma membrane bulges inward as it envelops a substance outside of the cell, eventually breaking free to become a closed vesicle within the cytoplasm. Compare *exocytosis*.

endoplasmic reticulum (ER) An organelle composed of many interconnected membrane sacs and tubes. The ER is the main site for the synthesis of lipids, and certain types of proteins, in eukaryotic cells.

energy carrier A molecule that can store energy and donate it to another molecule or a chemical reaction. ATP is the most commonly used energy carrier in living organisms.

enzyme A macromolecule, usually a protein, that acts as a catalyst, speeding the progress of chemical reactions. Nearly all chemical reactions in living organisms are catalyzed by enzymes.

epistasis (pl. epistases) A gene interaction in which the phenotypic effect of the alleles of one gene depends on which alleles are present for another, independently inherited gene.

ER See *endoplasmic reticulum*.

estuary An ecosystem of the marine biome defined as a region where a river empties into the sea.

ETC See *electron transport chain*.

eugenics movement An effort to breed "better humans" by encouraging the reproduction of people with certain genetic characteristics and discouraging the reproduction of people with other genetic characteristics.

Eukarya One of the three domains of life, encompassing the eukaryotes. Four kingdoms are included: Animalia, Plantae, Fungi, and Protista. Compare *Archaea, Bacteria*.

eukaryote A single-celled or multicellular organism in which each cell has a distinct nucleus and cytoplasm. All organisms other than the Bacteria and the Archaea are eukaryotes. Compare *prokaryote*.

eukaryotic flagellum (pl. flagella) A hairlike structure found in eukaryotes that propels the cell or organism through a whiplike action, with waves passing from the base to the tip of the flagellum. Compare *bacterial flagellum, cilium*.

eutrophication A process in which enrichment of water by nutrients (often from sewage or runoff from fertilized agricultural fields) causes bacterial populations to increase and oxygen concentrations to decrease.

evolution Change in the genetic characteristics of a population over time. See also *macroevolution* and *microevolution*.

evolutionary innovation A new and key adaptation of a group that improves members' chances of living and reproducing successfully.

evolutionary tree A diagrammatic representation showing the order in which different lineages arose, with the lowest branches having arisen first.

exchange pool In reference to nutrients, a source such as the soil, water, or air where nutrients are available to producers.

exocytosis A process by which a vesicle approaches and fuses with the plasma membrane of a cell, thereby releasing its contents into the cell's surroundings. Compare *endocytosis*.

exon A DNA sequence within a gene that encodes part of a protein. Compare *intron*.

exoskeleton An external framework of stiff or hard material that surrounds the soft tissues of an animal.

experiment A controlled manipulation of nature designed to test a hypothesis.

exploitation An interaction between two species in which one species benefits (the consumer) and the other species is harmed (the food organism). Exploitation includes the killing of prey by predators, the eating of plants by herbivores, and the harming or killing of a host by a parasite or pathogen.

exploitative competition A type of competition in which species compete indirectly for shared resources, with each reducing the amount of a resource available to the other. Compare *interference competition*.

exponential growth A type of rapid population growth in which a population increases by a constant proportion from one generation to the next.

extracellular matrix (pl. matrices) A coating of nonliving material, released by the cells of multicellular organisms, that often helps hold those cells together.

extremophile An organism, such as many archaeans, that lives in extreme environments, such as in boiling hot geysers or on salted meat.

F₁ generation The first generation of offspring in a genetic cross. Compare *P generation*.

F₂ generation The second generation of offspring in a genetic cross. Compare *P generation*.

family In reference to biological classification systems, the level in the Linnaean hierarchy above genus and below order.

fat An informal name for saturated lipids, usually of animal origin, that are solid at room temperature. Most such lipids are triglycerides, which consist of glycerol linked to three fatty acids and are used by living organisms to store energy.

fatty acid An organic molecule with a long hydrocarbon chain and a hydrophilic head group. Fatty acids are found in phospholipids, glycerides such as triglycerides, and waxes.

fermentation A series of catabolic reactions that produce small amounts of ATP through glycolysis and can function without oxygen. In most fermentation pathways, pyruvate from glycolysis is converted to other organic molecules, such as ethanol and carbon dioxide, or lactic acid.

fertilization The fusion of two different haploid gametes (egg and sperm) to produce a diploid zygote (the fertilized egg).

first law of thermodynamics The law stating that energy can be neither created nor destroyed, but only transformed or transferred from one molecule to another.

fixation In genetics, the removal of all alleles within a population at a particular genetic locus except one. The allele that remains has a frequency of 100 percent.

flagellum (pl. flagella) A long extension from the cell that is lashed or rotated to enable that cell to move.

flower A specialized reproductive structure that is characteristic of the plant group known as the angiosperms, or flowering plants.

flowering plants See *angiosperms*.

fluid mosaic model The concept of the plasma membrane as a phospholipid bilayer containing a variety of other lipids and embedded proteins, some of which can move laterally in the plane of the membrane.

food chain A single sequence of feeding relationships describing who eats whom in a community. Compare *food web*.

food web A summary of the movement of energy through a community. A food web is formed by connecting all of the *food chains* in the community to one another.

fossil Preserved remains or an impression of a formerly living organism. Fossils document the history of life on Earth, showing that many organisms from the past were unlike living forms, that many organisms have gone extinct, and that life has evolved through time.

founder effect A genetic bottleneck that results when a small group of individuals from a larger source population establishes a new population far from the original population.

frameshift In genetics, the large change in coding information that results when the deletion or insertion in a gene sequence is not a multiple of three base pairs (a codon). The amino acid sequence of the protein that is translated from such a gene is severely altered in most cases, and protein function is typically lost.

functional group A specific cluster of covalently bonded atoms, with distinctive chemical properties, that forms a discrete subgroup in a variety of larger molecules.

Fungi The kingdom of Eukarya that is made up of absorptive heterotrophs (consumers that absorb their food after digesting it externally). The kingdom includes mushroom-producing species, yeasts, and molds, most of which make their living as decomposers.

G₀ phase A resting state, when the cell withdraws from the cell cycle before the S phase. G₀ cells are not competent to divide, unless they receive and respond to signals that direct them to reenter the cell cycle and proceed through the S phase.

G₁ phase The stage of the cell cycle following mitosis and before S phase. The cell grows in size during G₁ phase and makes a commitment to enter S phase if it receives and responds to cell division signals.

G₂ phase The stage of the cell cycle following S phase and before mitosis. G₂ phase serves as a checkpoint ensuring that mitosis will not be launched under inappropriate conditions (such as inadequate nutrient supply, DNA damage, or incomplete DNA replication).

gamete A haploid sex cell that fuses with another sex cell during fertilization. Eggs and sperm are gametes.

gap junction A cytoplasmic channel, created by a narrow cylinder of proteins, that directly connects two animal cells and allows the passage of ions and small molecules between them.

GDP See *gross domestic product*.

gel electrophoresis A procedure in which DNA fragments are placed in a gelatin-like substance (a gel) and subjected to an electrical charge, which causes the fragments to move through the gel. Small DNA fragments move farther than large DNA fragments, causing the fragments to separate by size.

gene The smallest unit of DNA that governs a genetic characteristic and contains the code for the synthesis of a protein or an RNA molecule. Genes are located on chromosomes.

gene cascade A process in which the protein products of different genes interact with one another and with signals from the environment, thereby turning on other sets of genes in some cells, but not in other cells. Organisms use gene cascades to control how genes are expressed during development.

gene expression The creation of a functional product, such as a specific protein or RNA, using the coding information stored in a gene. Gene expression is the means by which a gene influences the cell or organism in which it is found.

gene flow The exchange of alleles between different populations.

gene therapy A treatment approach that seeks to correct genetic disorders by repairing the genes responsible for the disorder.

genetic bottleneck A drop in the size of a population that results in low genetic variation or causes harmful alleles to reach a frequency of 100 percent in the population.

genetic code The code that specifies how information in mRNA is translated to create the specific sequence of amino acids found in the protein encoded by that mRNA. The genetic code consists of all possible three-base combinations of each of the four nitrogenous bases found in RNA. Of the 64 possible three-base combinations (codons), 60 specify a particular amino acid, 3 serve as a "stop translation" signal, and 1 (AUG) acts as a "start translation" signal.

genetic cross A controlled mating experiment, usually performed to examine the inheritance of a particular characteristic.

genetic drift The natural process in which chance events cause certain alleles to increase or decrease in a population. The genetic makeup of a population undergoing genetic drift changes at random over time, rather than being shaped in a nonrandom way by natural selection.

genetic engineering The process in which a DNA sequence (often a gene) is isolated, modified, and inserted back into an individual of the same or a different species. Genetic engineering is commonly used to change the performance of the genetically modified organism, as when a crop plant is engineered to resist attack from an insect pest.

genetic linkage The situation in which different genes that are located close to one another on the same chromosome do not follow Mendel's law of independent assortment.

genetic recombination The creation of new groupings of alleles through the breaking and rejoining of of different DNA segments, as in the crossing-over that takes place between paired homologues during meiosis I.

genetic screening The examination of an individual's genes to assess current or future health risks and status.

genetic trait An inherited characteristic of an organism, such as its size, color, or behavior.

genetic variation The allelic differences among the individuals of a population.

genetically modified organism (GMO) An individual into which a modified gene or other DNA sequence has been inserted, typically with the intent of improving some aspect of the recipient organism's performance.

genetics The scientific study of the inheritance of characteristics encoded by DNA.

genome All the DNA of an organism, including all its genes; in eukaryotes, the term refers to the DNA in a haploid set of chromosomes, such as that found in a sperm or egg.

genomics The study of the structure and expression of entire genomes and how they change during evolution.

genotype The allelic makeup that is responsible for a particular *phenotype* displayed by an individual.

genotype frequency The proportion (percentage) of a particular genotype in a population.

genus (pl. genera) The level in the Linnaean hierarchy above species and below family.

geographic isolation The physical separation of populations from one another by a barrier such as a mountain chain or a river. Geographic isolation often causes the formation of new species, as when populations of a single species become physically separated from one another and then accumulate so many genetic differences that they become reproductively isolated from one another. Compare *reproductive isolation*.

global change Worldwide change in the environment. There are many causes of global change, including climate change caused by the movement of continents and changes in land and water use by humans.

global warming A worldwide increase in temperature. Earth appears to be entering a period of global warming caused by human activities—specifically, by the release of large quantities of greenhouse gases such as carbon dioxide into the atmosphere.

glucose A monosaccharide that is the primary metabolic fuel in most cells.

glycogen The storage carbohydrate in animals; it is found primarily in the liver and skeletal muscles in humans. Glycogen is a polysaccharide and is structurally similar to starch, the storage carbohydrate of plants.

glycolysis A series of catabolic reactions that split glucose to produce pyruvate, which is then used in either oxidative phosphorylation or fermentation. The glycolytic breakdown of one molecule of glucose yields two molecules of the energy carrier ATP.

GMO See *genetically modified organism*.

Golgi apparatus An organelle composed of flattened membrane sacs that routes proteins and lipids to various parts of the eukaryotic cell.

granum (pl. grana) A structure made up of a stack of membrane sacs, called thylakoids, that is part of the interconnected internal membrane system within a chloroplast.

grassland A terrestrial biome dominated by grasses. Grasslands occur in relatively dry regions, often with cold winters and hot summers.

green roof On a building, a 2- to 4-inch-thick "living" rooftop that has a layer of soil or other material in which plants grow, under which there are one or more layers that absorb water and prevent roots and water from damaging the underlying roof structure.

greenhouse effect The increase in Earth's temperature when absorbed heat that is reemitted by greenhouse gases becomes trapped because it lacks the energy to escape into outer space.

greenhouse gas Any of several gases in Earth's atmosphere that let in sunlight but trap heat.

gross domestic product (GDP) A traditional measure of a nation's economic output that records the value of goods produced but does not consider the social and environmental costs that result from producing those goods. Compare *index of sustainable economic welfare*.

groundwater Water from an underground source, such as an aquifer or belowground river.

growth factors A class of signaling molecules that play an especially important role in initiating and maintaining cell proliferation in the human body.

gymnosperms One of four major groups of plants. Represented by conifers such as pine or spruce trees, gymnosperms have vascular tissues and seeds but lack flowers and fruits. Compare *angiosperms*.

habitat A characteristic place or type of environment in which an organism lives.

haploid Of or referring to a cell or organism that has only one set (n) of homologous chromosomes, such that only one member of each homologous pair (either the paternal or maternal member) is represented in that set. Compare *diploid, polyploid*.

Hardy–Weinberg equation An equation ($p^2 + 2pq + q^2 = 1$) that predicts the genotype frequencies in a population that is not evolving.

herbivore A consumer that relies on living plant tissues for nutrients.

heterozygote An individual that carries one copy of each of two different alleles (for example, an *Aa* individual). Compare *homozygote*.

HGP See *Human Genome Project*.

histone proteins Proteins around which the DNA double helix winds to form a bead-on-a-string structure and that help create a more compact form of chromatin.

homeostasis The process of maintaining appropriate and constant conditions inside cells.

homeotic gene A master-switch gene that plays a key role in the control of gene expression during development. Each homeotic gene controls the expression of a series of other genes whose protein products direct the development of an organism.

hominid Any of a group of primates that encompasses humans and our now extinct humanlike ancestors.

homologous Of or referring to a characteristic shared by two groups of organisms because of their descent from a common ancestor. Compare *analogous*.

homologous chromosome pair The two members of a specific chromosome pair found in diploid cells, one of which comes from the individual's mother and the other from its father.

homologue One member, either the paternal or maternal partner, of a pair of homologous chromosomes.

homozygote An individual that carries two copies of the same allele (for example, an *AA* or an *aa* individual). Compare *heterozygote*.

horizontal gene transfer The movement of genes from one organism or group of organisms to another, not vertically within a lineage via reproduction, but horizontally to another lineage altogether by some other means.

hormone A signaling molecule released in very small amounts into the circulatory system of an animal, or into a variety of tissues in a plant, that affects the functioning of target tissues.

host The individual, or organism, in which a particular parasite or pathogen lives.

housekeeping gene A gene that has an essential role in the maintenance of cellular activities and is expressed by most cells in the body.

Human Genome Project (HGP) A publicly funded effort on the part of an international consortium created by the U.S. National Institutes for Health and the U.S. Department of Energy to determine the sequence of the human genome.

hybrid An offspring that results when two different species, or two different varieties or genotypic lines, are mated.

hybridize To cause hybrid offspring to be produced.

hydrogen bond A weak electrical attraction between a hydrogen atom that has a slight positive charge and another atom with a slight negative charge. Compare *covalent bond, ionic bond*.

hydrophilic Of or referring to substances, both salts and molecules, that interact freely with water. Hydrophilic molecules dissolve easily in water but not in fats or oils. Compare *hydrophobic*.

hydrophobic Of or referring to molecules or parts of molecules that do not interact freely with water. Hydrophobic molecules dissolve easily in fats and oils but not in water. Compare *hydrophilic*.

hypertonic solution A solution that has a higher solute concentration than the cytosol of a cell, causing more water to flow out of the cell than into it. Compare *hypotonic solution, isotonic solution*.

hypha (pl. hyphae) In fungi, a threadlike absorptive structure that grows through a food source. Mats of hyphae form mycelia, the main bodies of fungi.

hypothesis (pl. hypotheses) A possible explanation of how a natural phenomenon works. A hypothesis must have logical consequences that can be proved true or false.

hypotonic solution A solution that has a lower solute concentration than the cytosol of a cell, causing more water to flow into the cell than out of it. Compare *hypertonic solution, isotonic solution*.

incomplete dominance The situation in which heterozygotes (*Aa* individuals) are intermediate in phenotype between the two homozygotes (*AA* and *aa* individuals) for a particular gene. Compare *codominance*.

independent assortment of chromosomes The random distribution of maternal and paternal chromosomes into gametes during meiosis.

index of sustainable economic welfare A measure of economic output that includes a wider range of the benefits and costs of economic activities than does the traditional *gross domestic product*.

individual A single organism, usually physically separate and genetically distinct from other individuals.

induced defense A defensive response in plants that is directly stimulated by attacking herbivores.

induced fit model A model of substrate–enzyme interaction stating that as a substrate enters the active site, the parts of the enzyme shift about slightly to allow the active site to mold itself around the substrate.

inner cell mass The cluster of cells inside the blastocyst that eventually develops into the embryo and some of the membranes that surround a mammalian embryo and fetus.

insect Any of a group of six-legged arthropods that includes grasshoppers, beetles, ants, and butterflies. Insects are the most species-rich group of animals on Earth.

insertion In genetics, a mutation in which one or more nucleotides are inserted into the DNA sequence of a gene. Compare *deletion, substitution*.

interference competition A type of competition in which one organism directly excludes another from the use of resources. Compare *exploitative competition*.

intermediate filament One of a diverse class of ropelike protein filaments that serve as structural reinforcements in the cytoskeleton.

intermembrane space The space between the inner and outer membranes of a chloroplast or a mitochondrion.

interphase The period of time between two successive mitotic divisions, during which the cell increases in size and prepares for cell division.

intertidal zone The part of the coastal region that is closest to the shore, where the ocean meets the land. It extends from the highest tide mark to the lowest tide mark.

introduced species A species that does not naturally live in an area but has been brought there either accidentally or on purpose by humans.

intron A sequence of nitrogenous bases within a gene that does not specify part of the gene's final protein or RNA product. Enzymes in the nucleus must remove introns from mRNA, tRNA, and rRNA molecules for these molecules to function properly. Compare *exon*.

invasive species An introduced species that proliferates rapidly and becomes a major pest in its new environment.

inversion In genetics, a mutation in which a fragment of a chromosome breaks off and returns to the correct place on the original chromosome, but with the genetic loci in reverse order.

ion An atom or group of atoms that has either gained or lost electrons and therefore has a negative or positive charge.

ionic bond A chemical linkage between two atoms based on the electrical attraction between positive and negative charges. Compare *covalent bond, hydrogen bond*.

irregular fluctuations In reference to natural populations, a pattern of population growth in which the number of individuals in the population changes over time in an irregular manner.

isotonic solution A solution that has the same solute concentration as the cytosol of a cell, resulting in an equal amount of water flowing into the cell and out of it. Compare *hypertonic solution, hypotonic solution*.

isotope A variant form of a chemical element that differs in its number of neutrons, and therefore in its atomic mass number, from the most common form of that element.

J-shaped curve A pattern of population growth in which the number of individuals in the population rises rapidly over time, as in exponential growth. Compare *S-shaped curve*.

jumping gene See *transposon*.

karyotype A display of the specific number and shapes of chromosomes found in the diploid cells of a particular individual, or of a species in general.

keystone species A species that, relative to its own abundance, has a large effect on the presence and abundance of other species in a community.

kinetic energy The energy of motion.

kinetochore A patch of protein on the centromere of a chromosome where spindle microtubules attach during mitosis and meiosis.

kingdom In reference to biological classification systems, the highest taxonomic category in the Linnaean hierarchy. Generally six kingdoms are recognized: Bacteria, Archaea, Protista, Plantae, Fungi, and Animalia.

Krebs cycle Also called *citric acid cycle*. The second major phase of cellular respiration. This series of enzyme-driven oxidation reactions takes place in the mitochondrial matrix and yields many molecules of NADH (and a few of ATP and $FADH_2$).

Kyoto Protocol An international agreement designed to reduce global warming by reducing CO_2 emissions.

lake An ecosystem of the freshwater biome defined as a standing body water of variable size, ranging up to thousands of square kilometers.

land transformation Changes made by humans to the land surface of Earth that alter the physical or biological characteristics of the affected regions. Compare *water transformation*.

law of independent assortment Mendel's second law, which states that when gametes form, the separation of alleles of one gene is independent of the separation of alleles of other genes. We now know that this law does not apply to genes that are linked.

law of segregation Mendel's first law, which states that the two copies of a gene separate during meiosis and end up in different gametes.

lichen A symbiosis between an alga (kingdom Protista) and a fungus (kingdom Fungi).

life cycle engineering An approach in which a business seeks to document (and reduce, as necessary) the total environmental impact of its products, from manufacture to disposal.

ligase An enzyme that can connect two DNA fragments. Ligases are used in DNA technology when a gene from one species is inserted into the DNA of another species.

light reactions The series of chemical reactions in photosynthesis that harvest energy from sunlight and use it to produce energy-rich compounds such as ATP and NADPH. The light reactions occur at the thylakoid membranes of chloroplasts and produce O_2 as a by-product. Compare *Calvin cycle*.

lineage A group of closely related individuals, species, genera, or the like, depicted as a branch on an evolutionary tree.

linked genes See *genetic linkage*.

Linnaean hierarchy The classification scheme used by biologists to organize and name organisms. Its seven levels—from the most inclusive to the least—are kingdom, phylum, class, order, family, genus, and species.

lipid A hydrophobic molecule of biological origin. Lipids are a key component of cell membranes. See also *phospholipid bilayer*.

locus (pl. loci) The physical location of a gene on a chromosome.

lumen The space enclosed by the membrane of an organelle, or the cavity inside an organ.

lysosome A specialized vesicle with an acidic lumen containing enzymes that break down macromolecules.

macroevolution The rise and fall of major taxonomic groups due to evolutionary radiations that bring new groups to prominence and mass extinctions in which groups are lost; the history of large-scale evolutionary changes over time. Compare *microevolution*.

macromolecule A large organic molecule formed by the bonding together of small organic molecules.

malignant tumor A cancerous growth that begins as a single benign tumor and then spreads to other tissues in the body with life-threatening consequences. Compare *benign tumor*.

mass extinction A event during which large numbers of species become extinct throughout most of Earth.

maternal homologue In a homologous pair of chromosomes, the one that comes from the mother. Compare *paternal homologue*.

meiosis A specialized process of cell division in eukaryotes during which diploid cells divide to produce haploid cells. Meiosis has two division cycles, and in animals it occurs exclusively in cells that produce gametes. Compare *mitosis*.

meiosis I The first cycle of cell division in meiosis, in which the members of each homologous chromosome pair are separated into different daughter cells. Meiosis I produces haploid daughter cells, each with half of the chromosome set found in the diploid parent cell.

meiosis II The second cycle of cell division in meiosis, in which the sister chromatids of each duplicated chromosome are separated into different daughter cells. Meiosis II is essentially mitosis, but in a haploid cell.

messenger RNA (mRNA) A type of RNA that specifies the order of amino acids in a protein.

metabolic pathway A series of enzyme-controlled chemical reactions in a cell in which the product of one reaction becomes the substrate for the next.

metabolism All the chemical reactions in a cell that involve the acquisition, storage, or use of energy.

metaphase The stage of cell division during which chromosomes become aligned at the equator of the cell.

microevolution Changes in allele or genotype frequencies in a population over time; the smallest scale at which evolution occurs. Compare *macroevolution*.

microfilament A protein fiber composed of actin monomers. Microfilaments are part of a cell's cytoskeleton and are important in cell movements.

microtubule A cylinder of protein, composed of tubulin monomers. Microtubules are part of the cell's cytoskeleton.

mismatch error The insertion of an incorrect nitrogenous base during DNA replication that is not detected and corrected.

mitochondrion (pl. mitochondria) An organelle with a double membrane that is the site of cellular respiration in eukaryotes. Mitochondria break down simple sugars to produce ATP in an oxygen-dependent (aerobic) process.

mitosis The process of cell division in eukaryotes that produces two daughter nuclei, each with the same chromosome number as the parent nucleus. Compare *meiosis*.

mitotic spindle An football-shaped array of microtubules that guides the movement of chromosomes during mitosis.

molecule An association of atoms in which two or more of the atoms are linked through covalent bonds.

monomer A molecule that can be linked with other related molecules to form a larger *polymer*.

monosaccharide A simple sugar that can be linked to other sugars, forming a *disaccharide* or *polysaccharide*. Glucose is the most common monosaccharide in living organisms.

morphology The form and structure of an organism.

morula The initial multicellular ball of cells produced by division of the zygote during the first stage of vertebrate development.

most recent common ancestor The ancestral organism from which a group of descendants arose.

motor protein A protein that uses the chemical energy of ATP to produce mechanical work, such as the movement of organelles along microtubules.

mRNA See *messenger RNA*.

multicellular organism An organism made up of more than one cell.

multipotent Of or referring to a cell that can differentiate into only a relatively narrow range of cell types. Compare *pluripotent*, *totipotent*.

multiregional hypothesis A hypothesis stating that anatomically modern humans evolved from *Homo erectus* populations scattered throughout the world. According to this idea, worldwide gene flow caused different human populations to evolve modern characteristics simultaneously and to remain a single species. Compare *out-of-Africa hypothesis*.

mutagen A substance or energy source that alters DNA.

mutation A change in the sequence of an organism's DNA. New alleles arise only by mutation, so mutations are the original source of all genetic variation.

mutualism An interaction between two species in which both species benefit.

mutualist An organism that interacts with another organism to the mutual benefit of both.

mycelium (pl. mycelia) The main body of a fungus, composed of threadlike hyphae.

mycorrhiza (pl. mycorrhizae) A mutualism between a fungus and a plant, in which the fungus provides the plant with mineral nutrients while receiving organic nutrients from the plant.

NADH An energy carrier molecule that acts as a reducing agent in the catabolic reactions (cellular respiration) that produce ATP from the breakdown of sugars into water and carbon dioxide.

NADPH An energy carrier molecule that acts as a reducing agent in photosynthesis.

natural selection An evolutionary mechanism in which those individuals in a population that possess particular inherited characteristics survive and reproduce at a higher rate than other individuals in the population because of those characteristics. Natural selection is the only evolutionary mechanism that consistently improves the survival and reproduction of the organism in its environment.

negative growth regulators A variety of external and internal signals and regulatory proteins that control the cell cycle by halting cell division. Compare *positive growth regulators*.

net primary productivity (NPP) The amount of energy that producers capture by photosynthesis, minus the amount lost as metabolic heat. NPP is usually measured as the amount of new biomass produced by photosynthetic organisms per unit of area during a specified period of time. Compare *secondary productivity*.

neutron A particle, found in the nucleus of an atom, that has no electrical charge.

nitrogen fixation The process by which nitrogen gas (N_2), which is readily available in the atmosphere but cannot be used by plants, is converted to ammonium (NH_4^+), a form of nitrogen that can be used by plants. Nitrogen fixation is accomplished naturally by bacteria and by lightning, and by humans in industrial processes such as the production of fertilizer.

nitrogenous base Any of the five nitrogen-rich compounds found in nucleotides. The four nitrogenous bases found in DNA are adenine (A), cytosine (C), guanine (G), and thymine (T); in RNA, uracil (U) replaces thymine.

noncoding DNA A segment of DNA that does not encode proteins or RNA. Introns and spacer DNA are two common types of noncoding DNA.

noncovalent bond Any chemical linkage between two atoms that does not involve the sharing of electrons. Hydrogen bonds and ionic bonds are examples of noncovalent bonds.

nonpolar molecule A molecule that has an equal distribution of electrical charge across all its constituent atoms. Nonpolar molecules do not form hydrogen bonds and tend not to dissolve in water. Compare *polar molecule*.

NPP See *net primary productivity*.

nuclear envelope The double membrane that forms the outer boundary of the nucleus, an organelle found only in eukaryotic cells.

nuclear pore One of many openings in the nuclear envelope that allow selected molecules, including specific proteins and RNA, to move into and out of the nucleus.

nucleic acid A polymer made up of nucleotides. There are two kinds of nucleic acids: DNA and RNA.

nucleolus (pl. nucleoli) A region of the nucleus that specializes in churning out large quantities of ribosomal RNA.

nucleotide Any of a class of organic molecules that serve as energy carriers and as the chemical building blocks of nucleic acids (DNA and RNA). A nucleotide is made up of a phosphate group, a five-carbon sugar, and one of four nitrogenous bases (see *nitrogenous base*). Nucleotides are linked together to form a single strand of DNA or RNA.

nucleus (pl. nuclei) The organelle in a eukaryotic cell that contains the genetic blueprint in the form of DNA.

nutrient In an ecosystem context, an essential element required by a producer.

nutrient cycle The cyclical movement of a nutrient between organisms and the physical environment. There are two main types of nutrient cycles: *atmospheric cycles* and *sedimentary cycles*.

observations With respect to the scientific method, knowledge gained by watching, measuring, recording, and analyzing aspects of natural processes. Facts learned in this manner are subsequently used to formulate hypotheses.

oceanic region An ecosystem of the marine biome defined as the part of the ocean beginning about 40 miles offshore, where the continental shelf, and therefore the coastal region, ends.

oncogene A mutated gene that promotes excessive cell division, leading to cancer.

operator In prokaryotes, a regulatory DNA sequence that controls the transcription of a gene or group of genes.

opposable In primates, of or referring to a thumb (or big toe) that moves freely and can be placed opposite other fingers (or toes).

order In reference to biological classification systems, the level in the Linnaean hierarchy above family and below class.

organ A self-contained collection of different types of tissues, usually of a characteristic size and shape, that is organized for a particular set of functions.

organ system A group of organs of different types that work together to carry out a common set of functions.

organelle A discrete cytoplasmic structure with a specific function. Some cell biologists use the term only for membrane-enclosed cytoplasmic compartments; others include other cytoplasmic structures, such as ribosomes, in the definition.

organic molecule A molecule containing at least one carbon covalently bonded to one or more hydrogen atoms. Before modern chemistry, organic molecules on Earth were exclusively of biological origin, but now chemists can create many organic molecules artificially (synthetically).

osmoregulation The process of maintaining an internal water concentration that supports biological processes.

osmosis The passive movement of water across a selectively permeable membrane.

out-of-Africa hypothesis A hypothesis stating that anatomically modern humans evolved in Africa within the past 200,000 years and then spread throughout the rest of the world. According to this idea, as they spread from Africa, modern humans completely replaced older forms of *Homo sapiens*, including advanced forms such as the Neandertals. Compare *multiregional hypothesis*.

oxidation The loss of electrons by one atom or molecule to another. Compare *reduction*.

oxidative phosphorylation The shuttling of electrons down an electron transport chain in mitochondria that results in the production of ATP.

P generation The parent generation in a genetic cross. Compare F_1 *generation*, F_2 *generation*.

Pangaea An ancient supercontinent that contained all of the world's landmasses. Pangaea formed 250 million years ago and began to break apart 200 million years ago, ultimately yielding the continents we know today.

parasite An organism that lives in or on another organism (its host) and obtains nutrients from that organism. Parasites harm and may eventually kill their hosts but do not kill them immediately.

passive carrier protein A protein in the plasma membrane of a cell that, without the input of energy, changes its shape to transport a molecule across the membrane from the side of higher concentration to the side of lower concentration. Compare *active carrier protein*.

passive transport Movement of a substance from areas of higher concentration to areas of lower concentration without the expenditure of energy. Compare *active transport*.

paternal homologue In a pair of homologous chromosomes, the one that comes from the father. Compare *maternal homologue*.

pathogen An organism or virus that infects a host and causes disease, harming and in some cases killing the host.

PCR See *polymerase chain reaction*.

pedigree A chart that shows genetic relationships among family members over two or more generations of a family's history.

peptide bond A covalent bond between the amino group of one amino acid and the carboxyl group of another. Peptide bonds link amino acids together.

permafrost Permanently frozen soil that is found below the surface layers and may be a quarter of a mile deep.

Permian extinction The largest mass extinction in the history of life on Earth; it occurred 250 million years ago, driving up to 95 percent of the species in some groups to extinction.

persistent organic pollutant (POP) A long-lived organic molecule of synthetic origin that bioaccumulates in organisms, and that can have harmful effects. Some of the most damaging and widespread POPs are PCBs (polychlorinated biphenyls) and dioxins.

PGD See *preimplantation genetic diagnosis*.

pH The concentration of hydrogen ions in a solution. The pH scale runs from 1 to 14. A pH of 7 is neutral; values below 7 indicate acids, and values above 7 indicate bases.

phagocytosis A form of endocytosis by which a cell engulfs a large particle, such as another cell; "cell eating." Compare *pinocytosis*.

phenotype The specific version of a genetic trait that is displayed by a given individual. For example, black, brown, red, and blond are phenotypes of the hair color trait in humans. Compare *genotype*.

phosphate group A functional group consisting of a phosphate atom and four oxygen atoms.

phospholipid A lipid consisting of two fatty acids, a glycerol, and a phosphate as part of the hydrophilic head group. Phospholipids are the main component in all biological membranes.

phospholipid bilayer A double layer of phospholipid molecules arranged so that their hydrophobic "tails" lie sandwiched between their hydrophilic "heads." A phospholipid bilayer forms the basic structure of all biological membranes.

phosphorylation The addition of a phosphate group to an organic molecule.

photoautotroph An organism that obtains its energy from sunlight and derives its carbon from carbon dioxide in the air. Examples include cyanobacteria, green algae, and plants. Compare *photoheterotroph*, *chemoautotroph*.

photoheterotroph An organism that obtains its energy from sunlight and derives its carbon from carbon-containing compounds found mainly in other organisms. All photoheterotrophs are prokaryotes. Compare *photoautotroph*, *chemoheterotroph*.

photosynthesis A process by which organisms capture energy from sunlight and use it to synthesize sugars from carbon dioxide and water.

photosystem A large complex of proteins and chlorophyll that captures energy from sunlight. Two distinct photosystems (I and II) are present in the thylakoid membranes of chloroplasts.

photosystem I The photosystem that is primarily responsible for the production of NADPH.

photosystem II The photosystem in which light energy is used to initiate an electron flow along the electron transport chain, resulting in ATP synthesis and the release of oxygen gas (O_2) as a by-product.

phylum (pl. phyla) The level in the Linnaean hierarchy above class and below kingdom.

pinocytosis A form of nonspecific endocytosis by which cells take in fluid; "cell drinking." Compare *phagocytosis*.

Plantae The kingdom made up of plants.

plasma membrane The phospholipid bilayer that forms the outer boundary of any cell.

plasmid A small circular segment of DNA found naturally in bacteria. Plasmids are involved in natural gene transfers among bacteria and can be used as vectors in genetic engineering.

plasmodesma (pl. plasmodesmata) A tunnel-like channel between two plant cells that provides a cytoplasmic connection allowing the flow of small molecules and water between them.

pleiotropy The situation in which a single gene influences a variety of different traits.

pluripotent Of or referring to a cell that can differentiate into any of the cell types in the adult body. Compare *multipotent*, *totipotent*.

point mutation A mutation in which only a single base is altered.

polar molecule A molecule that has an uneven distribution of electrical charge. Polar molecules can easily interact with water molecules and are therefore soluble. Compare *nonpolar molecule*.

pollinator An animal that carries pollen grains from the stamens of one flower to the stigmas (see *carpel*) of other flowers of the same species.

polygenic Of or referring to inherited traits that are determined by the action of more than one gene.

polymer A large organic molecule composed of many *monomers* linked together.

polymerase chain reaction (PCR) A method of DNA technology that uses the DNA polymerase enzyme to make multiple copies of a targeted sequence of DNA.

polypeptide A polymer consisting of covalently linked linear chains of amino acids.

polyploid Of or referring to a cell or organism that has three or more complete sets of chromosomes (rather than the usual two complete sets). Populations of polyploid individuals can rapidly form new species without geographic isolation. Compare *diploid, haploid*.

polysaccharide A polymer composed of many linked *monosaccharides*. Examples include starch and cellulose. Compare *disaccharide*.

POP See *persistent organic pollutant*.

population A group of interacting individuals of a single species located within a particular area.

population cycle A pattern in which the population sizes of two species increase and decrease together in a tightly linked cycle; this pattern can occur when at least one of the two species involved is very strongly influenced by the other.

population density The number of individuals in a population, divided by the area covered by the population.

population ecology A branch of science concerned with questions that relate to how many organisms live in a particular environment, and why.

population size The total number of individuals in a population.

positive growth regulators A variety of internal signals, including growth factors, hormones, and regulatory proteins, that control the cell cycle by stimulating cell division. Compare *negative growth regulators*.

potential energy Stored energy.

predator An organism that kills other organisms (called *prey*) for food.

predictions With respect to the scientific method, statements about logical consequences that should be observed if a hypothesis is correct.

preimplantation genetic diagnosis (PGD) A procedure used in *in vitro* fertilization in which

one or two cells are removed from a developing embryo and tested for genetic disorders; embryos that are free of genetic disorders are then implanted into the mother's uterus.

prey Animals that *predators* kill and eat.

primary consumer An organism that eats a producer. Compare *secondary consumer, tertiary consumer*.

primary structure In reference to proteins, the sequence of amino acids in a protein. Compare *secondary structure, tertiary structure, quaternary structure*.

primary succession Ecological succession that occurs in newly created habitat, as when an island rises from the sea or a glacier retreats, exposing newly available bare ground. Compare *secondary succession*.

primate An order of mammals whose living members include lemurs, tarsiers, monkeys, humans, and other apes. Primates share characteristics such as flexible shoulder and elbow joints, opposable thumbs or big toes, forward-facing eyes, and brains that are large relative to body size.

producer An organism that uses energy from an external source, such as sunlight, to produce its own food without having to eat other organisms or their remains. Compare *consumer*.

product A substance that is formed by a chemical reaction. Compare *reactant*.

productivity (in ecology) The mass of living matter (biomass) that can be produced in a given area from the available nutrients and sunlight by the producers in a given ecosystem.

prokaryote A single-celled organism that does not have a nucleus. All prokaryotes are members of the domains Bacteria or Archaea. Compare *eukaryote*.

promoter In genetics, the DNA sequence in a gene to which RNA polymerase binds to begin transcription, and that therefore controls gene expression at the transcriptional level.

prophase The stage of mitosis or meiosis during which chromosomes first become visible under the microscope.

protein A polymer of amino acids that are linked together in a specific sequence. Most proteins are folded into complex three-dimensional shapes.

Protista The oldest eukaryotic kingdom, made up of a diverse collection of mostly single-celled but some multicellular organisms.

proton A positively charged particle found in atoms. Each atom contains a characteristic number of protons. Compare *electron*.

proton gradient An imbalance in the concentration of protons across a membrane.

proto-oncogene A gene that promotes cell division in response to growth signals as part of its normal cellular function.

protostome Any of a group of animals, including insects, worms, and snails, in which the first opening to develop in the early embryo becomes the mouth. Compare *deuterostome*.

pseudopodium (pl. pseudopodia) A dynamic protrusion of the plasma membrane that enables some cells to move. The extension of pseudopodia depends on actin filaments inside the cell.

Punnett square A diagram in which all possible genotypes of male and female gametes are listed on two sides of a square, providing a graphical way to predict the genotypes of the offspring produced in a genetic cross.

pyruvate A three-carbon molecule produced by glycolysis that is processed in the mitochondria to generate ATP.

quaternary structure In reference to proteins, the three-dimensional arrangement of two or more separate chains of amino acids into a functional protein complex. Compare *primary structure, secondary structure, tertiary structure*.

radioisotope An unstable, radioactive form of an element that releases energy as it decays to more stable forms at a constant rate over time.

rain shadow An area on the side of a mountain facing away from moist prevailing winds where little rain or snow falls.

rainforest A forest that receives high amounts of rainfall.

reactant A substance that undergoes a chemical reaction. Compare *product*.

reaction center A cluster of chlorophyll molecules within an antenna complex whose electrons become excited, and are passed to an electron transport chain, when the pigment molecules absorb light energy.

receptors See *receptor proteins*.

receptor-mediated endocytosis A form of endocytosis in which receptor proteins embedded in the plasma membrane of a cell recognize certain surface characteristics of materials to be brought into the cell by endocytosis.

receptor proteins Also called simply *receptors*. Proteins in the plasma membrane or cytoplasm of a target cell that bind signaling molecules, allowing those molecules to indirectly affect processes inside the cell.

recessive allele An allele that does not have a phenotypic effect when paired with a *dominant allele* in a heterozygote.

recombinant DNA An artificial assembly of genetic material created by the enzyme-mediated linking of DNA fragments.

recombination In reference to chromosomes or DNA, a collective term for the emergence of new combinations of genes or their alleles, for example, through crossing-over in meiosis.

redox reaction A chemical reaction in which electrons are transferred from one molecule or atom to another.

reduction The gain of electrons by one atom or molecule from another. Compare *oxidation*.

regulatory DNA A DNA sequence that can increase, decrease, turn on, or turn off the expression of a gene or a group of genes. Regulatory DNA sequences interact with regulatory proteins to control gene expression.

regulatory protein A protein that signals whether or not a particular gene or group of genes should be expressed. Regulatory proteins interact with regulatory DNA to control gene expression.

replicate In experimental design, an independent run or performance of an experiment.

repressor protein A protein that prevents the expression of a particular gene or group of genes.

reproductive cloning A technology used to produce an offspring that is an exact genetic copy (a clone) of another individual. The first two steps in reproductive cloning are the same as those in *therapeutic cloning*, but stem cells are not removed from the embryo. Instead, the embryo is transferred to the uterus of a surrogate mother, where, if all goes well, the birth of a healthy offspring ultimately results; this offspring is genetically identical to the individual who provided the donor nucleus.

reproductive isolation A condition in which barriers to reproduction prevent or strongly limit two or more populations from reproducing with one another. Many different kinds of reproductive barriers can result in reproductive isolation, but it always has the same effect: no or few genes are exchanged between the reproductively isolated populations. Compare *geographic isolation*.

respiration See *cellular respiration*.

restriction enzyme Any of a number of enzymes that cut DNA molecules at a specific target sequence; a key tool of DNA technology.

RFLP analysis A method of DNA technology in which restriction enzymes are used to cut an organism's genome into small pieces, which are sorted by size using gel electrophoresis. Next, a DNA probe is used to form a profile, whose pattern depends on the number and size of the fragments that can bind to the probe. *RFLP* stands for *restriction fragment length polymorphism*.

ribonucleic acid See *RNA*.

ribosomal RNA (rRNA) A type of RNA that is an important component of ribosomes.

ribosome A particle composed of proteins and RNA at which new proteins are synthesized. Ribosomes can be either attached to the endoplasmic reticulum or free in the cytosol.

ring species A species whose populations loop around a geographic barrier (such as a mountain chain) and in which the populations at the two ends of the loop are in contact with one another, yet cannot interbreed.

river An ecosystem of the freshwater biome defined as a body of water that moves continuously in a single direction.

RNA Ribonucleic acid; a polymer of nucleotides that is necessary for the synthesis of proteins in living organisms.

RNA polymerase The key enzyme in DNA transcription, which links together the nucleotides of the RNA molecule specified by a gene.

RNA splicing The process by which mRNA introns are snipped out and the remaining pieces of mRNA re-joined.

root system The branched, nongreen underground organ system that is specialized for absorbing water and mineral nutrients and for anchoring a plant in soil.

rough ER A region of the endoplasmic reticulum that has attached ribosomes and that specializes in protein synthesis. Compare *smooth ER*.

rRNA See *ribosomal RNA*.

rubisco The enzyme that catalyzes the first reaction of carbon fixation in photosynthesis.

S phase The stage of the cell cycle during which the cell's DNA is replicated.

S-shaped curve A pattern of population growth in which the number of individuals in a population at first increases at a rate similar to exponential growth; as the number of individuals increases, however, the growth rate gradually decreases and the population stabilizes at the

size that can be supported indefinitely by the environment. Compare *J-shaped curve*.

salt A compound consisting of ions held together by the mutual attraction between their opposite electrical charge.

saturated Of or referring to a lipid that has no double bonds between the carbon atoms in its hydrocarbon backbone. Compare *unsaturated*.

science A method of inquiry that provides a rational way to discover truths about the natural world.

scientific method A series of steps in which the investigator develops a hypothesis, tests its predictions by performing experiments, and then changes or discards the hypothesis if its predictions are not supported by the results of the experiments.

scientific name The unique two-part name given to each species that consists of, first, a Latin name designating the genus and, second, a Latin name designating that species.

second law of thermodynamics The law stating that all systems, such as a cell or the universe, tend to become more disordered, and that the creation and maintenance of order in a system requires the transfer of disorder to the environment.

secondary consumer An organism that eats a primary consumer. Compare *primary consumer*, *tertiary consumer*.

secondary productivity The rate of new biomass production by consumers per unit of area. Compare *net primary productivity*.

secondary structure In reference to proteins, the patterns of local three-dimensional form in segments of a protein. Spiral shapes and pleated sheets are common forms in the secondary structure of many proteins. Compare *primary structure, tertiary structure, quaternary structure*.

secondary succession Ecological succession that occurs as communities recover from disturbance, as when a forest grows back when a field ceases to be used for agriculture. Compare *primary succession*.

sedimentary cycle A type of nutrient cycle in which the nutrient does not enter the atmosphere easily. Compare *atmospheric cycle*.

seed A structure produced by a plant in which a plant embryo is encased in a protective covering.

selectively permeable membrane A membrane that controls which materials can pass through it. An example is the plasma membrane.

sex chromosome Either of a pair of chromosomes that determines the sex of an individual. Compare *autosome*.

sex-linked Of or referring to genes located on a sex chromosome. See also *X-linked* and *Y-linked*.

sexual reproduction The type of reproduction in which genes from two individuals are combined to give rise to a new individual, known as the offspring. Compare *asexual reproduction*.

sexual selection A type of natural selection in which individuals that differ in inherited characteristics differ, as a result of those characteristics, in their ability to get mates.

shared derived feature A feature unique to a common ancestor that is passed down to all of its descendants, clearly defining them as a group.

signal cascade A stepwise sequence of events within a cell, commonly involving a series of protein activations, that amplify a small stimulus and thereby provoke a large response.

signal transduction pathway A series of cellular events that relay receipt of a signal from protein receptors on the plasma membrane to the cytoplasm.

signaling molecule A molecule produced and released by one cell that affects the activities of another cell (referred to as a *target cell*). Signaling molecules enable the cells of a multicellular organism to communicate with one another and coordinate their activities.

single nucleotide polymorphism (SNP) A difference in a single base pair among the genomes of individuals.

sister chromatids A pair of identical double helices that are produced by the replication of DNA during the cell cycle.

smooth ER A region of the endoplasmic reticulum that is specialized for lipid synthesis. It is "smooth" because it does not have attached ribosomes. Compare *rough ER*.

SNP See *single nucleotide polymorphism*.

soluble Of or referring to a chemical that will dissolve (mix) in water.

solute A dissolved substance. Compare *solvent*.

solution Any combination of a solute and a solvent.

solvent A liquid (in biological systems, usually water) into which a *solute* has dissolved.

somatic cell Any cell in a multicellular organism that is not a gamete or part of a gamete-making tissue.

somatic stem cell See *adult stem cell*.

spacer DNA A region of noncoding DNA that separates two genes. Spacer DNA is abundant in eukaryotes, but not in prokaryotes.

specialist In reference to living organisms, a species that requires very specific conditions to survive, such as an insect that can eat only one kind of plant, as opposed to being able to eat and survive on many different kinds of plants.

speciation The process by which one species splits to form two or more species that are reproductively isolated from one another.

species (pl. species) A group of interbreeding natural populations that is reproductively isolated from other such groups.

spore The reproductive cell of a fungus, produced directly through meiosis and typically encased in a protective coating that shields it from drying or rotting.

SRY gene A gene, located on the Y chromosome, that functions as a master switch, committing the sex of a developing embryo to "male." *SRY* is short for "sex-determining *region* of *Y*."

stabilizing selection A type of natural selection in which individuals with intermediate values of an inherited characteristic have an advantage over other individuals in the population, as when medium-sized individuals produce offspring at a higher rate than do small or large individuals. Compare *directional selection, disruptive selection*.

start codon A three-nucleotide sequence on an mRNA molecule (usually the codon AUG) that signals where translation should begin. Compare *stop codon*.

stem cell An undifferentiated cell that can renew itself through cell division, theoretically indefinitely. Some of the daughter cells generated by stem cells may differentiate into a specialized cell type.

steroid hormone Any of a class of hydrophobic signaling molecules that can pass through the plasma membrane of a target cell.

sterols A group of lipids whose fundamental structure consists of four hydrocarbon rings fused to each other.

stop codon A three-nucleotide sequence on an mRNA molecule that signals where translation should end. Compare *start codon*.

stroma The space enclosed by the inner membrane of the chloroplast, in which the thylakoid membranes are situated.

subsidence Sinking of the land surface when groundwater levels have dropped and hence there is less water underground to support the land above.

substitution In genetics, a mutation in which one nitrogenous base is replaced by another at a single position in the DNA sequence of a gene. Compare *deletion, insertion*.

substrate In reference to enzymes, the particular substance on which an enzyme acts. Only the substrate will bind to the active site of the enzyme.

succession A process by which species in a community are replaced over time. For a given location, the order in which species are replaced is fairly predictable.

sugar A simple carbohydrate, generally a monosaccharide or a disaccharide, that has the general chemical formula $(CH_2O)n$, where n is less than 7 for monosaccharides and about 12 for most disaccharides.

sustainable Of or referring to an action or process that can continue indefinitely without using up resources or causing serious damage to ecosystems.

symbiosis A relationship in which two or more organisms of different species live together in close association.

sympatric speciation The formation of new species from populations that are not geographically isolated from one another. Compare *allopatric speciation*.

systematist A scientist who studies evolutionary relationships among organisms and constructs evolutionary trees.

taiga See *boreal forest*.

target cell A cell that receives and responds to a signaling molecule.

taxon (pl. taxa) A group defined within the Linnaean hierarchy—for example, a species or a kingdom.

telophase The stage of mitosis or meiosis during which chromosomes arrive at the opposite poles of the cell and new nuclear envelopes begin to form around each set of chromosomes.

temperate deciduous forest A terrestrial biome dominated by trees and shrubs that grow in regions with cold winters and moist, warm summers.

template strand In reference to gene transcription, the strand of DNA (of the two strands in a DNA molecule) that is copied into RNA and which is therefore complementary to the RNA synthesized of it.

terminator In reference to bacterial gene transcription, a DNA sequence that, when reached by RNA polymerase, causes transcription to end and the newly formed mRNA molecule to separate from its DNA template.

tertiary consumer An organism that eats a secondary consumer. Compare *primary consumer*, *secondary consumer*.

tertiary structure In reference to proteins, the overall three-dimensional form of a protein, created and stabilized by chemical interactions between distantly placed segments of the protein. Compare *primary structure, secondary structure, quaternary structure*.

therapeutic cloning A technology aimed at creating embryonic stem cells of a desired genotype (bearing a specific patient's genome, for example). The (haploid) nucleus of an unfertilized egg cell is replaced with the (diploid) nucleus of a nonreproductive donor cell, such as a skin cell. Next, chemicals are used to stimulate the egg to divide so that it begins to form an embryo. Finally, stem cells are removed from the developing blastocyst and stimulated to grow into a wide range of cell types. These goals have been accomplished with experimental animals, such as mice, but not with human cells. Compare *reproductive cloning*.

thylakoid One of a series of flattened, interconnected membrane sacs that lie one on top of another within a chloroplast in stacks called grana.

thylakoid membrane The membrane that encloses the thylakoid space inside a chloroplast. The thylakoid membrane houses both photosystems and their associated electron transport chains.

thylakoid space The space enclosed by the thylakoid membrane inside a chloroplast; the innermost compartment of the chloroplast.

tight junction A structure made up of rows of proteins associated with the plasma membranes of adjacent cells. Neighboring cells are held together tightly by the interlocking of the membrane proteins of adjacent cells. Cells connected by tight junctions form impermeable sheets that do not permit ions and molecules to pass from one side of the sheet to the other by slipping between the cells.

tissue A collection of coordinated and specialized cells that together fulfill a particular function for the organism.

totipotent Of or referring to a cell that can differentiate into any human cell type. Compare *multipotent, pluripotent*.

transcription In reference to genes, the synthesis of an RNA molecule from a DNA template. Transcription is the first of the two major steps in the process by which genes specify proteins; it produces mRNA, tRNA, and rRNA,

all of which are essential in the production of proteins. Compare *translation*.

transfer RNA (tRNA) A type of RNA that transfers the amino acid specified by mRNA to the ribosome during protein synthesis.

transformation In genetics, a change in the genotype of a cell as a result of the incorporation of external DNA by that cell.

translation In genetics, the conversion of a sequence of nitrogenous bases in an mRNA molecule to a sequence of amino acids in a protein. Translation is the second of the two major steps in the process by which genes specify proteins; it occurs at the ribosomes. Compare *transcription*.

translocation In genetics, a mutation in which a segment of a chromosome breaks off and is then attached to a different, nonhomologous chromosome.

transport vesicle A vesicle that specializes in moving substances from one location to another within the cytoplasm and to and from the exterior of the cell.

transposon A DNA sequence that can move from one position on a chromosome to another, or from one chromosome to another; known informally as a "jumping gene."

triglyceride A lipid in which all three hydroxyl groups in glycerol are bonded to a fatty acid. Animals store most of their surplus energy in the form of triglycerides.

trisomy In diploid organisms, the condition of having three copies of a chromosome (instead of the usual two).

tRNA See *transfer RNA*.

trophic level A level or step in a food chain. Trophic levels begin with producers and end with predators that eat other organisms but are not fed on by other predators.

tropical forest A terrestrial biome dominated by a rich diversity of trees, vines, and shrubs that grow in warm, rainy regions.

tubulin The protein monomer that makes up microtubules.

tumor A solid cell mass formed by the inappropriate proliferation of cells.

tumor suppressor gene A gene that inhibits cell division under normal conditions.

tundra A terrestrial biome dominated by low-growing shrubs and nonwoody plants that can tolerate extreme cold.

unsaturated Of or referring to a lipid, such as a fatty acid, that has one or more double bonds

between the carbon atoms in its hydrocarbon backbone. Compare *saturated*.

upwelling The stirring of bottom sediments in a body of water by air or water currents, or shifts in water temperature (as in northern lakes during fall and spring).

vacuole A large water-filled vesicle found in plant cells. Vacuoles help maintain the shape of plant cells and can also be used to store various molecules, including nutrients and antiherbivory chemicals.

vascular system The tissue system in plants that is devoted to internal transport.

vertebrates A group of animals with backbones. Vertebrates include fishes, amphibians, mammals, birds, and reptiles.

vesicle A small, membrane-enclosed sac found in the cytosol of eukaryotic cells.

vestigial organ A structure or body part that served a purpose in an ancestral species but is currently of little or no use to the organism that has it.

virus An infectious particle consisting of nucleic acids and proteins. A virus cannot reproduce on its own, and must instead use the cellular machinery of its host to reproduce.

water transformation Changes made by humans to the waters of Earth that alter their physical or biological characteristics. Compare *land transformation*.

weather Temperature, precipitation, wind speed, humidity, cloud cover, and other physical conditions of the lower atmosphere at a specific place over a short period of time. Compare *climate*.

wetland An ecosystem of the freshwater biome defined as standing water shallow enough that rooted plants emerge above the water surface. Bogs (stagnant, acidic, and oxygen-poor), marshes (grassy), and swamps (dominated by trees and shrubs) are all wetlands.

X-linked Of or referring to sex-linked genes located on an X chromosome. Compare *Y-linked*.

Y-linked Of or referring to sex-linked genes located on the Y chromosome. Compare *X-linked*.

zygote The diploid (2*n*) cell formed by the fusion of two haploid (*n*) gametes; a fertilized egg.

Credits

Photography Credits

Chapter 1 p. 3: Courtesy Dr. Philippa Uwins, University of Queensland; **1.1**: Strauss/Curtis/Corbis; **p. 5** (top two): Dr. JoAnn Burkholder; **1.3a**: Photo by the NC Division of Marine Fisheries; **1.3b**: Juvenile Atlantic Menhaden, Pamlico Estuary, NC; photo by H. Glasgow; **1.4a**: Yorgos Nikas/Stone/Getty Images; **1.4b**: Visuals Unlimited/Corbis; **1.4c**: Prof S. Cinti; **1.4** (diver): Patrik Giardino/Corbis; **1.4d**: Purestock/Getty Images; **1.4e**: Dr. John D. Cunningham/Visuals Unlimited; **1.4f**: Dr. John D. Cunningham/Visuals Unlimited; **1.7**: F. Stuart Westmorland/Photo Researchers, Inc; **1.8**: Momatiuk-Eastcott/Corbis; **1.9**: David Vintiner/zefa/Corbis; **p. 12**: Getty Images; **p. 13** (bottom): Courtesy of the Library Dept., American Museum of Natural History; **p. 16**: Bettmann/Corbis; **1.12**: Jeffrey L. Rotman/Corbis; **p. 19**: Courtesy Dr. Philippa Uwins, University of Queensland.

Chapter 2 p. 25: Vienna Report Agency/Sygma/Corbis: **2.1a** (all): Tim Graham/Getty Images; **2.3a**: Keren Su/Corbis; **2.3b**: Jane Sanders Miller/Wellstar Enterprises, LLC; **2.3c**: Goran Tomasevic/Reuters/Corbis; **2.6** (left): Bettmann/Corbis; **2.10**: Getty Images; **p. 37**: Vienna Report Agency/Sygma/Corbis; **p. 38**: AP Photo.

Chapter 3 p. 43: AP Photo/Tsunemi Kubodera of the National Science Museum of Japan, HO; **3.2** (top left): Photo Researchers, Inc; **3.2** (top right): Biology Media/Science Source/Photo Researchers, Inc; **3.2** (center right): Dr. Kari Lounatamaa/SPL/Science Source/Photo Researchers, Inc; **3.2** (bottom left): David M. Phillips/Science Source/Photo Researchers, Inc; **3.2** (bottom right): Dr. Jeremy Burgess/SPL/Science Source/Photo Researchers, Inc; **3.4**: Jerome Wexler/Photo Researchers, Inc; **p. 48** (top): Courtesy of Niles Eldredge; **p. 48** (bottom): Courtesy the Library, American Museum of Natural History; **3.5**: Krafft/Hoa-qui/Photo Researchers, Inc; **3.6** (left): James Marshall/Corbis; **3.6** (inset): Dr. Dennis Kunkel/Visuals Unlimited; **3.7** (top right): Science VU/Visuals Unlimited; **3.7** (bottom left): Dennis Kunkel Microscopy, Inc; **3.7** (bottom center): Biophoto Associates/Science Source/Photo Researchers, Inc; **3.7** (center right): Dennis Kunkel Microscopy, Inc; **3.7** (bottom right): Eye of Science/Photo Researchers, Inc; **3.8** (top left): Greg Vaughn/Tom Stack & Associates; **3.8** (top right): Greg Vaughn/Tom Stack & Associates; **3.8** (bottom left): Michael P. Gadomski/National Audubon Society Collection/Photo Researchers, Inc; **3.8** (bottom center): Compost/Visage/Peter Arnold, Inc; **3.8** (center right): Rod Planck/Photo Researchers, Inc; **3.12a**: Inga Spence/Tom Stack & Associates; **3.12b**: Ron Goulet/Dembinsky Photo Associates; **3.13** (top right): Sharon Cummings/Dembinsky Photo Associates; **3.13** (bottom left): Dennis Kunkel Microscopy, Inc; **3.13** (bottom right): Carolina Biological Supply Co./Visuals Unlimited; **3.15**: Michael & Patricia Fogden/Minden Pictures; **3.16**: Ed Ross; **3.17**: Stephen Sharnoff; **3.18** (p. 62, top): Newman & Flowers/National Audubon Society Collection/Photo Researchers, Inc; **3.18** (p. 62, bottom left): Thomas Zuraw/Animals Animals; **3.18** (p. 62, bottom center): Brian Parker/Tom Stack & Associates; **3.18** (p. 62, bottom right): Tom Stack/Tom Stack & Associates; **3.18** (p. 63 top left): Susan Blanchet/Dembinsky Photo Associates; **3.18** (p. 63, top center): F. Stuart Westmorland/National Audubon Society/Photo Researchers, Inc; **3.18** (p. 63, top right): Gladden William Willis/Animals Animals; **3.18** (p. 63, center left): Michael Fogden/Oxford Scientific Films; **3.18** (p. 63, center right): Andrew J. Martinez/National Audubon Society Collection/Photo Researchers, Inc; **3.18** (p. 63, bottom left): John W Banagan/Getty Images; **3.18** (p. 63, bottom right): Zigmund Leszczynski/Animals Animals; **3.23a**: Dorling Kindersley/Getty Images; **3.23b**: Michael Neveux; **p. 67**: AP Photo/Tsunemi Kubodera of the National Science Museum of Japan, HO.

Interlude A A.1: Michael & Patricia Fogden/Corbis; **A.2**: © Mark Moffet/Minden Pictures; **A.5**: Gregg Vaughn/Tom Stack & Associates; **A.6a**: E.S Ross; **A.6b**: E.S Ross; **A.7a**: David McIntyre; **A.7b**: Jim Hemenway; **A.8a**: Michael & Patricia Fogden/Corbis; **A.8b**: Corbis; **A.9** (both): Harald Pauli; **p. A10** (top): Getty Images; **p. A10** (bottom): Hal Lott/Corbis; **A.10**: Denis Scott/Corbis; **A.11**: Getty Images; **A.12a**: Dr. Till Eggers/Ecotron, NERC Centre for Population Biology; **A.12b**: Courtesy of David Tilman, University of Minnesota; **A.13a**: Eric and David Hosking/Corbis; **A.13b**: Dr. E. F. Anderson/Visuals Unlimited; **A.13c**: Kennan Ward/Corbis; **A.13d**: Hal Horwitz/Corbis.

Chapter 4 p. 73: ESO, European Southern Observatory; **4.2**: CNRI/Science Photo Library. Biology on the Job: Courtesy of Kyle Waggener; **4.4**: Joseph Sohm/Photo Researchers, Inc; **4.5**: Charles Falco/Photo Researchers, Inc; **4.8b** (center): Biophoto Associates/Photo Researchers, Inc; **4.8c**: Gary Gaugler/Visuals Unlimited; **p. 95**: ESO, European Southern Observatory.

Chapter 5 p. 101: Courtesy Justin Skoble and Daniel A. Portnoy; **p. 104** (left): CNRI/Photo Researchers, Inc; **p. 104** (right): Biological Photo Services; **5.3**: Dennis Kunkel Microscopy; **5.4** (top): Omikron/Photo Researchers, Inc; **5.4** (bottom): Don W. Fawcett/Photo Researchers, Inc; **5.6**: Dennis Kunkel Microscopy; **5.7**: David M. Philips/Visual Unlimited; **5.8**: Biophoto Associates/Science Source/Photo Researchers, Inc; **5.9**: Bill Longcore/Science Source/Photo Researchers, Inc; **5.10**: Dr. Karl Lounatmaa/SPL/Science Source/Photo Researchers, Inc; **5.11a**: Dr. Gopal Murti/Visuals Unlimited; **5.11b**: Dr. Mark McNiven; **5.11d**: Dr. Gopal Murti/Visuals Unlimited; **5.11e**: Dr. Gopal Murti/Visuals Unlimited; **5.12**: Louise Cramer; **5.13a**: Aaron Bell/Visuals Unlimited; **5.13b**: Eye of Science/Photo Researchers, Inc; **5.13c**: Science VU/Visuals Unlimited; **5.14a**: Dr. Gopal Murti/SPL/Photo Researchers, Inc; **p. 119**: Courtesy Justin Skoble and Daniel A. Portnoy.

Chapter 6 p. 125: Tom Grill/Iconica/Getty Images; **6.6a** (left): Susumu Nishinaga/Photo Researchers, Inc; **6.6** (center & right): David M. Phillips/Photo Researchers, Inc; **6.7e**: SPL/Photo Researchers, Inc; **6.7f**: Biology Media/Photo Researchers, Inc; **p. 133**: Corbis; **p. 138**: Tom Grill/Iconica/Getty Images.

David Denning, BioMedia Associates; **p. 372:** Stan Honda/AFP/Getty Images; **18.4a:** Bill Beatty/Visuals Unlimited; **18.4b:** Robert Lubeck/Animals Animals; **18.5:** Merlin D. Tuttle/Bat Conservation/Photo Researchers, Inc; **18.6a:** Rod Planck/Photo Researchers, Inc; **18.6b:** Jeff Lepore/Photo Researchers; **18.7** (left): Robert Sivinski; **18.7** (right): Doug Sokell/Visuals Unlimited; **p. 381:** Mark Smith/Photo Researchers, Inc.

Chapter 19 p. 387: © William Stott; **19.1a:** Ken Lucas/Visuals Unlimited; **19.1b:** Niles Eldridge; **19.1c:** B. Miller/Biological Photo Services; **19.1d:** Louie Psihoyos/Corbis; **19.1e:** David Grimaldi, American Museum of Natural History; **19.1.f:** Charlie Ott/Photo Researchers, Inc; **19.10a:** Rodolfo Coria; **19.10b:** Jason Edwards/National Geographic Image Collection; **p. 402** (left): DK Limited/Corbis; **p. 402** (right): Martin Harvey/Corbis; **p. 404** (top): William Stott; **p. 404** (bottom left): Adam Jones/Dembinsky Photo Associates; **p. 404** (bottom right): Gerald & Buff Corsi/Visuals Unlimited.

Interlude D D.1: Publiphoto/Photo Researchers, Inc; **D.3** (left): Leah Warkentin/Design Pics/Corbis; **D.3** (right): Cyril Ruoso/Peter Arnold Inc; **D.4b:** John Reader/SPL/Photo Researchers, Inc; **D.4c:** Dembinsky Photo Associates; **p. D4** (top right): Ken Lucas/Visuals Unlimited; **D.5a:** AFP/Getty Images; **D.5b:** Publiphoto/Photo Researchers, Inc; **D.7:** Copyright 1996, David L. Brill, from American Museum of Natural History; **D.8:** © Jay Matternes; **D.9b** (right): Tom McHugh/Science Source/Photo Researchers, Inc; **D.12:** David Perrett and Duncan Rowland, University of St. Andrews/SPL/Photo Researchers, Inc; **D.14a:** Ed Wargin/Corbis; **D.14b:** James L. Amos/Corbis; **D.15:** American Philosophical Society.

Chapter 20 p. 409 (top): Peter Yates/Photo Researchers, Inc; **p. 409** (inset): Ted Kinsman/Photo Researchers, Inc; **20.1a:** NASA; **20.1b:** NASA/Corbis; **p. 412** (dingo): Martin Harvey/Corbis; **20.2** (middle): Frans Lanting/Corbis; **20.4:** Medford Taylor/National Geographic/Getty Images; **20.5:** Alamy; **20.9:** Alamy; **20.10:** Alamy; **20.11:** Michael P. Gadomski/Earth Sciences/Animals Animals; **20.12a:** ionsofAmerica.com/Joe Sohm/The Image Bank/Getty Images; **20.12b:** Willard Clay/Dembinsky Photo Associates; **20.12c:** Mark De Fraeye/SPL/Photo Researchers, Inc; **20.13:** Ken Lucas/Visuals Unlimited; **20.14:** Willard Clay/Dembinsky Photo Associates; **20.15:** Gerry Ellis/Mindon Pictures; **20.17a:** Carr Clifton/Mindon Pictures; **20.17b:** Terry Donnelly/Tom Stack & Associates; **20.17c:** Scott T. Smith/Dembinsky Photo Associates; **20.18:** Alamy; **20.20a:** Jim Zipp/National Audubon Society Collection/Photo Researchers, Inc; **20.20b:** C. Van Dover/OAR/NURP/College of William and Mary/NOAA/Photo Researchers, Inc; **p. 426:** Ted Kinsman/Photo Researchers, Inc.

Chapter 21 p. 431: Art Wolfe/Stone/Getty Images; **21.1** (top): Jeremy Burgess/National Audubon Society Collection/Photo Researchers, Inc; **21.1** (bottom): Volker Staeger/SPL/Science Source/Photo Researchers, Inc; **21.2:** Frans Lanting/Minden Pictures; **21.4a:** Reproduced with permission of the Department of Natural Resources, Queensland, Australia. **21.4b:** Reproduced with permission of the Department of Natural Resources, Queensland, Australia; **21.4b** (top insert): Susan Ellis, USDA APHIS PPQ, Bugwood.org; **21.5:** Scott Camazine/National Audubon Society Collection/Photo Researchers, Inc; **21.6:** Laguna Design/Science Picture Library/Photo Researchers, Inc; **21.7:** Ecoscene/Corbis; **21.8:** Mark Newman/Tom Stack & Associates; **21.9:** G: A. Matthews/SPL/Photo Researchers, Inc; **21.10:** Alan G. Nelson/Dembinsky Photo Associates; **p. 439:** Frans Lanting/Minden

Pictures; **21.12:** Craig Mayhew and Robert Simmon, NASA GSFC; **p. 442** (top): Art Wolfe/Stone/Getty Images; **21.13:** NASA/Corbis.

Chapter 22 p. 447 (a): Gregory Dimijian/National Audubon Society Collection/Photo Researchers, Inc; **p. 447.**(b): Inga Spence/Tom Stack & Associates; **p. 447** (c): Science VU/Visuals Unlimited; **22.2** (left): Willard Clay/Dembinsky Photo Associates; **22.2** (right): Ken Wagner/Visuals Unlimited; **22.3:** courtesy of Mycorrhizal Applications Inc. www.mycorrhizae.com; **22.4:** F. Bravendam/Peter Arnold. Inc; **22.6a:** Mike Bacon/Tom Stack & Associates; **22.6b** (both): Dr. Lincoln Brewer; **22.7:** Fred Breummer/Peter Arnold, Inc; **22.9:** Courtesy of Dr. Charles Mitchell, University of North Carolina at Chapel Hill; **p. 454:** Mauro Fermariello/Photo Researchers, Inc; **22.11** (bottom): J.K. Clark/University of California, Davis; **22.13:** Forest & Kim Starr (USGS); **p. 458** (top): Gregory Dimijian/National Audubon Society Collection/Photo Researchers, Inc; **22.14a and b:** Reprinted by permission from *Nature,* vol. 406, pp. 255-256. © MacMillan Magazines, Ltd.

Chapter 23 p. 463: Greg Vaughn/Tom Stack & Associates; **23.4** (both): Courtesy Dr. Robert T. Paine; **23.5:** Dennis MacDonald /Photolibrary; **23.6a:** Stan Osolinski/Dembinsky Photo Associates; **23.6b:** Howard Garrett/Dembinsky Photo Associates; **23.6c:** Walt Anderson/Visuals Unlimited; **23.8a:** Charles E. Rotkin/Corbis; **23.8b:** B. Anthony Stewart/National Geographic/Getty Images; **23.9** (left): Ken Lucas/Visuals Unlimited/Getty Images; **23.9** (right): David Wrobel/Visuals Unlimited; **23.10a:** courtesy Robert Gibbens, Jornada Experimental Range, USDA, **23.10b:** courtesy Robert Gibbens, Jornada Experimental Range, USDA; **p. 472:** John W. Bova/Photo Researchers, Inc; **23.11:** Corbis; **p. 474** (top): Greg Vaughn/Tom Stack & Associates; **23.12** (left): © Douglas Peebles/CORBIS; **23.12** (top right, bottom right): Dr. Gerald D. Carr.

Chapter 24 p. 479: Joseph Sohm; ChromoSohm Inc./Corbis; **24.2a:** Martin Harvey/Corbis; **24.2b:** Simon Fraser/Photo Researchers, Inc; **24.4a:** Kevin Magee/Tom Stack & Associates; **24.4b:** Darryl Leniuk/Digital Vision/Getty Images; **p. 486:** Science Photo Library; **24.13a:** Courtesy of the Hubbard Brook Archives; **24.14:** Richard Packwood/Oxford Scientific Films; **24.15** (top): Adam Hart-Davis/Photo Researchers, Inc; **24.15** (inset): Mauro Fermariello/Photo Researchers, Inc; **p. 495:** Joseph Sohm; ChromoSohm Inc./Corbis; **24.16a:** AP Photo; **24.16b:** Bruce Ely/Landov.

Chapter 25 p. 501: Carr Clifton/Minden Pictures; **p. 501** (inset): Alamy; **25.1a:** James P. Blair/National Geographic/Getty Images; **25.1b:** Doug Sokell/Tom Stack & Associates; **25.1c and d:** USGS; **25.2** (inset): Jacques Jangoux/Photo Researchers, Inc; **p. 506:** Ralph Lee Hopkins/Photo Researchers, Inc; **25.6a and b:** Courtesy of David Tilman, University of Minnesota; **25.8:** Courtesy Joy Ward and Anne Hastley; **p. 513** (top): Carr Clifton/Minden Pictures; **25.12:** Jeffrey L. Rotman/Corbis; **25.C:** Brownie Harris/Corbis.

Interlude E E.1: Courtesy Rudolf B. Husar, Washington University at St. Louis; **E.2:** Corbis; **E.4a:** Raymond Gehman/Corbis; **E.4b:** Photo by Doug Gouzie, 2006 USGS; **E.5** (left): Gerry Ellis/Minden Pictures; **E.6:** Jeffrey Lepore/Photo Researchers, Inc; **E.7** (left): Courtesy of Mark Farina/City of Chicago; **E.7** (middle): Reuters/Corbis; **E.7** (right): Image copyright ZinCo, further information available at www.ZinCo.de; **E.8:** Courtesy of Eric Shriner; **E.9** (left): William Campbell/Corbis; **E.9** (right): Radu Sigheti Reuters/Corbis; **E.10:** Image courtesy of NASA Landsat Project Science

Office and USGS National Center for EROS; **E.11a**: NASA; **E.11b**: NASA; **E.12**: Owaki-Kulla/Corbis; **E.14**: Courtesy of DuPont.

"Biology in the News" Credits

Chapter 2 "Sasquatch sighting reported in Yukon," Canadian Broadcasting Corporation News, July 13, 2005. Reprinted by permission.

Chapter 4 "Trans Fat Banned in N.Y. Eateries; City Health Board Cites Heart Risks" by Annys Shin. From *The Washington Post,* Financial Section, 12/6/2006, p. D1, © 2006 The Washington Post. All rights reserved. Used by permission and protected by the Copyright Laws of the United States. The printing, copying, redistribution, or retransmission of the Material without express written permission is prohibited. www.washingtonpost.com

Chapter 5 "Science File; Without Gene, Mice Don't Know Meaning of Fear" by Alex Raksin, *Los Angeles Times,* Nov. 19, 2005, p. A12. Copyright 2005, Los Angeles Times. Reprinted with permission.

Chapter 7 "Doctors Warned About Common Drugs For Pain" by Shankar Vedantam. From *The Washington Post,* 2/27/2007, p. A8, © 2007 The Washington Post. All rights reserved. Used by permission and protected by the Copyright Laws of the United States. The printing, copying, redistribution, or retransmission of the Material without express written permission is prohibited. www.washingtonpost.com

Chapter 8 "Biofuels powering town's vehicles: Vegetable oils reduce emissions," *The Worcester Telegram & Gazette,* July 13, 2007, p. B1. Copyright © 2007 The Worcester Telegram & Gazette. Republished by permission.

Chapter 9 "New Type of Stem Cells May Help Regenerate Heart Tissues" by Adrienne Law, DAILY BRUIN, May 5, 2008.

Chapter 12 "Chernobyl wildlife baffles biologists; Animals are returning to area near meltdown, but scientist are split on their long-term fates" by Douglas Birch, *Toronto Star,* June 8, 2007. Used with permission of The Associated Press Copyright © 2008. All rights reserved.

Chapter 13 "Researchers delve into 'gene for speed'," Agence France-Presse, September 10, 2007. Copyright 2007 by Agence France-Presse. Reproduced with permission of Agence France-Presse, via Copyright Clearance Center, Inc.

Chapter 14 "Nobel Winners' Duel Draws Merck, Novartis in Genetic Drug Race" by John Lauerman, originally published June 7, 2007. © 2007 Bloomberg L.P. All rights reserved. Reprinted with permission.

Chapter 15 "Mutants or saviors? Rabbit genes create trees that eat poisons; UW scientists create transgenic poplars that neutralize toxins quickly" by Lisa Stiffler, *Seattle Post-Intelligencer,* Oct. 16, 2007. Reprinted by permission.

Chapter 16 "The 6th Annual Year in Ideas; Empty-Stomach Intelligence" by Christopher Shea, *New York Times Magazine,* December 10, 2006. © 2006, Christopher Shea. Reprinted by permission.

Chapter 17 "Antibiotic Runoff," *The New York Times,* Editorial Section, 9/18/2007, p. A26. © 2007 The New York Times. All rights reserved. Used by permission and protected by the Copyright Laws of the United States. The printing, copying, redistribution, or retransmission of the Material without express written permission is prohibited. www.nytimes.com

Chapter 18 "Study Says Bones Found in Far East Are of a Distinct Species" by John Noble Wilford, *The New York Times,* 9/21/07. © 2007 The New York Times. All rights reserved. Used by permission and protected by the Copyright Laws of the United States. The printing, copying, redistribution, or retransmission of the Material without express written permission is prohibited. www.nytimes.com

Chapter 20 Ralph Mitchell, "Invasion of the Cuban treefrogs," *Herald Sun,* May 27, 2007, excerpted from Steve A. Johnson, "The Cuban Treefrog (Osteopilus septentrionalis) in Florida," WEC-218, University of Florida Institute of Food and Agricultural Sciences, www.edis.ifas.ufl.edu/UW259, 3,738 words. Reprinted by permission.

Chapter 21 "Mosquitoes bring West Nile virus" by Mary K. Reinhart, *East Valley/Scottsdale Tribune,* May 5, 2007. Copyright © 2007 Freedom Communications/Arizona. Reprinted by permission.

Chapter 22 "River Parasite Eats At Children: Neglected Scourge in Africa is Cheap and Easy to Treat, But the 'Pennies' Cannot Be Found" by Colleen Mastony, *Chicago Tribune,* March 19, 2007. Reprinted with permission of the Chicago Tribune; copyright Chicago Tribune; all rights reserved.

Diligent efforts have been made to contact the copyright holders of all of the photographs and articles used in this book. Rights holders of any pieces not properly credited should contact W. W. Norton & Company, Inc., 500 Fifth Avenue, New York, NY 10110, in order for a correction to be made in the next reprinting of our work.

Index

Page numbers in *italics* refer to illustrations and tables.